Methods in Molecular Biology

Series Editor
John M. Walker
School of Life and Medical Sciences
University of Hertfordshire
Hatfield, Hertfordshire AL10 9AB, UK

For further volumes:
http://www.springer.com/series/7651

NADPH Oxidases

Methods and Protocols

Edited by

Ulla G. Knaus

Conway Institute, School of Medicine, University College Dublin, Dublin, Ireland

Thomas L. Leto

Laboratory of Clinical Immunology and Microbiology, National Institute of Allergy and Infectious Diseases, National Institutes of Health, Bethesda, MD, USA

Humana Press

Editors
Ulla G. Knaus
Conway Institute, School of Medicine
University College Dublin
Dublin, Ireland

Thomas L. Leto
Laboratory of Clinical Immunology
and Microbiology
National Institute of Allergy and Infectious
Diseases, National Institutes of Health
Bethesda, MD, USA

ISSN 1064-3745 ISSN 1940-6029 (electronic)
Methods in Molecular Biology
ISBN 978-1-4939-9423-6 ISBN 978-1-4939-9424-3 (eBook)
https://doi.org/10.1007/978-1-4939-9424-3

This Humana imprint is published by the registered company Springer Science+Business Media, LLC part of Springer Nature.
The registered company address is: 233 Spring Street, New York, NY 10013, U.S.A.

Preface

NADPH oxidases are the only enzymes solely dedicated to reduce molecular oxygen to superoxide and, under some circumstances, to hydrogen peroxide (H_2O_2). NADPH oxidases, also termed NOX or DUOX (in the presence of an additional peroxidase-like extracellular N-terminal domain), are expressed widely in prokaryotes and eukaryotes. The NOX/DUOX family consists of several structurally and functionally related members such as five NOX and two DUOX isoforms in humans. Although the general mechanism of electron transport from NADPH via FAD and two membrane-embedded low-potential hemes to molecular oxygen is conserved in all NOX enzymes, many differences exist in the activation mechanisms of oxidases. Most NOX/DUOX enzymes are activated in a tightly controlled fashion in response to signals transmitted by extracellular cues or intracellular metabolic needs. Reactive oxygen species (ROS) generation by these enzymes involves in many cases multiple cytosolic partner proteins and several molecular events that include second messenger involvement, transcriptional regulation, posttranslational modifications, protein-protein and protein-lipid associations, and subcellular translocation of oxidase components. H_2O_2, produced directly or indirectly by all NOX isozymes, is considered a key second messenger in signaling networks, thereby initiating redox relays. In addition, genetic diseases have causally linked NOX/DUOX-derived ROS to crucial steps in host defense and thyroid hormone biosynthesis. Finally, a plethora of observations have connected dysregulated ROS overproduction by these enzymes to pathophysiology, giving rise to drug development programs targeting these ROS generators.

The aim of *NADPH Oxidases: Methods and Protocols* is to provide an overview of our knowledge of this family of enzymes in human physiology and genetic disease, in which early discoveries represent prime examples of the finest translational "from bed to bench and back" studies (Parts I and VI). These authoritative reviews from leaders in the field are accompanied by detailed descriptions of methods to test assembly and function of multi-component oxidase complexes (Part II) and to analyze ROS generation in different systems by various means while addressing pitfalls of ROS probes currently being used (Part III). Parts IV and V provide protocols on NADPH oxidase regulation and their function in cells. Part VI includes overviews on genetic diseases caused or linked to loss-of-function mutations in NOX or DUOX isoforms and provides practical advice and protocols on diagnostics, disease modeling, and gene editing for chronic granulomatous disease (CGD), the phagocyte oxidase disorder associated with defective NOX2 complex components.

We thank everybody in the NOX/DUOX field who has contributed their outstanding chapters, and John M. Walker, Series Editor, and Springer Nature, in particular David C. Casey, for the opportunity, advice, and support for the publication of this volume.

Dublin, Ireland — *Ulla G. Knaus*
Bethesda, MD, USA — *Thomas L. Leto*

Contents

Contributors

FIONA AUGSBURGER • *Faculty of Medicine, Department of Pathology and Immunology, University of Geneva, Geneva, Switzerland*

GABRIELLA AVIELLO • *The Rowett Institute, University of Aberdeen, Aberdeen, UK*

LAURA BACIOU • *Laboratoire de Chimie Physique, UMR8000 CNRS, Université Paris-Sud, Université Paris-Saclay, Orsay, France*

SYLVAIN BEAUMEL • *Centre Diagnostic et Recherche CGD (CDiReC), Pôle Biologie, CHU Grenoble Alpes, Grenoble, France*

SAHRA A. BELAMBRI • *Stress Oxydatif et Inflammation, Laboratoire de Biochimie Appliquée, Département de Biochimie, Faculté des Sciences de la Nature et de la Vie, Université Ferhat Abbes 1, Sétif, Algeria*

VSEVOLOD V. BELOUSOV • *Shemyakin-Ovchinnikov Institute of Bioorganic Chemistry, Moscow, Russia; Pirogov Russian National Research Medical University, Moscow, Russia; Georg August University of Göttingen, Göttingen, Germany*

KAREN BERNARD • *Division of Pulmonary, Allergy and Critical Care Medicine, Department of Medicine, University of Alabama at Birmingham, Birmingham, AL, USA*

TANIA BIZOUARN • *Laboratoire de Chimie Physique, UMR8000 CNRS, Université Paris-Sud, Université Paris-Saclay, Orsay, France*

HOWARD E. BOUDREAU • *Laboratory of Clinical Immunology and Microbiology, National Institute of Allergy and Infectious Diseases, National Institutes of Health, Bethesda, MD, USA*

AICHA BOURAOUI • *Laboratoire de Chimie Physique, UMR8000 CNRS, Université Paris-Sud, Université Paris-Saclay, Orsay, France*

BILLY BOURKE • *Conway Institute, School of Medicine, University College Dublin, Dublin, Ireland; Department of Paediatric Gastroenterology, Our Lady's Children's Hospital Crumlin, Dublin, Ireland*

JULIE BRAULT • *Centre Diagnostic et Recherche CGD (CDiReC), Pôle Biologie, CHU Grenoble Alpes, Grenoble, France*

SUSAN BRUMFIELD • *Department of Plant Sciences and Plant Pathology, Montana State University, Bozeman, MT, USA*

HÉLÈNE BUVELOT • *Faculty of Medicine, Department of Pathology and Immunology, University of Geneva, Geneva, Switzerland*

YVES CAMBET • *Faculty of Medicine, Department of Pathology and Immunology, University of Geneva, Geneva, Switzerland; READS Unit, Faculty of Medicine, University of Geneva, Geneva, Switzerland*

BINDU CHANDRASEKHARAN • *Department of Pathology, Emory University School of Medicine, Atlanta, GA, USA*

GANG CHENG • *Department of Biophysics, Medical College of Wisconsin, Milwaukee, WI, USA; Free Radical Research Center, Medical College of Wisconsin, Milwaukee, WI, USA*

UIMOOK CHOI • *Laboratory of Clinical Immunology and Microbiology, Division of Intramural Research, National Institute of Allergy and Infectious Diseases, National Institutes of Health, Bethesda, MD, USA*

EUGENIA CIFUENTES-PAGANO • *Department of Pharmacology and Chemical Biology, Vascular Medicine Institute, University of Pittsburgh School of Medicine, Pittsburgh, PA, USA*

ROBERT A. CLARK • *Institute for Integration of Medicine and Science and Department of Medicine, University of Texas Health Science Center, and South Texas Veterans Healthcare System, San Antonio, TX, USA*
EMER CONROY • *Conway Institute, University College Dublin, Dublin, Ireland*
PHAM MY-CHAN DANG • *Centre de Recherche sur l'Inflammation (CRI), INSERM-U1149, CNRS-ERL8252, Laboratoire d'Excellence Inflamex, Université Paris Diderot-Sorbonne Paris Cité, Faculté de Médecine, Site Xavier Bichat, Paris, France*
XAVIER DE DEKEN • *Faculté de Médecine, Institut de Recherche Interdisciplinaire en Biologie Humaine et Moléculaire (IRIBHM), Université Libre de Bruxelles (ULB), Brussels, Belgium*
SUK SEE DE RAVIN • *Laboratory of Clinical Immunology and Microbiology, Division of Intramural Research, National Institute of Allergy and Infectious Diseases, National Institutes of Health, Bethesda, MD, USA*
RIAD ARABI DERKAWI • *Centre de Recherche sur l'Inflammation (CRI), INSERM-U1149, CNRS-ERL8252, Laboratoire d'Excellence Inflamex, Université Paris Diderot-Sorbonne Paris Cité, Faculté de Médecine, Site Xavier Bichat, Paris, France*
BECKY A. DIEBOLD • *Department of Pathology, Emory University, Atlanta, GA, USA*
MARY C. DINAUER • *Department of Pediatrics, Washington University in St. Louis School of Medicine, St. Louis, MO, USA; Department of Pathology and Immunology, Washington University in St. Louis School of Medicine, St. Louis, MO, USA*
ÁGNES P. DONKÓ • *Laboratory of Clinical Immunology and Microbiology, National Institute of Allergy and Infectious Diseases, National Institutes of Health, Bethesda, MD, USA*
JAMES H. DOROSHOW • *Developmental Therapeutics Branch, Center for Cancer Research, National Cancer Institute, National Institutes of Health, Bethesda, MD, USA; Division of Cancer Treatment and Diagnosis, Center for Cancer Research, National Cancer Institute, National Institutes of Health, Bethesda, MD, USA*
SOPHIE DUPRÉ-CROCHET • *LCP, CNRS UMR 8000, Université Paris-Sud, Université Paris-Saclay, Orsay, France*
CHRISTOPHER M. DUSTIN • *Department of Pathology and Laboratory Medicine, College of Medicine, University of Vermont, Burlington, VT, USA*
JAMEL EL-BENNA • *Centre de Recherche sur l'Inflammation (CRI), INSERM-U1149, CNRS-ERL8252, Laboratoire d'Excellence Inflamex, Université Paris Diderot-Sorbonne Paris Cité, Faculté de Médecine, Site Xavier Bichat, Paris, France*
JOHN F. ENGELHARDT • *Department of Anatomy and Cell Biology, Carver College of Medicine, University of Iowa, Iowa City, IA, USA; Department of Internal Medicine, Carver College of Medicine, University of Iowa, Iowa City, IA, USA; Center for Gene Therapy, Carver College of Medicine, University of Iowa, Iowa City, IA, USA*
BALÁZS ENYEDI • *Faculty of Medicine, Department of Physiology, Semmelweis University, Budapest, Hungary; MTA-SE Lendület Tissue Damage Research Group, Hungarian Academy of Sciences and Semmelweis University, Budapest, Hungary*
MARIE ERARD • *LCP, CNRS UMR 8000, Université Paris-Sud, Université Paris-Saclay, Orsay, France*
YULIA G. ERMAKOVA • *Shemyakin-Ovchinnikov Institute of Bioorganic Chemistry, Moscow, Russia; European Molecular Biology Laboratory, Heidelberg, Germany*
E. LIANA FALCONE • *Laboratory of Clinical Immunology and Microbiology (LCIM), National Institute of Allergy and Infectious Diseases (NIAID), National Institutes of Health (NIH), Bethesda, MD, USA*

ALEKSANDRA FILIPPOVA • *Faculty of Medicine, Department of Pathology and Immunology, University of Geneva, Geneva, Switzerland*
DAVID J. R. FULTON • *Vascular Biology Center, Medical College of Georgia at Augusta University, Augusta, GA, USA*
MIKLÓS GEISZT • *Faculty of Medicine, Department of Physiology, Semmelweis University, Budapest, Hungary*
MICAEL HARDY • *Aix Marseille Université, CNRS, ICR, UMR 7273, Marseille, France*
STEVEN M. HOLLAND • *Laboratory of Clinical Immunology and Microbiology (LCIM), National Institute of Allergy and Infectious Diseases (NIAID), National Institutes of Health (NIH), Bethesda, MD, USA*
CHANTAL HOUÉE-LEVIN • *Laboratoire de Chimie Physique, UMR8000 CNRS, Université Paris-Sud, Université Paris-Saclay, Orsay, France*
MILENA HRISTOVA • *Department of Pathology and Laboratory Medicine, College of Medicine, University of Vermont, Burlington, VT, USA*
VINCENT JAQUET • *Faculty of Medicine, Department of Pathology and Immunology, University of Geneva, Geneva, Switzerland; READS Unit, Faculty of Medicine, University of Geneva, Geneva, Switzerland*
MARK JELCIC • *Cell Biology Program, Memorial Sloan Kettering Cancer Center, New York, NY, USA; Louis V. Gerstner, Jr. Graduate School of Biomedical Sciences, Memorial Sloan Kettering Cancer Center, New York, NY, USA*
ALGIRDAS J. JESAITIS • *Department of Microbiology and Immunology, Montana State University, Bozeman, MT, USA*
BALARAMAN KALYANARAMAN • *Department of Biophysics, Medical College of Wisconsin, Milwaukee, WI, USA; Free Radical Research Center, Medical College of Wisconsin, Milwaukee, WI, USA; Cancer Center, Medical College of Wisconsin, Milwaukee, WI, USA*
GILDA KARIMI • *Laboratoire de Chimie Physique, UMR8000 CNRS, Université Paris-Sud, Université Paris-Saclay, Orsay, France*
ULLA G. KNAUS • *Conway Institute, School of Medicine, University College Dublin, Dublin, Ireland*
AGNIESZKA KORZENIOWSKA • *Laboratory of Clinical Immunology and Microbiology, National Institute of Allergy and Infectious Diseases, National Institutes of Health, Bethesda, MD, USA*
KARL-HEINZ KRAUSE • *Faculty of Medicine, Department of Pathology and Immunology, University of Geneva, Geneva, Switzerland*
DOUGLAS B. KUHNS • *Leidos Biomedical Research, Inc., Frederick National Laboratory for Cancer Research, Frederick, MD, USA*
J. DAVID LAMBETH • *Department of Pathology, Emory University, Atlanta, GA, USA*
THOMAS L. LETO • *Laboratory of Clinical Immunology and Microbiology, National Institute of Allergy and Infectious Diseases, National Institutes of Health, Bethesda, MD, USA*
HARRY L. MALECH • *Laboratory of Clinical Immunology and Microbiology, Division of Intramural Research, National Institute of Allergy and Infectious Diseases, National Institutes of Health, Bethesda, MD, USA*
RAWAND MASOUD • *Laboratoire de Chimie Physique, UMR8000 CNRS, Université Paris-Sud, Université Paris-Saclay, Orsay, France*
JAMES W. MCCOY • *Department of Pathology, Emory University, Atlanta, GA, USA*
JENNIFER L. MEITZLER • *Developmental Therapeutics Branch, Center for Cancer Research, National Cancer Institute, National Institutes of Health, Bethesda, MD, USA*

RANDALL K. MERLING • *Laboratory of Clinical Immunology and Microbiology, Division of Intramural Research, National Institute of Allergy and Infectious Diseases, National Institutes of Health, Bethesda, MD, USA*
REIKO MINAKAMI • *Department of Biochemistry, Kyushu University Graduate School of Medical Sciences, Fukuoka, Japan*
FRANÇOISE MIOT • *Faculté de Médecine, Université Libre de Bruxelles (ULB), Institut de Recherche Interdisciplinaire en Biologie Humaine et Moléculaire (IRIBHM), Brussels, Belgium*
NATALIYA M. MISHINA • *Shemyakin-Ovchinnikov Institute of Bioorganic Chemistry, Moscow, Russia*
KEI MIYANO • *Department of Biochemistry, Kyushu University Graduate School of Medical Sciences, Fukuoka, Japan*
STANISLAS MORAND • *Laboratory of Clinical Immunology and Microbiology, National Institute of Allergy and Infectious Diseases, National Institutes of Health, Bethesda, MD, USA; L'Oreal Advanced Research, Aulnay-Sous-Bois, Paris, France*
YOKO NAKANO • *Department of Medicine, Inflammation Program, Roy J. and Lucille A. Carver College of Medicine, University of Iowa, Iowa City, IA, USA*
WILLIAM M. NAUSEEF • *Department of Medicine, Inflammation Program, Roy J. and Lucille A. Carver College of Medicine, University of Iowa, Iowa City, IA, USA*
ANDREW S. NEISH • *Department of Pathology, Emory University School of Medicine, Atlanta, GA, USA*
PHILIPP NIETHAMMER • *Cell Biology Program, Memorial Sloan Kettering Cancer Center, New York, NY, USA*
OLIVER NÜβE • *LCP, CNRS UMR 8000, Université Paris-Sud, Université Paris-Saclay, Orsay, France*
SHARON O'NEILL • *Conway Institute, School of Medicine, University College Dublin, Dublin, Ireland*
PATRICK J. PAGANO • *Department of Pharmacology and Chemical Biology, Vascular Medicine Institute, University of Pittsburgh School of Medicine, Pittsburgh, PA, USA*
EDGAR PICK • *Department of Clinical Microbiology and Immunology, Sackler School of Medicine, Tel Aviv University, Tel Aviv, Israel*
HOUSSAM RAAD • *Centre de Recherche sur l'Inflammation (CRI), INSERM-U1149, CNRS-ERL8252, Laboratoire d'Excellence Inflamex, Université Paris Diderot-Sorbonne Paris Cité, Faculté de Médecine, Site Xavier Bichat, Paris, France*
BALÁZS RADA • *Department of Infectious Diseases, College of Veterinary Medicine, University of Georgia, Athens, GA, USA*
DELPHINE RASTI • *Faculty of Medicine, Department of Pathology and Immunology, University of Geneva, Geneva, Switzerland*
MARCIA RIESSELMAN • *Department of Microbiology and Immunology, Montana State University, Bozeman, MT, USA*
DIRK ROOS • *Sanquin Research, and Landsteiner Laboratory, Academic Medical Center, University of Amsterdam, Amsterdam, The Netherlands*
BEJAN J. SAEEDI • *Department of Pathology, Emory University School of Medicine, Atlanta, GA, USA*
CASPAR SCHIFFERS • *Department of Pathology and Laboratory Medicine, College of Medicine, University of Vermont, Burlington, VT, USA*
CARSTEN SCHULTZ • *European Molecular Biology Laboratory, Heidelberg, Germany; Oregon Health and Science University, Portland, OR, USA*

Xavier Serfaty • *Laboratoire de Chimie Physique, UMR8000 CNRS, Université Paris-Sud, Université Paris-Saclay, Orsay, France*

Weam S. Shahin • *Department of Anatomy and Cell Biology, Carver College of Medicine, University of Iowa, Iowa City, IA, USA*

Hager Souabni • *Laboratoire de Chimie Physique, UMR8000 CNRS, Université Paris-Sud, Université Paris-Saclay, Orsay, France*

Marie José Stasia • *Centre Diagnostic et Recherche CGD (CDiReC), Pôle Biologie, CHU Grenoble Alpes, Grenoble, France; Universite Grenoble Alpes, CNRS, CEA, Institut de Biologie Structurale, Grenoble, France*

Emily Stenke • *Conway Institute, School of Medicine, University College Dublin, Dublin, Ireland*

Hideki Sumimoto • *Department of Biochemistry, Kyushu University Graduate School of Medical Sciences, Fukuoka, Japan*

Colin L. Sweeney • *Laboratory of Clinical Immunology and Microbiology, Division of Intramural Research, National Institute of Allergy and Infectious Diseases, National Institutes of Health, Bethesda, MD, USA*

Ross M. Taylor • *Department of Microbiology and Immunology, Montana State University, Bozeman, MT, USA; Universal Cells, Seattle, WA, USA*

Victor J. Thannickal • *Division of Pulmonary, Allergy and Critical Care Medicine, Department of Medicine, University of Alabama at Birmingham, Birmingham, AL, USA*

Asma Tlili • *Centre de Recherche sur l'Inflammation (CRI), INSERM-U1149, CNRS-ERL8252, Laboratoire d'Excellence Inflamex, Université Paris Diderot-Sorbonne Paris Cité, Faculté de Médecine, Site Xavier Bichat, Paris, France*

Albert van der Vliet • *Department of Pathology and Laboratory Medicine, College of Medicine, University of Vermont, Burlington, VT, USA*

Bénédicte Vigne • *Centre Diagnostic et Recherche CGD (CDiReC), Pôle Biologie, CHU Grenoble Alpes, Grenoble, France*

S. Garrett Wilder • *Department of Pathology, Emory University, Atlanta, GA, USA*

Yerun Zhu • *Department of Pathology, Emory University, Atlanta, GA, USA*

Jacek Zielonka • *Department of Biophysics, Medical College of Wisconsin, Milwaukee, WI, USA; Free Radical Research Center, Medical College of Wisconsin, Milwaukee, WI, USA; Cancer Center, Medical College of Wisconsin, Milwaukee, WI, USA*

Monika Zielonka • *Department of Biophysics, Medical College of Wisconsin, Milwaukee, WI, USA; Free Radical Research Center, Medical College of Wisconsin, Milwaukee, WI, USA*

Part I

The NADPH Oxidase Family: Overviews

Chapter 1

Intersecting Stories of the Phagocyte NADPH Oxidase and Chronic Granulomatous Disease

William M. Nauseef and Robert A. Clark

Abstract

Neutrophils serve as the circulating cells that respond early and figure prominently in human host defense to infection and in inflammation in other settings. Optimal oxidant-dependent antimicrobial activity by neutrophils relies on the ability of stimulated phagocytes to utilize a multicomponent NADPH oxidase to generate oxidants. The frequent, severe, and often fatal infections experienced by individuals with chronic granulomatous disease (CGD), an inherited disorder in which one of the NADPH oxidase components is absent or dysfunctional, underscore the link between a functional phagocyte NADPH oxidase and robust host protection against microbial infection.

The history of the discovery and characterization of the normal neutrophil NADPH oxidase and the saga of recognizing CGD and its underlying causes together illustrate how the observations of astute clinicians and imaginative basic scientists synergize to forge new understanding of both basic cell biology and pathogenesis of human disease.

In this chapter, we review the events in the stepwise evolution of our understanding of the phagocyte NADPH oxidase, both in the context of normal human neutrophil function and in the setting of CGD. The phagocyte oxidase complex employs a heterodimeric transmembrane protein composed of gp91*phox* and p22*phox* to relay electrons from NADPH to molecular oxygen, while other cofactors contribute to localization and regulation of the activity of the assembled oxidase. The b-type cytochrome gp91*phox*, also known as NOX2, serves as the catalytic component of this multicomponent enzyme complex. Although many of the features of the composition and regulation of the phagocyte oxidase may apply as well to NOX2 expressed in non-phagocytes and to other members of the NOX protein family, exceptions exist and pose special challenges to investigators exploring the biology of NADPH oxidases.

Key words Phagocyte NADPH oxidase, NOX2, gp91*phox*, p22*phox*, Cytochrome b_{558}, p47*phox*, p67*phox*, p40*phox*, Chronic granulomatous disease (CGD)

1 Introduction

It may surprise some readers to learn that recognition of the wide distribution of NOX protein family members throughout both plant and animal kingdoms is relatively recent. For decades after discovery of the "respiratory burst" of neutrophils, the superoxide-generating NADPH oxidase was believed to be an attribute unique

Ulla G. Knaus and Thomas L. Leto (eds.), *NADPH Oxidases: Methods and Protocols*, Methods in Molecular Biology, vol. 1982, https://doi.org/10.1007/978-1-4939-9424-3_1,

to phagocytic cells (i.e., neutrophils, eosinophils, monocytes, and macrophages). Detection of superoxide production by non-phagocytic cells (reviewed in [1]) and the initial identification of NOX2 homologues [2, 3] revised this parochial view and introduced a new field for exploration.

The study of the NADPH oxidase of phagocytes reveals many principles that apply as well to other members of the NOX protein family. Many features of the neutrophil NADPH oxidase have been recently reviewed in detail [4], but we focus in this review on the more general aspects that characterize the NADPH oxidase in phagocytes, as we believe that an overview of the evolution of our understanding of the phagocyte system may provide insights that can be applied to the study of the cell and molecular biology of other NOX proteins. With that goal in mind, we review some of the milestones in the history of two intertwined tales of discovery, namely, those of the phagocyte NADPH oxidase and of chronic granulomatous disease (CGD) (see Fig. 1). The cross-fertilization between the biochemistry performed in laboratories around the world and the observations of astute clinicians exemplify biomedical research at its best. In a way, the history of discovery and elucidation of NOX2 in phagocytes represents only the *Book of Genesis* in what in time will be the Bible of the NOX protein family.

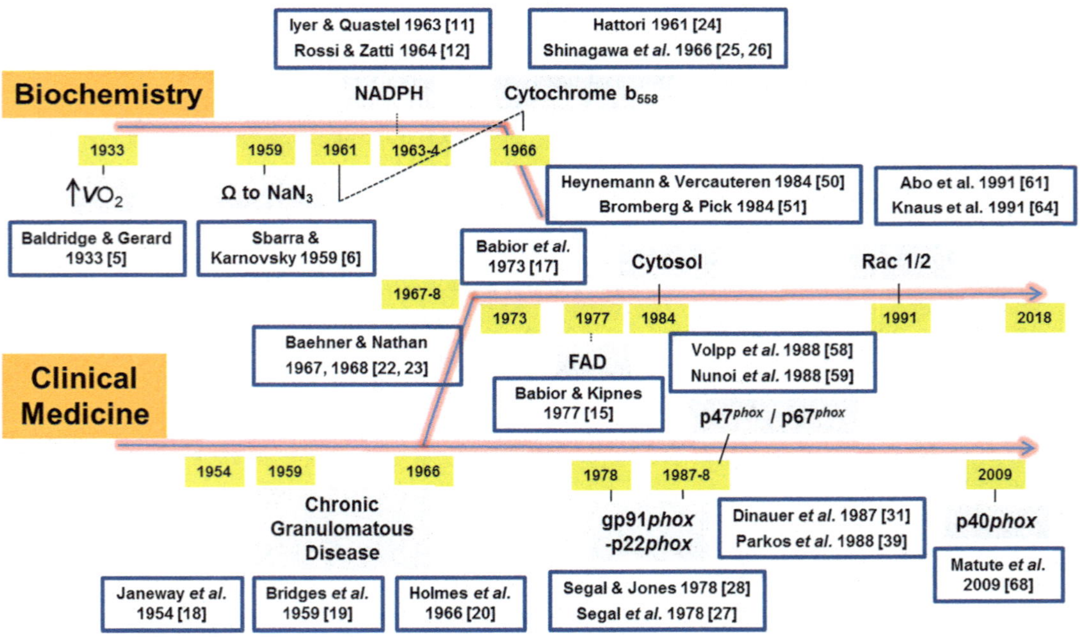

Fig. 1 The history of the discovery and characterization of the phagocyte respiratory burst oxidase is shown in the context of the evolution of our understanding of chronic granulomatous disease (CGD), emphasizing the interplay between analysis of the biochemistry of NADPH oxidase (NOX) and studies of neutrophils from CGD patients. Key advances are listed by author(s), publication year, and citation number from the chapter reference list

2 The "Respiratory Burst"

In the decades that followed Metchnikoff's initial description of phagocytosis, many investigators focused their attention on the mechanisms by which phagocytic cells engulf particles and the concomitant metabolic changes. Using Warburg manometers, Baldridge and Gerard detected increased oxygen consumption by dog leukocytes fed with a suspension of the Gram-positive bacterium *Sarcina lutea* [5]. The "burst of extra respiration" began immediately upon mixing leukocytes and bacteria and persisted for 10 to 15 min. The authors attributed the burst in oxygen consumption to "the excess energy liberation of engulfment and digestion." Subsequent work confirmed the observed increase in oxygen consumption during phagocytosis and extended the acute metabolic changes to include glucose utilization, activation of the hexose monophosphate shunt, and lactate production [6]. Unexpectedly, phagocytosis-triggered oxygen consumption was not inhibited by KCN or antimycin, suggesting that mitochondria were not responsible for the respiratory burst, as had been initially thought. Furthermore, phagocytosis was accompanied by production of hydrogen peroxide [7] and by oxidation of NADPH [8], as well as NADH [9, 10]. The relative roles of NADPH versus NADH fueled substantial debate among investigators, with some reporting that the oxidase activity was "much more active towards NADPH than towards NADH" [8, 11] and others reporting the opposite [9, 10]. The origins of contradictory results reflected in large part the confounding contribution of myeloperoxidase-mediated oxidation, as many investigators used leukocyte lysate that included granule proteins. Rossi and Zatti demonstrated that phagocytosis increases oxidation of both NADH and NADPH but that the oxidation of the latter by intact granules drives activation of the hexose phosphate shunt [12, 13]. Subsequent studies using subcellular particulate or soluble superoxide-generating fractions from stimulated neutrophils demonstrated a preference for NADPH over NADH, with a K_M of 33 μM versus 930 μM, respectively [14]. Furthermore, the loss of activity observed during preparation of soluble enzyme from neutrophils could be rescued by the addition of FAD, indicating that the NADPH oxidase of phagocytes is a flavin-dependent enzyme system [14–16].

Critical for rigorous study of the biochemistry of the phagocyte oxidase was the development of an accurate assay to quantitate the activity of the NADPH oxidase. Babior et al. met that need with the development of a spectrophotometric assay to quantitate superoxide anion as the superoxide dismutase-inhibitable reduction of ferricytochrome C [17]. With a consensus that phagocytes possess an agonist-elicited, flavin-dependent NADPH oxidase that reduces molecular oxygen to generate superoxide anion, investigators turned their attention to defining the composition and regulation of the novel enzyme.

3 Essential and Recurrent Role of Clinicians

At the May 1954 meeting of the American Pediatric Society, Dr. Charles Janeway reported the chance identification of five patients with severe and recurrent infections associated with splenomegaly, hepatomegaly, and generalized lymphadenopathy [18]. His presentation prompted Robert Good from the University of Minnesota to comment that "we have been struggling with a similar group of patients who fit the criteria outlined here this morning." The Good lab subsequently reported in greater detail their patients with "a fatal granulomatous disease of childhood" and speculated that "it could represent an alteration in immunological response turned against the reticuloendothelial system, whose functional and anatomical alteration and eventual destruction results in the clinical picture presented" [19]. In short order, physician-scientists demonstrated that neutrophils from affected individuals are unable to kill ingested staphylococci and lack oxidase activity [20–23]. At this point, the biochemistry of phagocytes gained clinical relevance in two complementary contexts. The studies established that *normal neutrophils* possess an NADPH oxidase that contributes to innate host defense against infection. Furthermore, *abnormal neutrophils* from patients with CGD lack oxidase activity and exhibit defective microbicidal activity, thus rendering the affected individual susceptible to frequent, severe, and often fatal infection. As illustrated repeatedly, clinical observations figure prominently in discoveries related to the phagocyte NADPH oxidase, often providing clues about both constituents and underlying mechanisms.

4 Cytochrome b_{558}

Two Japanese laboratories reported the presence of a novel b-type cytochrome in membranes from horse and rabbit neutrophils [24–26]. Although the authors explicitly linked this novel cytochrome with the neutrophil NADPH oxidase, cytochrome b_{558} did not merit attention until its rediscovery in human neutrophils and, most poignantly, recognition of its absence from phagocytes of patients with X-linked CGD by Segal and Jones [27, 28]. Although the coincidence of absent cytochrome b_{558} and the lack of NADPH oxidase provided circumstantial evidence that cytochrome b_{558} may transport electrons from NADPH to molecular oxygen, some skepticism remained because of a disparity between the kinetics of superoxide production and the rate of the cytochrome b_{558} reduction [29, 30]. The application of reverse genetics by Dinauer et al. resolved the dispute, however, with demonstration that the protein missing from the neutrophils of patients with X-linked CGD is cytochrome b_{558} and is encoded by *CYBB* [31].

Early studies of cytochrome b_{558} purified from neutrophils yielded molecular weight estimates ranging from 11,000 to 127,000 daltons [32–37]. Using several complementary analytical techniques, Parkos et al. demonstrated that cytochrome b_{558} exists as a heterodimeric membrane protein, composed of a 91 kDa glycosylated peptide (later named gp91*phox* [*ph*agocyte *ox*idase] and eventually NOX2) and a non-glycosylated 22 kDa peptide (p22*phox*) [38]. The gene encoding p22*phox*, *CYBA*, was cloned and sequenced [39]. Whereas RNA for p22*phox* was detected in all cell types examined, protein was stably expressed only in cells that also expressed gp91*phox*; no p22phox protein was detected in HepG2 cells, X-CGD neutrophils, K562 cells, or HeLa cells [39].

Studies of patients with CGD once again provided critical clues to understanding this complex enzyme system. The identification of patients with X-linked CGD, an inherited defect in the gene encoding gp91*phox*, and others with a rare autosomal recessive variant of CGD, both types lacking cytochrome b_{558} spectra in their neutrophils, suggested that two genetic defects resulted in the same biochemical phenotype, namely, absent oxidase activity and cytochrome b_{558} [40–43]. Patients with X-linked CGD also lack expression of p22*phox* [37, 44], and, conversely, neutrophils with mutations in p22*phox* also lack *both* proteins [45]. Taken together, these data suggested that heterodimer formation was required for stable expression of both gp91*phox* and p22*phox*. Subsequent studies of cytochrome b_{558} biosynthesis by myeloid precursors confirmed this interpretation [46].

5 Cytoplasmic Components of the Phagocyte Oxidase

With the identification of cytochrome b_{558} as the plasma membrane protein essential for electron transport from NADPH to oxygen, investigators turned their attention to a new question, likewise prompted in part by clinical observations. Clinicians recognized that whereas X-linked CGD was secondary to absent or inactive cytochrome b_{558} and accounted for approximately 60% of patients with CGD, the molecular basis for cytochrome b-positive autosomal CGD was unknown at that time. Segal and Jones observed that neutrophils from patients with autosomally inherited CGD express a normal cytochrome b_{558} but fail to reduce cytochrome b_{558} upon stimulation [47], raising the possibility that a different defect in electron transport may underlie the lack of superoxide generation in some forms of CGD. Support for this possibility came from studies using somatic cell hybridization, wherein fusion of monocytes from genetically different forms of CGD complement each other and yield heterologous hybrids that have a functional NADPH oxidase [48].

Speculation about cofactors that might regulate or possibly associate with cytochrome b_{558} led to investigation of oxidase activation in cell-free systems. Prompted by the novel observation that the dialysis of a granule fraction from unstimulated neutrophils stimulates oxidase activity [49], Heyneman and Vercauteren demonstrated that cytosol is required to drive superoxide generation in a cell-free system derived from equine neutrophils [50]. In studies of the role of unsaturated fatty acids as agonists in a cell-free system from guinea pig macrophages, Bromberg and Pick likewise demonstrated the requirement for the presence of cytosol to achieve activity [51]. The two systems required exogenous lipids (e.g., arachidonate, oleate) in addition to cytosol to convert the NADPH oxidase to an active state. Similar broken cell systems developed independently by other investigators confirmed that the transformation of the latent oxidase activity of resting phagocytes into a NADPH-dependent superoxide-generating system requires both membrane, as a source of cytochrome b_{558}, and something from cytosol that might serve to regulate or modify electron transport [52–54].

The addition of guanosine triphosphate (GTP) analogues or fluoride increases superoxide generation in the broken cell system, whereas guanosine diphosphate analogues decrease activity [55–57], observations that raised the possibility that the unknown cytosolic factor might be a GTP-binding protein. Using GTP-agarose affinity chromatography to isolate fractions from the cytosol of unstimulated human neutrophils, Volpp et al. recovered fractions enriched for the ability to support oxidase activity in a cell-free system that contained a suboptimal amount of cytosol [58]. Rabbit antiserum raised against the active fractions recognized only two proteins in neutrophil cytosol, with apparent molecular weights of 47-kDa and 67-kDa. At the same time, Nunoi et al. demonstrated the cytosol from neutrophils of patients with cytochrome b_{558}-positive autosomal CGD, i.e., individuals whose cytosol was defective in the cell-free system had complementary defects; the cytosol from one patient restored the defective cytosol activity of seven other cytochrome b_{558}-positive patients [59]. Immunoblots of cytosol from these patients with cytochrome b_{558}-positive autosomal CGD revealed the absence of either the 47-kDa or the 67-kDa protein as the basis for their defective NADPH oxidase.

Taken together, the complementation studies [59] and immunochemical analysis [58] established that p47*phox* and p67*phox* are two of the cytosolic components of the assembled phagocyte NADPH oxidase and that the absence of one or the other accounts for most patients with autosomal CGD [60]. Left unanswered at that time, however, was the identity of a third cytosolic factor, referred to as NCF3, that was essential for optimal activity of the cell-free system [59]. Although neither p47*phox* nor p67*phox* directly binds GTP, the importance of GTP for optimal activity of the broken cell system inspired the discovery of small-molecular-weight GTPases Rac1 (in macrophages) [61–63] and Rac2 (in neutrophils) [64] as the required GTP-binding protein in phagocyte cytosol.

Combining recombinant p47*phox*, p67*phox*, and Rac1 with neutrophil membrane [65], purified cytochrome b_{558} [66] or recombinant cytochrome b_{558} [67] fully reconstituted NADPH oxidase activity, thus suggesting that all oxidase components had been identified. However, a patient's clinical problems once again prompted a refinement in understanding of the phagocyte oxidase. Matute et al. described a boy who presented with severe granulomatous colitis and abnormal phagocyte NADPH oxidase activity [68]. Neutrophils from the affected patient exhibited normal oxidase activation in response to soluble agonists such as phorbol myristate acetate or formyl-methionylleucylphenylalanine but failed to generate superoxide anion within phagosomes when fed particles. Defective phagosomal oxidants compromised antimicrobial action, as affected neutrophils failed to kill ingested *Staphylococcus aureus*. The patient expressed normal cytochrome b_{558}, p47*phox*, p67*phox*, and Rac2 in neutrophils but exhibited compound heterozygosity for two mutations in the gene encoding p40*phox*.

p40*phox* copurifies with p47*phox* and p67*phox* [69], incorporated into the ternary complex by virtue of its interaction with p67*phox* [69–71]. Like p47*phox*, p40*phox* possesses a PX (*phox* homology) domain, a protein motif that mediates association with phosphoinositides [72–74]. However, the PX domains of p47*phox* and p40*phox* preferentially bind to different lipids; p40*phox* binds to phosphatidyl inositol-3-phosphate [74–76], whereas the PX domain of p47*phox* associates with phosphatidic acid and phosphatidyl inositol [77]. The patient described by Matute et al. [68] has two different mutant alleles of p40*phox*; one results in a premature stop codon and lack of protein, whereas the other causes a missense mutation at a residue in the PX domain required for phospholipid binding. Thus, the failure of the oxidase to assemble on and deliver oxidants to the phagosome reflects differential binding of p40*phox* to distinct phospholipids in phagosomal membranes.

It is likely that the current list of factors that contribute to phagocyte NADPH oxidase activity is incomplete. Additional proteins copurify with cytochrome b_{558} or with the assembled oxidase, including Rap1a [66, 78–80] and 6-phosphofructo-2-kinase/fructose-2,6-bisphosphate [81, 82], and have been implicated in modulating oxidase activity, including cytoplasmic phospholipase 2Aα [83–86] and the calcium-binding proteins S100A8 and S100A9 [81, 87–91]. Additional study will determine how best to integrate these elements, as well as those yet to be identified, into our understanding of the mechanisms of phagocyte NADPH oxidase activity.

6 Regulation of the Phagocyte NADPH Oxidase

Translocation of the cytosolic components to cytochrome b_{558} in the plasma or phagosomal membrane converts a latent NADPH oxidase into an active enzyme [reviewed in [92]]. Segregation of the electron transporter in membranes and the two multicomponent protein complexes, Rac2-RhoGDI and p47*phox*-p67*phox*-p40*phox*, in cytoplasm maintains the oxidase unassembled and inactive in resting neutrophils. Upon exposure of neutrophils to agonists, the cytoplasmic complexes translocate to membrane, associate with cytochrome b_{558}, and complete assembly of a functional oxidase [93]. In studies in which phagocytosis is synchronized, translocation occurs within 30 s after the onset of phagocytosis [94]. Although the p47*phox*-p67*phox*-p40*phox* complex initially docks at the membrane in neutrophils from X-linked CGD patients lacking cytochrome b_{558}, sustained association at the plasma membrane requires normal cytochrome b_{558} [94]. Furthermore, p67*phox* fails to translocate in p47*phox*-deficient neutrophils, underscoring the role of p47*phox* as an adaptor protein [94]. Membrane association of cytosolic proteins requires heterodimeric cytochrome b_{558}, as p22*phox* expressed alone in the membrane of Epstein Barr-transformed B cells is insufficient [95].

Agonist-dependent phosphorylation of individual cytosolic oxidase components triggers conformational changes in the complexes, thereby exposing otherwise cryptic interactive domains that then associate with targets in the membrane. Phosphorylation of membrane components likewise contributes to assembly and stability of the active oxidase [reviewed in [96]]. Furthermore, changes in the lipid environment around cytochrome b_{558} alter its electron-transporting capacity and thus influence superoxide anion generation [87, 88, 97–100].

The biochemical changes that accompany oxidase assembly are many, complex, and beyond the scope of this brief review. However, the overall principles most relevant to the NOX protein family are two. First, p47*phox* and p40*phox* serve as *organizing* adaptor proteins that lack intrinsic catalytic activity but rather target the cytoplasmic ternary complex to plasma or phagosomal membrane, respectively. Second, p67*phox* serves as an *activating* factor that regulates FAD-mediated reduction of NADPH [101]. The homologues of p47*phox* and p67*phox* that interact with some of the other NOX protein family members are named NOXO1 and NOXA1, respectively, indicating their presumed contribution to the assembly (organization) and activation of their associated NOX proteins [102–104].

7 Summary

Unraveling the complexities of the neutrophil NADPH oxidase benefited from at least three features of human neutrophil biology. First, human neutrophils are abundant in circulation, are easily isolated from venous blood, express large amounts of cytochrome b_{558} in their membranes (6 to 10 pmols heme/10^6 neutrophils [105]), and generate readily measurable amounts of superoxide anion extracellularly when stimulated. For example, stimulated neutrophils can produce 10 nmols of superoxide anion/min/10^6 cells, which requires ~10^9 electrons/s/cell or a current of ~16 pA [106]. Taken together, these conditions have made isolation and characterization of the neutrophil NADPH oxidase a tractable biochemical undertaking. Second, human neutrophils express only one NOX protein family member, NOX2, and, moreover, possess few mitochondria. Consequently, there is little, if any, ambiguity as to the biochemical basis for oxygen consumption and superoxide anion production by stimulated human neutrophils. Third, perhaps as a corollary of having only one NOX protein, the absence of NOX2 from phagocytes creates a distinct clinical phenotype. With the exception of NOX3 deficiency from inner ear hair cells and vestibular dysfunction [107] and DUOX2 mutations and congenital hypothyroidism [108], examples of clinical manifestations of a deficiency of other NOX proteins in non-phagocytic cells, in humans or experimental models, are limited. However, regardless of the circumstances that favored discoveries related to the phagocyte NOX2, aspects of the neutrophil oxidase story will continue to prove instructive in the exploration of the biology and biochemistry of NOX protein family members in non-phagocytic cells, as illustrated, for example, by the recognition of NOXO1 and NOXA1. The challenges are to distinguish promising clues from misdirection and to forge new trails where the phagocyte system proves not to be a useful model.

Acknowledgments

The authors have been supported by grants from the National Institutes of Health [R01- AI132335 (WMN); R01-AI16546 (WMN); UL1-TR002645 (RAC); P30-AG044271 (RAC); R01-AI020866 (RAC)] and the Veterans Health Administration [I01 BX000513 (WMN); I01-BX000117 (RAC); I01-BX003157 (RAC)].

References

1. Cross AR, Jones OT (1991) Enzymic mechanisms of superoxide production. Biochim Biophys Acta 1057(3):281–298
2. Suh YA, Arnold RS, Lassegue B, Shi J, Xu X, Sorescu D, Chung AB, Griendling KK, Lambeth JD (1999) Cell transformation by the superoxide-generating oxidase Mox1. Nature 401:79–82
3. Banfi B, Maturana A, Jaconi S, Arnaudeau S, Laforge T, Sinha B, Ligeti E, Demaurex N, Krause KH (2000) A mammalian H + channel generated through alternative splicing of the NADPH oxidase homolog NOH-1. Science 287:138–142
4. Nauseef WM (2018) The Neutrophil NADPH Oxidase. In: Vissers M, Hampton MB, Kettle AJ (eds) Hydrogen Peroxide Metabolism in Health and Disease. CRC Press, Boca Raton, FL, pp 237–277
5. Baldridge CW, Gerard RW (1933) The extra respiration of phagocytosis. Am J Phys 103:235–236
6. Sbarra AJ, Karnovsky ML (1959) The biochemical basis of phagocytosis. I. Metabolic changes during the ingestion of particles by polymorphonuclear leukocytes. J Biol Chem 234:1355–1362
7. Iyer GYN, Islam DMF, Quastel JH (1961) Biochemical aspects of phagocytosis. Nature 192:535–541
8. Roberts J, Quastel JH (1964) Oxidation of reduced triphosphopyridine nucleotide by guinea pig polymorphonuclear leucocytes. Nature 202:85–86
9. Evans WH, Karnovsky ML (1961) A possible mechanism for the stimulation of some metabolic functions during phagocytosis. J Biol Chem 236:Pc30–Pc32
10. Karnovsky ML (1962) Metabolic basis of phagocytic activity. Physiol Rev 42:143–168
11. Iyer GYN, Quastel JH (1963) NADPH and NADH oxidation by guinea pig polymorphonuclear leukocytes. Can J Biochem Physiol 41:427–434
12. Rossi F, Zatti M (1964) Biochemical aspects of phagocytosis in polymorphonuclear leucocytes. NADH and NADPH oxidation by the granules of phagocytizing cells. Experientia 20:21–23
13. Rossi F, Zatti M (1964) Changes in the metabolic pattern of polymorphonuclear leucocytes during phagocytosis. Br J Exp Pathol 45:548–559
14. Gabig TG, Babior B (1979) The O_2^- -forming oxidase responsible for the respiratory burst in human neutrophils. J Biol Chem 254(18):9070–9074
15. Babior BM, Kipnes RS (1977) Superoxide-forming enzyme from human neutrophils: evidence for a flavin requirement. Blood 50:517–524
16. Gabig TG, Kipnes RS, Babior BM (1978) Solubilization of the O_2^--forming activity responsible for the respiratory burst in human neutrophils. J Biol Chem 253 (19):6663–6665
17. Babior BM, Kipnes RS, Curnutte JT (1973) Biological defence mechanisms. The production by leukocytes of superoxide, a potential bactericidal agent. J Clin Invest 52:741–744
18. Janeway CA, Craig J, Davidson M, Downey W, Gitlin D, Sullivan JC (1954) Hypergammaglobulinemia associated with severe recurrent and chronic non-specific infection. Am J Dis Child 88:388–392
19. Bridges RA, Berendes H, Good RA (1959) A fatal granulomatous disease of childhood. Am J Dis Child 97:387–408
20. Holmes B, Quie PG, Windhorst DB, Good RA (1966) Fatal granulomatous disease of childhood. Lancet 1:1225–1228
21. Holmes B, Page AR, Good RA (1967) Studies of the metabolic activity of leukocytes from patients with a genetic abnormality of phagocytic function. J Clin Invest 46:1422–1432
22. Baehner RL, Nathan DG (1968) Quantitative nitroblue tetrazolium test in chronic granulomatous disease. N Engl J Med 278:971–976
23. Baehner RL, Nathan DG (1967) Leukocyte oxidase: defective activity in chronic granulomatous disease. Science 155:835–836
24. Hattori H (1961) Studies on the labile, stable Nadi oxidase and peroxidase staining reactions in the isolated particles of horse granulocyte. Nagoya J Med Sci 23:362–378
25. Shinagawa Y, Tanaka C, Teroaka A, Shinagawa Y (1966) A new cytochrome in neutrophilic granules of rabbit leucocyte. J Biochem 59:622–624
26. Shinagawa Y, Tanaka C, Teroaka A (1966) Electron microscopic and biochemical study of the neutrophilic granules from leucocytes. J Electron Microsc 15:81–85
27. Segal AW, Jones OT, Webster D, Allison AC (1978) Absence of a newly described cytochrome b from neutrophils of patients with

chronic granulomatous disease. Lancet 2:446–449
28. Segal AW, Jones OTG (1978) Novel cytochrome b system in phagocytic vacuoles of human granulocytes. Nature 276:515–517
29. Gabig TG, Scbervish EW, Santinga JT (1982) Functional relationship of the cytochrome b to the superoxide-generating oxidase of human neutrophils. J Biol Chem 257 (8):4114–4119
30. Cross A (1999) The participation of the hemes of flavocytochrome b245 in the electron transfer process in NADPH oxidase. Blood 93:4449–4449
31. Dinauer MC, Orkin SH, Brown R, Jesaitis AJ, Parkos CA (1987) The glycoprotein encoded by the X-linked chronic granulomatous disease locus is a component of the neutrophil cytochrome b complex. Nature 327:717–720
32. Pember SO, Heyl BL, Kinkade JM Jr, Lambeth JD (1984) Cytochrome b558 from (bovine) granulocytes. Partial purification from triton X-114 extracts and properties of the isolated cytochrome. J Biol Chem 259:10590–10595
33. Bellavite P, Papini E, Zeni L, Della Bianca V, Rossi F (1985) Studies on the nature and activation of $O_2(-)$-forming NADPH oxidase of leukocytes. Identification of a phosphorylated component of the active enzyme. Free Radic Res Commun 1(1):11–29
34. Harper AM, Dunne MJ, Segal AW (1984) Purification of cytochrome b-245 from human neutrophils. Biochem J 219:519–527
35. Harper AM, Chaplin MF, Segal AW (1985) Cytochrome b-245 from human neutrophils is a glycoprotein. Biochem J 227:783–788
36. Lutter R, van Schaik MLJ, van Zwieten R, Wever R, Roos D, Hamers MN (1985) Purification and partial characterization of the b-type cytochrome from human polymorphonuclear leukocytes. J Biol Chem 260:2237–2244
37. Segal AW (1987) Absence of both cytochrome b-245 subunits from neutrophils in X-linked chronic granulomatous disease. Nature 326:88–91
38. Parkos CA, Allen RA, Cochrane CG, Jesaitis AJ (1988) The quaternary structure of the plasma membrane b-type cytochrome of human granulocytes. Biochim Biophys Acta 932(1):71–83
39. Parkos CA, Dinauer MC, Walker LE, Allen RA, Jesaitis AJ, Orkin SH (1988) Primary structure and unique expression of the 22-kilodalton light chain of human neutrophil cytochrome b. Proc Natl Acad Sci U S A 85:3319–3323
40. Segal AW, Cross AR, Garcia RC, Borregaard N, Valerius NH, Soothill JF, Jones OTG (1983) Absence of cytochrome b-245 in chronic granulomatous disease. A multicenter European evaluation of its incidence and relevance. N Engl J Med 308:245–251
41. Bohler MC, Seger RA, Mouy R, Vilmer E, Fischer A, Criscelli C (1986) A study of 24 patients with chronic granulomatous disease: a new classification by correlating respiratory burst, cytochrome b, and flavoprotein. J Clin Immunol 6:136–145
42. Ohno Y, Buescher ES, Roberts R, Metcalf JA, Gallin JI (1986) Reevaluation of cytochrome b and flavin adenine dinucleotide in neutrophils from patients with chronic granulomatous disease and description of a family with probable autosomal recessive inheritance of cytochrome b deficiency. Blood 67 (4):1132–1138
43. Curnutte JT, Berkow RL, Roberts RL, Shurin SB, Scott PJ (1988) Chronic granulomatous disease due to a defect in the cytosolic factor required for nicotinamide adenine dinucleotide phosphate oxidase activation. J Clin Invest 81:606–610
44. Parkos CA, Allen RA, Cochrane CG, Jesaitis AJ (1987) Purified cytochrome b from human granulocyte plasma membrane is comprised of two polypeptides with relative molecular weights of 91,000 and 22,000. J Clin Invest 80:732–742
45. Heyworth PG, Cross AR, Curnutte JT (2003) Chronic granulomatous disease. Curr Opin Immunol 15:578–584
46. DeLeo FR, Burritt JB, Yu L, Jesaitis AJ, Dinauer MC, Nauseef WM (2000) Processing and maturation of flavocytochrome b 558 include incorporation of heme as a prerequisite for heterodimer assembly. J Biol Chem 275(18):13986–13993
47. Segal AW, Jones OTG (1980) Absence of cytochrome b reduction in stimulated neutrophils from both female and male patients with chronic granulomatous disease. FEBS Lett 110:111–114
48. Hamers MN, deBoer M, Meerhof LJ, Weening RS, Roos D (1984) Complementation in monocyte hybrids revealing genetic heterogeneity in chronic granulomatous disease. Nature 307:553–555
49. DeChatelet LR, McCall CE, Shirley PS (1980) Activation of dialysis of NAD(P)H

oxidase(s) from human neutrophils. J Reticuloendothel Soc 28(6):533–545
50. Heynemann RA, Vercauteren RE (1984) Activation of a NADPH-dependent oxidase from horse polymorphonuclear leukocytes in a cell-free system. J Leukoc Biol 36:751–759
51. Bromberg Y, Pick E (1984) Unsaturated fatty acids stimulate NADPH-dependent superoxide production by cell-free system derived from macrophages. Cell Immunol 88:213–221
52. Curnutte JT (1985) Activation of human neutrophil nicotinamide adenine dinucleotide phosphate, reduced (triphosphopyridine nucleotide, reduced) oxidase by arachidonic acid in a cell-free system. J Clin Invest 75:1740–1743
53. McPhail LC, Shirley PS, Clayton CC, Snyderman R (1985) Activation of the respiratory burst enzyme from human neutrophils in a cell-free system: evidence for a soluble cofactor. J Clin Invest 75:1735–1739
54. Clark RA, Leidal KG, Pearson DW, Nauseef WM (1987) NADPH oxidase of human neutrophils: subcellular localization and characterization of an arachidonate-activatable superoxide- generating system. J Biol Chem 262:4065–4074
55. Seifert R, Rosenthal W, Schultz G (1986) Guanine nucleotides stimulate NADPH oxidase in membranes of human neutrophils. FEBS Lett 205(1):161–165
56. Gabig TG, English D, Akard LP, Schell MJ (1987) Regulation of neutrophil NADPH oxidase activation in a cell-free system by guanine nucleotides and fluoride. J Biol Chem 262:1685–1690
57. Ligeti E, Doussiere J, Vignais PV (1988) Activation of the $O_2(^-)$-generating oxidase in plasma membrane from bovine polymorphonuclear neutrophils by arachidonic acid, a cytosolic factor of protein nature, and nonhydrolyzable analogues of GTP. Biochemistry 27(1):193–200
58. Volpp BD, Nauseef WM, Clark RA (1988) Two cytosolic neutrophil NADPH oxidase components absent in autosomal chronic granulomatous disease. Science 242:1295–1298
59. Nunoi H, Rotrosen D, Gallin JI, Malech HL (1988) Two forms of autosomal chronic granulomatous disease lack distinct neutrophil cytosol factors. Science 242:1298–1301
60. Clark RA, Malech HL, Gallin JI, Nunoi H, Volpp BD, Pearson DW, Nauseef WM, Curnutte JT (1989) Genetic variants of chronic granulomatous disease: prevalence of deficiencies of two cytosolic components of the NADPH oxidase system. N Engl J Med 321:647–652
61. Abo A, Pick E, Hall A, Totty N, Teahan CG, Segal AW (1991) Activation of the NADPH oxidase involves the small GTP-binding protein p21 rac1. Nature 353:668–670
62. Pick E, Kroizman T, Abo A (1989) Activation of the superoxide-forming NADPH oxidase of macrophages requires two cytosolic components-one of them is also present in certain nonphagocytic cells. J Immunol 143 (12):4180–4187
63. Sha'ag D, Pick E (1990) Nucleotide binding properties of cytosolic components required for expression of activity of the superoxide generating NADPH oxidase. Biochim Biophys Acta 1037(3):405–412
64. Knaus UG, Heyworth PG, Evans T, Curnutte JT, Bokoch GM (1991) Regulation of phagocyte oxygen radical production by the GTP-binding protein Rac 2. Science 254:1512–1515
65. Fuchs A, Dagher MC, Jouan A, Vignais PV (1994) Activation of the $O_2(-)$-generating NADPH oxidase in a semi-recombinant cell-free system. Assessment of the function of Rac in the activation process. Eur J Biochem 226 (2):587–595
66. Abo A, Boyhan A, West I, Thrasher AJ, Segal AW (1992) Reconstitution of neutrophil NADPH oxidase activity in the cell-free system by four components: p67- phox, p47-phox, p21 rac 1, and cytochrome b −245. J Biol Chem 267:16767–16770
67. Rotrosen D, Yeung CL, Katkin JP (1993) Production of recombinant cytochrome b558 allows reconstitution of the phagocyte NADPH oxidase solely from recombinant proteins. J Biol Chem 268:14256–14260
68. Matute JD, Arias AA, Wright NAM, Wrobel I, Waterhouse CCM, Li XJ, archal CC, Stull ND, Lewis DB, Steele M, Kellner JD, Yu W, Meroueh SO, Nauseef WM, Dinauer MC (2009) A new genetic subgroup of chronic granulomatous disease with autosomal recessive mutations in p40 phox and selective defects in neutrophil NADPH oxidase activity. Blood 114(15):3309–3315
69. Wientjes FB, Hsuan JJ, Totty NF, Segal AW (1993) p40phox, a third cytosolic component of the activation complex of the NADPH oxidase to contain src homology 3 domains. Biochem J 296:557–561
70. Someya A, Nagaoka I, Yamashita T (1993) Purification of the 260 kDa cytosolic complex involved in the superoxide production of

guinea pig neutrophils. FEBS Lett 330:215–218

71. Tsunawaki S, Mizunari H, Nagata M, Tatsuzawa O, Kuratsuji T (1994) A novel cytosolic component, p40 phox, of respiratory burst oxidase associates with p67 phox and is absent in patients with chronic granulomatous disease who lack p67 phox. Biochem Biophys Res Commun 199:1378–1387
72. Ponting CP (1996) Novel domains in NADPH oxidase subunits, sorting nexins, and PtdIns 3-kinases: Binding partners of SH3 domains. Protein Sci 5:2353–2357
73. Wishart M, Taylor G, Dixon J (2001) Phoxy lipids: revealing PX domains as phosphoinositide binding modules. Cell 105:817–820
74. Bravo J, Karathanassis D, Pacold CM, Pacold ME, Ellson CD, Anderson KE, Butler PJ, Lavenir I, Perisic O, Hawkins PT, Stephens L, Williams RL (2001) The crystal structure of the PX domain from p40(phox) bound to phosphatidylinositol 3-phosphate. Mol Cell 8(4):829–839
75. Ago T, Takeya R, Hiroaki H, Kuribayashi F, Ito T, Kohda D, Sumimoto H (2001) The PX domain as a novel phosphoinositide-binding module. Biochem Biophys Res Commun 287 (3):733–738
76. Ellson C, Gobert-Gosse S, Anderson K, Davidson K, Erdjument-Bromage H, Tempst P, Thuring J, Cooper M, Lim ZY, Holmes A, Gaffney P, Chilvers E, Hawkins P, Stephens L (2001) PtdIns(3)P regulates the neutrophil oxidase complex by binding to the PX domain of p40 phox. Nat Cell Biol 3:679–682
77. Karathanassis D, Stahelin RV, Bravo J, Perisic O, Pacold CM, Cho WW, Williams RL (2002) Binding of the PX domain of p47 phox to phosphatidylinositol 3,4-bisphosphate and phosphatidic acid is masked by an intramolecular interaction. EMBO J 21(19):5057–5068
78. Quinn MT, Parkos CA, Walker L, Orkin SH, Dinauer MC, Jesaitis AJ (1989) Association of a ras-related protein with cytochrome b of human neutrophils. Nature 342:198–200
79. Knoller S, Shpungin S, Pick E (1991) The membrane-associated component of the amphiphile-activated, cytosol-dependent superoxide-forming NADPH oxidase of macrophages is identical to cytochrome b 559. J Biol Chem 266:2795–2804
80. Li Y, Yan J, De P, Chang HC, Yamauchi A, Christopherson KW 2nd, Paranavitana NC, Peng X, Kim C, Munugalavadla V, Kapur R, Chen H, Shou W, Stone JC, Kaplan MH, Dinauer MC, Durden DL, Quilliam LA (2007) Rap1a null mice have altered myeloid cell functions suggesting distinct roles for the closely related Rap1a and 1b proteins. J Immunol 179(12):8322–8331
81. Paclet MH, Berthier S, Kuhn L, Garin J, Morel F (2007) Regulation of phagocyte NADPH oxidase activity: identification of two cytochrome b 558 activation states. FASEB J 21:1244–1255
82. Baillet A, Hograindleur MA, El Benna J, Grichine A, Berthier S, Morel F, Paclet MH (2016) Unexpected function of the phagocyte NADPH oxidase in supporting hyperglycolysis in stimulated neutrophils: key role of 6-phosphofructo-2-kinase. FASEB J 31 (2):663–673. https://doi.org/10.1096/fj.201600720R
83. Dana R, Leto TL, Malech HL, Levy R (1998) Essential requirement of cytosolic phospholipase A2 for activation of the phagocyte NADPH oxidase. J Biol Chem 273 (1):441–445
84. Li Q, Cathcart MK (1997) Selective inhibition of cytosolic phospholipase A 2 in activated human mononcytes. J Biol Chem 272 (4):2404–2411
85. Bae Y, Kim Y, Kim J, Lee T, Kim Y, Suh P, Ryu S (2000) Independent functioning of cytosolic phospholipase A 2 and phospholipase D 1 in Trp-Lys-Tyr-Met-Val-D-Met induced superoxide generationin human monocytes. J Immunol 164:4089–4096
86. Rubin BB, Downey GP, Koh A, Degousee N, Ghomashchi F, Nallan L, Stefanski E, Harkin DW, Sun C, Smart BP, Lindsay TF, Cherepanov V, Vachon E, Kelvin D, Sadilek M, Brown GE, Yaffe MB, Plumb J, Grinstein S, Glogauer M, Gelb MH (2005) Cytosolic phospholipase A 2- α is necessary for platelet-activating factor biosynthesis, efficient neutrophil-mediated bacterial killing, and the innate immune response to pulmonary infect. cPLA2- α does not regulate neutrophil nadph oxidase activity. J Biol Chem 280(9):7519–7529
87. Berthier S, Nguyen MVC, Baillert A, Hograindleur MA, Paclet MH, Polack B, Morel F (2012) Molecular interface of S100A8 with cytochrome b 558 and NADPH oxidase activation. PLoS One 7(7): e40277
88. Berthier S, Paclet MH, Lerouge S, Roux F, Vergnaud S, Coleman AW, Morel F (2003) Changing the conformation state of cytochrome b558 initiates NADPH oxidase activation. J Biol Chem 278(28):25499–25508

89. Doussiere J, Bouzidi F, Vignais PV (2002) The S100A8/A9 protein as a partner for the cytosolic factors of NADPH oxidase activation in neutrophils. Eur J Biochem 269:3246–3255
90. Kerkhoff C, Nacken W, Benedyk M, Dagher MC, Sopalla C, Doussiere J (2005) The arachidonic acid-binding protein S100A8/A9 promotes NADPH oxidase activation by interaction with p67phox and Rac-2. FASEB J 19(3):467–469
91. Taylor RM, Riesselman MH, Lord CI, Gripentrog JM, Jesaitis AJ (2012) Anionic lipid-induced conformational changes in human phagocyte flavocytochrome b precede assembly and activation of the NADPH oxidase complex. Arch Biochem Biophys 52 (1–2):24–31
92. Nauseef WM (2004) Assembly of the phagocyte NADPH oxidase. Histochem Cell Biol 122(4):277–291
93. Heyworth PG, Curnutte JT, Nauseef WM, Volpp BD, Pearson DW, Rosen H, Clark RA (1991) Neutrophil NADPH oxidase assembly. Membrane translocation of p47- phox and p67-phox requires interaction between p47-phox and cytochrome b558. J Clin Invest 87:352–356
94. Allen LAH, DeLeo FR, Gallois A, Toyoshima S, Suzuki K, Nauseef WM (1999) Transient association of the nicotinamide adenine dinucleotide phosphate oxidase subunits p47 phox and p67 phox with phagosomes in neutrophils from patients with X-linked chronic granulomatous disease. Blood 93:3521–3530
95. Dusi S, Nadalini KA, Donini M, Zentilin L, Wientjes FB, Roos D, Giacca M, Rossi F (1998) Nicotinamide-adenine dinucleotide phosphate oxidase assembly and activation in EBV-transformed B lymphoblastoid cell lines of normal and chronic granulomatous disease patients. J Immunol 161:4968–4974
96. Belambri SA, Rolas L, Raad H, Hurtado-Nedelec M, Dang PM, El-Benna J (2018) NADPH oxidase activation in neutrophils: role of the phosphorylation of its subunits. Eur J Clin Investig 48(Suppl 2):e12951. https://doi.org/10.1111/eci.12951
97. Brechard S, Plancon S, Tschirhart EJ (2013) New insights into the regulation of neutrophil NADPH oxidase activity in the phagosome: a focus on the role of lipid and Ca(2+) signaling. Antioxid Redox Signal 18(6):661–676. https://doi.org/10.1089/ars.2012.4773
98. Foubert TR, Burritt JB, Taylor RM, Jesaitis AJ (2002) Structural changes are induced in human neutrophil cytochrome b by NADPH oxidase activators, LDS, SDS, and arachidonate: intermolecular resonance energy transfer between trisulfopyrenyl-wheat germ agglutinin and cytochrome b 558. Biochim Biophys Acta 78380:1–11
99. Shao DM, Segal AW, Dekker LV (2003) Lipid rafts determine efficiency of NADPH oxidase activation in neutrophils. FEBS Lett 550 (1–3):101–106
100. Jin S, Zhou F, Katirai F, Li PL (2011) Lipid raft redox signaling: molecular mechanisms in health and disease. Antioxid Redox Signal 15 (4):1043–1083
101. Nisimoto Y, Motalebi S, Han CH, Lambeth JD (1999) The p67 phox activation domain regulates electron flow from NADPH to flavin in flavocytochrome b 558. J Biol Chem 274(33):22999–23005
102. Bánfi B, Clark RA, Steger K, Krause KH (2003) Two novel proteins activate superoxide generation by the NADPH oxidase NOX1. J Biol Chem 278(6):3510–3513
103. Cheng G, Ritsick D, Lambeth JD (2004) Nox3 regulation by NOXO1, $p47^{phox}$ and $p67^{phox}$. J Biol Chem 279(33):34250–34255
104. Geiszt M, Lekstrom K, Witta J, Leto TL (2003) Proteins homologous to $p47^{phox}$ and $p67^{phox}$ support superoxide production by NAD(P)H oxidase 1 in colon epithelial cells. J Biol Chem 278(22):20006–20012
105. Segal AW, Harper AM, Cross AR, Jones OT (1986) Cytochrome b-245. Methods Enzymol 132:378–394
106. Murphy R, DeCoursey TE (2006) Charge compensation during the phagocyte respiratory burst. Biochim Biophys Acta 1757:996–1011
107. Paffenholz R, Bergstrom RA, Pasutto F, Wabnitz P, Munroe RJ, Jagla W, Heinzmann Y, Marquardt A, Bareiss A, Laufs J, Russ A, Stumm G, Schimenti JC, Bergstrom DE (2004) Vestibular defects in head-tilt mice result from mutations in Nox3, encoding an NADPH oxidase. Genes Dev 18:486–491
108. Moreno JC, Bikker H, Kempers MJE, Van Trotsenburg ASP, Baas F, De Vijlder JJM, Vulsma T, Ris-Stalpers C (2002) Inactivating mutations in the gene for thyroid oxidase 2 (thox2) and congenital hypothyroidism. N Engl J Med 347(2):95–102

Chapter 2

Mammalian NADPH Oxidases

Hélène Buvelot, Vincent Jaquet, and Karl-Heinz Krause

Abstract

Reactive oxygen species (ROS) are highly reactive oxygen derivatives. Initially, they were considered as metabolic by-products (of mitochondria in particular), which consistently lead to aging and disease. Over the last decades, however, it became increasingly apparent that virtually all eukaryotic cells possess specifically ROS-producing enzymes, namely, NOX NADPH oxidases. In most mammals, there are seven NOX isoforms: three closely related isoforms, NOX1, 2, 3, which are activated by cytoplasmic subunits; NOX4, which appears to be constitutively active; and the EF-hand-containing Ca^{2+}-activated isoforms NOX5 and DUOX1 and 2. Loss-of-function mutations in *NOX* genes can lead to serious human disease. NOX2 deficiency leads to primary immune deficiency, while DUOX2 deficiency presents as congenital hypothyroidism. Nox-deficient mice provide important tools to explore the physiological functions of various NADPH oxidases as a loss of function in Nox2, Nox3, and Duox2 leads to a spontaneous phenotype. The genetic absence of Nox1, Nox4, and Duox1 does not result in an obvious mouse phenotype (the *NOX5* gene is absent in rodents and can therefore not be studied using knockout mice). Since the discovery of the NOX family at the turn of the millennium, much progress in understanding the biochemistry and the physiology of NOX has been made; however many questions remain unanswered to date. This chapter is an overview of our present knowledge on mammalian NOX/DUOX enzymes.

Key words NADPH oxidase, Reactive oxygen species, Redox signaling, Genetic deficiency, Mouse models

1 Introduction

In 1908, Otto Warburg investigated oxygen consumption during fertilization of sea urchin eggs [1]. He observed an increase in oxygen consumption and assumed that it was due to increased mitochondrial respiration. With the benefit of retrospection, we now know that the observed respiratory burst was not due to mitochondrial activity but rather to Udx1, an NADPH oxidase that generates hydrogen peroxide (H_2O_2) necessary for the cross-linking of the proteins found throughout the fertilization envelope. A similar extra-respiratory burst was also observed by Baldridge and Gerard in phagocytes [2] and by MacLeod in spermatocytes [3]. Few years later, Iyer, Islam, and Quastel

Ulla G. Knaus and Thomas L. Leto (eds.), *NADPH Oxidases: Methods and Protocols*, Methods in Molecular Biology, vol. 1982, https://doi.org/10.1007/978-1-4939-9424-3_2,

showed that phagocytosis results in the generation of H_2O_2 [4]. Finally, Rossi and Zatti identified that this H_2O_2-producing system used NADPH as a substrate [5]. In 1978, Segal and Jones discovered a membrane cytochrome *b* incorporated in phagocytic vacuoles that was missing in four patients suffering from X-linked chronic granulomatous disease (CGD) [6]. A purified cytochrome *b* complex showed functional oxidase activity, suggesting that it is a key component of the phagocyte NADPH oxidase [7]. The gene coding for this phagocyte NADPH oxidase was cloned for the first time in 1986, based on its chromosomal location on the X chromosome [8].

Nowadays, all human sequences of NADPH oxidases have been identified. They represent a family of seven different enzymatic isoforms, namely, NOX1–5 and DUOX1–2. These enzymes use NADPH as an electron donor to reduce oxygen to superoxide radical anion ($O_2{\bullet}^-$) according to the reaction below:

$$\mathrm{NADPH} + 2\mathrm{O}_2 \leftrightarrow \mathrm{NADP}^+ + 2\mathrm{O}_2{}^{\bullet-} + \mathrm{H}^+$$

$O_2{\bullet}^-$ is in turn converted to H_2O_2 either spontaneously or enzymatically [9]. The catalytic core, which contains six (seven for DUOX1–2) transmembrane α-helical domains, shares structural similarities between all isoforms with NADPH-binding sites and a FAD-binding site at the cytosolic COOH terminus [7]. NOX family members differ in their NH_2-terminal structure, subunits, and regulatory proteins (Fig. 1). Reactive oxygen species (ROS) generated by NADPH oxidases are not a by-product of metabolism but serve specific biochemical functions, as exemplified by the crosslinking of the proteins in the fertilization envelope of the sea urchin. It has now become clear that NADPH oxidases are a family of enzymes with a wide distribution throughout all families of eukaryotic organisms [10], and recent results suggest that NADPH oxidases might even be found in bacteria [11].

2 The NOX Family Members

2.1 NOX2

The phagocyte NADPH oxidase NOX2 is the first identified and most studied NOX isoform. Our knowledge about the topography and the structure of the other members of NOX family is mostly derived from studies on NOX2.

Descriptions of the phagocyte NADPH oxidase dates back to the middle of the twentieth century. Because of its high level of expression in neutrophils and macrophages, NOX2 is referred to as "phagocyte NADPH oxidase." However, NOX2 is also found in other cells such as cardiomyocytes [12], skeletal muscle cells [13], hepatocytes [14], endothelial cells [15–17], and different types of stem cells [18, 19]. The functional phagocyte NADPH oxidase is composed of six subunits (Fig. 1a), namely, gp91^{phox} (also referred to as NOX2), p22^{phox}, p47^{phox}, p67^{phox}, p40^{phox}, and Rac. The gp91^{phox} subunit represents the catalytic core of the enzyme and is directly involved in the electron transport, while p22^{phox} plays the role of stabilizer of NOX2 and binds to the organizer subunit, p47^{phox}. The p67^{phox} subunit acts as an activator subunit, p40^{phox} is a regulator subunit, and Rac is a GTP-binding protein [10]. In phagocytes, NOX2 is present in both intracellular organelles and plasma membrane. In resting neutrophils, the enzymatic complex is in an inactive form with gp91^{phox} and p22^{phox} located in the membrane of secondary granules, whereas p47^{phox}, p67^{phox}, and p40^{phox} are complexed in the cytosol, similar to the cytosolic Rac-GDP/RhoGDI complex [20]. Upon activation during phagocytosis, the NOX2 complex is assembled through membrane fusion of the granules with the phagosome and association of the cytosolic subunits. Rac is activated and translocates to the phagosome [21], where it binds to p67^{phox} in a GTP-dependent manner [22]. The assembly of the complex leads to the production of $O_2 \bullet^-$. In case of mutation in any of the subunit of the NOX2 complex, the production of ROS in the phagosome is decreased, and patients are suffering from CGD. CGD is a rare hereditary phagocyte dysfunction and is characterized by severe and recurrent infections with catalase-positive pathogens (*Staphylococcus aureus* and *Aspergillus* are most common) [23]. Patients suffering from CGD also demonstrate a hyperinflammatory state that can manifest with aseptic granulomas, colitis, or periodontitis [24].

The severe infections in CGD patients demonstrate that NOX2 is crucial for host defense. ROS participate in the defense against infection by two different mechanisms. The first mechanism is a direct role of ROS in killing. Indeed, high concentrations of ROS are toxic for cells and destroy microorganisms by targeting DNA, proteins, and lipids [25]. ROS damage DNA through oxidation of nucleotides, DNA-protein cross-linking, and single- and double-strand rupture [26]. Disruption of essential iron-sulfur clusters by ROS affects enzymatic activity [27] and leads to bacterial death [28]. The second mechanism is an indirect role of ROS in killing. NOX2 is also involved in many immune responses and in inflammation including regulation of the phagosomal pH [24, 29], antigen presentation by dendritic cells [30, 31], NETosis [32, 33], neutrophil apoptosis [34], cytokine release [35], and antibacterial autophagy [36, 37]. An indirect role of ROS in bacterial killing might be explained by a modification of phagocyte intracellular signaling [38].

Different types of Nox2 complex mutations exist in mice. Huang et al. identified mice with a spontaneous loss of function of p47^{phox} [39]. Mutant mice with loss of function in p22^{phox} have also been described [40]. Homologous recombination was used to generate knockout in gp91^{phox} [41], p40^{phox} [42], p47^{phox} [43],

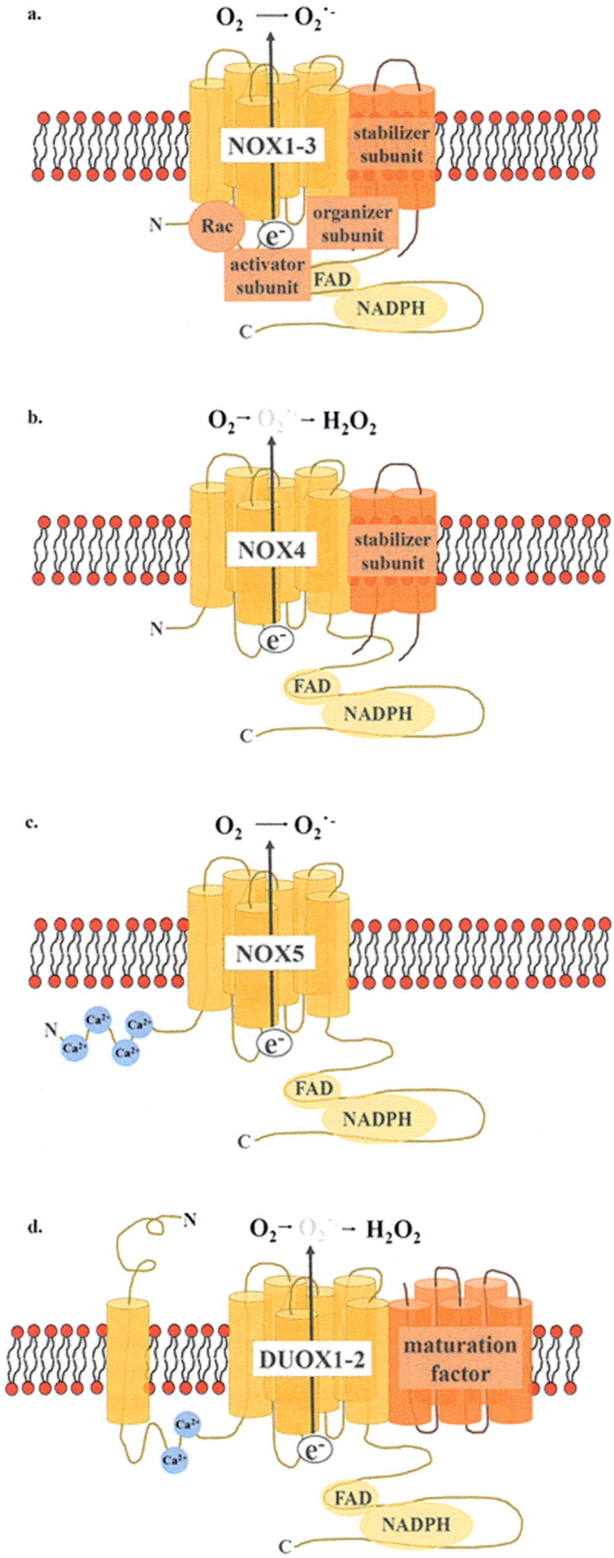

Fig. 1 NADPH oxidase isoforms and associated subunits. The catalytic core of NOX isoforms contains six transmembrane domains and a cytosolic C-terminal dehydrogenase domain that contains FAD cofactor and NADPH-binding sites. (**a**)

and p67^{phox} [44]. Mouse models of CGD show severe infections [41, 45, 46] and hyperinflammatory lesions [47] and thus have a good correlation with the human pathology. In contrast, rats with loss of function in p22^{phox} and in p47^{phox} have demonstrated a different phenotype. Indeed, p22^{phox} mutant rats have systemic hypereosinophilia, neutrophilia, and IgM and IgA hypergammaglobulinemia [48] but have no increased sensitivity to infection with *Staphylococcus aureus.* On the other hand, p47^{phox} mutant rats show increased sensitivity to collagen-induced arthritis (infection susceptibility has not been investigated in this model) [49].

The role of NOX2 in host defense is well described. Yet, the underlying mechanisms are not fully understood. The killing of microorganism by NOX2 is most likely not simply due to the toxicity of ROS but also due to the modification of intracellular signaling in phagocytes by NOX2-derived ROS. There is also increasing evidence that NOX2 acts as an important immune modulator. NOX2 influences both innate and adaptive immune responses. Indeed, CGD patients present a higher level of pro-inflammatory cytokines and IgG than the general population. Therefore, NOX2 is crucial not only for microbial killing but also for the cross talk between innate and adaptive immunities in order to generate an efficient immune response.

2.2 NOX1

The NOX1 isoform exhibits ~60% sequence identity at the protein level with NOX2 and is also located on the X chromosome. NOX1 is mainly expressed in colon epithelium [50–52] but can also be found in other tissues and cell types, such as smooth muscle cells [53], endothelial cells [54, 55], and tissues such as the uterus [51], placenta [56], and prostate [51]. NOX1 is composed of five subunits (Fig. 1a), i.e., NOX1, p22^{phox}, NOXO1, NOXA1, and Rac. Like the phagocyte NADPH oxidase, NOX1 plays the role of the

Fig. 1 (continued) NOX1–3 activity is dependent on a stabilizer subunit (p22^{phox}), organizer subunits (p47^{phox} or NOXO1), activator subunits (p67^{phox} or NOXA1), as well as a GTP-binding protein (Rac1 or Rac2). The NOX2 regulator subunit, p40^{phox}, which is absent in NOX1 and NOX3 complexes, is not represented in the figure. (**b**) NOX4 activity does not depend on cytosolic subunits and is composed of only two proteins, the catalytic NOX4 unit and the stabilizer subunit p22^{phox}. NOX4 activity is predominantly detected as H_2O_2. (**c**) NOX5 activation does not require a cytosolic subunit and is directly dependent on the binding of intracellular calcium on EF-hand domains. (**d**) DUOX isoforms are composed of seven transmembrane domains, and their activation is dependent on the binding of intracellular calcium on EF-hand domains. DUOX1–2 enzymes require maturation factors (DUOXA1–2) for proper translocation from the endoplasmic reticulum to the plasma membrane. Unlike other NOXs, DUOX isoforms contain an extracellular peroxidase-like domain in the N-terminal region. As in the case of NOX4, H_2O_2 is formed by DUOX1–2 catalytic core for electron transport, p22^{phox} acts as a stabilizer and binding subunit, NOXO1 is the organizer subunit, NOXA1 is the activator subunit, and Rac serves again as activating GTP-binding protein [50].

Several factors can induce the expression of NOX1 in the vascular smooth muscle, such as platelet-derived growth factor (PDGF), prostaglandin $F_{2\alpha}$, and angiotensin II [57–59]. In contrast, in renal mesangial cells, the expression of NOX1 is down-regulated by nitric oxide [60]. Nox1-deficient mice were first documented by Matsuno and colleagues [61] and Gavazzi et al. [62].

Following infusion with angiotensin II, Nox1 knockout mice showed a markedly attenuated blood pressure response [61, 62]. When studying a large number of mice, Gavazzi et al. also observed an effect on basal blood pressure [62]. Nox1-deficient mice do not present with an obvious specific spontaneous phenotype. However, Coant et al. demonstrated that progenitor cells in the colon of Nox1-deficient mice were converted into goblet cells in a higher proportion than that in wild-type mice and the number of colonocytes was lower [63]. Further, they proposed that Nox1-derived ROS in the colon might influence cell fate through a redox regulation of phosphatase and tensin homolog (PTEN) and NF-κB [63].

The presence of NOX1 in the colon suggests a role in host defense [64], and loss of function in children predisposes to very early-onset inflammatory bowel disease (VEOIBD) [65, 66], an inflammatory bowel disease that occurs in children less than 6 years old. In contrast to humans, no obvious signs of intestinal inflammation were observed in Nox1-deficient mice.

NOX1-derived ROS are likely to contribute to cell signaling. A key ROS target is the family of peroxiredoxins, which are cellular thiol peroxidases containing cysteine residues that are highly sensitive to H_2O_2 oxidation. Peroxiredoxins play a role in elimination of intracellular H_2O_2 produced in the cell, but they also act as a H_2O_2 signal receptor and transmitter in redox regulation of transcription factors [67]. Peroxiredoxins enable protein thiol oxidation by relaying H_2O_2-derived oxidizing equivalents to other proteins [68]. Peroxiredoxin 6 regulates NOX1 activity by acting as a stabilizer of NOXA1 [69]. In addition, NOX1 was shown to generate H_2O_2, controlling the activity of peroxiredoxin 1 [70].

A role for NOX1 in the intestinal immune response was initially suspected given the very high level of expression of NOX1 in the gut epithelia and the high concentration of bacteria in the gut lumen. However, to our knowledge there are no reports suggesting increased rate or severity of intestinal infections in Nox1-deficient mice. An alternative hypothesis would be the involvement of NOX1 in the homeostasis of the gut microbiota. A recent study demonstrating that H_2O_2 released by the intestinal epithelial cells alters tyrosine signaling in gut bacteria provides first evidence for such a concept [71].

2.3 NOX3

NOX3 shares ~56% of amino acid identity with NOX2. This enzymatic complex is highly expressed in the inner ear, including the cochlear and vestibular sensory epithelia [72]. As for NOX1, the NOX3 enzymatic complex is composed of NOX3, p22^{phox}, NOXO1, NOXA1, and Rac (Fig. 1a). Unlike NOX2, which is completely inactive in the absence of stimulation, NOX3 has a constitutive activity in the absence of regulators. However, this activity is dependent on p22^{phox} and can be enhanced by the organizer subunit NOXO1 [73]. Yet, the reason why constitutive ROS generation in the inner ear is required is not clear, and more in vivo studies are needed to clarify this.

The role of Nox3 was identified by reverse genetics on "head-tilt" mutant mice with a vestibular defect which had an underlying mutation in the *Nox3* gene [74]. A mutation in this gene leads to the lack of otoconia formation and to vestibular dysfunction. Head-tilt mutant mice demonstrated severe balance and spatial orientation defect [74]. As expected, mice with a mutation in Noxo1 also have a head-slant phenotype, and p22^{phox}-mutant mice exhibit balance defect [40, 75]. Because of its presence in other organs, the p22^{phox} mutation leads also to severe immunodeficiency due to the lack of Nox2 activity [40]. Loss of function in the Noxa1 subunit does not cause balance defects [76], suggesting that this subunit might not be essential for the activation of Nox3 in mice.

The role of Nox3 in the inner ear is well defined in rodents. Although mutations affecting the Nox3 system in mice have a clear impact on inner ear function, no such mutations have been detected in humans. Mutation in the gene coding for p22^{phox} exists in humans and should lead to an inactive NOX3 complex, but it is not linked to balance defects. It has been suggested that this might be due to sensorial compensation by the central nervous system [40]. Mutations in the Nox3 system do not affect normal hearing in mice as hearing tests did not show any impairment in head-tilt mutant mice [74], neither in p22^{phox} nor Noxo1 mutants [40, 75]. Yet, Lavinsky et al. reported that Nox3 mutants have a greater susceptibility to noise-induced hearing loss [77].

2.4 NOX4

NOX4 shares ~39% of amino acid identity with NOX2. NOX4 is highly expressed in the kidney [78, 79] but can also be found in osteoclasts [80, 81], endothelial cells [54, 82, 83], smooth muscle cells [84–88], hematopoietic stem cells [19], keratinocytes [89], and melanoma cells [90]. The NOX4 enzymatic complex is composed of only two proteins (Fig. 1b), namely, NOX4 and p22^{phox}, and its activity is not dependent on cytosolic subunits [91, 92]. Some evidence suggests that another protein (Poldip 2) may regulate NOX4 activation [93]. NOX4 activity is predominantly detected as H_2O_2 [94]. It has been proposed that an extracellular loop of NOX4 contains superoxide dismutase-like activity [95] and accelerates the spontaneous dismutation of $O_2 \bullet^-$ to form H_2O_2. This is supported by the fact that Nisimoto et al. observed that the electron flux through purified NOX4 forms 10% $O_2 \bullet^-$ and 90% H_2O_2 [91].

Several stimuli lead to upregulation of NOX4 expression, such as hypoxia [96], hyperoxia [97], hyperglycemia [98], shear stress [99], or transforming growth factor β1 (TGFβ1) [100]. There are conflicting data about the impact of NOX4 in disease models. Under certain conditions, NOX4 has a protective effect, for example, by promoting endothelial angiogenesis during ischemia [101, 102]. On the other hand, an overproduction of NOX4-derived ROS is associated with atherosclerosis [103] and the severity of stroke [104] and favors aortic aneurysm in Marfan syndrome [105]. To date, no spontaneous NOX4 loss-of-function mutations in humans or other mammals have been described.

When NOX4 is expressed, it generates H_2O_2 constitutively and has been implicated in cell death and cellular senescence. As discussed above, ROS can damage DNA, lipids, and proteins, leading to cellular death. ROS can induce cell apoptosis through other mechanisms, such as activation of MAP kinase [106], or, at higher concentration, ROS can inhibit caspases, which will lead to a switch from apoptosis to necrosis [107, 108]. Yet, in some circumstances, NOX-derived ROS may act as anti-apoptotic signal through activation of the Akt/ASK1 pathway [109]. In conclusion, ROS is most often associated with cell death, but, in some cases, it may act as an anti-apoptotic agent. Several reasons can explain these apparently conflicting observations including (1) the magnitude and duration of the ROS signal, (2) the subcellular localization of the NOX isoform, (3) the set of redox-sensitive signaling targets expressed in the cell, and (4) the metabolism of $O_2 \bullet^-$ (possibly anti-apoptotic) versus H_2O_2 (pro-apoptotic through conversion to hydroxyl anion or similar reactive species) [110].

Cellular senescence [111], i.e., the exit of aging cells from the cell cycle, is a cellular response to oxidative stress. Overexpression of NOX4 leads to cellular senescence [78]. Cellular senescence is not accompanied by cell death but refers to a state of nondividing cells which are rather apoptosis-resistant [112, 113]. Although the role of cellular senescence in the aging of mammalian organisms is still a matter of debate, it should be noted that there is ample evidence that ROS are involved in the aging process [114–116].

Oxygen sensing is a complex mechanism crucial for cellular function. NOX, in particular NOX4, might be involved in oxygen sensing, because NOX4 has a high Km for oxygen similar to the values of known oxygen-sensing enzymes [91]. Indeed, H_2O_2 leads to an increase in the HIF1α protein, a key sensor of hypoxia

[117]. An involvement of NOX4 in this process is somehow counterintuitive, as hypoxic conditions will limit the oxygen substrate necessary for NOX-derived catalysis of molecular oxygen. However, there is evidence that NOX4 activity can be increased under conditions of moderate (but not complete) hypoxia [96]. HIF1α can bind to the NOX4 promoter and thereby increase NOX4 expression, raising the possibility of a feed-forward mechanism [117].

The redox potential within the lumen of the endoplasmic reticulum is a crucial determinant of posttranslational protein processing, in particular for disulfide bridge formation. NOX4 is mostly found in the endoplasmic reticulum [118], where it forms a macromolecular complex with calnexin [119]. NOX4 can also associate with protein disulfide isomerase in smooth muscle cells [86]. Thus, NOX-derived ROS might contribute to the regulation of formation of protein disulfide bonds.

NOX4 is mostly expressed in the kidney, but Nox4-deficient mice do not show any detectable abnormality in their kidneys [120, 121]. However, Nox4-deficient mice develop more severe interstitial fibrosis and tubular apoptosis after obstruction when compared with wild-type mice [122]. Nox4 is also present in the heart and in endothelial cells and seems to have a protective role in blood vessels [102]. In fact, Nox4-deficient mice demonstrated lower formation of new blood vessels in case of ischemia [102] and an acceleration of atherosclerosis [123]. Furthermore, Nox4-deficient mice display higher bone density with a reduced number of osteoclasts [124].

Altogether, trying to summarize the present state of knowledge concerning NOX4 is a frustrating endeavor. NOX4 is probably the most widely distributed NOX, and no loss-of-function mutations in mammalian populations have been reported. Nox4 knockout mice were generated by different laboratories, but none of them show any conspicuous spontaneous phenotype, and published results using pathology models are often contradictory. Thus, key elements in the understanding of NOX4 are still missing. In the absence of more profound understanding of NOX4 function, we can only formulate a preliminary working hypothesis that might at least in part explain the situation. NOX4 may be involved in the fine-tuning of several essential cellular purposes or key developmental functions in the tissues where it is expressed, such as the kidney, vessels, bones, and cartilage. As essential cellular functions are typically regulated by redundant mechanisms, it is challenging to identify NOX4 function under standard laboratory conditions. Innovative experimental approaches will be needed to confirm or to discard this hypothesis.

2.5 NOX5

NOX5 is expressed in a variety of tissues, such as the testis, spleen, lymph nodes, vascular smooth muscle, bone marrow, pancreas, placenta, ovary, uterus, and stomach and in various fetal tissues

[125–127]. Unlike other isoforms, NOX5 activity is independent of cytosolic organizer or activator subunits [125], of p22^{phox}, and of Rac (Fig. 1c). NOX5 activation is directly dependent on free intracellular calcium [128] and is regulated by calmodulin. Indeed, calmodulin binds to a calmodulin-binding domain located in the NOX5 C-terminus and increases the calcium sensitivity of NOX5 [129]. In addition, there is a functionally relevant phosphorylation site in the C-terminus of NOX5 [92, 130] and two conserved polybasic domains located in both N- and C-termini [131].

The physiological function of NOX5 is not fully understood, but this oxidase seems to contribute to vascular oxidative stress in coronary lesions [132], hypertension, and stroke [133]. Also, in certain human populations, single nucleotide polymorphisms (SNPs) lead to partial NOX5 loss of function [134]. As of today no homozygous individuals were described, and no disease associated with NOX5 deficiency has ever been described [135].

In mammals, NOX5 is the least studied of all NOX isoforms. Indeed, the gene is missing in rodents [135] which explains in part our lack of knowledge on this NOX isoform. NOX5 is the only isoform that is not universally expressed across mammals. It implies that other models have to be developed such as the generation of transgenic rabbits and development of a NOX5 inhibitor.

2.6 DUOX1 and DUOX2

Both DUOX enzymes share ~50% of amino acid identity with NOX2, and DUOX1 shares ~83% similarity with DUOX2 [136]. DUOX1 and DUOX2 are NADPH oxidases that are highly expressed in the thyroid gland. They were initially referred to as thyroid oxidases. DUOX1 and DUOX2 are also expressed in airway epithelia [137–139], in the gastrointestinal tract [140], and in the prostate [141]. The main defining feature of DUOX enzymes, as compared to other NOX isoforms, is the presence of a seventh transmembrane domain and a peroxidase homology domain in their extracytosolic N-terminal regions (Fig. 1d). Like NOX5, DUOX isoforms are not dependent on other subunits for their activation, but they require the maturation factors DUOXA1 and DUOAX2, and their H_2O_2 generation is dependent on calcium [142] and phosphorylation [143].

The exact function of DUOX1 is poorly understood because a mutation in *DUOX1* does not affect the biosynthesis of thyroid hormones. Duox1-deficient mice do not have hypothyroidism [142] or any other spontaneous phenotype. However, DUOX1 may compensate for DUOX2 deficiency as digenic mutations in *DUOX1* and *DUOX2* lead to a more severe hypothyroidism phenotype [144]. Indeed, loss of function of DUOX2 (and DUOXA2) leads to congenital hypothyroidism.

DUOX2 is also found in the ducts of the salivary gland [138], rectal mucosa [138], and all along the gastrointestinal tract [145, 146]. Mutations in *DUOX2* are associated with a higher predisposition to develop VEOIBD [65, 147]. In terms of cellular function, the role of DUOX2 in the thyroid is well understood because of loss-of-function mutations in humans that preclude the synthesis of thyroid hormone [136, 148, 149]. Indeed, DUOX2 generates the H_2O_2 substrate for thyroid peroxidase in order to catalyze the oxidation of iodide which is necessary for the iodination of the thyroglobulin tyrosyl residue [150]. Interestingly, giant pandas have a mutation in the *DUOX2* gene that allows them to have a lower metabolic rate through a low level of thyroid hormones synthesis [151]. Indeed, giant pandas have a very low energy diet (mostly composed of bamboos) despite their carnivoran alimentary tract. Therefore, this mutation seems to be a survival advantage in these animals. Therefore, H_2O_2 derived from DUOX-type enzymes appears to play a particularly important role in biosynthesis processes. Mice with a spontaneous missense *Duox2* mutation were described and suffered from congenital hypothyroidism [152] similar to combined DuoxA1/DuoxA2 knockout mice [140]. Duox2 and Duox1 seem also to be involved in the defense against the gastric infection with *Helicobacter felis* in mice [140].

One of the particularities of DUOX isoforms is that H_2O_2 is the only ROS detected upon DUOX activity. This has led to a heated debate whether DUOX1–2 enzymes first generate H_2O_2 or whether generation of H_2O_2 occurs via an O_2• intermediate [153–156]. From a biochemical point of view, all NOX isoforms (including DUOX) are thought to be mono-electron transporters that deliver a single electron at a time to extracellular oxygen, and, hence, all NADPH oxidases should generate $O_2^{\bullet -}$ as a primary product. However, for DUOX1–2 this primary product is undetectable. As proposed by Takac et al. for NOX4 [95], DUOX1–2 might contain a structural region that accelerates spontaneous dismutation of $O_2^{\bullet -}$ to H_2O_2. Even if biochemically unlikely, one cannot entirely exclude the hypothesis that some NOX isoforms might be bi-electron rather than mono-electron transporters which could lead to direct generation of H_2O_2.

3 Conclusion

Before the discovery of the NADPH oxidase family members, ROS were considered metabolic by-products, toxic for the cell. During the last 30 years, our perception of ROS has changed, and it is now appreciated that they play a crucial role in the regulation of numerous physiological pathways. Consequently, loss-of-function mutations in NOX can lead to severe disease, as best documented for NOX2 and DUOX2. Experiments on gene-deleted animals have helped to understand and confirm the physiological function of some NOX isoforms. Yet, for several NOX isoforms, such as

NOX1, NOX4, NOX5, and DUOX1, their main function is still not fully understood, even in organs where the respective NOX isoforms are highly expressed. It is known that ROS are important for the regulation of many cellular signaling pathways, but the detailed involvement of NOX in this process remains unclear. The best-understood ROS targets are redox-sensitive cysteine residues. Through a reversible reaction with H_2O_2, cysteine residues are oxidized and thereby change the function of the respective protein, as, for example, for protein phosphatases. Protein tyrosine phosphatases control the phosphorylation of various signal transduction proteins and are involved in the regulation of cell proliferation, differentiation, survival metabolism, and motility [157]. The catalytic region of these phosphatases contains cysteine residues, which can be reversibly inactivated by oxidation [158].

Inhibition by ROS is also observed for lipid phosphatases, such as PTEN. PTEN acts as a tumor suppressor, and its inhibition leads to increased levels of phosphatidylinositol [3–5]-trisphosphate (PIP3), which in turn leads to activation of Akt signaling, thereby linking NOX/ROS signaling to the mTOR pathway [159]. Studies have made a connection between PTEN and NOX1 [160] and NOX2 [161], but the underlying mechanisms are unknown and further investigations are needed.

The role of NOX in the regulation of many physiological functions remains poorly understood. The use of NOX inhibitors may help to highlight how NOX regulates these pathways, but NOX inhibitors have also a potential therapeutic purpose. Indeed, while the physiological activity of NOX is crucial for health, oxidative stress due to excessive activity of NOX may lead to pathology, such as fibrotic diseases [162, 163], neurodegenerative disorders [164, 165], or cardiovascular diseases [120, 166]. Thus, ROS are essential for physiological function, but in excess (i.e., oxidative stress), they become toxic for the cell. Research for treatments for oxidative stress-induced diseases combined with an improved understanding of NOX has led to efforts to develop NOX inhibitors.

First-generation NOX inhibitors were only suited for cellular and animal experiments. Indeed, most of these first-generation inhibitors are non-specific (e.g., DPI) or need to be used at unreasonably high concentrations because they rather act as antioxidants (e.g., apocynin). Results obtained with these compounds should therefore be taken with caution. However, despite the lack of NOX specificity, DPI which can inhibit other flavin-dependent enzymes remains the most useful pharmacological tool for in vitro studies. Second-generation NOX inhibitors such as GKT136901 and GKT137831 were developed for their drug-like properties. The GKT compounds are reported to preferentially inhibit NOX1 and NOX4 activity, but they also have effects on other NOX isoforms [98, 167–170]. Their mode of action is unclear and is most likely not restricted to direct NOX inhibition [171]. Nevertheless, GKT compounds show remarkable activity in several animal models of oxidative stress-induced disease, and GKT137831 is currently in a phase II clinical trial for the treatment of primary biliary cholangitis (www.ClinicalTrials.gov Identifier: NCT03226067).

A first example of a third-generation NOX inhibitor is GSK2795039. This compound (1) specifically inhibits NOX2 and not other NOX isoforms and (2) has been validated by a panel of experiments showing NOX2 selectivity. GSK2795039 has a protective effect in the cerulean mouse model of pancreatitis [172]; however no information about a possible clinical development is available.

To conclude, there has been an amazing explosion in our knowledge about NOX over the last 20 years. Yet, a lot of work remains to be done. The structure of most NOX isoforms is only partially solved, their physiological role is—at least for some isoforms—only poorly understood, and the development of NOX inhibitors is in rather early stages. Given the progress of the last 20 years, we are optimistic that many of these issues will be solved over the next decade.

Acknowledgment

This research was supported by the Swiss National Science Foundation program.

References

1. Warburg O (1908) Beobachtungen über die Oxydationsprozesse im Seeigelei. Hoppe-Seyler´s Z Für Physiol Chem 57:1–16
2. Baldridge CW, Gerard RW (1932) The extra respiration of phagocytosis. Am J Physiol Content 103:235–236
3. MacLeod J (1943) The role of oxygen in the metabolism and motility of human spermatozoa. Am J Physiol Content 138:512–518
4. Iyer G, Islam M, Quastel J (1961) Biochemical Aspects of Phagocytosis. Nature 192:535–541
5. Rossi F, Zatti M (1964) Biochemical aspects of phagocytosis in polymorphonuclear leucocytes. NADH and NADPH oxidation by the granules of resting and phagocytizing cells. Experientia 20:21–23
6. Segal AW, Jones OT (1978) Novel cytochrome b system in phagocytic vacuoles of human granulocytes. Nature 276:515–517
7. Segal AW, West I, Wientjes F et al (1992) Cytochrome b-245 is a flavocytochrome containing FAD and the NADPH-binding site of the microbicidal oxidase of phagocytes. Biochem J 284:781–788
8. Royer-Pokora B, Kunkel LM, Monaco AP et al (1986) Cloning the gene for the inherited disorder chronic granulomatous disease on the basis of its chromosomal location. Cold Spring Harb Symp Quant Biol 51 (Pt 1):177–183
9. Lambeth J (2004) NOX enzymes and the biology of reactive oxygen. Nat Rev Immunol 4:181–189
10. Bedard K, Krause K-H (2007) The NOX family of ROS-generating NADPH oxidases: physiology and pathophysiology. Physiol Rev 87:245–313
11. Hajjar C, Cherrier MV, Mirandela GD et al (2017) The NOX family of proteins is also present in bacteria. mBio 8:e01487–e01417
12. Heymes C, Bendall JK, Ratajczak P et al (2003) Increased myocardial NADPH

oxidase activity in human heart failure. J Am Coll Cardiol 41:2164–2171

13. Javesghani D, MAGDER SA, Barreiro E et al (2002) Molecular characterization of a superoxide-generating NAD (P) H oxidase in the ventilatory muscles. Am J Respir Crit Care Med 165:412–418
14. Reinehr R, Becker S, Eberle A et al (2005) Involvement of NADPH oxidase isoforms and Src family kinases in CD95-dependent hepatocyte apoptosis. J Biol Chem 280:27179–27194
15. Görlach A, Brandes RP, Nguyen K et al (2000) A gp91phox containing NADPH oxidase selectively expressed in endothelial cells is a major source of oxygen radical generation in the arterial wall. Circ Res 87:26–32
16. Jones SA, O'Donnell VB, Wood JD et al (1996) Expression of phagocyte NADPH oxidase components in human endothelial cells. Am J Physiol-Heart Circ Physiol 271: H1626–H1634
17. Li J-M, Shah AM (2002) Intracellular localization and preassembly of the NADPH oxidase complex in cultured endothelial cells. J Biol Chem 277:19952–19960
18. Nayernia Z, Colaianna M, Robledinos-Antón N et al (2017) Decreased neural precursor cell pool in NADPH oxidase 2-deficiency: from mouse brain to neural differentiation of patient derived iPSC. Redox Biol 13:82–93
19. Piccoli C, Ria R, Scrima R, Cela O, D'April-e A, Boffoli D et al (2005) Characterization of mitochondrial and extra-mitochondrial oxygen consuming reactions in human hematopoietic stem cells: novel evidence of the occurrence of NAD(P)H OXIDASE activity. J Biol Chem 280:26467–26476
20. Borregaard N, Heiple JM, Simons ER et al (1983) Subcellular localization of the b-cytochrome component of the human neutrophil microbicidal oxidase: translocation during activation. J Cell Biol 97:52–61
21. Heyworth PG, Bohl BP, Bokoch GM et al (1994) Rac translocates independently of the neutrophil NADPH oxidase components p47phox and p67phox. Evidence for its interaction with flavocytochrome b558. J Biol Chem 269:30749–30752
22. Koga H, Terasawa H, Nunoi H et al (1999) Tetratricopeptide repeat (TPR) motifs of p67 (phox) participate in interaction with the small GTPase Rac and activation of the phagocyte NADPH oxidase. J Biol Chem 274:25051–25060
23. Buvelot H, Posfay-Barbe KM, Linder P et al (2017) Staphylococcus aureus, phagocyte NADPH oxidase and chronic granulomatous disease. FEMS Microbiol Rev 41:139–157
24. Cachat J, Deffert C, Hugues S et al (2015) Phagocyte NADPH oxidase and specific immunity. Clin Sci 128:635–648
25. Cabiscol E, Tamarit J, Ros J (2000) Oxidative stress in bacteria and protein damage by reactive oxygen species. Int Microbiol Off J Span Soc Microbiol 3:3–8
26. Bandyopadhyay U, Das D, Banerjee RK (1999) Reactive oxygen species: oxidative damage and pathogenesis. Curr Sci 77:658–666
27. Fontecave M, Ollagnier-de-Choudens S (2008) Iron-sulfur cluster biosynthesis in bacteria: mechanisms of cluster assembly and transfer. Arch Biochem Biophys 474:226–237
28. Imlay JA (2014) The mismetallation of enzymes during oxidative stress. J Biol Chem 289:28121–28128
29. Levine AP, Duchen MR, de Villiers S et al (2015) Alkalinity of neutrophil phagocytic vacuoles is modulated by HVCN1 and has consequences for myeloperoxidase activity. PLoS One 10:e0125906
30. Elsen S, Doussière J, Villiers CL et al (2004) Cryptic O_2^--generating NADPH oxidase in dendritic cells. J Cell Sci 117:2215–2226
31. Mantegazza AR, Savina A, Vermeulen M et al (2008) NADPH oxidase controls phagosomal pH and antigen cross-presentation in human dendritic cells. Blood 112:4712–4722
32. Fuchs TA, Abed U, Goosmann C et al (2007) Novel cell death program leads to neutrophil extracellular traps. J Cell Biol 176:231–241
33. Stoiber W, Obermayer A, Steinbacher P et al (2015) The Role of Reactive Oxygen Species (ROS) in the formation of Extracellular Traps (ETs) in humans. Biomolecules 5:702–723
34. Yamamoto A, Taniuchi S, Tsuji S et al (2002) Role of reactive oxygen species in neutrophil apoptosis following ingestion of heat-killed Staphylococcus aureus. Clin Exp Immunol 129:479–484
35. Kobayashi SD, Voyich JM, Braughton KR et al (2004) Gene expression profiling provides insight into the pathophysiology of chronic granulomatous disease. J Immunol 172:636–643
36. Huang J, Canadien V, Lam GY et al (2009) Activation of antibacterial autophagy by NADPH oxidases. Proc Natl Acad Sci U S A 106:6226–6231

37. de Luca A, Smeekens SP, Casagrande A et al (2014) IL-1 receptor blockade restores autophagy and reduces inflammation in chronic granulomatous disease in mice and in humans. Proc Natl Acad Sci U S A 111:3526–3531
38. Iles KE, Forman HJ (2002) Macrophage signaling and respiratory burst. Immunol Res 26:95–105
39. Huang C-K, Zhan L, Hannigan MO et al (2000) P47phox -deficient NADPH oxidase defect in neutrophils of diabetic mouse strains, C57BL/6J-m db/db and db/+. J Leukoc Biol 67:210–215
40. Nakano Y, Longo-Guess CM, Bergstrom DE et al (2008) Mutation of the Cyba gene encoding p22phox causes vestibular and immune defects in mice. J Clin Invest 118:1176–1185
41. Pollock JD, Williams DA, Gifford MAC et al (1995) Mouse model of X-linked chronic granulomatous disease, an inherited defect in phagocyte superoxide production. Nat Genet 9:202–209
42. Ellson CD, Davidson K, Ferguson GJ et al (2006) Neutrophils from p40phox-/- mice exhibit severe defects in NADPH oxidase regulation and oxidant-dependent bacterial killing. J Exp Med 203:1927–1937
43. Jackson SH, Gallin JI, Holland SM (1995) The p47phox mouse knock-out model of chronic granulomatous disease. J Exp Med 182:751–758
44. Jacob CO, Yu N, Yoo D-G et al (2017) Haploinsufficiency of NADPH oxidase subunit neutrophil cytosolic factor 2 is sufficient to accelerate full-blown lupus in NZM 2328 mice. Arthritis Rheumatol 69:1647–1660
45. Messina CGM, Reeves EP, Roes J et al (2002) Catalase negative Staphylococcus aureus retain virulence in mouse model of chronic granulomatous disease. FEBS Lett 518:107–110
46. Pizzolla A, Hultqvist M, Nilson B et al (2012) Reactive oxygen species produced by the NADPH oxidase 2 complex in monocytes protect mice from bacterial infections. J Immunol 188:5003–5011
47. Deffert C, Carnesecchi S, Yuan H et al (2012) Hyperinflammation of chronic granulomatous disease is abolished by NOX2 reconstitution in macrophages and dendritic cells. J Pathol 228:341–350
48. Mori M, Li G, Hashimoto M et al (2009) Pivotal advance: eosinophilia in the MES rat strain is caused by a loss-of-function mutation in the gene for cytochrome b(-245), alpha polypeptide (Cyba). J Leukoc Biol 86:473–478
49. Hultqvist M, Holmdahl R (2005) Ncf1 (p47phox) polymorphism determines oxidative burst and the severity of arthritis in rats and mice. Cell Immunol 233:97–101
50. Bánfi B, Clark RA, Steger K et al (2003) Two novel proteins activate superoxide generation by the NADPH oxidase NOX1. J Biol Chem 278:3510–3513
51. Suh YA, Arnold RS, Lassegue B et al (1999) Cell transformation by the superoxide-generating oxidase Mox1. Nature 401:79–82
52. Szanto I, Rubbia-Brandt L, Kiss P et al (2005) Expression of NOX1, a superoxide-generating NADPH oxidase, in colon cancer and inflammatory bowel disease. J Pathol 207:164–176
53. Lassègue B, Sorescu D, Szöcs K et al (2001) Novel gp91(phox) homologues in vascular smooth muscle cells: nox1 mediates angiotensin II-induced superoxide formation and redox-sensitive signaling pathways. Circ Res 88:888–894
54. Ago T, Kitazono T, Kuroda J et al (2005) NAD(P)H oxidases in rat basilar arterial endothelial cells. Stroke 36:1040–1046
55. Kobayashi S, Nojima Y, Shibuya M et al (2004) Nox1 regulates apoptosis and potentially stimulates branching morphogenesis in sinusoidal endothelial cells. Exp Cell Res 300:455–462
56. Cui X-L, Brockman D, Campos B et al (2006) Expression of NADPH oxidase isoform 1 (Nox1) in human placenta: involvement in preeclampsia. Placenta 27:422–431
57. Katsuyama M, Fan C, Yabe-Nishimura C (2002) NADPH oxidase is involved in prostaglandin F2alpha-induced hypertrophy of vascular smooth muscle cells: induction of NOX1 by PGF2alpha. J Biol Chem 277:13438–13442
58. Lee YS (2005) Role of NADPH oxidase-mediated generation of reactive oxygen species in the mechanism of apoptosis induced by phenolic acids in HepG2 human hepatoma cells. Arch Pharm Res 28:1183–1189
59. Wingler K, Wünsch S, Kreutz R et al (2001) Upregulation of the vascular NAD(P)H-oxidase isoforms Nox1 and Nox4 by the renin-angiotensin system in vitro and in vivo. Free Radic Biol Med 31:1456–1464
60. Plesková M, Beck K-F, Behrens MH et al (2006) Nitric oxide down-regulates the expression of the catalytic NADPH oxidase subunit Nox1 in rat renal mesangial cells. FASEB J 20:139–141

61. Matsuno K, Yamada H, Iwata K et al (2005) Nox1 is involved in angiotensin II-mediated hypertension: a study in Nox1-deficient mice. Circulation 112:2677–2685
62. Gavazzi G, Banfi B, Deffert C et al (2006) Decreased blood pressure in NOX1-deficient mice. FEBS Lett 580:497–504
63. Coant N, Ben Mkaddem S, Pedruzzi E et al (2010) NADPH oxidase 1 modulates WNT and NOTCH1 signaling to control the fate of proliferative progenitor cells in the colon. Mol Cell Biol 30:2636–2650
64. Rokutan K, Kawahara T, Kuwano Y et al (2006) NADPH oxidases in the gastrointestinal tract: a potential role of Nox1 in innate immune response and carcinogenesis. Antioxid Redox Signal 8:1573–1582
65. Hayes P, Dhillon S, O'Neill K et al (2015) Defects in NADPH oxidase genes NOX1 and DUOX2 in very early onset inflammatory bowel disease. Cell Mol Gastroenterol Hepatol 1:489–502
66. Schwerd T, Bryant RV, Pandey S et al (2018) NOX1 loss-of-function genetic variants in patients with inflammatory bowel disease. Mucosal Immunol 11(2):562–574
67. Sobotta MC, Liou W, Stöcker S et al (2015) Peroxiredoxin-2 and STAT3 form a redox relay for H_2O_2 signaling. Nat Chem Biol 11:64–70
68. Stöcker S, Maurer M, Ruppert T et al (2018) A role for 2-Cys peroxiredoxins in facilitating cytosolic protein thiol oxidation. Nat Chem Biol 14:148–155
69. Kwon J, Wang A, Burke DJ et al (2016) Peroxiredoxin 6 (Prdx6) supports NADPH oxidase1 (Nox1)-based superoxide generation and cell migration. Free Radic Biol Med 96:99–115
70. Woo HA, Yim SH, Shin DH et al (2010) Inactivation of peroxiredoxin I by phosphorylation allows localized H_2O_2 accumulation for cell signaling. Cell 140:517–528
71. Alvarez LA, Kovačič L, Rodríguez J et al (2016) NADPH oxidase-derived H_2O_2 subverts pathogen signaling by oxidative phosphotyrosine conversion to PB-DOPA. Proc Natl Acad Sci U S A 113:10406–10411
72. Bánfi B, Malgrange B, Knisz J et al (2004) NOX3, a Superoxide-generating NADPH Oxidase of the Inner Ear. J Biol Chem 279:46065–46072
73. Ueno N, Takeya R, Miyano K et al (2005) The NADPH oxidase Nox3 constitutively produces superoxide in a p22phox-dependent manner: its regulation by oxidase organizers and activators. J Biol Chem 280:23328–23339
74. Paffenholz R, Bergstrom RA, Pasutto F et al (2004) Vestibular defects in head-tilt mice result from mutations in Nox3, encoding an NADPH oxidase. Genes Dev 18:486–491
75. Kiss PJ, Knisz J, Zhang Y et al (2006) Inactivation of NADPH oxidase organizer 1 results in severe imbalance. Curr Biol CB 16:208–213
76. Flaherty JP, Spruce CA, Fairfield HE et al (2010) Generation of a conditional null allele of NADPH oxidase activator 1 (NOXA1). Genesis 48:568–575
77. Lavinsky J, Crow AL, Pan C et al (2015) Genome-wide Association Study identifies Nox3 as a critical gene for susceptibility to noise-induced hearing loss. PLoS Genet 11: e1005094
78. Geiszt M, Kopp JB, Várnai P et al (2000) Identification of renox, an NAD(P)H oxidase in kidney. Proc Natl Acad Sci U S A 97:8010–8014
79. Shiose A, Kuroda J, Tsuruya K et al (2001) A novel superoxide-producing NAD(P)H oxidase in kidney. J Biol Chem 276:1417–1423
80. Yang S, Madyastha P, Bingel S et al (2001) A new superoxide-generating oxidase in murine osteoclasts. J Biol Chem 276:5452–5458
81. Yang S, Zhang Y, Ries W et al (2004) Expression of Nox4 in osteoclasts. J Cell Biochem 92:238–248
82. Hu T, Ramachandrarao SP, Siva S et al (2005) Reactive oxygen species production via NADPH oxidase mediates TGF-beta-induced cytoskeletal alterations in endothelial cells. Am J Physiol Renal Physiol 289:F816–F825
83. Van Buul JD, Fernandez-Borja M, Anthony EC et al (2005) Expression and localization of NOX2 and NOX4 in primary human endothelial cells. Antioxid Redox Signal 7:308–317
84. Ellmark SHM, Dusting GJ, Fui MNT et al (2005) The contribution of Nox4 to NADPH oxidase activity in mouse vascular smooth muscle. Cardiovasc Res 65:495–504
85. Hoidal JR, Brar SS, Sturrock AB et al (2003) The role of endogenous NADPH oxidases in airway and pulmonary vascular smooth muscle function. Antioxid Redox Signal 5:751–758
86. Janiszewski M, Lopes LR, Carmo AO et al (2005) Regulation of NAD(P)H oxidase by associated protein disulfide isomerase in vascular smooth muscle cells. J Biol Chem 280:40813–40819

87. Laude K, Cai H, Fink B et al (2005) Hemodynamic and biochemical adaptations to vascular smooth muscle overexpression of p22phox in mice. Am J Physiol Heart Circ Physiol 288:H7–H12
88. Pedruzzi E, Guichard C, Ollivier V et al (2004) NAD(P)H oxidase Nox-4 mediates 7-ketocholesterol-induced endoplasmic reticulum stress and apoptosis in human aortic smooth muscle cells. Mol Cell Biol 24:10703–10717
89. Chamulitrat W, Stremmel W, Kawahara T et al (2004) A constitutive NADPH oxidase-like system containing gp91phox homologs in human keratinocytes. J Invest Dermatol 122:1000–1009
90. Brar SS, Kennedy TP, Sturrock AB et al (2002) An NAD(P)H oxidase regulates growth and transcription in melanoma cells. Am J Physiol Cell Physiol 282:C1212–C1224
91. Nisimoto Y, Diebold BA, Constentino-Gomes D et al (2014) Nox4: a hydrogen peroxide-generating oxygen sensor. Biochemistry (Mosc) 53:5111–5120
92. Serrander L, Cartier L, Bedard K et al (2007) NOX4 activity is determined by mRNA levels and reveals a unique pattern of ROS generation. Biochem J 406:105–114
93. Lyle AN, Deshpande NN, Taniyama Y et al (2009) Poldip2, a novel regulator of Nox4 and cytoskeletal integrity in vascular smooth muscle cells. Circ Res 105:249–259
94. Martyn KD, Frederick LM, von Loehneysen K et al (2006) Functional analysis of Nox4 reveals unique characteristics compared to other NADPH oxidases. Cell Signal 18:69–82
95. Takac I, Schröder K, Zhang L et al (2011) The E-loop is involved in hydrogen peroxide formation by the NADPH oxidase Nox4. J Biol Chem 286:13304–13313
96. Mittal M, Roth M, König P et al (2007) Hypoxia-dependent regulation of nonphagocytic NADPH oxidase subunit NOX4 in the pulmonary vasculature. Circ Res 101:258–267
97. Pendyala S, Moitra J, Kalari S et al (2011) Nrf2 regulates hyperoxia-induced Nox4 expression in human lung endothelium: identification of functional antioxidant response elements on the Nox4 promoter. Free Radic Biol Med 50:1749–1759
98. Sedeek M, Callera G, Montezano A et al (2010) Critical role of Nox4-based NADPH oxidase in glucose-induced oxidative stress in the kidney: implications in type 2 diabetic nephropathy. Am J Physiol Renal Physiol 299:F1348–F1358
99. Cunningham KS, Gotlieb AI (2005) The role of shear stress in the pathogenesis of atherosclerosis. Lab Invest 85:9–23
100. Piera-Velazquez S, Makul A, Jiménez SA (2015) Increased expression of NAPDH oxidase 4 in systemic sclerosis dermal fibroblasts: regulation by transforming growth factor β. Arthritis Rheumatol 67:2749–2758
101. Craige SM, Chen K, Pei Y et al (2011) NADPH oxidase 4 promotes endothelial angiogenesis through endothelial nitric oxide synthase activation. Circulation 124:731–740
102. Schröder K, Zhang M, Benkhoff S et al (2012) Nox4 is a protective reactive oxygen species generating vascular NADPH oxidase. Circ Res 110:1217–1225
103. Lozhkin A, Vendrov AE, Pan H et al (2017) NADPH oxidase 4 regulates vascular inflammation in aging and atherosclerosis. J Mol Cell Cardiol 102:10–21
104. Casas AI, Geuss E, Kleikers PWM et al (2017) NOX4-dependent neuronal autotoxicity and BBB breakdown explain the superior sensitivity of the brain to ischemic damage. Proc Natl Acad Sci U S A 114:12315–12320
105. Jiménez-Altayó F, Meirelles T, Crosas-Molist E, Sorolla MA, Del Blanco DG, López-Luque J et al (2018) Redox stress in Marfan syndrome: dissecting the role of the NADPH OXIDASE NOX4 in aortic aneurysm. Free Radic Biol Med 118:44–58
106. Irani K (2000) Oxidant signaling in vascular cell growth, death, and survival: a review of the roles of reactive oxygen species in smooth muscle and endothelial cell mitogenic and apoptotic signaling. Circ Res 87:179–183
107. Hampton MB, Orrenius S (1997) Dual regulation of caspase activity by hydrogen peroxide: implications for apoptosis. FEBS Lett 414:552–556
108. Hampton MB, Fadeel B, Orrenius S (1998) Redox regulation of the caspases during apoptosis. Ann N Y Acad Sci 854:328–335
109. Mochizuki T, Furuta S, Mitsushita J et al (2006) Inhibition of NADPH oxidase 4 activates apoptosis via the AKT/apoptosis signal-regulating kinase 1 pathway in pancreatic cancer PANC-1 cells. Oncogene 25:3699–3707
110. Laurent A, Nicco C, Chéreau C et al (2005) Controlling tumor growth by modulating endogenous production of reactive oxygen species. Cancer Res 65:948–956

111. Colavitti R, Finkel T (2005) Reactive oxygen species as mediators of cellular senescence. IUBMB Life 57:277–281
112. Hannken T, Schroeder R, Stahl RA et al (1998) Angiotensin II-mediated expression of p27Kip1 and induction of cellular hypertrophy in renal tubular cells depend on the generation of oxygen radicals. Kidney Int 54:1923–1933
113. Imanishi T, Hano T, Nishio I (2005) Estrogen reduces angiotensin II-induced acceleration of senescence in endothelial progenitor cells. Hypertens Res 28:263–271
114. Beckman KB, Ames BN (1998) The free radical theory of aging matures. Physiol Rev 78:547–581
115. Harman D (1956) Aging: a theory based on free radical and radiation chemistry. J Gerontol 11:298–300
116. Krause K-H (2007) Aging: a revisited theory based on free radicals generated by NOX family NADPH oxidases. Exp Gerontol 42:256–262
117. Diebold I, Flügel D, Becht S et al (2010) The hypoxia-inducible factor-2alpha is stabilized by oxidative stress involving NOX4. Antioxid Redox Signal 13:425–436
118. Chen K, Kirber MT, Xiao H et al (2008) Regulation of ROS signal transduction by NADPH oxidase 4 localization. J Cell Biol 181:1129–1139
119. Prior K-K, Wittig I, Leisegang MS et al (2016) The endoplasmic reticulum chaperone calnexin is a NADPH oxidase NOX4 interacting protein. J Biol Chem 291:7045–7059
120. Kleinschnitz C, Grund H, Wingler K et al (2010) Post-stroke inhibition of induced NADPH oxidase type 4 prevents oxidative stress and neurodegeneration. PLoS Biol 8: e1000479
121. Zhang M, Brewer AC, Schröder K et al (2010) NADPH oxidase-4 mediates protection against chronic load-induced stress in mouse hearts by enhancing angiogenesis. Proc Natl Acad Sci U S A 107:18121–18126
122. Nlandu Khodo S, Dizin E, Sossauer G et al (2012) NADPH-oxidase 4 protects against kidney fibrosis during chronic renal injury. J Am Soc Nephrol 23:1967–1976
123. Schürmann C, Rezende F, Kruse C et al (2015) The NADPH oxidase Nox4 has anti-atherosclerotic functions. Eur Heart J 36:3447–3456
124. Goettsch C, Babelova A, Trummer O, et al NADPH oxidase 4 limits bone mass by promoting osteoclastogenesis. https://www.jci.org/articles/view/67603/pdf
125. Bánfi B, Molnár G, Maturana A et al (2001) A Ca(2+)-activated NADPH oxidase in testis, spleen, and lymph nodes. J Biol Chem 276:37594–37601
126. Cheng G, Cao Z, Xu X et al (2001) Homologs of gp91phox: cloning and tissue expression of Nox3, Nox4, and Nox5. Gene 269:131–140
127. Salles N, Szanto I, Herrmann F et al (2005) Expression of mRNA for ROS-generating NADPH oxidases in the aging stomach. Exp Gerontol 40:353–357
128. Bánfi B, Tirone F, Durussel I et al (2004) Mechanism of Ca2+ activation of the NADPH oxidase 5 (NOX5). J Biol Chem 279:18583–18591
129. Tirone F, Cox JA (2007) NADPH oxidase 5 (NOX5) interacts with and is regulated by calmodulin. FEBS Lett 581:1202–1208
130. Jagnandan D, Church JE, Banfi B et al (2007) Novel mechanism of activation of NADPH oxidase 5. calcium sensitization via phosphorylation. J Biol Chem 282:6494–6507
131. Kawahara T, Lambeth JD (2008) Phosphatidylinositol (4,5)-bisphosphate modulates Nox5 localization via an N-terminal polybasic region. Mol Biol Cell 19:4020–4031
132. Guzik TJ, Chen W, Gongora MC et al (2008) Calcium-dependent NOX5 nicotinamide adenine dinucleotide phosphate oxidase contributes to vascular oxidative stress in human coronary artery disease. J Am Coll Cardiol 52:1803–1809
133. Kleikers PWM, Dao VT, Göb E et al (2014) SFRR-E Young Investigator AwardeeNOXing out stroke: identification of NOX4 and 5as targets in blood-brain-barrier stabilisation and neuroprotection. Free Radic Biol Med 75 (Suppl 1):S16
134. Wang Y, Chen F, Le B et al (2014) Impact of Nox5 polymorphisms on basal and stimulus-dependent ROS generation. PLoS One 9: e100102
135. Bedard K, Jaquet V, Krause K-H (2012) NOX5: from basic biology to signaling and disease. Free Radic Biol Med 52:725–734
136. De Deken X, Wang D, Many MC et al (2000) Cloning of two human thyroid cDNAs encoding new members of the NADPH oxidase family. J Biol Chem 275:23227–23233
137. Forteza R, Salathe M, Miot F et al (2005) Regulated hydrogen peroxide production by Duox in human airway epithelial cells. Am J Respir Cell Mol Biol 32:462–469

138. Geiszt M, Witta J, Baffi J et al (2003) Dual oxidases represent novel hydrogen peroxide sources supporting mucosal surface host defense. FASEB J 17:1502–1504
139. Schwarzer C, Machen TE, Illek B et al (2004) NADPH oxidase-dependent acid production in airway epithelial cells. J Biol Chem 279:36454–36461
140. Grasberger H, El-Zaatari M, Dang DT et al (2013) Dual oxidases control release of hydrogen peroxide by the gastric epithelium to prevent Helicobacter felis infection and inflammation in mice. Gastroenterology 145:1045–1054
141. Wang D, De Deken X, Milenkovic M et al (2005) Identification of a novel partner of duox: EFP1, a thioredoxin-related protein. J Biol Chem 280:3096–3103
142. Ameziane-El-Hassani R, Morand S, Boucher J-L et al (2005) Dual oxidase-2 has an intrinsic Ca2+-dependent H_2O_2-generating activity. J Biol Chem 280:30046–30054
143. Rigutto S, Hoste C, Grasberger H et al (2009) Activation of dual oxidases Duox1 and Duox2: differential regulation mediated by camp-dependent protein kinase and protein kinase C-dependent phosphorylation. J Biol Chem 284:6725–6734
144. Aycan Z, Cangul H, Muzza M et al (2017) Digenic DUOX1 and DUOX2 mutations in cases with congenital hypothyroidism. J Clin Endocrinol Metab 102:3085–3090
145. Dupuy C, Pomerance M, Ohayon R et al (2000) Thyroid oxidase (THOX2) gene expression in the rat thyroid cell line FRTL-5. Biochem Biophys Res Commun 277:287–292
146. El Hassani RA, Benfares N, Caillou B et al (2005) Dual oxidase2 is expressed all along the digestive tract. Am J Physiol Gastrointest Liver Physiol 288:G933–G942
147. Parlato M, Charbit-Henrion F, Hayes P et al (2017) First identification of biallelic inherited DUOX2 inactivating mutations as a cause of very early onset inflammatory bowel disease. Gastroenterology 153:609–611.e3
148. Dupuy C, Ohayon R, Valent A et al (1999) Purification of a novel flavoprotein involved in the thyroid NADPH oxidase. Cloning of the porcine and human cdnas. J Biol Chem 274:37265–37269
149. Moreno JC, Bikker H, Kempers MJE et al (2002) Inactivating mutations in the gene for thyroid oxidase 2 (THOX2) and congenital hypothyroidism. N Engl J Med 347:95–102
150. Carvalho DP, Dupuy C (2017) Thyroid hormone biosynthesis and release. Mol Cell Endocrinol 458:6–15
151. Nie Y, Speakman JR, Wu Q et al (2015) ANIMAL PHYSIOLOGY. Exceptionally low daily energy expenditure in the bamboo-eating giant panda. Science 349:171–174
152. Johnson KR, Marden CC, Ward-Bailey P et al (2007) Congenital hypothyroidism, dwarfism, and hearing impairment caused by a missense mutation in the mouse dual oxidase 2 gene, Duox2. Mol Endocrinol Baltim Md 21:1593–1602
153. Dupuy C, Virion A, Ohayon R et al (1991) Mechanism of hydrogen peroxide formation catalyzed by NADPH oxidase in thyroid plasma membrane. J Biol Chem 266:3739–3743
154. Leseney AM, Dème D, Legué O et al (1999) Biochemical characterization of a Ca2+/NAD(P)H-dependent H_2O_2 generator in human thyroid tissue. Biochimie 81:373–380
155. Nakamura Y, Ohtaki S, Makino R et al (1989) Superoxide anion is the initial product in the hydrogen peroxide formation catalyzed by NADPH oxidase in porcine thyroid plasma membrane. J Biol Chem 264:4759–4761
156. Nakamura Y, Makino R, Tanaka T et al (1991) Mechanism of H_2O_2 production in porcine thyroid cells: evidence for intermediary formation of superoxide anion by NADPH-dependent H_2O_2-generating machinery. Biochemistry (Mosc) 30:4880–4886
157. Hunter T (2000) Signaling--2000 and beyond. Cell 100:113–127
158. Denu JM, Tanner KG (1998) Specific and reversible inactivation of protein tyrosine phosphatases by hydrogen peroxide: evidence for a sulfenic acid intermediate and implications for redox regulation. Biochemistry (Mosc) 37:5633–5642
159. Leslie NR, Bennett D, Lindsay YE et al (2003) Redox regulation of PI 3-kinase signalling via inactivation of PTEN. EMBO J 22:5501–5510
160. Cui W, Matsuno K, Iwata K et al (2011) NOX1/nicotinamide adenine dinucleotide phosphate, reduced form (NADPH) oxidase promotes proliferation of stellate cells and aggravates liver fibrosis induced by bile duct ligation. Hepatol Baltim Md 54:949–958
161. Hervera A, De Virgiliis F, Palmisano I et al (2018) Reactive oxygen species regulate axonal regeneration through the release of exosomal NADPH oxidase 2 complexes into injured axons. Nat Cell Biol 20:307–319

162. Carnesecchi S, Deffert C, Donati Y et al (2011) A key role for NOX4 in epithelial cell death during development of lung fibrosis. Antioxid Redox Signal 15:607–619
163. Lan T, Kisseleva T, Brenner DA (2015) Deficiency of NOX1 or NOX4 prevents liver inflammation and fibrosis in mice through inhibition of hepatic stellate cell activation. PLoS One 10:e0129743
164. Nayernia Z, Jaquet V, Krause K-H (2014) New insights on NOX enzymes in the central nervous system. Antioxid Redox Signal 20:2815–2837
165. Sorce S, Stocker R, Seredenina T et al (2017) NADPH oxidases as drug targets and biomarkers in neurodegenerative diseases: what is the evidence? Free Radic Biol Med 112:387–396
166. Di Marco E, Gray SP, Chew P et al (2016) Differential effects of NOX4 and NOX1 on immune cell-mediated inflammation in the aortic sinus of diabetic ApoE-/- mice. Clin Sci 130:1363–1374
167. Aoyama T, Paik Y-H, Watanabe S et al (2012) Nicotinamide adenine dinucleotide phosphate oxidase in experimental liver fibrosis: GKT137831 as a novel potential therapeutic agent. Hepatology 56:2316–2327
168. Garrido-Urbani S, Jemelin S, Deffert C et al (2011) Targeting vascular NADPH oxidase 1 blocks tumor angiogenesis through a PPARα mediated mechanism. PLoS One 6: e14665
169. Musset B, Clark RA, DeCoursey TE et al (2012) NOX5 in human spermatozoa: expression, function, and regulation. J Biol Chem 287:9376–9388
170. Strengert M, Jennings R, Davanture S et al (2014) Mucosal reactive oxygen species are required for antiviral response: role of Duox in influenza a virus infection. Antioxid Redox Signal 20:2695–2709
171. Schildknecht S, Weber A, Gerding HR et al (2014) The NOX1/4 inhibitor GKT136901 as selective and direct scavenger of peroxynitrite. Curr Med Chem 21:365–376
172. Hirano K, Chen WS, Chueng ALW et al (2015) Discovery of GSK2795039, a novel small molecule NADPH oxidase 2 inhibitor. Antioxid Redox Signal 23:358–374

Part II

Biochemistry and Model Systems

Chapter 3

Enhanced Immunoaffinity Purification of Human Neutrophil Flavocytochrome B for Structure Determination by Electron Microscopy

Algirdas J. Jesaitis, Marcia Riesselman, Ross M. Taylor, and Susan Brumfield

Abstract

Determination of the structure of human neutrophil (PMN) flavocytochrome b (Cytb) is a necessary step for the understanding of the structure-function essentials of NADPH oxidase activity. This understanding is crucial for structure-driven therapeutic approaches addressing control of inflammation and infection. Our work on purification and sample preparation of Cytb has facilitated progress toward the goal of structure determination. Here we describe exploiting immunoaffinity purification of Cytb for initial examination of its size and shape by a combination of classical and cryoelectron microscopic (EM) methods. For these evaluations, we used conventional negative-stain transmission electron microscopy (TEM) to examine both detergent-solubilized Cytb as single particles and Cytb in phosphatidylcholine reconstituted membrane vesicles as densely packed random, partially ordered, and subcrystalline arrays. In preliminary trials, we also examined single particles by cryoelectron microscopy (cryoEM) methods. We conclude that Cytb in detergent and reconstituted in membrane is a relatively compact, symmetrical protein of about 100 Å in maximum dimension. The negative stain, preliminary cryoEM, and crude molecular models suggest that the protein is probably a heterotetramer of two p22^{phox} and gp91^{phox} subunits in both detergent micelles and membrane vesicles. This exploratory study also suggests that high-resolution 2D electron microscopic approaches may be accessible to human material collected from single donors.

Key words immunoaffinity purification, NOX2, flavocytochrome b, electron microscopy, cryoEM, 2D membrane reconstitution, negative stain, mAb CS9

1 Introduction

The balancing act performed by neutrophils (PMNs) and other phagocytes in deploying their biochemical oxidative arsenal [1] is central to our understanding of both normal immune function and the pathogenesis of inflammatory disease states [2]. These cells on one hand play a central role in the body's defense against infection, but when unregulated, are primary cellular mediators of inflammatory tissue damage [3]. The microbicidal activity of phagocytes

Ulla G. Knaus and Thomas L. Leto (eds.), *NADPH Oxidases: Methods and Protocols*, Methods in Molecular Biology, vol. 1982, https://doi.org/10.1007/978-1-4939-9424-3_3,

depends on their production of superoxide anion ($O_2^{\cdot -}$) [4] by the flavocytochrome b (Cytb) component of the NADPH oxidase complex. Although relatively nontoxic by itself [5], in different biological settings, $O_2^{\cdot -}$ serves as a precursor for the production of a variety of injury-producing and injury-signaling agents, such as hydrogen peroxide [5, 6], hypochlorous acid [5], hydroxyl radical [5, 7], peroxynitrite [5, 8], aldehydes [9], and chloramines [10], which appear to be involved in a myriad of chronic and acute injurious inflammatory conditions. Although the phagocyte oxidant production appears to be localized under certain conditions such as phagocytic killing of microbes [11] or while attached to tissue surfaces [12], reactive oxidants find their way to surrounding tissues where they have the capacity to inflict damage. One of the possible consequences of the inappropriate exposure of tissue to toxic oxygen species is an increased probability of mutation, DNA damage, and malignant transformation [13, 14]. Indeed, oxidants have been implicated as the damaging agent responsible for the high incidence of malignancies associated with inflammatory diseases such as chronic osteomyelitis, ulcerative colitis, chronic hepatitis, and pneumonitis [15]. Neutrophil oxidants also play an important direct role in many other inflammatory diseases, such as adult respiratory distress syndrome [16], atherosclerosis [17], rheumatoid arthritis [18], and ischemia reperfusion injury [19].

Due to the central role of $O_2^{\cdot -}$ in human health and the pathogenesis of inflammatory diseases, it is important to study aspects of the regulation of oxidant production by PMNs. Structure analysis of the core electron transferase of the superoxide-generating system, Cytb, is central to this understanding but has proven difficult for this relatively low-abundance integral membrane protein [20–22]. Structure determination of Cytb should provide a detailed structural map of key functional sites on the molecular surface of the protein and facilitate understanding of the molecular basis of superoxide production. In previous studies our group has generated and exploited Cytb-specific monoclonal antibodies to help reveal structures of sites on the surface of Cytb in functionally important and neutral regions [23–27], thus providing candidates for structure-based drug design and manipulation of the protein. We also have devised several strategies for purifying Cytb that have exploited the unique properties of the Cytb-specific mAbs [28–30].

Our previous technical publications in this series [29, 31] described the immunoaffinity purification of human neutrophil Cytb and its application for following the conformational dynamics of the protein. We showed that by using fluorescently tagged monoclonal antibodies and fluorescence resonance energy transfer quenching by intrinsic hemes, we could determine distance changes between the heme centers and surface epitopes of the protein. This system identified conformational changes in the protein, when it was activated by anionic phospholipids like phosphatidic acid or when bound by the NADPH oxidase-enhancing adaptor component $p47^{phox}$ SH3 domain [32].

The current work describes generation of material by an improved method of immunoaffinity purification with antibody CS9 that recognizes a region near the C-terminus of $p22^{phox}$ (162-PEARKKPSE-169) [33]. CS9-immunoaffinity purification of Cytb is ideally suited for TEM examination of the purified Cytb in detergent, reconstituted membranes, and frozen thin films, where only small amounts of protein are needed. Following rapid immunopurification on CS9 affinity matrices, sufficient Cytb from the membranes prepared from 5×10^9 PMN can be purified in 1 day. This cell number makes structural TEM methods applied to Cytb accessible to samples prepared from a single donor (usually 3–4 units of blood), thus minimizing the influence of genetic and individual polymorphisms on structure analysis, especially as it applies for 2D electron crystallography. Our study provides groundwork for more extensive structural analysis by describing methods for the purification and use of human PMN Cytb for conventional negative-stain TEM, 2D electron microscopic crystallography [34], and cryoelectron microscopy (cryoEM) [35, 36]. Taken together our initial studies suggest that, in both reconstituted membranes and detergent extracts, purified human neutrophil Cytb most probably exists as a heterotetramer with $2 \times gp91^{phox}$ and $2 \times p22^{phox}$ subunits.

2 Materials

2.1 Preparation of the mAb CS9 Affinity Purification Matrix

1. Antibody-binding matrix: protein G-Sepharose or Gamma-Bind Plus-Sepharose.
2. Hybridoma culture supernatant containing the $p22^{phox}$-specific mAb, CS9, is generated in-house by standard hybridoma culture technology and stored at 4 °C.
3. 250 mL polypropylene centrifuge tubes.
4. Disposable 5″ polystyrene chromatography columns.
5. Phosphate-buffered saline (PBS): 150 mM NaCl, 10 mM sodium phosphate pH 7.4 is made by dissolving 0.31 g $NaH_2PO_4 \cdot H_2O$ and 1.09 g of anhydrous Na_2HPO_4 up to 900 mL with 150 mM NaCl, titrating pH to 7.4 with small volumes of HCl and NaOH and bringing to a final volume of 1 L with 150 mM NaCl.
6. Cyanogen bromide-activated Sepharose 4B.
7. 1 M Tris Base, pH ~10.
8. 0.5 M acetic acid, pH 3.

9. 1 mM HCl (8.6 μL concentrated HCl (11.65 M) per 100 mL dH_2O).
10. 500 mM NaCl and 100 mM $NaHCO_3$-HCl, pH 8.3 (bead bicarbonate buffer).
11. 500 mM NaCl and 100 mM Tris–HCl, pH 8.3.
12. Clinical and ultracentrifuge with tubes and rotors.
13. Concentrators: Centricon YM-50 (MWCO 50) and YM-100 (MWCO 100) for compatibility with buffers containing different detergents (*see* Subheadings 2.2, **items 5** and **6**, and 2.3, **item 1**).
14. Dialysis tubing or cassettes (MW cutoff = 12–14,000).

2.2 Immunoaffinity Purification of Cytb

1. Human neutrophil membrane fractions are prepared as previously described [37] and stored at −70 °C prior to use.
2. A 5 M NaCl stock is prepared in dH_2O and stored at room temperature.
3. Hand-operated 15 mL glass homogenizer (Kimble/Kontes, Vineland, NJ).
4. Relax buffer: 100 mM KCl, 10 mM NaCl, 10 mM Hepes-HCl, pH 7.4; Relax (+) buffer, Relax buffer with 1 mM EDTA added.
5. The nonionic detergent octyl β-D-glucopyranoside (OG) is prepared as a 20% stock (w/v) in dH_2O at 20 °C and stored at −20 °C or for no more than 1 month at 4 °C.
6. n-Dodecyl-β-D-maltopyranoside (DDM) can be substituted for some procedures (*see* **Note 1**).
7. Protease inhibitors: PMSF is prepared as 200 mM stock in absolute ethanol and stored at 4 °C; mammalian protease inhibitor cocktail (P8340; Sigma-Aldrich) is purchased as a 1000-fold concentrated stock in DMSO and stored at −20 °C.
8. Dithiothreitol is prepared fresh as a 100 mM stock in dH_2O.
9. Membrane dispersion is conducted with a model 50 Sonic Dismembrator probe sonicator and/or Dounce homogenizer.
10. The CS9 antibody elution peptide Ac-AEARKKPSEEEAA-$CONH_2$ is stored lyophilized at −20 °C.
11. Absorption measurements are performed on an Agilent 8453 spectrophotometer with quartz microcuvettes.

2.3 Concentration and Desalting of Cytb

1. Concentration of purified Cytb samples in OG is conducted in a Sorvall RC-5B centrifuge with the SS34 rotor at 4 °C using Centricon YM-50 (MW cutoff 50,000) concentration devices. Cytb samples in DDM are concentrated with Centricon YM-100 concentrators.

2. Econo-Pac 10 DG desalting columns (30 × 10 mL) are equilibrated in Relax (+) buffer containing 1.2% OG at 4 °C for the removal of elution peptide from immunoaffinity-purified Cytb samples.
3. Relax (+) buffer and the OG stock are prepared as in Subheading 2.2.

2.4 Reconstitution of Cytb in 99% Phosphatidylcholine Vesicles by Dialysis

1. Cytb dialysis buffer: 10 mM Tris–HCl, pH 8.5, 100 mM KCl, 10 mM NaCl, 1 mM EDTA, 0.02% NaN_3, and 20% glycerol (w/v).
2. 1,2-Dimyristoyl-sn-glycero-3-phosphocholine (DMPC, Avanti Polar Lipids, Inc.) and 1,2-dioleoyl-*sn*-glycero-3-phosphocholine (DOPC, Avanti) are prepared as 10× dilution stocks, ~10–30 mg/mL depending on lipid-protein ratio (LPR) required (*see* Subheading 3.3). Stock solutions are prepared in Cytb dialysis buffer with added 1.2% OG and 20% glycerol.
3. 80% stock of glycerol in water.
4. 10 or 50 μL microdialysis cassettes with MW cutoff 10,000 filters.

2.5 Negative-Stain Transmission Electron Microscopy (TEM) of Detergent-Solubilized Cytb

1. EMS CF400-CU-50 carbon-coated grids (Electron Microscopy Sciences, Hatfield, PA).
2. Uranyl formate (0.75% w/v) made fresh and uranyl acetate stock solution (1% w/v).
3. Parafilm, micro-tweezers, and Whatman 20 paper filters.
4. Leo 912AB transmission electron microscope (Carl Zeiss, Inc.), 2 K × 2 K CCD camera (ProScan Imaging, LLC), and Soft Imaging System analysis software.
5. Vacuum Coating System (Agilent Technologies) for glow discharge.

2.6 Cryoelectron Microscopy (CryoEM)

1. QUANTIFOIL® R 1.2/1.3 A foil holey carbon grids.
2. Frozen thin vitreous ice films containing Cytb are generated by a Vitrobot freeze plunger with liquid ethane containment apparatus.
3. Gatan cryo specimen holder.
4. Philips Tecnai 20—SCX Cryoelectron Microscope equipped with TVIPS 2 K × 2 K Peltier-cooled CCD camera and Philips Compustage run at 120 kV, 2.0 mm spherical aberration, 50,000× magnification, 15 μm scan step, and 1.63 Å/pixel.

3 Methods

The following methods outline the preparation of purified, detergent-solubilized Cytb for direct visualization by uranyl formate or uranyl acetate negative-stain TEM. Also described is Cytb reconstitution into DMPC or DOPC phospholipid vesicles at different molar lipid-protein ratios to generate vesicles of different size and Cytb packing density. Such vesicles allow direct visualization of the Cytb membrane-incorporated form and provide for a starting point and experimental platform for 2D crystallization trials. All of these methods are suitable for obtaining low-resolution structural information about the size and shape of the molecule in detergent and in the membrane environment. Several modifications to the original protocol for immunoaffinity purification of Cytb [28] that result in improved preparations have been described in our prior publication in this series [31] and are used with minor modifications as described here, including the (1) washing of the affinity matrix in a column (rather than batch) format; (2) elution of Cytb at detergent concentrations near the critical micelle concentration; (3) concentration of eluted Cytb samples using a 50 kDa cutoff membrane to minimize the simultaneous concentration of detergent; and (4) rapid removal of elution peptide and/or buffer exchange on a desalting column. The affinity purification procedure can be readily conducted in a single day and generates ample material for conducting electron microscopic studies on fresh material for direct visualization. Additionally, new methods were utilized to reconstitute Cytb in liposomes at intermediate and high lipid-protein ratios that allow examination of Cytb in a fluid state, in closely packed arrangements, and in structural arrays in the membrane. All these applications serve to provide information about the size and shape of the protein in a membrane environment. Finally, we describe the results of preliminary experiments to assess feasibility of these preparations for Cytb structure determination by cryoEM [38, 39] and 2D protein electron microscopic crystallography [34]. These latter studies support the negative-stain structural estimate of the size and shape of the protein complex.

3.1 Preparation of the CS9 mAb Affinity Matrix

1. Prior to the Cytb purification, mAb CS9 is purified by passing 1 L hybridoma culture supernatant twice over 1 mL Gamma-Bind Plus-Sepharose in fritted plastic columns with gravity flow.
2. Column is washed with PBS until A_{280} returns to background level.
3. CS9 is eluted two times with 5 mL of 0.5 M acetic acid, pH 3, and neutralized immediately as it is collected with appropriate premeasured volume of 1 M Tris base, pH 10 (*see* **Note 2**).

4. Absorbance of fractions is measured, and fractions are pooled that have significant A_{280} absorbance. The pool is then centrifuged at 100,000 × *g* for 20 min to clarify and remove aggregates, if any.
5. The pool is concentrated to approximately 2.5–3.5 mg/mL using a Centricon 50 centrifugal concentrator and then dialyzed against 100–1000-fold higher volumes of PBS with 0.02% NaN_3 for at least 48 h and with four changes. Purified antibody is stored at −70 °C.
6. Dried CNBr-activated Sepharose (0.3 g) is resuspended in 10 mL of 1 mM HCl in a 50 mL plastic clinical centrifuge tube, allowed to swell (~1 mL settled bead volume), and washed extensively with 75–100 mL for 15 min with cold 1 mM HCl to activate functional groups. The washing steps can be carried out in 15 or 50 mL plastic tubes in a clinical centrifuge at 200 × *g*. The matrix is then washed once in bead bicarbonate buffer equilibrated briefly for 3–5 min and drained. Prior to or during equilibration, 2 mg CS9 mAb is diluted to 7.5 mL in the bead bicarbonate buffer, and the absorbance A_{280} of the diluted CS9 is measured. The antibody is then added and mixed with beads. The vessel is capped and gently tumbled to mix for 30 min at room temperature.
7. The bead/mAb mixture, as a slurry of 50–75% settled bead volume, is deposited into a fritted glass or disposable plastic column of convenient volume and additions, dilutions, and washes carried out by gravity flow or by peristaltic pump, always making sure the beads remain in the buffer.
8. The flow-through fractions are collected and A_{280} measured to confirm that mAb has coupled to beads. Coupling should be complete and stable as assessed by absorbance analysis of flow-through and wash fractions.
9. The column is then washed with five volumes of bead bicarbonate buffer.
10. The unreacted column coupling sites are blocked with 10 mL 500 mM NaCl, 100 mM Tris–HCl, pH 8.0 for 1 h at room temperature. The column can be capped and mixed periodically (3–4 times).
11. Beads are then washed with 5–10 volumes of PBS plus 0.02% NaN_3 and can be stored as a 50% slurry at 4 °C (*see* **Note 2**). 1.5 mL of beads (50% slurry) are used for immunoaffinity purification of 5×10^9 cell equivalents of neutrophil membrane extract (*see* Subheading 3.2).
12. Prior to immunoaffinity capture of Cytb in the membrane extracts, 350 μL of the mAb CS9 affinity matrix (2 mg CS9/mL beads) are first washed for 1 h with 10 mL 0.5% DDM in

Relax (+) buffer at 4 °C, followed by 10 mL of 1.2% OG in Relax (+) at 4 °C. The beads are then resuspended in 1.2% OG in Relax (+) buffer and tumbled overnight at 4 °C followed by warming to ambient temperature while tumbling for 2 h and a final draining and washing just prior to use, in order to equilibrate with 10 mL of the fresh RT ambient buffer of the same composition. This last washing and equilibration assure the removal of non-bound CS9 under conditions of purification (*see* also Subheading 3.2, **step 10**).

3.2 Immunoaffinity Purification of Cytb

1. 10 mL of neutrophil membranes (~5×10^9 cell equivalents) are homogenized and sonicated, bringing the suspension to 1 M NaCl (by addition from a 5 M NaCl stock). The mixture is centrifuged at 114,000 × *g* for 30 min at 4 °C.
2. The resulting membrane pellet is homogenized in 9 mL of Relax buffer (+), 0.1 mM DTT, 2 mM PMSF, and 10 μL/mL P8340 (*see* **Note 3**).
3. Nonionic detergent OG is added to 1.2% final concentration from a 20% stock and dispersed.
4. The extract is then briefly and gently sonicated (3 × 5 s, setting 3) to disperse particulates without generating too much foam.
5. The extract is then tumbled or gently mixed at 4 °C for 40 min.
6. The detergent extract of the membranes is centrifuged at 114,000 × *g* for 30 min at 4 °C to collect the soluble fraction and analyze the Cytb content by absorption spectroscopy (200–700 nm). The Cytb content is determined using $\varepsilon_{414} = 131{,}000\ M^{-1}\ cm^{-1}$ for the heme Soret band with the assumption of two heme prosthetic groups per Cytb heterodimer. $[Cytb] = A_{414}/2\varepsilon_{414}$ M.
7. The soluble fraction of the membrane extract is then added to 1.5 mL of the CS9 affinity matrix equilibrated in Relax (+) and 1.2% OG and tumbled for 1 h at 4 °C.
8. The above slurry can be poured into a disposable column, and post-binding flow fractions can be assessed by both absorption spectroscopy of the oxidized heme Soret band [30] and immunoblot analysis of the $p22^{phox}$ subunit [33] for Cytb retained by the affinity matrix.
9. The packed beads are washed with Relax (+) buffer and 1.2% OG until the absorption spectrum (200–700 nm) returns to baseline.
10. The bound CS9 affinity matrix is then eluted with 1.2% OG containing 200 μM elution peptide Ac-AEARKKPSEEEAA-$CONH_2$ in Relax (+) buffer plus 1.2% OG. Alternatively, 0.05–0.1% DDM can be substituted for OG in the elution buffer. Fractions of 0.5–1 mL are collected allowing the affinity

matrix to soak in the peptide elution buffer for 5–10 min between collections. Cytb should elute within 1–4 mL of elution volume. Assess the yield by absorbance spectroscopy and by immunoblot analysis of p22^{phox} (*see* **Note 4**).

11. To remove the elution peptide, the concentrated Cytb sample is passed over an Econo-Pac 10 DG desalting column at 4 °C with 0.5 mL fractions collected and analyzed for Cytb content by absorption spectroscopy.
12. Following desalting and pooling of fractions, Cytb is concentrated and diluted (*see* **Note 4**) to the desired protein and detergent concentration using Centricon YM-50 protein concentrators for OG samples and Centricon YM-100 for DDM samples. Both types of samples can be stored on ice or at 4 °C prior to application to grids. A good concentration range for single particles is 50–100 nM Cytb in 0.03% DDM as shown in Fig. 1.
13. The yield per 5 × 10^9 cell equivalents of starting membranes is about 160 μg or 1.9 nmol of Cytb (MW 86,163) in about 1 mL (*see* **Notes 5–7**).
14. For lipid reconstitutions, 100 μL Cytb is concentrated in a Centricon YM-50 rinsed with 100 μL Cytb dialysis buffer plus 1.2% OG to increase the Cytb concentration and to exchange Cytb buffers. This MW cutoff allows free OG micelles pass through while retaining Cytb. This is done for 3 min at 13,000 × *g* three times with addition of Cytb dialysis buffer plus 1.2% OG to bring the solution back to original 100 μL volume. After the third spin, volume should be approximately 10–15 μL with a Cytb concentration of 13–20 μM. Check filtrate flow-through by absorbance to ensure that Cytb is being retained.

3.3 Reconstitution of Cytb in Phosphatidylcholine Membrane Vesicles by Dialysis: DOPC:Cytb or DMPC:Cytb Molar LPR from 100 to 300

The reconstitution of Cytb into liposomes at various densities and molecular order is accomplished by slow removal of detergent by dialysis from detergent, lipid, and Cytb (protein) mixtures. For EM, only small amounts of material are required, i.e., 2 μL per grid, and so we dialyze 10 μL amounts in special dialysis cassettes that are floated in dialysis buffer with gentle stirring for up to 13 days. Detergent removal can be monitored by examining drop size and shape with time but requires cassette disassembly, so parallel experiments should initially be carried out to determine the minimum time required to obtain drop shape matching detergent-free mixtures. We have determined that 4–13 days of dialysis is required to remove sufficient detergent to allow formation of different types of membrane vesicles with incorporated Cytb. Vesicle size and protein packing are controlled by the LPR of the mixture and other parameters. Our studies were aimed at

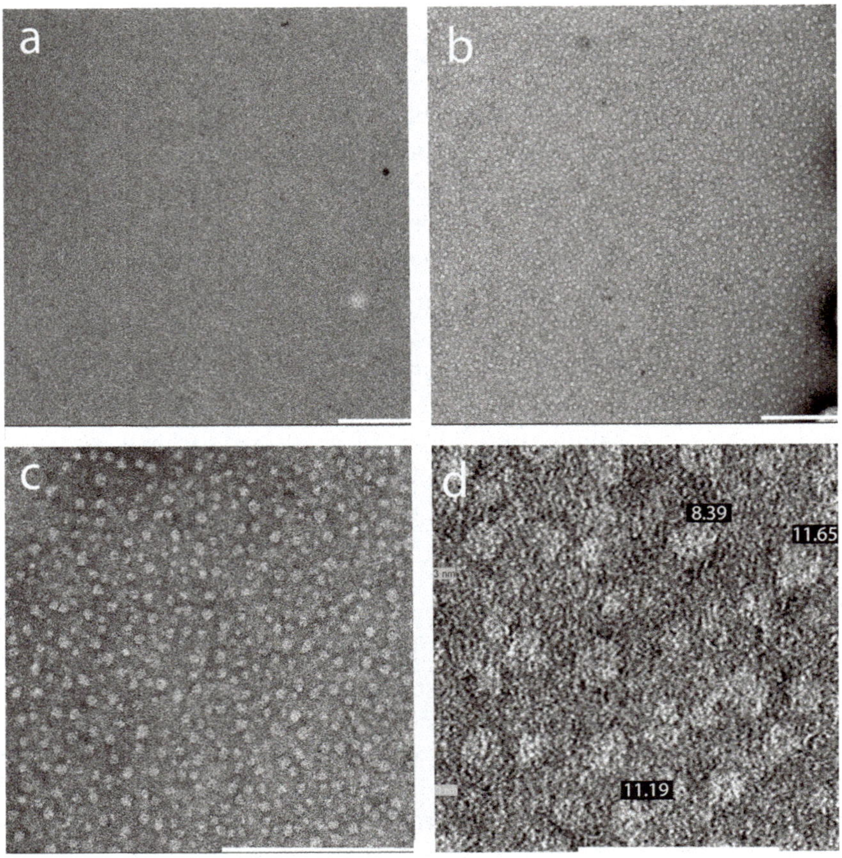

Fig. 1 CS9-immunopurified and DDM-solubilized Cytb for single-particle negative-stain TEM. Single-particle analysis was conducted on DDM-solubilized Cytb following immunoaffinity purification. Cytb purification was conducted as described in Subheading 3.2 but in 0.1% DDM and Relax (+) buffer, eluting with CS9 elution peptide, followed by the removal of peptide by desalting and concentration to a final Cytb concentration of 0.46 μM. The prepared Cytb is then diluted to 100 nM in Relax (+) to achieve a concentration of 0.03% DDM. Samples were then spotted on glow-discharged, carbon-filmed EMS CF400-Cu grids. Staining was conducted with uranyl formate as described in Subheading 3.4. Samples were examined using a LEO 912 AB TEM, equipped with cryostage, CCD camera, and Soft Imaging System analysis software. White bars indicate scale and are 200 nm for images in (**a**) Cytb-free control and Cytb containing samples at magnification 10,000× (**b**) and 20,000× (**c**). The highest magnification was 50,000× for the image shown in (**d**) which was further enlarged digitally to show three particles whose diameters were measured using the camera software. The white bar in (**d**) is 50 nm. In this preparation, the average maximum visual Cytb dimension was 110 ± 24 Å (σ, $n = 40$). In another preparation, using uranyl acetate for staining, the mean maximum dimension was (110 ± 11 Å; σ, $n = 40$). Similar results were obtained on multiple preparations

obtaining ordered arrays of Cytb in giant vesicles (>1 μm) in preparation for 2D crystallization trials. We examined a range of lipid:Cytb ratios, from 30 to 300, under various regimes, including activators such as phosphatidic acid, cofactors, and stabilizing agents such as FAD and Mg^{2+}, pH, temperature, and freeze-thawing, all of which could affect ordering and stability of the vesicle. Figure 2a shows a smaller vesicle from a heterogeneous population which ranged in size from approximately 0.1–0.5 μm and was obtained after 5 days of 4 °C dialysis of a mixture prepared at a DOPC:Cytb molar LPR of 100 with freeze-thaw prior to examination. Ordered linear structures of Cytb in a similar size range of vesicles were obtained at a DOPC:Cytb molar LPR of 300 and dialysis temperature of 20 °C and are shown in Fig. 2b. Formation of domains of semi-ordered linear arrays of Cytb in giant vesicles of as large as 4 μm is obtained at a DMPC:Cytb LPR of 233 and is shown in Fig. 3. These vesicles often break open and lie flat, facilitating EM 2D crystallography. Other relevant individual conditions are provided in the figure legends.

1. Cytb at 16 μM or 1 mg/mL is prepared in Cytb dialysis buffer supplemented with 1.2% OG (*see* Subheading 3.2, **step 14**).
2. 200 μL of the phospholipid, DOPC, is prepared at a concentration of 27.3 mg/mL (10× stock) solubilized in Cytb dialysis buffer plus OG as follows: dry 218 μL of the 25 mg/mL ≥99% purity DOPC chloroform stock in a glass tube under a stream of argon while vortexing. To the dried lipid, add 15 μL of Cytb dialysis buffer with vortexing and gentle sonication (2 × 3 s bursts at setting 2), producing a uniform cloudy suspension with minimal foaming. Add in 10 μL increments 20% OG to clarify solution, taking note how much was added. Add then Cytb dialysis buffer to bring the volume to 200 μL, and vortex and sonicate the mixture.
3. To make 40 μL of LPR 300 DOPC:Cytb solution, 28.2, 4, and 3.84 μL of 16 μM Cytb, Cytb dialysis buffer, and 10× DOPC, respectively, are mixed and incubated at ambient room temperature for 20 min.
4. To stabilize Cytb during dialysis, 2.92 μL of Relax (+) buffer containing 2% DDM is added to the mixture to bring the DDM concentration to 0.14%. The mixture is then allowed to incubate at ambient room temperature for 20 min.
5. 9.5 μL is then loaded into 10 μL spin dialysis chambers and dialyzed, with the chambers inverted and floating, using gentle mixing against 250 mL Cytb dialysis buffer. Dialysis buffer is replaced with fresh solution at 24 h intervals (*see* **Note 8**) (Fig. 2a, b).

3.4 Negative Staining Procedure for Single-Particle DDM-Solubilized Cytb (See Subheading 3.2, Step 14)

1. Freshly dissolve uranyl formate in dH_2O to 7.5 mg/mL and neutralize with 1/100 volume of 1 N NaOH, filter through a Whatman 0.22 μm filter, and wrap in aluminum foil to protect from light (*see* **Note 9**).
2. CF400-CU-50 carbon-coated grids are glow discharged for 30 s in an Agilent Vacuum Coating System bell jar.

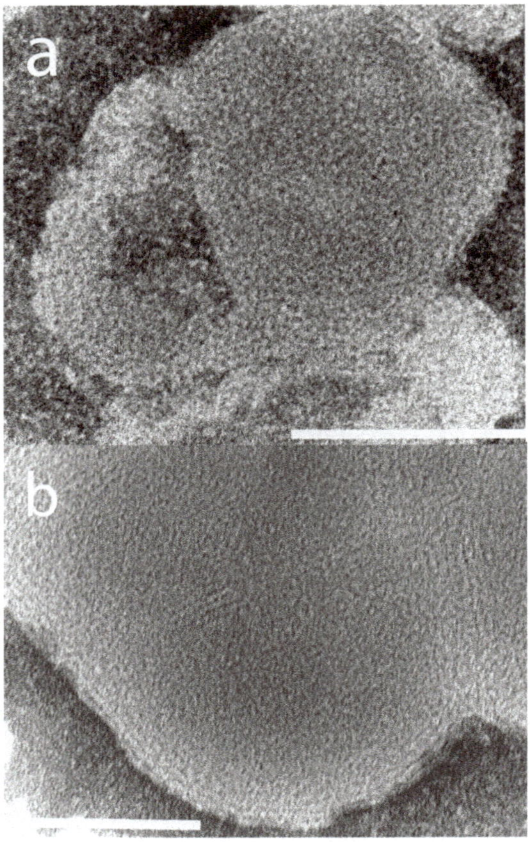

Fig. 2 Negative-stained EM images of immunoaffinity-purified Cytb reconstituted into liposomes at 100:1 and 300:1 DOPC:Cytb molar LPRs. (**a**) DOPC:Cytb ratio of 100. For reconstitution, 9 μL of 1.2 mg/mL Cytb was mixed with 9 μL of 10.5 mg/mL DOPC solution in Cytb dialysis buffer containing 1.2% OG and placed on ice for 20 min after which 0.75 μL of 2% DDM was added (0.14% DDM final). After an additional 20 min on ice, the mixture was deposited into a 10 μL spin dialysis cassette for dialysis against 250 mL of Cytb dialysis buffer for 5 days replacing Cytb dialysis buffer at ~24 h intervals. Prior to application to grid, 6 μL of dialysate was frozen in a 1 mL polyethylene tube dipped into liquid N_2 for 30 s. The sample tube was allowed to thaw for 1 min, and 2 μL was applied to the grid and allowed to equilibrate for 1 min. The sample was negatively stained as described in Subheading 3.5. In this micrograph, a broken or collapsed vesicle is shown with tightly packed Cytb particles spaced about 60–120 Å apart showing two types of faces. One face appears to be a larger roundish structure with a dense central hole and lighter surrounding regions of about 30 Å in dimension. There are also regions devoid of these structures but showing only the smaller lighter density regions but spaced about 60–120 Å apart. One interpretation of this micrograph is that the vesicles present a mixture of cytoplasmic and extracellular Cytb orientations with possible visible substructure. Scale bar is 108 nm. (**b**) DOPC:Cytb ratio of 300. For reconstitution, 28 μL of 16 μM Cytb, 3.84 μL of 34.8 mM DOPC, and 4 μL Cytb dialysis buffer plus OG were mixed and incubated for 20 min at ambient temperature. The mixture was supplemented with 2.9 μL of 2% DDM and incubated at

3. Attach parafilm strips to a clean benchtop.
4. 8 × 50 μL drops of double distilled water are lined up in a row.
5. 2–4 μL of DDM-solubilized Cytb at a concentration of 50–100 nM are pipetted directly on grid for 30 s.
6. Grids are then immediately serially transferred and touched or floated onto eight successive 50 μL washing drops of water. The last wash is about 20 s in duration.
7. Washed grids are then touched to a 50 μL drop of neutralized uranyl formate solution that is freshly made daily.
8. Grids are then blotted and touched to uranyl formate again, then blotted, and allowed to dry in air. They are examined by EM the same day.
9. Negative staining procedures follow those recommended by Walz and coworkers [40].

3.5 Negative Staining Procedures for Cytb Reconstituted in Phospholipid Vesicles

1. Dissolve uranyl acetate to 10 mg/mL in dH_2O and filter.
2. CF400-CU-50 carbon-coated grids (EMS, Hatfield, PA) are glow discharged for 30 s in Agilent Vacuum Coating System bell jar.
3. Attach parafilm strips to a clean benchtop.
4. 2 μL of dialysis sample is pipetted directly on grid for 30 s–1 min.
5. Grid is blotted on Whatman 20 filter paper.
6. Grid is touched to a 50 μL drop of uranyl acetate.
7. Grid is blotted and allowed to dry at ambient temperature (Fig. 3).

Fig. 2 (continued) ambient room temperature for another 20 min. A 9.5 μL volume of the final mixture was then placed in a 10 μL dialysis chamber and dialyzed for 6 days at constant temperature of 20 °C against Cytb dialysis buffer supplemented with 10 μM FAD and 1 μM 10:0 phosphatidic acid. EM grid preparation, sample application, and negative staining were conducted as described in Subheading 3.5. We interpret this micrograph to represent the ordering of Cytb particles at what appear to be boundaries between rounded vesicular regions. The ordered structure appears to be linear with a spacing of 50–55 Å between midlines of higher density. In the upper right, the two linear regions are seen to intersect adjacent to areas where spacing appears more random. Individual round structures of ~100 Å with a central higher density "hole" are observed that are also seen in (**a**). When these round structures are lined up, they appear to form the linear structures observed. The dark lines may represent boundaries formed by lipid and cavities of the molecules as they become organized. These structures may correspond to one of the two types of structures, hypothesized to represent right side-out and inside-out Cytb orientations in the membrane vesicles as observed in (**a**). Scale bar is 100 nm

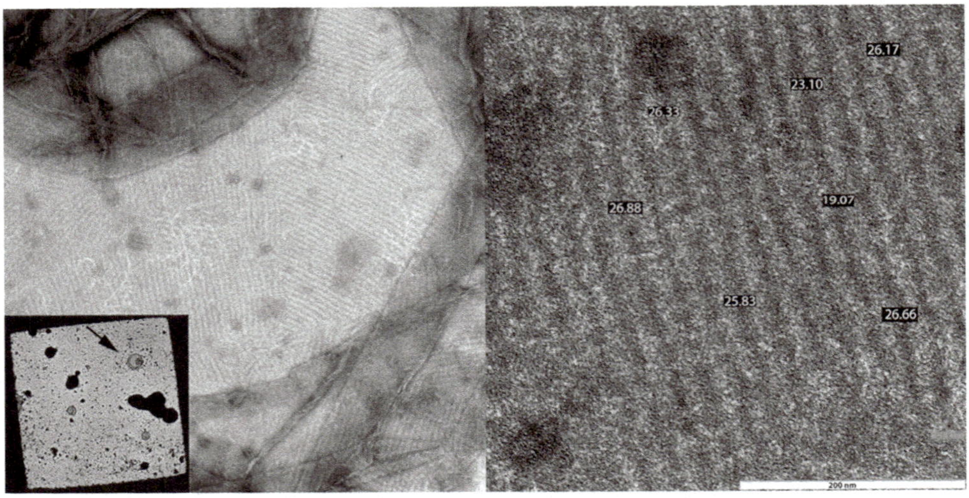

Fig. 3 Negative-stained TEM images of CS9-purified Cytb reconstituted into giant liposomes at intermediate LPR. In order to facilitate eventual, high-resolution structure determination by cryoelectron crystallography [34], our studies attempted to produce two-dimensional crystals of purified Cytb in reconstituted giant phospholipid vesicles at a DMPC:Cytb LPR of 233 at an elevated temperate. Such conditions would allow more mobility of the Cytb in the membrane needed for lateral organization. For these sample preparations, dialysis proceeded in 10 μL dialysis cassettes for 13 days against 250 mL Cytb dialysis buffer supplemented with 50 mM $MgCl_2$ and 10 μM FAD. Final concentrations of Cytb and DDM in the mixture prior to dialysis were 1 mg/mL and 0.31%, respectively. *Left panel*: Domains of ordered two-dimensional alternating linear bands of light and dark density represent Cytb and lipid arrays, respectively, which are not seen in pure lipid controls (not shown). The dimension of the edge of this micrograph is 2.35 μm observed at 10,000×. For scale and vesicle abundance, the inset shows the square, ~37 μm grid openings in a low magnification view (200×) of the ~3.5 μm diameter giant vesicle observed under examination and indicated by the arrow. Note several other smaller vesicles nearby. *Right panel*: A portion of one of the ordered domains of Cytb in the left panel is shown at higher magnification (50,000×, white bar 200 nm). Ordered two-dimensional linear stripes of Cytb can be seen with several measurements of the center-to-center distances between light bands shown in nm in the black rectangles. We interpret the image to represent loosely packed, semi-ordered Cytb with a spacing of about 250 ± 30 Å ($n = 7$), equally split between darkly stained lipid regions and lightly stained protein regions. These results may suggest that the prepared Cytb from pooled (and likely polymorphic) donors either may be too heterogeneous or that conditions are still not permissive for ordered 2D crystalline arrays of the protein. However, the dimensions are consistent with linear semi-order of ~100 Å structures separated by bands of lipids

3.6 Cryoelectron Microscopy (cryoEM) Procedures

CryoEM analysis involves automated collection of digitized images containing single-molecule particles capable of generating electron microscopic contrast using equipment provided as described in Subheading 2.6 (*see* **Note 10**).

1. Detergent-solubilized Cytb samples are first negatively stained and evaluated by conventional TEM to confirm that Cytb particles are observed at an appropriate density as described in Subheading 3.4.

2. Particles of Cytb are captured in thin frozen amorphous ice cryofilms on QUANTIFOIL® holey grids using a Vitrobot sample plunge freezing apparatus. The ice films are between 10 and 100 nm on grids in liquid ethane and capture Cytb in all its orientations as dissolved.
3. The frozen grids are then transferred into liquid N_2 for storage or placed on a Gatan cryo specimen holder for loading into a Philips Tecnai 20—SZCX cryoelectron microscope equipped with a TVIPS 2 K × 2 K Peltier-cooled CCD camera and a Philips Compustage.
4. Regions of particles are identified visually, and the microscope is programmed for automated photography of particle-rich regions far from focus configuration.
5. Digitized images are individually corrected for microscope contrast transfer function using EMAN ctfit.
6. Particles are selected and boxed by EMAN either manually or automatically as is shown in Fig. 4 upper panel.
7. Particles are then further analyzed by the program EMAN [41] which generates class averages, an initial model which is then refined as is shown in Fig. 4 lower panel.
8. Noise is removed by repeated refinement using EMAN refine. These preliminary efforts provided a low resolution (30–40 Å) and confirm the results observed by negative-stain TEM (Fig. 4).

4 Notes

1. The nominal micellar molecular weight of DDM is 72,000; thus concentration may require MW cutoff of 100,000. Studies from our group have indicated improved Cytb stability in DDM [30]. The critical micelle concentration (CMC) of DDM is 0.15 mM or 0.008%. At concentrations near or below the CMC, Cytb may aggregate irreversibly.
2. Following the mAb CS9 purification, the protein G-Sepharose beads can be regenerated for use in multiple preparations by washing with 2–3 bed volumes of cleaning 1 M acetic acid, pH 2.5, followed by re-equilibration with 2–3 bed volumes of binding buffer. For longer periods of storage, GammaBind G-Sepharose should be kept at 4–8 °C in a bacteriostatic solution such as 20% ethanol. The affinity matrix cannot be frozen after washing with PBS. Storage at 4 °C after 0.5 M acetic acid pH 3.0 elution and washing in PBS is possible, if the slurry is supplemented with 0.02% NaN_3.

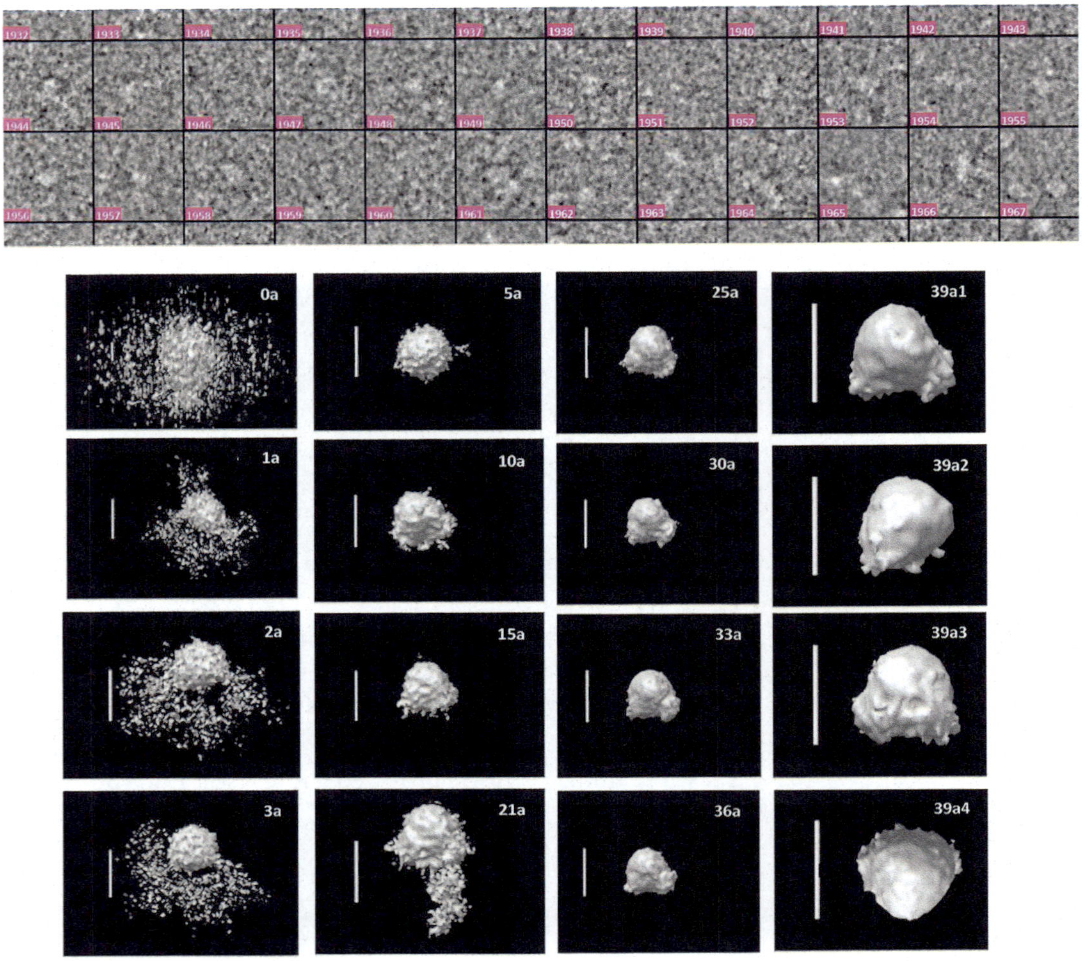

Fig. 4 CryoEM analysis of immunoaffinity-purified, DDM-solubilized Cytb in thin frozen films. *Upper panel:* A small sample of the 16,683 particles (2 runs of 5045 and 11,638) selected by the Tecnai automated cryoelectron microscope are shown as two rows of small square micrographs selected by the "boxer" function of EMAN with index number from 1932 through 1967 in the lower left of each box (*see* Subheading 3.6). These were computationally analyzed for reconstruction of a three-dimensional image from the collection of its class-averaged two-dimensional projections. It is notable that some of these images (10–20%) seem to represent complexes that are seen in higher abundance in the left half of the bottom row. The observed complexes were significantly larger than earlier crude hydrodynamic estimates of Cytb size in nonionic detergents [22]. For these cryoEM reconstructions, it was difficult to identify individual particles because of their relatively featureless and smaller size and because they were similar to some of the amorphous irregularities in the film. We believe the contaminating small percentage (10–20%) of immune complexes of Cytb and mAb CS9 used for immunoaffinity purification (*see* **Note 4**) may have supported the classification and orientation of unassociated particles, since their orientation relative to the easily visible binding Fab arm of the mAb would be fixed and thus allow development of classes shared between the unassociated and associated particles. The frozen films were produced and then examined at the National Resource for Automated Molecular Microscopy in La Jolla CA, where the far from focus particle images were collected. *Lower panel*: Cytb structure analysis using the particle images described in the upper panel. For these reconstructions, a variety of structure classes of Cytb alone, as well as some of the putative Cytb-CS9 antibody complexes, were apparent during refinement analysis in EMAN. The evaluation of these structure classes permitted orientation of the free, detergent-solubilized and putative Cytb-CS9 complexes around

3. Following the 1 M NaCl wash, membrane fractions can vary in the ease with which they are dispersed. To increase the extraction efficiency, this homogenization step should be carried out until the membranes appear optimally dispersed without visible chunks or aggregates.
4. Following peptide elution from the affinity matrix, mAb CS9 is occasionally observed as a minor contaminant of Cytb preparations (as confirmed by Western blot analysis and MALDI peptide mass mapping). This leakage probably occurs because of the incomplete removal of non-covalently bound CS9 from the CS9 affinity matrix. The contaminant antibody then elutes as non-covalently coupled CS9 that reassociates after the removal of peptide or as a Cytb-CS9 complex which does not dissociate under peptide elution conditions. To minimize this problem, the affinity matrix can be thoroughly washed and equilibrated overnight, followed by repeat washing under the conditions of binding and elution provided in Subheading 3.2, **step 10**. The final washing can also be done at elevated temperature of 20–25 °C without affecting antibody effectiveness. Alternatively, to remove any contaminating CS9-Cytb complexes from this preparation, an HPLC-size exclusion chromatography step using Agilent Zorbax GF 450 columns equilibrated in 0.1–0.05% DDM in Relax buffer may be used to separate the contaminating complexes of CS9 and Cytb. The size-separated antibody-free fractions can then be concentrated and used as described herein.
5. Concentration of purified Cytb samples by the MW 50,000 cutoff membranes proves critical for control of final OG levels as free OG micelles have been shown to pass through the membrane and do not concentrate with purified protein-detergent complex.
6. Our studies have indicated efficient elution of Cytb in DDM from the affinity matrix at room temperature without significant loss of spectral activity. The $t_{1/2}$ for spectral stability at

Fig. 4 (continued) common features. Thirty nine cycles of refinement were carried out and are so labeled in the lower set of images. Lower case letters correspond to different views of the structure. Ultimately, it became clear that one common feature appeared to be attached to an elongated cigar-shaped element as a knob at its terminus (21a). Upon further refinement, the elongated component disappeared, and what remained was the roundish shape with an axis of symmetry perpendicular to the round face of the featureless knob that nearly all images contained (25a–36a). We continued refinement until the Fourier shell correlation (FSC) converged (refinement cycle 39). Estimated resolution is 35–40 Å or better. The rightmost column shows four orthogonal views of the EMAN three-dimensional reconstruction (bar 100 Å). Such a shape was compatible with the crystal structure of dimeric cytochrome b6f and a crude model of the Cytb heterotetramer (2 × p22^{phox} and 2 × gp91^{phox}) considering that an annulus of detergent would span the transmembrane region as noted in Fig. 5

room temperature in OG is 3 h, while for DDM it is more than 10 h [30].

7. The absorbance spectrum of detergent-solubilized Cytb is indistinguishable from the original material following a single freeze-thaw cycle. However, one should also assess by immunoblot analysis or SDS-PAGE that proteolytic fragments have not been produced and that higher-order aggregates are not observed. OG-solubilized Cytb loses spectral activity slowly with time at 4 °C ($t_{1/2}$~1–2 days). To maintain structural and functional integrity, storage at −20 °C with 20% glycerol is recommended. In DDM, the protein has much higher stability at 4 °C and does not show loss of spectral activity after >1 week.
8. This procedure will allow comparison of 4 × 10 μL conditions of dialysis, e.g., neat, + 50 mM $MgCl_2$, 10 μM FAD, and 1 μM 10:0 phosphatidic acid. For Fig. 2b, the retentate was removed and examined after 6 days. Variables of this procedure include Cytb-stabilizing ligands FAD, NADPH, and Mg^{2+} and dialysis temperature.
9. For reconstituted membranes, uranyl acetate solution was substituted for uranyl formate with no washes.
10. CryoEM reconstructions represent preliminary studies using equipment and software that are no longer current. Technical advances in microscopy and software allow much higher resolution to be obtained [35, 36]. These preliminary studies were included, because they support the negative-stain data and because they represent an important future for the structural biology of Cytb (Fig. 5).

Acknowledgments

We thank the Montana State University College of Letters and Sciences for funding the Sabbatical leave during the 2007–2008 academic year for AJJ during which the initial phases of this work were carried out. We also acknowledge the PHS grant 5R01AI26711 and the ARRA supplement for support during the subsequent period. Thanks also go to J. Quispe, B. Carragher, and C. Potter of the National Resource for Automated Molecular Microscopy at the Scripps Research Institute and S. Ludtke of the Baylor College of Medicine for help in obtaining and processing the cryoEM images. Lastly, special thanks go to the Thermal Biology Institute, NSF EPSCOR, and The Montana Nanotechnology Facility (MONT) for funding the negative-stain electron microscopy time.

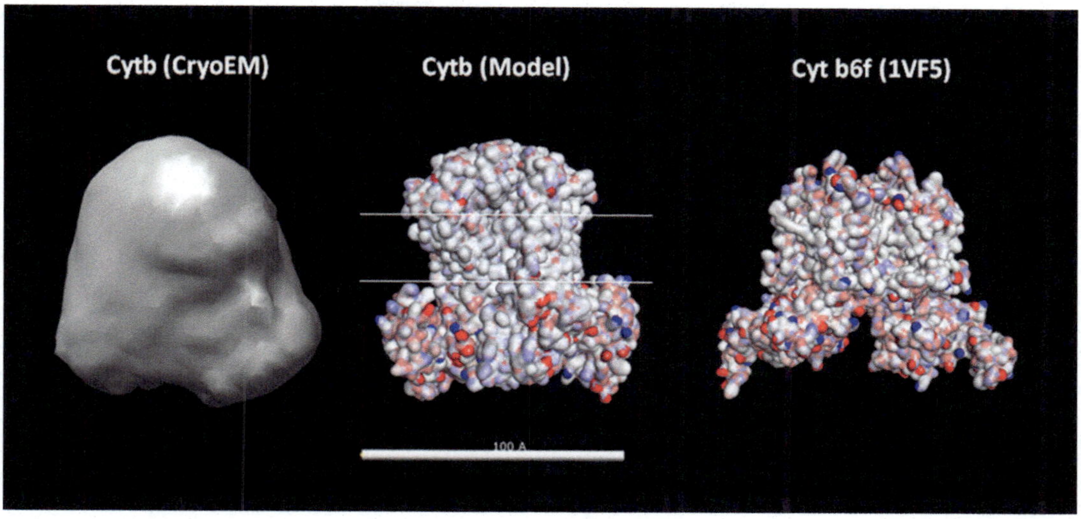

Fig. 5 Molecular models of Cytb are consistent with electron microscopy. A side view of the smoothed surface model derived from cryoEM image analysis using the program EMAN (*see* **Note 10**) is shown on the left and in our interpretation may represent a view along the membrane plane of the detergent-solubilized Cytb molecule. The model could be interpreted to suggest a relatively featureless structure with a rounded putative extracellular face and two cytoplasmic domains as shown with the extracellular face pointing up. This cryoEM-generated surface model is consistent with Cytb as a dimer of heterodimers of 2 $\times$ p22^{phox} and 2 $\times$ gp91^{phox} subunits shown in the center with the same orientation and double horizontal lines showing the placement of the membrane. This Cytb model was constructed using a Linux version of the Discovery 3.1 (Accelrys Software Inc., San Diego), a commercial molecular modeling program, by homology to two integral membrane b-type cytochromes, alignments suggested by Phyre protein fold recognition server (http:/www.sbg.bio.ic.ac.uk/phyre/html/), and by membrane minimization. It is useful mainly to group extracellular, intracellular, and transmembrane domains into a compact structure of a dimer of two p22^{phox}/gp91^{phox} heterodimers (A. Jesaitis, unpublished). The Cytb molecular model omits the detergent annulus that would necessarily fill in the membrane core of the protein [42] and result in a structure more closely resembling the EMAN reconstruction. As a comparison, the surface map derived from the crystal structure of cytochrome b6f of spinach chloroplasts (1VF5) is shown on the right. It is stripped of accessory polypeptide chains E–H and R–U and has a molecular size of 1588 amino acid residues, similar to that of Cytb which has 1530 amino acid residues as shown in the center. The two models of the two integral membrane b-type cytochromes support the size estimate from the EMAN three-dimensional reconstruction and negative-stain data and are displayed normalized for size (bar 100 Å)

References

1. Sumimoto H (2008) Structure, regulation and evolution of Nox-family NADPH oxidases that produce reactive oxygen species. FEBS J 275 (13):3249–3277
2. Dinauer MC (2016) Primary immune deficiencies with defects in neutrophil function. Hematology Am Soc Hematol Educ Program 2016 (1):43–50. https://doi.org/10.1182/asheducation-2016.1.43
3. Azcutia V, Parkos CA, Brazil JC (2017) Role of negative regulation of immune signaling pathways in neutrophil function. J Leukoc Biol. https://doi.org/10.1002/jlb.3mir0917-374r
4. Morgenstern DE, Gifford MA, Li LL, Doerschuk CM, Dinauer MC (1997) Absence of respiratory burst in X-linked chronic granulomatous disease mice leads to abnormalities in both host defense

and inflammatory response to Aspergillus fumigatus. J Exp Med 185(2):207–218

5. Klebanoff SJ (1999) Oxygen metabolites from phagocytes. In: Gallin JI, Goldstein IM, Snyderman R (eds) Inflammation: basic principle and clinical correlates. Lippincott Wiliams and Wilkins, Philadelphia, PA, pp 721–768
6. Schieber M, Chandel NS (2014) ROS function in redox signaling and oxidative stress. Curr Biol 24(10):R453–R462. https://doi.org/10.1016/j.cub.2014.03.034
7. Miller RA, Britigan BE (1997) Role of oxidants in microbial pathophysiology. Clin Microbiol Rev 10:1–18
8. Beckman JS, Chen J, Ischiropoulos H, Crow JP (1994) Oxidative chemistry of peroxynitrite. Methods Enzymol 233:229–240
9. Anderson MM, Hazen SL, Hsu FF, Heinecke JW (1997) Human neutrophils employ the myeloperoxidase-hydrogen peroxide-chloride system to convert hydroxy-amino acids amino acids into glycoaldehyde, 2-hydroxy-propanol, and acrolein. A mechanism for the generation of highly reactive alpha-hydroxy and alpha-beta unsaturated aldehydes by phagocytes at site of inflammation. J Clin Invest 99(424):432
10. Marcinkieowicz J (1997) Neutrophil chloramines the missing link between innate and aquired immunity. Immunol Today 18:577–580
11. Nauseef WM (2007) How human neutrophils kill and degrade microbes: an integrated view. Immunol Rev 219:88–102
12. McDonald B, Kubes P (2012) Neutrophils and intravascular immunity in the liver during infection and sterile inflammation. Toxicol Pathol 40(2):157–165
13. Weitzman S, Weitberg AB, Clark EP, Stossel TP (1985) Phagocytes as carcinogens: malignant transformation produced by human neutrophils. Science 227:1231–1233
14. Weitzman SA, Stossel TP (1981) Mutation caused by human phagocytes. Science 212:546–547
15. Davies KJ (1995) Oxidative stress: the paradox of aerobic life. Biochem Soc Symp 61:1–31
16. Ward PA (2010) Oxidative stress: acute and progressive lung injury. Ann N Y Acad Sci 1203:53–59
17. Soehnlein O (2012) Multiple roles for neutrophils in atherosclerosis. Circ Res. 110 (6):875–88
18. Quinonez-Flores CM, Gonzalez-Chavez SA, Del Rio Najera D, Pacheco-Tena C (2016) Oxidative stress relevance in the pathogenesis of the rheumatoid arthritis: a systematic review. Biomed Res Int 2016:6097417. https://doi.org/10.1155/2016/6097417
19. Walder CE, Green SP, Darbonne WC, Mathias J, Rae J, Dinauer MC, Curnutte JT, Thomas GR (1997) Ischemic stroke injury is reduced in mice lacking a functional NADPH oxidase. Stroke 28(11):2252–2258
20. Dinauer MC, Curnutte JT, Rosen H, Orkin SH (1989) A missense mutation in the neutrophil cytochrome b heavy chain in cytochrome-positive X-linked chronic granulomatous disease. J Clin Invest 84:2012–2016
21. Parkos CA, Dinauer MC, Walker LE, Allen RA, Jesaitis AJ, Orkin SH (1988) Primary structure and unique expression of the 22-kilodalton light chain of human neutrophil cytochrome b. Proc Natl Acad Sci U S A 85:3319–3323
22. Parkos CA, Allen RA, Cochrane CG, Jesaitis AJ (1988) The quaternary structure of the plasma membrane b-type cytochrome of human granulocytes. Biochim Biophys Acta 932:71–83
23. Burritt JB, Quinn MT, Jutila MA, Bond CW, Jesaitis AJ (1995) Topological mapping of neutrophil cytochrome b epitopes with phage- display libraries. J Biol Chem 270 (28):16974–16980
24. Burritt JB, Foubert TR, Baniulis D, Lord CI, Taylor RM, Mills JS, Baughan TD, Roos D, Parkos CA, Jesaitis AJ (2003) Functional epitope on human neutrophil flavocytochrome b558. J Immunol 170(12):6082–6089
25. Burritt JB, Busse SC, Gizachew D, Siemsen DW, Quinn MT, Bond CW, Dratz EA, Jesaitis AJ (1998) Antibody imprint of a membrane protein surface. Phagocyte flavocytochrome b. J Biol Chem 273(38):24847–24852
26. Burritt JB, DeLeo FR, McDonald CL, Prigge JR, Dinauer MC, Nakamura M, Nauseef WM, Jesaitis AJ (2001) Phage display epitope mapping of human neutrophil flavocytochrome b558. Identification of two juxtaposed extracellular domains. J Biol Chem 276 (3):2053–2061. https://doi.org/10.1074/jbc.M006236200
27. Ramaraj T, Angel T, Dratz EA, Jesaitis AJ, Mumey B (2012) Antigen-antibody interface properties: Composition, residue interactions, and features of 53 non-redundant structures. Biochim Biophys Acta 1824(3):520–532
28. Lord CI, Riesselman MH, Gripentrog JM, Burritt JB, Jesaitis AJ, Taylor RM (2008) Single-step immunoaffinity purification and functional reconstitution of human phagocyte flavocytochrome b. J Immunol Methods 329 (1–2):201–207

29. Taylor RM, Jesaitis AJ (2007) Immunoaffinity purification of human phagocyte flavocytochrome b and analysis of conformational dynamics. Methods Mol Biol 412:429–437. https://doi.org/10.1007/978-1-59745-467-4_26
30. Taylor RM, Burritt JB, Foubert TR, Snodgrass MA, Stone KC, Baniulis D, Gripentrog JM, Lord C, Jesaitis AJ (2003) Single-step immunoaffinity purification and characterization of dodecylmaltoside-solubilized human neutrophil flavocytochrome b. Biochim Biophys Acta 1612(1):65–75
31. Riesselman M, Jesaitis AJ (2014) Affinity purification and reconstitution of human phagocyte flavocytochrome B for detection of conformational dynamics in the membrane. Methods Mol Biol 1124:413–426. https://doi.org/10.1007/978-1-62703-845-4_24
32. Taylor RM, Lord CI, Riesselman MH, Gripentrog JM, Leto TL, McPhail LC, Berdichevsky Y, Pick E, Jesaitis AJ (2007) Characterization of surface structure and p47*phox* SH3 domain-mediated conformational changes for human neutrophil flavocytochrome b. Biochemistry 46(49):14291–14304
33. Taylor RM, Burritt JB, Baniulis D, Foubert TR, Lord CI, Dinauer MC, Parkos CA, Jesaitis AJ (2004) Site-specific inhibitors of NADPH oxidase activity and structural probes of flavocytochrome b: characterization of six monoclonal antibodies to the p22*phox* subunit. J Immunol 173(12):7349–7357
34. Kuhlbrandt W (1994) Two-dimensional crystallization of membrane proteins: a practical guide. In: Hunte CVJG, Schragger H (eds) Membrane protein purification and crystalization. Academic Press, San Diego, pp 253–280
35. Chiu W, Downing KH (2017) Editorial overview: cryo electron microscopy: exciting advances in CryoEM herald a new era in structural biology. Curr Opin Struct Biol 46: iv–viii. https://doi.org/10.1016/j.sbi.2017.07.006
36. Boisset N, Penczek P, Pochon F, Frank J, Lamy J (1993) Three-dimensional architecture of human alpha 2-macroglobulin transformed with methylamine. J Mol Biol 232(2):522–529. https://doi.org/10.1006/jmbi.1993.1408
37. Parkos CA, Allen RA, Cochrane CG, Jesaitis AJ (1987) Purified cytochrome b from human granulocyte plasma membrane is composed of two polypeptides with relative molecular weights of 91,000 and 22,000. J Clin Invest 80:732–742
38. Carragher B, Kisseberth N, Kriegman D, Milligan RA, Potter CS, Pulokas J, Reilein A (2000) Leginon: an automated system for acquisition of images from vitreous ice specimens. J Struct Biol 132(1):33–45. https://doi.org/10.1006/jsbi.2000.4314
39. Potter CS, Chu H, Frey B, Green C, Kisseberth N, Madden TJ, Miller KL, Nahrstedt K, Pulokas J, Reilein A, Tcheng D, Weber D, Carragher B (1999) Leginon: a system for fully automated acquisition of 1000 electron micrographs a day. Ultramicroscopy 77(3–4):153–161
40. Ohi M, Li Y, Cheng Y, Walz T (2004) Negative staining and image classification – powerful tools in modern electron microscopy. Biol Proced Online 6:23–34. https://doi.org/10.1251/bpo70
41. Ludtke SJ (2010) 3-D structures of macromolecules using single-particle analysis in EMAN. Methods Mol Biol 673:157–173. https://doi.org/10.1007/978-1-60761-842-3_9
42. Chen PC, Hub JS (2015) Structural properties of protein-detergent complexes from SAXS and MD simulations. J Phys Chem Lett 6(24):5116–5121. https://doi.org/10.1021/acs.jpclett.5b02399

Chapter 4

Purification and Characterization of DUOX Peroxidase Homology Domains (PHDs)

Jennifer L. Meitzler

Abstract

The dual oxidase (DUOX) enzymes (DUOX1 and DUOX2) are unique hydrogen peroxide (H_2O_2)-producing members of the NADPH oxidase (NOX) family, structurally distinguished from their related NOX isoforms by the presence of an additional N-terminal extracellular domain. This region has significant sequence and predicted structural homology to mammalian peroxidases, including myeloperoxidase (MPO) and lactoperoxidase (LPO), therefore justifying the nomenclature of the peroxidase homology domain (PHD). Obtaining detailed structural information and defining a function for this appended region are both critical for elucidation of the uncharacterized mechanism of H_2O_2 production by DUOX proteins. Purification strategies focused on isolated sections of each DUOX enzyme are a logical means to further characterization, particularly as isolation of the complete membrane-bound enzyme in significant quantities remains unachievable. In this chapter, a reproducible method for production of the homology domain applicable to both human DUOX isoforms is described. The approach utilizes a baculovirus expression vector in insect cell culture to produce secreted recombinant PHD; an appended C-terminal His_6 affinity tag was found to be crucial for structural stability. Finally, initial characterization of the activity of the purified PHDs is also described.

Key words Dual oxidase, Mammalian peroxidase, Baculovirus, Hydrogen peroxide

1 Introduction

Oxidative stress derived from endogenous sources has been implicated in a wide variety of human pathologies, fostered in part by a family of enzyme complexes that generate ROS (both superoxide and H_2O_2) and are known as the NADPH oxidases (NOXs). NOX enzymes are now recognized as contributors to the maintenance of cellular homeostasis through deliberate generation of ROS. All known NADPH oxidase enzymes are transmembrane (TM) proteins, with six helical TM sequences ligating two heme molecules that facilitate intramolecular electron transport from NADPH to oxygen for ROS production. The dual oxidase enzymes (DUOX1 and DUOX2) are unique calcium-activated members of the NOX family, first identified as H_2O_2 sources involved in iodide

Ulla G. Knaus and Thomas L. Leto (eds.), *NADPH Oxidases: Methods and Protocols*, Methods in Molecular Biology, vol. 1982, https://doi.org/10.1007/978-1-4939-9424-3_4, © Springer Science+Business Media, LLC, part of Springer Nature 2019

organification in the mammalian thyroid gland and distinguished by the additional presence of an N-terminal extracellular peroxidase homology domain (PHD) [1, 2]. To be functionally active, DUOX enzymes require interaction with maturation partners (DUOXA), providing stable cell heterodimers and co-translocation to the plasma membrane surface [3, 4]. Through studies with chimeric protein constructs, it is apparent that the COOH-terminus of DUOXA1 is critical for DUOX1-dependent H_2O_2 generation, and the NH_2-terminal tail of DUOXA2 defines the type of ROS released by DUOX2, either H_2O_2 or superoxide [5].

The size of each human DUOX (hDUOX) isoform PHD was defined through identification of the first transmembrane domain, utilizing the TMHMM server version 2.0 transmembrane helix algorithm. The resulting outer cellular regions, constituting the PHDs, were identified for hDUOX1 as amino acids 1–593 ($hDUOX1_{1-593}$) and amino acids 1–599 for hDUOX2 ($hDUOX2_{1-599}$), with sequence alignments indicating ~80% identity between the two domains [6, 7]. Both PHDs display significant sequence homology with the mammalian peroxidases thyroperoxidase (TPO) and LPO. Both isoforms have a calcium-binding site in the N-terminal peroxidase domain that mimics that found in the mammalian peroxidases (Fig. 1A) [6, 8]. Mutation of the cysteine residues in the PHD of hDUOX disrupts targeting and extracellular H_2O_2 production, indicating that these residues are critical for proper maturation and function [9, 10]. Alternatively, tissue-specific co-expression of DUOX with one or more of the mammalian peroxidases has also been observed [11, 12]. This is consistent with the perception that the N-terminal PHD is a site for mammalian peroxidase binding rather than having a catalytic function.

The ultimate goal of cell culture engineering is to obtain an active product in reliable quantities to support investigative efforts. Unfortunately, a general problem for heterologous protein expression is the low fraction of soluble and/or correctly assembled protein that is often achieved. Baculovirus expression systems have been successfully utilized to obtain significant portions of soluble protein with similar post-translational modifications (including oligomerization, phosphorylation, glycosylation, acylation, disulfide bond formation, and proteolytic cleavage) as those produced in mammalian cells [13]; examples include human kinases MKK1 or MEK1 [14, 15], members of the adenosine triphosphate (ATP)-binding cassette superfamily [16–18], cytochrome P450/oxidoreductase complexes [19], and lactoperoxidase (LPO) [20]. To establish whether covalent heme attachment occurred by an enzyme-dependent process for LPO, recombinant expression of the bovine LPO homolog was achieved through a baculovirus system [20]. This in vitro derivative was indistinguishable from the native protein, except for the degree of covalent heme modification, indicating that recombinant expression of similar proteins might be achieved through this system.

A.

```
hDUOX1   96  tvlgvffgyhvlsdlvs / 163 rdpanqvtgwldgsaiygsshsws
hDUOX2   102 tvlgvffgyhvlsdvvs / 169 rdlanqvtgwldgsaiygsshsws
hTPO     107 sdllmawgqyidhdiaf / 194 rqqmngltsfldastvygsspale
bLPO     94  sllfmqwgqivdhdldf / 174 reqinavtsfldaslvygsepsla
hMPO     109 slmfmqwgqlldhdldf / 187 rnqinaltsfvdasmvygseepla
hEPO     101 almfmqwgqfidhdldf / 179 rnqinaltsfvdasmvygsevsls
             : : : :*  :  *:          *:  * :*.::*.* :***.
```

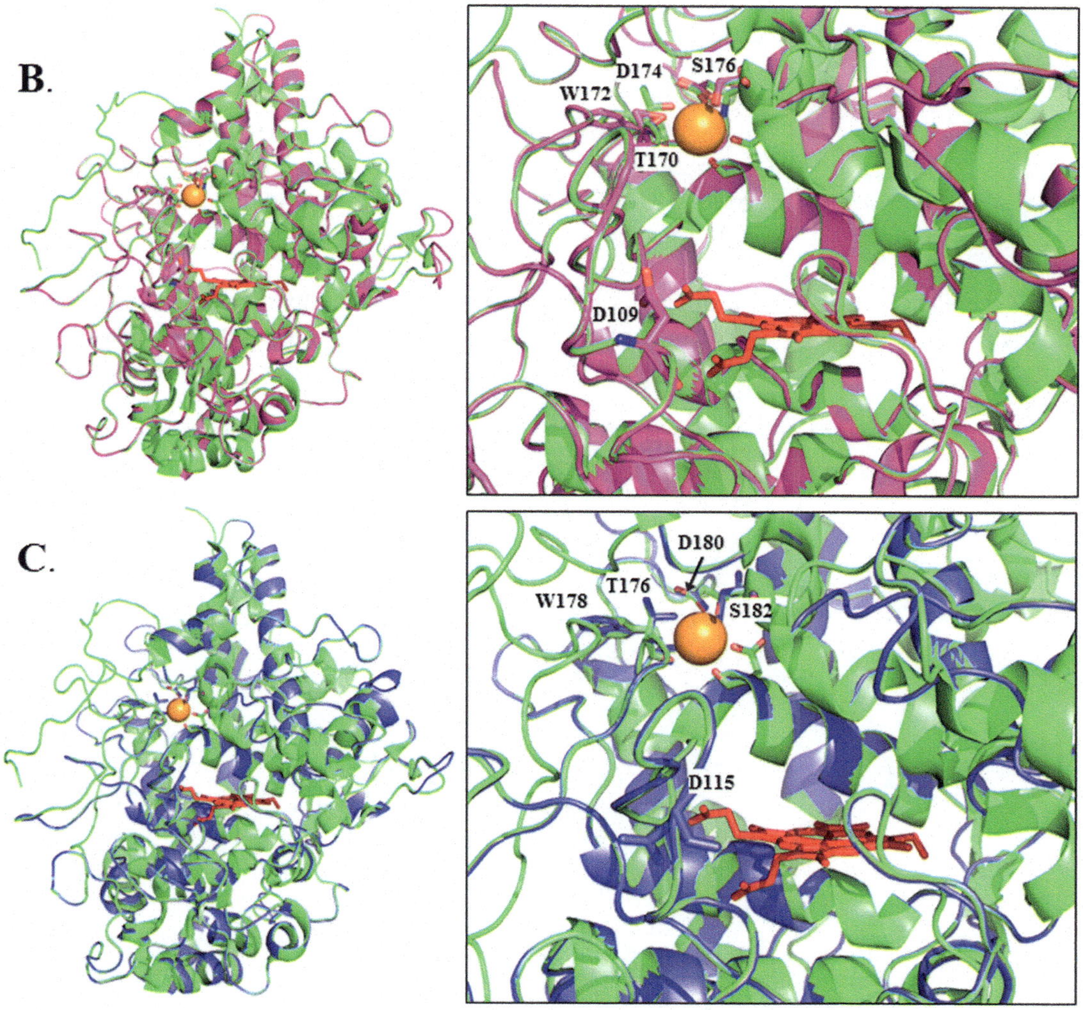

Fig. 1 Sequence and structural comparison of a conserved calcium-binding domain. (**A**) Sequence alignment of classical peroxidase domains with the DUOX proteins. Highlighted residues (red) focus on residues responsible for direct association with calcium. *hDUOX1*, human dual oxidase 1; *hDUOX2*, human dual oxidase 2; *hTPO*, human thyroid peroxidase; *bLPO*, bovine lactoperoxidase; *hMPO*, human myeloperoxidase; *hEPO*, human eosinophil peroxidase. (**B**) Structural model of hDUOX1$_{1-593}$ (purple) overlaid on the structure of LPO (PDB: 2IKC, green) (left). The calcium-binding domain residues are highlighted in stick representation with amino acid labeling for hDUOX1 (right) demonstrating their structural conservation surrounding calcium (orange) (*see* **Note 12**). (**C**) Structural model of hDUOX2$_{1-599}$ (blue) overlaid on the structure of LPO (PDB: 2IKC, green) (left). The calcium-binding domain residues are highlighted in stick representation with amino acid labeling for hDUOX2 (right). Heme mapped with the structure of LPO is highlighted in red

Previous success with a baculovirus system to stably express a structurally homologous protein (LPO) justified its use as a starting point for DUOX PHD expression and purification attempts. Herein a detailed protocol for preparing purified human PHD protein is described. A method for performing tyrosine ethyl ester and superoxide dismutase assays to characterize the function of these controversial regions is also summarized.

2 Materials

Prepare all solutions using ultrapure water and cell culture-grade reagents. Follow all MSDS instructions regarding handling of reagents and appropriate waste disposal guidelines when disposing of waste materials.

2.1 Generation of Viral Particles to Achieve PHD Expression

1. Human $duox1_{1\text{–}593}$ –His_6 (isolated from Quick-clone cDNA from the human lung, Clontech) and human $duox2_{1\text{–}599}$–His_6 (synthesized by GeneArt Inc.) were ligated into the specialized baculovirus expression vector pAcGP67-b (BD Biosciences), which affords a secreted, recombinant protein after viral incorporation, expansion, and infection (*see* **Note 1**) [6, 7].
2. BD Baculogold Transfection Kit with linearized Baculogold DNA (*see* **Note 1**).
3. Sf9 cells (Invitrogen) grown in ExCell® 420 insect medium (SAFC Biosciences) supplemented with glutamine (2.7 g/L) (*see* **Note 2**).
4. High Five™ (H5) cells grown in Express Five™ SFM medium (Invitrogen) supplemented with glutamine (2.7 g/L) and 10% fetal bovine serum (*see* **Note 2**).
5. LM agarose solution: 500 mg low melting point plaque assay agarose +5 mL H_2O, pre-autoclaved in a 100 mL bottle.
6. 60 mm sterile plates.
7. 27 °C incubator.
8. Microwave.
9. Penicillin-streptomycin solution (100×).
10. Neutral red dye solution: 100 mg neutral red dye dissolved in 10 mL H_2O, filtered through a 0.22 μm Millipore Express PES membrane syringe filter.

2.2 PHD Purification

1. 5-Aminolevulinic acid (ALA) (*see* **Note 3**).
2. Penicillin-streptomycin solution (10,000 U/mL penicillin, 10,000 μg/mL streptomycin, 100×).
3. Fungizone (250 μg/mL solution).

4. Spiral-wound regenerated cellulose membrane (molecular mass, 10 kDa cutoff; Millipore) connected to an Amicon TCF10 peristaltic pump equipped with a 2l RA2000S reservoir.
5. Phenylmethylsulfonyl fluoride (PMSF): 10 mL saturated solution in isopropanol.
6. Centrifuge (for large volumes).
7. Ultracentrifuge.
8. Nickel-nitrilotriacetic acid (Ni-NTA) agarose and Ni-NTA Superflow Cartridge, 5 mL (Qiagen).
9. Amersham Biosciences 900 Series FPLC with fraction collector.
10. Equilibration buffer: 20 mM phosphate buffer, 400 mM NaCl, pH 8.0 (4 °C).
11. Wash buffer 1: 20 mM imidazole added to equilibration buffer, adjusted to pH 8.0.
12. Wash buffer 2: 40 mM imidazole added to equilibration buffer, adjusted to pH 8.0.
13. Elution buffer: 200 mM imidazole added to equilibration buffer, adjusted to pH 8.0.
14. Dialysis cassette (20,000 MW cutoff).
15. YM-50 Centriprep Centrifugal Filter Unit (EMD Millipore).

2.3 Characterization of PHD Activity (See Note 4)

1. Hydrogen peroxide (H_2O_2) (for experiments utilizing H_2O_2, determine concentrations spectrophotometrically at 240 nm by using the extinction coefficient $\epsilon = 43.6\ M^{-1}\ cm^{-1}$).
2. Phosphate-buffered saline (PBS), pH 7.4.
3. Chelex-100 resin (Bio-Rad).
4. Tyrosine ethyl ester (TyrEE).
5. Lactoperoxidase (bovine, LPO) (Sigma-Aldrich).
6. Stir bar and quartz spectroscopy cuvette.
7. Fluorolog-3 spectrofluorometer (Horiba Jobin Yvon).
8. 50 × 2.1 mm C18 reversed-phase column (Waters XTerra MS).
9. 1200 Series HPLC (Agilent).
10. Cary 50 Bio UV-visible spectrophotometer (Varian).
11. Superoxide dismutase (SOD) assay kit (Fluka).
12. 50 mM HEPES buffer, 5% glycerol, pH 7.0.

3 Methods

It is important to mention that baculovirus expression vector systems, while often more complex to establish, garner significant advantages over prokaryotic expression systems. These advantages include expression of multiple proteins simultaneously in a single infection and production of multimeric proteins sharing functional similarity with their natural analogs; expression toxicity may also be avoided through secreted protein expression systems. Many kit-based protocols for co-transfection with viral DNA are commercially available and straightforward to follow [21]. However, a more difficult and crucial part of the baculovirus system process is the plaque assay, used both for generation of single viral populations (when the final viral stock may contain a mixture of parental and recombinant viruses) and to accurately infect cells for optimal protein overexpression at a specific multiplicity of infection (MOI). A method for this technique and subsequent PHD expression/purification are provided. All baculovirus generation steps should be carried out in a sterile environment, to prevent bacterial or fungal contamination.

3.1 Generation of Recombinant Baculoviruses for hDUOX PHD Expression: Plaque Assay Essentials

Plasmids for secreted expression of hDUOX1 and hDUOX2 PHD proteins must be co-transfected with viral DNA into Sf9 insect cells; commercial transfection kits are readily available (*see* **Note 1**). The resulting viruses are plaque-assayed, as outlined below, to generate high-titer recombinant baculovirus stocks before infection of High Five cells for PHD expression and purification.

1. Follow manufacturer instructions precisely for co-transfection of linearized baculovirus DNA and plasmid DNA (containing the hDUOX PHD genes). Transfection solutions will result from a mixture of baculovirus DNA and plasmid DNA in provided buffers (*see* **Note 5**). Incubate the resulting transfection solution(s) on Sf9 cells, pre-attached to 60 mm plates for 4 h at 27 °C.
2. Remove and discard the transfection solutions; wash cells gently with 3 mL ExCell 420 media; replace with 3 mL fresh ExCell 420 media; and incubate for 4 days at 27 °C in a humid environment.
3. Seed 60 mm plates with 2.5×10^6 Sf9 cells, and allow the cells to attach for 45 min at 27 °C.
4. Harvest hDUOX1 and hDUOX2 transfection solutions from plates, and place in Eppendorf tubes. Centrifuge at 8,600 × *g*, room temperature, for 2–3 min in a microcentrifuge.
5. Serially dilute each solution in ExCell 420 media from 10^{-1} to 10^{-5}. When creating diluted solutions, be sure to change pipette tips frequently and mix solutions thoroughly.

6. Verify Sf9 cell adhesion by microscope. Draw off media and replace with 1 mL fresh ExCell 420 media +100 μL viral serial dilution solution and incubate for 1 h at room temperature.
7. Prepare complete agarose solution for plaque assay addition: Heat the pre-autoclaved LM agarose solution bottle in the microwave until a thick liquid consistency of agarose is reached. Add 15 mL room temperature ExCell media and mix. If the agarose is not fully dissolved, reheat briefly. Add 30 mL room temperature media and 1 mL penicillin-streptomycin (100×) when the bottle is once again cool to the touch. Complete agarose solution is ready for addition to plates when close to room temperature, but not yet beginning to solidify.
8. Remove the diluted viral solution from each 60 mm plate carefully; each plate must be completely dry.
9. Tilt each plate and add 5 mL complete agarose solution to the edge; lay the plate flat for the solution to gently flow over the attached cells.
10. Place plates at 27 °C for 8 days in a humid environment (*see* **Note 6**).
11. Remove excess water from plate lids and the agarose layer in a sterile environment with a pipettor.
12. Add 1 mL of neutral red dye solution to 50 mL prepared complete agarose solution (as described above). Carefully pipette 4 mL agarose solution containing dye on top of the previous agarose layer (make sure the agarose being added is cool enough so as not to melt the first layer). Store plates overnight at 27 °C.
13. Plaques are now visible as white or light red circles surrounded by red stain.

3.2 Expression of hDUOX PHD Constructs

Single viral plaques, which demonstrate viral activity, are scaled up through infections of successively larger numbers of Sf9 cells, until significant volumes (>300 mL) of viral solution are achieved with a high titer (>1.0 × 10^8 pfu/mL) for optimized production of PHD construct. Once established, this high-titer stock is used to infect H5 insect cells for the secretion of the PHD proteins to be harvested and studied (*see* **Note 7**).

1. Infect 1 L of H5 cells (2.0 × 10^6 cells/mL) in ALA (250 μM) supplemented media with high-titer viral stock (hDUOX1_{1-593}-His_6 or hDUOX2_{1-599}-His_6 derived) at a multiplicity of infection of 5 for 3 days at 27 °C (*see* **Note 8**).
2. After 72 h of infection, centrifuge cell suspensions at 4200 × *g* for 20 min in a microcentrifuge.
3. The supernatant, containing the secreted hDUOX1 or hDUOX2 PHD, is processed for purification as outlined in the next section.

3.3 Purification of hDUOX PHD Protein

In a typical overexpression, ~0.7 mg/L of pure protein can be obtained following the procedure outlined below subsequent to 72 h of protein expression. All purification steps should be performed at 4 °C.

1. Concentrate the supernatant, containing the secreted peroxidase domain, to approximately 200 mL by ultrafiltration over a spiral-wound regenerated cellulose membrane (molecular mass, 10 kDa cutoff) connected to an Amicon TCF10 peristaltic pump equipped with a 2l RA2000S reservoir.
2. Add 800 mL equilibration buffer, and reconcentrate the mixture to ~200 mL volume.
3. Remove the concentrated protein solution from the reservoir, and adjust the protein solution containing the secreted PHD to pH 8.0 (*see* **Note 9**).
4. Clarify the protein solution by ultracentrifugation for 15 min at 45,000 × *g*.
5. Equilibrate a 10 mL nickel-nitrilotriacetic acid (Ni-NTA)-agarose resin loaded column with two column volumes of equilibration buffer.
6. Load the supernatant onto the Ni-NTA column at a flow rate of ~1 mL/min (save 50 μL of supernatant sample for final SDS-PAGE analysis). This can be done by a simple peristaltic pump or FPLC unit.
7. Wash the column with ten column volumes of equilibration buffer, five column volumes wash buffer 1 (20 mM imidazole), and two column volumes of wash buffer 2 (40 mM imidazole). Elute any remaining bound protein with 1.5 column volumes of elution buffer (200 mM imidazole).
8. Place the eluted protein into a 20 kDa dialysis cassette, and dialyze the eluted protein overnight in 1 L of pre-chilled equilibration buffer.
9. After dialysis, apply the protein solution onto a 5 mL Ni-NTA column attached to an FPLC unit. The Ni-NTA column should be pre-equilibrated with two column volumes of equilibration buffer prior to protein addition. The FPLC unit should be prepared with the following buffers in separate channels: equilibration buffer and elution buffer.
10. Elute the PHD protein by gradually increasing the imidazole concentration over a 0–55% elution buffer gradient at 1 mL/min for 2 h. Monitor the protein elution at 280 nm (Fig. 2A).
11. Evaluate fractions collected by SDS-PAGE for purity (Fig. 2B, C). Pool the purified fractions, and concentrate to ~2 mL volume by centrifugation in a YM-50 Centriprep. Dialyze the resulting protein solution overnight in dialysis buffer. Create a

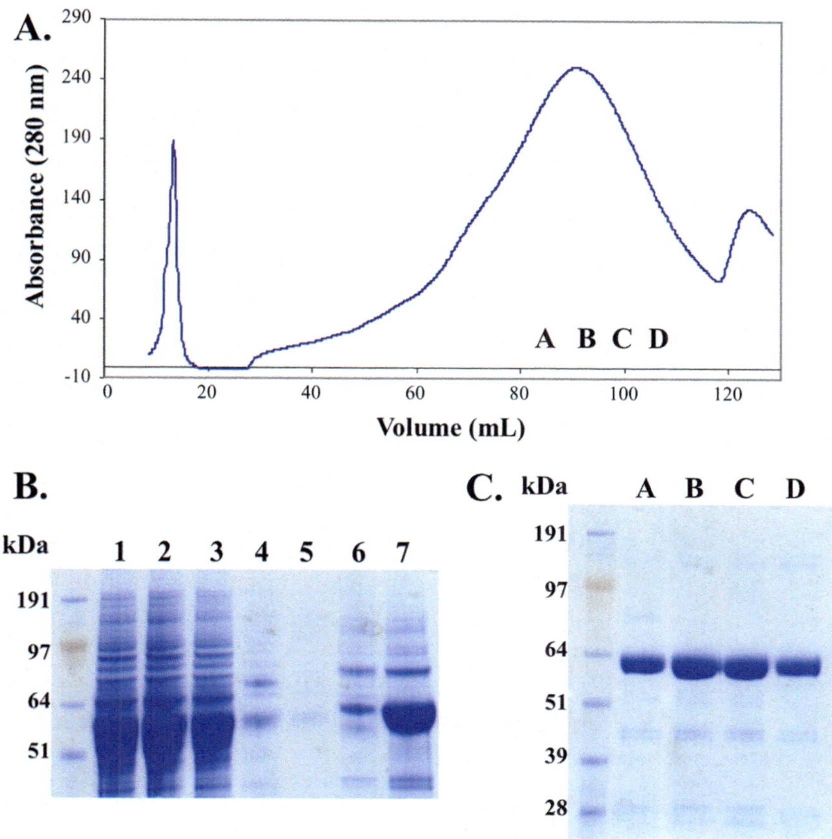

Fig. 2 Baculovirus-infected H5 cell suspension culture expression: DUOX1$_{1-593}$. (**A**) FPLC trace of DUOX1$_{1-593}$ elution due to increasing imidazole concentration (imidazole concentration at 100 mL elution volume = 100 mM). (**B**) SDS-PAGE analysis of initial Ni-NTA (imidazole step gradient) purification of DUOX1$_{1-593}$ 1. Concentrated media, 2. Flow-through from Ni-NTA column, 3. First wash: equilibration buffer (no imidazole), 4. Second wash: wash buffer 1 (10 mM imidazole), 5. Third wash: wash buffer 2 (20 mM imidazole), 6. Early elution volume (60 mL), 7. Late elution volume (100 mL). (**C**) SDS-PAGE analysis of the Ni-NTA purification of DUOX1$_{1-593}$ by FPLC, fractions A through D

final protein stock for storage at −20 °C by addition of glycerol to reach a final concentration of 25%.

12. Perform a Bradford or BCA assay on the stock solution to determine the protein stock concentration.

3.4 Activity Studies for Initial Characterization of a Purified PHD

Speculation surrounding the function of the PHD of DUOX proteins demands characterization of the purified protein in hand. The initial observation of the purified hDUOX PHD protein reveals a colorless product, demonstrating heme incorporation was not achieved while expression was undertaken in the presence of ALA (*see* **Note 10**). Preliminary characterization efforts should be undertaken to determine if this domain displays catalytic activity, regardless of heme incorporation. Outlined in this section are the protocols for two assay types: peroxidase activity measurements with tyrosine ethyl ester, undertaken due to speculation that DUOX H_2O_2 may support tyrosine cross-linking, and a superoxide dismutase (SOD) activity assay protocol, due to conjecture surrounding the nature of the ROS produced by DUOX enzymes (*see* **Note 11**).

3.4.1 Peroxidase Activity Assay with Tyrosine Ethyl Ester (TyrEE)

1. For buffer preparation, add Chelex-100 resin (5 g) to PBS (100 mL), and stir gently for 1 h. Filter the buffer from the resin and check the pH; adjust to pH 7.4 if required.
2. Make separate stock solutions of TyrEE and LPO in the Chelex resin treated PBS at a concentration of 2 mM.
3. Prepare a stock solution of H_2O_2 (2 mM) in ultrapure water and store on ice.
4. Thaw PHD protein stocks on ice. Add PHD or control protein (LPO) to an aliquot of PBS at a final concentration of 1 μM:1 mM TyrEE in a total volume of 1 mL in a quartz cuvette with stir bar. Begin stirring the reaction while monitoring fluorescence (Ex = 295 nm, Em = 414 nm, slit widths = 1 nm); choose a moderate rate of stirring to thoroughly mix the reaction while preventing data fluctuations.
5. Initiate the reaction by addition of H_2O_2 at a total reaction concentration of 100 μM.
6. Monitor the reaction fluorescence for dityrosine formation for 10 min, collecting data every 0.25 s at RT.
7. For further analysis, allow the reaction to proceed for a total of 1 h. After 1 h of reaction time, remove the solution from the cuvette and inject it onto a pre-equilibrated 50 × 2.1 mm C18 reversed-phase column attached to an Agilent 1200 series HPLC instrument. Pre-equilibrate the column with 9% acetonitrile/H_2O 0.1% formic acid for at least 10 min at a flow rate of 0.2 mL/min. Following injection, hold the equilibration condition constant for 5 min to elute unreacted starting material. Follow this isocratic step with a gradient of 9–10% acetonitrile/H_2O 0.1% formic acid from 5 to 15 min for dityrosine elution.
8. Monitor the reaction trace by absorbance (280 nm) and fluorescence intensity (Ex = 295 nm, Em = 414 nm). The expected fluorescent dityrosine assay product will elute around 8 min. Further confirmation of the product can be achieved by monitoring the trace by LC-MS and detecting the mass of the product peak (m/z 417).

3.4.2 SOD Activity Assay

1. Create a dilution series to evaluate PHD protein concentrations from 10 ng/mL to 120 ng/mL (150 μL total volume each) in HEPES buffer, 5% glycerol, pH 7.0.

2. Utilizing an SOD assay kit (Fluka, catalog number 19160), make up the required WST-1 working solution and enzyme working solution according to the kit protocol.
3. As an assay control, prepare solutions of SOD at concentrations of 25 ng/mL, 50 ng/mL, and 100 ng/mL (200 μL each) in HEPES buffer, 5% glycerol, pH 7.0.
4. Follow the manufacturer protocol using 20 μL of each sample at a final protein concentration of 1 μM, in triplicate, in the presence or absence of 100 μM $CaCl_2$ to initiate the reaction.
5. Record the absorbance at 450 nm of each sample after 20 min reaction time at 37 °C.
6. Plot the results as % inhibition versus enzyme concentration (ng/mL).

4 Notes

1. BD Biosciences no longer markets a baculovirus expression system; pAcGP67-b was formerly sold in conjunction with the BD Baculogold transfection kit, catalog #560129. Thermo Fisher Scientific Bac-to-Bac® and BaculoDirect™, Oxford Expression Technologies *Flash*BAC™, and Clontech Bac-PAK™ systems are currently available and provide an alternative for researchers to pursue baculovirus expression. Viral insect cell expression is essential for PHD expression, as *E. coli* cells are unable to produce a properly folded PHD product.
2. Both Sf9 and H5 cell lines were kept in suspension at 27 °C and maintained at densities between 0.5×10^6 and 2×10^6 cells/mL. To prevent excessive Sf9 cell aggregation when restarted from frozen stock, heparin (10 U/mL) was added for three passages.
3. ALA may not be necessary to include; heme was never found to bind directly to the hDUOX PHD constructs, but it is not detrimental to the purification procedure.
4. Currently, H_2O_2 production is most effectively monitored by Amplex red and is reliably used to validate DUOX1/2 activity in mammalian cells. This assay was not performed for the purified PHD domains, as related in publication [6, 7].
5. Many baculovirus expression technologies now rely on the production of a bacmid, through site-specific transposition in *E. coli*, rather than homologous recombination in insect cells. The provided protocol outlines homologous recombination; if a bacmid construct is utilized, after insect cell transfection, high-titer stocks can be directly generated without first

performing a plaque assay step to isolate single viral plaques. A plaque assay of this high-titer stock is still recommended to calculate a specific titer to optimize H5 infection conditions.

6. Place a beaker of water in the incubator, or seal the plates into a plastic bag with moist paper towels to help retain moisture.
7. High-titer viral stock concentrations are determined through plaque assay of a dilution series created from the stock solution. High-titer viral solutions may remain stable for up to 6 months when stored in the dark at 4 °C (sterile, amber glass bottles work well). Always freeze several individual 1–2 mL aliquots of high-titer virus at −80 °C for permanent storage. These stocks can be utilized directly for scale-up without returning to the initial co-transfection stage and will remain stable for years.
8. Utilizing H5 cells for production of the PHD is crucial; infection of Sf9 cells at the same scale and for the same time frame results in significantly less secreted PHD protein.
9. When adjusting the pH, a cloudiness will occur, due to precipitation of unrelated proteins from the growth media. This precipitation will clarify upon ultracentrifugation.
10. While neither hDUOX protein PHD revealed any heme incorporation, interestingly the purified *C. elegans* DUOX1 PHD revealed covalent heme binding [6, 22]; this clearly appears to be due to the presence of critical glutamate (responsible for covalent bond formation to the heme cofactor) and proximal histidine residues which are absent in the human counterparts. Analysis of PHD DUOX sequences from other lower organisms suggests that heme binding may be present, perhaps to support tyrosine cross-linking, which human DUOX proteins have evolved away from [22]. It remains to be determined if in the scope of the full-length native hDUOX protein, heme incorporation is somehow achieved, supported through delivery and/or stability of accessory proteins which are absent in a recombinant insect system.
11. Please keep in mind that the outlined purification protocol and activity assays may be applied to research regarding DUOX PHDs purified from other organisms. Homologous DUOX proteins have been identified in such organisms as *M. musculus*, *D. melanogaster*, *A. gambiae*, and *D. rerio* [22]. Previous research has demonstrated that *C. elegans* DUOX1 is a self-contained catalyst that mediates tyrosine cross-linking, with H_2O_2 generated by the NADPH oxidase domain being utilized by a heme-bound PHD to form tyrosine cross-links [6, 22]. This mechanism of action contrasts with that of the human DUOX system, which, as isolated, does not bind heme and requires several regulatory proteins to control enzymatic function. Further investigation of DUOX and the PHD from

these lower organisms may be of interest to researchers attempting to decipher the role of the PHD in humans.

12. Sequence alignment demonstrates conservation of D109 of hDUOX1 and D115 of hDUOX2 with a known aspartic acid residue that completes the calcium-binding region within mammalian peroxidases (Fig. 1A). However, as shown in Fig. 1B, the models generated for the PHD of each hDUOX protein place this residue far from calcium, while the remaining conserved calcium-associated residues surround and almost overlay their counterparts in LPO. It is possible that this suggests this aspartic acid does not associate with the bound calcium in hDUOX enzymes or that another residue participates in its place. It is more likely that as a model structure, the loop region containing D109 or D115 has been predicted incorrectly to be further from the calcium molecule; loop regions show significant variability in modeling relative to α-helices and β-sheets.

Acknowledgments

Many thanks are extended to Paul Ortiz de Montellano for supporting the PHD research and method developments discussed herein.

References

1. De Deken X, Wang D, Many MC, Costagliola S, Libert F, Vassart G, Dumont JE, Miot F (2000) Cloning of two human thyroid cDNAs encoding new members of the NADPH oxidase family. J Biol Chem 275:23227–23233
2. Dupuy C, Ohayon R, Valent A, Noel-Hudson MS, Deme D, Virion A (1999) Purification of a novel flavoprotein involved in the thyroid NADPH oxidase. Cloning of the porcine and human cdnas. J Biol Chem 274:37265–37269
3. Grasberger H, Refetoff S (2006) Identification of the maturation factor for dual oxidase. Evolution of an eukaryotic operon equivalent. J Biol Chem 281:18269–18272
4. Morand S, Ueyama T, Tsujibe S, Saito N, Korzeniowska A, Leto TL (2009) Duox maturation factors form cell surface complexes with Duox affecting the specificity of reactive oxygen species generation. FASEB J 23:1205–1218
5. Hoste C, Dumont JE, Miot F, De Deken X (2012) The type of DUOX-dependent ROS production is dictated by defined sequences in DUOXA. Exp Cell Res 318:2353–2364
6. Meitzler JL, Ortiz de Montellano PR (2009) Caenorhabditis elegans and human dual oxidase 1 (DUOX1) "peroxidase" domains: insights into heme binding and catalytic activity. J Biol Chem 284:18634–18643
7. Meitzler JL, Ortiz de Montellano PR (2011) Structural stability and heme binding potential of the truncated human dual oxidase 2 (DUOX2) peroxidase domain. Arch Biochem Biophys 512:197–203
8. Shin K, Hayasawa H, Lonnerdal B (2001) Mutations affecting the calcium-binding site of myeloperoxidase and lactoperoxidase. Biochem Biophys Res Commun 281:1024–1029
9. Meitzler JL, Hinde S, Banfi B, Nauseef WM, Ortiz de Montellano PR (2013) Conserved cysteine residues provide a protein-protein interaction surface in dual oxidase (DUOX) proteins. J Biol Chem 288:7147–7157
10. Carre A, Louzada RA, Fortunato RS, Ameziane-El-Hassani R, Morand S, Ogryzko V, de Carvalho DP, Grasberger H,

Leto TL, Dupuy C (2015) When an Intramolecular Disulfide Bridge Governs the Interaction of DUOX2 with Its Partner DUOXA2. Antioxid Redox Signal 23:724–733
11. Geiszt M, Witta J, Baffi J, Lekstrom K, Leto TL (2003) Dual oxidases represent novel hydrogen peroxide sources supporting mucosal surface host defense. FASEB J 17:1502–1504
12. Song Y, Ruf J, Lothaire P, Dequanter D, Andry G, Willemse E, Dumont JE, Van Sande J, De Deken X (2010) Association of duoxes with thyroid peroxidase and its regulation in thyrocytes. J Clin Endocrinol Metab 95:375–382
13. Sokolenko S, George S, Wagner A, Tuladhar A, Andrich JM, Aucoin MG (2012) Co-expression vs. co-infection using baculovirus expression vectors in insect cell culture: benefits and drawbacks. Biotechnol Adv 30:766–781
14. Smith CK, Carr D, Mayhood TW, Jin W, Gray K, Windsor WT (2007) Expression and purification of phosphorylated and non-phosphorylated human MEK1. Protein Expr Purif 52:446–456
15. Dent P, Chow YH, Wu J, Morrison DK, Jove R, Sturgill TW (1994) Expression, purification and characterization of recombinant mitogen-activated protein kinase kinases. Biochem J 303 (. Pt 1:105–112
16. Bakos E, Evers R, Szakacs G, Tusnady GE, Welker E, Szabo K, de Haas M, van Deemter L, Borst P, Varadi A, Sarkadi B (1998) Functional multidrug resistance protein (MRP1) lacking the N-terminal transmembrane domain. J Biol Chem 273:32167–32175
17. Gao M, Loe DW, Grant CE, Cole SP, Deeley RG (1996) Reconstitution of ATP-dependent leukotriene C4 transport by Co-expression of both half-molecules of human multidrug resistance protein in insect cells. J Biol Chem 271:27782–27787
18. Qian YM, Qiu W, Gao M, Westlake CJ, Cole SP, Deeley RG (2001) Characterization of binding of leukotriene C4 by human multidrug resistance protein 1: evidence of differential interactions with NH_2- and COOH-proximal halves of the protein. J Biol Chem 276:38636–38644
19. Lee CA, Kadwell SH, Kost TA, Serabjit-Singh CJ (1995) CYP3A4 expressed by insect cells infected with a recombinant baculovirus containing both CYP3A4 and human NADPH-cytochrome P450 reductase is catalytically similar to human liver microsomal CYP3A4. Arch Biochem Biophys 319:157–167
20. DePillis GD, Ozaki S, Kuo JM, Maltby DA, Ortiz de Montellano PR (1997) Autocatalytic processing of heme by lactoperoxidase produces the native protein-bound prosthetic group. J Biol Chem 272:8857–8860
21. Liu F, Wu X, Li L, Liu Z, Wang Z (2013) Use of baculovirus expression system for generation of virus-like particles: successes and challenges. Protein Expr Purif 90:104–116
22. Meitzler JL, Brandman R, Ortiz de Montellano PR (2010) Perturbed heme binding is responsible for the blistering phenotype associated with mutations in the *Caenorhabditis elegans* dual oxidase 1 (DUOX1) peroxidase domain. J Biol Chem 285:40991–41000

Chapter 5

A Close-Up View of the Impact of Arachidonic Acid on the Phagocyte NADPH Oxidase

Tania Bizouarn, Hager Souabni, Xavier Serfaty, Aicha Bouraoui, Rawand Masoud, Gilda Karimi, Chantal Houée-Levin, and Laura Baciou

Abstract

The NADPH oxidase NOX2 complex consists of assembled cytosolic and redox membrane proteins. In mammalian cells, natural arachidonic acid (*cis*-AA), released by activated phospholipase-A2, plays an important role in the activation of the NADPH oxidase, but the mechanism of action of *cis*-AA is still a matter of debate. In cell-free systems, *cis*-AA is commonly used for activation although its structural effects are still unclear. Undoubtedly *cis*-AA participates in the synergistic multi-partner assembly that can be hardly studied at the molecular level in vivo due to cellular complexity. The capacity of this anionic amphiphilic fatty acid to activate the oxidase is mainly explained by its ability to disrupt intramolecular bonds, mimicking phosphorylation events in cell signaling and therefore allowing protein-protein interactions. Interestingly the geometric isomerism of the fatty acid and its purity are crucial for optimal superoxide production in cell-free assays. Indeed, optimal NADPH oxidase assembly was hampered by the substitution of the *cis* form by the *trans* forms of AA isomers (Souabni et al., BBA-Biomembranes 1818:2314–2324, 2012). Structural analysis of the changes induced by these two compounds, by circular dichroism and by biochemical methods, revealed differences in the interaction between subunits. We describe how the specific geometry of AA plays an important role in the activation of the NOX2 complex.

Key words Cell-free system, Superoxide anion, Recombinant proteins, Thiol groups, Arachidonic acid, *Cis-trans* isomer fatty acid, Synchrotron radiation circular dichroism

1 Introduction

In phagocytes, the first line of innate immune defense against pathogens, the activity of the NADPH oxidase, relies on the formation of a macromolecular functional complex constituted of the membrane flavocytochrome b_{558} (Cytb_{558}) and cytosolic proteins (p47^{phox}, p67^{phox}, p40^{phox}, and the small G protein Rac1/2). The sole role of the assembled complex is to produce the primary product, $O_2^{\cdot -}$, by the NADPH-dependent one-electron reduction of molecular oxygen (O_2). This production is accomplished at the level of the membrane by the Cytb_{558}, comprised of tightly

Ulla G. Knaus and Thomas L. Leto (eds.), *NADPH Oxidases: Methods and Protocols*, Methods in Molecular Biology, vol. 1982, https://doi.org/10.1007/978-1-4939-9424-3_5, © Springer Science+Business Media, LLC, part of Springer Nature 2019

associated $p22^{phox}$ protein (22 kDa) and $gp91^{phox}$ glycoprotein, nowadays called NOX2 (91 kDa). The latter forms the redox core of the NADPH oxidase complex since it contains all the redox intermediates (two hemes, one flavin, and a NADPH-binding site) necessary for the transmembrane electron transfer leading to O_2 reduction. In vivo, the specific protein-protein interactions are triggered by biological stimuli (microorganisms or inflammatory mediators) initiating intracellular signaling pathways that lead to posttranslational protein modifications (phosphorylation) and fatty acid release. In order to decipher the functioning of the NOX2 NADPH oxidase complex, in vitro systems were developed by Pick's group (reviewed in Ref. [1]). This method of cell-free activation of the NADPH oxidase is an ingenious way to circumvent the complexity of cell signaling pathways and to be able to control the composition of the reaction mixture and the stoichiometry of each protein partner. The "canonical" cell-free assay uses native membrane fractions from bovine or human neutrophils or macrophages. The method has been adapted in several versions depending on the use of native or recombinant cytosolic proteins and of suitable additives. The protocols to obtain the "routine" semi-recombinant cell-free assay system have been described previously in details (see in Refs. [2, 3]). In more recent in vitro studies, the native Cytb_{558} was substituted by a recombinant Cytb_{558} produced by heterologous expression in yeast [4, 5]. This cell-free assay entirely comprising recombinant proteins was functional and was used for biophysical studies [6–8]. Cytosolic proteins are now routinely produced in *E. coli*, purified [9] and added as desired to permit strict quantification of each protein. NADPH is added to initiate the reaction, and the measurement of the generated rate of superoxide anion by the assembled NADPH oxidase complex is followed immediately by SOD-inhibitable cytochrome c reduction. The cell-free assay created from recombinant proteins gives detailed information on each step of assembly and on reaction steps that could not be easily observed in vivo and allows investigations of unusual component assembly. To illustrate this point, the assay was used to investigate the capacity of NOX2 to produce superoxide in the absence of its membrane partner $p22^{phox}$ [5].

In the resting phagocytes, $p47^{phox}$ and $p67^{phox}$ are known to interact through the C-terminal SH3 domain of $p67^{phox}$ and the C-terminal proline-rich region of $p47^{phox}$ [10–13], an interaction that is lost when the C-terminus of $p47^{phox}$ is deleted [14]. When activated, the cytosolic heterodimer $p47^{phox}$-$p67^{phox}$ probably separates in order to assemble properly with integral membrane proteins [15]. Arachidonic acid in its natural *cis*-polyunsaturated form ((5Z,8Z,11Z,14Z)-icosa-5,8,11,14-tetraenoic acid, *cis*-AA, Fig. 1) has been described to be an efficient activator of the enzyme in vivo and in vitro. *Cis*-AA induces conformational changes in the $p47^{phox}$ and $p67^{phox}$ subunits [16, 17] by perturbing the

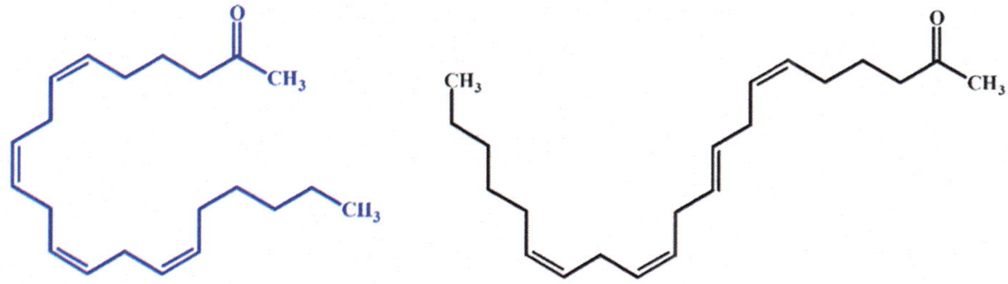

Fig. 1 Example of the geometrical differences between all-*cis*-AA (left, blue) and one of the four existing mono-*trans*-AA (right, black). Structure made with the software ACD/ChemSketch (Freeware) version 2017.2.1, Advanced Chemistry Development, Inc., Toronto, ON, Canada, www.acdlabs.com, 2018

intramolecular bonds in $p47^{phox}$ between the polybasic domain and the SH3 tandem and by mimicking phosphorylation events [18]. This leads to the unmasking of SH3 regions in $p47^{phox}$ [16, 19] and to interaction of $p67^{phox}$ and Rac [20] with $Cytb_{558}$. Moreover, AA interacts with the membrane fraction by eliciting conformational changes, which were proposed to be relevant for the activation of superoxide production [6, 21]. An important point is that the phagocyte NADPH oxidase is activated specifically by all-cis-AA isomer, which is the prevalent stereoisomer in eukaryotic cells, but not by the stereoisomer trans-AA (Fig. 1) [6]. How to explain that a mix of mono-*trans* isomers originating from the *cis-trans* isomerization of *cis*-AA is unable to activate the system and disturbs the ability of cytosolic proteins to translocate to the membrane partner? *Trans*-lipids are present in dietary compounds but are also produced by free radical reactions. Evidence was obtained that *cis* and *trans* binding have discrete functional effects on the NADPH oxidase through unique structural features, thereby allowing distinct residues on the protein surface to engage in interactions [22]. In this chapter we describe the methods used to determine the structural changes induced by AA and also draw attention to the role of the naturally occurring *cis-trans* fatty acid configuration in NADPH oxidase activation.

2 Materials

2.1 Bacterial Expression

1. The cDNA of human $p67^{phox}$ and $p47^{phox}$ were cloned into pET15b vector (Novagen) with the His tag sequence at the N-terminus.
2. The cDNA of human Rac1Q61L was cloned into pGEX2T vector (GE Healthcare) with the GST tag sequence at the N-terminus.
3. *Escherichia coli* bacterial strains BL21(DE3) and BL21(DE3) pLysS are used to overexpress $p47^{phox}$, $p67^{phox}$, and Rac1Q61L, respectively.

4. Ampicillin stock solution (100 mg/mL): 1 g ampicillin is dissolved in 10 mL deionized and sterilized water, filtered (0.22 μm pore size), aliquoted, and stored at −20 °C.
5. Isopropyl-1-thio beta-D-galactopyranoside (IPTG) stock solution (0.5 M): 2.38 g of IPTG is dissolved in 20 mL deionized water solution, filtered, aliquoted, and stored at −20 °C.
6. Agar plates are done by dissolving 40 g of LB agar in 1 L of distilled water and autoclaved. The medium is left to cool down to around 40 °C, then ampicillin is added, and the medium is poured into Petri dishes and left to solidify.
7. All media are sterilized using an autoclave before adding bacteria.
8. Scaling up of bacterial cultures is performed in 250 mL flasks containing 60 mL LB medium and 5 L baffled flasks containing 1.5 L Terrific Broth medium.

2.2 Protein Purification from Bacteria

When the solvent is not explicitly mentioned, all the solutions are made in 18.2 MΩ cm deionized filtered water.

1. Lysis buffer: 50 mM HEPES, 200 mM NaCl, 1mM ethylenediaminetetraacetic acid (EDTA), pH 7.5 adjusted with sodium hydroxide (NaOH) (*see* **Note 1**). Store at 4 °C.
2. Add just before use: 1 mg DNAse, 2 mM phenylmethylsulfonyl fluoride (PMSF), and 1 mM dithiothreitol (DTT).
3. PMSF stock (100 mM): dissolve 0.174 g of PMSF in 10 mL ethanol; can be stored at 4 °C for 2 weeks.
4. All steps should be performed in an ice bucket when possible or at temperatures below 5 °C using cold buffers to minimize as much as possible protease activities.
5. Buffer A: 20 mM HEPES, 100 mM NaCl, pH 7.5 adjusted with sodium hydroxide (NaOH). Store at 4 °C.
6. Packed SP-Sepharose FF (cation exchange) or Q-Sepharose FF (anion exchange) chromatography column (40 mL, GE Healthcare) equilibrated with buffer A supplemented with 1 mM PMSF.
7. Ni-Sepharose FF column (15–20 mL, GE Healthcare) equilibrated with buffer 1: 30 mM NaH_2PO_4, pH 7.5 adjusted with NaOH, 0.7 M NaCl, 30 mM imidazole, 1 mM PMSF (imidazole is added to limit non-specific adsorption of proteins on the resin).
8. Ni-Sepharose washing buffer (buffer 2): 30 mM NaH_2PO_4, pH 7.5 adjusted with sodium hydroxide (NaOH), 0.1 M NaCl, 10 mM imidazole, 1 mM PMSF.

9. Ni-Sepharose elution buffer (buffer 3): 30 mM NaH_2PO_4, pH 7.5 adjusted with sodium hydroxide (NaOH), 0.1 M NaCl, 300 mM imidazole, 1 mM PMSF.
10. Glutathione Sepharose HP chromatography column (10 mL, GE Healthcare) equilibrated with phosphate-buffered saline (PBS) supplemented with 1 mM PMSF.
11. Automated protein purification system (AKTA prime or similar).
12. Packed size-exclusion chromatography column (Superdex 75, GE Healthcare) with the running buffer: 30 mM NaH_2PO_4, pH 7.5 adjusted with NaOH, 0.1 M NaCl.
13. Pierce BCA assay: used for the determination of protein concentration of each sample with bovine serum albumin as standard.

2.3 Isolation of Human Neutrophil Membrane Fraction (for 400 to 500 mL Blood)

1. Sterile graduated cylinders of 1 L and 500 mL (*see* **Note 2**)
2. Four Falcon tubes of 15 mL containing 12 mL of sterile deionized water stored at 4 °C
3. Four Falcon tubes of 15 mL containing 4 mL of sterile KCl 0.6 M stored at 4 °C
4. 750 mL of sterile NaCl 0.9%, kept at room temperature
5. 50 mL of sterile Na-phosphate buffer (20 mM, pH 7.4, adjusted with NaOH) stored at 4 °C
6. 250 mL sterile Dulbecco's phosphate-buffered saline (DPBS), 50 mL stored at room temperature and 200 mL at 4 °C
7. Hank's Balanced Salt Solution stored at room temperature
8. Two 50 mL Falcon tubes containing 15 mL sterile Ficoll (GE Healthcare) at room temperature
9. 500 mL 2% dextran solubilized just before use in sterile 0.9% NaCl (*see* **Note 3**)
10. 30 mL breaking buffer (*see* **Note 4**): 20 mM sterile Na-phosphate buffer, pH 7.4 adjusted with NaOH, 340 mM saccharose, 7 mM $MgSO_4$, 200 μM leupeptin, 1 mM PMSF
11. 50 mL sterile polypropylene Falcon tubes
12. Sterile polypropylene pipettes of various volumes: 5, 10, 25 mL

2.4 Synchrotron Radiation Circular Dichroism (SRCD)

1. Protein concentrator with the molecular weight cutoff adapted to the size of the protein (*see* **Note 5**).
2. Determine the protein concentrations with a NanoDrop2000™ spectrophotometer (*see* **Note 6**) at 280 nm using the calculated absorption coefficient for each protein of interest (ExPASy, ProtParam free software).

2.5 Quantification of Free Thiol Groups

1. Prepare 10 mM 5-5′-dithiobis(2-nitrobenzoic acid) (DTNB, also known as Ellman's reagent; *see* **Note 7**) stock solution in Na-phosphate buffer (20 mM and adjust the pH to 8.0 with NaOH). Keep refrigerated.
2. Prepare 1 M Tris-HCl solution pH 8.0.
3. Prepare glutathione 600 μM in 10 mL (in deionized water).

2.6 Cell-Free Assays

1. Activity assay buffer: phosphate-buffered saline (PBS), pH 7.4 supplemented with 10 mM $MgSO_4$.
2. 20 mg/mL (65.7 mM) stock solution of *cis-* or *trans*-arachidonic acid (AA) is prepared in ethanol: dissolve 10 mg of AA in 500 μL absolute ethanol in ice bucket as fast as possible to avoid evaporation. Aliquot the solution and store at −80 °C (*see* **Note 8**).
3. Prepare a solution of NADPH 20 mM in deionized H_2O_2 (*see* **Note 9**).
4. Prepare equine cytochrome *c* (Cyt*c*) 5 mM stock solution in DPBS (*see* **Note 10**).
5. Double beam spectrophotometer (e.g., Uvikon XS Secomam).

3 Methods

3.1 Purification of Human Neutrophil Membrane Fractions

3.1.1 Isolation of Human Neutrophils from Blood

The purification of human neutrophil membrane fractions proceeds in three steps (*see* **Note 11**). First neutrophils are isolated from donor blood cells. Second, isolated neutrophils are broken, and granules and mitochondria are removed by differential centrifugation. Last, membrane fractions (plasma membrane, endoplasmic reticulum) are isolated from the cytosol by ultracentrifugation. The method used to purify human neutrophils was previously described in [23]. Modifications are as follows:

1. In a sterile 1 L graduated cylinder, pour the blood very gently (volume usually between 380 and 500 mL), and add immediately but slowly alongside the wall of the cylinder, the same volume of 2% dextran solution (instead of 3% dextran solution).
2. Close tightly the cylinder with a Parafilm® membrane, and mix three times by repeated very gentle inversion. The solution is kept at room temperature for 40 min (instead of 20 min) for sedimentation (*see* **Note 12**).
3. Using a sterile polypropylene 25 mL pipette fitted to a Pipetboy-type electric pipet, collect carefully the upper part containing the leukocytes into 50 mL Falcon tubes. Avoid taking the sedimented red blood cells. Usually 0.5 to 1 cm of the upper part are left behind. Usually 12 to 15 Falcon™ tubes of 50 mL are needed.

4. The tubes are centrifuged at 400 × *g* for 8 min at room temperature (instead of 500 × *g* 4 min).
5. This step has to be done as fast as possible to avoid cells to become dry but very gently in order to limit neutrophil activation. Take out the supernatant carefully. Leave a 3 mm layer of supernatant in each tube to avoid drying the cells. Resuspend the pellets of the two first tubes with the last drops by gently shaking the tubes, and add 5 mL of DPBS at room temperature in one of them, and immediately transfer into the second tube. Resuspend a third pellet, and add to the previous mixture. Do the same procedure with all pellets. At the same time, rinse with 5 mL DPBS the first empty tube, and then transfer to the second and so on. Rinse at least three times all the tubes. Transfer all the solutions into one tube. Complete the volume to 50 mL with DPBS and mix gently by inverting. If neutrophils are already activated, white aggregates will be visible in the suspension.
6. Add, very slowly, alongside the wall of a vertically held Falcon™ tube containing 15 mL Ficoll solution 25 mL of the cell suspension with a sterile 10 mL polypropylene pipette. The liquid has to flow slowly according to gravity, controlled via a manual pipette controller. This step can take 3–5 min for each tube in order to have a true biphasic system with the cell suspension as the top phase and the Ficoll® solution as the bottom one. The separation between the two phases is ideally clean without any mixing.
7. Centrifuge the biphasic solution at 400 × *g* for 30 min at room temperature.
8. During centrifugation, prepare 30 mL of breaking buffer in a sterile Falcon™ tube by solubilizing 3.5 g sucrose, 51.6 mg $MgSO_4$ heptahydrate, 2.8 mg leupeptin, and 5.22 mg PMSF at 4 °C.
9. Remove the tubes from the centrifuge and set the centrifuge to 4 °C for the following steps. Then, take out the supernatant of the first tube with a sterile 25 mL polypropylene pipette by carefully removing the lymphocytes, which are floating as a white turbid, thin section between the two phases. The pellet is only weakly attached, so when 1 cm of liquid is left, use a micropipette with a polypropylene sterile tip of 200 μL to remove all supernatant. Add immediately 200 μL of DPBS and resuspend the cells by gently stirring with this tip. Do not aspirate the cells inside the tip to avoid a loss of cells alongside the wall. Add quickly 12 mL of cold deionized 18.2 MΩ·cm water and mix immediately, then close the tube, and gently mix by inversion. Exactly 30 s after addition of H_2O, leave the tube on ice and 10 s later add quickly 4 mL of cold potassium

chloride 0.6 M, mix immediately, and complete the volume until 50 mL with cold DPBS. Do the same with the second tube and centrifuge 8 min at 400 × *g* at 4 °C. Remove the supernatant as described before, add 200 µL DPBS, and resuspend cells. Then, proceed as described previously. Two hemolysis steps are usually enough, but if, after this second centrifugation, red blood cells can be observed at the surface of the pellets, repeat this process just one more time.

10. Take out the supernatant, and resuspend the pellets in 200 µL DPBS as described before. With the same tip, put the cell suspension in one tube. Rinse the first tube several times with few volumes of DPBS by transferring each time the rinsed suspension into the second tube. Complete the volume to 10 mL with cold DPBS.
11. Count the cells on a hemocytometer after a 1:100 dilution with a solution of Trypan blue diluted four times in DPBS.
12. Centrifuge the cell suspension at 400 × *g* for 8 min at 4 °C. Remove the supernatant and resuspend the pellet in 200 µL cold breaking buffer, and extend the volume to reach 10^8 cells/mL (5–15 mL).

3.1.2 Preparation of the Neutrophil Membrane Fraction

The next step consists in breaking the cells and isolating the total membrane fraction. The purification of membrane fractions from guinea pig macrophages was described in details by Pick [2]. For the preparation of the membrane fraction from human neutrophils, we used the following protocol:

1. Cells in breaking buffer (~10^8 cells/mL) are sonicated as follows: 12 times 10 s with 30 s pause on ice between each 10 s sonication. The sonication pulses (30%) are composed of 300 ms of sonication and 700 ms pause. The power is around 3 on our device (Branson Sonifier 250).
2. Centrifuge the sonicated suspension at $9.0 \times 10^3 \times g$ for exactly 10 min, i.e., the phase of the speed increase is not counted in these 10 min. This step leads to a pellet containing granules, mitochondria, unbroken cells, and other organelles. The supernatant contains membrane fractions from cell membranes and other fragile membranes (such as endoplasmic reticulum membranes) in cytosol.
3. The supernatant (cytosol + membrane fractions) is then centrifuged for 90 min at $1.8 \times 10^4 \times g$ at 4 °C. This step leads to a pellet containing the membrane fractions and a supernatant containing the cytosol.
4. The supernatant can be eventually kept for further analysis. The pellet is first gently resuspended in 1 mL of buffer with a micropipette by aspiration/backflow. Then, aggregates are

broken by putting the tube in a bath sonicator with cycles of 30 s sonication and 30 s pause on ice and mixing after each cycle with the same yellow tip. Usually, five cycles are sufficient but sometimes more are needed.

5. When the homogeneity of the suspension is satisfactory, extend the volume with cold breaking buffer to 3 mL and make 40 μL aliquots, and keep at −80 °C. The concentration in Cytb_{558} reaches classically in the range of 0.7–1.4 μM.

3.2 Production and Purification of Cytosolic Proteins

Recombinant cytosolic protein purification protocols are rarely described in detail. Yet for some experiments, it is essential to optimize the purity of these proteins. This was indeed the case when studying structural modifications of p47^{phox} and p67^{phox} by circular dichroism in our laboratory. All steps must be performed at 4 °C.

3.2.1 Expression and Purification of His-p47^{phox}

1. Transformed bacteria (either freshly or from –80 °C glycerol stock) are plated onto LB Petri dishes containing ampicillin (100 μg/mL) and incubated overnight at 37 °C.
2. Preculture: a single colony is picked and used to inoculate 60 mL LB broth containing 100 μg/mL ampicillin. Bacteria are incubated overnight at 37 °C with shaking at 200 rpm in an orbital incubator.
3. Culture: 15 mL of preculture is added to 1.5 L of TB supplemented with 1.5 mL ampicillin in a baffled culture flask (5 L) during 4 h at 30 °C with shaking at 140 rpm. An absorbance close to 0.6 at 600 nm should be reached. Protein expression is then induced by adding 0.5 mM IPTG and growing cells at 20 °C overnight.
4. Harvest bacteria by centrifugation at 6200 × *g* for 15 min using Beckman centrifuge (Avanti J20) and rotor JLA8.1.0000, and store semidry pellets at −80 °C until use.
5. Resuspend bacterial pellets in 30 mL lysis buffer supplemented with 1 mg DNAse, 2 mM phenylmethylsulfonyl fluoride (PMSF), and 1 mM dithiothreitol (DTT) (*see* **Note 13**).
6. Bacteria are broken by sonication (*see* **Note 14**) in an ice-cooled beaker in 50% pulse mode, power setting 5. Four cycles of 2 min each are performed, pausing 2 min in between to reduce heat damage.
7. The bacterial lysate is centrifuged 90 min at 160,000 × *g* at 6 °C to pellet cell debris and membranes.
8. The purification protocol is designed for automated purification (e.g., using AKTA protein purification system) but could be easily adapted for batch and/or manual flow purification. The ultracentrifuge supernatant is filtered using a 0.22 μm pore

size filter and diluted twice with buffer A before loading on SP-Sepharose chromatography column (*see* **Note 15**).

9. Negatively charged proteins, DNA, and other molecules are eliminated by washing the column with buffer A (around 150 mL). $P47^{phox}$ is eluted in one step with buffer A supplemented with 500 mM NaCl.
10. Fractions containing $p47^{phox}$ are diluted twice with buffer 1, loaded on the Ni-Sepharose column (*see* **Note 16**), washed with buffer 2, and eluted with buffer 3.
11. Pool the fractions of the elution peak, dialyze the solution against the dialysis buffer (30 mM NaH_2PO_4, pH 7.5, 0.1 M NaCl), make aliquots, and store at −80 °C (*see* **Note 17**).
12. Fractions representative of each peak are analyzed biochemically in order to verify the protein purity. This is performed by sodium dodecyl sulfate-polyacrylamide gel electrophoresis (SDS-PAGE) using a standard Mini-Protean system from Bio-Rad (following manufacturer's instructions). The gel is stained with Coomassie blue.
13. The protein concentration is determined first by NanoDrop2000™ and then by BCA assay.

3.2.2 Expression and Purification of His-$p67^{phox}$

1. Follow the **steps 1–7** of the $p47^{phox}$ purification protocol. Then the protein solution is loaded on a Q-Sepharose (anion exchange column) chromatography column (*see* **Note 18**). Wash the column with buffer A, and elute in one step with buffer A supplemented with 500 mM NaCl.
2. The fractions containing the protein of interest are then loaded on a Ni-Sepharose column, and the proteins are separated with the same procedure described in **steps 10** and **13**.
3. The Q-Sepharose and Ni-Sepharose steps are illustrated in Fig. 2.

3.2.3 Purification of GST-Rac1Q61L

1. Follow the **steps 1–9** of the $p47^{phox}$ purification protocol with the exception that buffer A is adjusted to pH 7.0 with HCl.
2. The fractions containing the Rac GTPase are then loaded on a glutathione-Sepharose column and eluted according to the manufacturer's procedure.

3.2.4 Additional Purification Steps

If it is necessary to improve the purity of the proteins, load concentrated proteins in solution on an equilibrated Superdex 75 16/600 chromatography column, and elute the proteins with the running buffer.

3.3 Neutrophil NADPH Oxidase Cell-Free Assays

The NADPH oxidase cell-free assay consists of two steps. The first step allows the assembly of the different soluble and membrane components of the NADPH oxidase (the engine assembly). AA is

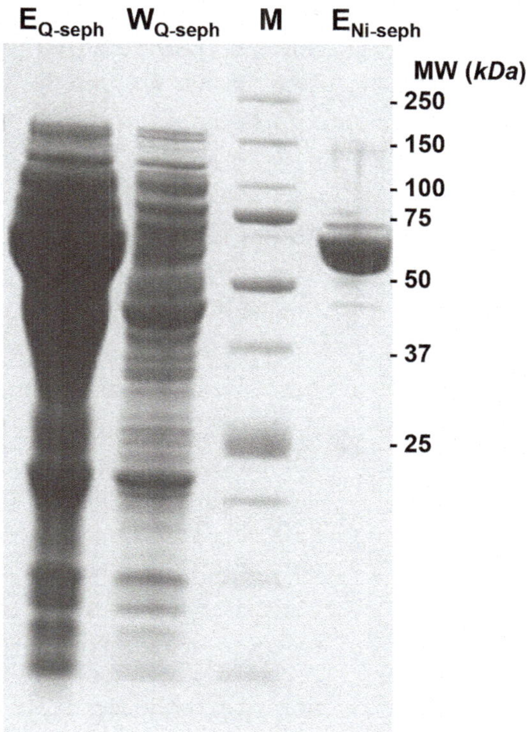

Fig. 2 SDS-PAGE analysis of the fractions from the p67^{phox} purification. $E_{Q\text{-seph}}$, Q-Sepharose elution pool; $W_{Q\text{-seph}}$, Q-Sepharose flow through; $E_{Ni\text{-seph}}$, Ni-Sepharose elution pool; M, protein weight maker (Precision Plus Protein standards, Bio-Rad). Gels are stained with Coomassie blue

important for this first stage which we call the activation step of the system by facilitating protein-protein interactions. AA used for NADPH oxidase activation in cell-free systems is the *cis* isomer. The second step is the start of the enzymatic reaction with the addition of the NADPH substrate. The functioning of the enzyme is followed indirectly by the reduction rate of cytochrome *c* (at 550 nm) by superoxide produced by the active complex (*see* **Note 19**). Thus, cell-free assays contain information on both: the activation level and the catalytic reaction. Since both are included in the same observable event, indirect studies should be performed to try to identify how AA impacts the system. Cell-free assays can be used in different configurations depending on the components that are mixed together. We commonly use the canonical assay which consists of mixing membrane fractions from neutrophils (*see* Subheading 3.1) with purified recombinant cytosolic proteins produced heterologously in *E. coli* (*see* Subheading 3.2) as follows (for a full recombinant cell-free system, *see* **Note 20**):

1. Mix gently in the cuvette a volume of neutrophil membrane fractions to have a final concentration of 2–4 nM of Cytb_{558} (*see*

Note 21) in the volume of activity buffer to a final volume of 600 mL. Usually 2 to 4 μL of the membrane fraction will be used. Avoid very thin tips which disrupt vesicles.

2. Add as fast as possible the purified cytosolic proteins ($p67^{phox}$, $p47^{phox}$, and Rac1Q61L (*see* **Note 22**)) and the desired volume of AA to have a final concentration of 200 nM of each protein.
3. Add as fast as possible AA and immediately close the cuvette with a Parafilm, and mix by inverting several times.
4. Incubate for 5 min at 25 °C to allow the assembly of the complex (*see* **Note 23**).
5. Add Cyt*c* to a final concentration of 70 μM (*see* **Note 24**).
6. Initiate the reaction with the addition of NADPH to obtain 200 μM final concentration; mix by inverting, and start immediately the acquisition of kinetic data by spectrophotometry at 550 nm. The trace represents the kinetic of superoxide Cyt*c* reduction. This is shown in Fig. 3 (black trace).
7. Calculate the amount of superoxide per unit of time generated (using the Cyt*c* molar extinction coefficient of $2.1 \times 10^4\ mM^{-1}\ cm^{-1}$ [24]), and normalize to the concentration of $Cytb_{558}$ (*see* **Note 21**) used into the sample (usually expressed in nM) to obtain the apparent turnover of the enzyme.

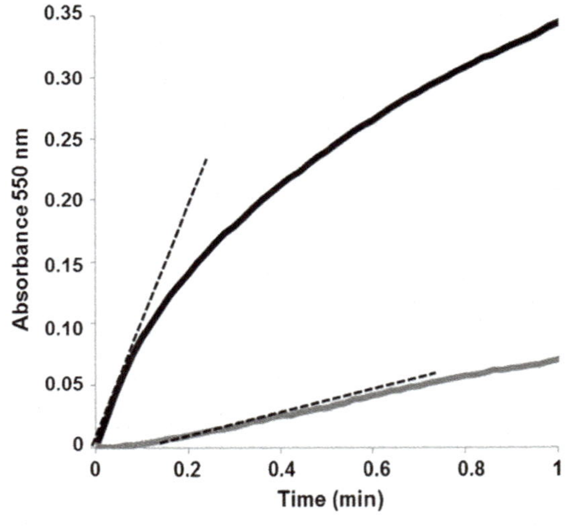

Fig. 3 Rate of formation of superoxide anions by NADPH oxidase measured by cytochrome *c* reduction kinetics using the canonical cell-free assay. The superoxide production was measured spectrophotometrically by following the reduction rate of Cyt*c* at 550 nm. The reaction was initiated by the addition of 200 μM NADPH in the presence of 50 μM cytochrome *c*. The reaction mixture contains membrane fractions with $Cytb_{558}$ (2.5 nM), purified recombinant cytosolic proteins and *cis*-arachidonic acid. The mixture was incubated 5 min at 25 °C (black trace). When the mixture was not preincubated (gray trace), the slope was less steep and followed generally a sigmoid pattern. The dashed lines correspond to the maximal rates at the origin

8. The preincubation time of 5 min is of particular importance for obtaining the maximal rate of superoxide anion production. All steps are performed at 25 °C. As shown in Fig. 3 (gray trace), the kinetic is different in the absence of incubation, and the rate of superoxide production measured as the initial slope of the curve (dashed line) is much higher when the preincubation is performed. It should be noted that in the absence of preincubation, the trace follows a sigmoid pattern, and in that case, the rate is calculated at the inflection point.

A more in-depth investigation of the NADPH oxidase activity versus the *cis*-AA concentration should be performed to determine the optimal concentration of AA to be added in the reaction mixture for obtaining the maximal rate of superoxide production. For that purpose, please note that (1) kinetics of superoxide production are measured spectrophotometrically for different *cis*-AA concentrations as abovementioned, (2) the rates are plotted versus the AA concentration, and (3) as illustrated in Fig. 4, the dose-response

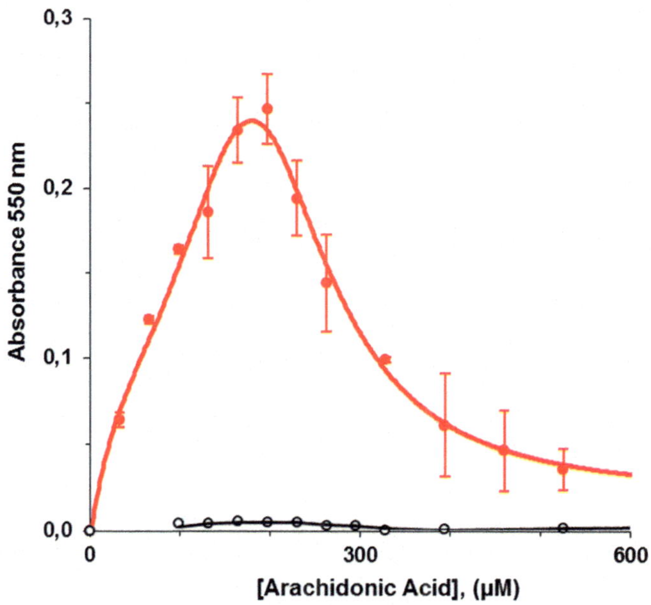

Fig. 4 Activation of the NADPH oxidase complex by different arachidonic acid isomers. The superoxide production was measured spectrophotometrically by the rate of cytochrome *c* reduction at 550 nm and plotted at different concentrations of *cis*- (red trace) and *trans*- (black trace) arachidonic acid. In the reaction mixture, membrane fractions from bovine neutrophils were incubated 5 min in the presence of recombinant cytosolic proteins and different concentrations of arachidonic acid. The reaction was initiated by the addition of 200 μM NADPH in the presence of 50 μM cytochrome *c*. Results are presented as the mean ± SD of four independent experiments. The visual fits of the curves were performed using KaleidaGraph (Synergy Software, Reading, PA, USA) (Modified from the original publication in [6])

curve of NADPH oxidase activity follows a bell-shaped pattern. Such curve is characteristic of phagocyte NADPH oxidase activation.

The optimal concentration of AA causing maximal activation has to be determined for each new membrane fraction from any blood donor. Depending on the origin of the membrane fraction (human blood donor, health state of the donor, different mammalian species), this optimal value may vary significantly. The differences are correlated to species variation (e.g., human, bovine), but the fundamental reason is parameters such as different ratio of membrane protein/lipids and protein and lipid composition (*see* **Note 25**).

3.4 Stereospecificity of NADPH Oxidase Activation by AA: the Trans-Acid Arachidonic Isomer

In eukaryotic cells, the prevalent geometry displayed by unsaturated fatty acids is the *cis* configuration. The change of the *cis* geometry to the geometrical *trans* isomer can happen in an exogenous manner with the intake of *trans* fatty acids through food containing partially hydrogenated or deodorized vegetable/fish oils or when food is processed at high temperature. Polyunsaturated fatty acids such as *cis*-AA are also targets for a variety of free radicals leading to generation of endogenous *trans* fatty acids by radical-induced *cis-trans* isomerization [25, 26]. Massive production of free radicals generated by the activated NADPH oxidase could contribute to the in situ concentration increase of *trans*-AA isomers.

The *trans*-AA isomer was synthesized and kindly provided by Drs C. Ferreri and C. Chatgilialoglu (ISOF, Bologna, Italy). Synthesis of *trans*-AA was performed following the protocol for free radical geometrical isomerization procedure [6, 27]. The *trans*-AA is a mixture of 4 *mono-trans* isomers in equimolar concentrations. We recorded the absorption spectra of the *trans*-AA (Fig. 5). *Trans*-AA displays a similar peak at 215 nm as *cis*-AA, but the peak at 235 nm is not resolved and resembles now a shoulder.

The superoxide production in the NADPH oxidase cell-free assay was measured spectrophotometrically following the same procedure as previously described (Fig. 6). In the reaction mixture, membrane fractions were incubated 5 min in the presence of recombinant cytosolic proteins and then with either *cis*- or *trans*-arachidonic acid. The superoxide production rate is highly decreased when *cis*-AA is replaced by *trans*-AA as illustrated by the slope being close to 0. The effect was investigated over a large *trans*-AA concentration range, and, as illustrated in Fig. 4 with black circles, the addition of *trans*-AA could not induce $O_2^{\cdot -}$ production in the cell-free system. This result indicates that *trans*-AA is unable to activate properly the NADPH oxidase complex although its negative charge is similar to the *cis* form. *Trans*-AA interferes with the NOX-activating capacity of all-*cis*-AA.

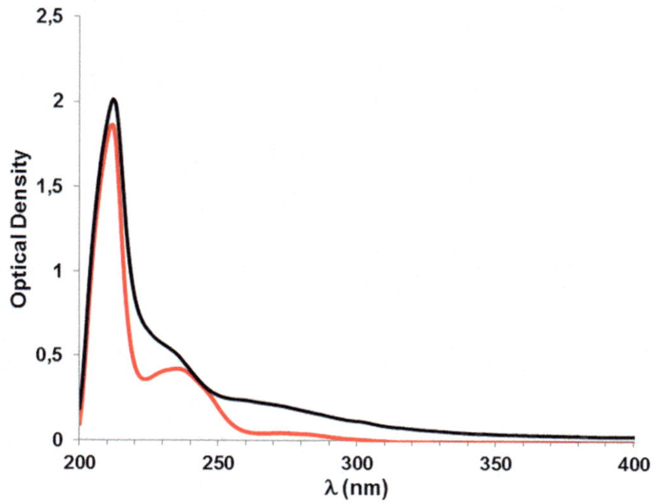

Fig. 5 Comparison of the absorption spectra of *cis*- and *trans*-arachidonic acid isomers. The red spectrum corresponds to *cis*-AA from Sigma-Aldrich (A9673). The *trans*-AA (black trace) is a mixture of four mono-*trans* isomers synthesized and kindly provided by Drs. C. Ferreri and C. Chatgilialoglu (ISOF, Bologna, Italy). Stock solutions of *cis*-AA and *trans*-AA are diluted with Na-phosphate buffer to 0.1 mg/mL

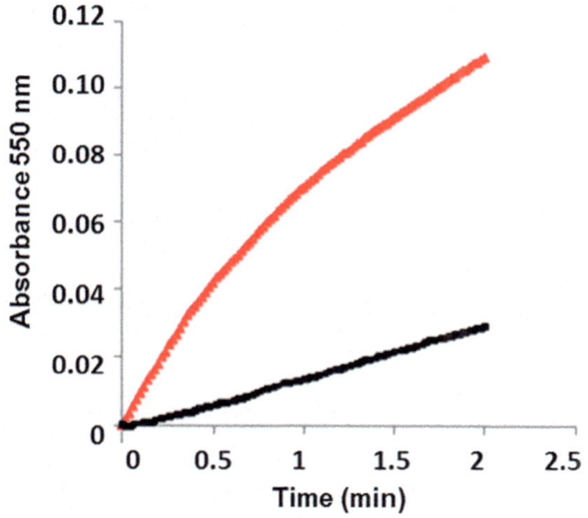

Fig. 6 Different effects of *cis*- and *trans*-AA on Cytb_{558} activity. The superoxide production was measured spectrophotometrically by the rate of cytochrome *c* reduction at 550 nm. In the reaction mixture, recombinant yeast membrane fractions (4 nM Cytb_{558}) were incubated 5 min in the presence of recombinant cytosolic proteins (200 nM each) and 450 μM of *cis*-arachidonic acid (red trace) or *trans*-arachidonic acid (black trace)

3.5 Colorimetric Free Thiol Quantification of p47^{phox} and p67^{phox} Proteins: Effect of AA on Thiol Accessibility

Besides the importance of AA on the membrane component of the oxidase, we evaluated its structural impact on the conformation of the cytosolic subunits p47^{phox} and p67^{phox}. We first investigated the accessibility of thiol groups. We used a standard protocol for the detection of free thiols [28]. This protocol works for small peptides and proteins where the thiol groups are accessible to 5,5′-dithiobis

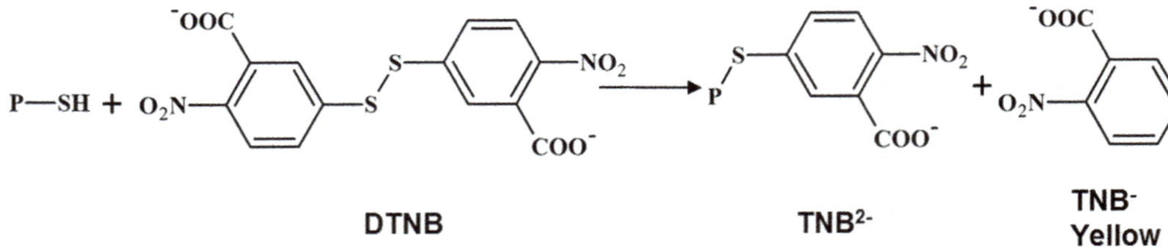

Fig. 7 Scheme of the protein thiol (PSH) reaction with DTNB

(2-nitrobenzoic acid) (DTNB) (Fig. 7). Ellman's reagent (DTNB) reacts with sulfhydryl groups to yield a yellow product, 5-thio-2-nitrobenzoic acid (TNB^{2-}). This compound is deprotonated at pH around 8. It is strongly yellow, providing a quantitative method to measure the amount of reduced cysteine and other free sulfhydryls in solution.

1. Set a standard glutathione (GSH) calibration curve with 0, 18, 36, 72, 90, and 103 μM by adding to 28 μL of the DTNB solution (400 μM final concentration in Na-phosphate pH 8.0), the appropriate GSH solution, and adjust to 700 μL final volume with 20 mM sodium phosphate buffer pH 8.0 at 25 °C. This permits a precise measurement of the extinction coefficient of TNB^{2-}.
2. Mix the solutions. The reaction with glutathione is immediate.
3. Measure the absorbance at 412 nm.
4. Plot the values against the GSH concentrations. This should be a straight line.
5. The extinction coefficient of the reagent is equal to the slope of the straight line. We obtained 13,800 M^{-1} cm^{-1} at 25 °C (*see* **Note 26**).
6. For free thiol detection in $p47^{phox}$ and $p67^{phox}$ proteins, the same procedure is followed except that the GSH solution is replaced by a known concentration of oxidase proteins (typically about 60 μM of DTNB and 1 μM of $p47^{phox}$ or $p67^{phox}$). Using these proteins, the reaction can be slow or inefficient when the groups are somewhat hidden (*see* **Note 27**).
7. Mix well, incubate for various times (depending on the protein), and record the spectra between 350 and 600 nm. Follow the absorbance at 412 nm versus time.
8. When the absorbance maximum is reached, calculate the molarity of (-SH) groups in the assay by dividing the absorbance maxima at 412 nm by 13,800 M^{-1} cm^{-1}.
9. The number of –SH groups per protein is calculated by dividing the molarity of –SH by that of the protein of interest. For

instance, the p67^{phox} concentration was 5 μM. After 20 min, the total amount of thiols detected was 47.5 μM. By dividing this amount by the protein concentration, one obtains the number of thiols detected per protein, e.g., 9.5 thiols per protein.

10. When the reaction is followed at 412 nm immediately after mixing and followed continuously until completion, it provides an indication of thiol accessibility as illustrated in Fig. 8. For p67^{phox}, nine thiol groups are present in the predicted amino acid sequence. All nine groups reacted with DTNB in about 25 min, four being immediately detected indicating that these four cysteines are completely solvent accessible. For p47^{phox}, four thiol groups are present in the predicted amino acid sequence. In contrast to p67^{phox}, no thiol groups are detected during the first 10 min. To detect the four thiol groups in p47^{phox}, a reaction time of more than 1 h is necessary. When *cis*-AA is added to the proteins, thiol accessibilities are modified as illustrated in Fig. 8 (inset): for p67^{phox}, while four thiol groups are still immediately measurable, only three additional cysteines remain accessible in about 20 min, and two cysteines are sequestered from the solvent consistently with structural changes in p67^{phox} being induced by *cis*-AA. For p47^{phox} one thiol is rapidly detected, while the other three cysteines are kept hidden. For this protein, it is likely that the addition of AA results in masking three SH functions. As shown in Fig. 8 (large panel), when both proteins are in the assay in equimolar

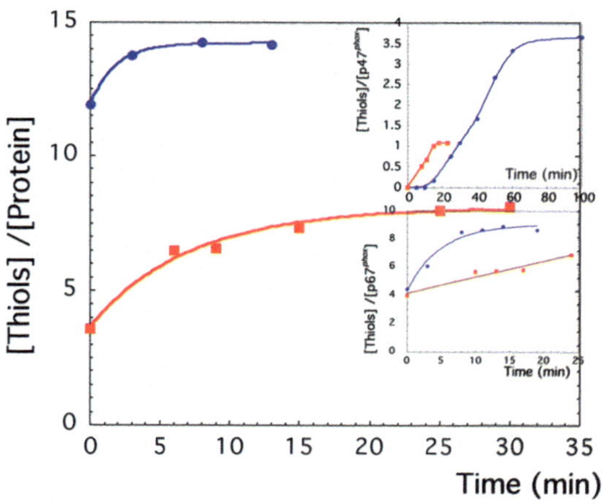

Fig. 8 Measurement of the amount of free thiol groups with Ellmann's reagent (DTNB 60 μM, Na-phosphate pH 8.0) in an equimolar mixture of p67^{phox} and p47^{phox} in the presence (red trace) and absence (blue trace) of *cis*-AA. Protein concentration 1 μM. In insets the same measurements on proteins alone in solution: top: 1 μM of p67^{phox}; bottom: 1 μM of p47^{phox}. Blue: no AA; Red: with 26 μM *cis*-AA. The measurements have been performed on two different protein preparations and each one was repeated 3 times

mix, the number of accessible thiol groups does not correspond to those obtained for proteins analyzed separately. This indicates that single addition versus mixed addition of $p47^{phox}$ and $p67^{phox}$ proteins will result in a different conformation of these proteins in the assay. Information deduced from thiol groups accessibility is an important aspect of the protein architecture.

3.6 Synchrotron Radiation Circular Dichroism (SRCD) Measurements

SRCD is in principle similar to CD, except that the intensity of light in the visible-UV region is much higher due to the synchrotron source. In the wavelength range investigated (175–280 nm), the absorption is mainly due to peptidic bonds. Hence, the spectrum reflects the secondary structure of a protein.

3.6.1 Sample Preparation for SRCD

1. Fill Slide-A-Lyzer MINI Dialysis Units [MW 30 K cutoff] with the purified protein of interest.
2. The filled Slide-A-Lyzer MINI Dialysis Units are stirred slowly in 100 mM NaF and buffered with a very low concentration of sodium phosphate buffer (10 mM sodium phosphate buffer, pH 7.4 adjusted with NaOH) overnight for buffer exchange (*see* **Note 28**).
3. The next day, recover the protein from the Slide-A-Lyzer MINI Dialysis Units, and keep 10 mL of the dialysate buffer for CD control analyses.
4. Briefly centrifuge the samples at 10,000 × *g* to remove any protein aggregates.

3.6.2 SRCD Spectroscopy

1. Collect SRCD spectra; in our case on the DISCO beamline at the synchrotron radiation SOLEIL, Gif-sur-Yvette, France [29].
2. Load 25 μL of the samples in CaF_2 cells (*see* **Note 29**) using volumes and optical path lengths of 33 and 50 μm depending on the protein concentrations (between 14 and 40 μM).
3. Spectra were measured from 280 to 170 nm (in this wavelength range, the absorption is due to the peptidic bonds), using the mid-height of the HT (high tension) as cutoff at 175 nm. The temperature is kept constant at 25 °C.
4. Spectra are expressed in delta epsilon units and calculated using mean residue weights of each measured protein.
5. Figure 9 shows the CD spectra obtained for $p47^{phox}$ at different *cis*-AA concentrations. Once the CD spectra have been acquired, structural studies can be performed by determining the secondary structure using CD spectra software. We use currently the recently developed software BeStSel (https://bestsel.elte.hu/index.php), but other free software is available such as DichroWeb or SELCON3. Figure 10 illustrates the distribution of secondary structure of $p47^{phox}$ and $p67^{phox}$ in the presence of different *trans*-AA concentrations obtained

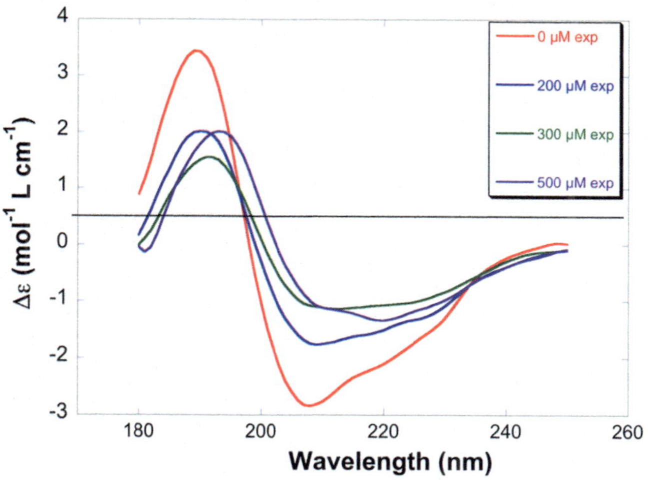

Fig. 9 Effect of *cis*-AA (0–500 μM) on the SRCD spectra of the p47^{phox} subunit. The p47^{phox} protein was purified, and its concentration was determined by NanoDrop measurements and the BCA method. Far UV spectra were acquired on the DISCO beamline at the synchrotron radiation SOLEIL, Gif-sur-Yvette, France. CD spectra were recorded in the absence and in the presence of increasing concentration of arachidonic acid. The acquisition was carried out at 25 °C with a 33 μm optical path length cell

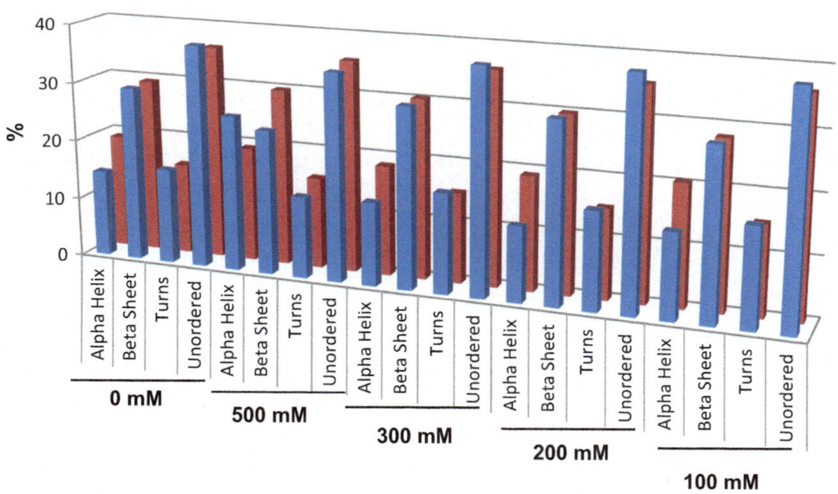

Fig. 10 Comparison of SRCD spectra of p47^{phox} and p67^{phox} treated with different *trans*-AA concentrations using SELCON3 (blue) and CONTIN3 (red) software

with two different analysis tools, SELCON and CONTIN. Even if variations in analyses of CD spectra are observed when using two different software programs, it is relevant to compare the trend upon increased AA concentration. This procedure will inform about differences in the behavior of p47^{phox} versus p67^{phox} under these conditions.

4 Notes

1. In the lysis buffer, NaCl is added for the ionic force that facilitates membrane separation; EDTA chelates metal ions and limits damage by protein oxidation; addition of DNAse avoids a slimy consistency of the suspension and facilitates sonication, centrifugation, and chromatography steps; PMSF is an efficient antiprotease to avoid protein degradation, and DTT reduces disulfide bonds.
2. Plastic tubes and pipettes in polypropylene are required to limit neutrophil activation. Large diameter cylinders are preferable.
3. Solubilization of dextran needs at least 1 hour of strong stirring in a large 500 mL sterile cylinder.
4. The breaking buffer should be prepared the same day.
5. For SRCD experiments even if the sample volume is very small (few μL), the protein solution should be very concentrated (>2.5–3 mg/mL).
6. NanoDrop2000™ spectrophotometer is a very fast and economical (only 2 μL of protein solution) instrument for the determination of the protein concentration, which is an important parameter for CD experiments, but the classical BCA method is more accurate.
7. The Ellman's reagent has a highly oxidizing disulfide bond which is stoichiometrically reduced by free thiols in an exchange reaction, forming a mixed disulfide and a molecule of 5-thio-2-nitrobenzoic acid ionized to the NTB^{2-} di-anion in basic medium. This reaction can be rapid and stoichiometric when the thiol groups are accessible, with one mole of thiol producing one mole of NTB^{2-}.
8. Small aliquot volumes of AA solution must be stored to avoid repetitive opening. Before activity experiments are performed, the AA stock solution is diluted ten times in ethanol (99.8%) to prepare a 2 mg/mL solution of arachidonic acid and aliquoted in several tubes to avoid repetitive opening. Commercial sources of *all-cis*-AA may not have the same degree of purity. AA from various manufacturers can be colorless to very faint yellow, as AA is not stable and can oxidize quickly. Absorption spectra of two different AA sources can display significant differences as illustrated in Fig. 11 where both utilized AAs have a spectrum with a major peak at 215 nm, but the AA from Sigma-Aldrich (black trace) displays an additional peak at 235 nm and a shoulder at about 280 nm. In this regard, the use of a specific AA may give different $O_2^{\cdot -}$ production values. Figure 12 illustrates the superoxide-induced Cyt*c* reduction rate measured at 550 nm with *cis*-arachidonic acid from two

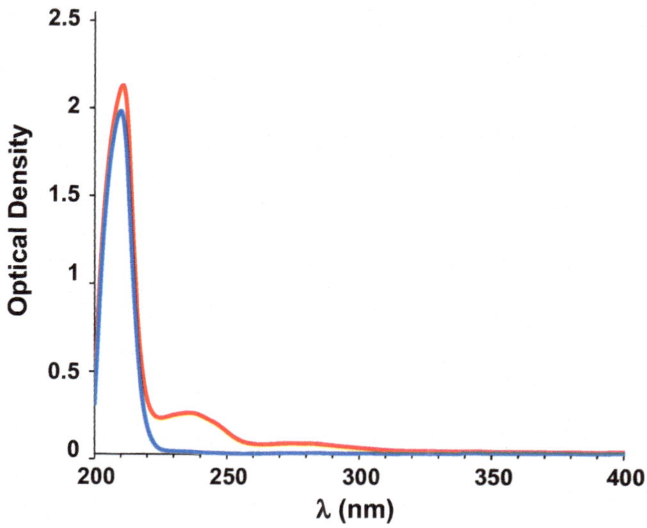

Fig. 11 Absorption spectra of *cis*-arachidonic acid from two different suppliers. Stock solutions of *cis*-AA are diluted with Na-phosphate buffer to 0.1 mg/mL. The red spectrum with additional peaks corresponds to *cis*-AA from Sigma-Aldrich (A9673)

different suppliers. In our hands, the *cis*-AA from Sigma-Aldrich (the red trace in Figs. 11 and 12; reference A9673) activates more efficiently the NADPH oxidase in cell-free assays as illustrated by a higher initial slope. Thus, just by changing the AA source or even, sometimes, the lot number, cell-free assays may lead to misinterpretations. Therefore, *cis*-AA from Sigma-Aldrich was used to perform all the NADPH oxidase activity tests, and we tried to keep the same lot number, and once the lot was emptied, the effectiveness of the new lot of AA was systematically compared.

9. NADPH solution must be prepared just before use. The solution is diluted 1000-fold to perform a spectrum. The NADPH concentration is calculated using the extinction coefficient of 6220 M^{-1} cm^{-1} at 340 nm.

10. Add 5 mL of PBS to the bottle containing 1 g Cyt*c* (MW 12384 Da), and solubilize as much as possible. Then transfer the solution into a 50 mL Falcon tube, and centrifuge for 5 min at 3200 × *g*. The supernatant is transferred into a new Falcon tube and centrifuged again for 10 min at 3200 × *g*. Repeat until the white pellet is completely dissolved. The supernatant is aliquoted (about 100 μL per aliquot) and stored at −80 °C. One aliquot is thawed to perform oxidized and reduced spectra. The Cyt*c* concentration is deduced from the difference in absorbance at 550 nm using the differential coefficient of 2.1×10^4 M^{-1} cm^{-1}.

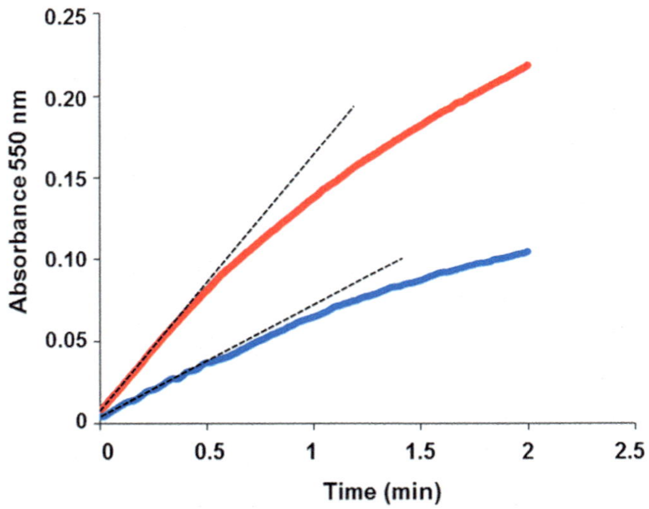

Fig. 12 Rate of formation of superoxide anions by NADPH oxidase measured by cytochrome *c* reduction kinetics using the canonical cell-free assay with *cis*-arachidonic acid from two different suppliers. The superoxide production rate was measured spectrophotometrically following the rate of cytochrome *c* reduction at 550 nm. In the reaction mixture, membrane fractions containing Cytb_{558} (2.5 nM) were incubated 5 min in the presence of recombinant cytosolic proteins and with the same concentration of *cis*-arachidonic acid from two different suppliers. The reaction was initiated by the addition of 200 μM NADPH in the presence of 50 μM cytochrome *c*. The two traces (blue and red) correspond to the kinetics obtained in the same experimental conditions except that *cis*-AA is from two different suppliers. In our hands, the *cis*-AA from Sigma-Aldrich (red trace) activates more efficiently the NADPH oxidase in cell-free assays as illustrated here by a higher slope at the beginning of the kinetic

11. Membrane purification of yeast expressing recombinant Cytb_{558}, detergent solubilization, and reconstitution into liposomes were previously described in [30].
12. Dextran causes the aggregation of red blood cells, which will sediment. In the lower part of the graduated cylinder which appears opaque and very dark red, one can find the majority of sedimented red cells. In the upper part, translucent and yellow orange, are leucocytes—including neutrophils—and some remaining red blood cells.
13. Cytosolic NADPH oxidase proteins are preferred targets for proteases. Several inhibitors such as leupeptin, glutamic acid, arginine, or a cocktail of inhibitors can be added. For the chromatography step, the most effective and cheap protease inhibitor remains PMSF. Keep all manipulation at low temperature (<6 °C), and perform the purification as quickly as possible.

14. The sonication breaks the cell membranes and releases the contents of the bacteria.
15. Filtration through 0.22 μm pore size filters eliminates contaminants or residual DNA.
16. Ni-Sepharose affinity chromatography is used to separate His_6-tagged $p47^{phox}$ from contaminating proteins.
17. The progress in achieving purity of proteins is followed after each purification step by SDS-PAGE and detection by Coomassie blue staining. For precise determinations of the purified protein concentration, BCA and NanoDrop measurements were systematically performed.
18. Purification of Rac1Q61L was previously described [9]. The isoelectric points of $p67^{phox}$ and Rac1Q61L are low (6.3 and 7.8, respectively). In the case of the Rac1Q61L purification, chromatography is performed at pH 7.0, which induces predominantly the positively charged form of the protein. Loading on an anion exchange column should be performed slowly in order to keep proteins as concentrated as possible.
19. Ferricytochrome *c* (Cyt*c*) traps the produced superoxide before it dismutates to hydrogen peroxide.
20. The protocol of cell-free systems using recombinant Cytb_{558} products in yeasts and proteoliposomes has been detailed in [30].
21. The Cytb_{558} concentration in the membrane fraction (MF) is determined by a Na-dithionite-reduced *minus* oxidized spectrum. First, isolated MFs are diluted (no more than twice for small sample quantity) with DPBS supplemented with 1% n-dodecyl β-D-maltoside (DDM) for a final volume of 200 μL. The solution is incubated at 22 °C for 70 min and mixed carefully every 10 min. The oxidized spectrum is recorded between 300 and 900 nm using an appropriate quartz cuvette. Then the reduced form is obtained by adding few grains of Na-dithionite, and the spectrum is recorded again. Both spectra are required to determine the reduced *minus* oxidized spectrum.

    One problem in such experimental conditions is turbidity which hinders the evaluation of the amplitude of the peaks in the Soret band. Posttreatments of the spectra can be performed to compensate for variation of the suspension turbidity:

    (a) MF spectra, oxidized and reduced, present a component characteristic of the turbidity. This component can be fitted by a nonlinear regression as a result of the function:

$$f(x) = a_0 + \frac{a_1}{x^{p_1}} + \frac{a_2}{x^{p_2}} + \ldots + \frac{a_i}{x^{p_i}} + \ldots + \frac{a_k}{x^{p_k}}$$

where $a_i \in \mathbb{R}$ are parameters and $p_i \in \mathbb{R}^+$ are exponent parameters. Indices are silent with $i \in \mathbb{N}$ and $0 < 1 < 2 < \ldots < i < \ldots < k$.

The nonlinear regression is usually done between 900 and 450 nm for both oxidized and reduced forms. The criteria are a regression coefficient of at least $r^2 = 0.99$ and a visual shape very close to the observed turbidity component of the experimental curve. Moreover, by extending the graphical representation of the function from 900 to 370 nm, the curve should be visually superimposed with experimental points, particularly before and after the characteristic bands at 411 nm for the oxidized form and at 427 nm for the reduced form (by keeping a scale adjusted from 370 to 900 nm for abscissa and a corresponding scale in the ordinate).

If the regression is difficult to realize, one has to change either the algorithm used to perform the fit (e.g., Simplex Nelder-Mead or Levenberg-Marquardt normalized or not), or the start value of the parameters, or the number of terms (and consequently the number of parameters) in the function. Usually, from two to eight terms can be used, but the more terms are used, the longer it will take to reach convergence. It is important that the only aim is to achieve the best possible fit for turbidity in order to remove it from the experimental spectra.

(b) After satisfying fits are obtained for both reduced and oxidized FM spectra, each spectrum is subtracted from the corresponding turbidity fit. The corrected difference spectrum can now be drawn using the adjusted reduced and oxidized FM spectra.

The Cytb_{558} heme amount is quantified by measuring the absorbance difference between the peak at 427 nm and the hollow at 411 nm using an extinction coefficient of $\Delta\varepsilon_{427\text{–}411} = 200\ \text{mM}^{-1}\ \text{cm}^{-1}$ [31]. Calculate the Cytb_{558} concentration [Cytb_{558}] (mol L^{-1}) in the membrane fraction suspension by using the equation:

$$[\text{Cyt}b_{558}] = F_D \cdot \frac{\Delta A_{427-411}}{\ell \cdot \Delta\varepsilon_{427-411}}$$

with F_D as the dilution factor of the FM suspension and ℓ as the optical path in centimeters.

22. In cell-free assays, the Rac1 mutant (Rac1Q61L) is often used as it represents a constitutively active GTP-bound form with increased affinity.

23. For reproducibility, all incubations and all the measurements should be done at the same temperature (here at 25 °C). Superoxide production rates showed significant temperature dependence [7].
24. The measurement of the generated superoxide anions by cytochrome c reduction is a classical technique, which can be employed to examine the catalytic parameters of enzymes implicated in oxidase redox reactions.
25. In the dose-response curves, the maximal activity (top of the bell curve) and its position (corresponding to optimal *cis*-AA concentrations) may vary. AA is an amphiphilic molecule that targets naturally membranes. Due to the different physicochemical properties (lipid chemical structure, membrane protein/lipids, and membrane proteins/total membrane protein ratio), the partitioning of the amphiphilic molecules within the bilayer is different. The differences that are observed in the literature concerning the optimal AA concentration values in cell-free systems correlate to the particular membrane composition that varies in species (e.g., human, bovine, recombinant yeast). The fundamental reasons for divergence are parameters like the ratio of membrane protein to lipids and protein and/or lipid composition. In the literature, depending on the type of human semi-recombinant cell-free system considered, the optimal AA concentration values range from 30 [16, 32, 33] to 200 μM [34]. In our hands, in the bovine cell-free system, the maximal $O_2^{\cdot-}$ production occurred at 150–200 μM of AA, using neutrophil bovine membrane fractions (containing in general >0.06 mg of Cytb_{558}/mg of total membrane proteins). This value reaches 750–800 μM with yeast membrane fraction containing bovine recombinant proteins (<0.01 mg of Cytb_{558}/mg of total membrane proteins), i.e., four times higher. This optimal value is highly variable, and it can be different even within the same species as we often observe with human neutrophil membrane fractions depending on the blood donor and even on the experimentalist. The transfer of AA from aqueous phase to the lipid membrane phase, i.e., the lipid membrane partitioning of AA into the membrane, has a significant influence on the mechanical properties of the lipid bilayer such as rigidity, bending, and the transmembrane barrier properties which are fundamental regarding the enzyme's catalytic reaction and the superoxide measurement.
26. The extinction coefficient for dilute solutions can be underestimated and should be reevaluated for the experimental condition of the target protein solution.
27. To measure the total amount of free thiol groups in proteins, one should add 6 M guanidinium chloride for denaturation.

28. The CD experiments require buffers without absorbance values at low wavelengths in the UV and without ingredients that may cause undesirable interactions. For example, it can be problematic using phosphate buffers, and NaF was added to maintain ionic strength (anions, such as fluoride, maintain ionic strength and preserve the structure of proteins).
29. Cells with different optical path lengths are available depending on the protein concentrations. For standard protein concentration of 2–3 mg/mL, a cell with an optical length of 100 μm is classically used.

Acknowledgments

Authors want to acknowledge Drs. M. Réfrégiers and F. Wien for SRCD measurements.

References

1. Dagher MC, Pick E (2007) Opening the black box: lessons from cell-free systems on the phagocyte NADPH-oxidase. Biochimie 89:1123–1132
2. Molshanski-Mor S, Mizrahi A, Ugolev Y, Dahan I, Berdichevsky Y, Pick E (2007) Cell-free assays: the reductionist approach to the study of NADPH oxidase assembly, or "all you wanted to know about cell-free assays but did not dare to ask". Methods Mol Biol 412:385–428
3. Pick E (2014) Cell-free NADPH oxidase activation assays: "in vitro veritas". Humana Press, Totowa, NJ
4. Ostuni MA, Lamanuzzi LB, Bizouarn T, Dagher MC, Baciou L (2010) Expression of functional mammal flavocytochrome b(558) in yeast: comparison with improved insect cell system. Biochim Biophys Acta 1798:1179–1188
5. Ezzine A, Souabni H, Bizouarn T, Baciou L (2014) Recombinant form of mammalian gp91(phox) is active in the absence of p220 (phox). Biochem J 462:337–345
6. Souabni H, Thoma V, Bizouarn T, Chatgilialoglu C, Siafaka-Kapadai A, Baciou L, Ferreri C, Houee-Levin C, Ostuni MA (2012) Trans Arachidonic acid isomers inhibit NADPH-oxidase activity by direct interaction with enzyme components. BBA-Biomembranes 1818:2314–2324
7. Souabni H, Wien F, Bizouarn T, Houee-Levin C, Refregiers M, Baciou L (2017) The physicochemical properties of membranes correlate with the NADPH oxidase activity. Biochim Biophys Acta 1861:3520–3530
8. Souabni H, Machillot P, Baciou L (2014) Contribution of lipid environment to NADPH oxidase activity: influence of sterol. Biochimie 107 (Pt A):33–42
9. Baciou L, Erard M, Dagher MC, Bizouarn T (2009) The cytosolic subunit p67phox of the NADPH-oxidase complex does not bind NADPH. FEBS Lett 583:3225–3229
10. Leto TL, Adams AG, Demendez I (1994) Assembly of the phagocyte NADPH oxidase – binding of src homology-3 domains to proline-rich targets. Proc Natl Acad Sci U S A 91:10650–10654
11. Lapouge K, Smith SJ, Groemping Y, Rittinger K (2002) Architecture of the p40-p47-p67phox complex in the resting state of the NADPH oxidase. A central role for p67phox. J Biol Chem 277:10121–10128
12. Iyer SS, Pearson DW, Nauseef WM, Clark RA (1994) Evidence for a readily dissociable complex of P47phox and P67phox in cytosol of unstimulated human neutrophils. J Biol Chem 269:22405–22411
13. Lee JH, Lee KS, Chung T, Park J (2000) C-terminal region of the cytosolic subunit p47(phox) is a primary target of conformational change during the activation of leukocyte NADPH oxidase. Biochimie 82:727–732
14. Hata K, Ito T, Takeshige K, Sumimoto H (1998) Anionic amphiphile-independent activation of the phagocyte NADPH oxidase in a cell-free system by p47phox and p67phox,

both in C terminally truncated forms. Implication for regulatory Src homology 3 domain-mediated interactions. J Biol Chem 273:4232–4236

15. Karimi G, Levin CH, Dagher MC, Baciou L, Bizouarn T (2014) Assembly of phagocyte NADPH oxidase: a concerted binding process? BBA-Gen Subjects 1840:3277–3283
16. Swain SD, Helgerson SL, Davis AR, Nelson LK, Quinn MT (1997) Analysis of activation-induced conformational changes in p47(phox) using tryptophan fluorescence spectroscopy. J Biol Chem 272:29502–29510
17. Park HS, Park JW (1998) Fluorescent labeling of the leukocyte NADPH oxidase subunit p47 (phox): evidence for amphiphile-induced conformational changes. Arch Biochem Biophys 360:165–172
18. Groemping Y, Lapouge K, Smerdon SJ, Rittinger K (2003) Molecular basis of phosphorylation-induced activation of the NADPH oxidase. Cell 113:343–355
19. Shiose A, Sumimoto H (2000) Arachidonic acid and phosphorylation synergistically induce a conformational change of p47(phox) to activate the phagocyte NADPH oxidase. J Biol Chem 275:13793–13801
20. Matono R, Miyano K, Kiyohara T, Sumimoto H (2014) Arachidonic acid induces direct interaction of the p67(phox)-rac complex with the phagocyte oxidase Nox2, leading to superoxide production. J Biol Chem 289:24874–24884
21. Doussiere J, Gaillard J, Vignais PV (1996) Electron transfer across the O-2(−) generating flavocytochrome b of neutrophils. Evidence for a transition from a low-spin state to a high-spin state of the heme iron component. Biochemistry 35:13400–13410
22. Bizouarn T, Karimi G, Masoud R, Souabni H, Machillot P, Serfaty X, Wien F, Refregiers M, Houee-Levin C, Baciou L (2016) Exploring the arachidonic acid-induced structural changes in phagocyte NADPH oxidase p47 (phox) and p67(phox) via thiol accessibility and SRCD spectroscopy. FEBS J 283:2896–2910
23. Nauseef WM (2007) Isolation of human neutrophils from venous blood. Methods Mol Biol 412:15–20
24. van GB, Slater EC (1962) The extinction coefficient of cytochrome c. Biochim Biophys Acta 58:593–595
25. Chatgilialoglu C, Ferreri C (2005) Trans lipids: the free radical path. Acc Chem Res 38:441–448
26. Ferreri C, Chatgilialoglu C (2005) Geometrical trans lipid isomers: a new target for lipidomics. Chembiochem 6:1722–1734
27. Anagnostopoulos D, Chatgilialoglu C, Ferreri C, Samadi A, Siafaka-Kapadai A (2005) Synthesis of all-trans arachidonic acid and its effect on rabbit platelet aggregation. Bioorg Med Chem Lett 15:2766–2770
28. Ellman GL (1958) A colorimetric method for determining low concentrations of mercaptans. Arch Biochem Biophys 74:443–450
29. Refregiers M, Wien F, Ta HP, Premvardhan L, Bac S, Jamme F, Rouam V, Lagarde B, Polack F, Giorgetta JL, Ricaud JP, Bordessoule M, Giuliani A (2012) DISCO synchrotron-radiation circular-dichroism endstation at SOLEIL. J Synchrotron Radiat 19:831–835
30. Souabni H, Ezzine A, Bizouarn T, Baciou L (2017) Functional assembly of soluble and membrane recombinant proteins of mammalian NADPH oxidase complex. Methods Mol Biol 1635:27–43
31. Light DR, Walsh C, O'Callagahn A, Goetzl E, Tauber A (1981) Characteristics of the cofactor requirements for the superoxide-generating NADPH oxidase of human polymorphonuclear leukocytes. Biochemistry 17:1468–1476
32. Curnutte JT (1985) Activation of human neutrophil nicotinamide adenine-dinucleotide phosphate, reduced (triphosphopyridine nucleotide, reduced) oxidase by arachidonic-acid in a cell-free system. J Clin Investig 75:1740–1743
33. Clark RA, Leidal KG, Pearson DW, Nauseef WM (1987) NADPH oxidase of human-neutrophils – subcellular-localization and characterization of an arachidonate-activatable superoxide-generating system. J Biol Chem 262:4065–4074
34. Ligeti E, Pizon V, Wittinghofer A, Gierschik P, Jakobs KH (1993) Gtpase activity of small Gtp-binding proteins in Hl-60 membranes is stimulated by arachidonic-acid. Eur J Biochem 216:813–820

Chapter 6

NOX5 Cell-Free Assay for the High-Throughput Screening of Small Molecules

Fiona Augsburger, Delphine Rasti, Yves Cambet, and Vincent Jaquet

Abstract

NADPH oxidases (NOX) are a family of transmembrane enzymes, which catalyze the formation of $O_2^{\cdot -}$ and H_2O_2. Membrane fractions of leukocytes are highly enriched in the phagocyte NOX isoform (NOX2). This feat has allowed the development of a complex NOX2 cell-free assay, which has been a key tool for the understanding of the mode of action of NOX2, its biochemistry, pharmacology, and identification of NOX2-specific inhibitors. In addition to NOX2, there are six other NOX isoforms in humans, but cell-free assays of non-phagocytic oxidases are infrequently used, and their specificity has recently been debated. Here we describe a NOX5 cell-free assay. We present a method to purify the membranous component of cells stably transduced with NOX5 and to measure $O_2^{\cdot -}$ in a high-throughput format (96-w or 384-w plates). The experimental description allows high-throughput screening of small molecules with limited cost.

Key words NOX5, Membrane purification, Cell-free assay, High-throughput screening

1 Introduction

NADPH oxidases (NOX) are a family of transmembrane enzymes catalyzing the formation of reactive oxygen species (ROS). In humans, there are seven NOX isoforms, each with a specific tissue localization and mode of action. As major ROS generators in pathological states, NOX are potential drug targets for a range of diseases [1], and several drug discovery programs have been undertaken for identification of small molecule therapeutic NOX inhibitors. NOX enzymology is essential for establishing robust, reliable, and appropriate assays for screening campaigns as well as determining the pharmacological mode of inhibition of small molecule NOX inhibitors [2]. For this purpose, purified cell membranes containing active NOX enzymatic systems (NOX cell-free assays) represent optimal systems. Historically, NOX2 cell-free systems were developed in the 1980s using purified leukocyte membranes and cytosolic preparation to which the NADPH and FAD cofactors were

Ulla G. Knaus and Thomas L. Leto (eds.), *NADPH Oxidases: Methods and Protocols*, Methods in Molecular Biology, vol. 1982, https://doi.org/10.1007/978-1-4939-9424-3_6, © Springer Science+Business Media, LLC, part of Springer Nature 2019

added. The NOX2 cell-free system was later improved by the addition of recombinant cytosolic components enabling pivotal advances in the mechanistic understanding of NOX2 kinetics [3]. Intriguingly, upon discovery of the other six NOX isoforms, few studies were undertaken to develop non-phagocytic NOX cell-free assays, possibly due to the difficulty in detecting endogenous non-phagocytic NOX activities and the challenges posed by detection systems of NOX-derived oxidants in tissue homogenates and raw membrane preparations [4].

NOX5 contains the typical NOX structural elements necessary for electron transfer with six transmembrane domains, two sets of heme spanning histidines in transmembrane domains 3 and 5, and a cytosolic C-terminus containing NADPH- and FAD-binding sites. However NOX5 differs from other NOX by the presence of polybasic regions, a serine-rich region containing specific phosphorylation sites, and a longer cytosolic N-terminus containing four EF-hand motifs. In terms of activity, NOX5 catalyzes primarily the formation of $O_2^{\cdot -}$, it is independent of $p22^{phox}$ for stability in the cell membrane, and its activity is independent from other known cytosolic subunits [5]. As anticipated by the presence of EF-hand domains, NOX5 activation is dependent on intracellular Ca^{2+}. Study of intramolecular interactions suggests that an increase in free Ca^{2+} induces conformational changes of the EF-hand domain, leading to its interaction with the catalytic COOH terminus [5]. The calcium sensitization occurs especially through phosphorylation, which results in a rapid response of the enzyme upon exposure of NOX5-expressing cells to both protein kinase-activating *phorbol* 12-myristate 13-acetate (PMA) and the calcium ionophore ionomycin [6].

The physiological function of NOX5 is still unclear – simply due to practical reasons as it is absent in rodents – which limits access to knockout animals and inexpensive animal models. Recent evidence points toward a role of increased NOX5 activity in cardiovascular pathologies, including atherosclerosis, hypertension, diabetes, and their complications [7, 8]. NOX5 is highly expressed in the endothelium, and in terms of mechanism, NOX5 requires increased cytosolic Ca^{2+} for activity, a known pathological hallmark of cardiovascular disease and stroke in particular. Thus, NOX5 is a novel pharmacological target of high importance for cardiovascular pathologies [9].

NOX5 inhibitors will be key to understand the physiological role of NOX5 and a potential treatment for cardiovascular diseases. Today NADPH oxidase isoform-specific inhibitors are scarce, and none is targeting NOX5. In this chapter, we describe a NOX5 membrane assay derived from cellular systems overexpressing human NOX5 cDNA [5]. This NOX5 cell-free assay allows detection of $O_2^{\cdot -}$ by the ferricytochrome C assay in the presence of NOX5, NADPH, and Ca^{2+} and was first described in Ref. [5].

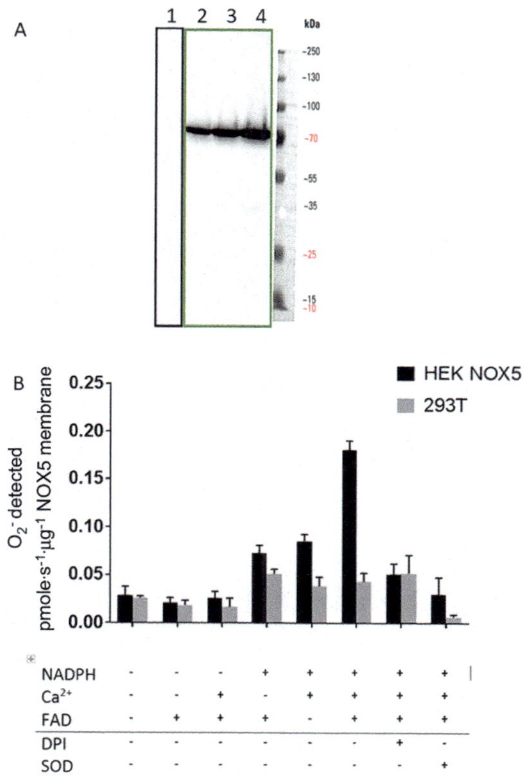

Fig. 1 (**A**) Representative image of a Western blot experiment of purified membranes from HEK293T (lane 1, 2.8 µg protein) and HEK293T stably expressing human NOX5 (lanes 2–4, 0.7, 1.4 and 2.8 µg protein). A NOX5-specific antibody [14] detected a specific band at approximately 70 kDa only in NOX5-expressing membranes. (**B**) Characterization of $O_2^{\cdot-}$ production by NOX5 in purified membranes. $O_2^{\cdot-}$ production was measured by the WST-1 reduction assay using NOX5-containing membranes in the presence of various cofactors (NADPH 200 µM, FAD 1 µM, and Ca^{2+} 700 µM) in 96-w plates. The signal is inhibited by 10 µM of the flavoprotein inhibitor diphenylene iodonium and 50 U of superoxide dismutase (V. Jaquet, internal data)

It was adapted for high-throughput screening using the $O_2^{\cdot-}$ detection probe WST-1 because of increased sensitivity [10] and compatibility to measure NOX-derived $O_2^{\cdot-}$ [11]. The WST-1 membrane assay allows detection of $O_2^{\cdot-}$ in presence of NOX5, FAD, NADPH, and Ca^{2+} (Fig. 1).

2 Materials

2.1 Preparation of Membranes from NOX5-Expressing Cells

1. HEK293T cells stably transfected to express human NOX5 cDNA [5] (Fig. 1).
2. Complete medium: Dulbecco's Modified Eagle's Medium, 10% fetal calf serum.

3. Dissociation buffer: EDTA disodium salt solution (0.05%) in Dulbecco's phosphate-buffered saline (DPBS) 1×.
4. Sonication buffer: DPBS 0.1× ($CaCl_2$- and $MgCl_2$-free), 11% sucrose, 120 mM NaCl, 1 mM EGTA, and 1 tablet/10 mL protease inhibitor cocktail (cOmplete™ Mini, Roche) in Milli-Q H_2O.
5. Sucrose solution 1: 40% sucrose in DPBS 1×.
6. Sucrose solution 2: 17% sucrose in DPBS 1×.
7. Ultracentrifuge tube (Ultra-Clear™ tubes, Beckman Coulter #344062) or similar.
8. SW60 Ti rotor (Beckman) or similar.
9. Ultracentrifuge (Beckman OptimaTM XL-100 K) or similar.
10. Protein Assay Dye Reagent Concentrate.
11. Bovine Serum Albumin (BSA).

2.2 Measurement of NOX5-Derived O_2

1. Dimethyl sulfoxide (DMSO).
2. 800 μM β-nicotinamide adenine dinucleotide 2′-phosphate reduced, tetrasodium salt (NADPH) 4× in Milli-Q H_2O.
3. 2-(4-Iodophenyl)-3-(4-nitrophenyl)-5-(2,4-disulfophenyl)-2H-tetrazolium, monosodium salt (WST-1) 100 mM in DMSO.
4. Flavin adenine dinucleotide disodium salt hydrate (FAD) 1 mM in Milli-Q H_2O.
5. Superoxide dismutase (SOD) from bovine erythrocytes 5000 U/mL in Milli-Q H_2O.
6. Diphenylene iodonium chloride (DPI) 10 mM in DMSO.
7. Chemical compounds to be tested are dissolved in DMSO at 10 mM (*see* **Note 1**).
8. Nitrilotriacetic acid (NTA) 50 mM in Milli-Q H_2O.
9. Ethylenediaminetetraacetic acid (EDTA) 50 mM in Milli-Q H_2O.
10. 1,2-bis(o-aminophenoxy)ethane-*N,N,N′,N′*-tetraacetic acid (BAPTA) 50 mM in Milli-Q H_2O.
11. Reaction mixture 2×: 100 mM HEPES pH 7.1 (adjusted using NaOH 1 M), 1 mM NTA, 1 mM EDTA, 1 mM BAPTA and 2 mM $MgCl_2$, 11 μM phosphatidic acid, 1400 μM $CaCl_2$, 2 μM FAD, 2 mM WST-1, and 60 ng/μL of purified NOX5-containing membranes (*see* **Note 2**) in Milli-Q H_2O (*see* **Notes 3** and **4**).
12. Controls.
13. DMSO 4× (0.4%) in Milli-Q H_2O.
14. SOD IC_{100} 4× (200 U/mL, 0.4% DMSO) in Milli-Q H_2O.

15. DPI IC_{100} 4× (40 μM, 0.4% DMSO) in Milli-Q H_2O (*see* **Note 5**).
16. 96/384-w Clear Flat Bottom Polystyrene low-binding surface microplates.

3 Methods

3.1 Preparation of Membrane from NOX5-Expressing Cells

3.1.1 Membrane Purification

1. Grow cells in ten dishes 150 cm^2 at approximately 90% confluency in complete medium at 37 °C, 5% CO_2.
2. Remove medium, then add 20 mL DPBS 1×, remove DPBS and add 10 mL pre-warmed dissociation buffer until cells are detached with smooth up and down pipetting.
3. Count cells.
4. Collect cells in 15 mL tubes and centrifuge at 400 × *g* for 10 min at room temperature. Discard supernatant.
5. Resuspend in DPBS 1× and centrifuge at 400 × *g* for 10 min. Discard supernatant.
6. Freeze the pellet at −80 °C for at least 30 min.
7. Resuspend the pellet in sonication buffer (70,000,000 cells/mL).
8. Insert the sonicating probe in cell preparation, and sonicate two times for 25 s (level 7, Branson Sonifier 250, duty cycle constant). Work on ice (*see* **Note 6**).
9. Remove cell debris by centrifugation at 800 × *g* for 10 min at 4 °C.
10. Prepare sucrose gradient. Add 1 mL of sucrose solution 1 (bottom layer), and add carefully 1 mL of sucrose solution 2 (upper layer) (Fig. 2).
11. Lay approximately 1–2 mL of sonicated supernatant on sucrose gradient. Tubes should be filled up to 2–3 mm from the top with extra sonication buffer if required.
12. Ultracentrifuge at 150,000 × *g* for 30 min at 4 °C (swing-out rotor).
13. Remove carefully the upper cytosolic fraction, and collect the cloudy membrane fraction (Fig. 2).
14. Add the collected membrane fraction into an ultracentrifuge tube, and fill it to 2–3 mm from the top with cold DPBS.
15. Ultracentrifuge at 200,000 × *g* for 1 h at 4 °C.
16. Resuspend pellet containing membranes in a minimum of 300 μL of DPBS 1×.
17. Measure protein concentration using Bradford assay with BSA as standard (*see* **Note 7**).
18. Prepare aliquots and store at −80 °C.

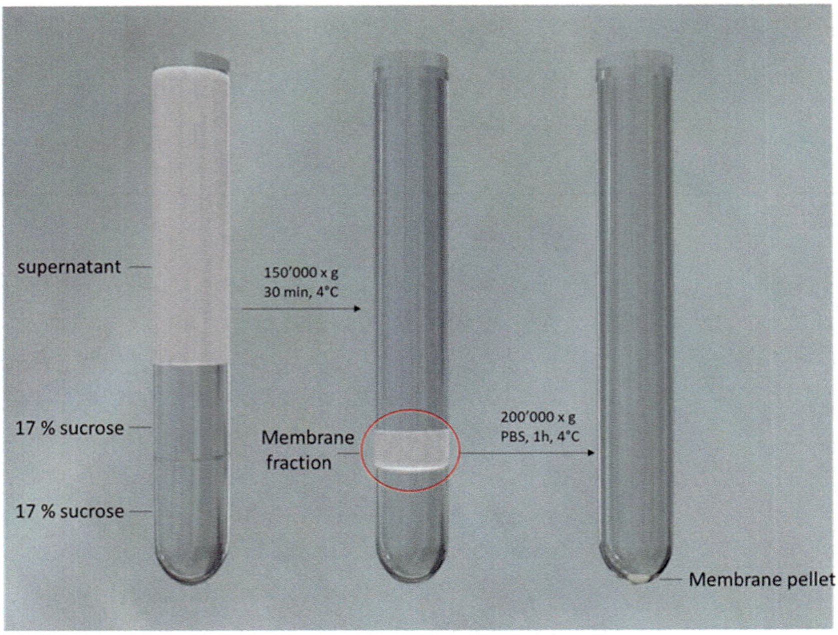

Fig. 2 Scheme illustrating the different steps of the protocol for membrane preparation

3.2 Measurement of O_2 with Membranes Using WST-1 in 96-w or 384-w Plates

The assay is performed in 100 μL/well for 96-w and 40 μL/well for 384-w plates. Add as defined in plate layout (Fig. 3) in the following order:

1. Add ¼ volume control solution 4× (25 μL for 96-w plate and 10 μL for 384-w plate) (columns 1 and 12 for 96-w plate and columns 1, 2, 23, 24 for 384-w plate) as defined in the plate layout using a multichannel pipette, or add ¼ volume chemical compounds (4×, 0.4% DMSO) to be tested in remaining empty wells (columns 2–11 for 96-w and 3–22 for 384-w plates).
2. ½ reaction mixture 2× (50 μL for 96-w plate and 20 μL for 384-w plate) (*see* **Note 8**).
3. ¼ volume NADPH 4× (25 μL for 96-w plate and 10 μL for 384-w plate) (*see* **Note 9**), except 25 μL Milli-Q H_2O for the following wells (A01-D01 for 96-w and A01-H02 for 384-w plates). This will initiate the reaction.

Measure absorbance at 440 nm immediately after addition of NADPH for at least 1 h at 37 °C in a plate reader in kinetic mode (*see* **Note 10**).

Superoxide equivalents were calculated from absorbance values using the following equation:

$$A = \varepsilon_{440\ \mathrm{nm}} \cdot l \cdot c$$

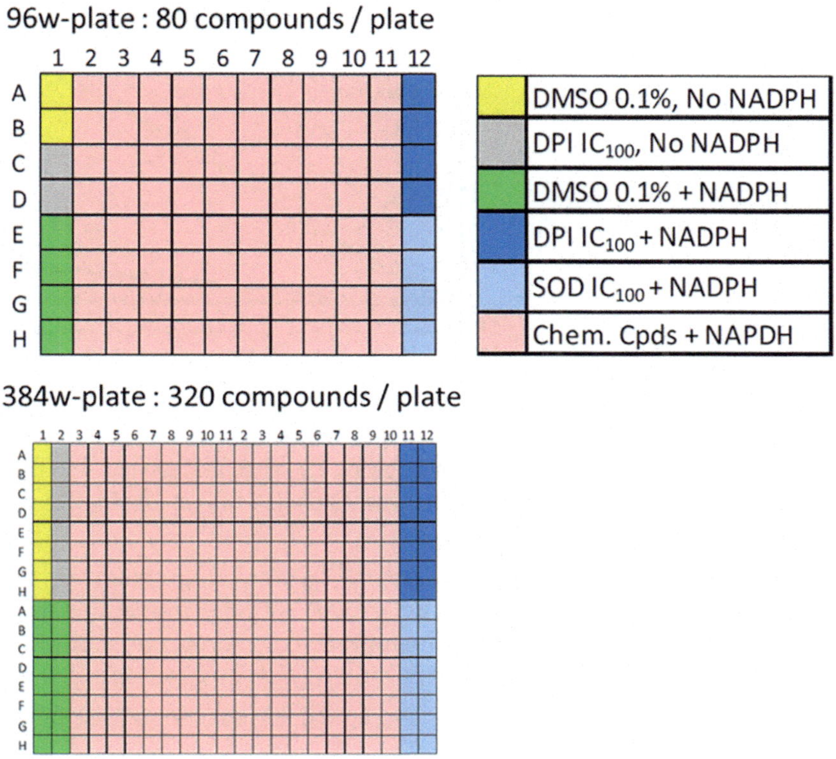

Fig. 3 Scheme illustrating the proposed layout for NOX5 membrane assay in 96-w plate (upper panel) and 384-w plate (below panel)

3.3 Data Analysis of O_2 Rate of Production from Absorbance Measurements Using WST-1

A: absorbance measured (AU).$\varepsilon_{440\ \mathrm{nm}}$: $37{\cdot}10^3\ \mathrm{M}^{-1}\ \mathrm{cm}^{-1}$ for WST-1. *l*: 0.4 cm. Focal distance corresponds to 0.1 cm/25 μL for 96-w plate and 0.1 cm/10 μL for 384-w plate.*c*: $O_2^{\cdot-}$ concentration in molar.

Figure 4 shows a representative image of the reaction rate represented as a rate of $O_2^{\cdot-}$ production per second. In order to identify inhibitors, data are conveniently reported as percentage of control. Potential inhibitors are further confirmed using 8–10 serial concentrations.

4 Notes

1. Chemical compounds are stored at 10 mM in DMSO. Final concentration for the screen is 10 μM corresponding to 0.1% DMSO. We observed that signal decreases only above 3% of DMSO.
2. The amount of NOX5-containing membranes depends on the batch. Optimally, the amount of NOX5 present in the membrane fraction should be monitored by measuring a redox absorbance spectrum in order to calculate the concentration of heme present in membranes as described in Refs.

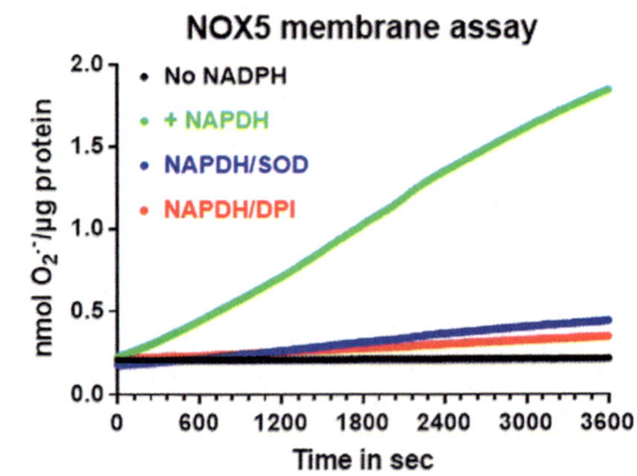

Fig. 4 Representative figure of WST-1 reduction kinetics in the presence of NOX5-containing cell membranes, complete reaction mix, and NADPH 200 μM. The signal is inhibited in presence of 50 U SOD or 10 μM DPI

[3, 12]. However, in HEK293, high mitochondrial-derived heme conceals the specific NOX5-heme-containing spectrum. Thus, different membrane concentrations should be tested for each batch. Typically 60 ng/μL was used for 384-w plate assay.

3. The mixture was shown to be stable for at least 90 min at room temperature.
4. Kinetics of the reaction were recorded using varying concentrations of NADPH and FAD in order to estimate their K_m using Michaelis-Menten plots. The following values were obtained: NADPH, $K_m = 176$ μM; FAD, $K_m = 96$ nM.
5. The NOX inhibitor DPI was used as control due to its high inhibitory potency. We calculated an IC_{50} of 200 nM for NOX5 using this assay.
6. After sonication, the supernatant should be cloudy. In case that the supernatant is clear, repeat the sonication step.
7. The yield is typically of 1–5 mg for 300,000,000 cells of starting material.
8. Prepare 5 mL of reaction mixture 2× per 96-w and 8 mL for 384-w plates.
9. Prepare 2.5 mL of NADPH 4× per 96-w plate and 4 mL for 384-w plate.
10. For HTS, measurements are optimally made at endpoint. Identified hits are confirmed in dose-responses. Other assays, such as NADPH consumption, are required to confirm bona fide NOX5 inhibitors as described in Ref. [13].

References

1. Bedard K, Whitehouse S, Jaquet V (2015) Challenges, progresses, and promises for developing future NADPH oxidase therapeutics. Antioxid Redox Signal 23(5):355–357. https://doi.org/10.1089/ars.2015.6450
2. Hirano K, Chen WS, Chueng AL, Dunne AA, Seredenina T, Filippova A, Ramachandran S, Bridges A, Chaudry L, Pettman G, Allan C, Duncan S, Lee KC, Lim J, Ma MT, Ong AB, Ye NY, Nasir S, Mulyanidewi S, Aw CC, Oon PP, Liao S, Li D, Johns DG, Miller ND, Davies CH, Browne ER, Matsuoka Y, Chen DW, Jaquet V, Rutter AR (2015) Discovery of GSK2795039, a novel small molecule NADPH oxidase 2 inhibitor. Antioxid Redox Signal 23(5):358–374. https://doi.org/10.1089/ars.2014.6202
3. Pick E (2014) Cell-free NADPH oxidase activation assays: "in vitro veritas". Methods Mol Biol 1124:339–403. https://doi.org/10.1007/978-1-62703-845-4_22
4. Rezende F, Prior KK, Lowe O, Wittig I, Strecker V, Moll F, Helfinger V, Schnutgen F, Kurrle N, Wempe F, Walter M, Zukunft S, Luck B, Fleming I, Weissmann N, Brandes RP, Schroder K (2017) Cytochrome P450 enzymes but not NADPH oxidases are the source of the NADPH-dependent lucigenin chemiluminescence in membrane assays. Free Radic Biol Med 102:57–66. https://doi.org/10.1016/j.freeradbiomed.2016.11.019
5. Banfi B, Tirone F, Durussel I, Knisz J, Moskwa P, Molnar GZ, Krause KH, Cox JA (2004) Mechanism of Ca2+ activation of the NADPH oxidase 5 (NOX5). J Biol Chem 279 (18):18583–18591. https://doi.org/10.1074/jbc.M310268200
6. Jagnandan D, Church JE, Banfi B, Stuehr DJ, Marrero MB, Fulton DJ (2007) Novel mechanism of activation of NADPH oxidase 5. calcium sensitization via phosphorylation. J Biol Chem 282(9):6494–6507. https://doi.org/10.1074/jbc.M608966200
7. Montezano AC, Tsiropoulou S, Dulak-Lis M, Harvey A, Camargo Lde L, Touyz RM (2015) Redox signaling, Nox5 and vascular remodeling in hypertension. Curr Opin Nephrol Hypertens 24(5):425–433. https://doi.org/10.1097/MNH.0000000000000153
8. Jha JC, Banal C, Okabe J, Gray SP, Hettige T, Chow BSM, Thallas-Bonke V, De Vos L, Holterman CE, Coughlan MT, Power DA, Skene A, Ekinci EI, Cooper ME, Touyz RM, Kennedy CR, Jandeleit-Dahm K (2017) NADPH oxidase Nox5 accelerates renal injury in diabetic nephropathy. Diabetes 66 (10):2691–2703. https://doi.org/10.2337/db16-1585
9. Chen F, Wang Y, Barman S, Fulton DJ (2015) Enzymatic regulation and functional relevance of NOX5. Curr Pharm Des 21 (41):5999–6008
10. Tan AS, Berridge MV (2000) Superoxide produced by activated neutrophils efficiently reduces the tetrazolium salt, WST-1 to produce a soluble formazan: a simple colorimetric assay for measuring respiratory burst activation and for screening anti-inflammatory agents. J Immunol Methods 238(1–2):59–68
11. Jaquet V, Rutter AR (2015) Response to pick. Antioxid Redox Signal 23(15):1251–1253. https://doi.org/10.1089/ars.2015.6565
12. Pick E (2015) Absolute and relative activity values in assessing the effect of NADPH oxidase inhibitors. Antioxid Redox Signal 23 (15):1250–1251. https://doi.org/10.1089/ars.2015.6470
13. Sorce S, Stocker R, Seredenina T, Holmdahl R, Aguzzi A, Chio A, Depaulis A, Heitz F, Olofsson P, Olsson T, Duveau V, Sanoudou D, Skosgater S, Vlahou A, Wasquel D, Krause KH, Jaquet V (2017) NADPH oxidases as drug targets and biomarkers in neurodegenerative diseases: what is the evidence? Free Radic Biol Med 112:387–396. https://doi.org/10.1016/j.freeradbiomed.2017.08.006
14. Serrander L, Jaquet V, Bedard K, Plastre O, Hartley O, Arnaudeau S, Demaurex N, Schlegel W, Krause KH (2007) NOX5 is expressed at the plasma membrane and generates superoxide in response to protein kinase C activation. Biochimie 89(9):1159–1167. https://doi.org/10.1016/j.biochi.2007.05.004

Chapter 7

Spectroscopy of NOX Protein Family Members

Yoko Nakano and William M. Nauseef

Abstract

All members of the NOX protein family contain a unique b-type cytochrome that mediates the electron transport that characterizes the activity of the multicomponent oxidase complexes. Referred to as cytochrome b558, because of its signature spectral absorbance at 558 nm in reduced-minus-oxidized difference spectroscopy, or cytochrome b(-245), because of its very low midpoint potential of −245 mV at pH 7.0, the protein possesses two stacked inequivalent hemes ligated by pairs of histidine residues in membrane helices h3 and h5. In a flavin-dependent fashion, cytochrome b558 shuttles electrons from cytoplasmic NADPH across membranes to molecular oxygen and thereby generates superoxide anion. By performing reduced-minus-oxidized difference spectroscopy and using the millimolar extinction coefficient, $E_{559-540\ nm} = 21.6\ cm^{-1}\ mM^{-1}$, one can calculate the amount of cytochrome b558 in intact cells or partially purified membrane preparations. Measurements in samples where cytochrome b558 is relatively high and the presence of unrelated heme-containing proteins low, as in neutrophils, are straightforward. However, low levels of cytochrome b558 expression combined with an abundance of mitochondria and other sources of heme proteins make spectral detection of cytochrome b558 in non-phagocytic cells extremely challenging.

Key words Cytochrome b558, Cytochrome b(-245), gp91^{phox}, NOX proteins, Dithionite-reduced-minus-oxidized spectroscopy

1 Introduction

Located in the plasma membrane and membranes of secretory vesicles and specific granules of neutrophils, NOX protein family members shuttle electrons across two inequivalent hemes, whose porphyrins are ligated by two pairs of histidines located in helices h3 and h5 [1, 2], from cytoplasmic NADPH to molecular oxygen, thereby generating superoxide anion. Spectroscopic characteristics of the membrane-associated phagocyte NADPH oxidase figure prominently in the story of the discovery of gp91^{phox}/p22^{phox}. Much of the early work on the newly recognized cyanide-resistant, flavin-dependent respiratory burst oxidase of neutrophils [3–5] focused on its dependence on NADH or NADPH as the source of electrons, its relationship to other hemeproteins in neutrophils (most notably myeloperoxidase), and its composition. Investigators

Ulla G. Knaus and Thomas L. Leto (eds.), *NADPH Oxidases: Methods and Protocols*, Methods in Molecular Biology, vol. 1982, https://doi.org/10.1007/978-1-4939-9424-3_7, © Springer Science+Business Media, LLC, part of Springer Nature 2019

rigorously analyzed the biochemical properties of superoxide-generating subcellular particles recovered from stimulated phagocytes.

In 1961, early in the exploration of the phagocyte oxidase, a Japanese investigator detected an absorbance band at 558 nm while performing spectroscopy on isolated granules from equine granulocytes [6]. Dr. Hattori interpreted the spectra as evidence of a b-type cytochrome and proposed that it was involved in the newly described cyanide- and antimycin-resistant burst in oxygen consumption by phagocytosing neutrophils. In two subsequent publications, Shinagawa et al. reported that reduced-minus-oxidized spectra of granules recovered from rabbit inflammatory exudates [7, 8] exhibit features typical of b-type cytochromes and are nearly identical to that of bacterial cytochrome b1. After identifying features that distinguished the granulocyte spectra from those of cytochrome b1, cytochrome b559, and peroxidase 559, the authors concluded that they had found a novel cytochrome, which they referred to as "cytochrome b558, neutrophils."

Several years later, cytochrome b558 in neutrophils was rediscovered. Segal and Owens reported the presence of a b-type cytochrome in neutrophils that was distinct from cytochrome P450 and mitochondrial cytochrome oxidase and that localized to the phagosome of human neutrophils fed IgG-opsonized latex beads [9]. Since the first report of patients with what would later be called chronic granulomatous disease (CGD) [10], the identification of fundamental aspects of normal phagocyte oxidase biology and the elucidation of the abnormalities in CGD neutrophils have evolved in parallel (see Chapter 1 in this Volume). Keeping with that tradition, the Segal laboratory reported that membranes of neutrophils from patients with X-linked CGD lack spectral evidence of cytochrome b558 [9, 11]. Over the ensuing decades, seminal studies by Andrew Cross and others described many of the biochemical and thermodynamic properties of cytochrome b558 and of the phagocyte NADPH oxidase [12–16].

One of the studies by Cross and Jones includes a survey of a variety of non-phagocyte cells for their capacity to support superoxide production [17]. Given what is now known about the far-reaching prevalence of NOX proteins throughout nature [18–20], it is likely that the findings summarized by Cross and Jones were early hints that the NADPH oxidase was not a unique biochemical property of phagocytes. Although spectral data were not included in the survey study, a contemporaneous study described plasma membranes from human glomerular mesangial cells that exhibited superoxide anion generation in response to calcium ionophore, immunoreactivity with a gp91^{phox}-specific monoclonal antibody [21, 22], and a spectra consistent with cytochrome b558 [23].

Cytochrome b558 exhibits three prominent peaks in reduced-minus-oxidized spectra, at 428, 529, and 558/559 nm, and the height of the peak at 559 nm relative to the low point at 540 nm reflects the amount of cytochrome b558 in the sample. Although spectra of cytochrome b558 closely resemble those of mitochondrial b cytochromes and hemoglobin [24], the extremely low oxidized-reduced midpoint potential at pH 7.0 (−245 mV [12]) distinguishes it from nearly all other mammalian b-type cytochromes. However, the latter measurement requires equipment more sophisticated than is typically present in a laboratory [24].

2 Materials

1. Dual beam spectrophotometer, ideally one equipped to study turbid samples (e.g. Lambda 35, Perkin Elmer).
2. Solid sodium dithionite ($Na_2S_2O_4$).
3. Phosphate buffered normal saline: 137 mM NaCl, 2.7 mM KCl, 10 mM Na_2HPO_4, 1.8 mM NaH_2PO_4, pH 7.0, or another buffer in which to suspend neutrophils that is free of pH indicators or other agents that would color the buffer.

3 Methods

1. Cell suspensions can be directly scanned for the presence of cytochrome b558. In the case of human neutrophils, cell suspensions of ~5×10^7/mL yield interpretable spectra.
2. Distribute cell suspensions into two cuvettes and place in a dual-beam spectrophotometer.
3. Scan from 400 to 600 nm to create a baseline.
4. Add a few grains of sodium dithionite ($Na_2S_2O_4$) to the sample cuvette, mix, and scan again (*see* **Note 1**).
5. The resultant scan will be a reduced-minus-oxidized difference spectrum (Fig. 1).
6. Measure the height of the peak at ~558 nm above the nadir at 540 nm (*see* **Note 2**).
7. Using the millimolar extinction coefficient, $E_{559\text{–}540\ \mathrm{nm}} = 21.6\ \mathrm{cm}^{-1}\ \mathrm{mM}^{-1}$ [13], calculate the concentration of cytochrome b558 in the sample (*see* **Note 3**). Typically, a sample of neutrophils at 2×10^7 cells/mL yields $A_{559\text{–}540\ \mathrm{nm}}$ of 0.002 and contains 100 pmol cytochrome b558 [24].
8. NOX3, the closest homologue of gp91*phox* (NOX2, 58%) [25], is apparently only expressed in cells in the inner ear [26]. To study the biosynthesis of NOX3, we expressed NOX3 and its

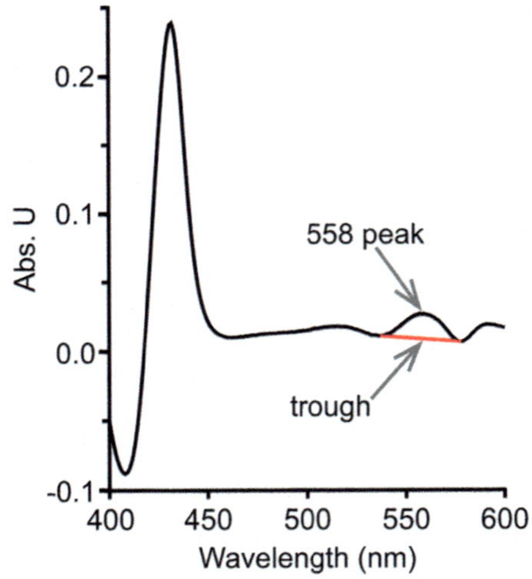

Fig. 1 Dithionite-reduced-minus-oxidized spectrum of human neutrophils. Neutrophil suspensions (8.5 × 10^7 cells/mL in phosphate-buffered saline, pH 7.0) were aliquoted in sample and reference cuvettes. After scanning 400 to 600 nm to obtain a baseline, a few grains of dithionite were added to the sample cuvette and the scan was repeated. The resultant reduced-minus-oxidized scan demonstrates peaks for cytochrome b558 at 428, 529, and 558 nm. The difference in absorbance between the peak at 558 nm, and the midpoint of a line connecting troughs at 540 nm and 570 nm is 0.0178. Using the millimolar extinction coefficient 21.6 cm^{-1} mM^{-1} yields a concentration of 96 pmol/10^7 neutrophils

essential cofactors NOXO1 and NOXA1 [27] in HEK293 cells [28]. Non-transfected HEK293 cells served as negative controls and gp91^{phox} transfectants as positive controls for cytochrome b558 expression. We used differential centrifugation to recover low speed (8000 × *g*, 10 min at 4 °C) P2 and high speed (114,000 × *g*, 60 min at 4 °C) P3 pellets from each of the three cell lines and washed the resultant membranes with high salt to deplete peripheral membrane proteins and adherent cytoplasmic proteins before analysis. Whereas P2 is enriched for the mitochondrial proteins porin and cytochrome c, neither protein is present in P3 (Fig. 2).

9. Spectra in Fig. 2 illustrate the difficulty in detecting the spectrum of cytochrome b558 in non-phagocytic cells. In panel b are spectra of dithionite-reduced-minus-oxidized tracings of P3 (5 × 10^8 cells/mL) of each cell type. The calculated amounts of cytochrome b are similar for the two transfectants (10.0 pmoles/10^8 cells or 40.8 pmol/mg protein for gp91^{phox} vs. 5.3 pmoles/10^8 or 18.8 pmol/mg protein for NOX3), but note the presence of a small peak near 558 nm in non-transfected HEK293 cells. Because these cells lack endogenous NOX proteins and do not generate superoxide

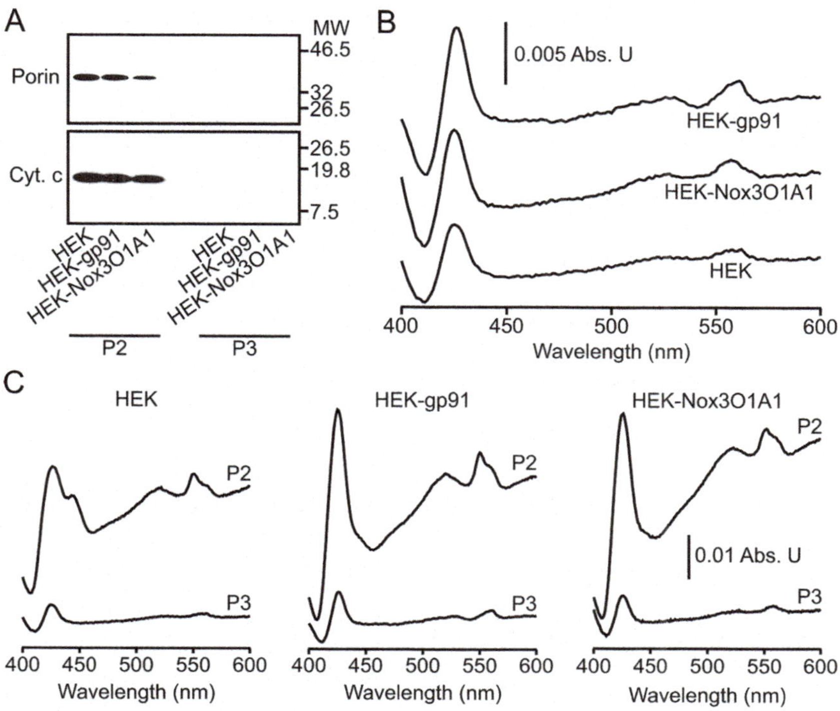

Fig. 2 Analyses of membranes from wild-type and transfected HEK293 cells. Membrane pellets from wild-type HEK293 cells (HEK) or from HEK transfectants expressing gp91^{phox} or NOX3/NOXO1/NOXA1 were (**A**) probed for the presence of mitochondrial proteins cytochrome c and porin and (**B**, **C**) subjected to dithionite-reduced-minus-oxidized spectroscopy. Note the different scales used in scans shown in panels **B** and **C**. Data reproduced from ref. 28

anion [28], this peak likely reflects the presence of an unrelated b-type cytochrome (e.g., cytochrome b5 [29]). Panel c includes spectra for P2 and P3 for each of the samples. The broad peaks from 540 to 560 nm in the spectra for the mitochondria-enriched P2 samples are the same for all, but note the relative magnitude of signals in P2 and P3: the former are >20-fold greater. Thus, spectra from mitochondrial and other heme proteins in cells that are not of phagocyte origin will obscure detection of the NOX proteins in samples that are not enriched for membranes of interest before study.

4 Notes

1. Because the distribution of cytochrome b558 in some cell types includes intracellular compartments, dithionite-reduced-minus-oxidized spectra of simple suspensions of intact cells may underestimate the total cellular amount of cytochrome b558. For example, only ~15% of total cellular cytochrome b558 in unstimulated human neutrophils resides in the plasma

membrane, and the bulk of the protein is expressed in membranes of secretory vesicles and peroxidase-negative granules intracellularly [30], compartments that are inaccessible to the added dithionite and will not be reduced optimally. Addition of 0.1% Triton X-100 will permeabilize the plasma membrane and allow dithionite to reduce intracellular cytochrome b558. However, permeabilization of neutrophils introduces a new confounding feature, the acquisition of the spectra of myeloperoxidase, which is stored in azurophilic granules [31]. Fortunately, the predominant peak for MPO at 473 nm does not alter the 558 nm signal from cytochrome b558.

2. Sometimes sample turbidity can create challenges to obtaining quality spectra. If the baseline on tracings is not horizontal, one can measure the distance between the peak at 559 nm and a line drawn between the valleys at 540 and 570 nm [12] (*see* Fig. 1).
3. Scans of nonmyeloid cells pose genuine challenges for two reasons. First, whereas neutrophils express large amounts of cytochrome b558 per cell (417 pmol/mg plasma membrane protein [13, 14]), making spectral detection relatively easy, expression of NOX protein family members in primary cells that are not phagocytes is orders of magnitude less, and detection by spectroscopy requires much more material to scan, which raises inherent challenges related to sample turbidity. Second, whereas neutrophils have few mitochondria and, with the exception of myeloperoxidase, lack other heme-containing proteins, non-phagocytes often have abundant mitochondria and other sources of heme proteins that can confound spectroscopy. Our work on NOX3 expressed in HEK293 cells illustrates the challenges of spectral analysis of NOX proteins in non-phagocytes.

Acknowledgments

The Nauseef lab is supported by National Institute of Health grants AI116546 and AI132335, a Merit Review award from the Veterans Affairs, and use of facilities at the Iowa City Department of Veterans Affairs Medical Center, Iowa City, IA.

References

1. Biberstine-Kinkade KJ, DeLeo FR, Epstein RI, LeRoy BA, Nauseef WM, Dinauer MC (2001) Heme-ligating histidines in flavocytochrome b 558. J Biol Chem 276:31105–31112
2. Magnani F, Nenci S, Millana Fananas E, Ceccon M, Romero E, Fraaije MW, Mattevi A (2017) Crystal structures and atomic model of NADPH oxidase. Proc Natl Acad Sci U S A 114(26):6764–6769. https://doi.org/10.1073/pnas.1702293114
3. Baldridge CW, Gerard RW (1933) The extra respiration of phagocytosis. Am J Phys 103:235–236

4. Sbarra AJ, Karnovsky ML (1959) The biochemical basis of phagocytosis. I. Metabolic changes during the ingestion of particles by polymorphonuclear leukocytes. J Biol Chem 234:1355–1362
5. Iyer GYN, Islam DMF, Quastel JH (1961) Biochemical aspects of phagocytosis. Nature 192:535–541
6. Hattori H (1961) Studies on the labile, stable Nadi oxidase and peroxidase staining reactions in the isolated particles of horse granulocyte. Nagoya J Med Sci 23:362–378
7. Shinagawa Y, Tanaka C, Teroaka A, Shinagawa Y (1966) A new cytochrome in neutrophilic granules of rabbit leucocyte. J Biochem 59:622–624
8. Shinagawa Y, Tanaka C, Teroaka A (1966) Electron microscopic and biochemical study of the neutrophilic granules from leucocytes. J Electron Microsc 15:81–85
9. Segal AW, Jones OTG (1978) Novel cytochrome b system in phagocytic vacuoles of human granulocytes. Nature 276:515–517
10. Berendes H, Bridges RA, Good RA (1957) A fatal granulomatosus of childhood; the clinical study of a new syndrome. Minn Med 40 (5):309–312
11. Segal AW, Jones OT, Webster D, Allison AC (1978) Absence of a newly described cytochrome b from neutrophils of patients with chronic granulomatous disease. Lancet 2:446–449
12. Cross AR, Jones OTG, Harper AM, Segal AW (1981) Oxidation-reduction properties of the cytochrome b found in the plasma-membrane fraction of human neutrophils. A possible oxidase in the respiratory burst. Biochem J 194:599–606
13. Cross AR, Higson FK, Jones OTG (1982) The enzymic reduction and kinetics of oxidation of cytochrome b of neutrophils. Biochem J 204:479–485
14. Cross AR, Jones OTG, Garcia R, Segal AW (1982) The association of FAD with the cytochrome b-245 of human neutrophils. Biochem J 208:759–763
15. Cross AR, Rae J, Curnutte JT (1995) Cytochrome b-245 of the neutrophil superoxide-generating system contains two nonidentical hemes. Potentiometric studies of a mutant form of gp91phox. J Biol Chem 270:17075–17077
16. Light DR, Walsh C, O'Callaghan AM, Goetzl EJ, Tauber AI (1981) Characteristics of the cofactor requirements for the superoxide-generating NADPH oxidase of human polymorphonuclear leukocytes. Biochemistry 20:1468–1476
17. Cross AR, Jones OT (1991) Enzymic mechanisms of superoxide production. Biochim Biophys Acta 1057(3):281–298
18. Bedard K, Krause KH (2007) The NOX family of ROS-generating NADPH oxidases: physiology and pathophysiology. Physiol Rev 87:245–313
19. Bedard K, Lardy B, Krause KH (2007) NOX family NADPH oxidases: not just in mammals. Biochemie 89:1107–1112
20. Aguirre J, Lambeth JD (2010) Nox enzymes from fungus to fly to fish and what they tell us about nox function in mammals. Free Radic Biol Med 49:1342–1353
21. Nakamura M, Murakami M, Koga T, Tanaka Y, Minakami S (1987) Monoclonal antibody 7D5 raised to cytochrome b 558 of human neutrophils: immunocytochemical detection of the antigen in peripheral phagocytes of normal subjects, patients with chronic granulomatous disease, and their carrier mothers. Blood 69 (5):1404–1408
22. Burritt JB, DeLeo FR, McDonald CL, Prigge JR, Dinauer MC, Nakamura M, Nauseef WM, Jesaitis AJ (2001) Phage display epitope mapping of human neutrophil flavocytochrome b558. Identification of two juxtaposed extracellular domains. J Biol Chem 276:2053–2061
23. Radeke HH, Cross AR, Hancock JT, Jones OT, Nakamura M, Kaever V, Resch K (1991) Functional expression of NADPH oxidase components (alpha- and beta-subunits of cytochrome b558 and 45-kDa flavoprotein) by intrinsic human glomerular mesangial cells. J Biol Chem 266(31):21025–21029
24. Segal AW, Harper AM, Cross AR, Jones OT (1986) Cytochrome b-245. Methods Enzymol 132:378–394
25. Cheng G, Cao Z, Xu X, Van Meir E, Lambeth J (2001) Homologs of gp91 phox: cloning and tissue expression of Nox3, Nox4, and Nox5. Gene 269:131–140
26. Paffenholz R, Bergstrom RA, Pasutto F, Wabnitz P, Munroe RJ, Jagla W, Heinzmann Y, Marquardt A, Bareiss A, Laufs J, Russ A, Stumm G, Schimenti JC, Bergstrom DE (2004) Vestibular defects in head-tilt mice result from mutations in nox3, encoding an NADPH oxidase. Genes Dev 18:486–491
27. Banfi B, Malgrange B, Knisz J, Steger K, Dubois-Dauphin M, Krause KH (2004) NOX3, a superoxide-generating NADPH oxidase of the inner ear. J Biol Chem 279:46065–46072

28. Nakano Y, Banfi B, Jesaitis AJ, Dinauer MC, Allen LA, Nauseef WM (2007) Critical roles for p22 phox in the structural maturation and subcellular targeting of Nox3. Biochem J 403 (1):97–108
29. Aoyama T, Nagata K, Yamazoe Y, Kato R, Matsunaga E, Gelboin HV, Gonzalez FJ (1990) Cytochrome b 5 potentiation of cytochrome P-450 catalytic activity demonstrated by a vaccinia virus-mediated in situ reconstitution system. Proc Natl Acad Sci U S A 87:5425–5429
30. Borregaard N, Tauber AI (1984) Subcellular localization of the human neutrophil NADPH oxidase. b-cytochrome and associated flavoprotein. J Biol Chem 259:47–52
31. Nauseef WM (2014) Myeloperoxidase in human neutrophil host defence. Cell Microbiol 16(8):1146–1155. https://doi.org/10.1111/cmi.12312

Chapter 8

Soluble Regulatory Proteins for Activation of NOX Family NADPH Oxidases

Hideki Sumimoto, Reiko Minakami, and Kei Miyano

Abstract

NOX family NADPH oxidases deliberately produce reactive oxygen species and thus contribute to a variety of biological functions. Of seven members in the human family, the three oxidases NOX2, NOX1, and NOX3 form a heterodimer with p22*phox* and are regulated by soluble regulatory proteins: p47*phox*, its related organizer NOXO1; p67*phox*, its related activator NOXA1; p40*phox*; and the small GTPase Rac. Activation of the phagocyte oxidase NOX2 requires p47*phox*, p67*phox*, and GTP-bound Rac. In addition to these regulators, p40*phox* plays a crucial role when NOX2 is activated during phagocytosis. On the other hand, NOX1 activation prefers NOXO1 and NOXA1, although Rac is also involved. NOX3 constitutively produces superoxide, which is enhanced by regulatory proteins such as p47*phox*, NOXO1, and p67*phox*. Here we describe mechanisms for NOX activation with special attention to the soluble regulatory proteins.

Key words p47*phox*, NOXO1, p67*phox*, NOXA1, p40*phox*, Rac, NOX1, NOX2, NOX3, p22*phox*

1 Introduction

NOX family NADPH oxidases deliberately produce reactive oxygen species (ROS), thereby contributing to a variety of functions such as host defense, signal transduction, and hormone synthesis [1–4]. The NOX oxidases contain two distinct hemes in the N-terminal transmembrane region and FAD- and NADPH-binding sites in the C-terminal cytoplasmic domain, thus forming a complete electron-transporting apparatus from NADPH to O_2 via FAD and hemes in a single polypeptide [1–4].

The human genome contains seven genes encoding NADPH oxidases: NOX1 through NOX5 and the distantly related oxidases DUOX1 and DUOX2. Among them, NOX5, DUOX1, and DUOX2 contain Ca^{2+}-binding EF-hand motifs in the N-terminal extension, and their activation involves elevation in intracellular Ca^{2+} concentrations. In contrast, NOX1 through NOX4 lack Ca^{2+}-binding motifs but form a heterodimer with the membrane-integrated p22*phox*, which stabilizes these NOX proteins and controls their

Ulla G. Knaus and Thomas L. Leto (eds.), *NADPH Oxidases: Methods and Protocols*, Methods in Molecular Biology, vol. 1982, https://doi.org/10.1007/978-1-4939-9424-3_8, © Springer Science+Business Media, LLC, part of Springer Nature 2019

activity by directly interacting with soluble regulatory proteins. NOX1, NOX2, and NOX3 become activated in stimulated cells, whereas NOX4 is constitutively active [1–4].

Mammalian NOX2 (aka gp91^{phox}) is highly expressed in phagocytes such as neutrophils and macrophages and forms a stable complex with p22^{phox}; the heterodimer is known as flavocytochrome b_{558}. In neutrophils, the flavocytochrome localizes to the plasma membrane or specific granules and is in a completely inactive state at both sites under resting conditions. NOX2 is activated during phagocytosis to produce superoxide, a precursor of microbicidal ROS. The membrane of neutrophil phagosomes is derived mainly from the plasma membrane but also delivered from specific granules by a process known as degranulation. As a result, the NOX2–p22^{phox} heterodimer is enriched in the phagosomal membrane and becomes activated via the assembly with the soluble proteins p47^{phox} (aka NOX organizer 2 (NOXO2)), p67^{phox} (aka NOX activator 2 (NOXA2)), and p40^{phox} as well as the small GTPase Rac; they translocate upon cell stimulation from the cytoplasm to the phagosome. The significance of the NOX2-based oxidase in host defense is evident, because recurrent and life-threatening infections occur in a patient with chronic granulomatous disease (CGD), whose neutrophils fail to kill pathogenic microbes due to genetic defects in NOX2, p22^{phox}, p47^{phox}, or p67^{phox} [1–4].

Superoxide production by NOX2 is triggered not only during phagocytosis but also in response to cell stimulants such as the chemoattractant formyl-methionyl-leucyl-phenylalanine (fMLP), arachidonic acid (AA), and phorbol 12-myristate 13-acetate (PMA), a potent activator of protein kinase C (PKC). In contrast to fMLP and PMA, AA is also capable of activating the phagocyte oxidase under cell-free conditions in the presence of p47^{phox}, p67^{phox}, and GTP-bound Rac.

The non-phagocytic oxidase NOX1, abundant in the plasma membrane of colon epithelial and vascular smooth muscle cells, is also dimerized with p22^{phox}. NOX1 is inactive in the absence of an organizer or an activator, but generates superoxide without stimulants such as PMA in cells expressing both the p47^{phox}-related protein NOXO1 (NOX organizer 1) and the p67^{phox}-related protein NOXA1 (NOX activator 1). Rac also participates in NOX1 activation via interacting with NOXA1. NOX3, a non-phagocytic oxidase for otoconia formation in the inner ear, forms a stable heterodimer with p22^{phox} and produces a substantial amount of superoxide even in the absence of an organizer or an activator. The superoxide-producing activity of NOX3 can be strongly enhanced by one of NOXO1, p67^{phox}, or p47^{phox} and also upregulated by Rac in the presence of p67^{phox} or NOXA1. Here we focus on the mechanisms whereby these soluble regulatory proteins activate NOX2, NOX1, and NOX3.

2 Regulators of NOX1–3

2.1 p47^{phox}, NOX Organizer 2 (NOXO2)

The phagocyte oxidase NOX2, in complex with p22^{phox}, is dormant in resting cells, but becomes activated during phagocytosis or in response to soluble cell stimulants via the coordination of regulatory proteins: p47^{phox}, p67^{phox}, p40^{phox}, and Rac. The soluble *phox* proteins p47^{phox}, p67^{phox}, and p40^{phox} form a trimeric complex with 1:1:1 stoichiometry in resting neutrophils. Upon cell stimulation or during phagocytosis, they translocate en bloc to the membrane to assemble with NOX2 and Rac (which is independently recruited from the cytoplasm), leading to superoxide production. Key to the assembly process is p47^{phox}, which directly binds to p22^{phox} upon activation and thus plays a central role in interaction of the p47^{phox}–p67^{phox}–p40^{phox} complex with the NOX2–p22^{phox} heterodimer on the membrane. The organizer protein p47^{phox}, containing 390 amino acids, has several modules for protein–protein or protein–lipid interaction: starting from the N-terminus, it harbors a PX domain, two SH3 domains in tandem, an autoinhibitory region (AIR; amino acids 286–340), and a proline-rich region (PRR) (Fig. 1). In the cytoplasmic *phox* ternary complex, p47^{phox} constitutively associates with p67^{phox} via interaction between the PRR and the p67^{phox} C-terminal SH3 domain. The tandem SH3 domains of p47^{phox} simultaneously interact with the PRR in the C-terminal cytoplasmic tail of p22^{phox}, an interaction which is essential for membrane translocation and subsequent NOX2 activation [1–4]. The p47^{phox} PX domain also contributes to membrane translocation by binding to phosphoinositides such as phosphatidylinositol 3,4-bisphosphate [PI(3,4)P_2], albeit with low affinity and broad specificity [5]. In neutrophils, PI(3,4)P_2 is predominantly formed from phosphatidylinositol 3,4,5-trisphosphate [PI (3,4,5)P_3], a product of type I phosphatidylinositol 3-kinase (PI3K) that is activated during phagocytosis and in response to chemoattractants. In contrast to the p47^{phox} PX domain, the module of p40^{phox} binds specifically and strongly to phosphatidylinositol 3-phosphate [PI(3)P], which is directly produced from phosphatidylinositol by type III PI3K.

Stimulus-triggered translocation of p47^{phox} requires both bis-SH3-mediated interaction with the transmembrane protein p22^{phox} and PX binding to membrane phosphoinositides [1–4]. However, the SH3 domains of p47^{phox} are normally masked via an intramolecular interaction with the AIR and thus incapable of interacting with p22^{phox} [6–9]. The lipid-binding activity of the PX domain is also negatively regulated via intramolecular interactions with the bis-SH3 domain and the AIR under resting conditions [10–13]. Upon cell stimulation, the AIR undergoes phosphorylation at multiple serine residues (e.g., Ser-303, Ser-304, and Ser-328 in human p47^{phox}) by protein kinases such as PKC and AKT. The

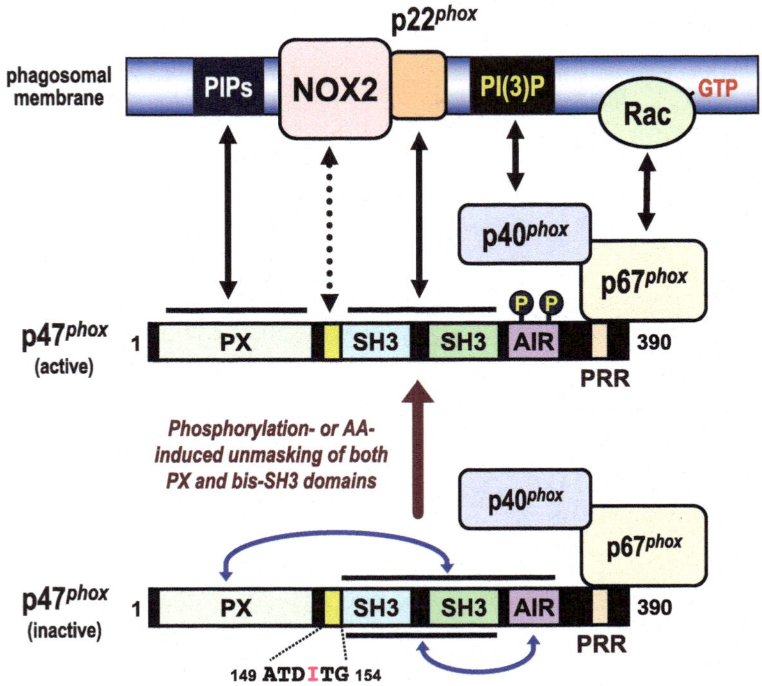

Fig. 1 Role of p47*phox* in activation of the phagocyte oxidase NOX2. The organizer protein p47*phox* forms a ternary complex with p67*phox* and p40*phox* in the cytoplasm in resting phagocytes; in the complex, bis-SH3 and PX domains of p47*phox* are normally masked via intramolecular interactions with an autoinhibitory region (AIR). During phagocytosis p47*phox* undergoes a conformational change, which allows the bis-SH3 and PX domains to associate with their respective membrane targets p22*phox* and phosphoinositides (PIPs). On the phagosomal membrane, p67*phox* and p40*phox* bind to GTP-bound Rac and phosphatidylinositol 3-phosphate [PI(3)P], respectively. Solid two-headed arrows in *black* show individual associations that participate in membrane translocation of p47*phox*. Solid two-headed arrows in *blue* indicate intramolecular interactions. A dotted line with bidirectional arrows denotes possible binding of the p47*phox* Ile-152-containing region to NOX2, which may be directly involved in enzymatic activation. A solid upward-pointing arrow indicates the conformational change of p47*phox*, which is induced by phosphorylation of the AIR or direct interaction with arachidonic acid (AA)

former is activated by diacylglycerol produced by phospholipase C (PLC), whereas the latter is dependent on PI(3,4,5)P_3 formed by stimulus-activated type I PI3K. It should be noted that activation of both PLC and PI3K occurs during phagocytosis and in response to chemoattractants. The phosphorylation of AIR acts as a regulatory switch to relieve its inhibition and facilitates the interaction of the bis-SH3 and PX domains with their respective membrane targets p22*phox* and phosphoinositides [10]. The in vitro NOX2 stimulant AA is also able to induce this conformational change by directly dissociating the bis-SH3 domain from the AIR [13]. Intriguingly,

AA synergizes with phosphorylation of p47^{phox} to promote its interaction with p22^{phox} for NOX2 activation [13]. It is thus possible that NOX2 is regulated also via AA, which is locally released by phospholipase A_2 (PLA_2) in stimulated cells. In this context, it seems interesting that, via the PX domain, p47^{phox} anchors cytosolic PLA_2 ($cPLA_2$) to the assembled oxidase on the plasma membrane in stimulated cells [14].

In p47^{phox}, the PX–SH3–SH3–AIR supercomplex adopts an autoinhibited conformation via intramolecular interactions, and is converted upon stimulation into an active state, which renders both PX and bis-SH3 domains accessible to their membrane targets. Since the supercomplex is connected with the PRR by a flexible linker, the stimulus-induced conformational change of p47^{phox} does not appear to largely affect or block the PRR-mediated constitutive interaction with p67^{phox}.

In addition to these modules in p47^{phox}, Ile-152 and Thr-153 in a region N-terminal to the bis-SH3 domain are involved in activation of the NOX2-based oxidase (Fig. 1). Whereas replacement of Thr-153 by methionine or alanine partially but significantly reduces NOX2-catalyzed superoxide production [15, 16], alanine substitution for Ile-152 results in an almost complete loss of NOX2 activation both in vivo and in vitro [16]. These residues participate neither in binding to p22^{phox} [16, 17] nor in intramolecular interaction with the AIR [8, 9]. The molecular mechanism whereby the residues Ile-152 and Thr-153 contribute to NOX2 activation is presently unknown. It seems possible that the Ile-152-containing stretch functions in a process after membrane translocation, because a mutant p47^{phox} with the I152A substitution appears to normally translocate in PMA-stimulated cells [16].

The significance of the p22^{phox}-binding bis-SH3 domain and the Ile-152-containing activation stretch agrees with findings showing that the two moieties of p47^{phox} are minimally required for cell-free activation of NOX2 [16, 18]. It is tempting to speculate that the bis-SH3 domain serves by directing its N-terminal, Ile-152-containing activation stretch toward NOX2 complexed with p22^{phox}, in addition to the generally accepted role of p47^{phox} in targeting the oxidase activator p67^{phox} (especially the activation domain) toward NOX2 via the en bloc membrane translocation.

2.2 NOXO1, a p47^{phox}-Related Oxidase Organizer

The non-phagocytic oxidase NOX1, forming a heterodimer with p22^{phox}, is preferentially activated by a protein combination consisting of the p47^{phox}-related oxidase organizer NOXO1 and the p67^{phox}-related oxidase activator NOXA1, in addition to GTP-bound Rac. NOXO1, consisting of 376 amino acids, shows a domain architecture similar to that of p47^{phox} except of the absence of an AIR (Fig. 2). In contrast to p47^{phox}, even in resting cells, NOXO1 localizes to the plasma membrane via interaction of the N-terminal PX and bis-SH3 domains with their respective

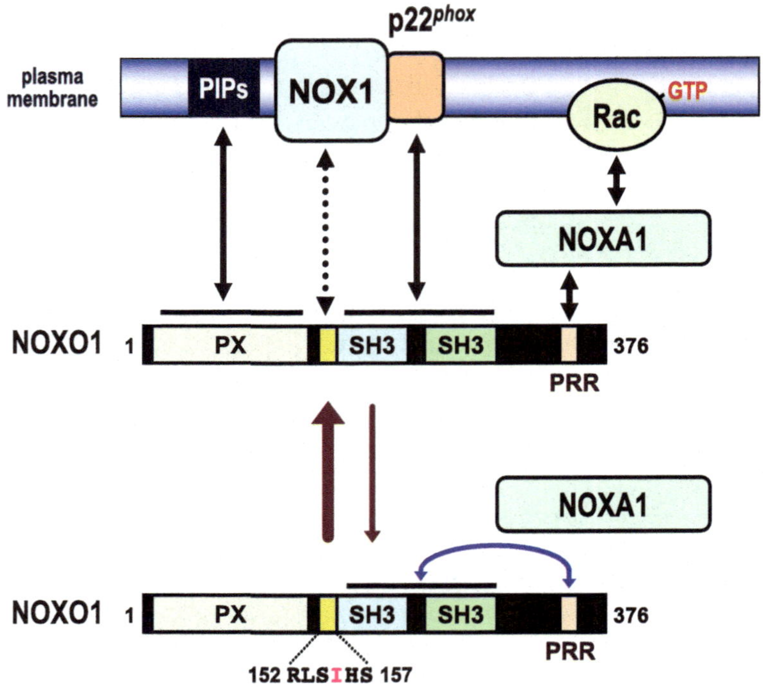

Fig. 2 Role of NOXO1 in activation of the non-phagocytic oxidase NOX1. Even in resting cells, the organizer protein NOXO1 localizes to the plasma membrane via bis-SH3 domain-mediated association with p22*phox* and PX domain binding to phosphoinositides (PIPs). Although NOXO1 simultaneously associates with p22*phox* and NOXA1 via the bis-SH3 domain and a C-terminal proline-rich region (PRR), respectively, these associations are partially prevented by an intramolecular interaction between the bis-SH3 domain and the PRR. Phosphorylation of NOXO1 disrupts the intramolecular interaction and thus enhances membrane localization of NOXA1. On the membrane, NOXA1 associates with GTP-bound Rac for activation of NOX1. Solid two-headed arrows in *black* show individual associations that participate in NOXO1-mediated assembly of the NOX1-based oxidase. A solid two-headed arrow in *blue* indicates an intramolecular interaction. A dotted two-headed arrow denotes possible interaction of the NOXO1 Ile-155-containing region with NOX1, which may be directly involved in enzymatic activation

targets such as phosphoinositides and the p22*phox* C-terminal PRR [19–22]. The C-terminal PRR of NOXO1 directly mediates a tail-to-tail interaction with NOXA1, which allows this activator protein to constitutively localize to the plasma membrane [19, 23, 24]. The three interactions are all required for superoxide production by NOX1 [19, 23, 24]. In addition, Ile-155, an invariant residue that localizes N-terminal to the bis-SH3 domain and corresponds to Ile-152 in p47*phox*, is involved in NOX1 activation [16].

In cells co-expressing NOXO1 and NOXA1, NOX1 constitutively produces superoxide. This is in sharp contrast with the fact that NOX2 is dormant even in the presence of p47*phox* and p67*phox*, and its activation further requires cell stimulants such as PMA, a potent PKC activator. Superoxide production by NOX1 in unstimulated cells is consistent with constitutive membrane localization of NOXO1 and NOXA1. This is at least partially due to the lack of AIR in NOXO1, a region which maintains an autoinhibited conformation in p47*phox*. NOX1-catalyzed superoxide production is enhanced by cell stimulation with PMA in a manner sensitive to the PKC inhibitor GF109203X [25], whereas this inhibitor does not affect the constitutive activity of NOX1. In NOXO1, its SH3 domains interact intramolecularly with the C-terminal PRR, a target of the NOXA1 SH3 domain, albeit with a low affinity [26, 27]. Because of this weak intramolecular interaction (compared with a much stronger interaction between the bis-SH3 domain and the AIR in p47*phox*), the bis-SH3 domain of NOXO1 is in a state partially accessible to the p22*phox* PRR, which permits constitutive superoxide production (Fig. 2). The accessibility may be enhanced by the presence of NOXA1, since its SH3 domain competes with the NOXO1 bis-SH3 domain for binding to the PRR of NOXO1. Thr-341 in a region C-terminal to the PRR of NOXO1 is phosphorylated by PKC in PMA-stimulated cells [25], and the phosphorylation disrupts the bis-SH3–PRR interaction in NOXO1 without preventing the binding of the PRR to the NOXA1 SH3 domain [25]. Thus, PKC-catalyzed phosphorylation of Thr-341 facilitates the two productive interactions (one between NOXO1-bis-SH3 and p22*phox*-PRR and one between NOXO1-PRR and NOXA1-SH3), which results in further activation of NOX1 as observed in PMA-stimulated cells [25]. In addition to Thr-341, Ser-154 is also phosphorylated [25, 28], which appears to positively regulate both constitutive and enhanced superoxide-producing activity of NOX1 [25]. It should be noted that Ser-154 exists immediately N-terminal to the invariant Ile-155 (Fig. 2), a crucial residue that participates in NOX1 activation, as described above [16].

NOXO1 is also an important regulator of NOX3. In contrast to NOX1 and NOX2, NOX3 is capable of constitutively producing a significant amount of superoxide in a manner independent of an organizer and an activator [29]. The constitutive production occurs even when NOX3 is complexed with a mutant p22*phox* (with substitution of glutamine for Pro-156 in the PRR), which fails to interact with NOXO1 and p47*phox* [29]. The superoxide-producing activity of NOX3 is strongly enhanced by NOXO1 and by p47*phox* without co-expression of an activator protein [24, 29]. The enhancement requires the bis-SH3 domain for the interaction with p22*phox* [24, 29] and the invariant isoleucine (Ile-155 in NOXO1; Ile-152 in p47*phox*) [16].

2.3 p67phox, NOX Activator 2 (NOXA2)

The NOX activator p67*phox* consists of 526 amino acids and contains several protein–protein interaction modules: an N-terminal domain that comprises four tetratricopeptide repeat (TPR) motifs,

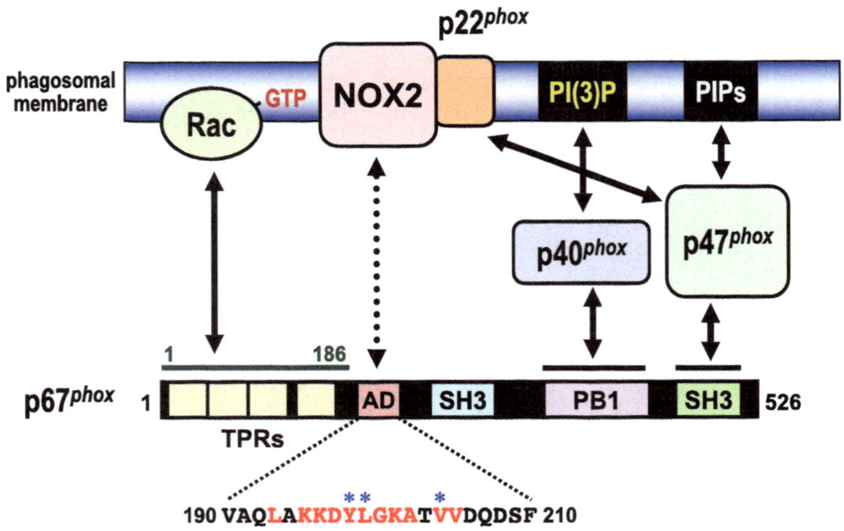

Fig. 3 Role of p67^{phox} in activation of the phagocyte oxidase NOX2. The activator protein p67^{phox} constitutively interacts via a PB1 domain and a C-terminal SH3 domain with p40^{phox} and p47^{phox}, respectively, in the cytoplasm in resting phagocytes. Translocation of p67^{phox} involves both p47^{phox}-mediated association with membrane components [p22^{phox} and phosphoinositides (PIPs)] and p40^{phox} binding to phosphatidylinositol 3-phosphate [PI(3)P], a phosphoinositide highly abundant in the phagosomal membrane. Phagocytosis also induces conversion of Rac into a GTP-bound form; Rac–GTP binds to a p67^{phox} N-terminal domain comprising four tetratricopeptide repeats (TPRs) on the phagosomal membrane. Subsequently, p67^{phox} directly interacts via an activation domain (AD) with NOX2, leading to superoxide production. Solid two-headed arrows show individual associations that participate in membrane translocation of p67^{phox}, whereas a dotted bidirectional arrow denotes an interaction that is directly involved in enzymatic activation of NOX2. Asterisks indicate most crucial residues in the activation domain

a central SH3 domain, a PB1 (Phox and Bem 1) domain, and a C-terminal SH3 domain (Fig. 3). In contrast to the presence of multiple intramolecular interactions in p47^{phox}, as described above, p67^{phox} per se adopts an elongated conformation with no or few apparent associations between the domains [30, 31]. The flexible structure likely allows p67^{phox} to constitutively form a ternary complex with p47^{phox} and p40^{phox} in the cytoplasm and to interact with Rac depending on the status of Rac but not that of p67^{phox}. Indeed Rac–GTP but not Rac–GDP is able to interact with both full-length p67^{phox} and an isolated p67^{phox} TPR domain [32, 33].

In the ternary cytoplasmic *phox* complex, p67^{phox} associates with p47^{phox} via a tail-to-tail interaction of the p67^{phox} C-terminal SH3 domain with the p47^{phox} PRR [34, 35] and stably binds to p40^{phox} via an interaction between PB1 domains [36, 37]. In general, a single SH3 domain interacts with a polyproline II (PPII) helix in a PRR with a low affinity (K_d value of about 1–10 μM). In the p47^{phox}–p22^{phox} interaction, the tandem SH3 domains of p47^{phox} sandwich a short PRR of p22^{phox} (amino acids 151–160), which increases the affinity as well as the specificity. On the other hand, the C-terminal SH3 domain of p67^{phox} specifically binds to the C-terminus of p47^{phox} at an unusual high affinity of 20 nM by simultaneously interacting with the PPII helix of the PRR (amino acids 362–369) and its C-terminally flanking region that comprises two α helices (amino acids 372–386) [34]. The p47^{phox}–p67^{phox} association constitutively occurs in resting neutrophils, but it is possible that p47^{phox} partially dissociates from p67^{phox} after cell stimulation: Ser-379 in the flanking region is known to undergo phosphorylation after cell stimulation, which likely attenuates p47^{phox} binding to p67^{phox} [35, 38] and thus negatively regulates phagocyte oxidase activation [35].

In the p47^{phox}–p67^{phox}–p40^{phox} complex, the constitutive association of p67^{phox} with p40^{phox} is mediated by stable interaction between the PB1 domains of p67^{phox} and p40^{phox} [36, 37]. The PB1 domains are composed of about 80 amino acid residues and are grouped into type I and type II (and also type I/II): the type I includes the PB1 domain of p40^{phox}, whereas p67^{phox} contains the type II PB1 domain [39]. Heterodimeric assembly occurs between type I and II PB1 domains mainly via specific electrostatic interactions between a conserved acidic DX(D/E)GD region of the OPCA motif from a type I PB1 domain (from Asp-289 to Asp-293 in human p40^{phox}) and an invariant lysine residue from a type II PB1 domain (Lys-355 in human p67^{phox}). The central SH3 domain is the most evolutionarily conserved region in p67^{phox} [2, 40]. However, its target protein remains to be identified, and its precise role is presently unknown. Although this SH3 domain is dispensable for the superoxide-producing activity of NOX2 both in vivo and in vitro, the module can positively regulate oxidase activation in vivo, possibly by increasing the affinity of p67^{phox} for the oxidase complex [30, 40, 41].

The N-terminal TPR domain of p67^{phox} specifically interacts with GTP-bound Rac albeit with low affinity [32, 33, 42]. This weak but specific interaction plays a crucial role in phagocyte oxidase activation, as the invariant residue Arg-102 near a β-hairpin insertion between the third and fourth TPR motifs makes direct contacts with Rac–GTP [33] and the R102E substitution results in a complete loss of interaction with Rac–GTP and of NOX2 activation both in vivo and in vitro [32]. Although the TPR domain (amino acids 1–186) is essential and sufficient for p67^{phox} binding to Rac–GTP, the domain by itself is insufficient for oxidase activation [43, 44]. In addition to the Rac-binding TPR domain, NOX activation also requires its C-terminal flanking region (amino acids 190–210), known as an activation domain [45]. Rac binding is considered to induce a conformational change of p67^{phox}, which allows the activation domain to interact with NOX2, leading to superoxide production. Earlier experiments using a cell-free system showed that a p67^{phox} fragment of 1–212 amino acids fully supports NOX2-catalyzed superoxide production, while a fragment of 1–199 amino acids is incapable of activating the oxidase [43] and that oxidase activation is severely impaired by alanine substitution of Val-204 in this region [44]. A later study revealed that, in addition to the Val-204-containing stretch (amino acids 200–210), an additional N-terminal region (spanning amino acids 190–199, especially Tyr-198 and Leu-199) is also required for oxidase activation [45] (Fig. 3). It should be noted that the protein–protein interaction modules for membrane translocation of p67^{phox} (i.e., the two SH3 domains and the PB1 domain) are dispensable for cell-free activation of the phagocyte oxidase, probably because of the presence of p67^{phox}, p47^{phox}, and Rac–GTP at much higher concentrations than those in intact cells. The dispensability suggests that the TPR and activation domains are directly involved in a conformational change of NOX2, which leads to superoxide production. Indeed Tyr-198, Leu-199, and Val-204 are crucial not only for oxidase activation but also for direct interaction of Rac-bound p67^{phox} with the NADPH-binding region of NOX2 [46].

2.4 NOXA1, a p67^{phox}-Related Oxidase Activator

The p67^{phox}-related oxidase activator NOXA1 is essential for activation of the non-phagocytic oxidase NOX1. Human NOXA1 of 476 amino acids contains an N-terminal Rac-binding TPR domain followed by an activation domain and a C-terminal SH3 domain (Fig. 4). In contrast to p67^{phox}, NOXA1 lacks a central SH3 domain

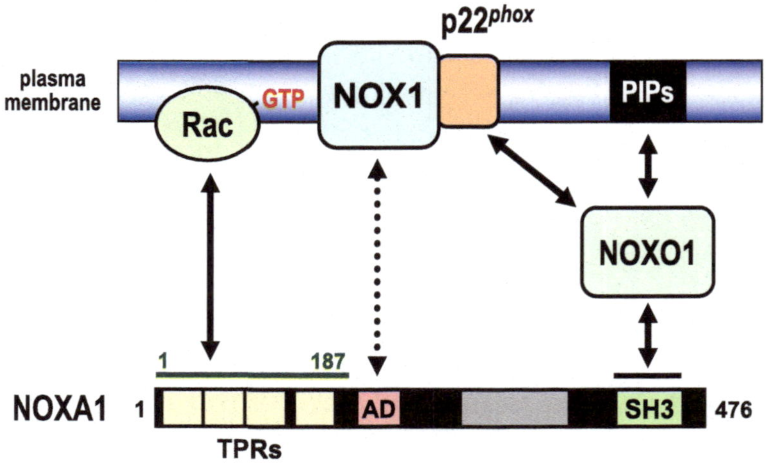

Fig. 4 Role of NOXA1 in activation of the non-phagocytic oxidase NOX1. The activator protein NOXA1 localizes to the plasma membrane via an N-terminal domain consisting of four tetratricopeptide repeats (TPRs) and a C-terminal SH3 domain, which associate with Rac–GTP and NOXO1, respectively: NOXO1 simultaneously binds to p22^{phox} and phosphoinositides (PIPs). An activation domain (AD) of NOXA1 likely interacts with NOX1, which results in superoxide production. Solid two-headed arrows show individual associations that participate in membrane translocation of NOXA1, whereas a dotted line with arrow denotes an interaction that is likely involved in enzymatic activation of NOX1

and a functional PB1 domain [23]. The absence of the latter seems to be reasonable, considering that NOXA1 is an activator of non-phagocytic oxidases. This is because the PB1 domain of p67*phox* is responsible for its association with p40*phox*, an adaptor protein that promotes NOX2 oxidase assembly at the phagosome.

NOXA1-mediated activation of NOX1 and NOX3 is dependent on interaction of its TPR domain with GTP-bound Rac: Arg-103, corresponding to Arg-102 in p67*phox*, participates in binding to Rac and oxidase activation [24, 47]. It has been reported that NOXA1 undergoes phosphorylation at Ser-171 in the TPR domain as well as at Ser-282, which is involved in negative regulation of NOX1 activation [48]. The activation domain of NOXA1 is also functional, as the domain is required for NOX1 activation and the critical residues are evolutionarily well conserved among NOXA1 and p67*phox* proteins [45]. It has been reported that PKC-catalyzed phosphorylation of NOX1 at Thr-429 facilitates its association with the activation domain of NOXA1 for oxidase complex assembly and ROS formation [49]. The C-terminal SH3 domain of NOXA1 strongly interacts with NOXO1 probably via simultaneously binding to the PRR and its C-terminally flanking region, similar to the interaction between p67*phox* and p47*phox* [2]. The association between NOXA1 and NOXO1 is indispensable for membrane localization of NOXA1 and activation of NOX1 [19, 24].

2.5 The Small GTPase Rac

In addition to p47*phox* and p67*phox*, GTP-bound Rac is an essential factor for NOX2 activation both in cells and in a cell-free system. In resting cells, Rac is sequestered in the cytoplasm as GDP-bound Rac, forming a heterodimer with Rho GDP dissociation inhibitor (RhoGDI), whereas p67*phox* is complexed with p47*phox* and p40*phox* as a ternary complex. Upon cell stimulation, Rac is activated to its GTP-bound state and translocates to the membrane in a manner independent of the en bloc translocation of the cytoplasmic *phox* complex (composed of p67*phox*, p47*phox*, and p40*phox*). Rac activation in neutrophils is controlled by guanine nucleotide exchange factors (GEFs), such as Vav1, Vav3, P-Rex1, DOCK2, and DOCK5, all of which are dependent on $PI(3,4,5)P_3$, a phosphoinositide produced by type I PI3K activated by cell stimulants [50–54]. After translocation, Rac–GTP binds to the TPR domain of p67*phox* at the membrane, which leads to a productive interaction of the Rac–p67*phox* complex with NOX2 via the activation domain of p67*phox* [45, 46]. Thus, the GTP–GDP exchange of Rac by GEFs, which are activated via a PI3K-mediated signal transduction pathway, functions as another switch for activation of NOX2, in addition to the signal-induced conformational change of p47*phox* as switch mechanism. Although Rac–GTP binds to p67*phox* with a low affinity in solution [32, 33, 42], they are considered to effectively associate at the membrane.

The human genome contains three genes that encode Rac GTPases: Rac1 is a ubiquitously expressed protein; Rac2 expression is mostly restricted to bone marrow-derived cells; and Rac3 is present in various cells and most abundant in the brain. These three Rac GTPases are all able to participate in activation of NOX1, NOX2, and NOX3 [55]. In contrast, the closely related GTPase Cdc42 neither binds to p67^{phox} nor activates oxidases [55, 56]. The strict specificity is likely due to the difference in only two residues between Rac and Cdc42: Ala-27 and Gly-30 are replaced by lysine and serine, respectively, in Cdc42 [56]. Interestingly, a doubly mutated Cdc42 with the K27A/S30G substitution is capable of effectively binding to the p67^{phox} TPR domain [33] and fully activating the NOX2-based oxidase [55].

The insert helix (a surface-exposed α helix) of Rac, a region unique to the Rho family among the Ras-related superfamily of small GTPases, is distant from the interface between Rac–GTP and the p67^{phox} TPR domain, and thus does not contribute to complex formation [33]. Although this region was previously postulated to directly interact with NOX2 for superoxide production, the insert helix of Rac1/2/3 is dispensable for NOX2 activation in cell-free and intact cell systems [55]; in addition, it does not play a role in activation of NOX1 and NOX3 [55]. On the other hand, it is still possible that a region other than the insert helix in Rac makes a direct contact of the Rac–p67^{phox} complex with NOX for enzyme activation, as does the p67^{phox} activation domain.

2.6 p40^{phox}

The adaptor protein p40^{phox} of 339 amino acids is predominantly expressed in cells of the hematopoietic lineage, including neutrophils and macrophages. It contains an N-terminal PI(3)P-binding PX domain and a central SH3 domain and ends with a PB1 domain for its constitutive interaction with p67^{phox} (Fig. 5). As expected from its unique expression profile, p40^{phox} is specifically involved in activation of the phagocyte oxidase NOX2 [57–60], which agrees with its ability to form a stable complex with p67^{phox}, but not with the non-phagocytic oxidase activator NOXA1 [23]. In PMA-stimulated cells or in cell-free systems, p40^{phox} is not required for NOX2 activation, although this adaptor protein can enhance superoxide production to some extent. In contrast, p40^{phox} plays a crucial role in activation of NOX2 during phagocytosis [57–60] and upon cell stimulation via the heterotrimeric GTPase G_i [37].

In phagocytosis-induced activation of NOX2, p40^{phox} translocates together with p47^{phox} and p67^{phox} to phagosomes largely via PX domain binding to PI(3)P, which is generated by activation of type III PI3K on phagosomes. It has been suggested that p40^{phox}-regulated translocation of the cytosolic *phox* trimeric complex to phagosomes is of great importance when p47^{phox} does not fully function [2, 61, 62]. The p40^{phox} PX domain is normally masked via an intramolecular interaction with the PB1 domain without

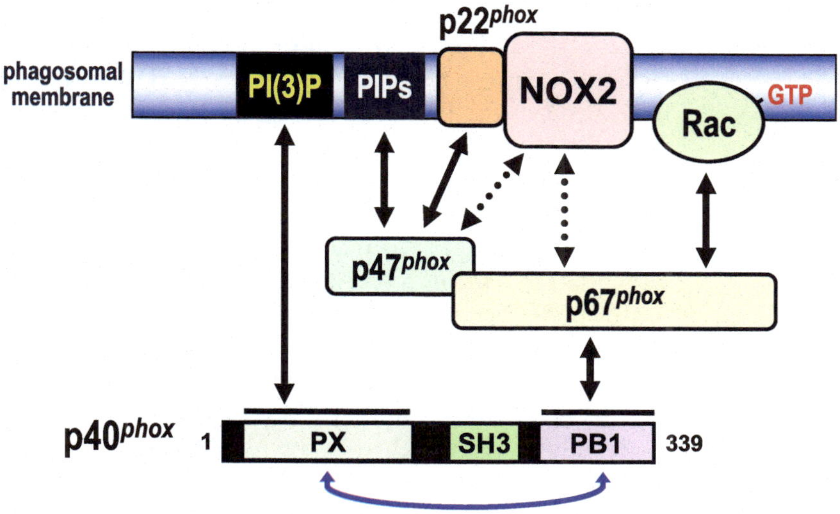

Fig. 5 Role of p40*phox* in activation of the phagocyte oxidase NOX2. In phagocytes, p40*phox* forms a constitutive trimer with p67*phox* and p47*phox* via a PB1 domain-mediated association with p67*phox*, which is not inhibited by an intramolecular interaction with an N-terminal PX domain. Translocation of the trimer from the cytoplasm to the phagosomal membrane requires binding of the p40*phox* PX domain to phosphatidylinositol 3-phosphate [PI (3)P], a phagosome-enriched phospholipid, in addition to p67*phox* interaction with Rac–GTP and p47*phox* association with both p22*phox* and phosphoinositides (PIPs). Solid two-headed arrows in *black* show individual associations that participate in phagocyte oxidase assembly, whereas dotted bidirectional arrows denote interactions involved in enzymatic activation of NOX2. A solid two-headed arrow in *blue* indicates an intramolecular interaction

preventing it from constitutively binding to the p67*phox* PB1 domain [63, 64]. The closed, autoinhibited conformation is likely disrupted during oxidase activation, which enables p40*phox* to interact with PI(3)P-enriched phagosomes. It is also possible that high-affinity binding of the PX domain to PI(3)P may overcome PX–PB1 autoinhibition in p40*phox* on phagosomal membranes which contain abundantly this phosphoinositide. The central SH3 domain of p40*phox* seems to participate in phagocytosis-induced activation of NOX2 [65, 66]. However, the physiological target of this SH3 domain remains obscure, although it is able to interact with p47*phox* very weakly [38, 67].

3 Discussion

We describe here a current view on soluble regulatory proteins for activation of NOX1 through NOX3 in mammals. Activation of the phagocyte oxidase NOX2 is initiated by two stimulus (or phagocytosis)-induced events: a conversion of Rac into a GTP-bound form and a conformational change of p47*phox* in the p47*phox*–p67*phox*–p40*phox* complex (Fig. 1). The trimer and Rac are independently targeted to the plasma (or phagosomal) membrane, where Rac–GTP binds to p67^{phox} for activation of NOX2. In this process, the activation domain of p67^{phox} interacts with NOX2 to facilitate electron transfer (e.g., from NADPH to FAD and/or from FAD to the proximal heme), leading to superoxide production (Fig. 3). However, the precise mechanism whereby the conformation of NOX2 is changed into a catalytically active form is largely unknown. The Ile-152-containing stretch in p47^{phox} may contribute to the conformational change of NOX2 [16], but its target region remains to be identified. In addition, a partner protein for the p67^{phox} central SH3 domain, a module that regulates NOX2 activation [30, 40, 41], is presently unclear.

In NOX2-based oxidase assembly on the phagosome, p40^{phox} plays a central role via specific binding of its PX domain to PI(3)P, a phosphoinositide enriched in the phagosomal membrane [57–60] (Fig. 5). The PI(3)P-binding activity of p40^{phox}, however, is normally suppressed via an intramolecular interaction of the PX domain with the PB1 domain [63, 64]. Although the p40^{phox} PX domain becomes accessible to PI(3)P during phagocytosis, its molecular mechanism remains to be clarified. Phagosomal oxidase assembly appears to be also regulated by the p40^{phox} SH3 domain [65, 66], a target of which should be identified.

Future studies, in which the above questions will be addressed, are expected to contribute to the improvement of our understanding of NOX regulation and the development of novel drugs to treat NOX-related diseases.

Acknowledgments

This work was supported in part by JSPS (Japan Society for the Promotion of Science: a Grant-in-Aid for Scientific Research on Innovative Areas "Oxygen Biology: a new criterion for integrated understanding of life" No. 26111009).

References

1. Lambeth JD (2004) NOX enzymes and the biology of reactive oxygen. Nat Rev Immunol 4:181–189
2. Sumimoto H (2008) Structure, regulation and evolution of Nox-family NADPH oxidases that produce reactive oxygen species. FEBS J 275:3249–3277
3. Leto TL, Morand S, Hurt D, Ueyama T (2009) Targeting and regulation of reactive oxygen species generation by Nox family NADPH oxidases. Antioxid Redox Signal 11:2607–2619
4. Nauseef WM, Borregaard N (2014) Neutrophils at work. Nat Immunol 15:602–611
5. Stampoulis P, Ueda T, Matsumoto M, Terasawa H, Miyano K, Sumimoto H, Shimada I (2012) Atypical membrane-embedded PI (3,4)P_2 binding site on p47^{phox} PX domain revealed by NMR. J Biol Chem 287:17848–17859
6. Ago T, Nunoi H, Ito T, Sumimoto H (1999) Mechanism for phosphorylation-induced activation of the phagocyte NADPH oxidase protein p47^{phox}. Triple replacement of serines 303, 304, and 328 with aspartates disrupts the SH3 domain-mediated intramolecular interaction in p47^{phox}, thereby activating the oxidase. J Biol Chem 274:33644–33653

7. Groemping Y, Lapouge K, Smerdon SJ, Rittinger K (2003) Molecular basis of phosphorylation-induced activation of the NADPH oxidase. Cell 113:343–355
8. Yuzawa S, Suzuki NN, Fujioka Y, Ogura K, Sumimoto H, Inagaki F (2004) A molecular mechanism for autoinhibition of the tandem SH3 domains of p47*phox*, the regulatory subunit of the phagocyte NADPH oxidase. Genes Cells 9:443–456
9. Yuzawa S, Ogura K, Horiuchi M, Suzuki NN, Fujioka Y, Kataoka M, Sumimoto H, Inagaki F (2004) Solution structure of the tandem Src homology 3 domains of p47*phox* in an autoinhibited form. J Biol Chem 279:29752–29760
10. Ago T, Kuribayashi F, Hiroaki H, Takeya R, Ito T, Kohda D, Sumimoto H (2003) Phosphorylation of p47*phox* directs phox homology domain from SH3 domain toward phosphoinositides, leading to phagocyte NADPH oxidase activation. Proc Natl Acad Sci U S A 100:4474–4479
11. Marcoux J, Man P, Petit-Haertlein I, Vivès C, Forest E, Fieschi F (2010) p47*phox* molecular activation for assembly of the neutrophil NADPH oxidase complex. J Biol Chem 285:28980–28990
12. Ueyama T, Nakakita J, Nakamura T, Kobayashi T, Kobayashi T, Son J, Sakuma M, Sakaguchi H, Leto TL, Saito N (2011) Cooperation of p40*phox* with p47*phox* for Nox2-based NADPH oxidase activation during Fcγ receptor (FcγR)-mediated phagocytosis: mechanism for acquisition of p40*phox* phosphatidylinositol 3-phosphate (PI(3)P) binding. J Biol Chem 286:40693–40705
13. Shiose A, Sumimoto H (2000) Arachidonic acid and phosphorylation synergistically induce a conformational change of p47*phox* to activate the phagocyte NADPH oxidase. J Biol Chem 275:13793–13801
14. Shmelzer Z, Karter M, Eisenstein M, Leto TL, Hadad N, Ben-Menahem D, Gitler D, Banani S, Wolach B, Rotem M, Levy R (2008) Cytosolic phospholipase $A_2\alpha$ is targeted to the p47*phox*-PX domain of the assembled NADPH oxidase via a novel binding site in its C2 domain. J Biol Chem 283:31898–31908
15. Olofsson P, Holmberg J, Tordsson J, Lu S, Akerström B, Holmdahl R (2003) Positional identification of Ncf1 as a gene that regulates arthritis severity in rats. Nat Genet 33:25–32
16. Taura M, Miyano K, Minakami R, Kamakura S, Takeya R, Sumimoto H (2009) A region N-terminal to the tandem SH3 domain of p47*phox* plays a crucial role in the activation of the phagocyte NADPH oxidase. Biochem J 419:329–338
17. Ogura K, Nobuhisa I, Yuzawa S, Takeya R, Torikai S, Saikawa K, Sumimoto H, Inagaki F (2006) NMR solution structure of the tandem Src homology 3 domains of p47*phox* complexed with a p22*phox*-derived proline-rich peptide. J Biol Chem 281:3660–3668
18. Mizrahi A, Berdichevsky Y, Casey PJ, Pick E (2010) A prenylated p47*phox*-p67*phox*-Rac1 chimera is a quintessential NADPH oxidase activator: membrane association and functional capacity. J Biol Chem 285:25485–25499
19. Miyano K, Ueno N, Takeya R, Sumimoto H (2006) Direct involvement of the small GTPase Rac in activation of the superoxide-producing NADPH oxidase Nox1. J Biol Chem 281:21857–21868
20. Cheng G, Lambeth JD (2004) NOXO1, regulation of lipid binding, localization, and activation of Nox1 by the Phox homology (PX) domain. J Biol Chem 279:4737–4742
21. Ueyama T, Lekstrom K, Tsujibe S, Saito N, Leto TL (2007) Subcellular localization and function of alternatively spliced Noxo1 isoforms. Free Radic Biol Med 42:180–190
22. Takeya R, Taura M, Yamasaki T, Naito S, Sumimoto H (2006) Expression and function of Noxo1γ, an alternative splicing form of the NADPH oxidase organizer 1. FEBS J 273:3663–3677
23. Takeya R, Ueno N, Kami K, Taura M, Kohjima M, Izaki T, Nunoi H, Sumimoto H (2003) Novel human homologues of p47*phox* and p67*phox* participate in activation of superoxide-producing NADPH oxidases. J Biol Chem 278:25234–25246
24. Ueyama T, Geiszt M, Leto TL (2006) Involvement of Rac1 in activation of multicomponent Nox1- and Nox3-based NADPH oxidases. Mol Cell Biol 26:2160–2174
25. Yamamoto A, Takeya R, Matsumoto M, Nakayama KI, Sumimoto H (2013) Phosphorylation of Noxo1 at threonine 341 regulates its interaction with Noxa1 and the superoxide-producing activity of Nox1. FEBS J 280:5145–5159
26. Yamamoto A, Kami K, Takeya R, Sumimoto H (2007) Interaction between the SH3 domains and C-terminal proline-rich region in NADPH oxidase organizer 1 (Noxo1). Biochem Biophys Res Commun 352:560–565
27. Dutta S, Rittinger K (2010) Regulation of NOXO1 activity through reversible interactions with p22phox and NOXA1. PLoS One 5:e10478

28. Debbabi M, Kroviarski Y, Bournier O, Gougerot-Pocidalo MA, El-Benna J, Dang PM (2013) NOXO1 phosphorylation on serine 154 is critical for optimal NADPH oxidase 1 assembly and activation. FASEB J 27:1733–1748
29. Ueno N, Takeya R, Miyano K, Kikuchi H, Sumimoto H (2005) The NADPH oxidase Nox3 constitutively produces superoxide in a p22*phox*-dependent manner: its regulation by oxidase organizers and activators. J Biol Chem 280:23328–23339
30. Yuzawa S, Miyano K, Honbou K, Inagaki F, Sumimoto H (2009) The domain organization of p67*phox*, a protein required for activation of the superoxide-producing NADPH oxidase in phagocytes. J Innate Immun 1:543–555
31. Durand D, Vivès C, Cannella D, Pérez J, Pebay-Peyroula E, Vachette P, Fieschi F (2010) NADPH oxidase activator p67phox behaves in solution as a multidomain protein with semi-flexible linkers. J Struct Biol 169:45–53
32. Koga H, Terasawa H, Nunoi H, Takeshige K, Inagaki F, Sumimoto H (1999) Tetratricopeptide repeat (TPR) motifs of p67*phox* participate in interaction with the small GTPase Rac and activation of the phagocyte NADPH oxidase. J Biol Chem 274:25051–25060
33. Lapouge K, Smith SJ, Walker PA, Gamblin SJ, Smerdon SJ, Rittinger K (2000) Structure of the TPR domain of p67phox in complex with Rac·GTP. Mol Cell 6:899–907
34. Kami K, Takeya R, Sumimoto H, Kohda D (2002) Diverse recognition of non-PxxP peptide ligands by the SH3 domains from p67*phox*, Grb2 and Pex13p. EMBO J 21:4268–4276
35. Mizuki K, Takeya R, Kuribayashi F, Nobuhisa I, Kohda D, Nunoi H, Takeshige K, Sumimoto H (2005) A region C-terminal to the proline-rich core of p47*phox* regulates activation of the phagocyte NADPH oxidase by interacting with the C-terminal SH3 domain of p67*phox*. Arch Biochem Biophys 444:185–194
36. Ito T, Matsui Y, Ago T, Ota K, Sumimoto H (2001) Novel modular domain PB1 recognizes PC motif to mediate functional protein-protein interactions. EMBO J 20:3938–3946
37. Kuribayashi F, Nunoi H, Wakamatsu K, Tsunawaki S, Sato K, Ito T, Sumimoto H (2002) The adaptor protein p40*phox* as a positive regulator of the superoxide-producing phagocyte oxidase. EMBO J 21:6312–6320
38. Massenet C, Chenavas S, Cohen-Addad C, Dagher MC, Brandolin G, Pebay-Peyroula E, Fieschi F (2005) Effects of p47*phox* C terminus phosphorylations on binding interactions with p40*phox* and p67*phox*. Structural and functional comparison of p40*phox* and p67*phox* SH3 domains. J Biol Chem 280:13752–13761
39. Sumimoto H, Kamakura S, Ito T (2007) Structure and function of the PB1 domain, a protein-interaction module conserved in animals, fungi, amoebas, and plants. Sci STKE 2007:re6
40. Maehara Y, Miyano K, Sumimoto H (2009) Role for the first SH3 domain of p67*phox* in activation of superoxide-producing NADPH oxidases. Biochem Biophys Res Commun 379:589–593
41. de Mendez I, Garrett MC, Adams AG, Leto TL (1994) Role of p67-*phox* SH3 domains in assembly of the NADPH oxidase system. J Biol Chem 269:16326–16332
42. Miyano K, Sumimoto H (2012) Assessment of the role for Rho family GTPases in NADPH oxidase activation. Methods Mol Biol 827:195–212
43. Hata K, Takeshige K, Sumimoto H (1997) Roles for proline-rich regions of p47*phox* and p67*phox* in the phagocyte NADPH oxidase activation in vitro. Biochem Biophys Res Commun 241:226–231
44. Han C-H, Freeman JL, Lee T, Motalebi SA, Lambeth JD (1998) Regulation of the neutrophil respiratory burst oxidase. Identification of an activation domain in p67*phox*. J Biol Chem 273:16663–16668
45. Maehara Y, Miyano K, Yuzawa S, Akimoto R, Takeya R, Sumimoto H (2010) A conserved region between the TPR and activation domains of p67*phox* participates in activation of the phagocyte NADPH oxidase. J Biol Chem 285:31435–31445
46. Matono R, Miyano K, Kiyohara T, Sumimoto H (2014) Arachidonic acid induces direct interaction of the p67*phox*–Rac complex with the phagocyte oxidase Nox2, leading to superoxide production. J Biol Chem 289:24874–24884
47. Miyano K, Sumimoto H (2007) Role of the small GTPase Rac in p22*phox*-dependent NADPH oxidases. Biochimie 89:1133–1144
48. Kroviarski Y, Debbabi M, Bachoual R, Périanin A, Gougerot-Pocidalo MA, El-Benna J, Dang PM (2010) Phosphorylation of NADPH oxidase activator 1 (NOXA1) on serine 282 by MAP kinases and on serine 172 by protein kinase C and protein kinase A prevents NOX1 hyperactivation. FASEB J 24:2077–2092
49. Streeter J, Schickling BM, Jiang S, Stanic B, Thiel WH, Gakhar L, Houtman JC, Miller FJ Jr (2014) Phosphorylation of Nox1 regulates association with NoxA1 activation domain. Circ Res 115:911–918

50. Kunisaki Y, Nishikimi A, Tanaka Y, Takii R, Noda M, Inayoshi A, Watanabe K, Sanematsu F, Sasazuki T, Sasaki T, Fukui Y (2006) DOCK2 is a Rac activator that regulates motility and polarity during neutrophil chemotaxis. J Cell Biol 174:647–652
51. Utomo A, Cullere X, Glogauer M, Swat W, Mayadas TN (2006) Vav proteins in neutrophils are required for FcγR-mediated signaling to Rac GTPases and nicotinamide adenine dinucleotide phosphate oxidase component p40(phox). J Immunol 177:6388–6397
52. Graham DB, Robertson CM, Bautista J, Mascarenhas F, Diacovo MJ, Montgrain V, Lam SK, Cremasco V, Dunne WM, Faccio R, Coopersmith CM, Swat W (2007) Neutrophil-mediated oxidative burst and host defense are controlled by a Vav-PLCγ2 signaling axis in mice. J Clin Invest 117:3445–3452
53. Lawson CD, Donald S, Anderson KE, Patton DT, Welch HC (2011) P-Rex1 and Vav1 cooperate in the regulation of formyl-methionyl-leucyl-phenylalanine-dependent neutrophil responses. J Immunol 186:1467–1476
54. Watanabe M, Terasawa M, Miyano K, Yanagihara T, Uruno T, Sanematsu F, Nishikimi A, Côté J-F, Sumimoto H, Fukui Y (2014) DOCK2 and DOCK5 act additively in neutrophils to regulate chemotaxis, superoxide production, and extracellular trap formation. J Immunol 193:5660–5667
55. Miyano K, Koga H, Minakami R, Sumimoto H (2009) The insert region of the Rac GTPases is dispensable for activation of superoxide-producing NADPH oxidases. Biochem J 422:373–382
56. Kwong CH, Adams AG, Leto TL (1995) Characterization of the effector-specifying domain of Rac involved in NADPH oxidase activation. J Biol Chem 270:19868–19872
57. Suh CI, Stull ND, Li XJ, Tian W, Price MO, Grinstein S, Yaffe MB, Atkinson S, Dinauer MC (2006) The phosphoinositide-binding protein $p40^{phox}$ activates the NADPH oxidase during FcγIIA receptor-induced phagocytosis. J Exp Med 203:1915–1925
58. Ellson CD, Davidson K, Ferguson GJ, O'Connor R, Stephens LR, Hawkins PT (2006) Neutrophils from $p40^{phox-/-}$ mice exhibit severe defects in NADPH oxidase regulation and oxidant-dependent bacterial killing. J Exp Med 203:1927–1937
59. Tian W, Li XJ, Stull ND, Ming W, Suh CI, Bissonnette SA, Yaffe MB, Grinstein S, Atkinson SJ, Dinauer MC (2008) FcγR-stimulated activation of the NADPH oxidase: phosphoinositide-binding protein $p40^{phox}$ regulates NADPH oxidase activity after enzyme assembly on the phagosome. Blood 112:3867–3877
60. Anderson KE, Boyle KB, Davidson K, Chessa TA, Kulkarni S, Jarvis GE, Sindrilaru A, Scharffetter-Kochanek K, Rausch O, Stephens LR, Hawkins PT (2008) CD18-dependent activation of the neutrophil NADPH oxidase during phagocytosis of *Escherichia coli* or *Staphylococcus aureus* is regulated by class III but not class I or II PI3Ks. Blood 112:5202–5211
61. Ueyama T, Kusakabe T, Karasawa S, Kawasaki T, Shimizu A, Son J, Leto TL, Miyawaki A, Saito N (2008) Sequential binding of cytosolic Phox complex to phagosomes through regulated adaptor proteins: evaluation using the novel monomeric Kusabira-Green System and live imaging of phagocytosis. J Immunol 181:629–640
62. Li XJ, Marchal CC, Stull ND, Stahelin RV, Dinauer MC (2010) $p47^{phox}$ Phox homology domain regulates plasma membrane but not phagosome neutrophil NADPH oxidase activation. J Biol Chem 285:35169–35179
63. Honbou K, Minakami R, Yuzawa S, Takeya R, Suzuki NN, Kamakura S, Sumimoto H, Inagaki F (2007) Full-length $p40^{phox}$ structure suggests a basis for regulation mechanism of its membrane binding. EMBO J 26:1176–1186
64. Ueyama T, Tatsuno T, Kawasaki T, Tsujibe S, Shirai Y, Sumimoto H, Leto TL, Saito N (2007) A regulated adaptor function of $p40^{phox}$: distinct $p67^{phox}$ membrane targeting by $p40^{phox}$ and by $p47^{phox}$. Mol Biol Cell 18:441–454
65. Bissonnette SA, Glazier CM, Stewart MQ, Brown GE, Ellson CD, Yaffe MB (2008) Phosphatidylinositol 3-phosphate-dependent and -independent functions of $p40^{phox}$ in activation of the neutrophil NADPH oxidase. J Biol Chem 283:2108–2119
66. Chessa TA, Anderson KE, Hu Y, Xu Q, Rausch O, Stephens LR, Hawkins PT (2010) Phosphorylation of threonine 154 in $p40^{phox}$ is an important physiological signal for activation of the neutrophil NADPH oxidase. Blood 116:6027–6036
67. Ito T, Nakamura R, Sumimoto H, Takeshige K, Sakaki Y (1996) An SH3 domain-mediated interaction between the phagocyte NADPH oxidase factors $p40^{phox}$ and $p47^{phox}$. FEBS Lett 385:229–232

Chapter 9

Insights into the NOX NADPH Oxidases Using Heterologous Whole Cell Assays

Mary C. Dinauer

Abstract

Assays based on ectopic expression of NOX NADPH oxidase subunits in heterologous mammalian cells are an important approach for investigating features of this family of enzymes. These model systems have been used to analyze the biosynthesis and functional domains of NOX enzyme components as well as their regulation and cellular activities. This chapter provides an overview of the basic principles and applications of heterologous whole cell assays in studying NOX NADPH oxidases.

Key words NOX, Flavocytochrome *b*, COS-7, Expression, *phox*

1 Introduction

The NOX NADPH oxidases are a family of multi-subunit membrane enzymes whose electron transferase activity is mediated by a transmembrane flavocytochrome referred to as "NOX." This mini-review will focus on the use of heterologous whole cell assays for the analysis of NOX NADPH oxidases. A major focus will be the NOX2 NADPH oxidase, which is highly expressed in leukocytes. The NOX2 oxidase generates large quantities of superoxide from molecular oxygen (O_2) on plasma or phagosome membranes upon activation by particulate or soluble inflammatory stimuli [1]. This is the best understood of the NOX family NADPH oxidases, both from the standpoint of structural and regulatory features as well as the biologic functions of derivative oxidants. The NOX2 oxidase is expressed in myeloid cells, including neutrophils, monocytes and macrophages, eosinophils, and dendritic cells. Derivative reactive oxygen species (ROS) play important roles in host defense against bacterial and fungal species as well as in limiting inflammation and autoimmunity by redox regulation of pro-inflammatory responses, degradation of ingested material, and antigen processing. The NOX2 oxidase is also expressed in B and perhaps T lymphocytes,

Ulla G. Knaus and Thomas L. Leto (eds.), *NADPH Oxidases: Methods and Protocols*, Methods in Molecular Biology, vol. 1982, https://doi.org/10.1007/978-1-4939-9424-3_9,

although its function in lymphocytes is not well understood. Inactivating mutations in the NOX2 NADPH oxidase lead to the primary immunodeficiency disorder, chronic granulomatous disease (CGD), and hypomorphic variants of enzyme subunits are associated with autoimmune diseases.

The NOX2 NADPH oxidase has four canonical subunits. The redox center of the enzyme is flavocytochrome *b*, a heterodimer of two multipass integral membrane proteins, the glycoprotein gp91^{phox} (*CYBB* or NOX2) and p22^{phox} (*CYBA*). Transfer of electrons from NADPH across the membrane through FAD and a pair of heme groups is activated upon binding of cytosolic regulatory subunits, p47^{phox} (*NCF1*) and p67^{phox} (*NCF2*). The NOX2 oxidase also employs a third regulatory subunit, p40^{phox} (*NCF4*), which is specialized to upregulate the oxidase in response to phosphatidyl inositol 3-phosphate (PI(3)P). These cytosolic subunits are in turn regulated by their phosphorylation and binding of phosphoinositides and other membrane-associated lipids. Importantly, GTP-bound Rac (either Rac1 or Rac2) is also required for enzyme activity, acting as a molecular switch to induce additional activating conformational changes in p67^{phox} and gp91^{phox}. NOX1, NOX3, and NOX4 also partner with p22^{phox}, with heterodimer formation important for stability, trafficking out of the endoplasmic reticulum and, for NOX1, NOX2, and NOX3, providing a key docking site for cytosolic regulatory subunits.

The analysis of the biosynthesis of NOX/p22^{phox}, function of specific oxidase subunits and their domains, regulated assembly of the enzyme complex, and cellular functions of the NOX NADPH oxidases have all benefited from studies in heterologous whole cell models. These models involve the use of transient or stable transfection of transgenes to achieve ectopic expression of NOX enzyme subunits in tissue culture cell lines that do otherwise not express them, or in some cases, only at very low levels. Central advantages of a whole cell model are the ability to pursue studies in the context of an intact cell environment while taking advantage of facile approaches for genetic manipulation in order to express either wild type or mutant versions of NADPH oxidase subunits and/or regulatory proteins. Analysis in intact cells is crucial for certain questions, such as those involving biosynthesis, trafficking of NOX/p22^{phox} heterodimers, receptor-mediated activation of enzyme activity, and oxidase regulation on different membrane compartments. The approach is also convenient in that expression of oxidase subunits in cells followed by functional analysis avoids the necessity of purifying recombinant proteins for cell-free assays.

2 Heterologous Cell Lines for Expression of NOX NADPH Oxidases

There are a number of cell lines that have been successfully employed for heterologous cell studies of NOX NADPH oxidases (*see* also Refs. [2, 3]). Most are adherent mammalian epithelial cell lines, including monkey kidney COS-7 cells, human embryonic kidney (HEK)-293 cells, human HeLa cells, human H661 lung carcinoma cells, or Chinese hamster ovary (CHO) cells. Another widely used assay system, particularly for the NOX2 oxidase, is based on K562 cells, a multipotent primitive myeloid cell line derived from a patient with myeloid leukemia, which grows in suspension culture. K562 cells express endogenous Fcγ receptor for IgG, thus enabling NOX2 NADPH oxidase activation through a physiologic ligand. Although not well characterized, K562 cells may also express blood cell-specific signaling proteins that offer theoretical advantages. However, the level of transiently expressed (*see* below) oxidase subunits in K562 cells is often not as high as what is achievable with epithelial cell lines. Epithelial cell lines are also often preferred for microscopy-based studies because of their adherent nature and greater cytoplasmic/nuclear ratio compared to K562 cells.

Heterologous cell lines can be used to express wild type or mutant NADPH oxidase subunits in various combinations, depending on the goal of the experiment. The high conservation of NOX oxidase subunits minimizes difficulties of assaying oxidase subunits across different species. The activity of NOX1, NOX2, or NOX3 has been assessed in various combinations with the classic NOX2 regulatory subunits, $p47^{phox}$ and $p67^{phox}$, or their homologues, NOXO1 and NOXA1 (e.g. [4, 5]). All of these cell lines express endogenous Rac1, but one can interrogate function of Rac1or Rac2 using overexpression strategies. Most of the cell lines express endogenous $p22^{phox}$, with the exception of H661 lung carcinoma cells and CHO cells, and thus the latter are useful to assess $p22^{phox}$ mutants. The abundance of $p22^{phox}$ is variable in COS-7 lines, and therefore a version that uses a stable $p22^{phox}$ transgene to enforce expression of a fixed amount of $p22^{phox}$ [6] is often employed (e.g. [7].). Transgenic overexpression of $p22^{phox}$ can also be used to interrogate the behavior of $p22^{phox}$ mutants in cell lines expressing endogenous $p22^{phox}$ at low levels. COS-7 or K562 lines with expression of the four canonical NOX2 NADPH oxidase subunits, $gp91^{phox}$ (NOX2), $p22^{phox}$, $p47^{phox}$, and $p67^{phox}$, have often been referred to as COS^{phox} or $K562^{phox}$ cells, respectively.

3 Approaches for Transgenic Expression of NADPH Oxidase Subunits

There are two general approaches for transgenic expression of NADPH oxidase subunits, which are often used in combination. First, cDNAs encoding oxidase subunits are cloned into expression plasmids designed for high-level transient expression in the cytoplasm. Such plasmids include pcDNA3 and pRK5, transfected into epithelial cell lines using DNA-complexing reagents such as calcium phosphate or liposome-based methods, whereas electroporation is typically employed for K562 cells. The transient expression approach offers rapid turnaround time, with assays performed within 24–72 h. A caveat is that subunits are often overexpressed relative to normal endogenous levels. As a second approach, transgenes encoding oxidase subunits can be stably transfected as chromosome-integrated transgenes, using either plasmid, retrovirus, or lentivirus-based expression vectors. In this case, either an antibiotic-resistant or fluorescent protein (FP) marker is used for selection of cells expressing the desired transgene. Fluorescent proteins can be either expressed from a linked gene or attached to the N- or C-terminus of oxidase subunits. The latter has been useful in monitoring intracellular localization and enzyme assembly in live imaging studies (e.g. [8, 9]). As expression levels of transgenes can have clone-to-clone variation, it is often desirable to isolate individual clones in order to identify those that have appropriate expression levels. Cell lines generated in this way offer the advantage of stability for repetitive analysis; lines with stable expression of only some of the oxidase subunits, for example, NOX2/p22^{phox} or p22^{phox}, can also be generated and used in transient transfection studies to study properties of the other subunits.

4 Activation of NOX NADPH Oxidases in Heterologous Cell Lines

Phorbol myristate acetate (PMA) is a pharmacologic stimulus that is well-known to induce NOX2 NADPH oxidase assembly in leukocytes. PMA also works well to induce oxidase activation in heterologous cells. PMA activates protein kinase C family enzymes, which phosphorylate cytosolic regulatory proteins to induce conformational changes and enable binding to the NOX/p22^{phox} complex. PMA also activates Rac guanine nucleotide exchange factors (GEFs), which is important to generate the GTP-bound Rac that provides an essential activating switch for p67^{phox}. Which exchange factors mediate GDP to GDP exchange on Rac for PMA-induced activation of the oxidase is not well understood. Arachidonic acid also can activate NADPH oxidase activity in heterologous cells by multiple mechanisms and is able to induce conformational changes in p47^{phox}, activate Rac, and promote the interaction of the p67^{phox}-

Rac-GTP module with NOX2 [10]. As mentioned, K562 cells express endogenous Fcγ receptors, thus enabling NADPH oxidase activation through physiologic stimuli such as IgG-opsonized particles or IgG immune complexes. Receptor transgenes can also be introduced into heterologous cells, e.g., FcγRIIA into COS-7 cells, or the formyl peptide receptor into either COS-7 [11] or K562 [2] cells. Endogenous receptors can permit uptake of microbes and concomitant NADPH oxidase activation in COS^{phox} cells [12]).

5 Applications of Heterologous Whole Cell Assays for the Analysis of NOX NADPH Oxidases

Heterologous whole cell assays have been used for a range of studies on NOX NADPH oxidases. Selected examples are described below.

5.1 NOX-p22^{phox} Biosynthesis, Heterodimer Formation, and Intracellular Trafficking

Studies using COS-7 or CHO cells expressing NOX isoforms with or without p22^{phox} have been used extensively to study the steps involved in biosynthesis, NOX glycosylation, and heterodimer formation between NOX and its p22^{phox} partner. Their association is important for stability of each subunit, heme incorporation into NOX, and transit out of the endoplasmic reticulum to the plasma membrane and/or intracellular membrane compartments. Our group has used this approach to characterize these steps in the formation of the NOX2-p22^{phox} heterodimer [6, 9, 13–15]. Studies using NOX2 co-expressed in balanced levels with a fluorescent protein-tagged p22^{phox} in CHO cells revealed that the assembled heterodimer resided not only in plasma membranes but also in endosomal compartments, including Rab11 recycling endosomes [9]. This distribution proved to be similar in murine macrophages, and intracellular vesicles appear to be a source of NOX2/p22^{phox} incorporated into phagosome membranes [9]. NOX4-p22^{phox} interactions and their subcellular localization were investigated by Knaus and colleagues [16, 17]. The NOX4/p22^{phox} heterodimer was found largely on the plasma membrane in COS-7 and H661 cells. In an elegant recent study, small fusion tags were placed on NOX4 and p22^{phox}, which generate luciferase activity when in close proximity. Derivatives were expressed in COS-7 cells and used to probe the role of specific amino acids in each subunit for heme incorporation, heterodimerization, and catalytic activity [17].

5.2 Functional Analysis of NOX NADPH Subunits and Contribution to Enzyme Activity

Numerous studies have taken advantage of heterologous whole cell assays to explore the function of modular domains within the oxidase subunits. Early studies examined the importance of the multiple SH3 domains and their proline-rich targets, p47^{phox} phosphorylation, and Rac-p67^{phox} interactions that mediate the assembly of the active oxidase complex, as, for example, in [18–20].

The COS^{phox} system was also key in helping establish the role of the enigmatic $p40^{phox}$ subunit of the NOX2 oxidase, which was not required for high-level oxidase activity in either cell-free systems or in heterologous whole cell oxidase models activated by PMA. It was eventually recognized that a PX motif in $p40^{phox}$ has high affinity for PI(3)P [21, 22], a phosphoinositide found in high concentrations on early endosomes and on phagosomes. Introduction of the FcγIIa receptor into COS^{phox} cells enabled phagocytosis of IgG-opsonized beads. However, NADPH oxidase activity on phagosomes was absent unless $p40^{phox}$ with a functional PI(3)P binding domain was co-expressed [8]. This finding was corroborated by studies in mice genetically engineered to lack $p40^{phox}$ or a form of $p40^{phox}$ mutated in the PI(3)P binding domain [23, 24]. The $p40^{phox}$ subunit thus plays a selective role in promoting NADPH oxidase activation on phagosomes via a PI(3)P signal. A variant form of CGD involving mutations in $p40^{phox}$ was since identified in humans, associated with inflammatory disease and mild infections [25, 26].

The $p47^{phox}$ subunit also has a PX domain, which has affinity for several other membrane phospholipids, which do not include PI(3)P (*see* [27]). Studies in $K562^{phox}$ cells established that the $p47^{phox}$ PX domain plays an important role in upregulating oxidase activity on plasma membranes [27]. Interestingly, a R90H missense variant in $p47^{phox}$, which disables $p47^{phox}$ binding to phospholipids, was recently shown to be strongly associated with multiple autoimmune diseases, including systemic lupus erythematosus [28, 29].

With the advent of large-scale genomic sequencing, NOX homologues to the NOX2 flavocytochrome and regulatory subunits NOXO1 and NOXA1 with homology to $p47^{phox}$ and $p67^{phox}$, respectively, were identified by homology searching. However, little was known about their biochemical activities or the identity and role in their endogenous location. Heterologous cell assays played a crucial role in the functional analysis of these "new kids on the block." These include studies on NOX1 [4, 5, 30, 31], NOX3 [32, 33], and NOX4 [34, 35]. In general, NOX1 requires not only heterodimer formation with $p22^{phox}$ but multiple interactions with cytosolic regulatory subunits and activated Rac, similar to NOX2. There is some cross-utilization of $p47^{phox}$ and NOXO1 by NOX1/$p22^{phox}$ in these heterologous systems and also possibly in cells that express these subunits endogenously, including vascular smooth muscle cells [36]. In contrast, NOX4-$p22^{phox}$ has constitutive activity, whereas NOX3 has both constitutive and regulated activity in the presence of cytosolic regulatory subunits and Rac.

5.3 Regulation of NOX NADPH Oxidase by Rac and Other Cellular Proteins

GTP-bound Rac is an essential component of the NADPH oxidase complex and binds to a target region in the N-terminus of $p67^{phox}$ to activate NOX electron transferase activity. The importance of Rac-GTP as an essential cofactor for the NOX2 NADPH oxidase was first shown in in vitro assays using proteins purified from neutrophils and macrophages [37, 38]. However, studies in heterologous cells were important for establishing this role in whole cells and dissecting the relevant functional domains in Rac and p67*phox*, e.g. [19, 31, 39]. Most recently, this approach was used to show that arachidonic acid induces a direct interaction of p67*phox*-Rac-GTP with the cytosolic C-terminus of NOX2, which is essential for activating superoxide production [10]. Thus, this interaction may act as a "third switch" to activate the oxidase, in addition to phosphorylation of p47*phox* that unmasks its p22*phox* binding domain and GDP to GTP exchange on Rac to activate p67*phox*.

The importance of the VAV1 guanine nucleotide exchange factor for GDP-GTP exchange on Rac in the context of the NOX2 oxidase also emerged from studies in heterologous whole cell models. Transient overexpression of constitutively active VAV1 enabled oxidase activation in COS*phox* cells in the absence of PMA. Moreover, constitutively active VAV1 was much more effective than constitutively active VAV2 or TIAM1 exchange factors, despite the even higher cellular levels of Rac-GTP with the latter two GEFs [40]. Direct binding of the p67*phox* subunit to VAV1 was next shown in studies in COS-7 and in myeloid cells. This interaction promoted GEF activity and a feed-forward loop for Rac-GTP-p67*phox* activation of the oxidase [41]. Subsequent genetic studies in mice established the importance of Vav1/Vav3 as GEF downstream of immunoreceptors, such as the Fcγ receptor, for oxidase activation in myeloid cells [42]. The region of p67*phox* that interacts with VAV1 was recently shown to be the site of a common missense variant, H389Q, that is associated with systemic lupus erythematosus [43]. When expressed in K562 cells, p67*phox* H389Q supported normal PMA-activated NOX2 oxidase activity, a stimulus for which Rac activation does not require VAV1. However, using IgG-beads to activate the oxidase, where Rac activation is mediated by VAV1, superoxide production was reduced by ≈50%. This result is consistent with the importance of an intact p67*phox* H389Q –VAV1 interaction for optimal oxidase activation.

The COS*phox* system was also used to interrogate signaling components important for activating the NOX2 NADPH oxidase downstream of the formyl peptide receptor (FPR), which detects the formylated peptide fMLF. PKC-delta was important for oxidase activation through FPR, but other PKC isoforms were not necessary [11]. Further mechanistic studies in the FPR-COS*phox* cells showed that fMLF induced phosphorylation of the D-loop of PKC-delta, which in turn was essential for PKC-delta to catalyze p47*phox* phosphorylation [44].

There has long been indirect evidence that association of *phox* regulatory subunits with the actin cytoskeleton is important for their translocation from the cytosol to NOX2/p22*phox* on neutrophil membranes [45]. Studies in heterologous cells have also supported this notion. These include a study in COS*phox* cells using a mutant form of $p40^{phox}$ that led to inappropriate cytoskeletal retention of cytosolic *phox* proteins [46]. Interestingly, angiotensin II-activation of the NOX1 oxidase in vascular smooth muscle cells via $p47^{phox}$ was found to be dependent on the ezrin-radixin-moesin-binding phosphoprotein 50 (EBP50) in mice with a genetic deletion of EBP50 [47]. This finding was corroborated in studies using COS-7-based assays. A bacterial effector protein conferring protection against NOX oxidase-produced ROS also appears to target the cytoskeleton, as described below.

5.4 Other Studies Employing Heterologous Whole Cells Expressing NOX NADPH Oxidase

Heterologous whole cell models expressing the NOX2 NADPH oxidase have contributed other insights into the regulation of NAPDH oxidase and its cellular targets. Some examples are summarized here.

The NADPH oxidase is electrogenic because it catalyzes electron transfer across the membrane and requires a compensating charge for optimal activity [1]. In the absence of a compensating charge, membrane depolarization inhibits NADPH oxidase activity. A study using COS^{phox} cells used an indirect but convincing approach to show that neutrophil proton (H^+) channels are responsible for charge compensation during neutrophil NAPDH oxidase activation [48]. Voltage-gated H^+ channels are inhibited by zinc, and COS-7 cells are one of the few cells known not to express H^+ channels. Superoxide production by COS^{phox} cells was insensitive to zinc inhibition, in contrast to neutrophils, and COS^{phox} cells must compensate charge by a different channel. Subsequently, the *HCVN1* gene encoding the voltage-gated H^+ channel was identified, and neutrophils from Hcvn1-deficient mice have markedly reduced NOX2 NADPH oxidase activity (reviewed in [1]).

Recruitment of the autophagy complex-associated protein LC3 to phagosome membranes, sometimes referred to as "LC3-associated phagocytosis" or LAP, was shown to require oxidants generated by the NOX2 NADPH oxidase in myeloid cells [49]. This finding was corroborated using COS^{phox} cells co-expressing the Fcγ receptor. Recruitment of LC3 to IgG-bead phagosomes occurred only if $p40^{phox}$ was also co-expressed in order to activate NADPH oxidase on phagosomes. This result highlights the importance of localized oxidant production by the NOX2 oxidase for targeting autophagy complex proteins to phagosomes. The presence of LC3 on phagosomes appears important for promoting phagosome membrane remodeling and acidification by recruitment of vacuolar-type H+ ATPases [50, 51].

The gastrointestinal pathogen *Vibrio parahaemolyticus*, the leading cause of food poisoning due to ingestion of raw seafood, is sensitive to epithelial cell production of NOX1-generated ROS, which induces bacterial filamentation and reduces its growth. The VopL protein produced by the *V. parahaemolyticus* type III secretion system counteracts this effect of ROS by inducing the assembly of actin into nonfunctional filaments, as discerned in studies using intestinal Caco-2 epithelial cells [12]. As expression of endogenous NADPH oxidase components and ROS production in Caco-2 cells is very low, the investigators turned to COS^{phox} cells to examine a link between this VopL effect and the NOX oxidase. Oxidant production was easily measured in COS^{phox} cells, and they could also track oxidase assembly by immunofluorescent staining for p67^{phox} recruitment to membranes. The results showed that VopL-induced rearrangement of actin reduced ROS and prevented NADPH oxidase assembly induced either by PMA or by *V. parahaemolyticus* [12]. This finding not only established the mechanism by which VopL allows *V. parahaemolyticus* to evade ROS but also provided additional evidence for the importance of an intact actin cytoskeleton for NADPH oxidase assembly.

6 Final Remarks

This brief review highlights how heterologous cell models can be used to examine many different facets of NOX NADPH oxidases. While valuable, one caveat is that oxidase subunits can be expressed at levels higher than endogenous and/or out of "balance" with each other, which could potentially confound interpretation of studies. For example, functional defects in hypomorphic variants could be masked by overexpression. Findings in these model systems should ideally also be confirmed in the corresponding native cell background. For the NOX2 oxidase, for example, approaches include myeloid lines that can be differentiated into neutrophil- or monocyte-like cells as well as primary cells. In order to facilitate analysis of mutant subunits in a "null" background, cell lines and primary cells can be genetically engineered to delete expression of endogenous NOX oxidase subunits by homologous recombination, RNA interference. or CRISPR/Cas9 technology.

Acknowledgments

MCD is funded by the Children's Discovery Institute at Washington University and St. Louis Children's Hospital and NIH grants R01HL045635 and R01AR072212. I thank Tina McGrath and Diane Jensen for assistance with preparation of the manuscript. I apologize to colleagues whose work could not be cited due to space limitations.

References

1. Nunes P, Demaurex N, Dinauer MC (2013) Regulation of the NADPH oxidase and associated ion fluxes during phagocytosis. Traffic 14(11):1118–1131. https://doi.org/10.1111/tra.12115
2. Leto TL, Lavigne MC, Homoyounpour N, Lekstrom K, Linton G, Malech HL, de Mendez I (2007) The K-562 cell model for analysis of neutrophil NADPH oxidase function. In: Quinn MT, DeLeo FR, Bokoch GM (eds) Neutrophil methods and protocols. Humana Press, Totowa, NJ, pp 365–383. https://doi.org/10.1007/978-1-59745-467-4_24
3. Miyano K, Sumimoto H (2012) Assessment of the role for rho family GTPases in NADPH oxidase activation. In: Rivero F (ed) Rho GTPases: methods and protocols. Springer New York, New York, NY, pp 195–212. https://doi.org/10.1007/978-1-61779-442-1_14
4. Banfi B, Clark RA, Steger K, Krause KH (2003) Two novel proteins activate superoxide generation by the NADPH oxidase NOX1. J Biol Chem 278(6):3510–3513
5. Takeya R, Ueno N, Kami K, Taura M, Kohjima M, Izaki T, Nunoi H, Sumimoto H (2003) Novel human homologues of p47phox and p67phox participate in activation of superoxide-producing NADPH oxidases. J Biol Chem 278(27):25234–25246
6. Yu L, Zhen L, Dinauer MC (1997) Biosynthesis of the phagocyte NADPH oxidase cytochrome b558. Role of heme incorporation and heterodimer formation in maturation and stability of gp91phox and p22phox subunits. J Biol Chem 272(43):27288–27294
7. von Lohneysen K, Noack D, Jesaitis AJ, Dinauer MC, Knaus UG (2008) Mutational analysis reveals distinct features of the Nox4-p22 phox complex. J Biol Chem 283 (50):35273–35282. https://doi.org/10.1074/jbc.M804200200
8. Suh CI, Stull ND, Li XJ, Tian W, Price MO, Grinstein S, Yaffe MB, Atkinson S, Dinauer MC (2006) The phosphoinositide-binding protein p40phox activates the NADPH oxidase during FcgammaIIA receptor-induced phagocytosis. J Exp Med 203(8):1915–1925. https://doi.org/10.1084/jem.20052085
9. Casbon AJ, Allen LA, Dunn KW, Dinauer MC (2009) Macrophage NADPH oxidase flavocytochrome B localizes to the plasma membrane and Rab11-positive recycling endosomes. J Immunol 182(4):2325–2339. https://doi.org/10.4049/jimmunol.0803476
10. Matono R, Miyano K, Kiyohara T, Sumimoto H (2014) Arachidonic acid induces direct interaction of the p67phox-Rac complex with the phagocyte oxidase Nox2, leading to superoxide production. J Biol Chem 289 (36):24874–24884. https://doi.org/10.1074/jbc.M114.581785
11. He R, Nanamori M, Sang H, Yin H, Dinauer MC, Ye RD (2004) Reconstitution of chemotactic peptide-induced nicotinamide adenine dinucleotide phosphate (reduced) oxidase activation in transgenic COS-phox cells. J Immunol 173(12):7462–7470
12. de Souza SM, Salomon D, Orth K (2017) T3SS effector VopL inhibits the host ROS response, promoting the intracellular survival of *Vibrio parahaemolyticus*. PLoS Pathog 13 (6):e1006438. https://doi.org/10.1371/journal.ppat.1006438
13. Biberstine-Kinkade KJ, DeLeo FR, Epstein RI, LeRoy BA, Nauseef WM, Dinauer MC (2001) Heme-ligating histidines in flavocytochrome b (558): identification of specific histidines in gp91(phox). J Biol Chem 276 (33):31105–31112. https://doi.org/10.1074/jbc.M103327200
14. Yu L, Quinn MT, Cross AR, Dinauer MC (1998) Gp91(phox) is the heme binding subunit of the superoxide-generating NADPH oxidase. Proc Natl Acad Sci U S A 95 (14):7993–7998
15. Biberstine-Kinkade KJ, Yu L, Stull N, LeRoy BA, Bennett S, Cross AR, Dinauer M (2002) Mutagenesis of p22phox Histidine 94. J Biol Chem 277(33):30368–30374
16. von Löhneysen K, Noack D, Wood MR, Friedman JS, Knaus UG (2010) Structural insights into Nox4 and Nox2: motifs involved in function and cellular localization. Mol Cell Biol 30 (4):961–975. https://doi.org/10.1128/mcb.01393-09
17. O'Neill S, Mathis M, Kovačič L, Zhang S, Reinhardt J, Scholz D, Schopfer U, Bouhelal R, Knaus UG (2018) Quantitative interaction analysis permits molecular insights into functional NOX4 NADPH oxidase heterodimer assembly. J Biol Chem 293 (23):8750–8760. https://doi.org/10.1074/jbc.RA117.001045
18. Leto TL, Adams AG, de Mendez I (1994) Assembly of the phagocyte NADPH oxidase: binding of Src homology 3 domains to proline-rich targets. Proc Natl Acad Sci U S A 91 (22):10650–10654

19. Koga H, Terasawa H, Nunoi H, Takeshige K, Inagaki F, Sumimoto H (1999) Tetratricopeptide repeat (TPR) motifs of p67(phox) participate in interaction with the small GTPase Rac and activation of the phagocyte NADPH oxidase. J Biol Chem 274(35):25051–25060
20. Ago T, Nunoi H, Ito T, Sumimoto H (1999) Mechanism for phosphorylation-induced activation of the phagocyte NADPH oxidase protein p47(phox). Triple replacement of serines 303, 304, and 328 with aspartates disrupts the SH3 domain-mediated intramolecular interaction in p47(phox), thereby activating the oxidase. J Biol Chem 274(47):33644–33653
21. Kanai F, Liu H, Field SJ, Akbary H, Matsuo T, Brown GE, Cantley LC, Yaffe MB (2001) The PX domains of p47phox and p40phox bind to lipid products of PI(3)K. Nat Cell Biol 3 (7):675–678
22. Ellson CD, Gobert-Gosse S, Anderson KE, Davidson K, Erdjument-Bromage H, Tempst P, Thuring JW, Cooper MA, Lim ZY, Holmes AB, Gaffney PR, Coadwell J, Chilvers ER, Hawkins PT, Stephens LR (2001) PtdIns (3)P regulates the neutrophil oxidase complex by binding to the PX domain of p40(phox). Nat Cell Biol 3(7):679–682
23. Ellson CD, Davidson K, Ferguson GJ, O'Connor R, Stephens LR, Hawkins PT (2006) Neutrophils from p40phox−/− mice exhibit severe defects in NADPH oxidase regulation and oxidant-dependent bacterial killing. J Exp Med 203(8):1927–1937
24. Ellson C, Davidson K, Anderson K, Stephens LR, Hawkins PT (2006) PtdIns3P binding to the PX domain of p40phox is a physiological signal in NADPH oxidase activation. EMBO J 25(19):4468–4478. https://doi.org/10.1038/sj.emboj.7601346
25. Matute JD, Arias AA, Wright NA, Wrobel I, Waterhouse CC, Li XJ, Marchal CC, Stull ND, Lewis DB, Steele M, Kellner JD, Yu W, Meroueh SO, Nauseef WM, Dinauer MC (2009) A new genetic subgroup of chronic granulomatous disease with autosomal recessive mutations in p40 phox and selective defects in neutrophil NADPH oxidase activity. Blood 114(15):3309–3315. https://doi.org/10.1182/blood-2009-07-231498
26. van de Geer A, Nieto-Patlán A, Kuhns DB, Tool ATJ, Arias AA, Bouaziz M, de Boer M, Franco JL, Gazendam RP, van Hamme JL, van Houdt M, van Leeuwen K, Verkuijlen PJH, van den Berg TK, Alzate JF, Arango-Franco CA, Batura V, Bernasconi AR, Boardman B, Booth C, Burns SO, Cabarcas F, Cerf Bensussan N, Charbit-Henrion F, Corveleyn A, Deswarte C, Esnaola Azcoiti M, Foell D, Gallin JI, Garcés C, Guedes M, Hinze CH, Holland SM, Hughes SM, Ibañez P, Malech HL, Meyts I, Moncada-Velez M, Moriya K, Neves E, Oleastro M, Perez L, Rattina V, Oleaga-Quintas C, Warner N, Muise AM, Serafin López J, Trindade E, Vasconselos J, Vermeire S, Wittkowski H, Worth A, Abel L, Dinauer MC, Arkwright PD, Roos D, Casanova J-L, Kuijpers TW, Bustamante J (2018) Inherited p40phox deficiency differs from classic chronic granulomatous disease. J Clin Invest 128(9):3957–3975. https://doi.org/10.1172/JCI97116. [Epub ahead of print]
27. Li XJ, Marchal CC, Stull ND, Stahelin RV, Dinauer MC (2010) p47phox Phox homology domain regulates plasma membrane but not phagosome neutrophil NADPH oxidase activation. J Biol Chem 285(45):35169–35179. https://doi.org/10.1074/jbc.M110.164475
28. Zhao J, Ma J, Deng Y, Kelly JA, Kim K, Bang SY, Lee HS, Li QZ, Wakeland EK, Qiu R, Liu M, Guo J, Li Z, Tan W, Rasmussen A, Lessard CJ, Sivils KL, Hahn BH, Grossman JM, Kamen DL, Gilkeson GS, Bae SC, Gaffney PM, Shen N, Tsao BP (2017) A missense variant in NCF1 is associated with susceptibility to multiple autoimmune diseases. Nat Genet 49 (3):433–437. https://doi.org/10.1038/ng.3782
29. Olsson LM, Johansson AC, Gullstrand B, Jonsen A, Saevarsdottir S, Ronnblom L, Leonard D, Wettero J, Sjowall C, Svenungsson E, Gunnarsson I, Bengtsson AA, Holmdahl R (2017) A single nucleotide polymorphism in the NCF1 gene leading to reduced oxidative burst is associated with systemic lupus erythematosus. Ann Rheum Dis 76 (9):1607–1613. https://doi.org/10.1136/annrheumdis-2017-211287
30. Geiszt M, Lekstrom K, Witta J, Leto TL (2003) Proteins homologous to p47phox and p67phox support superoxide production by NAD(P)H oxidase 1 in colon epithelial cells. J Biol Chem 278(22):20006–20012. https://doi.org/10.1074/jbc.M301289200
31. Miyano K, Ueno N, Takeya R, Sumimoto H (2006) Direct involvement of the small GTPase Rac in activation of the superoxide-producing NADPH oxidase Nox1. J Biol Chem 281(31):21857–21868. https://doi.org/10.1074/jbc.M513665200
32. Cheng G, Ritsick D, Lambeth JD (2004) Nox3 regulation by NOXO1, p47phox, and p67phox. J Biol Chem 279 (33):34250–34255. https://doi.org/10.1074/jbc.M400660200
33. Ueno N, Takeya R, Miyano K, Kikuchi H, Sumimoto H (2005) The NADPH oxidase

Nox3 constitutively produces superoxide in a p22phox-dependent manner: its regulation by oxidase organizers and activators. J Biol Chem 280(24):23328–23339

34. Martyn KD, Frederick LM, von Loehneysen K, Dinauer MC, Knaus UG (2006) Functional analysis of Nox4 reveals unique characteristics compared to other NADPH oxidases. Cell Signal 18(1):69–82. https://doi.org/10.1016/j.cellsig.2005.03.023

35. von Löhneysen K, Noack D, Hayes P, Friedman JS, Knaus UG (2012) Constitutive NADPH oxidase 4 activity resides in the composition of the B-loop and the penultimate C terminus. J Biol Chem 287(12):8737–8745. https://doi.org/10.1074/jbc.M111.332494

36. Al Ghouleh I, Khoo NK, Knaus UG, Griendling KK, Touyz RM, Thannickal VJ, Barchowsky A, Nauseef WM, Kelley EE, Bauer PM, Darley-Usmar V, Shiva S, Cifuentes-Pagano E, Freeman BA, Gladwin MT, Pagano PJ (2011) Oxidases and peroxidases in cardiovascular and lung disease: new concepts in reactive oxygen species signaling. Free Radic Biol Med 51(7):1271–1288. https://doi.org/10.1016/j.freeradbiomed.2011.06.011

37. Abo A, Pick E, Hall A, Totty N, Teahan CG, Segal AW (1991) Activation of the NADPH oxidase involves the small GTP-binding protein p21rac1. Nature 353(6345):668–670

38. Knaus UG, Heyworth PG, Evans T, Curnutte JT, Bokoch GM (1991) Regulation of phagocyte oxygen radical production by the GTP-binding protein Rac 2. Science 254(5037):1512–1515

39. Cheng G, Diebold BA, Hughes Y, Lambeth JD (2006) Nox1-dependent reactive oxygen generation is regulated by Rac1. J Biol Chem 281(26):17718–17726

40. Price MO, Atkinson SJ, Knaus UG, Dinauer MC (2002) Rac activation induces NADPH oxidase activity in transgenic COSphox cells, and the level of superoxide production is exchange factor-dependent. J Biol Chem 277(21):19220–19228. https://doi.org/10.1074/jbc.M200061200

41. Ming W, Li S, Billadeau DD, Quilliam LA, Dinauer MC (2007) The Rac effector p67phox regulates phagocyte NADPH oxidase by stimulating Vav1 guanine nucleotide exchange activity. Mol Cell Biol 27(1):312–323. https://doi.org/10.1128/MCB.00985-06

42. Graham DB, Stephenson LM, Lam SK, Brim K, Lee HM, Bautista J, Gilfillan S, Akilesh S, Fujikawa K, Swat W (2007) An ITAM-signaling pathway controls cross-presentation of particulate but not soluble antigens in dendritic cells. J Exp Med 204(12):2889–2897. https://doi.org/10.1084/jem.20071283

43. Jacob CO, Eisenstein M, Dinauer MC, Ming W, Liu Q, John S, Quismorio FP Jr, Reiff A, Myones BL, Kaufman KM, McCurdy D, Harley JB, Silverman E, Kimberly RP, Vyse TJ, Gaffney PM, Moser KL, Klein-Gitelman M, Wagner-Weiner L, Langefeld CD, Armstrong DL, Zidovetzki R (2012) Lupus-associated causal mutation in neutrophil cytosolic factor 2 (NCF2) brings unique insights to the structure and function of NADPH oxidase. Proc Natl Acad Sci U S A 109(2):E59–E67. https://doi.org/10.1073/pnas.1113251108

44. Cheng N, He R, Tian J, Dinauer MC, Ye RD (2007) A critical role of protein kinase C delta activation loop phosphorylation in formyl-methionyl-leucyl-phenylalanine-induced phosphorylation of p47(phox) and rapid activation of nicotinamide adenine dinucleotide phosphate oxidase. J Immunol 179(11):7720–7728

45. Nauseef WM (2004) Assembly of the phagocyte NADPH oxidase. Histochem Cell Biol 122(4):277–291

46. Chen J, He R, Minshall RD, Dinauer MC, Ye RD (2007) Characterization of a mutation in the Phox homology domain of the NADPH oxidase component p40phox identifies a mechanism for negative regulation of superoxide production. J Biol Chem 282(41):30273–30284. https://doi.org/10.1074/jbc.M704416200

47. Al Ghouleh I, Meijles DN, Mutchler S, Zhang Q, Sahoo S, Gorelova A, Henrich Amaral J, Rodríguez AI, Mamonova T, Song GJ, Bisello A, Friedman PA, Cifuentes-Pagano ME, Pagano PJ (2016) Binding of EBP50 to Nox organizing subunit p47phox is pivotal to cellular reactive species generation and altered vascular phenotype. Proc Natl Acad Sci U S A 113(36):E5308–E5317. https://doi.org/10.1073/pnas.1514161113

48. DeCoursey TE, Morgan D, Cherny VV (2003) The voltage dependence of NADPH oxidase reveals why phagocytes need proton channels. Nature 422(6931):531–534

49. Huang J, Canadien V, Lam GY, Steinberg BE, Dinauer MC, Magalhaes MA, Glogauer M, Grinstein S, Brumell JH (2009) Activation of antibacterial autophagy by NADPH oxidases. Proc Natl Acad Sci U S A 106(15):6226–6231. https://doi.org/10.1073/pnas.0811045106

50. Martinez J, Almendinger J, Oberst A, Ness R, Dillon CP, Fitzgerald P, Hengartner MO, Green DR (2011) Microtubule-associated

protein 1 light chain 3 alpha (LC3)-associated phagocytosis is required for the efficient clearance of dead cells. Proc Natl Acad Sci U S A 108(42):17396–17401. https://doi.org/10.1073/pnas.1113421108

51. Bagaitkar J, Huang J, Zeng MY, Pech NK, Monlish DA, Perez-Zapata LJ, Miralda I, Schuettpelz LG, Dinauer MC (2018) NADPH oxidase activation regulates apoptotic neutrophil clearance by murine macrophages. Blood 131(21):2367–2378. https://doi.org/10.1182/blood-2017-09-809004

Chapter 10

The X-CGD PLB-985 Cell Model for NOX2 Structure-Function Analysis

Sylvain Beaumel and Marie José Stasia

Abstract

Structure-function analysis of specific regions of NOX2 can be carried out after stable expression of site-directed mutagenesis-modified NOX2 in the X^0-CGD PLB-985 cell model. Indeed, the generation of this human cellular model by Prof. MC Dinauer's team gave researchers the opportunity to gain a deeper understanding of functional regions of NOX2. With this model cell line, the functional impact of X^+-CGD or of new mutations in NOX2 can be highlighted, as the biological material is not limited. PLB-985 cells transfected with various NOX2 mutations can be easily cultured and differentiated into neutrophils or monocytes/macrophages. Several measurements in intact mutated NOX2 PLB-985 cells can be carried out such as NOX2 expression, cytochrome b_{558} spectrum, enzymatic activity, and assembly of the NADPH oxidase complex. Purified membranes or purified cytochrome b_{558} from mutated NOX2 PLB-985 cells can be used for the study of the impact of specific mutations on NADPH oxidase or diaphorase activity, FAD incorporation, and NADPH or NADH binding in a cell-free assay system. Here, we describe a method to generate mutated NOX2 PLB-985 cells in order to analyze NOX2 structure-function relationships.

Key words PLB-985 cell line, Chronic granulomatous disease, NOX2 deficiency, NADPH oxidase activity, Cellular model, Structure-function analysis

1 Introduction

NOX2 (also called gp91^{phox}) is one of the subunits of the NADPH oxidase complex expressed mainly in phagocytic cells but also in endothelial cells, cardiomyocytes, platelets, neural cells, and particularly microglia [1]. NOX2 is the prototype of all NOX/DUOX family members and the most studied oxidase so far. NOX2 is a membrane protein composed of a transmembrane domain (TM) and a C-terminal cytosolic dehydrogenase domain (DH) as present in all NOXs. DH contains the binding sites for NADPH and FAD, and the TM is composed of six transmembrane helical domains; two intracytosolic loops, B and D; and three extracellular loops, A, C, and E. NOX1–4 forms a stable complex with membrane-bound p22^{phox}, and the complex NOX2-p22^{phox}

Ulla G. Knaus and Thomas L. Leto (eds.), *NADPH Oxidases: Methods and Protocols*, Methods in Molecular Biology, vol. 1982, https://doi.org/10.1007/978-1-4939-9424-3_10,

constitutes the cytochrome b_{558}. Like NOX1 and NOX3, NOX2 requires additional regulatory proteins termed p67*phox* and p47*phox* in an assembled complex to be activated, whereas NOX4 is constitutively active. NOX5 as well as DUOX1/2 possess additional calcium-binding sites compared to NOX1-4.

The importance of NOX2 for health is illustrated by its absence/defect in X-linked chronic granulomatous disease (X-CGD) (70% of cases), a rare inherited syndrome characterized by severe and recurrent infections in childhood [2]. X^+-CGD cases are extremely rare (<5% of cases), often due to missense mutations in the NOX2 sequence that are characterized by a normal expression of NOX2 in the absence of any catalytic activity in neutrophils. These mutations have enhanced our knowledge of functional regions in NOX2 [3]. An elegant way to gain a deeper understanding of the mechanism contributing to dysfunction of these mutants is to model X^+-mutations in the X-CGD PLB-985 cell line. The human PLB-985 cell line was established in 1987 from the peripheral blood of a patient with acute nonlymphocytic leukemia [4]. A few years later, a knockout for *CYBB* encoding NOX2 was developed in this cell line to produce an in vitro model of X-CGD. Superoxide formation in X-CGD PLB-985 cells after differentiation of granulocytes was absent but could be rescued by stable transfection and expression of wild-type NOX2 cDNA [5].

The first functional study of an X^+-mutation in the X-CGD PLB-985 cell model was an amino acid 488 to 497 deletion in the DH domain, which disrupts electron transfer to FAD [6]. Subsequently, several regions important for NOX2 function such as the DH domain [7–10], the TM domain [11], the D-loop [8, 12], the C terminus, and especially a small insertion sequence (NIS) [8, 13, 14] were identified by using this cellular model. Recently, due to obtaining the crystal structure of the DH and the TM domain of *Cylindrospermum stagnale* NOX5 (csNOX5), the involvement of both domains regulating the electron transfer from NADPH to molecular oxygen was clarified [15]. These studies confirmed the importance of certain NOX regions and amino acids in the electron transfer process as previously revealed by mutagenesis studies in X-CGD PLB-985 cells and HEK293 cells. Examples are the C-terminal residue (F693 in csNOX5 and F570 in NOX2) [10, 13], charged amino acids in the D-loop (R360 and K361 in csNOX5 and R198 and R199 in NOX2) [8, 12], the B-loop interfacing with the DH domain [16, 17], and the central role of R256 in csNOX5 or R54 in NOX2 [11]. Studies delineating the functional impact of NOX2 mutations expressed in the X-CGD PLB-985 cells were reinforced by three-dimensional models of NOX2 built from analogy with the ferredoxin-$NADP^+$ reductase (FNR), an FAD-containing enzyme with homology to the NOX DH domain [10, 14, 18]. A recent article revealed the mechanistic details of the NADPH to FAD hydride transfer step through high-resolution studies of the corn root FNR [19]. Sequence alignment of the corn root FNR with functional residues in NOX pointed out T341, G408, C537, and E568 as essential residues. When mutated in NOX2, these residues lead to X^+CGD cases that involved the electron transfer from NADPH to FAD, as demonstrated in the PLB-985 NOX2 KO cell model [10]. The X-CGD PLB-985 cell model also permits determining the impact of each single mutation when a double mutation is observed in X^+CGD, thereby confirming or refuting a new polymorphism in the *CYBB* gene [7]. In addition, X^--CGD cases can be useful when reproduced in the X-CGD PLB-985 cell model in order to identify residues involved in the stable expression of NOX2 protein [20]. With CRISPR-Cas9 technology, it is now feasible to generate any knockout or mutation in the PLB-985 cell line in order to carry out structure-function studies of any NADPH oxidase subunits. This technology is also promising for the development of new vectors for CGD gene therapy [21, 22].

Generation of X-CGD PLB-985 cells expressing mutated NOX2 consists of three principal steps (Fig. 1): (1) directed mutagenesis in NOX2 cDNA cloned in pBluescriptKS+, (2) mutated NOX2 cDNA subcloned into the plasmid vector pEF-PGKneo, and (3) nucleofection and selection of mutated NOX2 X-CGD PLB-985 cell clones by limited dilution and antibiotic selection. Two to three clones for each studied mutation are isolated and stored in nitrogen. These X-CGD PLB-985 cells harboring mutated NOX2 can be used to carry out functional and biochemical experiments to decipher the role of specific domains in NOX2 without any limitation in biological material.

2 Materials

Prepare all solutions and media using deionized water unless specified otherwise.

2.1 Solutions for Molecular Biology

1. Several plasmid DNA Miniprep solutions are prepared in sterilized deionized water. Solution I for suspension of bacteria: 1 M (w/v) Tris HCl pH 8, 250 mM Ethylene Diamine Tetraacetic Acid (EDTA), 50 mM (w/v) Glucose, autoclaved, keep at 4 °C. Solution II is 0.2 N NaOH, 1% SDS for alkaline lysis prepared extemporaneously and kept at room temperature. Solution III is 3 M (w/v) potassium acetate, 11.5% (v/v) acetic acid glacial, keep at 4 °C.
2. Tris-EDTA (TE) Buffer composed of 10 mM (w/v) Tris HCl, 1 mM EDTA, pH 8, keep at 4 °C.
3. JM109 competent cells (Promega™).
4. Bacterial culture medium: Lysogeny broth (LB) medium sterilized—10 g/L tryptone, 5 g/L yeast extract, 10 g/L NaCl, pH 7.0. LB can also contain 15 g/L agar and 100 μg/mL ampicillin.

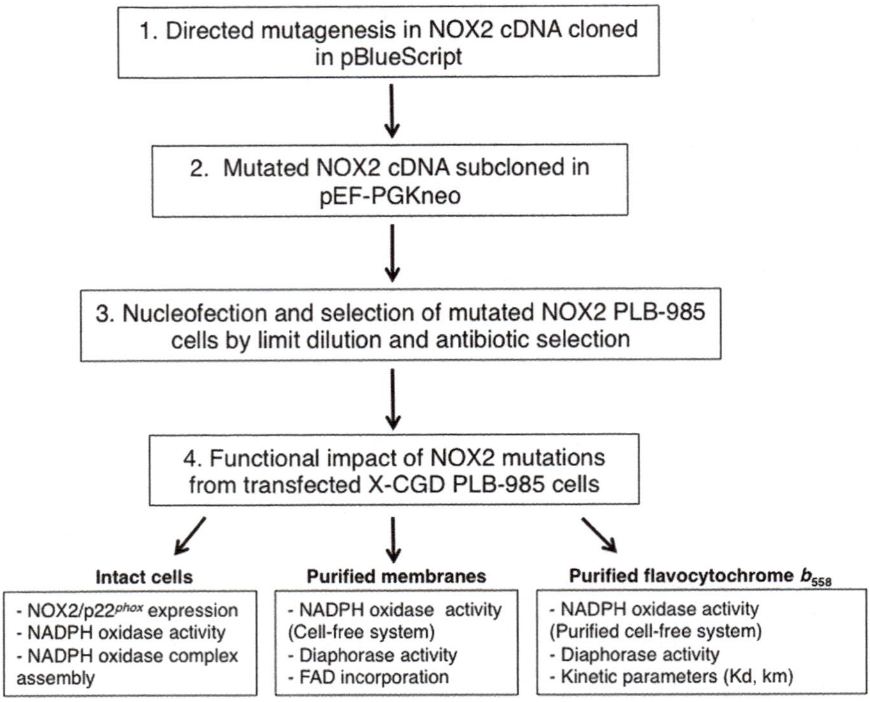

Fig. 1 Schematic representation of the method to generate mutated NOX2 PLB-985 cells in order to analyze NOX2 structure-function relationship

5. 0.8% or 1% (w /v) agarose gel.
6. BamH1 restriction enzyme.
7. Reagents to precipitate DNA: (1) Sodium acetate 3 M pH 5.2; (2) 100% and 70% (v/v) ethanol; and (3) 1 M $CaCl_2$—dissolve 7.35 g of $CaCl_2$ in 50 mL of deionized water, sterilize by filtration with a 0.22 μm membrane filter, and store at room temperature for 6 months. This stock solution must be diluted 1:20 extemporaneously with deionized water and sterilized by filtration with a 0.22 μm membrane filter before use.

2.2 Directed Mutagenesis of NOX2 cDNA Cloned into pBluescript

1. QuikChange II—Site-directed mutagenesis kit contains QuikChange II buffer, dNTP mixt, pfu Turbo DNA polymerase, and DPN1 (Agilent™).
2. NOX2 cDNA cloned in pBluescriptKS+, a generous gift from Prof. MC Dinauer, Washington University, USA.
3. Mix for PCR; 1 × Taq buffer, 100 μM dNTP, 1 μM β1 forward primer (5′cctctgccaccatggggaact 3′) and NOX2-End reverse primer (3′ggaagagacaagttagaagttttcc 5′) (*see* **Note 1**), 1,25 U pfu Turbo DNA polymerase.
4. 500 μg/mL RNase (DNase-free) from bovine pancreas.
5. Quick Gel Extraction Kit to purify DNA band from agarose gel.

2.3 NOX2 cDNA Subcloned in pEF-PGKneo

1. pEF-PGKneo plasmid, a generous gift from Prof. MC Dinauer, Washington University (*see* **Note 2**).
2. Shrimp alkaline phosphatase.
3. T4 DNA ligase.
4. EcoR1 restriction enzyme.

2.4 Nucleofection and Selection of NOX2 X-CGD PLB-985 Cells by Limited Dilution and Antibiotic Selection

1. Cell culture medium: RPMI 1640 medium is supplemented with 10 or 20% (v/v) heat-inactivated fetal bovine serum, 100 units/mL penicillin, 100 μg/mL streptomycin, and 2 mM L-glutamine with or without 1.5 mg/mL geneticin G418 (*see* **Note 3**).
2. Endo-free Plasmid Maxi Kit (Qiagen™).
3. Isopropanol.
4. Amaxa Nucleofector and Cell line Nucleofector V (Lonza™).

2.5 Analysis of Structural and Functional Impact of NOX2 Mutations

2.5.1 NOX2 Expression by Flow Cytometry

1. Phosphate-buffered saline (PBS).
2. PBS/BSA buffer: PBS with 0.02% (w/v) bovine serum albumin (BSA).
3. Monoclonal antibody specific for NOX2 (clone 7D5, MBL Med. Biol. Lab.) (*see* **Note 4**).
4. Mouse IgG1, isotype control.
5. Goat anti-mouse secondary antibody directed against F(ab′)2-IgG (H + L) labeled with phycoerythrin (PE).

2.5.2 NADPH Oxidase Activity by Chemiluminescence

1. Dimethylformamide (DMF).
2. PBS^+: PBS with 0.9 mM $CaCl_2$ and 0.5 mM $MgCl_2$ filtered through a 0.22 μm membrane filter.
3. 0.8 M glucose in PBS. Store aliquots at −20 °C.
4. 100 mM luminol in dimethyl sulfoxide (DMSO), keep for 1 week at 4 °C and in the dark. 2 mM luminol (working solution) is prepared extemporaneously by two successive dilutions and protected from light. First 1/10 dilution in PBS^+ (10 μL of luminol 100 mM + 90 μL of PBS^+), vortex, then dilute 1/5 (addition of 400 μL of PBS^+ in the previous dilution), and vortex vigorously for 1 min.
5. 500 U/mL horseradish peroxidase (HRP) in PBS^+, keep for 1 week at 4 °C and in the dark.
6. The chemiluminescence medium (total volume of 1 mL), prepared extemporaneously and protected from light, is composed of 10 μL of 2 mM luminol, 945 μL of PBS with 0.9 mM $CaCl_2$, and 0.5 mM $MgCl_2$ (PBS^+) filtered through a 0.22 μm membrane filter.

7. 2 mg/mL phorbol myristate acetate (PMA) in DMSO, keep for 1 year at −20 °C. This stock solution must be diluted 1:10 in DMSO (10 μL in 90 μL of DMSO) to obtain a second stock solution at 0.2 mg/mL, keep for 1 month at −20 °C. Working solution at 2 μg/mL is prepared extemporaneously by diluting the 0.2 mg/mL stock solution of PMA 1:100 in PBS^+ (10 μL in 990 μL of PBS^+). Vortex vigorously for 1 min just before use.

3 Methods

3.1 Directed Mutagenesis of NOX2 cDNA Cloned into pBluescript

Site-directed mutagenesis can reproduce mutations that cause X-type chronic granulomatous disease (*see* **Note 5**). Wild-type NOX2 cDNA (1.7 kb) cloned via BamH1 restriction sites in the plasmid pBluescriptKS+ is used as template strand [5].

3.1.1 Site-Directed Mutagenesis

First, the desired mutation is introduced by PCR. Hybridization reactions with "forward" and "reverse" primers are performed separately (*see* **Note 6**). One PCR is performed in a mix containing 5 μL of QuikChange buffer, 42.6 μL of dNTP mix, 0.4 μL of 2.5 U/μL PFU turbo, 1 μL of 100 ng/μL plasmid, 1 μL of 10 μM forward primer, and a second PCR with the same medium but containing 1 μL of 10 μM reverse primer. The PCR program comprises different phases: denaturation at 95 °C for 30 s, 3 cycles of amplification/elongation—30 s at 95 °C, 60 s at 55 °C, and 12 min at 68 °C—and a final elongation at 68 °C for 10 min. A second amplification is performed by mixing 25 μL of the previous "forward" reaction medium with 25 μL of the previous "reverse" reaction medium and using the same program as previously described except that 18 cycles are performed instead of 3 cycles. The non-mutated methylated template strands are specifically digested with 10 U of DpnI for 1 h at 37 °C.

3.1.2 Purification of Mutated cDNA

1. Purify mutated DNA by precipitation with 5 μL of 3 M sodium acetate (one-tenth the volume) and 150 μL of 100% ethanol (three volumes).
2. After 1 h at −80 °C, centrifuge the precipitate at 10,000 × *g* for 30 min at 4 °C, and wash the DNA pellet with 100 μL of ice-cold 70% ethanol.
3. After a further centrifugation at 10,000 × *g* for 15 min at 4 °C, the DNA pellet is air dried on the bench for at least 1 h and dissolved in 30 μL of TE buffer. The pellet must become white, and all traces of ethanol must have disappeared before adding TE buffer otherwise it is impossible to dissolve it correctly.
4. Analyze the mutagenesis product on a 1% agarose gel in order to verify that its size is close to 4.6 kb (2.9 kb for the size of the

plasmid and 1.7 kb for the size of the NOX2 cDNA). It is essential to check the mutagenesis product systematically by Sanger sequencing to ensure the presence of the desired mutation and the absence of unwanted mutations. Bacterial transformation is performed with 10 μL of the mutagenesis product.

3.1.3 Bacterial Transformation

1. Inoculate a starter culture of 2 mL LB medium with JM109 competent bacteria. Incubate overnight at 37 °C with vigorous shaking (approx. 300 rpm).
2. Take 50 μL of a 12 h JM109 bacterial culture, and mix with 4 mL of LB medium. Incubate with shaking at 37 °C to obtain a bacterial suspension with an optical density of 0.5 at 600 nm.
3. Centrifuge at 4000 × *g* for 5 min at 4 °C.
4. Discard supernatant and resuspend the pellet with 2 mL of ice-cold 50 mM $CaCl_2$ (half of the initial volume).
5. Incubate on ice for 30 min.
6. Centrifuge the bacteria at 4000 × *g* for 5 min at 4 °C.
7. Remove supernatant and resuspend the pellet in 400 μL of ice-cold 50 mM $CaCl_2$ (1/10 initial volume).
8. For each transformation, transfer 100 μL of the previous bacterial suspension into an ice-cold tube.
9. Add 10 μL of the plasmid (the directed mutagenesis product) (*see* **Note** 7).
10. Incubate on ice for 30 min, then rapidly and precisely at 42 °C for 45 s, and finally incubate on ice for 5 min.
11. Add 450 μL of LB medium preheated to 42 °C.
12. Incubate at 37 °C for 1 h with shaking.
13. Spread 50 μL on LB agar plate containing 100 μg/mL ampicillin.
14. In order to have a second inoculated LB agar plate with a higher bacterial concentration, harvest bacteria by centrifugation at 4000 × *g* for 5 min at 20 °C.
15. Discard 400 μL of supernatant, resuspend the bacterial pellet in remaining supernatant, and spread on LB agar containing 100 μg/mL ampicillin.
16. Incubate dishes overnight at 37 °C.

3.1.4 Verification of Bacterial Transformation

Selection of the positive clones containing the insert of interest is performed by PCR. β1 and NOX2-End primer, respectively, hybridizing 5′ and 3′ of NOX2 cDNA are used to amplify all of NOX2 cDNAs (control or mutated).

1. 8 to 10 of ampicillin-resistant colonies present on the agar plate are suspended in 20 μL of sterile water.

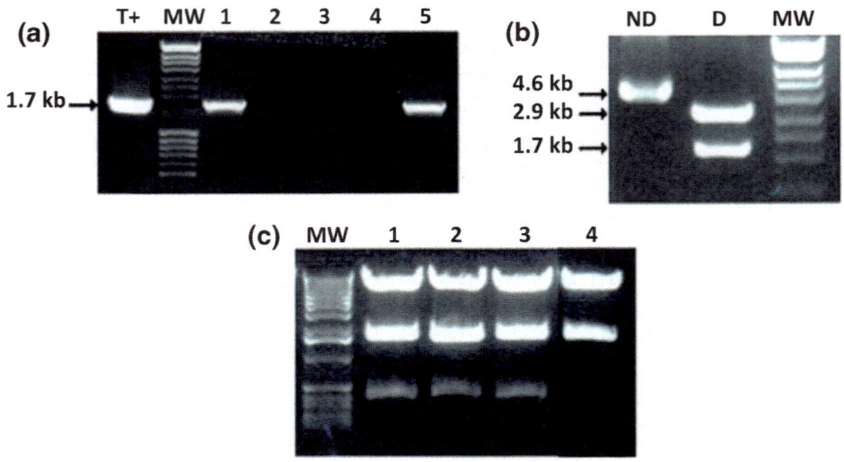

Fig. 2 Cloning of mutated NOX2 cDNA from bacteria to eukaryotic cells. (**a**) Verification of bacterial transformation by the NOX2 cDNA cloned in pBluescriptKS (+). The selection of positive clones of JM109 bacteria containing the insert of interest is performed by PCR as described in Methods. Positive clones reveal a 1.7 kb band corresponding to the NOX2 cDNA (clones 1 and 5) unlike clones 2 to 4. (**b**) Enzymatic digestion by BamH1 of NOX2 cDNA cloned in pBluescriptKS+. BamH1-digested product (D) reveals the presence of two bands at 2.9 kb and 1.7 kb, unlike the undigested control (ND). (**c**) Analyze the sense of cDNA integration in the pEF-PGKneo vector. DNA plasmid digested by EcoR1 enzyme reveals a correct cDNA sense of integration into pEF-PGKneo vector in well 4 (two bands) unlike in wells 1–3 (three bands)

2. 10 μL of this suspension is diluted in 2.5 mL of LB medium, 100 μg/mL ampicillin and incubated overnight with vigorous shaking at 37 °C.
3. The remaining 10 μL undergoes a heat shock (5 min at 95 °C, 2 min at 40 °C, 5 min at 95 °C) to release their intracellular contents.
4. PCR mix (1× Taq buffer, 100 μM dNTP, 1 μM β1 and NOX2-End primer, 1.25 U Taq polymerase) is added to 10 μL of heat-shocked bacterial suspension from each selected ampicillin-resistant colonies. A first denaturation step (30 s at 95 °C) is followed by an amplification of 30 cycles (1 min at 95 °C, 1 min at 55 °C, 3 min at 72 °C) with a final elongation (10 min at 72 °C).
5. PCR products are loaded on a 1% agarose gel. A 1.7 Kb band corresponding to the PCR-amplified NOX2 cDNA is visible under UV light for clones, which have integrated the mutagenesis product (Fig. 2a).

3.1.5 Plasmid DNA Extraction and BamH1 Digestion of Insert

From bacterial pre-cultures of clones that have integrated the mutagenesis product (Subheading 3.1.4, **step 2**), plasmid DNA preparation is performed by Miniprep (*see* **Note 8**).

1. Transfer 1.5 mL of bacterial pre-culture to an Eppendorf tube, centrifuge at 12,000 × *g* for 1 min at 4 °C, and discard the supernatant.

2. Resuspend the pellet with 100 μL of ice-cold Solution I and vortex.
3. Add 200 μL of Solution II, mix by inverting (5 times), and put on ice for 5 min.
4. Add 150 μL of ice-cold Solution III to the previous mixture and vortex for 10 s, and then incubate for 10 min on ice.
5. Centrifuge at 12,000 × *g* for 5 min at 4 °C, and transfer the supernatant into an Eppendorf tube (which corresponds to 1 volume of DNA solution).
6. Add 2 volumes of ice-cold 100% ethanol and 1/10 volume of 3 M sodium acetate pH 5.2, and incubate for 15 min at room temperature.
7. Centrifuge at 12,000 × *g* for 5 min at 4 °C and discard the supernatant; wash the yellowish diffuse pellet with 1 mL of ice-cold 70% ethanol while gently flushing.
8. Centrifuge at 12,000 × *g* for 5 min at 4 °C, aspirate the supernatant, and air-dry the pellet for at least 1 h.
9. Dissolve the pellet with 50 μL of TE buffer.
10. The DNA concentration is determined spectrophotometrically by the optical density at 260 nm.
11. 5 μg (about 5 μL) of DNA is digested with 15 U BamH1 restriction enzyme in the presence of 100 ng of RNase (DNase-free) and 1× reaction buffer for 1 h at 37 °C in a final volume of 10–20 μL.
12. After migration onto a 1% agarose gel, the presence of two bands (1.7 and 2.9 bp) corresponds to the NOX2 cDNA and the open pBluescriptS+ plasmid, respectively (Fig. 2b).
13. The 1.7 bp band is cut under UV light, and the DNA is extracted from the agarose gel using the Quick Gel Extraction Kit. The DNA concentration is estimated using the absorbance at 260 nm.

3.2 NOX2 cDNA Subcloned in pEF-PGKneo Plasmid

The nucleotide sequence of the pEF-PGKneo plasmid is about 7 kb and is not commercially available [5]. This plasmid contains a strong promoter, the human elongation factor 1α (EF-1α) (*see* **Note 9**), a phosphoglycerate kinase-1 (PGK) gene promoter enabling stable expression for long term, a geneticin resistance cassette (eukaryotic cell selection), and an ampicillin resistance cassette (prokaryotic cell selection). The restriction map and sequence of this plasmid are not completely known (data Prof. MC Dinauer, Washington University).

3.2.1 pEF-PGKneo Dephosphorylation

1. 20 μg of pEF-PGKneo plasmid is digested by BamH1 enzyme for 2 h at 37 °C. Then, the restriction enzyme is denatured at 60 °C for 15 min.
2. Once the plasmid is open, it is dephosphorylated by shrimp alkaline phosphatase (3 U) in 1× dephosphorylation buffer for 1 h at 37 °C. Then, the enzyme is inactivated at 60 °C for 15 min.
3. The dephosphorylated open plasmid migrates on a 0.8% (w/v) agarose gel, and the band is purified from the gel with the Quick Gel Extraction Kit.
4. The DNA concentration is determined by measuring the absorbance at 260 nm. The dephosphorylated vector can be stored at −20 °C until use.

3.2.2 NOX2 cDNA Ligation into Dephosphorylated pEF-PGKneo Plasmid and Bacterial Transformation of the Ligation Product

Ligation is performed for 2 h at room temperature with 100 ng of NOX2 cDNA, 100 ng of dephosphorylated pEF-PGKneo plasmid, T4 DNA ligase (3 U), and 1× ligase buffer in a final volume of 10 μL. The ligation product is used for transformation of competent JM 109 bacteria as described in Subheading 3.1.

3.2.3 Selection of Positive Clones After Bacterial Transformation

Positive clones that contain the desired insert are selected by PCR, and their DNA plasmid content is extracted by Miniprep as described (Subheading 3.1) (*see* **Note 8**).

1. In order to control the direction of insert integration into the pEF-PGKneo plasmid, digestion of 5 μg of purified plasmid DNA is performed with 15 U of the restriction enzyme EcoR1 in 1× reaction buffer for 1 h at 37 °C in a final volume of 10 μL (*see* **Note 2**).
2. After migration onto a 1% agarose gel, the presence of two bands reveals integration of the insert in the correct direction. The presence of three bands reveals integration in the wrong direction (Fig. 2c).
3. Perform Sanger sequencing of the entire mutated NOX2 to verify the presence of the mutation introduced by site-directed mutagenesis and the absence of other mutations that could have been created during genetic manipulations.

3.3 Nucleofection and Selection of NOX2 X-CGD PLB-985 Cells by Limited and Antibiotic Selection

The PLB-985 cell line can be differentiated into monocytes/macrophages or granulocytes by using chemical agents such as dimethylformamide (DMF), dimethyl sulfoxide (DMSO), or retinoic acid [4, 23]. Therefore, this cell line expresses all the NADPH oxidase subunits such as $p22^{phox}$, NOX2, $p67^{phox}$, $p47^{phox}$, and $p40^{phox}$ and is capable of performing the main phagocyte functions such as phagocytosis and superoxide anion production. However, neutrophils differentiated from PLB-985 cells are not totally mature with regard to some specialized functions such as adherence and chemotaxis. Although some protocols have been recommended for obtaining mature cells, secretory vesicles are difficult to identify [23, 24]. In addition, some immune receptors or signaling pathways seem to be defective [12]. Nevertheless, the X^0-CGD PLB-985 cell line, where *CYBB* encoding NOX2 was knocked out by homologous recombination, represents a powerful cell model for structure-function studies of NOX2 [5].

Cells are maintained in RPMI-1640 containing 10% (v/v) heat-inactivated fetal bovine serum, 2 mM L-glutamine, 100 U/mL penicillin, and 100 μg/mL streptomycin at 37 °C in a 5% CO_2 humidified atmosphere. This medium is changed biweekly (dilution 1:10). "Neutrophil-like" differentiation: To induce expression of endogenous NADPH oxidase components, PLB-985 cells (2×10^5 cells/mL) are differentiated for 6 days with 0.5% (v/v) DMF. This medium is partially changed at day 3 of differentiation (about half of the volume).

3.3.1 Extraction of Purification of NOX2 in pEF-PGKneo

DNA quality is essential for an efficient transfection into eukaryotic cells. The extracted DNA plasmid must be free of any bacterial toxins; therefore purify DNA with an Endo-free Plasmid Maxi Kit according to the manufacturer's instructions.

1. Pick a single colony from a freshly streaked selective plate, and inoculate a starter culture of 2–5 mL LB medium with 100 U/mL ampicillin. Incubate for approx. 8 h at 37 °C with vigorous shaking (approx. 300 rpm).
2. 200 μL of starter culture is diluted with 100 mL medium. Grow at 37 °C for 12–16 h with vigorous shaking (approx. 300 rpm).
3. Harvest the bacterial cells by centrifugation at 6000 × *g* for 15 min at 4 °C.
4. In order to purify bacterial DNA, refer to the manufacturer's instructions.
5. DNA concentration is determined by measuring its absorbance at 260 nm. Generally, 300–800 μg of DNA is recovered. A 260/280 ratio of ~1.8 or higher is generally accepted as "pure" for DNA.

3.3.2 Nucleofection

1. For transfection, cells must be in their logarithmic growth phase (cell concentration between 5×10^5 and 1.5×10^6 cells/mL).
2. Centrifuge 2×10^6 cells at 90 × g for 10 min at room temperature.
3. Discard supernatant completely so that no residual medium covers the cell pellet. Resuspend the pellet in 100 μL of room-temperature Nucleofector solution V supplemented.

4. Add 2 μg of plasmid and transfer to an Amaxa™ cuvette, and close the vial with its cap. Make sure that the sample covers the cuvette bottom, and avoid air bubbles while pipetting. Avoid storing the cell suspension for longer than 15 min in the nucleofection solution, as this reduces cell viability and gene transfer efficiency.
5. Install the cuvette in the Amaxa™ system, select program C023 (suitable for PLB-985 cells) (*see* **Note 10**), and start the program.
6. Immediately recover the cuvette, and add 500 μL of cell culture medium (as in 2.4), 20% FBS at 37 °C under the hood.
7. Using a sterile plastic pipette, aspirate the total cuvette contents in one go (do not flush), and transfer to a flask containing 4.5 mL of cell culture medium, 20% FBS at 37 °C.
8. Incubate for 48 h in a humidified 37 °C, 5% CO_2 incubator.
9. After 48 h, recover the transfected cells (5 mL), and transfer them to a large flask containing 45 mL of cell culture medium, 20% FBS, 1.5 mg/mL geneticin (dilution 1:10). This flask corresponds to the pool of transfected cells (*see* **Note 11**).
10. From the pool, dilute the cells 1:10 in a sterile 50 mL tube (5 mL of cell suspension to 45 mL of cell culture medium, 20% FBS, 1.5 mg/mL geneticin).
11. With this new dilution, carry out the limited cell dilution: 2× 24-well plates with 1 mL cells per well (*see* **Note 12**).
12. Incubate plates in a humidified 5% CO_2 incubator at 37 °C.

3.4 Structural and Functional Impact of NOX2 Mutations

3.4.1 Analysis of NOX2 Expression by Flow Cytometry

The wild-type (control NOX2) or mutated (site-directed mutagenesis NOX2) NOX2 protein expression level is evaluated by flow cytometry using the mouse monoclonal antibody 7D5 directed against two external epitopes of NOX2 [25] (*see* **Note 13**). In order to determine any nonspecific interactions, a control is performed in parallel with an irrelevant mouse IgG1 antibody.

1. For each labeling and control tube, 5×10^5 PLB-985 cells are centrifuged for 10 s at 12,000 × *g* and washed with 1 mL of sterile PBS.
2. After a further centrifugation of 10 s at 12,000 × *g*, cells are incubated for 15–30 min at room temperature (or 30 min to 1 h on ice) with 0.5 μg of primary antibody 7D5 or irrelevant mouse IgG1 diluted in 100 μL (final volume) of PBS with 0.2% (w/v) BSA.
3. Following a centrifugation of 10 s at 12,000 × *g*, cells are washed twice with 500 μL of PBS/BSA buffer.
4. A secondary anti-mouse IgG antibody coupled to a phycoerythrin (fluorochrome) is used. Cells are incubated with 0.25 μg

of secondary antibody diluted in 100 μL (final volume) PBS/BSA buffer for 15–30 min at room temperature (or 30 min to 1 h on ice) in the dark.

5. Cells are washed twice with 500 μL of PBS/BSA buffer to remove the excess of secondary antibody.
6. Finally, cells are analyzed by flow cytometry. Data are collected on 10,000 events and analyzed with software.

Initially, NOX2 expression analysis is performed on the entire transfected cell population after about 10 days of transfection (before starting the selection by limited dilution) and then, on selected clones that developed after limited dilution, in order to select several clones for each studied mutation. These clones will be used later for functional analyses. In general, three different clones for every studied mutation are kept in liquid nitrogen after expansion. Figure 3 shows the analysis of NOX2 expression in the pool of

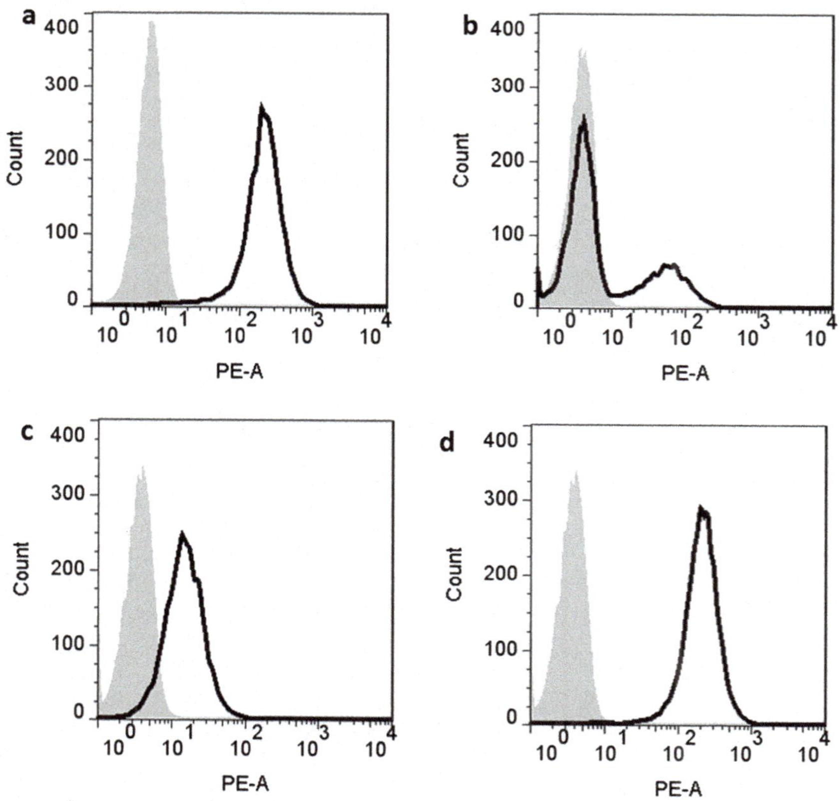

Fig. 3 Analysis of NOX2 protein expression by flow cytometry in transfected X-CGD PLB-985 cells. The NOX2 expression was analyzed with the 7D5 mouse monoclonal antibody and revealed by a secondary anti-mouse IgG antibody coupled to phycoerythrin (black curve). The isotype control was performed with an irrelevant mouse IgG1 antibody (filled gray curve). (**a**) Expression of NOX2 in wild-type PLB-985. (**b**) Wild-type NOX2 expression in the pool of transfected X-CGD PLB-985 cells before clonal selection by limited dilution. (**c**) An example of NOX2 expression when analyzing a X^--CGD mutation (W18C mutation in NOX2) in the transfected PLB-985 cell model. (**d**) An example of NOX2 expression when analyzing a X^+-CGD mutation (P415H mutation in NOX2) in the transfected PLB-985 cell model

transfected PLB-985 cells before selection by limited dilution. In Fig. 3b, the pool contains two populations of PLB-985 cells, one expressing NOX2 and the other one not (*see* **Notes 14** and **15**). This is an indication that the nucleofection experiment worked well and that the pool can be used to select positive clones expressing NOX2. Figure 3c shows low surface expression of the NOX2 W18C mutation (characteristic of a X^--CGD case) compared to endogenous NOX2 expression in wild-type PLB-985 cells (Fig. 3a). NOX2 with a X^+-CGD mutation (P415H) (Fig. 3d) has the same expression level as that of NOX2 in wild-type PLB-985 cells (Fig. 3a).

3.4.2 NADPH Oxidase Activity by Chemiluminescence

The impact of mutations on NADPH oxidase activity can be evaluated by chemiluminescence.

For all functional measurements, PLB-985 cells should be differentiated into "neutrophil-like" cells as previously described (Subheading 3.3). This technique enables measurement of reactive oxygen species (ROS), mainly hydrogen peroxide (H_2O_2) produced by dismutation of superoxide, which is generated by the NADPH oxidase of phagocytic cells after activation with soluble (PMA) or particulate (opsonized particles) stimuli. H_2O_2 reacts with luminol in the presence of a peroxidase (HRP) to emit photons detected by a chemiluminometer (Fig. 4). ROS production is optimal at 37 °C, and thus it is important to pre-warm the chemiluminometer. Measurement is performed in white 96-well plates with 5×10^5 differentiated PLB-985 cells/well. For each cell type, a duplicate of unstimulated cells and a triplicate of PMA stimulated cells are analyzed.

1. Count the day 6-differentiated PLB-985 cells (*see* Subheading 3.3), and collect the volume needed for 3×10^6 cells.
2. Centrifuge at 550 × *g* for 8 min at room temperature, discard the supernatant, and resuspend the cells with 1 mL of PBS (*see* **Note 16**).
3. Centrifuge at 550 × *g* for 8 min at room temperature, remove the supernatant, and resuspend the pellet with 300 μL of PBS (5×10^5 cells/50 μL).
4. Calculate the amount of chemiluminescence solution required for the experiment (200 μL/well), and prepare it extemporaneously, protected from light as described in Subheading 2.5.
5. Add 50 μL of cells per well and then 200 μL of chemiluminescence medium.
6. Incubate for 5 min at 37 °C, and stimulate the cells by adding 10 μL of 2 μg/mL PMA (prepared as described in Subheading 2.5).

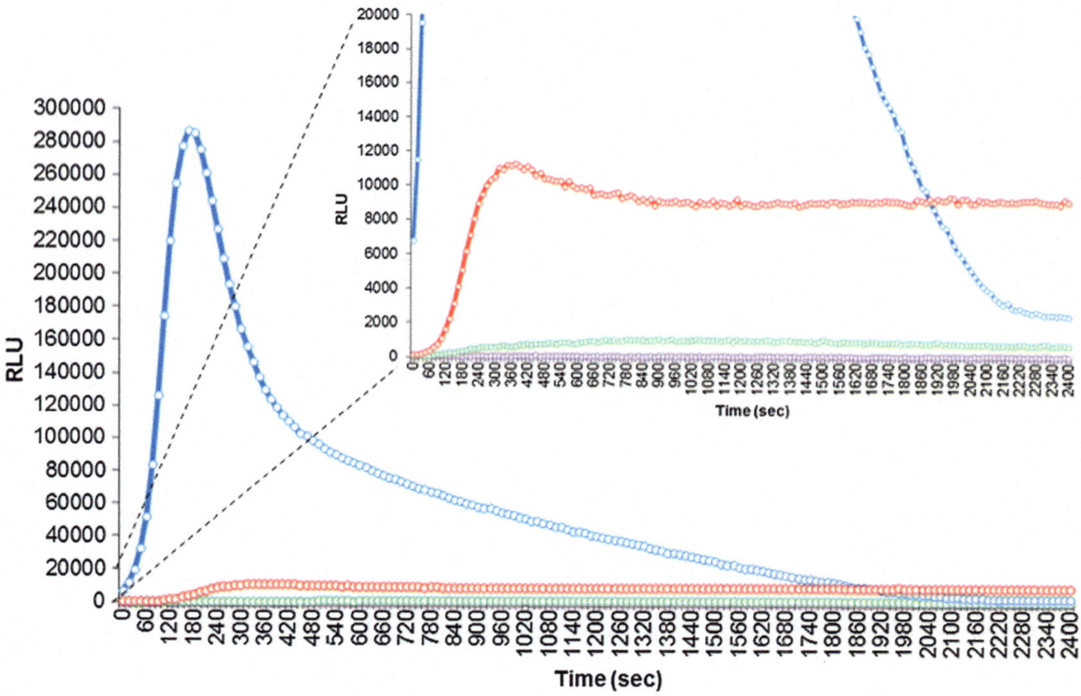

Fig. 4 NADPH oxidase activity of NOX2-transfected PLB-985 cells. The NADPH oxidase activity of differentiated wild-type NOX2-transfected PLB-985 cells (5×10^5 cells) was measured by chemiluminescence in resting state (green curve) and after stimulation with 80 ng/mL PMA (blue curve). This activity serves as reference for the comparison of the NADPH oxidase activity of mutated NOX2 PLB-985 cells. For example, the impact of a X^--CGD mutation (W18C) (red curve) on the NADPH oxidase activity of mutated NOX2-transfected PLB-985 cells after PMA stimulation can be evaluated. The insert represents the expansion of the previous kinetics and the total absence of the NADPH oxidase activity in X^+-CGD PLB-985 cells (P415H mutation in NOX2) after PMA activation (purple curve) can be observed

7. Emission kinetics are started for 60 min at 37 °C with a measurement made every 15 s. Results are in relative units of luminescence (RLU) and can be expressed by total activity (sum of RLU) measured during kinetics or by the maximum value (peak) of RLU during kinetics (Fig. 4).

Besides analysis of NOX2 protein expression and NADPH oxidase activity in intact NOX2-transfected PLB-985 cells, other functionalities and activities can be measured (e.g., NADPH oxidase complex assembly by confocal microscopy) [10]. In addition, several kinetic measurements can be carried out in purified membranes from these NOX2-transfected PLB-985 cells as well as in purified mutated cytochrome b_{558} [11, 26].

4 Notes

1. β1 and NOX2-End primers hybridize at the start codon and the stop codon of NOX2 cDNA, respectively, thus making it possible to amplify a fragment of approximately 1700 bp.
2. Plasmid or viral methods can be used for stable expression of NOX2, but we describe here the plasmid approach as it is accessible to most research laboratories. NOX2 cDNA is cloned into a unique BamH1 site, the Kozac sequence being located just after the promoter sequence; therefore the direction of the insert integration into the pEF-PGKneo vector must be verified.
3. Geneticin sensitivity of PLB-985 cells varies depending on the source and batch of the antibiotic. It is necessary to test a range of concentrations for each new batch of geneticin and for each new cell type.
4. This antibody is directed against external epitopes located on NOX2 extracellular C and E loops [25]. Thereby, the use of this antibody in flow cytometry does not require cell permeabilization.
5. Attempting to change a large number of bases at once can be difficult; otherwise site-directed mutagenesis is suitable for point mutations (1 to 5 bases). Choice of primers is essential for the success of site-directed mutagenesis, and three rules must be respected: Primer size must range between 25 and 45 bases; the residue to be mutated must be centered in the primers; primers must end with a G or a C.
6. This step makes it possible to avoid hybridization between forward and reverse primers and thus increases the efficiency of the next step.
7. It is important to perform a nucleofection with the plasmid encoding wild-type NOX2 at the same time. This control will determine the level of wild-type NOX2 expression, which will serve to quantify and compare expression of the mutated NOX2.
8. We describe the classical Miniprep method, but several commercial kits exist.
9. This promoter is the most efficient in PLB-985 cells; other conventional strong promoters such as CMV are much less efficient.
10. The C-023 program is best adapted to PLB-985 cells with a transfection efficiency of 60% and cell viability of 75%.
11. The cell pool provides a quick overview of transfection efficiency (7–10 days). It contains cell mixtures that either express mutated NOX2 protein or not. During each transfection, some

cells acquire a truncated plasmid which confers geneticin resistance, but does not enable full length NOX2 cDNA expression.

12. The purpose of limited cell dilution is to obtain a clonal population of transfected cells expressing mutated NOX2 protein from the transfected cell pool. Plates are incubated for 15–21 days in a humidified 37 °C/5% CO_2 incubator. Positive clones (resistant to geneticin) are detected due to a color change in the medium (from pink to yellow) illustrating cell proliferation (metabolic activity). Clones are maintained in complete RPMI medium with 1.5 mg/mL geneticin as described (Subheading 3.3). Once a clonal population has been obtained, the geneticin concentration can be decreased to 0.5 mg/mL.
13. Mutated NOX2 expression is regulated by the pEF-PGKneo promoter and is independent of cell differentiation. Therefore, to evaluate mutated NOX2 protein expression, it is not necessary to differentiate cells into a neutrophil-like type. However, expression of NOX2 with X^+-CGD mutations [10, 11] or with X^--CGD mutations [20] or de novo missense mutations [8] to be studied is always confirmed and compared with the expression of wild-type NOX2 expressed in NOX2 WT-transfected KO-PLB-985 cells after 6 days of differentiation.
14. If NOX2 protein expression is not observed in the pool, one needs to differentiate whether the lack of NOX2 expression is due to the impact of the mutation or to inefficient transfection. For this purpose, it is necessary to detect the presence of NOX2-specific messenger RNA by extraction of total RNAs, reverse transcription (RT), and specific PCR.
15. If limited dilution selection is not sufficiently effective, it is possible that despite testing 20 clones, two different populations still exist. In these rare cases, cells of interest can be obtained by flow cytometry sorting.
16. The objective is to stimulate cells with PMA. Thus, it is essential that cells remain in a resting state before stimulation. Cells should be treated as gently as possible (e.g., no vortex, gentle centrifugation), and sterile materials and reagents must be used.

References

1. Bedard K, Krause KH (2007) The NOX family of ROS-generating NADPH oxidases: physiology and pathophysiology. Physiol Rev 87:245–313
2. Van den Berg JM, van Koppen E, Ahlin A et al (2009) Chronic granulomatous disease: the European experience. PLoS 4:e5234
3. Stasia MJ, Li XJ (2008) Genetics and immunopathology of chronic granulomatous disease. Semin Immunopathol 30:209–235
4. Tucker KA, Lilly MB, Heck L Jr, Rado TA (1987) Characterization of a new human diploid myeloid leukemia cell line (PLB-985) with

granulocytic and monocytic differentiating capacity. Blood 70:372–378

5. Zhen L, King AA, Xiao Y, Chanock SJ, Orkin SH, Dinauer MC (1993) Gene targeting of X chromosome-linked chronic granulomatous disease locus in a human myeloid leukemia cell line and rescue by expression of recombinant gp91phox. Proc Natl Acad Sci U S A 90:9832–9836
6. Yu L, Cross AR, Zhen L, Dinauer MC (1999) Functional analysis of NADPH oxidase in granulocytic cells expressing a delta 488-497 gp91(phox) deletion mutant. Blood 94:2497–2504
7. Bionda C, Li XJ, van Bruggen R et al (2004) Functional analysis of two-amino acid substitutions in gp91 phox in a patient with X-linked flavocytochrome b558-positive chronic granulomatous disease by means of transgenic PLB-985 cells. Hum Genet 115:418–427
8. Li XJ, Grunwald D, Mathieu J, Morel F, Stasia MJ (2005) Crucial role of two potential cytosolic regions of Nox2, 191TSSTKTIRRS200 and 484DESQANHFAVHHDEEKD500, on NADPH oxidase activation. J Biol Chem 280:14962–14973
9. Li XJ, Fieschi F, Paclet MH et al (2007) Leu505 of Nox2 is crucial for optimal p67phox-dependent activation of the flavocytochrome b558 during phagocytic NADPH oxidase assembly. J Leukoc Biol 81:238–249
10. Debeurme F, Picciocchi A, Dagher MC et al (2010) Regulation of NADPH oxidase activity in phagocytes: relationship between FAD/NADPH binding and oxidase complex assembly. J Biol Chem 285:33197–33208
11. Picciocchi A, Debeurme F, Beaumel S et al (2011) Role of putative second transmembrane region of Nox2 protein in the structural stability and electron transfer of the phagocytic NADPH oxidase. J Biol Chem 286:28357–28369
12. Carrichon L, Picciocchi A, Debeurme F et al (2011) Characterization of superoxide overproduction by the D-Loop(Nox4)-Nox2 cytochrome b(558) in phagocytes-Differential sensitivity to calcium and phosphorylation events. Biochim Biophys Acta 1808:78–90
13. Zhen L, Yu L, Dinauer MC (1998) Probing the role of the carboxyl terminus of the gp91phox subunit of neutrophil flavocytochrome b558 using site-directed mutagenesis. J Biol Chem 273:6575–6581
14. Beaumel S, Picciocchi A, Debeurme F et al (2017) Down-regulation of NOX2 activity in phagocytes mediated by ATM-kinase dependent phosphorylation. Free Radic Biol Med 113:1–15
15. Magnani F, Nenci S, Millana Fananas E et al (2017) Crystal structures and atomic model of NADPH oxidase. Proc Natl Acad Sci U S A 114:6764–6769
16. Biberstine-Kinkade KJ, Yu L, Dinauer MC (1999) Mutagenesis of an arginine- and lysine-rich domain in the gp91(phox) subunit of the phagocyte NADPH-oxidase flavocytochrome b558. J Biol Chem 274:10451–10457
17. Jackson HM, Kawahara T, Nisimoto Y, Smith SM, Lambeth JD (2010) Nox4 B-loop creates an interface between the transmembrane and dehydrogenase domains. J Biol Chem 285:10281–10290
18. Taylor WR, Jones DT, Segal AW (1993) A structural model for the nucleotide binding domains of the flavocytochrome b-245 beta-chain. Protein Sci 2:1675–1685
19. Kean KM, Carpenter RA, Pandini V et al (2017) High resolution studies of hydride transfer in the ferredoxin: NADP(+) reductase superfamily. FEBS J 284:3302–3319
20. Beaumel S, Grunwald D, Fieschi F, Stasia MJ (2014) Identification of NOX2 regions for normal biosynthesis of cytochrome b558 in phagocytes highlighting essential residues for p22phox binding. Biochem J 464:425–437
21. Ding C, Kume A, Björgvinsdóttir H, Hawley RG, Pech N, Dinauer MC (1996) High-level reconstitution of respiratory burst activity in a human X-linked chronic granulomatous disease (X-CGD) cell line and correction of murine X-CGD bone marrow cells by retroviral-mediated gene transfer of human gp91phox. Blood 88:1834–1840
22. Wrona D, Siler U, Reichenbach J (2017) CRISPR/Cas9-generated p47(phox)-deficient cell line for Chronic Granulomatous Disease gene therapy vector development. Sci Rep 7:44187
23. Pedruzzi E, Fay M, Elbim C, Gaudry M, Gougerot-Pocidalo MA (2002) Differentiation of PLB-985 myeloid cells into mature neutrophils, shown by degranulation of terminally differentiated compartments in response to N-formyl peptide and priming of superoxide anion production by granulocyte-macrophage colony-stimulating factor. Br J Haematol 117:719–726
24. Pivot-Pajot C, Chouinard FC, El Azreq MA, Harbour D, Bourgoin SG (2010) Characterisation of degranulation and phagocytic capacity of a human neutrophilic cellular model, PLB-985 cells. Immunobiology 215:38–52

25. Yamauchi A, Yu L, Pötgens AJ et al (2001) Location of the epitope for 7D5, a monoclonal antibody raised against human flavocytochrome b558, to the extracellular peptide portion of primate gp91phox. Microbiol Immunol 45:249–257

26. Lord CI, Riesselman MH, Gripentrog JM, Burritt JB, Jesaitis AJ, Taylor RM (2008) Single-step immunoaffinity purification and functional reconstitution of human phagocyte flavocytochrome b. J Immunol Methods 329:201–207

Chapter 11

Functional Characterization of DUOX Enzymes in Reconstituted Cell Models

Agnieszka Korzeniowska, Ágnes P. Donkó, Stanislas Morand, and Thomas L. Leto

Abstract

Biosynthesis of active human dual oxidases (DUOX1 and DUOX2) requires maturation factors, a.k.a. DUOX activator proteins (DUOXA1 and DUOXA2), that form covalent complexes with DUOX; both chains together represent the mature catalytic unit that functions as a dedicated hydrogen peroxide-generating enzyme. Genetic defects in DUOX2 or DUOXA2 can result in congenital hypothyroidism, whereas partial defects in DUOX2 activity also have been associated with very early-onset inflammatory bowel disease. Our understanding of the links between DUOX dysfunction and these diseases remains incomplete. An important challenge in developing a better understanding of the pathogenic roles of DUOX defects requires robust and reliable DUOX reconstitution cell models to examine the functional consequences of candidate DUOX missense mutations and polymorphisms. Here, we describe methods for efficient heterologous DUOX/DUOXA co-expression and functional characterization, including detailed assessments of posttranslational processing and subcellular translocation of DUOX that accompanies the maturation of these enzymes into catalytically active NADPH oxidases.

Key words NADPH oxidases, Dual oxidases, DUOX, DUOX activator, DUOXA, NOX, Hydrogen peroxide, Cell surface targeting

1 Introduction

The dual oxidases, also known as DUOX enzymes, are dedicated hydrogen peroxide generators in the NOX family of NADPH oxidases that support the activities of extracellular heme peroxidases. DUOX1 and DUOX2 were first described as sources of hydrogen peroxide in thyroid tissue that were proposed to support thyroperoxidase-mediated iodine organification during the process of thyroid hormone biosynthesis [1, 2]. In subsequent years, several groups described molecular variants of DUOX2 associated with congenital hypothyroidism (CH) (*see* reviews Refs. [3–6]).

Agnieszka Korzeniowska and Ágnes P. Donkó contributed equally to this work.

Ulla G. Knaus and Thomas L. Leto (eds.), *NADPH Oxidases: Methods and Protocols*, Methods in Molecular Biology, vol. 1982, https://doi.org/10.1007/978-1-4939-9424-3_11, © Springer Science+Business Media, LLC, part of Springer Nature 2019

Early studies aimed at establishing a direct causal role for these DUOX2 variants in CH were hampered by the inability to reconstitute DUOX activity in heterologous whole cell systems. Breakthrough studies then demonstrated that the DUOX enzymes require specific maturation factors, a.k.a. DUOX activators (DUOXA), to permit faithful processing and transport of the active enzyme to the cell surface [7]. Although initially described as maturation factors required to allow the escape of DUOX from the endoplasmic reticulum (ER), it is now clear that the DUOXA proteins form covalent heterodimer complexes with DUOX at the plasma membrane, representing the mature, active form of these oxidases [8–10]. These developments allowed investigators to explore the functional consequences of candidate DUOX2 or DUOXA2 mutations on oxidase activity in reconstituted cells. Despite these advances, our understanding of links between DUOX or DUOXA structural variants and congenital hypothyroidism remains incomplete. For example, some patients with transient hypothyroidism were described with biallelic DUOX2 or DUOXA2 defects, while others with permanent CH present with only monoallelic DUOX2 defects [11, 12]. These observations are further complicated by other factors that can influence thyroid hormone production, including the possible compensatory role for DUOX1, also expressed in thyroid tissue, as well as dietary, environmental, and autoimmune factors that can affect thyroid insufficiency. Many studies described DUOX2 variants that are predicted to be deleterious without supporting functional data, while others provided only a partial functional characterization, such as hydrogen peroxide output in a variety of cell models, some of which may not recapitulate the specifics of thyrocyte function. Several DUOX2 variants exhibiting reduced NADPH oxidase activities have also been associated with very early-onset inflammatory bowel disease in patients with no evidence of thyroid hormone deficiency [13, 14]. However, one recent large patient-based study reported an increased risk of inflammatory bowel disease in patients with transient congenital hypothyroidism, where partial defects in DUOX2 activity are commonly observed [15]. The above observations highlight the need for reliable test cell systems and clearer guidelines for investigating the functional consequences of any DUOX or DUOXA structural variant or mutant protein.

In this chapter, we provide detailed protocols for testing the functional reconstitution of DUOX and DUOXA in transfected cells. Figure 1 outlines key events in the biosynthesis of active DUOX, including the requisite interaction of DUOX and DUOXA allowing their exit from the ER, glycosyl modifications of both polypeptide chains as they transit through the Golgi apparatus, and translocation of the active heterodimeric complex to the plasma membrane. This chapter provides methods for monitoring each of the biosynthetic events: (1) hydrogen peroxide release

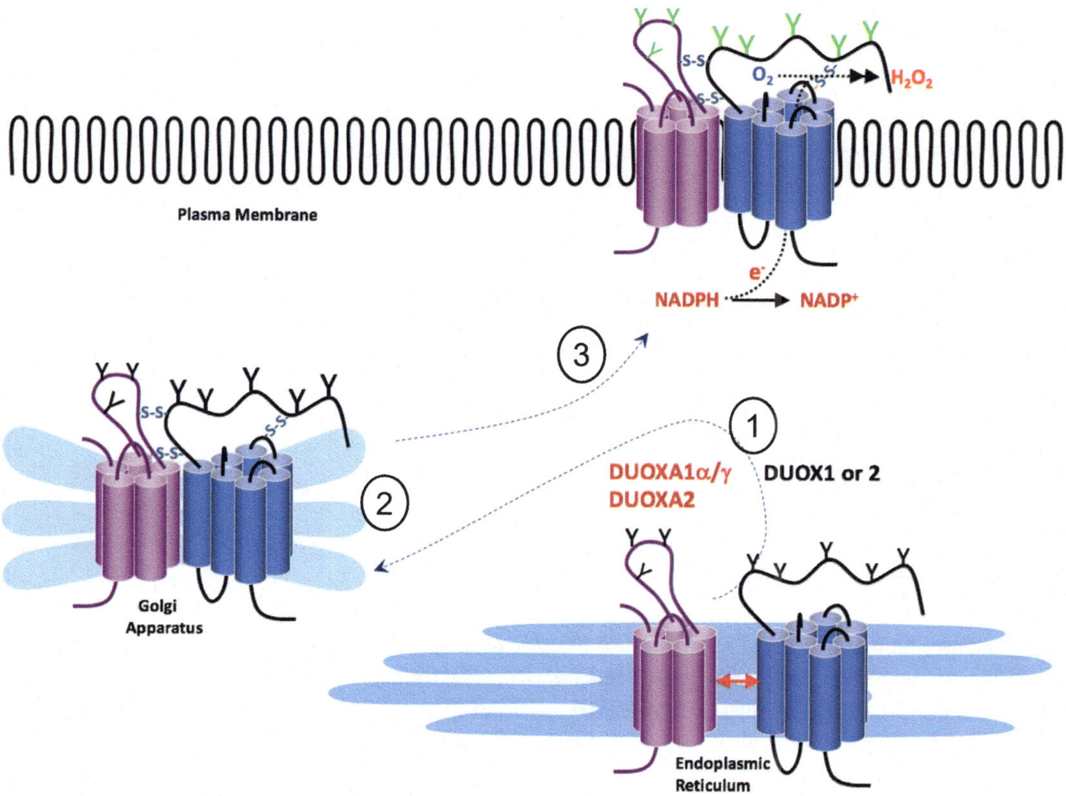

Fig. 1 Schematic representation of the biosynthesis of the DUOX/DUOXA oxidase complex, illustrating stages of (1) stable heterodimer formation required for export from the endoplasmic reticulum, (2) Golgi apparatus-based N-linked glycosyl modifications of both chains, and (3) accumulation of the mature, active enzyme that releases hydrogen peroxide into the extracellular milieu. Both DUOXA1α (isoform 3) and γ (isoform 1) splice variants support DUOX1 maturation, although the DUOX1/DUOXA1γ complex also accumulates in a vesicular compartment [8]

assays are described to detect DUOX NADPH oxidase activity in reconstituted cells, (2) Western blotting methods are described for detecting production of epitope-tagged versions of both DUOX and DUOXA proteins, (3) Endoglycosidase H assays are described to detect Golgi apparatus-based modifications that endow resistance of both proteins to deglycosylation, and (4) immunostaining methods for monitoring translocation of both proteins to the cell surface are described using immunofluorescence microscopy or flow cytometry. We have confirmed the proper biosynthesis and activities of several peptide epitope-tagged versions of DUOX and DUOXA proteins (*see* Fig. 2) that enable their detection on the cell surface or by Western blotting [8]. Together, these methods provide a coherent strategy and rigorous means for delineating functional consequences of structural variations in DUOX or DUOXA proteins that may compromise their biosynthesis and processing into active hydrogen peroxide-generating NADPH oxidases.

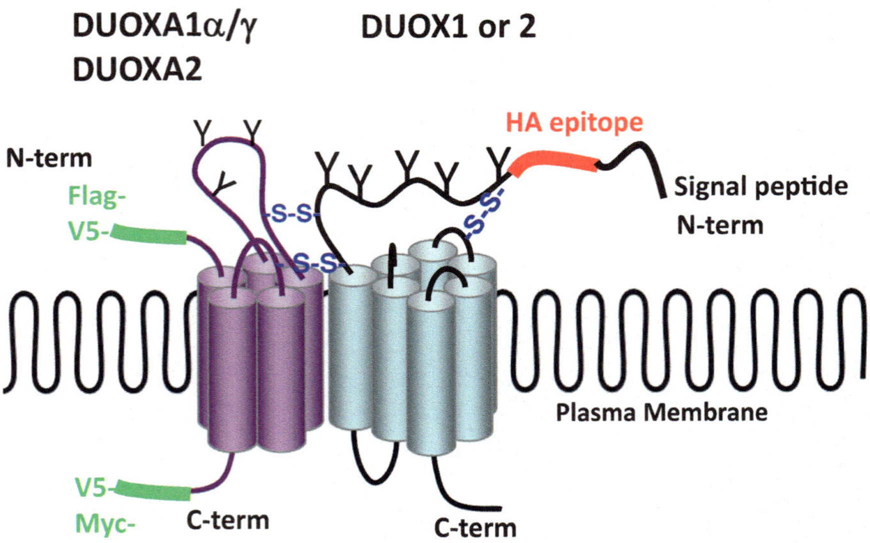

Fig. 2 Design of peptide epitope-tagged versions of recombinant DUOX and DUOXA proteins used to track expression and subcellular targeting of DUOX/DUOXA complexes in reconstituted cell models [8]. DUOXA proteins shown here, tagged at either the amino or carboxyl terminus, support DUOX reconstitution, whereas the DUOX HA epitope tag must be inserted immediately following the signal peptide sequence in order to retain NADPH oxidase function [7–9]. Only N-terminal V5- and FLAG-tagged DUOXA proteins are detected on the cell surface in non-permeabilized cells when co-expressed with active DUOX [8]

2 Materials

1. Flp-In 293 cells (ThermoFisher) or stable isogenic DUOX or DUOXA Flp-In 293 cloned cell lines [8] (*see* **Note 1**).
2. Minimum essential medium-α (MEM-α) complete medium: supplement with 10% fetal bovine serum (FBS), 100 U/mL penicillin, and 100 μg/mL streptomycin (P/S). Include 50 μg/mL hygromycin for retention of transgenes in stable cloned lines.
3. OPTI-MEM medium.
4. Trypsin/EDTA.
5. FuGENE6 or other suitable lipofection reagent (*see* **Note 2**).
6. pcDNA5/FRT plasmids encoding the human-tagged ^{HA}DUOX1 or ^{HA}DUOX2 and ^{V5}DUOXA1α or ^{V5}DUOXA2 cDNAs (*see* **Note 3**).
7. Hank's balanced salt solution with Ca^{2+} and Mg^{2+} ($HBSS^{++}$).
8. Phosphate-buffered saline (PBS), without calcium or magnesium.
9. HVA/HRP hydrogen peroxide assay reagent (2× stock): 2 mM homovanillic acid (HVA), 40 U/mL horseradish peroxidase (HRP) in $HBSS^{++}$, prepared fresh on the day of assay.

10. H_2O_2 standards prepared using H_2O_2 (30% w/v solution) diluted in $HBSS^{++}$ (see below).
11. Luminol/HRP peroxide assay reagent (2× stock): 2 mM luminol, 40 U/mL HRP, $HBSS^{++}$, prepared fresh on the day of assay.
12. NP-40 lysis buffer (NP40/PIC/PMSF): 50 mM Tris–HCl pH 8.0, 150 mM NaCl, 1% NP-40, supplemented with 1% (v/v) Protease Inhibitor Cocktail (Sigma-Aldrich) and 1% phenylmethanesulfonyl fluoride.
13. Bicinchoninic acid (BCA) protein assay kit.
14. 4× LDS-loading buffer (ThermoFisher): Supplement 10 mL initially with 2 mL mercaptoethanol as a reducing agent.
15. Prestained protein marker: Novex Sharp Pre-Stained Protein Standard (ThermoFisher).
16. 4–12% polyacrylamide gradient precast NuPAGE Bis-Tris gels.
17. 1× NuPAGE transfer buffer: 20× NuPAGE Transfer Buffer (ThermoFisher) diluted with distilled water and supplemented with 10% methanol and 0.04% SDS.
18. TBS-T wash buffer: 1× Tris-buffered saline supplemented with 0.1% Tween 20.
19. Western blot (WB) blocking buffer: 5% nonfat dry milk in TBS-T.
20. Immunofluorescence (IF) blocking buffer: 5% BSA in PBS.
21. FACS blocking buffer: 2% fetal bovine serum in PBS.
22. Antibodies (diluted in the respective blocking buffers):

    Mouse anti-HA (Covance): 1:2000 (WB), 1:100 (IF) 1:100 (FACS).

    Mouse anti-HSP90 (SantaCruz): 1:2000 (WB).

    Mouse anti-V5 (ThermoFisher): 1:5000 (WB), 1:200 (IF), 1:100 (FACS).

    Mouse anti-Calnexin (Santa Cruz): 1:50 (IF).

    Secondary antibodies (diluted in the respective blocking buffers w/o normal serum):

    Anti-mouse-HRP (GE Healthcare) WB 1:3000.

    Goat anti-mIgG1-Alexa647 (ThermoFisher): 1:1000 (FACS).

    Goat anti-mIgG2a-Alexa488 (ThermoFisher): 1:1000 (FACS).

    Goat anti-mouse Alexa Fluor®-488, –594, and – 647 (ThermoFisher): 1:1000 (IF).

23. DAPI: 4′, 6-diamidino-2-phenylindole, blue-fluorescent DNA stain, 1:10,000 (IF).

24. ECL Plus detection reagent (GE Healthcare).
25. N-glycosidase F (New England Biolabs).
26. Endoglycosidase H (New England Biolabs).
27. Collagenl Cellware 10 cm dishes *or* 6-well plates (*optional*) (Corning BioCoat).
28. Collagenl-coated 35 mm glass-bottom dishes (MatTek).
29. Permeabilization buffer: 0.3% Triton X-100 in PBS.
30. Luminescence reading plates: 96-well, white opaque.
31. Fluorescence reading plates: 96-well, black with clear bottoms.
32. Protein slab gel electrophoresis apparatus.
33. Protein gel electrophoretic transblotting apparatus.
34. Plate-reading fluorometer.
35. Plate-reading luminometer.
36. iBright CL1000 Imaging System (ThermoFisher) or comparable system.
37. Laser-scanning fluorescent microscope (e.g., Leica SP2 NA 1.4 or Zeiss LSM 880).
38. FACSort flow cytometer (BD Biosciences) or comparable system.

3 Methods

The DUOX reconstitution systems described here are based on pcDNA 5/FRT expression vectors [8]. We have created stably transfected clonogenic Flp-In 293 cell lines (a derivative of HEK-293 cells) expressing either the DUOXA or DUOX component from a single integrated copy of either cDNA. This approach minimizes variability in experimental outcomes as it requires transient transfection of only a single vector encoding the complementary component under investigation (e.g., candidate mutant or SNP variants). The methods below can be adapted for transient co-transfection with both the DUOX and DUOXA expression vectors together in other lines.

3.1 Cell Culture and Transfections

1. Propagate stable Flp-In 293-DUOXA (or Flp-In 293-DUOX) clone cell line in complete MEM-α medium supplemented with 10% FBS and antibiotics, and incubate at 37 °C and 5% CO_2. The Flp-In 293 or HEK-293 cells grow best on collagenl-coated plastic.
2. Passage cells every 3–4 days when nearly confluent by washing with PBS, followed by trypsinization in 2 mL Trypsin-EDTA at 37 °C. Split culture back at a 1:3 to 1:6 ratio.

3. Seed cells at 1.2×10^6 cells suspended in 12 mL per one 10 cm collagen1-coated dishes in complete MEM-α medium, and allow to attach for 24 h at 37 °C and 5% CO_2. Transfection of one 10 cm dish is preferred because it will provide enough cells to perform the NADPH oxidase assays, Western blotting, and flow cytometry on the same lot of transfected cells at once. For smaller-scale pilot transfections, *see* **Note 4**.
4. Transfect cells the next day once they reach ~60–75% confluence. Remove culture medium, and replace with 12 mL (10 cm dish) or 2 mL/well (6-well plate) of fresh MEM-α medium containing 10% FBS and P/S for overnight transfections.
5. Prepare DNA/Fugene transfection mixes in Eppendorf tubes using a ratio of 6 μL Fugene6/1 μg plasmid DNA, as follows: For 10 cm plates, mix 600 μL serum-free OPTI-MEM with 36 μL of Fugene6. After 5 min, add 6 μg plasmid DNA, mix gently, and incubate at RT for 20 min. Then dispense this mix onto the cell culture by adding dropwise while gently swirling the plate.
6. Incubate overnight (14–16 h), and then replace the medium with fresh MEM-α complete medium. Cells are then harvested at 48 h post-transfection by trypsinization and divided into separate lots for processing in the three assays described below: [1] H_2O_2 release by Methods below in Subheading 3.2, [2] DUOX and DUOXA Western blotting by Subheading 3.3, and [3] detection of cell surface-exposed DUOX and DUOXA proteins by flow cytometry by Subheading 3.4.

3.2 Measurement of Extracellular H_2O_2 Production

The HVA oxidation assay provides a fluorescence-based means for quantitative determination of H_2O_2 release by DUOX/DUOXA-expressing cells. Although cell surface-expressed DUOX enzymes have N-terminal extracellular peroxidase-like domains, the assay requires addition of an exogenous heme-peroxidase, HRP, since the ectodomains of mature, active mammalian DUOX proteins lack demonstrable endogenous peroxidase activity [8]. Other oxidizable substrates of HRP (i.e., Amplex Red or tyrosine [16]) can be used in place of HVA, generating fluorescent products that can be quantified based on H_2O_2 standards.

3.2.1 HVA/HRP Assay

1. Obtain transfected cells from **step 6** (Subheading 3.1), trypsinize, wash twice, and gently resuspend in $HBSS^{++}$ at the desired cell densities (typically, 2.0×10^6 viable cells/mL). Distribute 100 μL suspended cells into each well of a black/clear-bottom 96-well fluorescence reading plate. Plan on running triplicate well assays for every unique transfection or experimental condition (+/− cell stimulation, inhibitors, etc.). For other experimental design details, including appropriate negative controls, *see* **Note 5**.

2. Prepare fresh 2× HVA/HRP hydrogen peroxide assay reagent for injection from the reservoir (100 μL/ well) at the start of the assay. Separately, prepare 10× stock of cell activator of choice that will be dispensed (20 μL/well) into designated wells after the assay reagent (*see* **Note 6**). Prime the injection lines for delivery of each reagent.
3. Select the plate-reading fluorometer filters to monitor extracellular H_2O_2 production by measuring formation of oxidized HVA (excitation, 320 nm; emission, 405 nm). Then program the instrument to monitor kinetics of fluorescence emission for reactions performed at 37 °C, at 60 s intervals over the desired time course, typically 30 min.
4. Load the plate containing the trypsinized cell suspensions, and allow 5–10 min for cells to equilibrate at 37 °C before starting the injection of assay/activator reagents and kinetic fluorescence readings.
5. At the end of the reading cycle, process data to obtain relative reaction rates from slopes of fluorescence increases over time. Alternatively, 30-min total fluorescence endpoint assays can be used to determine absolute H_2O_2 yields per unit cell number/ well, based on a calibration curve obtained from H_2O_2 concentration standards assayed in parallel in the same plate (*see* Fig. 3a and **Note** 7). The yields of ionomycin-stimulated H_2O_2 are linear with respect to the number of DUOX2/ DUOXA2-expressing Flp-In 293 cells over the range of 50,000–300,000 cells/reaction (Fig. 3b). Based on these findings, routine assays typically use 200,000 cells/200 μL assay.

3.2.2 Luminol/HRP Assay

Chemiluminescence of luminol, when oxidized by H_2O_2 in the presence of HRP, provides a convenient and sensitive means to monitor real-time H_2O_2 generation by DUOX/DUOXA-reconstituted cells. The method described here has been validated as being completely dependent on co-expression of DUOX and DUOXA proteins, as there is no background luminescence observed with the parent cell line, in the absence of either DUOX or DUOXA chain or in the absence of exogenous HRP. Total integrated chemiluminescence observed over a 30-min time course following cell stimulation is linear with respect to transfected DUOX or DUOXA plasmid amounts or reconstituted cell numbers routinely used in the fluorescence-based assays (Fig. 3c).

1. Obtain transfected cells from **step 6** (Subheading 3.1), trypsinize, wash twice, and gently resuspend in $HBSS^{++}$ at the desired cell densities (typically, 2.0 × 10^6 viable cells/mL). Distribute 100 μL suspended cells into each well of a white opaque 96-well luminescence reading plate. Plan on running triplicate well assays in parallel; for other experimental design details, including negative controls, *see* **Note 5**.

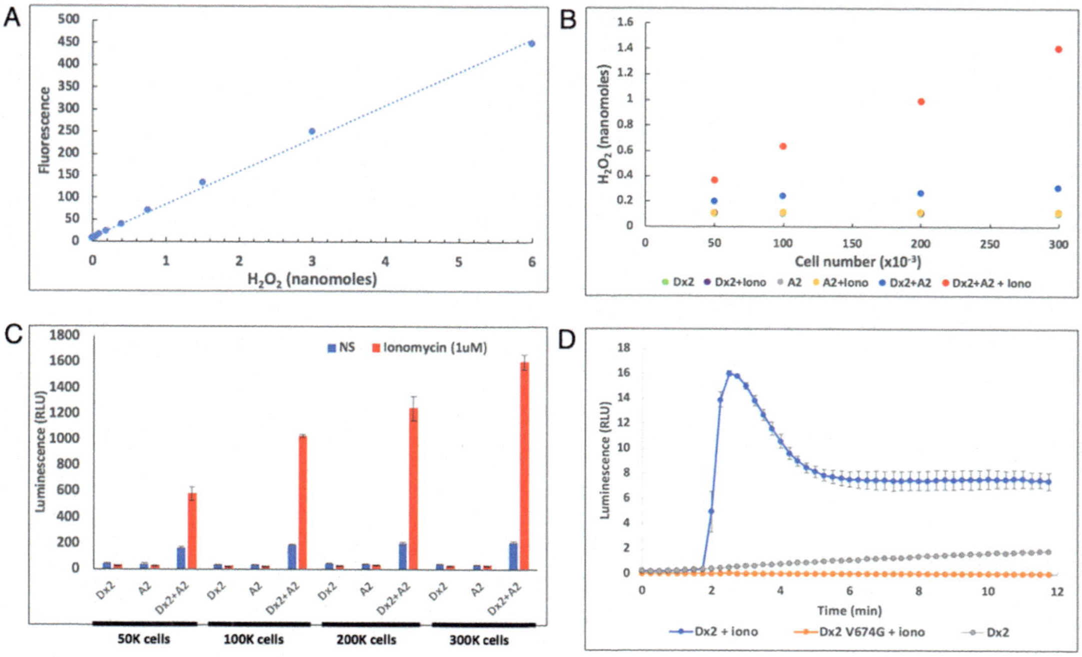

Fig. 3 Detection of H_2O_2 release by DUOX2/DUOXA2-reconstituted Flp-In 293 cells. (**A**) Standard calibration curve for detection of H_2O_2 by the HVA/HRP-based fluorimetric assay. (**B**) Cell concentration-dependent detection of H_2O_2 release by ^{HA}DUOX2/^{V5}DUOXA2-transfected Flp-In 293 detected by the HVA/HRP-based fluorimetric assay. (**C**) Cell concentration-dependent detection of H_2O_2 release by ^{HA}DUOX2/^{V5}DUOXA2-transfected Flp-In 293 detected by the luminol/HRP-based chemiluminescence assay. (**D**) Kinetic luminol/HRP chemiluminescence assays demonstrating the loss of NADPH oxidase activity in stable ^{V5}DUOXA2-expressing cells transiently transfected with ^{HA}DUOX2 V674G mutant cDNA in comparison with cells transfected with wild-type ^{HA}DUOX2. DUOX2 (Dx2), DUOXA2 (A2), ionomycin (Iono). Data in Fig. 3D reproduced from Ref. [17]

2. Prepare fresh 2× luminol/HRP hydrogen peroxide assay reagent for injection (100 μL), as well as the 10× activator reagent (20 μL). *See* **Note 6** regarding choice of DUOX activating agents. Load each into separate reservoirs and prime the injection lines. Then set up the plate template to command injection of activator/assay reagents into designated wells at the beginning of the 30-min kinetic assay.
3. Set up the plate-reading luminometer to monitor kinetics of luminescence for reactions performed at 37 °C, at 60 s intervals over the entire time course.
4. Load the plate containing the trypsinized cell suspensions into the instrument, and allow 5–10 min for cells to equilibrate at 37 °C before starting the injection of activator/assay reagent and kinetic luminescence readings.
5. At the end of 30-min reading cycle, process data to obtain the integrated relative luminescence over the 12 or 30 min reading cycle. Figure 3d shows that most of the ionomycin-stimulated

luminescence occurs within a few minutes following cell stimulation, whereas PMA-stimulated DUOX2 activation begins at a slower rate and is sustained for a longer time course. Thus, the real-time kinetic luminescence readings may detect subtle functional differences between DUOX or DUOXA structural variants.

3.3 Detection of DUOX and DUOXA Proteins

Any indications of compromised H_2O_2 generation observed with candidate DUOX or DUOXA structural variants must be investigated further by Western blotting methods described below to confirm normal protein expression or determine whether the defect relates to any structural instabilities. For example, defects in the interaction of DUOX and DUOXA or their inability to transit beyond the ER can lead to reduced stability of both polypeptide chains [17]. The addition of various peptide epitope tags illustrated in Fig. 2 provides a means for sensitive immunoblotting-based detection of both proteins.

3.3.1 Western Blotting

1. Prepare protein lysates from 2 × 10^6 DUOX/DUOXA-expressing cells from **step 6** (Subheading 3.1): Trypsinize, count, and wash once with HBSS, and then harvest by centrifugation at 200 × *g*, 3 min at 4 °C. Suspend cell pellet in 150 μL NP40/PIC/PMSF buffer by vortexing briefly, and then lyse for 15 min at 4 °C by incubating on a rotating platform (40 rpm). Recover solubilized proteins by centrifugation for 10 min at 10,000 × *g* at 4 °C, and transfer the supernatant into a new 1.5 mL tube. Samples can be stored at −80 °C or loaded onto SDS-PAGE gels.
2. Measure protein concentrations of lysates by BCA method (*see* **Note 8**).
3. Mix 50 μg of protein samples in 1.5 mL tubes with appropriate amount of 4× SDS-loading buffer before loading proteins onto SDS-PAGE gels. DUOX protein containing lysates should not be heated, as this can cause protein aggregation and prevent DUOX proteins from entering the gel. Separate protein samples on 4–12% gradient precast gels at 120 V (up to 170 V for faster migration) in 1× MOPS-SDS running buffer. Include 5 μL of pre-stained protein marker in a reference lane.
4. Electrophorese proteins onto a nitrocellulose membrane by wet transfer at 4 °C for 1 h 30 min at 100 V or (or 12 h at 45 V) in 1× NuPAGE transfer buffer. Confirm protein transfer to nitrocellulose membrane by staining with Ponceau S solution for 3 min under constant agitation. Rinse the membrane vigorously with water to visualize protein bands. If the quality of transfer is good (i.e., no bubbles, smearing, or presence of high molecular weight proteins), then cut the membrane into three pieces corresponding to M_r ranges of 30–60 kDa, 60–110 kDa, and 110–260 kDa.

5. Block membranes in TBS-T for 1 h at room temperature (RT), and then probe the appropriate membranes for 2 h at RT with anti-HA (1:2000), anti-HSP90 (1:2000), or anti-V5 (1:5000) antibodies.
6. Wash membranes six times, 5 min in TBS-T, and incubate at RT with HRP-conjugated goat secondary antibodies for 1 h (1:3000).
7. Wash membranes again six times, 5 min in TBS-T.
8. Remove blots, and incubate for 2 min at RT in 4 mL of ECL Plus detection reagent, and then visualize immune complexes on blots using a luminescence imaging system (i.e., iBright) or by exposure to X-ray films.

3.3.2 Detection of Modifications of Glycosylated Intermediates

During proper protein biosynthesis and subcellular transport, the *N*-glycosylated DUOX enzymes exit the ER in complexes with their activator (DUOXA) proteins and undergo Golgi apparatus-based carbohydrate modifications, in a process that renders both proteins resistant to endoglycosidase H digestion [8]. This maturation appears to be necessary to form an active enzyme complex at the cell surface. Endoglycosidase H treatment cleaves only unmodified high-mannose-type carbohydrates, characteristic of premature forms synthesized in the ER, resulting in a shift to higher mobility bands (lower M_r forms). The endoglycosidase H-resistant forms remain intact, exhibiting M_r's of 180kDa (DUOX2) and 40 kDa (DUOXA2). A typical experiment investigating proper DUOX-DUOXA posttranslational processing and Golgi apparatus-based glycosyl modifications that endow endoglycosidase H resistance is illustrated in Fig. 4. N-glycosidase F treatment provides a useful reference, as it removes all asparagine-linked carbohydrates from DUOX and DUOXA proteins, resulting in downshifted, smaller M_r bands of app. 170 kDa (DUOX1 and 2) and 33 kDa (DUOXA1α and DUOXA2). Insufficient protein processing, due to disrupted DUOX-DUOXA interactions, causes their retention in the ER and prevents Golgi apparatus transport and preservation of higher M_r, endoglycosidase H-resistant forms. For example, the loss-of-function DUOX2 V674G mutant exhibits a complete endoglycosidase H sensitivity indicated by the appearance of the downshifted, lower M_r forms (Fig. 4b).

1. For deglycosylation assays, prepare protein lysate from 2×10^6 DUOX- and DUOXA-expressing cells (from Subheading 3.1) in cold NP-40 buffer supplemented with PIC by the same methods as detailed above for Western blotting (Subheading 3.3.1, **step 1**).
2. In separate tubes digest, or not, equal amounts of proteins, usually 50 μg, with 1000 U of N-glycosidase F or endoglycosidase H at 37 °C for 1 h, and then stop reactions by adding 4× SDS-loading buffer. Proceed to SDS-PAGE and WB, as described above (Subheading 3.3.1).

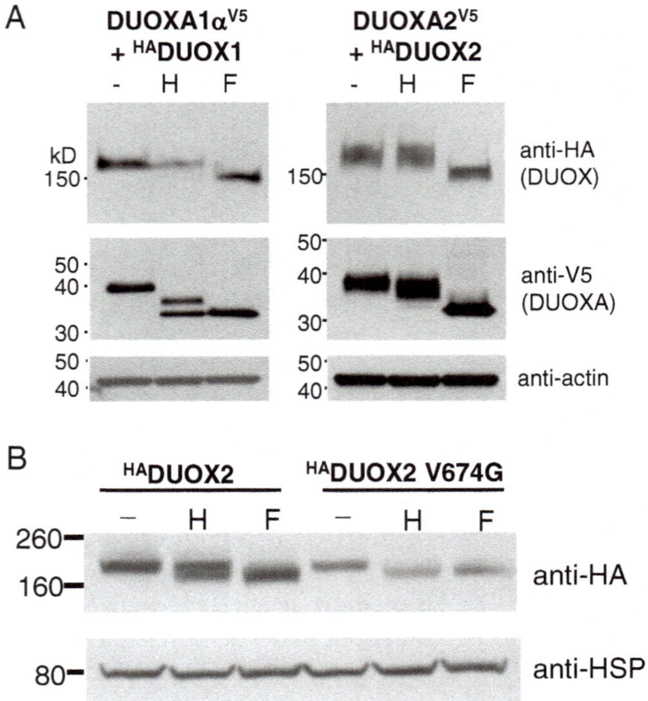

Fig. 4 Detection of endoglycosidase H-resistant forms of DUOX and DUOXA proteins indicative of Golgi apparatus-based modifications of *N*-linked glycosylated chains on DUOX and DUOXA. (**A**) Higher molecular weight bands represent species resistant to endoglycosidase H treatment (H) when active DUOX and DUOXA are co-expressed. Migration of N-glycosidase F-treated protein lysates (F), shown as a reference of the deglycosylated (low molecular weight) species. No treatment lane denoted by (–), showing the fully glycosylated species. (**B**) Absence of Golgi-based glycosyl modifications indicated by no Endo H resistance (shift to higher mobility species), in the case of mutant DUOX2 (V674G) that is associated with congenital hypothyroidism in mice

3.4 Detection of Cell Surface-Expressed DUOX and DUOXA Proteins

Alterations in cell surface expression of DUOX/DUOXA variants in reconstituted cells can be detected by two methods described below: immunofluorescence (IF) microscopy of fixed adherent stained cells and flow cytometry (FACS). Both methods reveal whether any putative DUOX or DUOXA mutation affects plasma membrane trafficking by immunostaining of the cell surface-exposed N-terminal epitope tags in non-permeabilized cells (Fig. 5). IF staining of detergent permeablized cells can provide insight on whether diminished H_2O_2 generation results from formation of aggregates of misfolded proteins that are retained within intracellular compartments. We observed this phenomenon with expression of the V674G DUOX2 mutant protein that causes

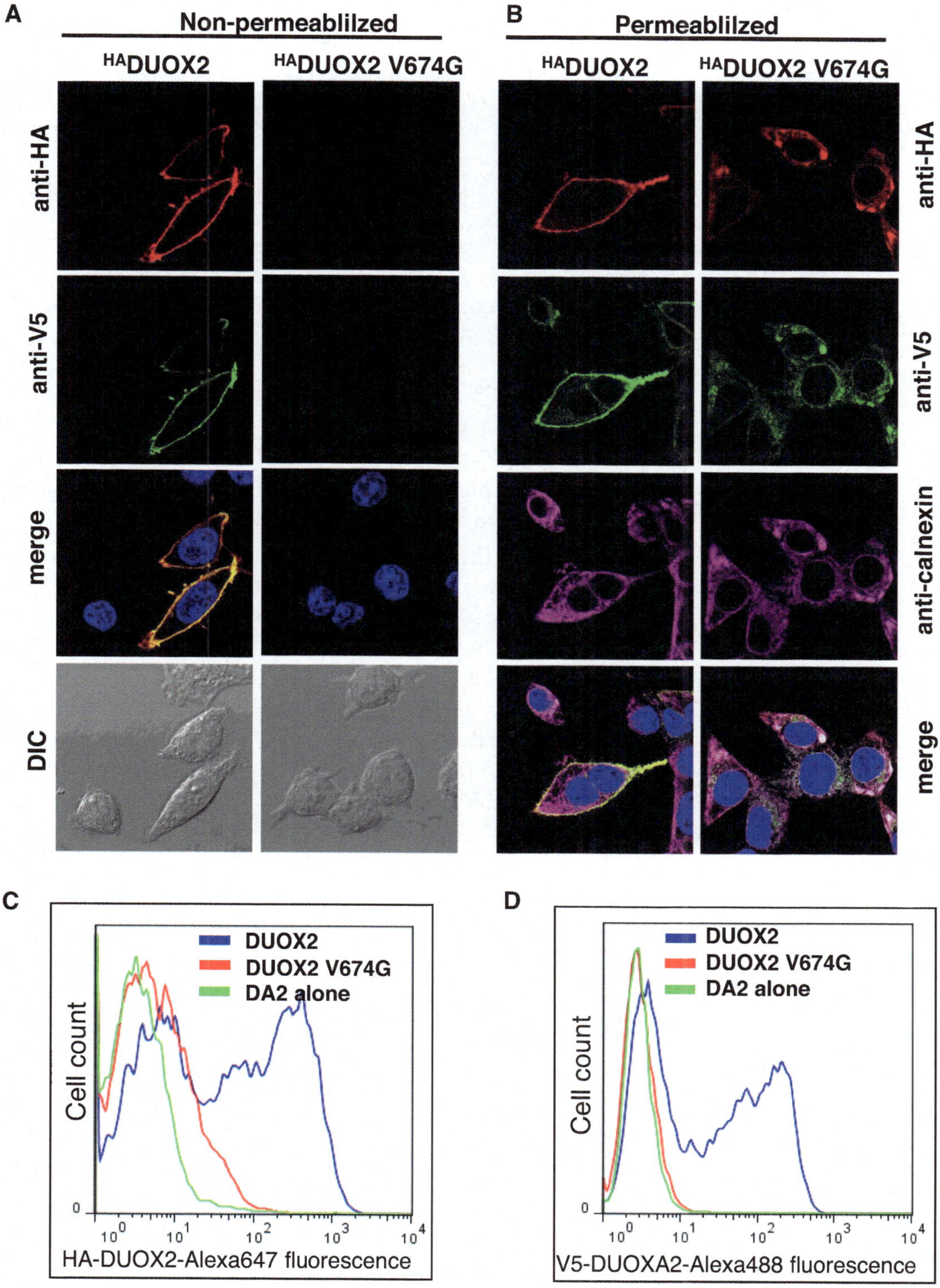

Fig. 5 Immunofluorescence detection of disrupted cell surface targeting and expression of missense mutant HA-DUOX2 V674G associated with congenital hypothyroidism in mice (**A**). The mutated protein is not detected in non-permeabilized cells (**A**) but is stained in detergent-permeabilized cells in a pattern resembling ER staining detected by the marker calnexin (**B**). Flow cytometry of non-permeabilized cells also fails to detect cell surface expression of HA-DUOX2 V674G (**C**) or the co-expressed DUOXA2 (DA2) partner in the same cells (**D**). Reproduced from Ref. [17]

congenital hypothyroidism in mice (Fig. 5a, b). Here we observed the co-staining of DUOXA2 and mutated DUOX2 aggregates in permeabilized cells overlapping with staining patterns for the ER marker calnexin.

While confocal imaging provides only a qualitative assessment of surface expression, the flow cytometric method allows precise quantification of cell surface expression of variants in non-permeabilized cells, usually expressed as a percentage relative to the wild-type proteins. A typical experiment analyzing cell surface expression by flow cytometry is shown in Fig. 5c, d.

3.4.1 Detection by Immunofluorescence Microscopy

1. Plate Flp-In 293 cells onto collagen1-coated 35 mm glass-bottom dishes, and transfect them with DUOX enzymes according to the above-described transfection guidelines.
2. After 24 h of expression, rinse cells once gently with 1 mL PBS, and then fix with 4% paraformaldehyde in PBS for 10 min.
3. Depending on the location of your DUOX and DUOXA epitope tag (N- vs. C-terminal) and on the experimental design, permeabilize cells (or not) with permeabilization buffer for 10 min, and then block with IF blocking buffer supplemented with 5% normal goat serum 1 h at room temperature.
4. Dilute primary antibodies (i.e., anti-HA, anti-V5, or anti-calnexin, depending on the DUOX/DUOXA tags used) in blocking solution, and incubate for 2 h at room temperature. Calnexin is a 67 kDa calcium-binding chaperon protein widely used as an ER marker.
5. Wash cells six times with PBS.
6. To reveal staining, dilute Alexa Fluor®-594-, Alexa Fluor®-488-, and Alexa Fluor®-647-labeled secondary goat antibodies (1:1000) in IF blocking buffer, and incubate for 1 h at RT.
7. Wash cells six times with PBS. During one of the washing steps, counterstain nuclei with DAPI (1:10000) for 3 min at RT.
8. Collect fluorescence images using a confocal laser-scanning fluorescent microscope equipped with a 63× oil immersion objective.

3.4.2 Detection by Flow Cytometry

1. Obtain transfected cells from Subheading 3.1, **step 6** following trypsinization, and then count and resuspend in HBSS. Transfer 1×10^6 into a 5 mL polystyrene tube from each DUOX or DUOXA variants and one additional tube for double staining of (+) control wild-type DUOX/DUOXA-expressing cells, as well as three additional tubes for (−) control cell types: one for no antibody staining and others for each (−) antibody isotype control staining. A typical experimental setup looks as follows:

DUOX/ DUOXA	Variant/X	Variant/Y	Variant/Z	(–) ctrl cell	(–) ctrl cell	(–) ctrl cell
Double staining	Double staining	Double staining	Double staining	Isotype IgG1 + 647	Isotype IgG2a + 488	No staining

2. Centrifuge cells for 3 min 200 × *g* at 4 ° C, and block with 100 μL FACS blocking buffer for 15 min.
3. Sequentially supplement FACS blocking buffer with anti-HA (1:100), anti-V5 (1:100), mouse IgG1 (1:10), or mouse IgG2a (1:25) isotype primary antibodies, and incubate on ice for 30 min.
4. Wash three times with 4 mL cold PBS followed by centrifugation for 3 min 200 × *g* at 4 °C. Each time resuspend the cell pellet in two steps, add first 1 mL cold PBS, gently shake, and then add the remaining 3 mL of cold PBS.
5. After the third wash, aspirate the liquid completely, and add 100 μL of cold 2% FBS in PBS containing goat α-mIgG1-Alexa647 and/or goat α-mIgG2a-Alexa488 secondary antibodies at 1:1000 dilutions. Gently shake to resuspend the cells and incubate on ice for 15 min.
6. Wash three times with 4 mL cold PBS, and harvest cells by centrifugation for 3 min 200 × *g* at 4 °C. Each time first resuspend the cell pellet in 1 mL cold PBS, gently vortex, and then add the remaining 3 mL of cold PBS.
7. After the third wash, add 500 μL cold PBS, and vortex gently to resuspend the pellet. Cells are now ready for FACS analysis.
8. Set flow cytometer background using cells without staining, and then monitor fluorescence of DUOX/DUOXA-expressing cells on a FACSort flow cytometer or comparable system by acquiring 10,000 events for each condition.
9. Data analysis can be performed with FlowJo or similar software. For comparisons of surface expression of different variant DUOX- or DUOXA-expressing cells, populations exhibiting fluorescence intensities greater than the peak observed from cells expressing DUOX or DUOXA alone are gated, and plasma membrane exposure of the unique variants is expressed as the percent of that observed with wild-type DUOX/DUOXA-transfected cells.

4 Notes

1. We favor the clonal FLP-In 293 lines stably expressing DUOXA (or DUOX) to test activities of DUOX (or DUOXA) variants, although we have used COS-7 cells transiently transfected with equal amounts of DUOX and DUOXA [8, 17]. The latter are large cells that work well for microscopic imaging. Other commonly used lines include HeLa cells or H661 respiratory epithelial cells stably expressing DUOXA2 [7, 9]. While not well developed, thyrocyte lines may be better suited to test candidate DUOX2 and DUOXA2 functional variants associated with congenital hypothyroidism.
2. High transfection rates (50–70%) have been achieved with Fugene6 or Lipofectamine LTX/Plus transfection reagents, which are desirable for cell trafficking experiments.
3. The activities of N-terminally tagged V5- or FLAG-DUOXA and signal peptide-fused HA DUOX proteins shown in Fig. 1 compare well with the untagged native versions of DUOXA and DUOX; these epitope tags are detected on the cell surface of DUOX-reconstituted HEK-293 or Flp-In 293 cells [8]. C-terminal peptide fusions on DUOX block its NADPH oxidase activity, as with other NOX enzymes. C-terminal Myc-tagged versions of DUOXA have been used to reconstitute DUOX activity, although C-terminal EGFP/Myc-tagged DUOXA does not appear to exit the ER [7, 9]. C-terminal V5- or Myc-tagged DUOXA translocates to the plasma membrane when co-expressed with DUOX, but is not detected without cell permeabilization [8, 9].
4. For smaller-scale transfections in 6-well plates, seed 2×10^5 cells/well in 2 mL medium 1 day in advance. Scale down transfection reagent volumes in **step 5** by 1/6 per well (1 μL Fugene6/ 1 μg DNA in 200 μL).
5. Include negative control assays of transfected cells lacking DUOX or DUOXA, along with other assays lacking HRP, or containing catalase (H_2O_2 scavenger) or the oxidase/flavoenzyme inhibitor, diphenylene iodonium (10 μM).
6. Although DUOX1 and DUOX2 exhibit low levels of constitutive activity, assays performed in response to DUOX activators are needed for a complete functional study. While we routinely use the calcium ionophore ionomycin (1 μM) in calcium-containing medium, other calcium-mobilizing agonists may be used, including thapsigargin or ATP. DUOX2 also responds to PKC activators, whereas DUOX1 responds to PKA activation [18].
7. The H_2O_2 standard calibration curve and its use in converting fluorescence into cellular yields of H_2O_2: Prepare triplicate wells of twofold serial dilutions of hydrogen peroxide standards

in HBSS^{++}, i.e., 10 dilutions over the range of 0–60 μmolar concentrations in 100 μL. These dilutions of 60, 30, 15, 7.5, 3.75, etc. μM of H_2O_2 correspond to 6, 3, 1.5, 0.75, 0.375, etc. nanomoles of H_2O_2 per well. Then add 100 μL of 2× HVA/HRP assay buffer to each well, and incubate 30 min at 37 °C in parallel with the other kinetic cellular assays of H_2O_2 release in the same plate. Use the 30-min endpoint fluorescence values from these standards to obtain a standard calibration curve (e.g., Fig. 3a) from which the cellular yield of H_2O_2 can be calculated. Yields can be expressed as nanomoles of H_2O_2 per unit cell number.

8. BCA assay: Use BSA to prepare a standard calibration reference curve. A typical 6-point series of calibration standards ranges as follows: 0, 0.5, 1.0, 2.0, 4.0, and 8.0 mg/mL BSA dissolved in the same NP40/PIC/PMSF buffer. Load 5 μL of samples in triplicate (standards and unknowns) into a transparent 96-well plate, add 200 μL of BCA mixture, and incubate at 37 °C for 30 min on a rocker platform (180 rpm). Read the absorbance at 562 nm on a spectrophotometer, and calculate protein concentrations using the standard calibration curve.

Acknowledgments

This work was supported by the Intramural Research Program of the National Institute of Allergy and Infectious Diseases, NIH.

References

1. Dupuy C, Ohayon R, Valent A et al (1999) Purification of a novel flavoprotein involved in the thyroid NADPH oxidase. Cloning of the porcine and human cdnas. J Biol Chem 274:37265–37269
2. De Deken X, Wang D, Many MC et al (2000) Cloning of two human thyroid cDNAs encoding new members of the NADPH oxidase family. J Biol Chem 275:23227–23233
3. Moreno JC, Visser TJ (2007) New phenotypes in thyroid dyshormonogenesis: hypothyroidism due to DUOX2 mutations. Endocr Dev 10:99–117
4. Grasberger H (2010) Defects of thyroidal hydrogen peroxide generation in congenital hypothyroidism. Mol Cell Endocrinol 322:99–106
5. Ohye H, Sugawara M (2010) Dual oxidase, hydrogen peroxide and thyroid diseases. Exp Biol Med 235:424–433
6. Muzza M, Fugazzola L (2017) Disorders of H_2O_2 generation. Best Pract Res Clin Endocrinol Metab 31:225–240
7. Grasberger H, Refetoff S (2006) Identification of the maturation factor for dual oxidase. Evolution of an eukaryotic operon equivalent. J Biol Chem 281:18269–18272
8. Morand S, Ueyama T, Tsujibe S et al (2009) Duox maturation factors form cell surface complexes with Duox affecting the specificity of reactive oxygen species generation. FASEB J 23:1205–1218
9. Luxen S, Noack D, Frausto M et al (2009) Heterodimerization controls localization of Duox-DuoxA NADPH oxidases in airway cells. J Cell Sci 122:1238–1247
10. Carre A, Louzada RA, Fortunato RS et al (2015) When an Intramolecular Disulfide Bridge Governs the Interaction of DUOX2 with Its Partner DUOXA2. Antioxid Redox Signal 23:724–733

11. Maruo Y, Takahashi H, Soeda I et al (2008) Transient congenital hypothyroidism caused by biallelic mutations of the dual oxidase 2 gene in Japanese patients detected by a neonatal screening program. J Clin Endocrinol Metab 93:4261–4267
12. Zamproni I, Grasberger H, Cortinovis F et al (2008) Biallelic inactivation of the dual oxidase maturation factor 2 (DUOXA2) gene as a novel cause of congenital hypothyroidism. J Clin Endocrinol Metab 93:605–610
13. Hayes P, Dhillon S, O'Neill K et al (2015) Defects in NADPH oxidase genes NOX1 and DUOX2 in very early onset inflammatory bowel disease. Cell Mol Gastroenterol Hepatol 1:489–502
14. Parlato M, Charbit-Henrion F, Hayes P et al (2017) First identification of biallelic inherited DUOX2 inactivating mutations as a cause of very early onset inflammatory bowel disease. Gastroenterology 153:609–611
15. Grasberger H, Noureldin M, Kao TD et al (2018) Increased risk for inflammatory bowel disease in congenital hypothyroidism supports the existence of a shared susceptibility factor. Sci Rep 8:10158
16. Donko A, Orient A, Szabo PT et al (2009) Detection of hydrogen peroxide by lactoperoxidase-mediated dityrosine formation. Free Radic Res 43:440–445
17. Donko A, Morand S, Korzeniowska A et al (2014) Hypothyroidism-associated missense mutation impairs NADPH oxidase activity and intracellular trafficking of Duox2. Free Radic Biol Med 73:190–200
18. Rigutto S, Hoste C, Grasberger H et al (2009) Activation of dual oxidases Duox1 and Duox2: differential regulation mediated by camp-dependent protein kinase and protein kinase C-dependent phosphorylation. J Biol Chem 284:6725–6734

Chapter 12

Guidelines for the Detection of NADPH Oxidases by Immunoblot and RT-qPCR

Becky A. Diebold, S. Garrett Wilder, Xavier De Deken, Jennifer L. Meitzler, James H. Doroshow, James W. McCoy, Yerun Zhu, and J. David Lambeth

Abstract

The identification of NADPH oxidase (NOX) isoforms in tissues is essential for interpreting experiments and for next step decisions regarding cell lines, animal models, and targeted drug design. Two basic methods, immunoblotting and reverse transcriptase-quantitative polymerase chain reaction (RT-qPCR), are important to monitor NOX protein and messenger RNA (mRNA) levels, respectively, for a range of investigations from understanding cell signaling events to judging NOX inhibitor efficacies. For many other genes that are expressed in high abundance, these methods may seem rather simple. However, detecting the low expression levels of endogenous NOX/DUOX is difficult and can be frustrating, so some guidelines would be helpful to those who are facing difficulties. One reason why detection is so difficult is the limited availability of vetted NOX/DUOX antibodies. Many of the commercial antibodies do not perform well in our hands, and dependable antibodies, often generated by academic laboratories, are in limited supply. Another problem is the growing trend in the NOX literature to omit end-user validation of antibodies by not providing appropriate positive and negative controls. With regard to NOX mRNA levels, knockdown of *NOX/DUOX* has been reported in cell lines with very low endogenous expression (C_q values ≥ 30) or in cell lines devoid of the targeted NOX isoform (e.g., *NOX4* expression in NCI-60 cancer cell panel cell line 786–0). These publications propagate misinformation and hinder progress in understanding NOX/DUOX function. This chapter provides overdue guidelines on how to validate a NOX antibody and provides general methodologies to prepare samples for optimal detection. It also includes validated methodology to perform RT-qPCR for the measurement of NOX mRNA levels, and we suggest that RT-qPCR should be performed prior to embarking on NOX protein detection.

Key words Immunoblot, NOX antibodies, NOX qPCR primers

1 Introduction

NADPH oxidases (NOX) are the major producers of purpose-driven ROS production. Unlike other enzymes that produce ROS as a side reaction during the course of catalysis, the NADPH oxidase enzymes use NADPH to reduce molecular oxygen to form superoxide and hydrogen peroxide as their sole known enzymatic function. NOX enzyme expression is complex and dependent

Ulla G. Knaus and Thomas L. Leto (eds.), *NADPH Oxidases: Methods and Protocols*, Methods in Molecular Biology, vol. 1982, https://doi.org/10.1007/978-1-4939-9424-3_12,

upon factors such as species, tissue, cell type, and physiological or pathological status. Thus, the cataloguing of NOX expression has often been reported by specific tissue type. While immunoblotting and PCR are seemingly simple methods, NOX detection presents a special case requiring knowledge of certain precautions and preparations prior to embarking on these methods. Membrane proteins, in general, are more difficult to detect, and NOX proteins have adorned themselves with an "invisibility cloak" [a la Harry Potter] which may be due to their low abundance, glycosylation, resistance to standard extraction methods, susceptibility to proteolytic degradation, and perhaps a variety of other factors. Many manufacturers market "anti-NOX isoform antibodies," but the majority of their validation data (and/or literature referenced in their product data sheets) are incomplete (e.g., lack of appropriate positive and negative controls, no information about isoform specificity). In our hands many of these antibodies fail to identify the claimed NOX isoform. Experts in the field are frustrated by the poor results obtained when using such antibodies. Astonishingly, a large number of manuscripts are accepted without proper antibody validation. Thus, guidelines are needed in order to properly evaluate immunoblot data to avoid misleading new investigators in this field who are searching for appropriate reagents.

A number of validated NOX antibodies have emerged from academic laboratories. Some of them are listed in Tables 5, 6, 7, 8, 9, 10, and 11, and additional ones are described in [1]. NOX and DUOX antibodies listed in Tables 5, 6, 7, 8, 9, 10, and 11 are not all-inclusive, and the reader is advised to judge other antibodies in the literature based on the methods of validation described in these guidelines. Exact dilutions for each antibody are not provided since it will vary depending on the characteristics of the sample.

Given the lack of dependable commercial antibodies, many researchers have relied upon PCR as a surrogate for the detection of NOX expression. Yet even with the specificity of NOX primers, conflicting results can be obtained regarding the presence of specific NOX isoforms in identical tissues under basal conditions. Thus, this chapter also includes a section describing one method for RT-qPCR. The basal levels of NOX mRNA are often low, indicated by quantification cycle (C_q) values in the range of 30–32. Thus, relying solely on reverse transcriptase PCR (RT-PCR) without performing reverse transcriptase-quantitative PCR (RT-qPCR) does not provide a secure sense of the quantity of NOX mRNA present. In addition, the high C_q values, indicating very low mRNA quantities, should heighten the level of concern about false-positive results arising from trace contamination from genomic DNA (gDNA), NOX-bearing plasmids, and other sources. Thus, the inclusion of appropriate negative controls is crucial, and the handling of samples requires rigorous protocols to avoid contamination. In addition, several NOX isoforms have variants due to alternative splicing. Thus, an added layer of precision is required in designing primers that detect the full-length form. We recommend that RT-qPCR be performed prior to immunoblotting to obtain an indication of whether the level of protein is likely to be detectable by immunoblot. Here, we share what we have learned and hope that these guidelines will improve the validity of published NOX immunoblots and PCR data in the future.

2 Materials

2.1 RNA Isolation and Reverse Transcriptase-Quantitative PCR (RT-qPCR)

1. RNase*Zap*™ RNase decontamination solution.
2. Decontaminated workspace and equipment free of NOX plasmids.
3. Nuclease-free water.
4. Nuclease-free pipette tips with barrier filters.
5. RNeasy Plus Kit containing gDNA eliminator columns.
6. 70% ethanol prepared in DEPC-treated water.
7. DEPC-treated water.
8. β-mercaptoethanol.
9. 1.5 mL micro-centrifuge tubes.
10. Phosphate-buffered saline (PBS).
11. Cell scrapers.
12. 200-proof ethanol (100%) to prepare 70% ethanol in DEPC-treated water.
13. 1 mL syringes and 21-gauge needles for homogenization or Qiashredder columns.
14. qScript XLT™ cDNA SuperMix.
15. 0.2 mL PCR tubes.
16. A cold 4 °C cube designed to hold small tubes and 96-well plates for PCR setup.
17. Mini-centrifuge capable of briefly spinning down 0.2 mL PCR tubes.
18. RT-qPCR instrument capable of accepting 96-well plates such as Bio-Rad's CFX96.
19. 96-well PCR plate suggested by the manufacturer of the PCR instrument.
20. Adhesive sealing film for PCR plates.
21. A paddle to apply the adhesive sealing film.
22. Primers for the reference gene and for the specific NOX isoform(s).

23. PerfeCta® SYBR® Green PCR SuperMix or similar.
24. A centrifuge capable of spinning 96-well plates.

2.2 SDS-PAGE and Immunoblotting

1. SDS-PAGE, appropriate percentage or gradient gels.
2. Standard running buffer: 0.025 M Tris-base, do not adjust pH, 0.192 M glycine, 0.1% SDS.
3. Power supply capable of attaining 200 volts.
4. Pre-stained MW protein ladder.
5. Polyvinylidene fluoride (PVDF) membrane, filter papers, sponges, and transfer cassettes and tanks (standard wet transfer).
6. Standard transfer buffer: 0.025 M Tris-base, do not adjust pH, 0.192 M glycine, 20% (v/v) of 100% methanol, chilled to 4 °C.
7. 100% methanol.
8. TBST: 10 mM Tris–HCl, pH 7.4, 0.15 M NaCl. 0.1% Tween-20.
9. Blocking buffer: 5% nonfat dry milk in TBST. Store at 4 °C.
10. Primary antibodies: *See* Tables 5, 6, 7, 8, 9, 10, and 11 for recommendations.
11. Secondary antibodies: HRP-conjugated goat anti-rabbit, HRP-goat anti-mouse, or fluorescent-labeled antibodies.
12. ECL reagent such as SuperSignal West Pico Chemiluminescent Substrate.
13. Plastic sheets to encase the immunoblot.
14. Autoradiography cassette equipped with intensifying screens on each side.
15. Autoradiography film and processor (or other film-less detection system).
16. Optional: *LI-COR* Odyssey fluorescent near-infrared (IR) scanner.
17. Optional: *LI-COR* Odyssey Blocking buffer (TBS).
18. Optional: *LI-COR* Odyssey IRDye 680LT (red)-labeled secondary antibody.
19. Optional *LI-COR* Odyssey IRDye 800Cw (green)-labeled secondary antibody.
20. Optional *LI-COR* Odyssey fluorescent IRDye-labeled protein ladder.
21. Plasmids for positive control antigens: Mammalian expression plasmid containing the NOX cDNA sequence for a specific NOX isoform.
22. Negative control plasmid: Same plasmid as above without the NOX sequence.

23. Human embryonic kidney cell line (HEK 293 cells) or any other cell line with very low or no endogenous expression of NOX/DUOX enzymes.
24. DMEM medium, high glucose (4.5 g/L), containing 10% FBS, or similar.
25. Cell line to be tested, appropriate culture medium.
26. Transfection reagent, such as X-tremeGENE 9, Lipofectamine, etc.
27. OptiMEM medium without any supplements.
28. RIPA buffer: 50 mM Tris–HCl, pH 7.5, 150 mM NaCl, 0.5% sodium deoxycholate, 0.1% SDS, 1.0% Triton X-100.
29. Protease and phosphatase inhibitor cocktail or tablets.
30. Sonicator equipped with a pencil-tip probe or 1 mL syringes and 21-gauge 1-in. needles for homogenization.
31. Micro-centrifuge, cooled to 4 °C.
32. 4× Laemmli's sample buffer: 200 mM Tris–HCl, pH 6.8, 20% glycerol, 0.4% SDS, 4% β-mercaptoethanol.
33. Protein determination kit, such as the BCA protein determination kit.

3 Methods

3.1 RNA Isolation

1. Use a PCR-clean workspace and tools that have been washed with soap and have been decontaminated from DNA (e.g., NOX plasmids and amplicons) by treatment with UV light. Spray the workspace, tools, and your gloves with ZAP spray. Wipe off any excess with a paper towel (*see* **Note 1**, Fig. 1).
2. Use barrier-filtered, PCR-clean pipet tips for all the following steps. Spray the surface of the pipette tip boxes with RNase Zap spray, especially where the thumb is placed.
3. Follow the protocol from Qiagen RNeasy Plus Kit described briefly below for cell lines. Refer to the manual for isolation of RNA from other sources such as tissues. Be sure to wipe the interior and exterior surfaces of the micro-centrifuge with Zap spray, especially knobs, dials, buttons, and the lid where the thumb is placed. Clean any rings of precipitated material from previous spins from the interior walls of the micro-centrifuge.
4. Wash adherent cells twice with PBS (*see* **Note 2**).
5. Add 600 μL of resuspension buffer containing fresh 5% β-mercaptoethanol to a 10-cm plate of rinsed cells (about 2.5×10^6 cells).
6. Scrape the cells off the plate, and transfer to the next plate and scrape again. A total of 1×10^7 cells can be processed in 600 μL. Transfer to a 1.5 mL micro-centrifuge tube.

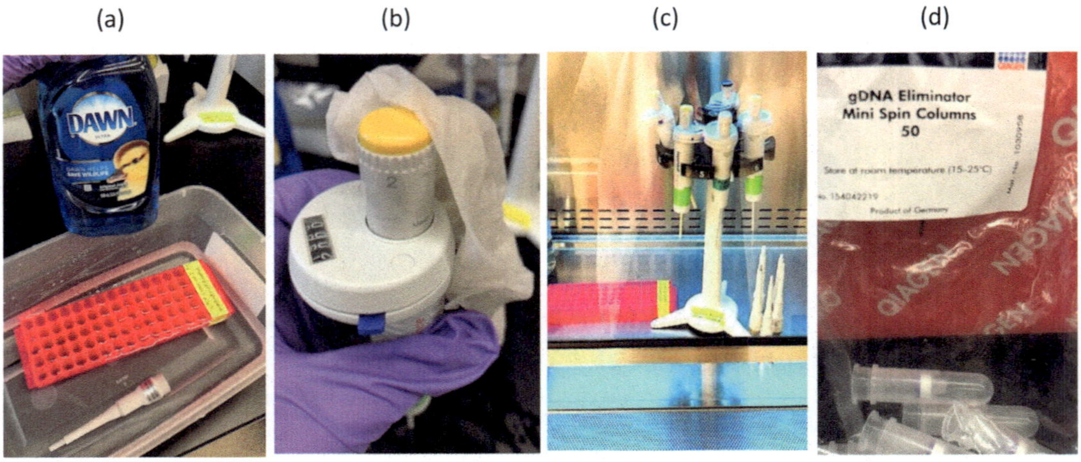

Fig. 1 Preventing contamination in RT-qPCR. (**a**) Wash the bench and surfaces of tools and equipment with a paper towel soaked in diluted detergent and water. If possible unscrew the lower parts of the pipette barrel and wash in soapy water. (**b**) The dial and top of the pipette where the thumb is placed is a large source of contamination. (**c**) Treat the washed pipettes and test tube racks with UV light in a biosafety cabinet or in a UV light box for 1 h. (**d**) Use a gDNA eliminator column twice to ensure gDNA removal

7. Homogenize the cells using a Qiashredder column (Qiagen), or manually homogenize by using a 1-mL syringe and 21-gauge needle, 5 passes. Take 600 μL and add it to the gDNA eliminator column. Spin 10,000 rpm (8000 × *g*) for 30 s in a micro-centrifuge. Collect the flow through. Dispose of the used gDNA column, and reapply the flow through to a fresh gDNA column, and spin again to ensure that the gDNA is removed (*see* **Note 3**).
8. Precipitate the RNA by adding an equal volume (600 μL) of 70% ethanol prepared in DEPC-treated water.
9. Take 700 μL of the 1200 μL sample and add it to the RNA-binding column. Spin as before. Discard the flow through. Add the remainder of the sample to the same column and spin again.
10. Discard the flow through, and wash the column with the wash buffers as directed by the kit.
11. Dry the column by performing the optional spin at 13,200 rpm (15,000 × *g*), 1 min without buffer added to the column. Elute with 30–50 μL of nuclease-free water to keep the RNA concentrated. Determine the RNA concentration using a Nanodrop spectrophotometer or by other standard methods (*see* **Note 4**).
12. Store RNA samples at −70 to −80 °C or proceed to cDNA synthesis on the same day (*see* **Note 5**).

3.2 cDNA First-Strand Synthesis

1. On the day of cDNA first-strand synthesis, thaw the RNA and keep it on ice. Place 2 μg of RNA into each 0.2 mL PCR tube (or well of a 96-well PCR plate for many samples). Prepare the negative control (Tube 1) first by adding nuclease-free water to the RNA to 20 μL total. To the test sample (Tube 2), add nuclease-free water to the RNA for a final volume of 16 μL. Add 4 μL of 5× qScript XLT cDNA SuperMix only to Tube 2. Each tube should have 20 μL total volume. Of these two tubes, 10 μL will be used for one NOX PCR master mix of 100 μL (4 replicate wells) and 10 μL will be used for one reference gene PCR master mix (4 replicate wells). Prepare enough cDNA first strand (and negative controls) if additional conditions are intended. Spin the PCR tubes briefly in a mini-centrifuge to bring the liquid to the bottom of the tubes before placing them in the thermocycler. If using a 96-well plate for many cDNA first-strand samples, seal the plate and centrifuge the plate at 400 × *g* for 5 min at 4 °C. Place in a standard thermocycler (it is not necessary to use a real-time PCR instrument at this step), and set the program according to the manufacturer's instructions. If using the qScript XLT cDNA SuperMix, set the program for 5 min at 25 °C, 50 min at 42 °C, 5 min at 85 °C, and hold at 4 °C (*see* **Note 6**).
2. Return the RNA to the −70 °C freezer. Plan the experiment so that the first-strand cDNA can be used immediately, but if necessary freeze the first-strand cDNA tubes at −20 °C overnight (*see* **Note 7**).

3.3 Setting Up the RT-qPCR Master Mix and 96-Well Plate

1. Set up the following master mix RT-qPCR reactions (100 μL total volume in 1.5 mL tubes or in a 96-well plate if many samples are being processed). Avoid working on ice since ice may accidently contaminate the samples. Instead use a 4 °C test tube rack (cube) designed to hold PCR tubes or plates. If using a SuperMix containing a hot-start DNA polymerase, then the setup may be performed at 25 °C. Each 100 μL of master mix reaction will be pipetted into 4 replicate wells of the PCR plate (20 μL/well plus 20 μL extra).
2. Always start with the no template control (NTC) tubes first, then the negative control (NC) tubes, and finally the unknown (UNK) tubes (*see* **Note 8**).
3. No template controls (NTC): Add 50 μL of 2× PerfeCtA® SYBR® Green SuperMix (*see* **Note 9**). Add 30 μL of nuclease-free water. Add 10 μL of 10× (2 μM) NOX forward primer and 10 μL of 10× (2 μM) NOX reverse primer to NTC-1. The NOX primers should be isoform-specific and able to detect the full-length variant. Table 1 lists some of the variants of human NOX isoforms, their main differences, and their accession numbers. Table 2 lists the human NOX primers that we use.

Table 1
Human NOX variants, accession numbers, and main characteristics

NOX isoform	Variant number (V) and NCBI accession #	Notes about the variant (see accession # for full details)
NOX1	(V1) NM_007052.4 (V2) NM_013955.2 (V3) NM-001271815	Full length Deletion of 1503–1610 (exon 10) Deletion of 459–543 (exon 4)
NOX2 (CYBB)	NM_000397.3	Full length
NOX3	NM_015718.2	Full length
NOX4	(V1) NM_016931.4 (V2) NM_001143836.2 (V3) NM_001143837 (V4) NM_001291926 (V5) NM_001291927.1 (V6) NM_001291929.1 (V7) NR_120406.1 (V8) NM_001300995.1	Full length Deletion of 1546–1656 (exon 14) Deletion of 1–375 (exon 1) Deletion of 472–475 (part of exon 3) Deletion of 1–375 (exon 1) Deletion of 1–471 (exon 1 and 2) Non-coding Deletion of 1–375 (exon 1) and of nucleotides corresponding to exon 14 of full length
NOX5 [8, 9]	(α) NM_001184779.1 (β) NR_033672.1 (γ) NM_024505.3 (δ) NR_033671.2 (ε) AF317889 (ζ) NM_001184780.1	NOX5α; 18 aa extension at the start of the N-terminus (exon 3); deletion of 27 aa in exon 6; predominant isoform in spleen NOX5β; deletion of 27 aa in exon 6; predominant isoform in testis NOX5γ complete exon 6 without the 18 aa extension of the N-terminus NOX5δ complete exon 6 with the 18 aa extension of the N-terminus NOX5ε (a.k.a. NOX5-S); truncated from exon 3 through most of exon 7 NOX5ζ: contains an extra 25 aa at the start of the N-terminus
DUOX1	NM_017434.4	Full length
DUOX2	NM_014080.4	Full length

All of them detect the full-length isoform as well as all the variants known to date (*see* **Note 10**). Table 3 lists mouse NOX primers from a reference [2]. Add 10 μL of 10× (2 μM) reference gene forward primer and 10 μL of 10× (2 μM) reference reverse primer to NTC-2. Do not add any cDNA or RNA to these NTC tubes. Table 4 lists a selection of human primers for reference genes that have been suggested by Integrated DNA Technologies' RT-qPCR tutorial manual available on their website (*see* **Note 11**). Primers for other species can be purchased from Integrated DNA Technologies or other companies and validated as we have done for human NOX RT-qPCR primers.

Table 2
Primers used in the detection of human NOX isoforms by RT-qPCR

NOX isoform	Human sequence of primer	Exon location and T_m	Integrated DNA technologies Cat. #
NOX1	GGTCAACACGAGGAGAGC CAAGGATCCACTTCCAAGACTC	E7F, 60.7 E8R, 61.4	Hs.PT.49a.435525.4
NOX2 (CYBB)	CCCAATCCCTCAGTTTGCT CCTTCTGTTGAGATCGCCAA	E7F, 61.1 E8R, 61.2	Hs.PT.51.3983106
NOX3	ACCTTCTGTAGAGACCGCTAT TCACATGCATACAAGACCACA	E7F, 61.2 E8R, 61.1	Hs.PT.51.3065606
NOX4	CTGTGGTGTTACTATCTGTA TTTTCTC CTTGCTGCATTCAGTTCAACA	E4F, 60.7 E5R, 60.8	Hs.PT.51.22695360.g
NOX5	GCCAGTGCCTCAACTTCG CCACTACCACGTAGCCCATA	E7F, 62.1 E8R, 62.1	Hs.PT.58.40874887
DUOX1	GCGTCTACATGAGAAATGCCA GCAGCAGTGCATCCACAT	E10F, 61.8 E12R, 61.9	Hs.PT.51.21346453
DUOX2	CGCCACCTACCAGAACATC GGTAGAGAAGAACTGC TCAGAG	E7F, 61.1 E9R, 61.3	Hs.PT.51.20690545

Primers from PrimeTime™ qPCR Pre-designed Assays (without probes) were purchased from Integrated DNA Technology (IDT) and may have been modified slightly to obtain desired Tm. Primers are listed in the 5′ to 3′ direction (R = reverse and F = forward). The exon number may be different for different accession numbers. The T_m values were determined using IDT's OligoAnalyzer tool (http://www.idtdna.com/calc/analyzer) using the following final concentrations in the qPCR reaction: [primers] = 200 nM, [Na^+] = 55 mM, [Mg^{2+}] = 1.5 mM, and [dNTP] = 0.2 mM. The actual annealing temperatures were determined by performing a gradient RT-qPCR for each primer set using the methods described in the text. These primers detect variant 1 (the full-length variant). To determine whether these primers detect other variants, the user should check the accession number of the variant, some of which are provided in Table 1

4. Negative controls (NC): Add 50 μL of 2× PerfeCtA® SYBR® Green SuperMix. Add 20 μL of nuclease-free water. Add 10 μL 10× of forward and 10 μL of 10× reverse NOX primers to NC-1. Add 10 μL (1 ug RNA) of the negative control first-strand cDNA synthesis tube prepared during the cDNA synthesis step. Those tubes only contain RNA and no cDNA. Each tube will now have 100 μL total volume. Continue with NC-2, which will have the reference gene primers instead of NOX primers.
5. Unknowns (UNK's): Add 50 μL of 2× PerfeCtA® SYBR® Green SuperMix. Prepare one UNK tube for each set of primers to be tested. Add 20 μL of water, 10 μL of 10× forward and 10 μL of 10 × reverse primer, and 10 μL of cDNA (1 ug RNA equivalents) that was synthesized in the presence of first-strand cDNA synthesis SuperMix. The total volume is 100 μL.

Table 3
Primers used in the detection of mouse Nox isoforms by RT-qPCR

Nox isoform	Mouse primer sequence	Exon location and T_m	Accession no.
Nox1	CCCAGCAGAAGGTCGTGATT GCTAAAGCCTCGCTTCCTCAT	E8F, 63.2 E9R, 63.5	NM_172203.2
Nox2 (Cybb)	CAGGAACCTCACTTTCCATAAGAT AACGTTGAAGAGATGTGCAATTGT	E4F 61.5 E5R 63.0	NM_007807.5
Nox3	CGACGAATTCAAGCAGATTGC AAGAGTCTTTGACATTGCTTTGG	E12F. 61.3 E13R, 61.0	NM_198958.2
Nox4	CCGGACAGTCCTGGCTTATCT TGCTTTTATCCAACAATCTTCTTGTT	E3F, 64.3 E4R, 61.7	NM_015760.5
Duox1	CACCAGGAACGGATTGTTCT CCTGCAAGCCAAAAGAAGAC	E12–13F, 61.2 E13R, 60.7	NM_001099297.1
Duox2	TGCAACAGCTACTGGATTCG ATCCTGTCCTCCAGCTCTGA	E9F, 61.4 E10R, 63.3	NM_001362755.1

These primers were selected from [2]. Primers are listed in the 5′ to 3′ direction (R = reverse and F = forward). The exon number may be different for different accession numbers. The T_m values were determined using IDT's OligoAnalyzer tool (http://www.idtdna.com/calc/analyzer) using the following final concentrations in the qPCR reaction: [primers] = 200 nM, [Na^+] = 55 mM, [Mg^{2+}] = 1.5 mM, and [dNTP] = 0.2 mM. These primers detect the full-length variant 1 sequence. The listed accession numbers contain details of existing variants

Table 4
Some suggested human reference primers

Gene	Primer sequence	Exon location	Accession # IDT Cat. #
HMBS	GCAACTGTACCTGACTGGA TCAGGGCCATCTTCATGC	E12–13F, 60.1 E14R, 60.7	NM_000190.3 Hs. PT.58.242481164
HPRT1	CATCAAAGCACTGAATAGAAATAGTGA CCAATTACTTTTATGTCCCCTGTT	E3F, 60.5 E4R, 61.1	NM_000194 Hs. PT.53a.20881146
HPRT1	AGGTATGCAAAATAAATCAAGGTCAT TTCTCCTGAGCAGTCAGC	E1F E2R	NM_000194 Hs. PT.53a.2145446
YWHAZ	TCTGATCCCCAATGCTTCAC TGGTATGCTTGTTGTGACTGA	E8F, 60.9 E9R, 60.9	NM_003406 Hs. PT.53a.20437075

A discussion of reference gene choice can be found in Refs. [3, 10]. Primer sequences are 5′ to 3′. Catalog numbers are for IDT's Prime Time Pre-Designed qPCR Assays. The exon number may be different for different accession numbers. The T_m values were determined using IDT's OligoAnalyzer tool using the following final concentrations in the qPCR reaction: [primers] = 200 nM, [Na^+] = 55 mM, [Mg^{2+}] = 1.5 mM, and [dNTP] = 0.2 mM. The actual annealing temperatures were determined by performing a gradient RT-qPCR for each primer set using the methods described in the text. Primers for mouse or rat can be obtained by entering the name of the gene and desired species on the IDT website

6. Use a clean, decontaminated vortex to gently vortex all the PCR master mix tubes for a few seconds, and then spin in a clean, decontaminated micro-centrifuge for 30 s at 10,000 rpm

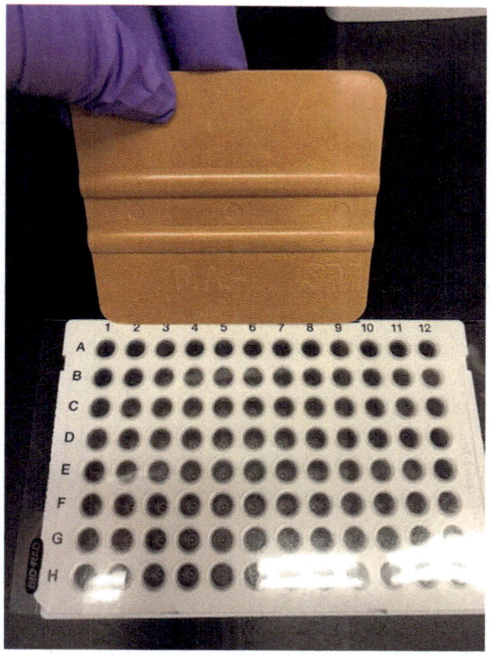

Fig. 2 Correct manual application of the adhesive seal to PCR plates. A plastic paddle is available commercially to assist in the proper application of the adhesive seal to PCR plates. Large standard deviations result from evaporation from individual wells that are not sealed completely

(8000 × *g*) to bring the liquid to the bottom of the tubes. Starting with the NTC tubes, pipet 20 μL into each of 4 wells of a 96-well PCR plate. There will not be enough volume for a fifth well. Proceed then with the NC tubes; pipet 20 μL into 4 wells. Pipet the UNK wells last, and then change gloves again.

7. Seal the PCR plate with a clear adhesive film made for PCR plates. Use a plastic PCR paddle, and drag it over the surface of the film to ensure that each well is sealed. Use the edge of the paddle and drag it in between column, rows, and all along the edge of the outer rows and columns to ensure that each individual well is sealed tightly (*see* **Note 12** and Fig. 2). Centrifuge the sealed plate at 400 × *g* (2000 rpm) for 5 min at 4 °C to bring all the liquid to the bottom of the wells.

3.4 Running the RT-qPCR Program and Obtaining the Data

1. Insert the 96-well plate into a RT-qPCR instrument, and enter the plate layout using the software of the instrument. Figure 3a shows an example of a 96-well PCR plate layout using Bio-Rad's CFX 96 instrument and CFX Manager software. Choose a two-step PCR program followed by a melt curve. An example of a two-step program that we use is 3 min at 95 °C for 1 cycle (hot start), then 40 cycles of 95 °C for 15 s and 55 °C for 30 s,

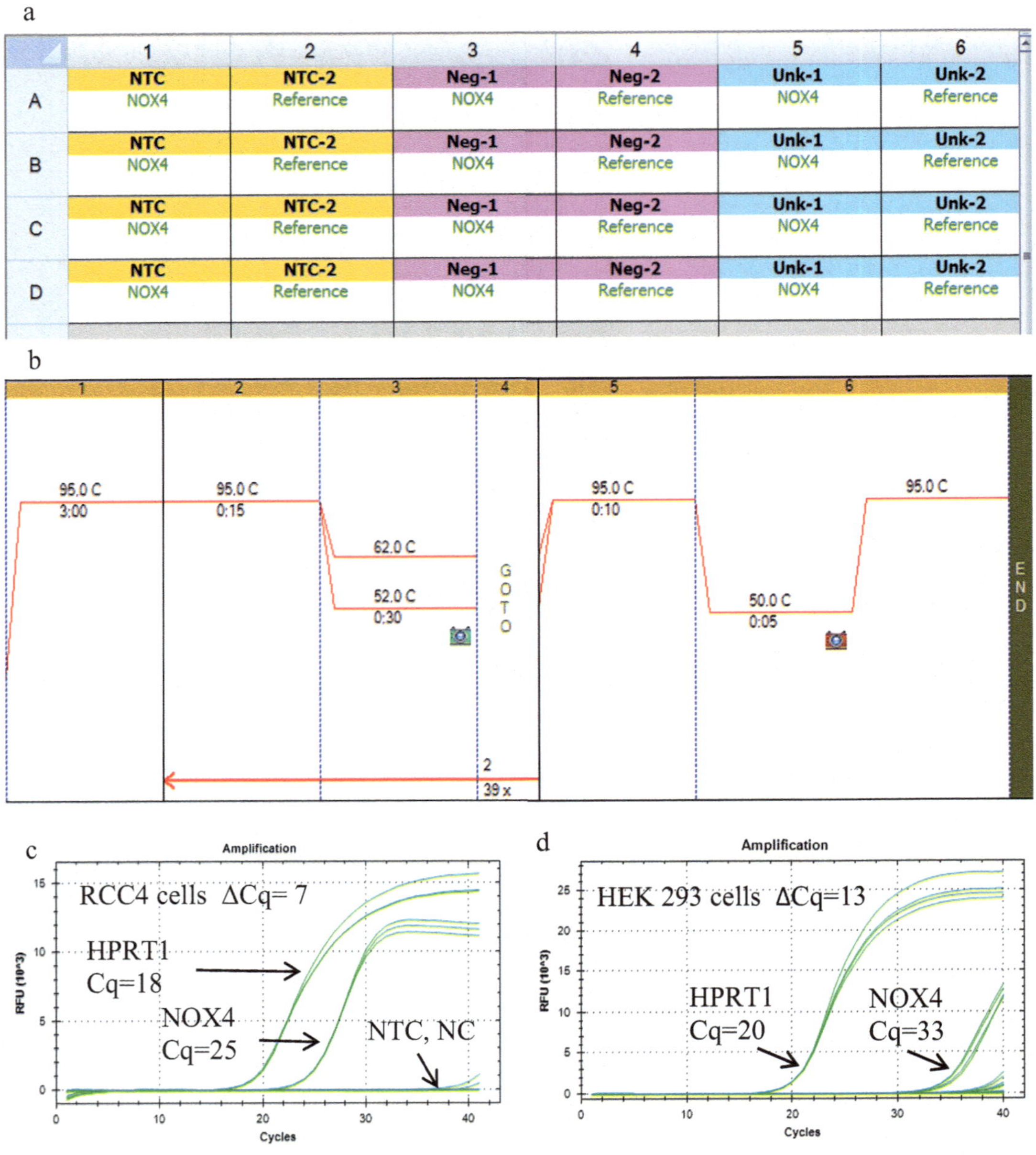

Fig. 3 Example of a basic qPCR setup using a Bio-Rad CFX 96 instrument and CFX manager software. (**a**) Plate layout of the no template control (NTC), negative control (NC) and unknown (UNK) wells, 4 replicate wells each for the NOX primers, and for the reference primers. (**b**) Example of a two-step PCR program with an annealing temperature gradient inserted. (**c**) Representative amplification curves (3–4 replicate wells per set of curves) for the reference gene (*HPRT1*, left set of curves) and *NOX4* (right set of curves) for untreated RCC4 clear cell renal cell carcinoma cells. (The cells were a kind gift from Dr. William Kaelin, Harvard University). (**d**) Amplification curves for untreated HEK 293 cells. Note that for the negative controls (NC) and no template controls (NTC) wells, the curves should resemble the baseline (flat lines) up to cycles 35–40. Primers (200 nM) for *NOX4* (E4F and E5R) from Table 2 and *HPRT1* (E3F and E4R) from Table 4 were used. Each UNK well contained 200 ng of cDNA (RNA equivalents) or 200 ng RNA in place of cDNA in the NC wells. The optimal annealing temperature was determined from a gradient RT-qPCR experiment

plus a melt curve (Fig. 3b). The optimal annealing temperature is determined by performing a gradient RT-qPCR as explained next.

2. To optimize the annealing temperature for each primer set, perform a gradient RT-qPCR by inserting a gradient step in the program such that each row represents a different annealing temperature (Fig. 3b). Then prepare a master mix for 25 identical reactions, and place 8 samples in 1 column of the plate, or if tubes, place the tubes in one column of the instrument. Replicates are recommended and should be placed in additional columns. Run the two-step program with the gradient step inserted. The annealing temperature yielding the lowest C_q value should be chosen. For the human primers in Table 2, the optimal annealing temperatures are within 55–58 °C using the reagents and instrument we used. Examples of the amplification curves obtained using the human NOX4 primer set E4F, E5R in Table 2 and reference gene HPRT1 primer set E3F, E4R in Table 4 for RCC4 and HEK 293 cells are shown in Fig. 3c, d, respectively.

3. The software of the instrument calculates the threshold cycle (formerly known as "C_t"). The correct terminology according to the MIQE rules is quantification cycle (C_q) [3]. The C_q values are dependent upon the value of the threshold level which is either set by the user or determined by the instrument. Consult the instrument's manual for directions. Preferably, obtain the C_q values by linear regression of the amplification curves which does not depend on the user's input of the threshold level. Consult the instrument's manual on how to choose this setting. Observe the software's calculated mean and standard deviation of the replicate wells for each sample. If necessary, eliminate one outlier out of the four replicates. If more than one well is an outlier, or the standard deviation is too large (>0.5), it is recommended that additional replicates be used and that the experiment repeated until technical errors (pipetting, mixing, sealing) are minimized as determined by the standard deviation. If a C_q value <35 is obtained in the NTC or NC wells, this indicates either a primer-dimer problem, or DNA contamination. In this case, the melting temperature of the amplicons of the NTC and NC wells should be examined to see if the values are due to NOX contamination or primer-dimers. For NOX DNA contamination, the melting temperature will be identical to that obtained from a positive control (plasmid containing the NOX isoform). The PCR products should be run on an agarose gel to check that the anticipated amplicon size is correct. Primer-dimers run less than 100 bp.

4. Sequencing of the PCR products from the UNK wells adds another layer of confidence that the products are correct.
5. Obtain the ΔC_q values for each well by subtracting the C_q value of the reference sample from the C_q value of the NOX sample. The mean ΔC_q value can be used to compare relative NOX mRNA levels between different samples if the reference gene C_q value does not differ significantly between samples. These values can also be used as an indication of whether or not protein detection will be feasible (*see* **Note 13**).
6. If two conditions are being compared, e.g., treated vs. untreated, then the software will also calculate the relative fold change, *R*, using the $\Delta\Delta C_q$ (Livak) method. If the efficiency of the NOX primers is determined to be quite different (10% or more) from the reference gene, then the Pfaffl method should be used as well and the results compared using the two methods. The primer efficiency values can be entered into the dialogue box (*see* **Notes 14** and **15**).
7. To determine the efficiency of the NOX and reference gene primer sets, generate a RT-qPCR standard curve using linearized plasmids that contain the NOX cDNA or the reference gene. Be careful not to use the PCR-designated bench and tools to make a series of ten 10-fold dilutions of the linearized plasmid in nuclease-free water starting with a known amount of the plasmid, about 1 μg/μL. Leave the dilutions at the non-PCR bench. Change gloves and go to the PCR-designated bench. Set up ten master mix PCR tubes for one primer set without cDNA. Add 10 μL of one of the plasmid dilutions per tube. Vortex gently and centrifuge the tubes briefly. Pipet four replicate wells per column of the PCR plate (40 wells). Seal the plate well. It is not necessary to mix the wells since centrifugation of the sealed plate (400 × *g*, 2000 rpm, 5 min, 4 °C) will induce mixing. Freeze the diluted stocks at −20 °C so that they can be used again. Run the qPCR program under the same conditions, but label each well as STD-1, STD-2, STD-3, etc., and enter the dilution-fold values associated with each set of wells. The software will generate a standard curve of C_q values vs. dilution-fold as shown in Fig. 4. The efficiency is calculated from the slope, which ideally would be −3.32 (the binary exponent required for a tenfold change in C_q). The standard curve can also be used to estimate copy number by converting the mass of the plasmid (plus insert) to moles and then to the number of molecules, which is the number of copies since each plasmid molecule contains one copy of the gene (*see* **Note 16**).

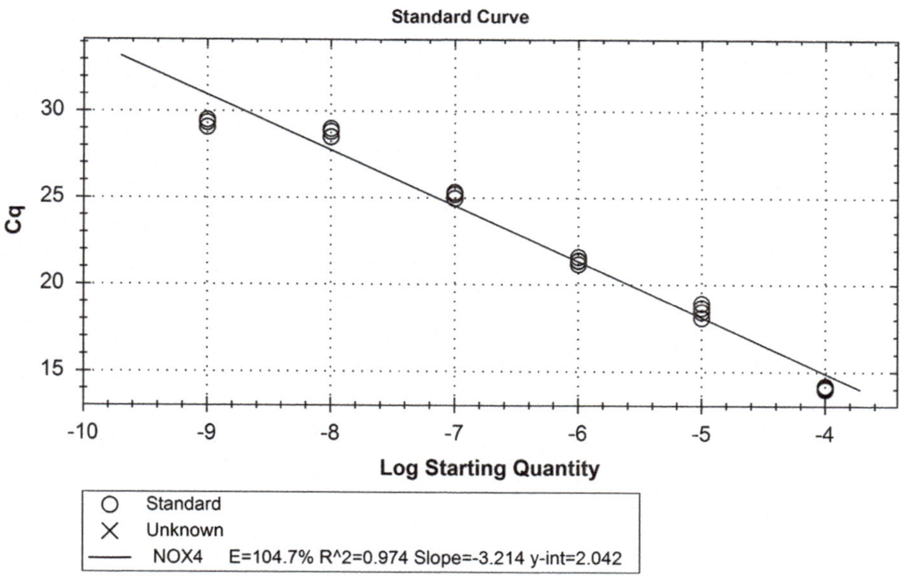

Fig. 4 Example of a RT-qPCR standard curve used for the determination of primer set efficiency. Ten tenfold serial dilutions were made from pcDNA3.1-*NOX4* variant 1 at a starting concentration of 1 μg/μL. Primers (200 nM) for *NOX4* (E4F and E5R) from Table 2 were used. The C_q values were plotted vs. log (dilution-fold), and the efficiency of the primer set was determined from the slope

3.5 Immunoblotting: Preparing Lysates of Positive and Negative Control Cells

1. The overexpression of NOX isoforms in a cell line having little or no NOX expression is used as a positive control, and the overexpression of the corresponding empty vector in the same cell line is used as the negative control. NOX1–4 require $p22^{phox}$ for maturation and expression, so use a cell line that expresses $p22^{phox}$ for these isoforms. We recommend HEK 293 cells because they contain sufficient endogenous $p22^{phox}$ protein to support stable NOX isoform expression [4], and they have a transfection efficiency close to 100%. If using a different cell line, the presence of $p22^{phox}$ protein should be confirmed. If the cell line does not express $p22^{phox}$ protein (e.g., COS7 cells), then the CYBA gene that codes for $p22^{phox}$ should be co-transfected along with NOX plasmids. DUOX1 and DUOX2 require the co-transfection of their maturation proteins, DUOXA1 and DUOXA2, respectively. Harvest HEK 293 cells from freshly seeded master plates prepared 1–2 days before the day of seeding. Seed the HEK 293 cells at 2×10^6 cells (high confluency for X-tremeGENE 9, ~80%) in 10 mL of DMEM containing 10% FBS per 10 cm dish without addition of antibiotics the day before transfection.

2. On the day of transfection, add a total of 6 μg of positive control plasmid (e.g., pcDNA3.1-NOX) to 600 μL of Opti-MEM in a 1.5 mL micro-centrifuge tube and vortex gently. The DNA concentration should be kept at 0.01 μg/μL.

Prepare a second tube in the same way for the negative control plasmid (e.g., pcDNA3.1 plasmid lacking the NOX cDNA as a negative control). If desired, a GFP plasmid (0.5 μg/10 cm plate) can be co-transfected to estimate transfection efficiency (*see* **Note 17**).

3. Transfect the cells following the manufacturer's directions. A number of transfection reagents are suitable such as Lipofectamine, but we use the X-tremeGENE 9 reagent because it does not require any media changes after transfection. If using X-tremeGENE 9, it should be at room temperature and gently vortexed for a few seconds just prior to use. For a 10-cm dish of cells containing 10 mL of media (no antibiotics), add 18 μL of X-tremeGENE 9 to 6 μg of DNA in 600 μL OptiMEM, keeping the ratio of X-tremeGENE 9:DNA at a ratio of 3 μL:1 μg DNA. Be sure not to touch the walls of the tube while depositing the transfection reagent directly into the liquid. Vortex the mixture gently, and spin briefly in a micro-centrifuge to collect all the liquid to the bottom of the tube. Incubate at 25 °C for 15 min to allow lipid micelles to incorporate the DNA.
4. Add the mixture dropwise to the 10 cm plate while swirling the plate. Swirl or rock the plate a few more times prior to placing the plate(s) back in the incubator. Incubate for 24–48 h for protein expression. No media change is required if using XtremeGENE 9 (*see* **Note 18**).
5. Remove the medium from the plates containing the transfected HEK293 cells. Wash the cells twice with PBS. Remove as much of the wash as possible. Place the plates on ice. To the first 10 cm plate, add a minimal volume (300–400 μL) of ice-cold RIPA buffer containing protease and phosphatase inhibitors. Scrape the plate, collect the scraped cells, and keep on ice. If multiple plates are used per condition, then add the collected cells from the first plate to the next plate, and scrape using the same 300–400 μL of RIPA buffer (*see* **Note 19**).
6. Sonicate or homogenize both the control and test sample lysates gently by placing the micro-centrifuge tube in a beaker of ice, and administer two 1 s bursts from a sonicator equipped with a pencil-tip probe. Avoid foam by using a low-output setting of 30%. If a sonicator is not available, homogenize 5 times by using a 1-mL syringe and a 21-gauge needle, avoiding foam formation.
7. Clarify the lysate by centrifuging the samples at 13,000 rpm (15,000 × *g*), 10 min, 4 °C.
8. Collect the clarified lysate, and determine the protein concentration using the BCA protein determination kit or similar methods.

9. Storage of samples: Do not use all of the lysate for SDS-PAGE gel sample preparation because the NOX proteins degrade within 1 week if stored in Laemmli's sample buffer. Prepare only as much gel sample as needed on the day of use. Aliquot and store the lysates at −20 °C for up to 2 weeks or for longer storage at −70 °C. If the protein precipitates after thawing, re-sonicate or homogenize, re-clarify, and determine the protein concentration again for positive and negative controls. For test samples, fresh lysates should be prepared each time and stored for no more than 1–2 weeks or as determined by each user.
10. To prepare concentrated SDS-PAGE gel samples, use a concentrated recipe for Laemmli's sample buffer, for example, 4× sample buffer. Take three parts of sample (e.g., 75 μL of the cell lysate), and add one part (25 μL) of 4× Laemmli's sample buffer. The protein concentration of the SDS-PAGE sample is now 75% of the original lysate. For a 4 mg/mL sample, it is now 3 mg/mL. For a 3 mg/mL sample, load 33 μL which is 100 μg of the test sample on the SDS-PAGE gel. For the positive and negative control lysates, load 0.1, 1, and 10 μg initially, and then determine the best amount to use in your conditions (exposure time, ECL reagent strength, etc.).
11. For NOX, do not boil the samples (*see* **Note 20**).

3.6 Preparing Lysates of Cells with Endogenous NOX

1. For the cells to be tested (e.g., RCC4 renal clear cell carcinoma cells), grow enough cells to obtain 2–3 × 10^7 cells/mL of lysis buffer for each test condition.
2. If desired, treat the cells with a stimulus that can induce or upregulate NOX isoform expression (e.g., TGFβ1, 5 ng/mL final in complete media for 24 and 48 h). Leave some plates untreated or treated with the vehicle.
3. Harvest the cells, and prepare the lysates from the test plates as done in **steps 5–11** in Subheading 3.5 above.

3.7 SDS-PAGE and Transfer

1. Use an appropriate percentage SDS-PAGE gel. A 1.5-mm-thick gel and comb will permit large sample volumes to be loaded (e.g., 100 μg protein in 30–50 μL) (*see* **Note 21**).
2. Load the pre-stained or fluorescently labeled protein ladder and 100 μg of test samples (30–50 μL). Also load the positive and negative control cell lysates (between 0.1 and 10 μg, determined by user). Run the 1.5-mm-thick mini-gels (about 8 × 10 cm plus the stacking gel) at 25 °C using constant amperage (30 mA/gel) for 15 min to allow stacking, and then increase the amperage to 40 mA/gel for about 1–1.5 h until the dye front reaches the very bottom edge of the gel to obtain optimal separation of the bands. After the run, rinse the

gels briefly by dipping them in transfer buffer, and place them on the PVDF membrane. The PVDF membrane should be pre-wetted in 100% methanol and equilibrated in transfer buffer. Remove any bubbles by smoothing with a gloved finger or roller tool.

3. Transfer the proteins to the PVDF membrane. For a standard wet transfer, set the voltage at 90 V for 1 h and transfer at room temperature or in a cold room, use chilled transfer buffer containing 20% methanol, add a small magnetic stir bar to the tank, and place the tank on a magnetic stir plate.
4. Block the PVDF membrane after the transfer for 1 h at RT on a rocking platform. Use 5% nonfat dry milk in TBST, or use *LI-COR* Odyssey Blocking Buffer without Tween-20. Incubations of up to 3 days (over the weekend) at 4 °C on a rocker should not result in high background if PVDF membranes are used. If using 5% nonfat dry milk in TBST as a blocking buffer, wash the blot with TBST 4 times (5 min each) before adding a primary antibody that is diluted in *LI-COR* Odyssey Blocking Buffer plus 0.1% Tween-20 (*see* **Note 22**).
5. For most antibodies, dilute the antibodies in 3–5% nonfat dry milk in TBST or use *LI-COR* Odyssey's Blocking Buffer (TBS) plus 0.1% Tween 20 which works for both fluorescent and non-fluorescent antibodies. A dilution of 1000-fold to 2000-fold (0.5–1 μg/mL final) is a good starting point for immunoblot optimization (*see* **Note 22**).
6. Incubate the blot with the primary antibody overnight or longer at 4 °C on a rocking platform. Detection of endogenous NOX proteins may require 72 h for detection (*see* **Note 23**).
7. Choice of primary antibody: Some recommended antibodies are listed in Tables 5, 6, 7, 8, 9, 10, and 11. Some commercial antibodies that did not pass validation under the conditions we used (HEK 293 cells overexpressing NOX isoforms) are also listed. Figures 5, 6, 7, and 8 show representative immunoblots using some of the selected antibodies from Tables 5, 6, 7, 8, 9, 10, and 11. Additional information about specific antibodies is provided in the footnotes of Tables 5, 6, 7, 8, 9, 10, and 11.
8. Recover and save the primary antibody and wash the blot 4 times, 5 min each with TBST.
9. Add the secondary antibody to the blot. Dilute HRP-conjugated goat anti-rabbit or mouse antibodies according to the manufacturer's instructions (*see* **Note 22**). Incubate the blot with the secondary antibody solution for 1 h at room temperature. If using *LI-COR* Odyssey's fluorescent secondary antibodies, dilute them 20,000-fold in *LI-COR* blocking buffer plus 0.1% Tween 20 and 0.01% SDS as recommended by *LI-COR*. Protect the solution from light during incubation.

Table 5
NOX1 and NOXA1 antibodies

NOX1 source/catalog #	Host and clonality	Immunogen (aa)	Is the target detected in WB (other applications)?	Species detected (human, mouse, rat)	Reference
Dr. J. Doroshow	Mouse Monoclonal Clone 22	224–564	Yes[a] (IF, IHC)	H	[11, 12]
Dr. C. Yabe-Nishimura	Rabbit Polyclonal	139–152 (mouse)	Yes[b] (IF, IHC)	M	[7]
Dr. C. Yabe-Nishimura	Rat Monoclonal Clone 31C1	139–152 (mouse)	Yes[b] (IF, IHC)	M	Unpublished
Dr. J. David Lambeth	Mouse Monoclonal Clone E39	3rd extracellular loop	Yes[c] (IF, IHC)	H	[13]
OriGene Technologies #TA349080	Rabbit Polyclonal	Proprietary	No[d]	H	
Novus #NBP1-31546	Rabbit Polyclonal	Proprietary	No[d]	H, M, R	
NOXA1 subunit Santa Cruz sc-398873	Mouse Monoclonal Clone H-6	1–200	Yes[e]	H, M, R	

Evaluation of academic and commercial NOX1 antibodies

[a]Specificity in cell lines is shown in Ref. [11]; specific in CHO cells overexpressing NOX1 (Dr. V. Jaquet, pers. communication)

[b]Authors stated that the polyclonal antibody detected Nox1 in mouse colon but not in mouse small intestine or lung. Their mAb detects Nox1 in mouse colon, small intestine, and lung (personal communication Dr. M. Matsumoto) (*see* Fig. 8)

[c]There was a limited quantity of this antibody available for WB verification (*see* Fig. 5d). This antibody was made by diaDexus (out of business). The specific epitope was not disclosed, only the location (3rd extracellular loop)

[d]Unable to differentiate between NOX1 stably overexpressed in HEK 293 cells and empty vector-transfected controls

[e]Product datasheet indicates that NOXA1 was detected in HEK 293T cells overexpressing NOXA1, but not in vector control

Incubate the blot with the secondary antibody solution for 1 h at room temperature.

10. Discard the used solution of HRP-conjugated secondary antibody. Save fluorescent secondary antibody solutions in a foil-wrapped tube at 4 °C, and reuse until the signal fades (~3 repeats).
11. Wash the blots 4 times with TBST, 5 min per wash, and then add the ECL reagent or proceed to *LI-COR* Odyssey scanning.

Table 6
NOX2 antibodies

NOX2 source/ catalog #	Host and clonality	Immunogen (aa)	Is the target detected in WB (other applications)?	Species detected (human, mouse, rat)	Reference
Dr. A. Jesaitis Abcam #ab80897 Santa Cruz #sc-130543	Mouse Monoclonal Clone 54.1	383–390	Yes[a] (IF)	H, M	[14]
Dr. D. Roos, LSbio #LS-C85347	Mouse Monoclonal Mo 48	Purified cytochrome b_{558}	Yes (IHC)	H	[17, 18]
Dr. A. Jesaitis Santa Cruz #sc-130549	Mouse Monoclonal Clone CL5	135–147	Yes[c]	H	[19]
BD Transduction Laboratories, #BD61141	Mouse Monoclonal Clone 53	450–556	Yes[d] (IF rat)	H, M, R	[2, 20]
LifeSpan Biosciences #LS-C179418	Mouse Monoclonal Clone 7D5	143–151	Yes[b] (Flow)	H	[15, 16]

Evaluation of academic and commercial NOX2 antibodies
[a]Cross-reactivity with NOX1-4 and GRP58
[b]Cross-reactivity with gelsolin
[c]Isoform specificity not tested. Other applications not tested
[d]According to the manufacturer, a 58 kDa band is detected in mouse macrophages

Table 7
NOX3 antibodies

NOX3 source/ catalog #	Host and clonality	Immunogen (aa)	Is the target detected in WB (other applications)?	Species detected (human, mouse, rat)	Reference
Bioss bs-3683R	Rabbit Polyclonal	70–120	Yes[a]	H, M,R	Unpublished

At this time there is little information on NOX3 antibodies and their validation. Several NOX3 antibodies are on the market (see www.abcam.com and www.antibodies-online.com), but none of these have, thus far, been validated nor tested for isoform specificity
[a]Worked well in detecting NOX3 overexpression in HEK 293 cells for one group but not for another (Dr. Ulla Knaus, Dr. Vincent Jaquet, pers. communication), and specificity over other isoforms has not been tested

12. Prepare the ECL reagent as directed by the manufacturer. For SuperSignal West Pico Enhanced Chemiluminescent Substrate, we do not dilute the reagent. For an 8 × 10 cm blot, add 2 mL of the SuperSignal West Pico Enhanced

Table 8
NOX4 antibodies

NOX4 source/ catalog #	Host and clonality	Immunogen (aa)	Is the target detected in WB (other applications)?	Species detected (human, mouse, rat)	Reference
Dr. A. Shah	Rabbit Polyclonal	556–568	Yes[a] (IF) (*see* Fig. 5a)	H, M	[21]
Dr. P. Jansen-Dürr	Rabbit Monoclonal (27-6)	335–578	Yes[b] (*see* Fig. 5f)	H, M	[22]
Dr. U. Knaus	Rabbit Polyclonal (7843)	251–266	Yes[c] (IF, Flow)	H	[23]
Dr. U. Knaus	Rabbit Monoclonal (20.1, 37.7)	251–266	Yes[c] (IF, Flow)	H	[23]
Dr. J. Doroshow	Rabbit Monoclonal (47-6)	242–262	Yes[c] (IHC) (*see* Fig. 5b)	H	[12]
Abcam ab133303	Rabbit Monoclonal	Unknown	No[d,e] (*see* Fig. 5h)	H, R	
Abcam ab109225	Rabbit Monoclonal	500–578	No[d,e]	H, R	
Millipore ABC459	Rabbit Polyclonal	C-terminus	No[d,f] (*see* Fig. 5g)	H, M	
Novus Biologicals #NB110-58851	Rabbit Polyclonal	500–578	No[d,g] (*see* Fig. 5c)	H, M, R	

Evaluation of academic and commercial NOX4 antibodies
[a]Excellent sensitivity for mouse and human NOX4 although many bands are detected, some of which may be NOX4 variants; does not cross-react with NOX1 in HEK 293 cells stably overexpressing NOX1 (*see* Fig. 5a)
[b]Detects human NOX4 in Trex cells overexpressing NOX4; detects Nox4 in mouse kidney cortex, multiple bands
[c]NOX4-specific (epitope is absent in other NOX isoforms) but only detects human NOX4
[d]Does not distinguish between HEK293 cells expressing an empty vector vs. HEK293 cells stably overexpressing NOX4
[e]Identical results were obtained by two independent laboratories (Dr. J.D. Lambeth and Dr. J. Doroshow)
[f]Website states that application is not for WB, but for IHC and ICC
[g]Unspecific in mouse; older lots may have been specific for NOX4 in overexpressing cells, but recent lots are not specific

Chemiluminescent Substrate, and leave it on for 5 min on a platform rocker (or rock occasionally manually).

13. Remove the blot from the ECL detecting solution, and blot lightly between two pieces of filter paper so that the ECL solution is not dripping. Tape the blot in a sheet protector that has been cut (with the folded edge intact) to the shape of a (used) autoradiography film. Open the sheet protector and tape the blots between the plastic sheets. Close the sheet

Table 9
NOX5 antibodies

NOX5 source/ catalog #	Host and clonality	Immunogen (aa)	Is the target detected in WB (other applications)?	Species detected (rodents lack NOX5 gene)	Reference
Dr. W. Nauseef	Rabbit Polyclonal	1–169	Yes[a] (IF, IP)	H	[6, 24, 25]
Dr. J. D. Lambeth	Rabbit Polyclonal (#8579)	491–506 and 577–588	Yes[b] (IHC)	H	[4]
Dr. J. Doroshow	Mouse Monoclonal	600–746	Yes[b] (IF, IHC)	H	[26]

Evaluation of NOX5 antibodies
[a]Reference [24] shows a MW of 75 kDa in K562 cells stably transfected with NOX5β. Detects a 70 kDa band and a smaller band at 43kDa (degradation product) in HEK cells expressing NOX5. Antibody works for immunoprecipitation
[b]Reference [4] shows a MW of 75kDa in HEK 293 cells transiently transfected with NOX5

protector over the blots, and insert it into the autoradiography cassette containing intensifying screens on both sides.

14. Expose the membrane for 1, 10, and 30 s and then 1, 3, 5, and 10 min. If no bands appear by 30 min, then remove the blots and place them in an ECL reagent with higher sensitivity such as SuperSignal West Dura solution. No rinsing of the blot is required before going to the next level of SuperSignal West ECL reagent. SuperSignal West Femto usually results in high background during long exposures.
15. For *LI-COR* Odyssey scanning, the blot can be wet or lightly dried between two filter papers. Place the wet blot facedown on the glass bed of the scanner. For dried blots, place a glass gel plate or similar weight on top of the membrane to keep it flat. Be sure to choose the membrane mode (169 μm thickness). Preview under an initial intensity of 3.0 for each IRDye wavelength. If the preview looks fine, then proceed to scan the membrane, and adjust the signal to noise ratio as desired using the settings of the instrument. Consult the instrument's manual for further instructions. If necessary, scan again at a higher intensity for increased sensitivity. The membranes can be re-scanned up to 6 months if stored in the dark.

4 Notes

1. To minimize contamination from plasmids containing NOX cDNA sequences, the surface of the workspace and all tools should be washed with diluted detergent. We use household

Table 10
DUOX antibodies

DUOX1, DUOX2, DUOXA1, DUOXA2 source/catalog #	Host and clonality	Immunogen (aa)	Is the target detected in WB (other applications)?	Species detected (human, mouse, rat)	Reference
DUOX1 and 2 Dr. U. Knaus	Rabbit Polyclonal	DUOX1 775–1026	Yes[a] (Flow, IHC)	H, M	[27]
DUOX2 specific Dr. U. Knaus	Rabbit Polyclonal	634–648	Yes (Flow, IHC)	H	[28]
DUOXA1 specific Dr. U. Knaus	Rabbit Polyclonal	280–299	Yes	H	[28]
DUOX1 and 2 Dr. X. De Deken	Rabbit Polyclonal I2	DUOX1 618–1004	Yes[a]	H, M	[29]
DUOX1 and 2 Dr. C. Dupuy	Rabbit Polyclonal	DUOX2 410–423	Yes[a]	H, M	[30]
DUOX1 specific Dr. C. Dupuy	Rabbit Polyclonal	DUOX1 988–1011	Yes	H, M	[31]
DUOX1 and 2 Dr. C. Dupuy	Mouse Monoclonal Clone	DUOX2 21–589	Yes[a]	H	[32]
DUOX1 and 2 Dr. J. Doroshow	Mouse Monoclonal Clone S-40	DUOX2 131–540	Yes[a]	H, M	[33]
DUOX1 Bioss bs-11431R	Rabbit Polyclonal	Unspecified	No[b]	H, M, R	
DUOX2 Santa Cruz (Y-15) sc-49939	Goat Polyclonal	Unspecified	Yes[a]	H, M, R	
DUOX2 Santa Cruz (E-8) sc-398681	Mouse Monoclonal	DUOX2 626–680	Not tested[c]	H, M, R	
DUOX1 Sigma-Aldrich HPA023544	Polyclonal	910–1011	Yes[a] (IHC)	H	
DUOXA1 specific Sigma-Aldrich HPA041578	Polyclonal	270–326	Yes[d] (IHC)	H	

Evaluation of DUOX1, DUOX2, DUOXA1, and DUOXA2 antibodies

[a]Cross-reactivity between DUOX1 and DUOX2. Dr. Dupuy's monoclonal antibody has ~tenfold preference for DUOX2, and in porcine tissues this antibody does not cross-react with Duox1 (*see* reference [32])

[b]Detects a non-specific band at a similar molecular weight as DUOX in non-transfected cells

[c]Recommended by Santa Cruz as a replacement for Santa Cruz (Y-15) sc-49939. See website for images

[d]Detects additional non-specific bands

Table 11
Antibodies for subunits (p47^{phox}, p67^{phox}) and for p22^{phox}

Subunits and p22^{phox} source/ catalog #	Host and clonality	Immunogen (aa)	Is the target detected in WB (other applications)?	Species detected (human, mouse, rat)	Reference
p47^{phox} Dr. M. Quinn Millipore 07-001	Rabbit Polyclonal	Full-length human p47^{phox}	Yes[a,b] (IHC, IP)	H, M, R	[34]
p47^{phox} Dr. M. Quinn	Mouse Monoclonal Clone 43.12	Full-length human p47^{phox}	Yes (others not tested)	H, M, R	Unpublished
p67^{phox} Dr. M. Quinn Millipore 07-002	Rabbit Polyclonal	Full-length human p67^{phox}	Yes[a,b] (IHC, IP)	H, M, R	[34]
p67^{phox} Dr. M. Quinn	Mouse Monoclonal Clone 81.1	Full-length human p67^{phox}	Yes (IHC)	H, M, R	Unpublished
p47^{phox} Santa Cruz sc-17844	Mouse Monoclonal Clone A-7	196–390	Yes[a] (others not tested)	H, M, R.	
p67^{phox} Santa Cruz sc-374510	Mouse Monoclonal Clone D-6	Full-length p67^{phox}	Yes[a] (others not tested)	H, M, R	
p40^{phox} Dr. M. Quinn	Mouse Monoclonal Clone 1.9	Full-length human p40^{phox}	Yes (IHC)	H	
p22^{phox} Dr. A. Jesaitis Dr. M. Quinn Santa Cruz sc-130550	Mouse Monoclonal Clone 44.1	29–33 and 182–188	Yes[a,b] (IP)	H, M, R	[14]

Many academic laboratories have published antibodies generated for these subunits. We have used antibodies generated in the laboratory of Dr. Mark Quinn

Also listed are commercial antibodies that worked in our conditions for immunoblots

[a]Tested with human neutrophil cytosolic fractions

[b]Tested in HEK 293 cells overexpressing cytosolic components

dish detergent. Do not spray the surfaces with ethanol, which can precipitate DNA. Rinse off any soap with a wet paper towel 2–3 times, dry, and spray with Zap spray. Wipe off the excess Zap spray solution without rinsing. Do this for all the items on the workspace such as pipettes, pipette aids, pipette racks, pipette tip boxes, micro-centrifuge, vortex, tube racks, pens, and calculator. We use pipettes and dual filter tips (Eppendorf) designated for RNA and PCR setup. No amplified PCR products (amplicons) should be brought to this workspace. After washing the surface of the pipettes, treat them with UV light for 1 h on a UV light box or in a biosafety cabinet with the UV light turned on. Other equipment (e.g., tube racks) can be treated overnight with UV light in the biosafety cabinet. It is especially important to clean the top of the pipette where the thumb is placed and the dial of the pipet. These areas should be exposed to the UV light. If it is easy to separate the pipette's lower barrel, wash and treat with UV light along with the other parts of the pipette. It is not necessary to do this washing protocol before every PCR experiment (although we do), but if contamination (C_q values ≤ 35) appears in the negative controls or no template controls, then washing and UV treatment is strongly recommended. If contamination is still evident, this may be due to traces of genomic cDNA (overloading of the gDNA eliminator column) or other non-NOX products such as primer-dimers. Thus, always include negative controls and a melting curve to check if the contamination is due to gDNA, primers, or an unknown source. If it is due to plasmid DNA, the melting temperature of the contaminating amplicon will be identical or close to that of the desired NOX amplicon.

2. Washing the cells twice removes phenol red which distorts the UV spectrum of RNA, thus interfering with the measurement of RNA concentration.
3. Removal of gDNA has been simplified by Qiagen's gDNA eliminator columns, which are only available in Qiagen's RNeasy Plus kits. This allows results to be more consistent since DNase I digestion is not performed by the user. It may be necessary to repeat this step twice (as we do) when processing large quantities of cells (5–10 million per column).
4. Some cell lines have a low yield of RNA, so we elute the first column with 50 μL and use the eluant to elute a second column. We aim for an RNA concentration of 0.5–1.0 μg/μ L. We check the quality of our RNA preparations by sending a small sample to a core facility to be analyzed by a bioanalyzer. The analyzer measures the sizes of the rRNA bands and determines an RNA Integrity Number (RIN) that can be used to normalize RNA samples. For cell lines, the RIN number is usually 10, indicating a very good quality.
5. The RNA preparation should be kept at −70 to −80 °C. It will deteriorate if stored at −20 °C.
6. We prefer qScript XLT cDNA SuperMix (Quanta Biosciences) because it features a unique blend of oligo (dT) and random primers that captures unbiased representation of all RNA

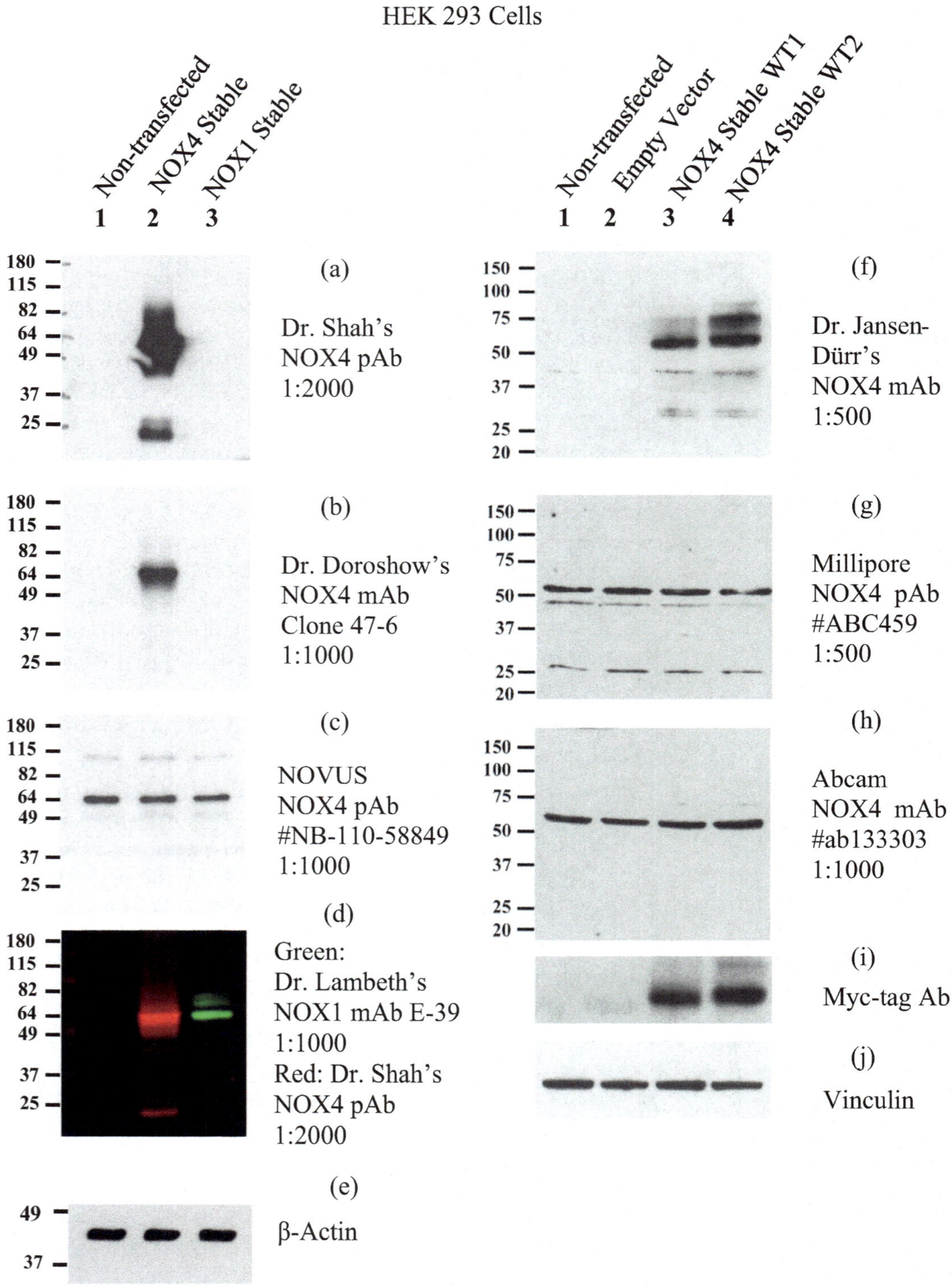

Fig. 5 Examples of using positive and negative control lysates for the validation of NOX4 antibodies. Two laboratories tested several NOX4 antibodies from Table 8. Left column: NOX4 antibodies tested in the laboratory of Dr. J. D. Lambeth. Right column: NOX4 antibodies tested in the laboratory of Dr. J. Doroshow. *Left column*: Identical PVDF immunoblots consisting of 10 μg of the following non-heated cell line lysates.

sequences into cDNA product (including the 5′end, 3′-end, or central regions of long RNAs). This reagent is highly sensitive and easy to use for two-step RT-PCR and RT-qPCR (up to 1 kb). The engineered M-MLV reverse transcriptase has reduced RNase H activity and improved activity and stability at higher temperatures and over a wide dynamic range of input RNA, with up to eightfold higher sensitivity than cDNA synthesis kits utilizing RNaseH(+) reverse transcriptase (RT). The 5× concentrated master mix provides all necessary components for first-strand synthesis including buffer, dNTPs, $MgCl_2$, primers, RNase inhibitor protein, qScript XLT reverse transcriptase, and stabilizers.

7. The best results (low standard deviation, clean negative controls, and lowest C_q values) are obtained if the cDNA is used on the day of synthesis. If the cDNA is stored overnight, store at −20 °C.

8. It is better to start with the NTC (no template control) tubes first, followed by the NC (negative control) tubes to avoid contaminating these tubes with cDNA from the UNK (unknown). The thumb used to open cDNA tubes is a large source of contamination. Change gloves after setting up the UNK tubes.

9. We use Quanta Biosciences PerfeCtA SYBR Green SuperMix because it provides a high level of specificity to reduce the occurrence of primer-dimers and other non-specific artifacts associated with the binding of SYBR Green I dye to any dsDNA generated. The AccuStart Taq DNA polymerase contains monoclonal antibodies that bind to the polymerase and keep it inactive prior to the initial PCR denaturation step.

Fig. 5 (continued) Lane (1) non-transfected HEK 293 cells, lane (2) HEK 293 cells stably transfected with pcDNA3.1-NOX4, untagged (a kind gift from Dr. R. Brandes (Goethe-University Frankfurt), lane (3) HEK 293 cells stably transfected with pcDNA3.1-NOX1, untagged (also a gift from Dr. Brandes). The blots were probed with the following NOX4 antibodies from Table 8 (**a**) Dr. A. Shah's NOX4 rabbit pAb, 1:2000 (a kind gift from Dr. A. Shah, University College London); (**b**) Dr. J. Doroshow's NOX4 rabbit mAb, clone 47–6, 1 μg/mL (1:1000); (**c**) NOVUS NOX4 rabbit pAb #NB110–58849 Lot F1 (1:1000, 1 μg/mL); (**d**) green, Dr. J. D. Lambeth's NOX1 E39 mAb, 1:1000, as a positive control for NOX1; red, Dr. Shah NOX4 rabbit pAb, 1:2000. (**e**) β-actin mouse mAb (Sigma-Aldrich), 1:10,000. Blocking buffer, 5% nonfat dry milk in TBST; antibody diluent, *LI-COR* Odyssey blocking buffer +0.1% Tween-20; primary antibodies incubated overnight; secondary antibodies (HRP-conjugated goat anti-rabbit or anti-mouse), 1 h. ECL reagent: SuperSignal West Pico, 1 s. *Right column*: identical nitrocellulose membranes consisting of 40 μg of the following cell lysates. Lane (1) non-transfected HEK 293 cells, lane (2) HEK 293 transfected with pCMV6-empty vector, lane (3) HEK 293 stable clone (WT1) overexpressing N-terminal tagged-Myc-DDK-NOX4, lane (4) HEK 293 stable clone (WT2) overexpressing N-terminal tagged-Myc-DDK-NOX4. The following antibodies from Table 8 were tested: (**f**) Dr. P. Jansen-Dürr's NOX4 mAb 1:500; (**g**) Millipore NOX4 pAb (#ABC549), 1:500; (**h**) Abcam NOX4 antibody (ab133303), 1:1000; (**i**) anti-Myc antibody; (**j**) anti-vinculin antibody. Conditions: similar to left panel, except the antibody was diluted in 5% milk in TBST and the ECL reagent was ECL Plus reagent (GE Healthcare)

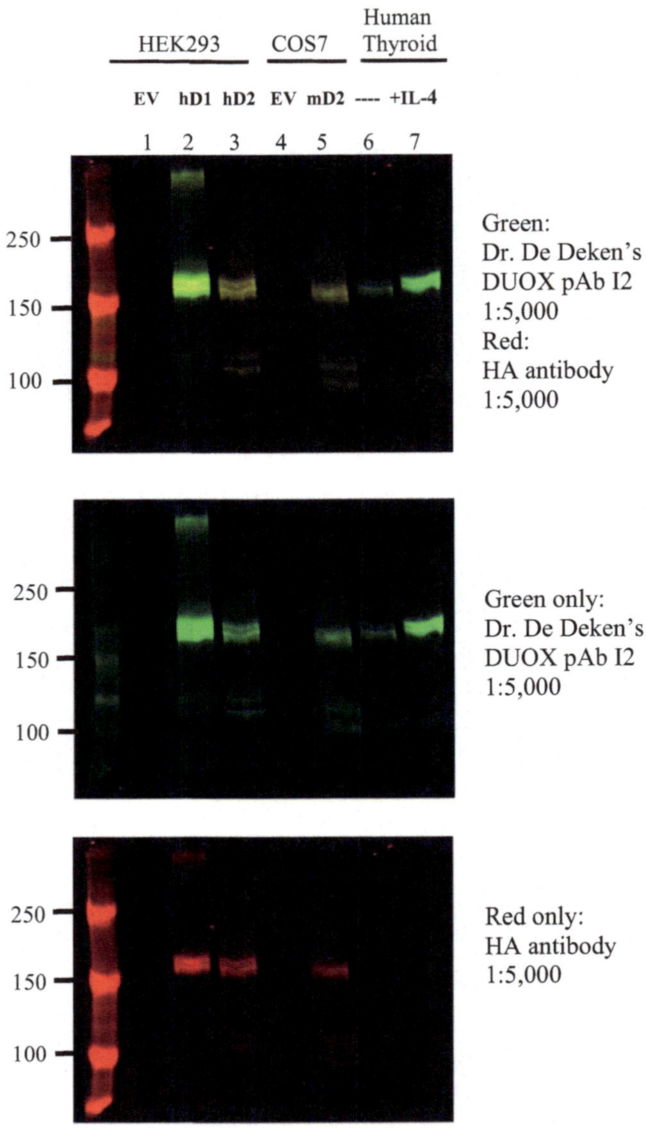

Fig. 6 Validation of DUOX antibodies. HEK293 cells were transfected with pcDNA3 HA- DUOX1(hD1) or DUOX2(hD2) and pcDNA3 V5-DUOXA1,2 constructs or empty vectors (EV) for the control condition. COS7 cells were transfected with pcDNA3 HA-mouse Duox2 (mD2) and pcDNA3 V5-mDuoxa2. Primary human thyroid cells were incubated 48 h with or without 14 ng/mL human IL-4. Lysis buffer: 50 mM phosphate buffer, pH 7.5, containing 20 mM EDTA, 1% Triton X-100 plus protease inhibitors (Complete; Roche Applied Science) and phosphatase inhibitors (1 mM vanadate, 100 mM NaF, and 50 nM okadaic acid). DUOX proteins (20–30 μg) in Laemmli buffer were heated at 80 °C for 5 min and were separated on a 6% gel and transferred to a low-fluorescence PVDF membrane (Immobilon-FL, Millipore, Billerica, USA). Blocking: *LI-COR* Odyssey blocking buffer, 2 h RT. Primary antibody: Dr. De Deken's anti-DUOX pAb I2 1:5000 or anti-HA-mAb 1:5000 (Covance) overnight at 4 °C. Secondary antibodies: anti-rabbit IRDye800 (green) and anti-mouse IRDye680 (red) from *LI-COR*, 1 h RT. This DUOX antibody recognizes both human DUOX1 and DUOX2 as well as mouse Duox2 in overexpressing cell lines. DUOX proteins were also detected in human thyrocyte extracts, with higher expression after IL-4 treatment

(a)

Primer Set	Ave Cq ± Std. Dev. (n=4 wells)	ΔCq (Cq_{NOX}-Cq_{HPRT1})
RCC4 Cells: Untreated		
NOX4	27.9 ± 0.2	6.2
HPRT1	20.9 ± 0.1	
RCC4 Cells + TGFβ1 24 hrs		
NOX4	25.0 ± 0.1	2.4
HPRT1	22.6 ± 0.1	
Fold-change	ΔΔCq = -3.8 (2-6)	R= $2^{-(-3.8)}$= 13.9-fold increase Relative to parent
RCC4 Cells + TGFβ1 48 hrs		
NOX4	24.0 ± 0.08	1.6 ± 0.1
HPRT1	22.4 ± 0.1	
Fold-change	ΔΔCq = -4.6 (2-6)	R= $2^{-(-4.6)}$= 24.2-fold increase Relative to parent

Fig. 7 Correlation between ΔC_q values and endogenous NOX4 protein detection in RCC4 cells. (**a**) C_q, ΔCq, and *R* (fold-change) values for untreated RCC4 parent cells and for TGFβ1-treated cells (5 ng/mL in complete medium). (**b**) NOX4 immunoblot probed with Dr. Shah's NOX4 antibody (1:2000 for 72 h at 4 °C), then developed with SuperSignal West Pico Chemiluminescent reagent, and exposed for 30 min. Lanes (1–4) lysates of parent RCC4 cells (100 μg) untreated or treated with TGFβ1 for 0–72 h as indicated. Lanes (5–8) lysates of a representative RCC4 stable NOX4 knockdown (KD) clone untreated or treated with TGFβ1 for 0–72 h as indicated. Lane (9) positive NOX4 control (1.0 μg) of HEK293 cells stably expressing NOX4. The ΔC_q values corresponding to each condition are shown above each lane, except for lane 4 (72 h) for which the C_q was not determined (n.d.). (**c**) Immunoblot using Dr. Doroshow's NOX4 rabbit mAb clone 47–6 (1:1, 000 for 72 h at 4 °C). Lane (1) 48 h, TGFβ1-treated parent RCC4 cells; lane (2) 48 h treated, representative RCC4 stable scramble shRNA clone; lane (3) 48 h treated, representative RCC4 stable NOX4 knockdown (KD) clone. The ΔC_q values corresponding to each condition are shown above each lane. NOX4 protein could not be detected in untreated RCC4 cells in these conditions using Dr. Doroshow's NOX4 antibody. (**e–g**) Amplification curves for untreated and 24–72 h TGFβ1-treated parent cells corresponding to the C_q values in panel a; (**h**) amplification curves for untreated parent and representative clones of scramble shRNA and NOX4 shRNA (KD) corresponding to the C_q values in panel d.The scramble shRNA plasmid had an effect of decreasing NOX4 mRNA expression by eightfold, but the effect on protein expression could not be evaluated without TGFβ1 treatment. When the scramble clone was treated with TGFβ1 for 48 h, the NOX4 protein expression was similar to that of the parent cells treated for 48 h (lane 1 vs. 2, panel c). The NOX4 primer sets were NOX4 E4F/E5R and HPRT1 E3F/E4R as indicated in Table 2. The NOX4 shRNA (KD, retroviral) plasmid (OriGene Technologies, #TG302911) contained the target sequence 5′CACTGAACTATGAGGTCAGCCTCTGCGTG. The scramble (Scr) shRNA (retroviral) plasmid (OriGene Technologies, #TR30013) contained the non-targeting sequence: 5′ GCACTACCAGAGCTAACTCAGATAGTACT 3′

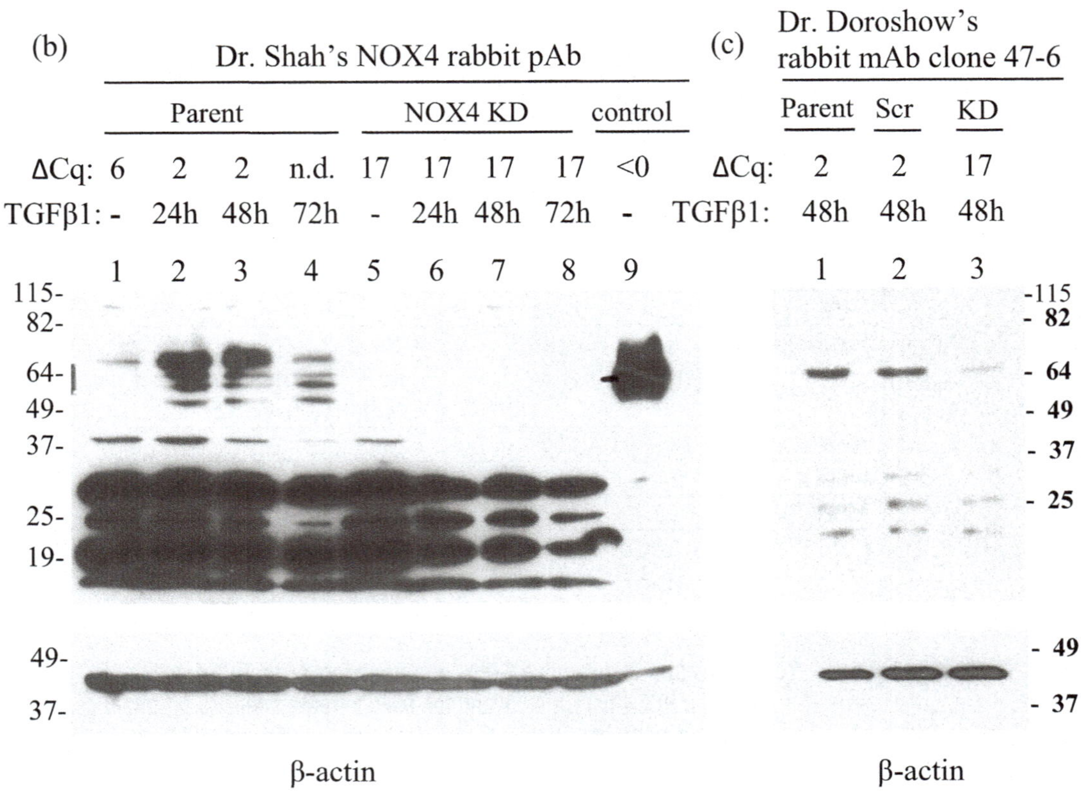

(d)

Primer	Ave Cq ± Std. Dev. (n=4 wells)	ΔCq (Cq_{NOX}-Cq_{HPRT1})
RCC4 parent cells (non-transfected) + TGFβ1 48 hrs		
NOX4	28 ± 0.2	7
HPRT1	21 ± 0.1	
RCC4 stably expressing scramble shRNA + TGFβ1 48 hrs		
NOX4	31 ± 0.2	10
HPRT1	21 ±0.2	
Fold-change	ΔΔCq= 3 (10-7)	R= $2^{-(3)}$= 0.125 8-fold decrease relative to parent
RCC4 stably expressing NOX4 shRNA) + TGFβ1 48 hrs		
NOX4	36 ± 0.1	17
HPRT1	19 ± 0.1	
Fold-change	ΔΔCq= 7 (17-10)	R= $2^{-(7)}$ = 0.0078 128-fold decrease Relative to scramble

Fig. 7 (continued)

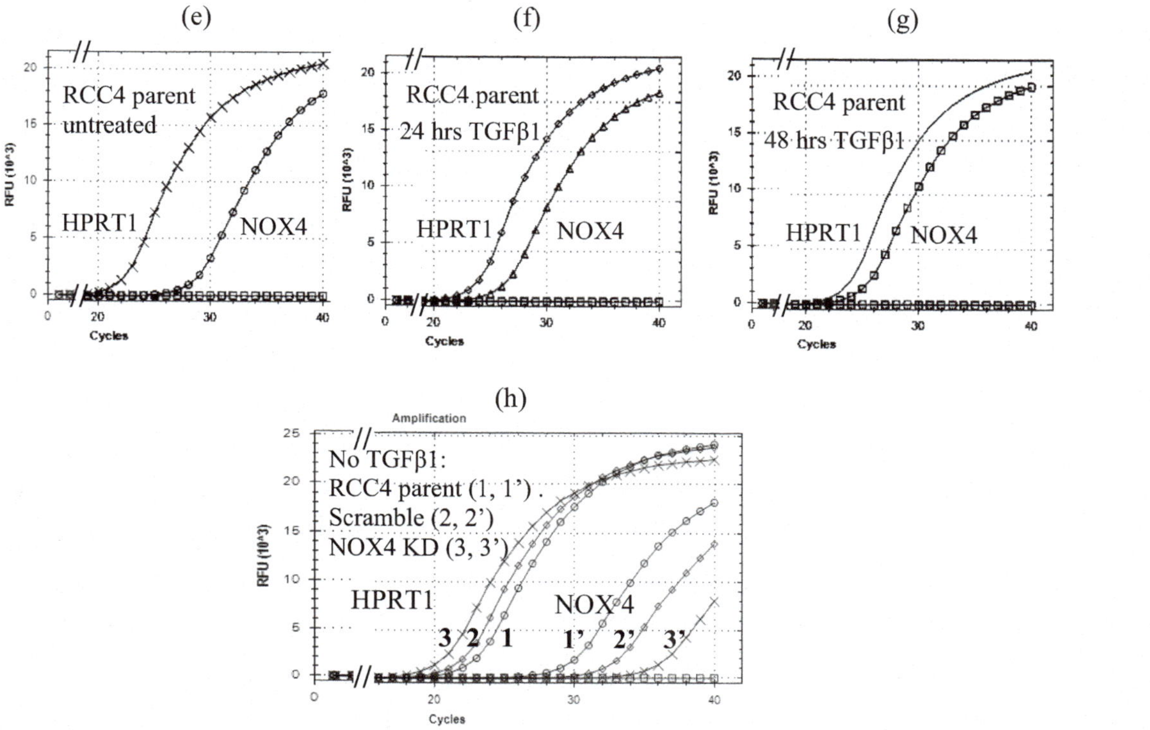

Fig. 7 (continued)

Upon heat activation (3 min at 95 °C), the antibodies denature irreversibly, releasing fully active, unmodified TaqDNA polymerase. This enables specific and efficient primer extension with the convenience of room temperature reaction assembly. We also found that Bio-Rad's Universal iTaq SYBR Green SuperMix performs equally well. To increase specificity, use a fluorescently labeled probe (instead of SYBR Green technology) that anneals to the target sequence and emits fluorescence due to FRET when displaced by the primer set. Consult your primer manufacturer's website for descriptions on the types of probe technologies available.

10. Upon receipt of freeze-dried primers, original primer stocks (10 μM or 100 μM) are prepared in TE buffer (10 mM Tris-Cl, pH 8, 0.1 mM EDTA). To avoid contamination of the original primer stocks, aliquots are prepared by one designated person and given to other users. The final primer concentration in the PCR reaction is 200 nM, so a 10× solution is prepared in nuclease-free water for setting up the PCR reactions. Higher final concentrations may be used such as 500 nM, but one should check that the C_q values of the NTC wells are not lower than 35, which is an indication of primer-dimer

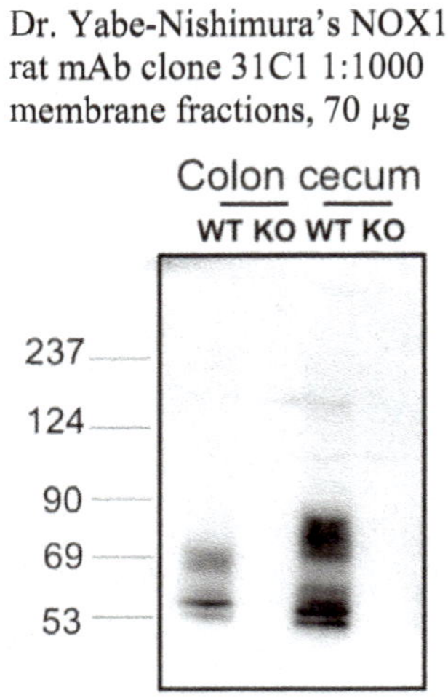

Fig. 8 Validation of a Nox1 rat monoclonal antibody. A rat monoclonal antibody against mouse Nox1 (clone 31C1) detects Nox1 in wild-type (WT) mouse colon (lane 1) and cecum (lane 3) membrane fractions (70 μg) solubilized in 2% *n*-dodecyl β-D-maltoside in 10 mM Hepes pH 7.4, 10 mM NaCl, 100 mM KCl, 1 mM EDTA, and 1 mM PMSF. This monoclonal antibody is more specific and sensitive than a previous polyclonal antibody to the same epitope (*see* Table 5) [7]. The Nox1-specific bands are observed diffusely at 56–80 kDa due to differently glycosylated forms of Nox1 (lane 1). The specificity for Nox1 was confirmed using solubilized membrane fractions from the colon from Nox1-deficient mice (KO) as a negative control (lane 2, 4). This rat monoclonal antibody also detects Nox1 in solubilized membrane fractions of mouse small intestine and mouse lung (unpublished, personal communication from Dr. M. Matsumoto). However, Nox1 protein was not detected in other mouse tissues, possibly due to its low abundance

formation. We purchase pre-designed primers from IDT (Integrated DNA Technology) and verify their exon docking site, variant specificity as well as isoform specificity, and anticipated amplicon sizes using appropriate accession numbers available in the National Center for Biotechnology Information (NCBI) (*see* Table 1). For example, for human NOX4 exon 14 is present in variant 1, the full-length form, but is absent in variant 2.

11. There are several reference genes that are often used in the literature. However, we do not recommend using β-actin or GAPDH because there are about 60 pseudogenes for actin and about 197 for GAPDH [5]. However, we emphasize that with any reference gene, one should confirm that the expression level (C_q value) does not vary with the experimental conditions, although it may vary by 1–2 cycles due to the accuracy of the amount of cDNA pipetted into the well. We used HPRT1

(hypoxanthine-guanine phosphoribosyltransferase (HGPRT), which is an enzyme encoded by the *HPRT1* gene. In mouse, it only has one pseudogene and in humans three, but there are several regions in the exon sequences that are not homologous in the pseudogenes.

12. Applying the adhesive seal is tricky and if not done well can lead to large standard deviation values. We noticed variation in the standard deviation values of the mean C_q values of identical plates when different students applied the adhesive seal. We discovered that each student had a different technique to do this. An automated plate sealer is better for consistency if available. However, a paddle that is supplied with the adhesive seals (Bio-Rad) (Fig. 2) is adequate if used correctly. First drag the paddle on its edge across the entire surface of the seal. Then, drag it between each column and row and all along the perimeter of the outer wells to ensure that each well is sealed to avoid evaporation and cross-contamination between wells. This greatly decreases the standard deviation values. Be sure to centrifuge the plate as directed before proceeding.
13. Figure 7 shows that the detection of endogenous NOX4 protein in untreated RCC4 cells ($\Delta C_q \sim 6$), using the methods in this chapter, requires extreme conditions such as 72 h of incubation with the primary antibody (Dr. Shah's NOX4 antibody, 1:2000 and SuperSignal West Chemiluminescent Substrate, 30 min) (Fig. 7a, b). It is possible that lower dilutions might shorten the incubation time. This antibody detects several lower MW bands (likely variants or degradation products) as major bands. Treatment with TGFβ1 (5 ng/mL in complete medium) for 24–72 h induced NOX4 expression ($\Delta C_q \sim 2$), and this treatment can be helpful when judging NOX4 protein knockdown. Dr. Doroshow's NOX4 rabbit mAb (1:1000, SuperSignal West Chemiluminescent Substrate, 2 min exposure) also requires about 72 h of incubation (Fig. 7c). It is less sensitive than Dr. Shah's antibody since endogenous NOX4 in RCC4 cells could not be detected without TGFβ1 treatment. However, it appears to be more specific since less of the lower MW bands were observed. Importantly, this example shows that endogenous NOX4 protein from a cell line with a high basal level of mRNA expression ($C_q \sim 27$ and $\Delta C_q \sim 6$) requires extreme conditions to be detected by immunoblot, suggesting that for cell lines with lower mRNA expression levels ($C_q > 27$), detection of NOX protein may be very difficult or impossible given our current tools. Here, RCC4 cells were used because they have a high basal level of NOX4 mRNA ($C_q \sim 27$), possibly due to a mutation of the von Hippel Lindau gene, VHL, and the subsequent impaired protein degradation of hypoxia inducible factor-1-alpha, HIF1-α. Figure 9 shows

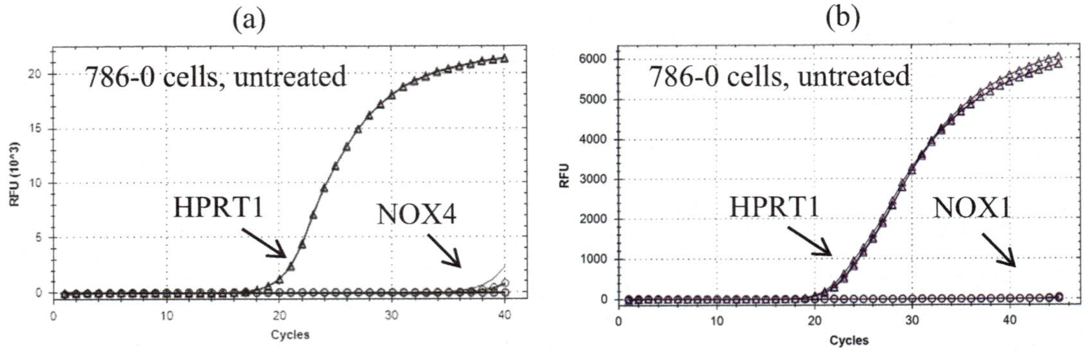

(a)

Primer Set	Ave Cq ± Std. Dev. (n=4 wells)
NOX4 E4F/E5R	>40
HPRT1 E3F/E4R	19.4 ± 0.04

(b)

Primer Set	Ave Cq ± Std. Dev. (n=4 wells)
NOX1 E7F/E8R	>40
HPRT1 E3F/E4R	22.1 ± 0.30

Fig. 9 786–0 cells do not express NOX4 and NOX1. 786–0 cells have been used as a model for studying NOX4 and NOX1, but this cell line does not express NOX4 mRNA (**a**) or NOX1 mRNA (**b**) using the methods in this chapter. The *y*-axis scales differ in (**a**) and (**b**) due to the use of two different commercial SYBR Green reagents from Quanta in (**a**) and Bio-RAD in (**b**)

that untreated 786–0 cells, one of the NCI-60 panel cell lines (renal clear cell carcinoma cell line), do not express NOX4 (or NOX1) mRNA. Thus, 786–0 cells should not be used to determine the effects of deleting NOX4 (or NOX1) in renal cancer.

14. Several companies such as IDT DNA Technologies and Bio-Rad have RT-qPCR tutorials in PDF format available on their websites with full explanations of qPCR methods and show examples of how to calculate relative and absolute changes in mRNA levels. For the $\Delta\Delta C_q$ method, first the ΔC_q of the control condition is subtracted from the ΔCq of the experimental condition:

$$\Delta\Delta C_q = \left(C_{q\text{NOX}} - C_{q\text{REF}}\right)_{\text{experimental}} - \left(C_{q\text{NOX}} - C_{q\text{REF}}\right)_{\text{control}}$$

Then, the relative fold change, R, of NOX mRNA levels with respect to the reference mRNA levels is calculated by inserting the $\Delta\Delta C_q$ value into the following equation:

$R = 2^{-(\Delta\Delta Cq)}$. If there is an increase in the experimental NOX mRNA levels (a decrease in the experimental ΔC_q), then $\Delta\Delta C_q$ will be a negative number. If there is a one cycle decrease ($\Delta\Delta C_q = -1$), then R is twofold. Thus, it is important to take into consideration whether the standard deviations of the mean C_q values are much less than 1 when interpreting qPCR data that report twofold changes. Even for a tenfold change in mRNA levels, the change between the experimental and control values for C_q ($\Delta\Delta C_q$) is only 3.32 cycles. Thus, the C_q values and standard deviations should be included when reporting RT-qPCR data, especially for low fold changes.

15. A plot of C_q values vs. concentration (or fold dilution) generates a standard curve, and the efficiency of the primer set is calculated from the slope. Efficiency is usually within 90–110%. If the efficiency of the reference gene primers is very different than that of the NOX primers, the Pfaffl equation should be also be used and compared to the Livak method since it takes into account the difference in efficiencies. In the Pfaffl equation, the relative fold change in mRNA levels, R, is

$$R = \frac{(1 + E_{NOX})^{\Delta C_{qNOX}}}{(1 + E_{REF})^{\Delta C_{qREF}}}$$

where E is the efficiency of the primer set, and ΔC_{qNOX} and ΔC_{qREF} are the differences between the Cq values for the control condition and the experimental condition for the NOX and reference primers, respectively. The efficiencies in the percentage format are to be converted to a decimal before entering their values into this equation, e.g., for 90%, $1 + E = 1.9$.

16. From the total number of base pairs of the plasmid plus insert that is used to generate the standard curve, it is possible to estimate the molecular weight of the plasmid using the average MW of a base pair (650 g/mol) and hence the molarity of the initial 1 μg/μL stock of linearized plasmid. From the number of moles, the number of molecules (copies) of plasmid can be calculated using Avogadro's number. The copy number can be estimated by plotting copy number rather than dilution-fold on the x-axis. Thus the standard curve could be used as an estimation of copy number.

17. Using a C-terminal tagged NOX/DUOX isoform for transfection will abolish catalytic activity, while N-terminal tags may be tolerated (certain tags for NOX4, DUOX) or decrease activity (NOX1). Epitope tags (e.g., Myc-DDK-NOX4) may also distort localization. Thus, tagged NOX/DUOX constructs should not be used when NOX activity and/or localization will be investigated or need to be validated stringently. However, for purposes of testing antibodies using lysates, a tagged NOX isoform may be used.

18. NOX1–5 require 24 h for initial expression, but 48 h yield higher expression levels when activity measurements are intended. DUOX1 and DUOX2 require co-transfection of DUOXA1 and DUOXA2, respectively, and at least 48 h for optimal expression in HEK 293 cells.

19. Remove as much of the rinse solution as possible so that the rinse solution does not add volume to the RIPA buffer. This is important to keep the protein concentration high for endogenous proteins (2–3 × 10^7 cells/mL; 4–6 mg/mL) as generally 100 μg is loaded in a volume of about 30 μL. For overexpressed NOX proteins, it is not as important, and a protein concentration of about 1 mg/mL (5 × 10^6 cells/mL) should be sufficient to load 10–20 μg. Hold the plate at an angle for 10–15 s and aspirate any remaining rinse solution. Repeat for all the plates before adding RIPA buffer. For the positive control (overexpressed NOX) and empty vector negative controls, one 10 cm plate of transfected cells should yield abundant control protein. However, for cell lines to be tested for endogenous NOX protein, at least 3–4 full plates of cells should be scraped in a total volume of 400 μL to obtain a concentrated protein sample of about 4 mg/mL. After dilution with 4× Laemmli's sample buffer, the protein concentration will be 3 mg/mL, and 33 μL is loaded for 100 μg of protein.
20. Boiling samples compromises detection of integral membrane proteins, including NOX proteins. Boiling NOX proteins will result in bands below 25 kDa or in a high MW band as insoluble material that cannot enter the gel. We find it unnecessary to heat NOX samples, but some laboratories apply heat. For example, heating NOX5 samples at 37 °C for 15 min improves solubility without adverse effects when using Dr. Nauseef's NOX5 antibody (Dr. W. Nauseef, pers. communication) [6]. For DUOX samples, Dr. De Deken applies 80 °C for 5 min due to the high number of glycosylation sites (Fig. 6). It is best to test whether heating will improve your results or whether it is unnecessary. Also a number of reducing agents can be tested in addition to β-mercaptoethanol, such as DTT. Smeared high MW bands may be avoided by preparing lysates fresh and adding reducing agent immediately (Dr. U. Knaus, pers. communication).
21. In our hands, freshly poured gels give better resolution than pre-cast gels. However, this may depend on the manufacturer, the ratio of acrylamide and BIS, and buffer system.
22. We prefer *LI-COR* blocking buffer (TBS-based) without Tween-20 and also 5% nonfat dry milk in TBST from Sigma-Aldrich as blocking buffers. We normally compare both buffers for new antibodies. We also prefer to dilute our antibodies (non-fluorescent as well as fluorescent) in the *LI-COR* blocking buffer +0.1% Tween 20 because the antibody performs well for over a year in this buffer, whereas antibodies stored in milk lose sensitivity due to degradation. We prepare fresh dilutions of antibody in milk after 2 weeks of storage. When using the *LI-COR* Odyssey blocking buffer, add 0.1% Tween when using

it as antibody diluent, and also add 0.01% SDS for secondary antibodies as recommended by the manufacturer.

23. For NOX antibodies, an overnight incubation is adequate for 10 μg of overexpressed NOX proteins; for endogenous NOX protein detection, incubate for 24–72 h at 4 °C while rocking (to be determined by the end user; depends on tissue, antibody dilution of primary and secondary antibodies, and ECL reagent). For secondary antibodies and antibodies used for loading controls (e.g., β-actin), an incubation period of 1 h at 25 °C is sufficient.

Acknowledgments

We thank for their generous support of this work: Dr. Ajay Shah and Dr. Pidder Jansen-Dürr for the NOX4 polyclonal rabbit antibodies; Dr. William Kaelin for the RCC4 parent cell line; Dr. Ralf Brandes for HEK 293 cells stably expressing NOX4 or NOX1; and Dr. Vincent Jaquet, Dr. Mark Quinn, Dr. Corinne Dupuy, Dr. Chihiro Yabe-Nishimura, Dr. Misaki Matsumoto, Dr. Ulla Knaus, and Dr. William Nauseef for contributing specific information for some of the NOX/DUOX antibodies and reviewing the manuscript. We also thank the editors, Dr. Ulla Knaus and Dr. Tom Leto, for the invitation to contribute a chapter to this book.

This chapter is dedicated to the late Dr. Gary Bokoch (The Scripps Research Institute, La Jolla, CA, USA), who made many contributions to the field of NADPH oxidases and served as a post-doctoral mentor of both Dr. Becky Diebold and Dr. Ulla Knaus.

This work was made possible by funding from the National Institute of Health, USA, and the American Heart Association. XDD is supported by the "Fonds de la Recherche Scientifique" (FRS-FNRS), the "Fonds Docteur J.P. Naets" managed by the "Fondation Roi Baudouin," and the "Fondation Tournay-Solvay."

References

1. Altenhofer S, Kleikers PW, Radermacher KA, Scheurer P, Rob Hermans JJ, Schiffers P, Ho H, Wingler K, Schmidt HH (2012) The NOX toolbox: validating the role of NADPH oxidases in physiology and disease. Cell Mol Life Sci 69(14):2327–2343. https://doi.org/10.1007/s00018-012-1010-9

2. Seredenina T, Nayernia Z, Sorce S, Maghzal GJ, Filippova A, Ling SC, Basset O, Plastre O, Daali Y, Rushing EJ, Giordana MT, Cleveland DW, Aguzzi A, Stocker R, Krause KH, Jaquet V (2016) Evaluation of NADPH oxidases as drug targets in a mouse model of familial amyotrophic lateral sclerosis. Free Radic Biol Med 97:95–108. https://doi.org/10.1016/j.freeradbiomed.2016.05.016

3. Bustin SA, Benes V, Garson JA, Hellemans J, Huggett J, Kubista M, Mueller R, Nolan T, Pfaffl MW, Shipley GL, Vandesompele J, Wittwer CT (2009) The MIQE guidelines: minimum information for publication of quantitative real-time PCR experiments. Clin Chem 55(4):611–622. https://doi.org/10.1373/clinchem.2008.112797

4. Kawahara T, Lambeth JD (2008) Phosphatidylinositol (4,5)-bisphosphate modulates Nox5 localization via an N-terminal polybasic region.

Mol Biol Cell 19(10):4020–4031. https://doi.org/10.1091/mbc.E07-12-1223

5. Sun Y, Li Y, Luo D, Liao DJ (2012) Pseudogenes as weaknesses of ACTB (Actb) and GAPDH (Gapdh) used as reference genes in reverse transcription and polymerase chain reactions. PLoS One 7(8):e41659. https://doi.org/10.1371/journal.pone.0041659
6. Dho SH, Kim JY, Kwon ES, Lim JC, Park SS, Kwon KS (2015) NOX5-L can stimulate proliferation and apoptosis depending on its levels and cellular context, determining cancer cell susceptibility to cisplatin. Oncotarget 6 (36):39235–39246. https://doi.org/10.18632/oncotarget.5743
7. Matsumoto M, Katsuyama M, Iwata K, Ibi M, Zhang J, Zhu K, Nauseef WM, Yabe-Nishimura C (2014) Characterization of N-glycosylation sites on the extracellular domain of NOX1/NADPH oxidase. Free Radic Biol Med 68:196–204. https://doi.org/10.1016/j.freeradbiomed.2013.12.013
8. Bedard K, Jaquet V, Krause KH (2012) NOX5: from basic biology to signaling and disease. Free Radic Biol Med 52(4):725–734. https://doi.org/10.1016/j.freeradbiomed.2011.11.023
9. Chen F, Wang Y, Barman S, Fulton DJ (2015) Enzymatic regulation and functional relevance of NOX5. Curr Pharm Des 21 (41):5999–6008
10. Jacob F, Guertler R, Naim S, Nixdorf S, Fedier A, Hacker NF, Heinzelmann-Schwarz V (2013) Careful selection of reference genes is required for reliable performance of RT-qPCR in human normal and cancer cell lines. PLoS One 8(3):e59180. https://doi.org/10.1371/journal.pone.0059180
11. Doroshow JH, Juhasz A, Ge Y, Holbeck S, Lu J, Antony S, Wu Y, Jiang G, Roy K (2012) Antiproliferative mechanisms of action of the flavin dehydrogenase inhibitors diphenylene iodonium and di-2-thienyliodonium based on molecular profiling of the NCI-60 human tumor cell panel. Biochem Pharmacol 83 (9):1195–1207. https://doi.org/10.1016/j.bcp.2012.01.022
12. Meitzler JL, Makhlouf HR, Antony S, Wu Y, Butcher D, Jiang G, Juhasz A, Lu J, Dahan I, Jansen-Durr P, Pircher H, Shah AM, Roy K, Doroshow JH (2017) Decoding NADPH oxidase 4 expression in human tumors. Redox Biol 13:182–195. https://doi.org/10.1016/j.redox.2017.05.016
13. Laurent E, McCoy JW 3rd, Macina RA, Liu W, Cheng G, Robine S, Papkoff J, Lambeth JD (2008) Nox1 is over-expressed in human colon cancers and correlates with activating mutations in K-Ras. Int J Cancer 123 (1):100–107. https://doi.org/10.1002/ijc.23423
14. Burritt JB, Quinn MT, Jutila MA, Bond CW, Jesaitis AJ (1995) Topological mapping of neutrophil cytochrome b epitopes with phage-display libraries. J Biol Chem 270 (28):16974–16980
15. Nakamura M, Murakami M, Koga T, Tanaka Y, Minakami S (1987) Monoclonal antibody 7D5 raised to cytochrome b558 of human neutrophils: immunocytochemical detection of the antigen in peripheral phagocytes of normal subjects, patients with chronic granulomatous disease, and their carrier mothers. Blood 69 (5):1404–1408
16. Kawai C, Yamauchi A, Kuribayashi F (2018) Monoclonal antibody 7D5 recognizes the R147 epitope on the gp91(phox), phagocyte flavocytochrome b558 large subunit. Microbiol Immunol 62(4):269–280. https://doi.org/10.1111/1348-0421.12584
17. Verhoeven AJ, Bolscher BG, Meerhof LJ, van Zwieten R, Keijer J, Weening RS, Roos D (1989) Characterization of two monoclonal antibodies against cytochrome b558 of human neutrophils. Blood 73(6):1686–1694
18. Seredenina T, Sorce S, Herrmann FR, Ma Mulone XJ, Plastre O, Aguzzi A, Jaquet V, Krause KH (2017) Decreased NOX2 expression in the brain of patients with bipolar disorder: association with valproic acid prescription and substance abuse. Transl Psychiatry 7(8):e1206. https://doi.org/10.1038/tp.2017.175
19. Taylor RM, Maaty WS, Lord CI, Hamilton T, Burritt JB, Bothner B, Jesaitis AJ (2007) Cloning, sequence analysis and confirmation of derived gene sequences for three epitope-mapped monoclonal antibodies against human phagocyte flavocytochrome b. Mol Immunol 44(4):625–637. https://doi.org/10.1016/j.molimm.2005.10.022
20. Sorce S, Nuvolone M, Keller A, Falsig J, Varol A, Schwarz P, Bieri M, Budka H, Aguzzi A (2014) The role of the NADPH oxidase NOX2 in prion pathogenesis. PLoS Pathog 10(12):e1004531. https://doi.org/10.1371/journal.ppat.1004531
21. Anilkumar N, Weber R, Zhang M, Brewer A, Shah AM (2008) Nox4 and nox2 NADPH oxidases mediate distinct cellular redox signaling responses to agonist stimulation. Arterioscler Thromb Vasc Biol 28(7):1347–1354. https://doi.org/10.1161/ATVBAHA.108.164277
22. Holl M, Koziel R, Schafer G, Pircher H, Pauck A, Hermann M, Klocker H, Jansen-Durr P, Sampson N (2016) ROS signaling by

NADPH oxidase 5 modulates the proliferation and survival of prostate carcinoma cells. Mol Carcinog 55(1):27–39. https://doi.org/10.1002/mc.22255

23. von Lohneysen K, Noack D, Jesaitis AJ, Dinauer MC, Knaus UG (2008) Mutational analysis reveals distinct features of the Nox4-p22 phox complex. J Biol Chem 283 (50):35273–35282. https://doi.org/10.1074/jbc.M804200200
24. El Jamali A, Valente AJ, Lechleiter JD, Gamez MJ, Pearson DW, Nauseef WM, Clark RA (2008) Novel redox-dependent regulation of NOX5 by the tyrosine kinase c-Abl. Free Radic Biol Med 44(5):868–881. https://doi.org/10.1016/j.freeradbiomed.2007.11.020
25. Marzaioli V, Hurtado-Nedelec M, Pintard C, Tlili A, Marie JC, Monteiro RC, Gougerot-Pocidalo MA, Dang PM, El-Benna J (2017) NOX5 and p22phox are 2 novel regulators of human monocytic differentiation into dendritic cells. Blood 130(15):1734–1745. https://doi.org/10.1182/blood-2016-10-746347
26. Antony S, Wu Y, Hewitt SM, Anver MR, Butcher D, Jiang G, Meitzler JL, Liu H, Juhasz A, Lu J, Roy KK, Doroshow JH (2013) Characterization of NADPH oxidase 5 expression in human tumors and tumor cell lines with a novel mouse monoclonal antibody. Free Radic Biol Med 65:497–508. https://doi.org/10.1016/j.freeradbiomed.2013.07.005
27. Pacquelet S, Lehmann M, Luxen S, Regazzoni K, Frausto M, Noack D, Knaus UG (2008) Inhibitory action of NoxA1 on dual oxidase activity in airway cells. J Biol Chem 283(36):24649–24658. https://doi.org/10.1074/jbc.M709108200
28. Luxen S, Noack D, Frausto M, Davanture S, Torbett BE, Knaus UG (2009) Heterodimerization controls localization of Duox-DuoxA NADPH oxidases in airway cells. J Cell Sci 122. (Pt 8:1238–1247. https://doi.org/10.1242/jcs.044123
29. De Deken X, Wang D, Many MC, Costagliola S, Libert F, Vassart G, Dumont JE, Miot F (2000) Cloning of two human thyroid cDNAs encoding new members of the NADPH oxidase family. J Biol Chem 275 (30):23227–23233. https://doi.org/10.1074/jbc.M000916200
30. Caillou B, Dupuy C, Lacroix L, Nocera M, Talbot M, Ohayon R, Deme D, Bidart JM, Schlumberger M, Virion A (2001) Expression of reduced nicotinamide adenine dinucleotide phosphate oxidase (ThoX, LNOX, Duox) genes and proteins in human thyroid tissues. J Clin Endocrinol Metab 86(7):3351–3358. https://doi.org/10.1210/jcem.86.7.7646
31. Ameziane-El-Hassani R, Talbot M, de Souza Dos Santos MC, Al Ghuzlan A, Hartl D, Bidart JM, De Deken X, Miot F, Diallo I, de Vathaire F, Schlumberger M, Dupuy C (2015) NADPH oxidase DUOX1 promotes long-term persistence of oxidative stress after an exposure to irradiation. Proc Natl Acad Sci U S A 112(16):5051–5056. https://doi.org/10.1073/pnas.1420707112
32. El Hassani RA, Benfares N, Caillou B, Talbot M, Sabourin JC, Belotte V, Morand S, Gnidehou S, Agnandji D, Ohayon R, Kaniewski J, Noel-Hudson MS, Bidart JM, Schlumberger M, Virion A, Dupuy C (2005) Dual oxidase 2 is expressed all along the digestive tract. Am J Physiol Gastrointest Liver Physiol 288(5):G933–G942. https://doi.org/10.1152/ajpgi.00198.2004
33. Wu Y, Antony S, Hewitt SM, Jiang G, Yang SX, Meitzler JL, Juhasz A, Lu J, Liu H, Doroshow JH, Roy K (2013) Functional activity and tumor-specific expression of dual oxidase 2 in pancreatic cancer cells and human malignancies characterized with a novel monoclonal antibody. Int J Oncol 42(4):1229–1238. https://doi.org/10.3892/ijo.2013.1821
34. De Leo FR, Ulman KV, Davis AR, Jutila KL, Quinn MT (1996) Assembly of the human neutrophil NADPH oxidase involves binding of p67phox and flavocytochrome b to a common functional domain in p47phox. J Biol Chem 271(29):17013–17020

Part III

Detection of ROS

Chapter 13

Methods for Detection of NOX-Derived Superoxide Radical Anion and Hydrogen Peroxide in Cells

Fiona Augsburger, Aleksandra Filippova, and Vincent Jaquet

Abstract

NADPH oxidases (NOX) are transmembrane enzymes, which catalyze the formation of reactive oxygen species (ROS). In humans and most mammals, the NOX family comprises seven members, namely, NOX1-5 and the dual oxidases DUOX1 and 2. The primary product of most NOX isoforms is the superoxide radical anion $O_2^{\bullet-}$, which is rapidly dismutated in hydrogen peroxide (H_2O_2), while NOX4 and DUOX mostly generate H_2O_2. ROS are multifunctional molecules in tissues, and NOX-derived ROS cellular functions are as diverse as microbial killing (NOX2), thyroid hormone synthesis (DUOX2), or otoconia formation in the inner ear (NOX3). NOX are potential pharmacological targets in numerous diseases such as diabetes, fibrosis, and brain ischemia, and NOX inhibitors are currently under development. Here we describe two cellular assays to detect extracellular $O_2^{\bullet-}$ and H_2O_2 in cells overexpressing specific NOX isoforms and their subunits.

Key words NADPH oxidase, Hydrogen peroxide, Superoxide radical anion, Peroxidase, Amplex red, WST-1, Cellular assay

1 Introduction

Reactive oxygen species (ROS) are chemically reactive molecules containing oxygen. ROS include oxygen radicals, such as superoxide radical anion ($O_2^{\bullet-}$), nitric oxide ($NO^{\bullet}$), hydroxyl ($^{\bullet}OH$), peroxyl ($RO_2^{\bullet}$), alkoxyl ($RO^{\bullet}$), and peroxynitrite (NO_3^{-}) and oxygen non-radicals, such as hydrogen peroxide (H_2O_2), ozone (O_3), singlet oxygen ($^1O_2$), and hypochlorous acid (HOCl). Presence of ROS was detected in plant and animal tissues using electron paramagnetic resonance (EPR) spectroscopy as early as the 1950s [1]. ROS are typically generated as by-products of biological reactions in mitochondria and peroxisomes, but also from enzymatic reactions, such as cytochrome P450 or xanthine oxidase. In order to limit potential cellular damage by excessive ROS, enzymatic systems and natural oxidant scavengers neutralize them. These include superoxide dismutases (SOD), catalase, glutathione peroxidases, peroxiredoxins,

Ulla G. Knaus and Thomas L. Leto (eds.), *NADPH Oxidases: Methods and Protocols*, Methods in Molecular Biology, vol. 1982, https://doi.org/10.1007/978-1-4939-9424-3_13,

glutathione, ascorbic acid (vitamin C), α-tocopherol (vitamin E), and melatonin, among others. However, since their discovery, the role of ROS as active intermediates in some metabolic processes has been progressively recognized (*see* review [2]).

The discovery of a whole family of enzymes (i.e., NOX) with the sole function of generating ROS has highlighted the physiological importance of ROS in cells and tissues. All NOX isoforms catalyze the transport of electrons across membranes. NOX use cytoplasmic NADPH as electron donor to reduce extra-cytosolic O_2, thereby generating $O_2^{\bullet -}$ as primary product. There are seven NOX isoforms (NOX1-5 and DUOX1 and 2) with specialized expression and modes of activation [2, 3]. NOX1, 2, 3, and 5 generate $O_2^{\bullet -}$ which is rapidly dismutated in H_2O_2 in solution, while only H_2O_2 is detected upon activity of NOX4 and DUOX (Fig. 1).

Genetic loss of function of NOX2 leads to chronic granulomatous disease, an immune deficiency characterized by the absence of functional NOX2. DUOX2 mutations can cause congenital hypothyroidism, while NOX3 loss of function leads to absence of otoconia in the inner ear and impaired balance in rodents. The exact

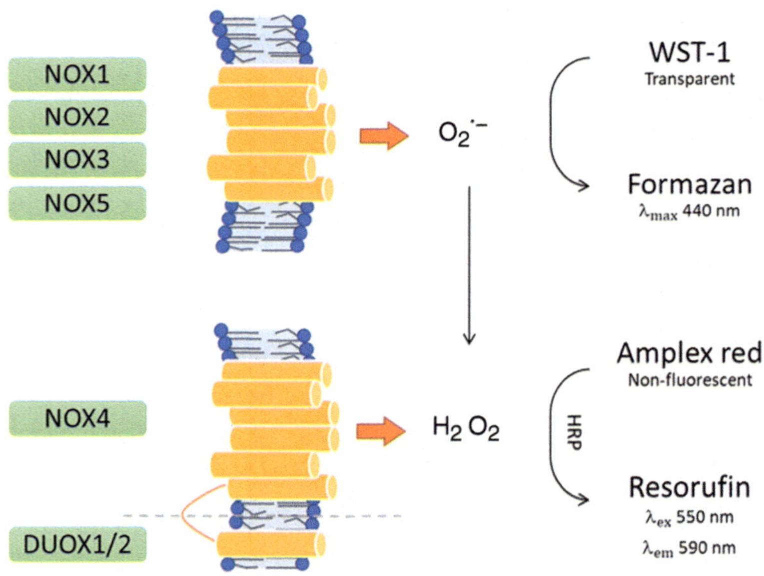

Fig. 1 Scheme showing the principle of detection of the ROS-generating activity of NOX isoforms. Extracellular $O_2^{\bullet -}$ generated by NOX1, NOX2, NOX3, and NOX5 is detected by reduction of WST-1, which produces a yellow soluble formazan detected by absorbance spectrophotometry (440 nm). NOX4 and DUOX activities are undetectable by WST-1 as they directly produce H_2O_2 extracellularly. Due to rapid dismutation of $O_2^{\bullet -}$, H_2O_2 is a product common to all NOX isoforms. H_2O_2 is detected by the conversion of the nonfluorescent, colorless Amplex red to a fluorescent product by HRP-mediated oxidation. Amplex red is oxidized by H_2O_2 in the presence of HRP to generate resorufin, which is colored (pink) and highly fluorescent at 587 nm

physiological role of other NOX isoforms is still cryptic, but they may regulate key redox-dependent signaling by oxidizing phosphatases, receptors, and transcription factors. As professional ROS producers, NOX are attractive pharmacological targets to reduce excessive ROS production in numerous disease states [4].

Detection of NOX activity is a complex matter as probes used to detect oxidants are artifact-prone [5]. Specificity of signal needs confirmation by appropriate controls: $O_2^{\bullet-}$ is mitigated by addition of SOD and H_2O_2 by addition of catalase, while all NOX isoforms are inhibited by the flavoprotein inhibitor diphenylene iodonium (DPI). Here we describe two methods to detect extracellular $O_2^{\bullet-}$ and H_2O_2 following specific activity of NOX isoforms (*see* **Note 1**).

2 Materials

2.1 Common Material for Both Assays

1. NOX1 cells: CHO cells transduced with human NOX1, NOXA1, NOXO1, and CYBA coding sequences (*see* **Note 2**).
2. NOX2 cells: Human PLB-985 cells differentiated with DMSO 1.25% for 3 days (*see* **Note 2**).
3. T-REx NOX3 cells: T-REx 293 cells transduced with human coding sequences of NOX3, NOXA1, NOXO1, and CYBA. A tetracycline-dependent promoter controls the expression of NOX3 (*see* **Note 2**).
4. T-REx NOX4 cells: T-REx 293 cells transduced with human coding sequence of NOX4. A tetracycline-dependent promoter controls the expression of NOX4 (*see* **Note 2**).
5. NOX5 cells: HEK293T cells stably transfected to express human NOX5 cDNA (*see* **Note 2**).
6. DUOX1 cells: HEK293T cells stably transfected with human DUOX1 and DUOXA1 (*see* **Notes 2** and **3**).
7. DUOX2 cells: HEK293T cells stably transfected with human DUOX2 and DUOXA2 (*see* **Notes 2** and **3**).
8. Complete medium 1: Dulbecco's Modified Eagle's medium, 10% fetal calf serum, penicillin (100 U/mL), and streptomycin (100 μg/mL).
9. Complete medium 2: Roswell Park Memorial Institute medium 1640, 10% fetal calf serum, penicillin (100 U/mL), and streptomycin (100 μg/mL).
10. Dulbecco's phosphate-buffered saline (DPBS) 1×.
11. Trypsin-EDTA (0.05%).
12. HBSS (Hanks' balanced salt solution), sodium bicarbonate.
13. Phorbol 12-myristate 13-acetate (PMA) 1 μM in HBSS (*see* **Note 4**).

14. Ionomycin calcium salt 10 μM in HBSS (*see* **Note 5**).
15. Superoxide dismutase (SOD) from bovine erythrocytes 500 U/mL in HBSS.
16. Catalase from bovine liver 5000 U/mL in HBSS.
17. Diphenylene iodonium chloride (DPI) 200 μM in HBSS (*see* **Note 6**).
18. Multiplate reader, such as FlexStation 3 Multi-Mode Microplate Reader (Molecular Devices) or FLUOstar OPTIMA Microplate Reader (BMG).

2.2 Amplex Red Assay

1. *N*-Acetyl-3,7-dihydroxyphenoxazine (Amplex red) 10 mM in DMSO. Avoid exposure to direct light as much as possible.
2. Horseradish peroxidase 0.05 U/mL in HBSS (*see* **Note 7**).
3. Hydrogen peroxide (H_2O_2) 1 mM in Milli-Q H_2O.
4. Black 96-well plate, flat bottom.

2.3 WST-1 Assay

1. 2-(4-Iodophenyl)-3-(4-nitrophenyl)-5-(2,4-disulfophenyl)-2H-tetrazolium, monosodium salt (WST-1) 100 mM in DMSO.
2. Clear 96-well plate, flat bottom.

3 Methods

3.1 Cell Preparation for Both Assays

1. NOX-expressing cells are grown in dishes 75 cm^2 at approximately 90% confluency in complete medium 1—except for PLB-985 cells, which are grown in suspension in complete medium 2—at 37 °C, 5% CO_2 (*see* **Note 8**).
2. Remove medium, add 10 mL DPBS 1×, remove DPBS 1x, and add 3 mL pre-warmed Trypsin-EDTA for 5 min. Stop reaction by adding 3 mL complete medium.
3. Collect cells in 15 mL tubes, and centrifuge at 400 × *g* for 10 min at room temperature. Discard supernatant.
4. Resuspend in HBSS 20 mL and count cells using a hemocytometer, dilute to 500,000 cells/mL in fresh HBSS, and keep on ice (*see* **Note 9**).

3.2 H_2O_2 Detection Using Amplex Red/HRP System (See Note 10)

1. Prepare black 96-well plate with control wells: dispense 10 μL of each control (500 U/mL SOD (final 25 U/mL), 5000 U/mL catalase (final 250 U/mL) and 200 μM DPI (final 10 μM), all in HBSS) in separate wells. Dispense 10 μL HBSS in all other wells.
2. Prepare H_2O_2 standard curve: dispense 100 μL/well of HBSS in an entire row of 96-well plate, except in the first well

dispense 200 μL. Add 2.5 μL of 1 mM H_2O_2 (12.5 μM final) in the first well, mix, and take 100 μL to the next well, mix, and take 100 μL to the next well, until before the last well. The last well should not contain H_2O_2.

3. Dispense 100 μL/well of cells in all wells, except for the H_2O_2 standard row (*see* **Note 9**).
4. For NOX3 and NOX4 cells and H_2O_2 standard curve, prepare Amplex red reaction mix 1 (*see* **Note 11**): 0.5 μL/well 10 mM Amplex red, 20 μL/well 0.05 U/mL HRP, and 69.5 μL/well HBSS (*see* **Note 12**).
5. For NOX1 and NOX2 cells, prepare Amplex red reaction mix 2 (PMA) (*see* **Note 11**): 0.5 μL/well 10 mM Amplex red, 20 μL/well 0.05 U/mL HRP, 20 μL/well 1 μM PMA (final 100 nM), and 49.5 μL/well HBSS (*see* **Note 12**).
6. For NOX5, DUOX1, and DUOX2 cells, prepare Amplex red reaction mix 3 (ionomycin) (*see* **Note 11**): 0.5 μL/well 10 mM Amplex red, 20 μL/well 0.05 U/mL HRP, 20 μL/well 1 μM PMA (final 100 nM), 20 μL/well 10 μM ionomycin (final 1 μM), and 29.5 μL HBSS (*see* **Note 12**).
7. Set fluorescence reader temperature at 37 °C.
8. Dispense 90 μL/well of reaction mixture according to corresponding cells and H_2O_2 standard in wells.
9. Measure the kinetic of fluorescence for 1 h with excitation wavelength at 550 nm and emission wavelength at 590 nm using a fluorescence reader. The Amplex red signal is inhibited by catalase and DPI, but not SOD. Results of Amplex red assay for each NOX isoform are shown in Fig. 2a.

3.3 $O_2^{\bullet-}$ Detection Using WST-1 (See Note 13)

1. Prepare clear 96-well plate with control wells: dispense 10 μL of each control (500 U/mL SOD (final 25 U/mL), 5000 U/mL catalase (final 250 U/mL), and 200 μM DPI (final 10 μM), all in HBSS) in separate wells.. Dispense 10 μL HBSS in other wells.
2. Dispense 100 μL/well of cells in all wells (*see* **Note 9**).
3. For NOX3 and NOX4 cells, prepare WST-1 reaction mix 1: 1 μL/well 100 mM WST-1 and 89 μL/well HBSS (*see* **Note 12**).
4. For NOX1 and NOX2 cells, prepare WST-1 reaction mix 2 (PMA): 1 μL/well 100 mM WST-1, 20 μL/well 1 μM PMA (final 100 nM), and 69 μL/well HBSS (*see* **Note 12**).
5. For NOX5, DUOX1, and DUOX2 cells, prepare WST-1 reaction mix 3 (ionomycin): 1 μL/well 100 mM WST-1, 20 μL/well 1 μM PMA (final 100 nM), 20 μL/well 10 μM ionomycin (final 1 μM), and 49 μL/well HBSS (*see* **Note 12**).
6. Set absorbance reader temperature at 37 °C.

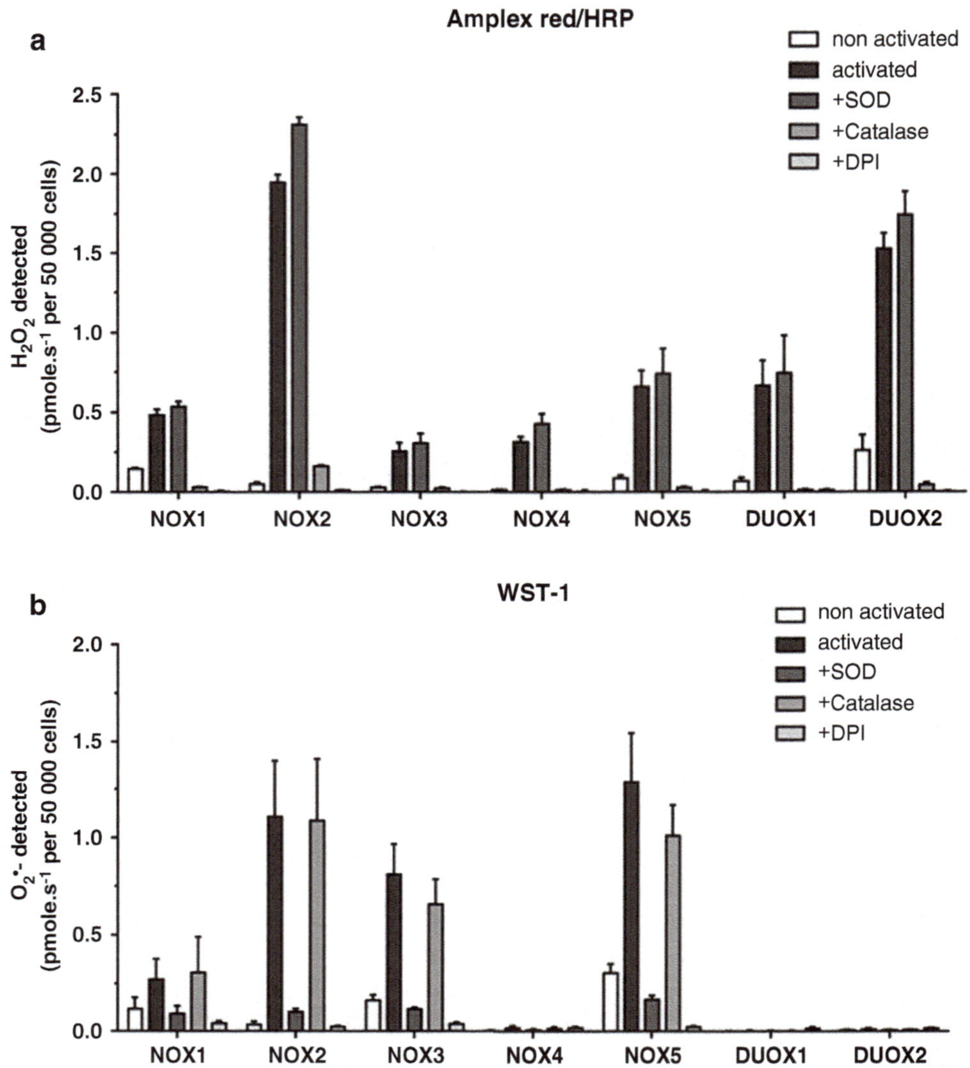

Fig. 2 ROS measurement in cells expressing different NOX isoforms. All required components for each NOX isoform were transduced in HEK or CHO cells except for NOX2, which is endogenously expressed in PLB-985 cells. Oxidant generation was induced as described. Specificity of the signals was controlled using SOD, catalase, and DPI. (**a**) Rate of H_2O_2 production as measured with the Amplex red/HRP system. (**b**) Rate of $O_2^{\bullet -}$ production as measured with WST-1 probe. The bars represent the mean ± standard deviation of three independent experiments performed in triplicates

7. Dispense 90 μL/well of reaction mixture.
8. Measure immediately the kinetic of absorbance at both 440 nm and 600 nm using an absorbance reader. The WST-1 signal is inhibited by SOD and DPI, but not catalase. NOX4, DUOX1, and DUOX2 activities are not detected using this probe. Results of WST-1 assay for each NOX isoform are shown in Fig. 2b.

3.4 Data Analysis

For Amplex red assay, hydrogen peroxide equivalents are calculated from fluorescence values using the standard curve with H_2O_2. Rates of H_2O_2 production are evaluated by calculating the slope in the linear part of the curve, usually at around 20 min and 40 min after activation of the reaction.

For WST-1 assay, rates of $O_2^{\bullet-}$ are evaluated by calculating the slope in the linear part of absorbance values using the following equation:

$$A = \varepsilon_{440\ \mathrm{nm}} \cdot l \cdot c$$

A: absorbance measured (A. U.)$\varepsilon_{440\ \mathrm{nm}}$: $37{\cdot}10^3\ \mathrm{M}^{-1}{\cdot}\mathrm{cm}^{-1}$ for WST-1l: 0.4 cm. focal distance corresponds to 0.1 cm/25 μL for 96-well plate$c$: $O_2^{\bullet-}$ concentration in molar

4 Notes

1. Countless probes are suitable for measuring ROS. Advantages and pitfalls of commonly used probes are reviewed in detail in [5].
2. Details of cell lines expressing NOX1-5 are found in [6]. Briefly, ROS generation by NOX1 and NOX2 is activated by 100 nM PMA. NOX3 and NOX4 are constitutively active; expression of NOX3 and NOX4 is induced by pre-treatment with tetracycline 1 μg/mL 24 h before the experiment; NOX5, DUOX1, and DUOX2 are activated by combining 100 nM PMA and 1 μM ionomycin. While PLB-985 cells express endogenously NOX2 and its subunits, all described cell lines are artificial systems based on heterologous NOX expression. Thus the relative ROS output measured cannot be interpreted as an indication of the inherent enzymatic kinetics of a specific NOX isoform to generate $O_2^{\bullet-}$ or H_2O_2. Expression levels of NOX and subunits are most likely the main determinants for ROS output in these artificial systems.
3. Cells stably transfected with DUOX1/DUOXA1 and DUOX2/DUOXA2 are generous gifts from William Nauseef, University of Iowa, USA.
4. PMA stock is kept as aliquots at 1 mM in DMSO at −20 °C. Working solution is prepared by adding 10 μL PMA stock in 10 mL HBSS on ice. Final concentration of PMA is 100 nM.
5. Ionomycin is kept as aliquots at 1 mM in DMSO at −20 °C. Working solution is prepared by adding 10 μL ionomycin stock in 1 mL HBSS on ice. Final concentration of ionomycin is 1 μM.

6. DPI stock is kept as aliquots at 10 mM in DMSO at −20 °C. Working solution is prepared by adding 20 μL DPI stock in 1 mL HBSS on ice. Final concentration of DPI is 10 μM.
7. HRP stock is kept as aliquots at 500 U/mL in PBS at −20 °C. Working solution is prepared by adding 2 μL HRP stock in 20 mL HBSS on ice. Final concentration of HRP is 0.005 U/mL.
8. Adherent cells grown to 80–90% confluence in a 75-cm^2 cell culture flask typically yield 10–25 × 10^6 cells.
9. In this protocol we describe measurement of extracellular H_2O_2 or $O_2^{\bullet -}$ using cells in suspension in order to better control cell number. Adherent cells may be used. This may possibly generate a different amount of ROS.
10. Amplex red is a quantitative method to measure H_2O_2 in kinetic mode with high fidelity. Amplex red can detect the activity of all NOX isoforms (Fig. 2a). The signal is stable for at least 1 h. A potential pitfall resides in the fact that it does not directly react with H_2O_2, as its oxidation depends on peroxidase activity. When searching for NOX inhibitors, the use of an Amplex red assay to measure NOX activity will increase the rate of false positive as a decreased Amplex red signal is often due to direct peroxidase inhibition (as shown in [6]).
11. Amplex red is optimally added in the reaction mixture immediately before measurement to avoid auto-oxidation. Avoid exposure to direct light as much as possible.
12. To prepare the mix, calculate at least one additional well to ensure that the amount of mix is sufficient.
13. WST-1 is formazan salt allowing quantitatively measuring $O_2^{\bullet -}$ in kinetic mode. WST-1 reacts directly with $O_2^{\bullet -}$. WST-1 can only detect the activity of NOX isoforms generating $O_2^{\bullet -}$ (Fig. 2b). The signal is stable for at least 1 h. Compared to the probe ferrocytochrome C commonly used to measure $O_2^{\bullet -}$, WST-1 shows very low background absorbance and generates a twofold greater signal. Furthermore, reduction of WST leads to a stable water-soluble formazan [7].

Acknowledgments

This work was supported by the European Community's Framework Programme (FP7/2007–2013) under grant agreement number 278611.

References

1. Commoner B, Townsend J, Pake GE (1954) Free radicals in biological materials. Nature 174 (4432):689–691
2. Hensley K, Floyd RA (2002) Reactive oxygen species and protein oxidation in aging: a look back, a look ahead. Arch Biochem Biophys 397 (2):377–383. https://doi.org/10.1006/abbi.2001.2630
3. Bedard K, Krause KH (2007) The NOX family of ROS-generating NADPH oxidases: physiology and pathophysiology. Physiol Rev 87 (1):245–313. https://doi.org/10.1152/physrev.00044.2005
4. Lambeth JD, Krause KH, Clark RA (2008) NOX enzymes as novel targets for drug development. Semin Immunopathol 30(3):339–363. https://doi.org/10.1007/s00281-008-0123-6
5. Maghzal GJ, Krause KH, Stocker R, Jaquet V (2012) Detection of reactive oxygen species derived from the family of NOX NADPH oxidases. Free Radic Biol Med 53(10):1903–1918. https://doi.org/10.1016/j.freeradbiomed.2012.09.002
6. Seredenina T, Chiriano G, Filippova A, Nayernia Z, Mahiout Z, Fioraso-Cartier L, Plastre O, Scapozza L, Krause KH, Jaquet V (2015) A subset of N-substituted phenothiazines inhibits NADPH oxidases. Free Radic Biol Med 86:239–249. https://doi.org/10.1016/j.freeradbiomed.2015.05.023
7. Tan AS, Berridge MV (2000) Superoxide produced by activated neutrophils efficiently reduces the tetrazolium salt, WST-1 to produce a soluble formazan: a simple colorimetric assay for measuring respiratory burst activation and for screening anti-inflammatory agents. J Immunol Methods 238(1–2):59–68

Chapter 14

HPLC-Based Monitoring of Oxidation of Hydroethidine for the Detection of NADPH Oxidase-Derived Superoxide Radical Anion

Jacek Zielonka, Monika Zielonka, and Balaraman Kalyanaraman

Abstract

Hydroethidine is a fluorogenic probe that in the presence of the superoxide radical anion is oxidized to a red fluorescent product, 2-hydroxyethidium. In cells, hydroethidine is also oxidized to other products, including red fluorescent ethidium. Thus, selective monitoring of 2-hydroxyethidium is required for specific detection of the superoxide radical anion. Here, we provide protocols for HPLC- and LC-MS-based quantitation of 2-hydroxyethidium, among other oxidation products. Also, a protocol for continuous sampling for real-time monitoring of superoxide production using rapid HPLC measurements of 2-hydroxyethidium is described.

Key words Superoxide radical anion, Hydroethidine, 2-hydroxyethidium, HPLC, LC-MS

1 Introduction

Superoxide radical anion ($O_2^{\bullet-}$) is a primary species formed upon one-electron reduction of molecular oxygen, a key step in the enzymatic function of nicotinamide adenine dinucleotide phosphate (NADPH) oxidases (NOX enzymes) [1, 2]. Because production of $O_2^{\bullet-}$ and hydrogen peroxide (H_2O_2, a product of dismutation of $O_2^{\bullet-}$) is regarded as a primary function of the members of the family of NADPH oxidases [1, 3], their detection is often used to monitor the activity of NOX enzymes in cell-free and cell-based assays [4]. Luminescent (chemiluminescent and fluorescent) probes are typically used for this purpose, as they allow for sensitive, nondestructive, and real-time measurement of $O_2^{\bullet-}$ or H_2O_2 production [5, 6]. The applicability of currently available chemiluminescent probes for specific detection of $O_2^{\bullet-}$ is rather limited, due to their inherent chemical reactivity and the reactivity of reaction intermediates on the pathways to chemiluminescence [5–7]. Among fluorescent probes for $O_2^{\bullet-}$, hydroethidine

Ulla G. Knaus and Thomas L. Leto (eds.), *NADPH Oxidases: Methods and Protocols*, Methods in Molecular Biology, vol. 1982, https://doi.org/10.1007/978-1-4939-9424-3_14,

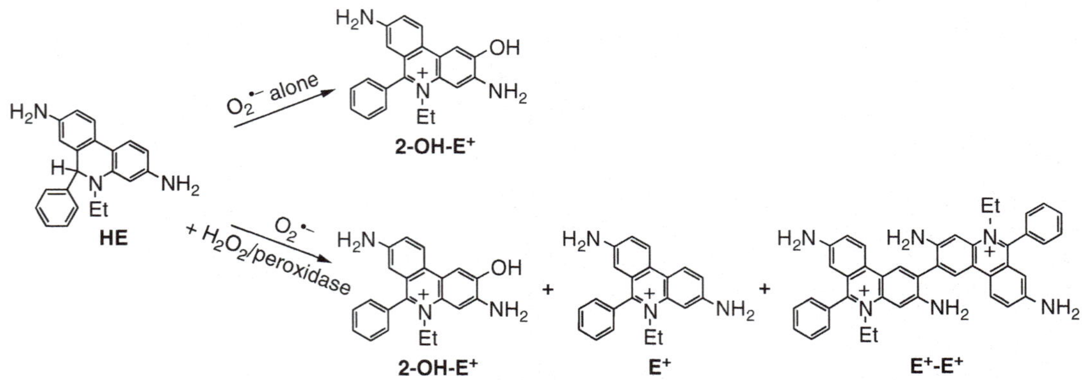

Fig. 1 HE oxidation scheme

(HE, also known as dihydroethidium) has proven to be most useful [8]. In the presence of $O_2^{\bullet-}$, HE is oxidized to a red fluorescent product, 2-hydroxyethidium (2-OH-E^+, Fig. 1). Because this product was demonstrated to be formed only by $O_2^{\bullet-}$, but not other biologically relevant oxidizing species, assays based on the conversion of HE into 2-OH-E^+ are currently regarded as a gold standard for $O_2^{\bullet-}$ detection. In cells, HE is also oxidized to other products, including ethidium and dimers. Because ethidium is also red fluorescent and is formed intracellularly typically at more than tenfold excess of 2-OH-E^+, selective detection of 2-OH-E^+ requires separation of the products by high-performance liquid chromatography (HPLC) or other techniques. In addition, one-electron oxidants (e.g., peroxidase + H_2O_2) have been shown to increase the yield of 2-OH-E^+ in the presence of a steady flux of $O_2^{\bullet-}$, as well as to produce both E^+ and E^+-E^+ (Fig. 1). Thus, profiling all oxidation products of HE, including dimers, is crucial for accurate interpretation of the changes in 2-OH-E^+ formation. Several papers reported the use of HPLC-based detection of 2-OH-E^+ for monitoring the activity of NADPH oxidases in cellular systems [9–15].

Here, we describe how profiling the oxidation products of the HE probe can be utilized to selectively monitor superoxide produced by NADPH oxidases, both intracellularly and in the extracellular medium. The method is based on HPLC or liquid chromatography-mass spectrometry (LC-MS)-based separation and quantification of the probe and the oxidation products, including 2-OH-E^+, E^+, and E^+-E^+. We describe the protocols for cell sample collection, processing, and HPLC and LC-MS-based analyses of the products formed.

2 Materials

Prepare all solutions using ultrapure water, endotoxin-free dimethyl sulfoxide (DMSO), and analytical-grade reagents. Samples and solutions should be protected from light and stored on ice (unless indicated otherwise) (*see* **Note 1**).

2.1 Probes and Standards of the Oxidation Products

1. Hydroethidine: prepare stock solution (20 mM) in DMSO, aliquot in brown or black tubes (20 μL per tube), and store at −80 °C. Discard any leftovers; do not reuse (*see* **Note 2**).
2. 2-Hydroxyethidium: prepare using a published protocol [16, 17]. Prepare a stock solution of 20 mM in DMSO and store at −80 °C.
3. Ethidium bromide: prepare a stock solution of 20 mM in DMSO and store at −80 °C.
4. Diethidium: prepare using a published protocol [16, 17]. Prepare a stock solution of 20 mM in DMSO and store at −80 °C.
5. Internal standard: 3,8-diamino-6-phenylphenanthridine (DAPP). Prepare a stock solution of 1 mM in DMSO and store at −20 °C.

2.2 Cell Incubation Components

1. Cell growth medium (*see* **Note 3**).
2. Assay medium (if different than cell growth medium) (*see* **Note 4**).
3. Dulbecco's phosphate-buffered saline (DPBS) for cell washing (2 × ~10 mL in 15 mL tubes per cell dish), stored on ice (*see* **Note 5**).

2.3 Cell Extraction Components

1. Cell lysis buffer: 4 mL of DPBS containing 0.1% Triton X-100, and 1 μM DAPP (internal standard), and place on ice.
2. 10 mL of ice-cold LC-MS-grade acetonitrile (MeCN), containing 0.1% (v/v) formic acid (FA).
3. 10 mL of ice-cold LC-MS-grade water, containing 0.1% (v/v) FA.
4. Protein assay reagent (e.g., Bradford reagent).
5. Bovine serum albumin (BSA) in the lysis buffer: 20 mg/mL BSA. Prepare a series of BSA solutions for a calibration curve for protein assay. Dilute the 20 mg/mL BSA solution in the lysis buffer to final concentrations of 0.5, 1, 1.5, 2, 3, 4, 5, 7, and 10 mg/mL. Keep all solutions on ice or refrigerated.

2.4 Medium Extraction Components

1. 10 mL of ice-cold LC-MS-grade MeCN, containing 0.1% (v/v) FA and 1 μM DAPP (internal standard).
2. 10 mL of ice-cold LC-MS-grade water, containing 0.1% (v/v) FA.

2.5 Stopping Solution for "Real-Time" Analyses

1 mL of a mixture of superoxide dismutase (SOD, 50×) and catalase (CAT, 50×): prepare a solution of SOD (5 mg/mL) and catalase (5 kU/mL) in water (*see* **Note 6**).

2.6 HPLC and LC-MS Analysis Components

1. HPLC mobile phase: 0.1% trifluoroacetic acid in water (mobile phase A) and 0.1% trifluoroacetic acid in MeCN (mobile phase B). LC-MS mobile phase: 0.1% FA in water (mobile phase A) and 0.1% FA in MeCN (mobile phase B) (*see* **Note 7**).
2. Solvent for standards: water-to-MeCN (3:1, v/v) mixture, containing 0.1% FA and 1 μM DAPP (internal standard).
3. Mixture of the HE oxidation products standards (2-OH-E^+, E^+, and E^+-E^+; 100 μM each) in the solvent for standards (prepared in the previous step). Prepare serial dilutions from 1 μM down to 1 nM.
4. Prepare serial dilutions of HE probe from 10 μM down to 10 nM. Use the solvent for standards prepared in **step 2**.

3 Methods

3.1 Cell Incubation with the Probe (Fig. 2)

1. Prepare the cells according to the required experimental conditions.
2. Add HE (from the 20 mM stock solution in DMSO) to obtain a final concentration in the medium of 10 μM (*see* **Note 8**).
3. Incubate the cells for 1 h at 37 °C.

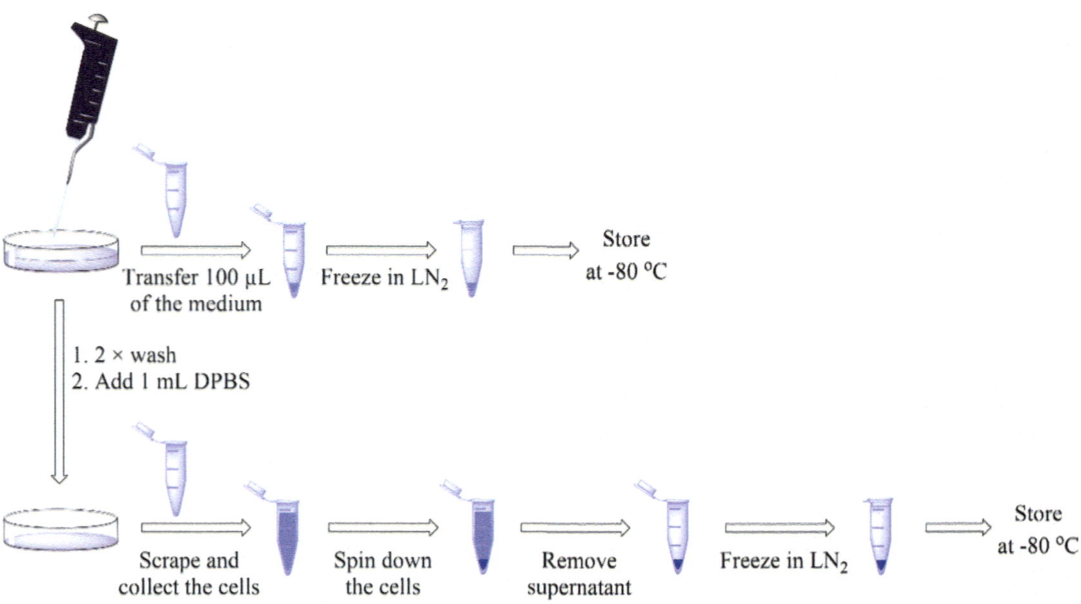

Fig. 2 Scheme showing collection of cells and media from cell experiment

4. Collect an aliquot of the medium (100 μL) in a 1.5 mL microcentrifuge tube and freeze immediately in liquid nitrogen.
5. Remove the rest of medium and wash the cells twice with ice-cold DPBS (2 × 10 mL in 15 mL conical tubes).
6. Add 1 mL of ice-cold DPBS and harvest the cells, transfer the cells into a 1.5 mL microcentrifuge tube, and spin down the cells by quick (30 s, 2000 × *g*) centrifugation. Discard the supernatant and freeze the cell pellet in liquid nitrogen.
7. Repeat **steps 4–6** for each sample (*see* **Note 8**).
8. Frozen cells and pellets can be stored at −80 °C for up to 1 week before analysis.

3.2 Extraction of the Products

3.2.1 Cell Pellets (Fig. 3)

1. Preload one set of 1.5 mL microcentrifuge tubes with 100 μL of MeCN containing 0.1% FA and place on ice.
2. Preload a second set of 1.5 mL microcentrifuge tubes with 100 μL of water containing 0.1% FA and place on ice.
3. Prepare a clear-bottom 96-well plate for protein assay and place on ice.
4. Place the tubes with frozen cell pellets on ice.
5. Add 150 μL of the lysis buffer containing DAPP (1 μM), and lyse the cells by 10 syringe strokes using a 0.5 mL insulin syringe with needle 28 G × 0.5 in (0.36 mm × 13 mm).
6. Immediately after lysing, transfer 100 μL of the cell lysate into a tube containing 100 μL MeCN with 0.1% FA, vortex for 10 s, and put back on ice. Transfer 3 × 2 μL aliquots of the cell lysate into three wells on a 96-well plate for the protein assay.

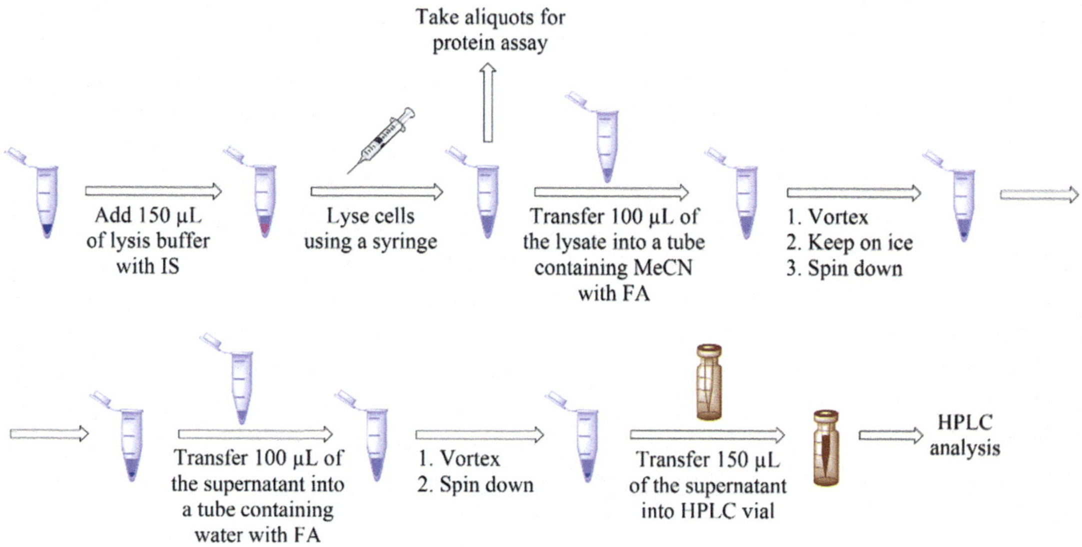

Fig. 3 Scheme showing preparation of the cell samples for HPLC analysis

7. Repeat **steps 5** and **6** for each sample (*see* **Note 9**).
8. Incubate the mixtures of cell lysates with MeCN for 30 min on ice.
9. During incubation, measure protein concentration in the lysates using the Bradford assay and a plate reader with absorption detection.
10. Vortex the tubes for 5 s and centrifuge for 30 min at 20,000 × *g* at 4 °C.
11. Place the tubes back on ice and transfer 100 μL of the supernatants into the second set of tubes containing 0.1% FA in water.
12. Vortex the tubes for 5 s and centrifuge for 15 min at 20,000 × *g* at 4 °C.
13. Transfer 150 μL of the supernatants into HPLC vials preloaded with conical inserts, seal the vials, and place on ice. After all solutions have been transferred, place the vials in an HPLC autosampler precooled to 4 °C.

3.2.2 Media (Fig. 4)

1. Preload one set of 1.5 mL microcentrifuge tubes with 100 μL of water containing 0.1% FA and place on ice.
2. Place the tubes with frozen media on ice.
3. Add 100 μL of the ice-cold MeCN, containing 1 μM DAPP (internal standard, IS), vortex for 10 s, and place back on ice.
4. Incubate the mixtures of media with MeCN for 30 min on ice.
5. Vortex the tubes for 5 s and centrifuge for 30 min at 20,000 × *g* at 4 °C.

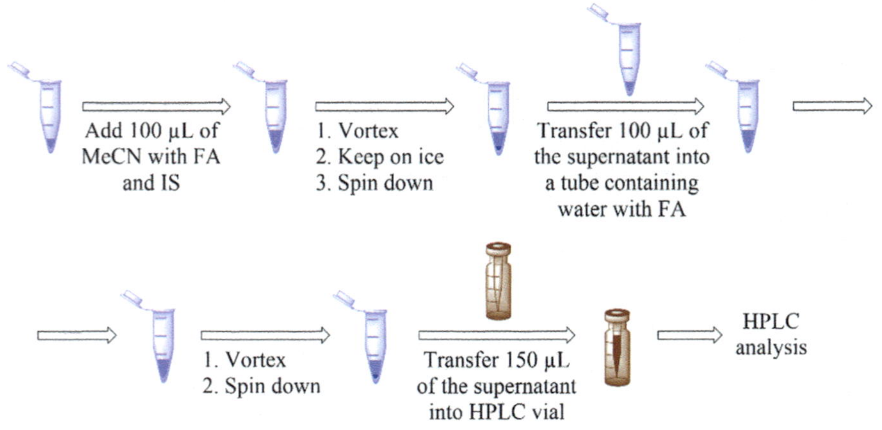

Fig. 4 Scheme showing preparation of the media samples for HPLC analysis

6. Place the tubes back on ice and transfer 100 μL of the supernatants into the second set of tubes containing 0.1% FA in water.
7. Vortex the tubes for 5 s and centrifuge for 15 min at 20,000 × *g* at 4 °C.
8. Transfer 150 μL of the supernatants into HPLC vials preloaded with conical inserts, seal the vials, and place on ice. After all solutions have been transferred, place the vials in an HPLC autosampler precooled to 4 °C.

3.3 HPLC Analyses of the Extracts

1. Install the column Kinetex C_{18} 100 mm × 4.6 mm, 2.6 μm (Phenomenex), in the HPLC system equipped with UV-Vis absorption and fluorescence detectors (HPLC-Abs/Fl). The column should be equipped with a UHPLC column filter or guard column to protect the column and extend its lifetime.
2. Equilibrate the column with the mobile phase (80% of mobile phase A and 20% of mobile phase B).
3. Set up the HPLC method and detection parameters according to Tables 1 and 2, respectively.
4. Test the system by three injections of a mix of standards (HE, 2-OH-E^+, E^+, E^+-E^+, and DAPP; 1 μM each) for the reproducibility of the retention times and peak intensities for all analytes and the internal standard, as shown in Fig. 5.
5. Set up the batch table including the calibration standards.
6. Run the analysis of the batch of samples.

Table 1
Gradient HPLC method parameters

Flow rate	1.5 mL/min		
Gradient	0 min	80% A	20% B
	4.5 min	44% A	56% B
	5.0 min	0% A	100% B
	6.5 min	0% A	100% B
	7.0 min	80% A	20% B
	9.0 min	80% A	20% B
ABS detector	290 nm	370 nm	
FL detector	0 min	Excitation at 358 nm Emission at 440 nm	
	2.3 min	Excitation at 490 nm Emission at 567 nm	

Table 2
UV-Vis absorption and fluorescence detection parameters

Analyte	Detector	Wavelength	Retention time (min)
DAPP	ABS	290 nm	2.5
HE	ABS FL	370 nm Exc. 358 nm Emi. 440 nm	1.6
2-OH-E^+	FL	Exc. 490 nm Emi. 567 nm	3.2
E^+	FL	Exc. 490 nm Emi. 567 nm	3.3
E^+-E^+	ABS	290 nm	4.3

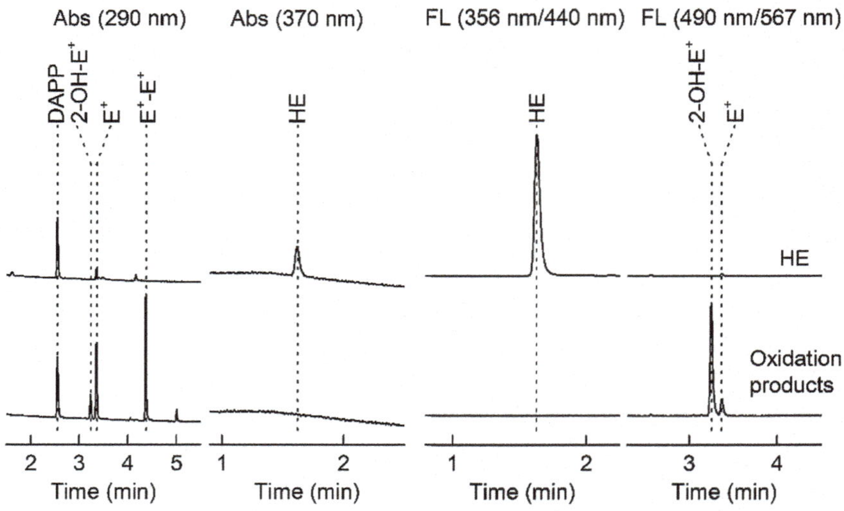

Fig. 5 HPLC-Abs/Fl chromatograms of standards of HE (2 μM, upper traces) and its oxidation products (2-OH-E^+, E^+, and E^+-E^+; 1 μM each, lower traces). Both samples contained DAPP (1 μM) as the internal standard. Samples were analyzed using the gradient elution method, as described in Subheading 3.3

7. Include the system and column wash with the water-to-methanol mixture (1:1) at the end of batch.
8. Quantify each analyte based on the detection parameters shown in Table 2 and calibration curves constructed in the concentration range relevant to levels detected in the samples analyzed, using an internal standard-based method.
9. When appropriate, normalize the concentrations of analytes to the protein levels in cell lysates, as determined by the Bradford method.
10. An increase in the peak intensity of 2-OH-E^+ but not E^+-E^+ indicates increased superoxide formation (*see* Fig. 6 as an example). A concomitant increase in 2-OH-E^+ and E^+-E^+ suggests an increase in peroxidatic activity.

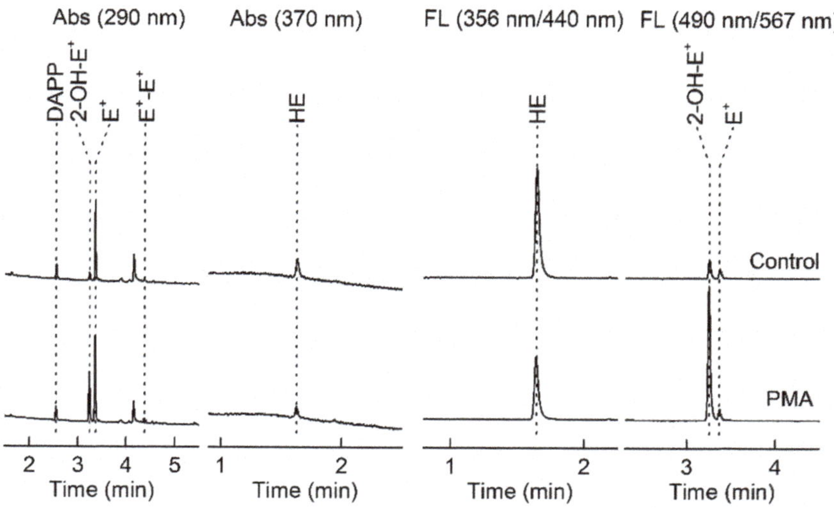

Fig. 6 HPLC-Abs/Fl chromatograms of the extracts from RAW 264.7 cells. The upper traces correspond to untreated (control) cells, and the lower traces correspond to cells stimulated with a phorbol ester (PMA) to activate superoxide production. Samples were analyzed using the gradient elution method, as described in Subheading 3.3

3.4 LC-MS Analyses of the Extracts

1. Install the column Titan C_{18} 100 mm × 2.1 mm, 1.9 μm (Supelco), in the LC-MS system equipped with a triple quadrupole (MS/MS) detector. The column should be equipped with an ultrahigh-performance liquid chromatography (UHPLC) column filter or guard column to protect the column and extend its lifetime.
2. Equilibrate the column with the mobile phase (80% of mobile phase A and 20% of mobile phase B).
3. Set up the HPLC-MS/MS method and detection parameters according to Tables 3 and 4, respectively.
4. Test the system by three injections of a mix of standards (HE, 2-OH-E^+, E^+, E^+-E^+, and DAPP; 1 μM each) for the reproducibility of the retention times and peak intensities for all analytes and the internal standard, as shown in Fig. 7.
5. Set up the batch table including the calibration standards.
6. Run the analysis of the batch of samples.
7. Include the system and column wash with the water-to-methanol mixture (1:1) at the end of batch.
8. Quantify each analyte based on the specific multiple reaction monitoring (MRM) transitions (Table 4) and calibration curves constructed in the concentration range relevant to the levels detected in the samples analyzed using an internal standard-based method.
9. When appropriate, normalize the concentrations of analytes to the protein levels in cell lysates, as determined by the Bradford method.

Table 3
LC-MS method parameters

Flow rate	0.5 mL/min		
Gradient	0 min	80% A	20% B
	7.0 min	50% A	50% B
	7.25 min	0% A	100% B
	7.75 min	0% A	100% B
	8.0 min	80% A	20% B
	9.0 min	80% A	20% B
Diverter valve	0 min	Waste	
	1.8 min	Detector	
	7.0 min	Waste	

Table 4
MS/MS detection parameters

Analyte	MRM transition	Retention time (min)
DAPP	286.0 > 208.0	2.2
HE	316.1 > 287.1	2.1
2-OH-E^+	330.1 > 300.0	2.7
E^+	314.1 > 284.0	2.8
E^+-E^+	313.1 > 298.9	3.5

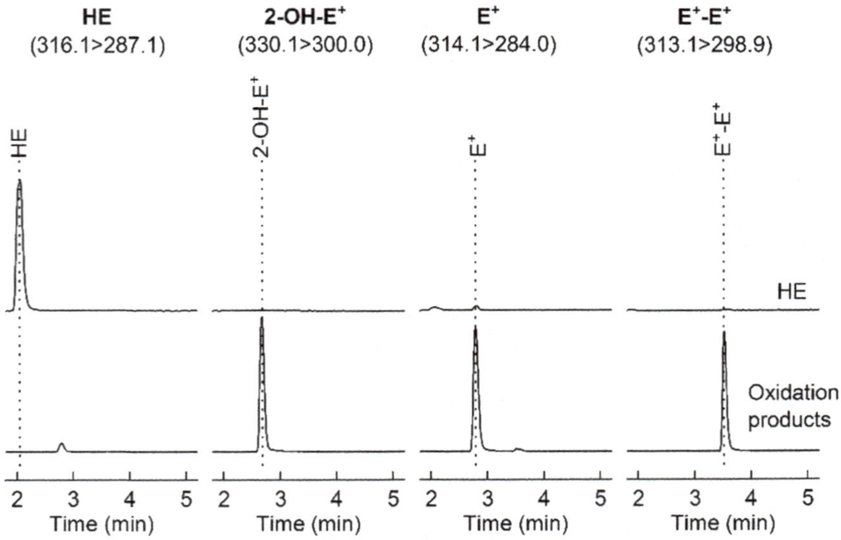

Fig. 7 LC-MS/MS chromatograms of standards of HE (2 μM, upper traces) and its oxidation products (2-OH-E^+, E^+, and E^+-E^+; 1 μM each, lower traces). Samples were analyzed using the gradient elution method, as described in Subheading 3.4

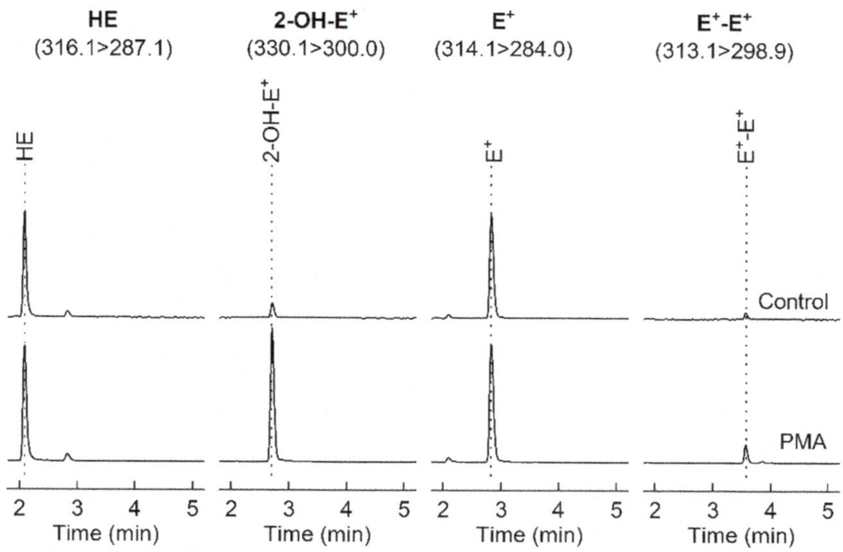

Fig. 8 LC-MS/MS chromatograms of the extracts from RAW 264.7 cells. The upper traces correspond to untreated (control) cells, and the lower traces correspond to cells stimulated with a phorbol ester (PMA) so as to induce superoxide production. Samples were analyzed using the gradient elution method, as described in Subheading 3.4

10. An increase in the peak intensity of 2-OH-E^+ but not E^+-E^+ indicates increased superoxide formation (Fig. 8). A concomitant increase in 2-OH-E^+ and E^+-E^+ suggests an increase in the peroxidatic activity.

3.5 Cell Incubation with the Probe Combined with Media Sampling for Real-Time Monitoring of Superoxide Released into Extracellular Medium (Fig. 9)

1. Prepare the cells according to the required experimental conditions (*see* **Note 10**).
2. Remove the medium and wash the cells with Hank's Balanced Salt Solution (HBSS) supplemented with HEPES buffer (25 mM) and DTPA (0.1 mM) (*see* **Note 4**).
3. Add the assay medium containing 10 μM HE and all necessary supplements for cell function (*see* **Notes 4** and **11**).
4. Incubate the cells at 37 °C.
5. Collect an aliquot of the medium (200 μL) into a 1.5 mL microcentrifuge tube containing 4 μL of SOD and CAT solution (50×). Add 2 μL of 100 μM DAPP (internal standard). Invert the tube three times for complete mixing.
6. Spin down any floating cells by quick (1 min, 2000 × *g*) centrifugation.
7. Transfer 150 μL of the supernatant into an HPLC vial preloaded with conical inserts, seal the vial, place it in the HPLC autosampler, and start the analysis.
8. Repeat **steps 5–7** at the intended time intervals.

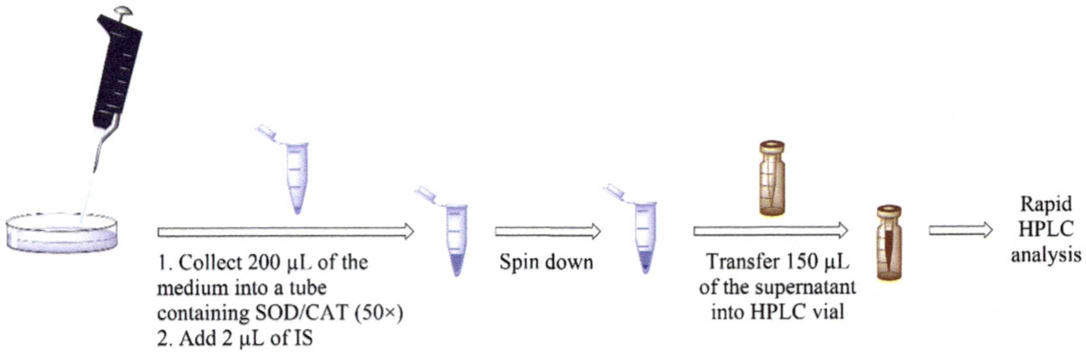

Fig. 9 Scheme showing sampling of the cell media samples for real-time monitoring of $O_2^{\bullet-}$ formation

Table 5
Rapid HPLC method parameters

Flow rate	2.0 mL/min		
Mobile phase		65% A	35% B
ABS detector	290 nm	370 nm	
FL detector		Excitation at 370 nm Emission at 565 nm	

3.6 Rapid HPLC Analyses

1. Install the column Ascentis Express Phenyl-Hexyl 50 mm × 4.6 mm, 2.7 μm (Supelco), in the HPLC system equipped with UV-Vis absorption and fluorescence detectors. The column should be equipped with a UHPLC column filter or guard column to protect the column and extend its lifetime (*see* **Note 12**).
2. Equilibrate the column with the mobile phase (65% of mobile phase A and 35% of mobile phase B) (*see* **Note 13**).
3. Set up the HPLC method and detection parameters according to Tables 5 and 6, respectively (*see* **Note 14**).
4. Test the system by three injections of a mix of standards (HE, 2-OH-E^+, E^+, E^+-E^+, and DAPP; 1 μM each) for the reproducibility of the retention times and peak intensities for all analytes and the internal standard, as shown in Fig. 10.
5. Run analyses of standards for calibration purposes.
6. Run the analysis of the batch of samples.
7. Include the system and column wash with the water-to-methanol mixture (1:1) at the end of batch.
8. Quantify each analyte based on the detection parameters shown in Table 6 and calibration curves constructed in the concentration range relevant to the levels detected in the samples analyzed, using an internal standard-based method.

Table 6
UV-Vis absorption and fluorescence detection parameters for the rapid HPLC method

Analyte	Detector	Wavelength	Retention time (s)
DAPP	ABS	290 nm	26
HE	FL	Exc. 370 nm Emi. 565 nm	19
2-OH-E^+	FL	Exc. 370 nm Emi. 565 nm	36
E^+	ABS	290 nm	41
E^+- E^+	ABS	290 nm	86

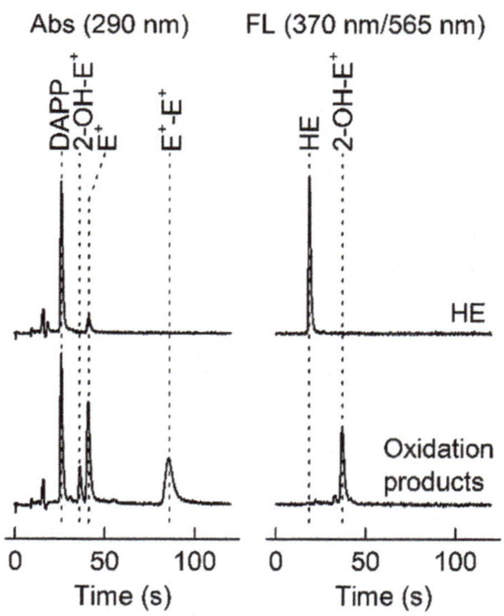

Fig. 10 HPLC-Abs/Fl chromatograms of standards of HE (1 μM) and its oxidation products (2-OH-E^+, E^+, and E^+-E^+; 1 μM each, lower traces). Both samples contained DAPP (1 μM) as the internal standard. Samples were analyzed using the rapid HPLC method with isocratic elution, as described in Subheading 3.6

9. When appropriate, normalize the concentrations of analytes to the number of cells.
10. An increase in the peak intensity of 2-OH-E^+ indicates increased superoxide formation and release into medium (*See* Fig. 11 for an example).

4 Notes

1. HE has been shown to be prone to photooxidation [18]. Also, 2-OH-E^+ may act as a photosensitizer, catalyzing the photooxidation process. Thus, it is important to minimize the light exposure of the samples at all stages of the experiment.

2. If the stock solution of HE is prepared and aliquoted under anaerobic conditions, the aliquots can be stored at −80 °C for at least 6 months without probe degradation.
3. The cell culture medium depends on the cell type and should be based on the requirements for normal cell growth, unless intended otherwise.
4. For most studies, the regular cell culture medium can be used during incubation of the cells with the HE probe. For real-time sampling of media combined with the rapid HPLC method (isocratic elution), DPBS or HBSS with necessary supplements (glucose, pyruvate, etc.) and buffer + metal chelator (e.g., 25 mM HEPES + 0.1 mM DTPA) are recommended. This is to avoid clogging the column with hydrophobic components of the regular cell growth medium while providing the bioenergetic substrates and preventing metal-catalyzed oxidation of the probe.
5. The 15 mL centrifuge tubes should be prefilled with DPBS and placed on ice before the experiment. That way, cells can be washed twice with ice-cold DPBS quickly (<30 s). DPBS should be poured on the dish gently to avoid cell detachment.
6. From our experience, different preparations of commercial catalase exhibit peroxidatic activity to different extents. To minimize the peroxidatic activity, we recommend using catalase from *Corynebacterium glutamicum* (Sigma, cat. no. 02071).
7. HPLC-grade solvents should be used for HPLC analyses. For LC-MS analyses, UHPLC/MS-grade solvents should be used. After preparation, the mobile phase should be passed through a 0.2 μm filter. The mobile phase should be used within 1 week after preparation and stored at 4 °C in between analyses.
8. The probe in DMSO should be added slowly, with a gentle shaking of the dish, to avoid exposing the cells to a local high concentration of DMSO. Alternatively, a large volume of the medium containing the probe can be prepared immediately before starting the experiment. For all samples, incubation with the probe should be started in time intervals of 3–5 min, depending on the predetermined time needed to harvest cells per a single sample. That way, cells and media are placed in liquid nitrogen as quickly as possible after stopping the incubation, and the time between stopping the incubation and freezing the pellet should be consistent for all samples. To decrease the time, two persons may be needed for sample collection.
9. **Steps 5–7** should be done for one sample at a time; i.e., after finishing **step 7** for one sample, start lysing another sample (**step 5**). To decrease the time, two persons may be needed for cell lysis and extraction.

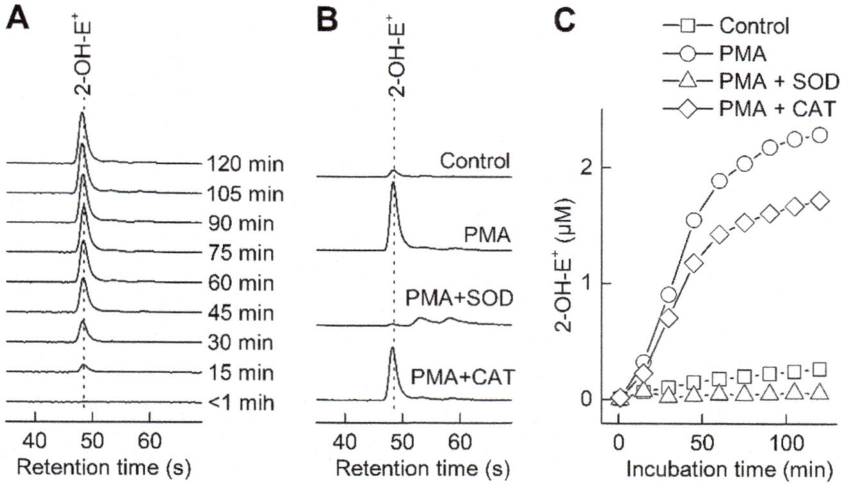

Fig. 11 Real-time monitoring of superoxide production by differentiated HL60 (*d*HL60) cells stimulated with PMA. (**A**) HPLC peak of 2-OH-E$^+$ as a function of incubation time of *d*HL60 cells with the HE probe. (**B**) Effect of PMA, SOD, and CAT on the HPLC peak of 2-OH-E$^+$ detected at 60 min of incubation time of *d*HL60 cells with the HE probe. (**C**) Effect of PMA, SOD, and CAT on the dynamics of 2-OH-E$^+$ formation during incubation of *d*HL60 cells with the HE probe. Samples were analyzed using the isocratic elution method, as described in Subheading 3.6, but using the column and mobile phase composition as detailed in **Note 12**

10. The protocol described is for adherent cells, but it also can be used for cells in suspension (e.g., in differentiated HL60 cells [13, 14], *see* Fig. 11).
11. The assay medium volume should be large enough to not significantly change upon sampling. The combined sampling volume per sample over whole experiment should be less than 10% of the initial volume.
12. Alternatively, the Kinetex Phenyl-Hexyl column (Phenomenex, 50 mm × 4.6 mm, 2.6 μm) can be used with isocratic elution at the mobile phase composition of 70% mobile phase A and 30% mobile phase B [14]. All other conditions are the same as described in Subheading 3.6. An example of data obtained with this column is shown in Fig. 11.
13. The same mobile phase (with trifluoroacetic acid) should be used as is used for a gradient method (described in Subheading 3.3).
14. The same HPLC method and detection parameters may also be used for simultaneous monitoring of superoxide and H_2O_2 production if hydroethidine and coumarin boronic acid probes are combined in the assay medium and both 2-OH-E$^+$ and 7-hydroxycoumarin products are monitored [13, 14].

Acknowledgment

This work was supported by NIH grants NCI U01 CA178960 and R01 AA022986 to B.K.

References

1. Lambeth JD (2004) NOX enzymes and the biology of reactive oxygen. Nat Rev Immunol 4:181–189
2. Winterbourn CC (2008) Reconciling the chemistry and biology of reactive oxygen species. Nat Chem Biol 4:278–286
3. Bedard K, Krause KH (2007) The NOX family of ROS-generating NADPH oxidases: physiology and pathophysiology. Physiol Rev 87:245–313
4. Maghzal GJ, Krause KH, Stocker R, Jaquet V (2012) Detection of reactive oxygen species derived from the family of NOX NADPH oxidases. Free Radic Biol Med 53:1903–1918
5. Wardman P (2007) Fluorescent and luminescent probes for measurement of oxidative and nitrosative species in cells and tissues: progress, pitfalls, and prospects. Free Radic Biol Med 43:995–1022
6. Zielonka J, Kalyanaraman B (2018) Small-molecule luminescent probes for the detection of cellular oxidizing and nitrating species. Free Radic Biol Med 128:3–22. https://doi.org/10.1016/j.freeradbiomed.2018.03.032
7. Zielonka J, Lambeth JD, Kalyanaraman B (2013) On the use of L-012, a luminol-based chemiluminescent probe, for detecting superoxide and identifying inhibitors of NADPH oxidase: a reevaluation. Free Radic Biol Med 65:1310–1314
8. Zielonka J, Kalyanaraman B (2010) Hydroethidine- and MitoSOX-derived red fluorescence is not a reliable indicator of intracellular superoxide formation: another inconvenient truth. Free Radic Biol Med 48:983–1001
9. Fernandes DC, Goncalves RC, Laurindo FR (2017) Measurement of superoxide production and NADPH oxidase activity by HPLC analysis of dihydroethidium oxidation. Methods Mol Biol (Clifton, NJ) 1527:233–249
10. Laurindo FR, Fernandes DC, Santos CX (2008) Assessment of superoxide production and NADPH oxidase activity by HPLC analysis of dihydroethidium oxidation products. Methods Enzymol 441:237–260
11. Fernandes DC, Wosniak J Jr, Pescatore LA, Bertoline MA, Liberman M, Laurindo FR, Santos CX (2007) Analysis of DHE-derived oxidation products by HPLC in the assessment of superoxide production and NADPH oxidase activity in vascular systems. Am J Physiol Cell Physiol 292:C413–C422
12. Zielonka J et al (2017) Recent developments in the probes and assays for measurement of the activity of NADPH oxidases. Cell Biochem Biophys 75:335–349
13. Zielonka J et al (2016) Mitigation of NADPH oxidase 2 activity as a strategy to inhibit peroxynitrite formation. J Biol Chem 291:7029–7044
14. Zielonka J et al (2014) High-throughput assays for superoxide and hydrogen peroxide design of a screening workflow to identify inhibitors of NADPH oxidases. J Biol Chem 289:16176–16189
15. Zielonka J et al (2012) Global profiling of reactive oxygen and nitrogen species in biological systems: high-throughput real-time analyses. J Biol Chem 287:2984–2995
16. Zielonka J, Hardy M, Kalyanaraman B (2009) HPLC study of oxidation products of hydroethidine in chemical and biological systems: ramifications in superoxide measurements. Free Radic Biol Med 46:329–338
17. Zielonka J, Vasquez-Vivar J, Kalyanaraman B (2008) Detection of 2-hydroxyethidium in cellular systems: a unique marker product of superoxide and hydroethidine. Nat Protoc 3:8–21
18. Zielonka J, Vasquez-Vivar J, Kalyanaraman B (2006) The confounding effects of light, sonication, and Mn(III)TBAP on quantitation of superoxide using hydroethidine. Free Radic Biol Med 41:1050–1057

Chapter 15

Visualization of Intracellular Hydrogen Peroxide with the Genetically Encoded Fluorescent Probe HyPer in NIH-3T3 Cells

Yulia G. Ermakova, Nataliya M. Mishina, Carsten Schultz, and Vsevolod V. Belousov

Abstract

Reactive oxygen species (ROS) are involved in regulating normal physiological cell functions as second messengers as well as nonspecific damage of biomolecules in a pathological process known as oxidative stress. The HyPer family of genetically encoded probes are a useful noninvasive tool for monitoring the real-time dynamics of ROS in individual cells or model organisms. HyPer, the first genetically encoded probe for detection of hydrogen peroxide (H_2O_2), is oxidized with high specificity and sensitivity by H_2O_2, leading to ratiometric changes in the fluorescence excitation spectrum of the probe. These changes can be detected with a wide range of commercial confocal and wide-field microscope systems. Here we describe a detailed protocol for ratiometric monitoring of H_2O_2 produced by D-amino acid oxidase (DAAO) or by NADPH oxidase (NOX) in NIH-3T3 cells using the HyPer probe.

Key words ROS, Hydrogen peroxide, HyPer, HyPerRed, PDGF, D-amino acid oxidase, DAAO, Ratiometric imaging, Microscopy

1 Introduction

Reactive oxygen species (ROS) such as superoxide anion radical, nitric oxide, hydroxyl radical, and hydrogen peroxide (H_2O_2) are produced in cells both spontaneously as nonspecific products during oxygen respiration and directly by specialized enzymes [1]. H_2O_2 is the most stable form of ROS and often plays the role of a second messenger in intracellular and extracellular transmission of signals [2, 3]. H_2O_2 at low (nanomolar) concentrations is able to play a regulatory function in cells, regulating the activity of enzymes in signaling and metabolic cascades by direct oxidation of specific ionized cysteine residues in proteins [4–6] or by indirect oxidation via redox relays [7]. This regulation is reversible: oxidized thiols are reduced by cellular thiol-exchange systems, and H_2O_2 is

Ulla G. Knaus and Thomas L. Leto (eds.), *NADPH Oxidases: Methods and Protocols*, Methods in Molecular Biology, vol. 1982, https://doi.org/10.1007/978-1-4939-9424-3_15,

removed by enzymes including catalase, glutathione peroxidases, or peroxiredoxins [8, 9]. Simultaneous operation of these enzyme systems determines the cellular redox status and links ROS and pools of oxidized and reduced glutathione and NAD(P)H/NAD(P)$^+$ [10, 11]. However, for many years after the discovery of the role of H_2O_2 in living systems, no specific methods existed to detect this molecule in living cells in real time.

Often information about ROS participation in biochemical processes is based on indirect evidence, for example, the level of oxidized DNA, lipids, and proteins. In case of imaging, the significance of the data obtained using synthetic small-molecule optical fluorescent probes for ROS detection is questionable because of the high risk of artifacts. These dyes can be oxidized by different types of ROS and reactive nitrogen species. Moreover, certain dyes are capable to produce ROS upon exposure to light, which leads to aberrant signal amplification [12]. Intracellular targeting of synthetic dyes requires a complex approach of linking them to specific proteins, for example, using a SNAP-tag, which makes them difficult to use in cellular compartments [13]. We have developed an alternative approach to H_2O_2 detection based on genetically encoded fluorescent indicators.

The HyPer biosensor is a chimeric protein, in which circularly permuted fluorescent protein (cpFP) is integrated into a regulatory domain of the bacterial transcription factor OxyR (Fig. 1). HyPer will be oxidized by the reaction with H_2O_2. This leads to the formation of a disulfide bond between cysteine residues 199 and 208 of the OxyR part of the sensor, resulting in conformational changes. These changes are transmitted through linker peptides SAG and GT to cpFP. Because of oxidation of the sensor, the fluorescent spectrum of HyPer changes [14].

Currently available HyPer versions have spectral properties of green FP (HyPer [14], HyPer-2 [15], HyPer-3 [16]) or red FP (HyPerRed [17]). Green HyPer probes are ratiometric and have two excitation peaks with maxima at 420 and 500 nm corresponding to neutral (protonated) and charged (deprotonated) forms of the cpYFP chromophore, respectively. After oxidation of HyPer, the 500 nm peak increases and the 420 nm peak decreases. The emission maximum for both forms of the chromophore is the same, i.e., 516 nm. Therefore, the imaging channel of the protonated form can be described as CFP excitation/GFP emission and of the charged form as GFP excitation/GFP emission. Using a ratio of two peaks is a more straightforward approach compared to single-channel fluorescence detection: the ratio does not depend on the expression level of the probe, which could differ between individual cells. This allows avoiding artifacts of diffusion or redistribution of the probe in the sample. The observed ratio value can be assigned to a particular concentration of H_2O_2 in an experimental system that allows using the ratiometric sensor as a

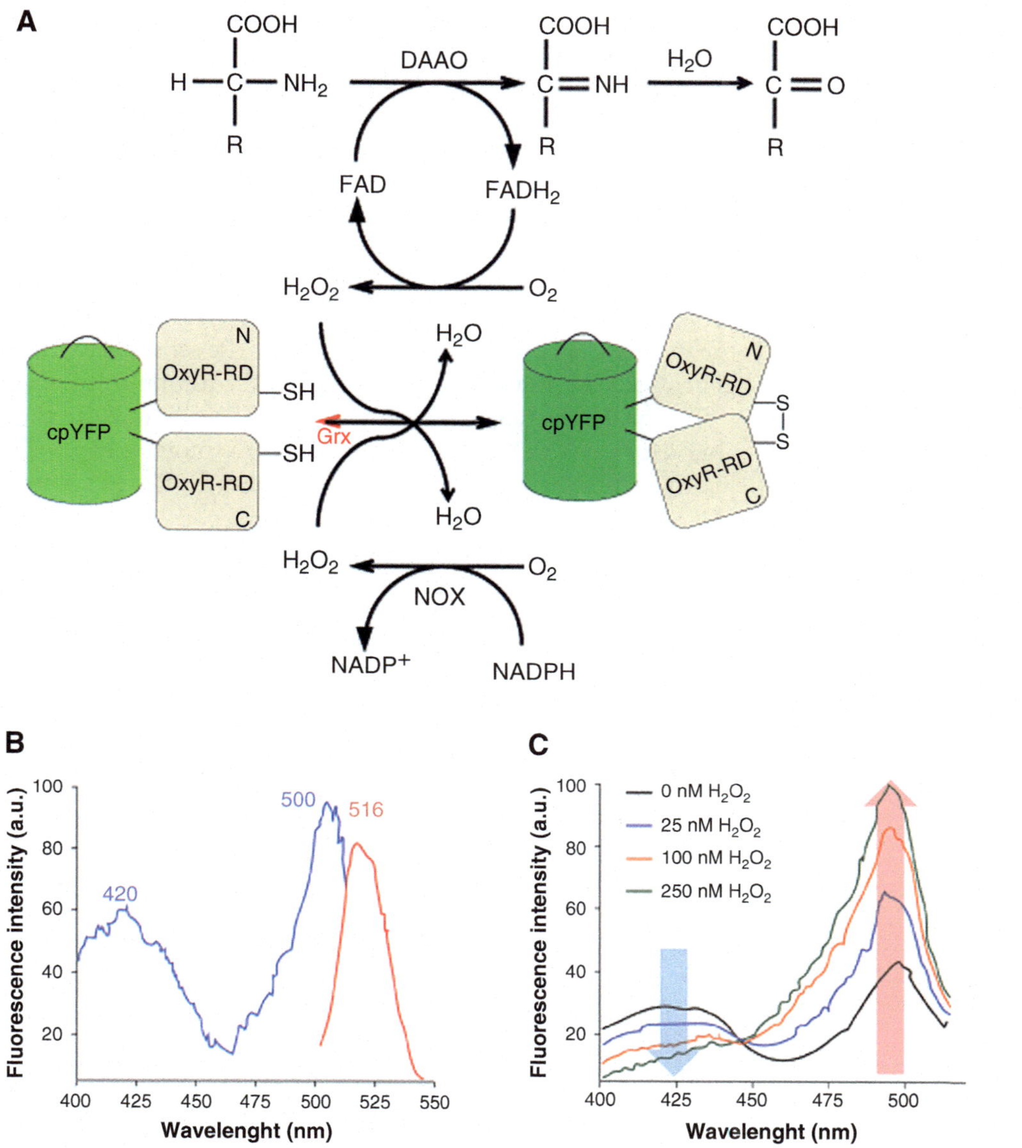

Fig. 1 Generation and detection of H_2O_2. (**A**) D-amino acid oxidase or NOX family enzymes produce H_2O_2 visualized using HyPer family probes. (**B**) Fluorescence excitation (blue) and emission (red) spectra of HyPer. (**C**) Fluorescence excitation spectrum of HyPer in the presence of various amounts of H_2O_2

semiquantitative tool for the detection of the oxidant. HyPerRed is based on the fluorescent protein cpmApple taken from the genetically encoded fluorescent sensor for calcium R-GECO1 [18]. HyPerRed is a single-channel probe having spectral properties of red fluorescent proteins and therefore can be used together with the green sensor Hyper (HyPer-2, HyPer-3) for the simultaneous detection of H_2O_2 in different compartments of a single cell or with other GFP-based genetically encoded probes such as roGFP [19], Perceval [20], SypHer [21], GCaMP [22], and many others. The performance of HyPerRed including its selectivity and sensitivity is similar to the ratiometric green variants.

The advantages of HyPer probes are high sensitivity and selectivity for H_2O_2, low cytotoxicity, and the possibility of targeted expression in different cellular compartments and organelles. The dynamics and sensitivity of HyPer allow detecting nanomolar concentrations of H_2O_2 in an experimental system. HyPer probes can also be used in nontargeted or organelle-targeted versions. The best spatial resolution will be achieved by fusing the sensor with proteins located just near the source of the oxidant or with at least the same localization sequence: the closer the sensor is to the H_2O_2 generator, the higher ratio change R/R_0 (or brightness change F/F_0 for HyPerRed) will be detected. This is due to the high abundance of thiol peroxidases efficiently removing H_2O_2. Bringing the probe closer to the generator allows HyPer to compete for H_2O_2 with the peroxidases.

Here we provide a protocol for detection of H_2O_2 using HyPer probes at two conditions of intracellular H_2O_2 production with a similar experimental design (Fig. 1). Example 1 shows DAAO-induced H_2O_2 production. DAAO is an enzyme that oxidizes the amino group of several D-amino acids, including D-Alanine, producing the corresponding α-keto acid and H_2O_2. The DAAO gene from red yeast *Rhodotorula gracilis* has been previously cloned, and the enzyme was characterized in vitro [23, 24]. Here DAAO has been used as a genetically encoded H_2O_2 generator coupled with H_2O_2 detection by HyPer following D-Ala addition [25]. Other D-amino acids can also be used to activate DAAO [26, 27]. Example 2 presents stimulation of NIH-3T3 cells with platelet-derived growth factor (PDGF). PDGF receptor activation has been shown to stimulate H_2O_2 generation via phosphoinositide 3-kinase activation, translocation, and activation of the small GTP-binding protein Rac1 and subsequent activation of NOX [28].

This protocol can be used with different cell lines, sections, and tissues without significant changes except of sample preparation. DAAO or NOX are used as examples for a ROS source, and the protocol can be adapted for other ROS-generating enzymes.

2 Materials

As there are several similar HyPer probes, the first question is as follows: how to choose the best HyPer probe for your experiment? Table 1 will help with this decision. Note that the HyPer signal is pH-dependent. The excitation ratio changes more than tenfold between pH 6.0 and 9.0. Therefore, in parallel one should perform a control experiment using SypHer, or it sequels SypHer-2 [29] or SypHer3s [30] to monitor pH changes in the cell system of choice. SypHer and SypHer-2 are C199S mutants of HyPer and HyPer-2, respectively, and due to this mutation, these probes are insensible to H_2O_2. pH dependency of SypHer, SypHer-2, and HyPerRed-C199S, the red analog of SypHer, is similar, while SypHer3s has more than double of the dynamic range compared to other SypHer versions. Addgene catalog numbers for available HyPer and SypHer probes are listed in Table 1. The provided protocol for microscopy or sample preparation is the same for both HyPer and SypHer, differing only in the plasmid DNA encoding the probe.

2.1 Reagents and Solutions

1. Dulbecco's Modified Eagle Medium (DMEM) supplemented with 10% fetal calf serum (FCS).
2. Opti-MEM without phenol red supplemented with 2.2 g/L sodium bicarbonate.
3. Minimum Essential Medium Eagle (MEM) or PBS with Ca^{2+} and Mg^{2+}.
4. HEPES, pH 7.4, adjusted with NaOH, 1 M stock in MQ water.
5. H_2O_2 (180 mM solution in MQ water, 1000 × stock). Can be prepared by 50 × dilution of 30% H_2O_2. Prepare fresh before the experiment; store in dark.
6. X-tremeGene 9 transfection reagent or any other lipid-based transfection reagent.
7. D-alanine (**Example 1**).
8. HyPer-DAAO fusion plasmid (used in experiment **Example 1**), HyPer plasmid DNA (used in **Example 2**), SypHer-2-plasmid DNA (both experiments, control). HyPer-DAAO plasmid is available upon request. HyPer plasmid can be purchased from Evrogen (alternatively, HyPer-2 or HyPer-3 can be used, available from Addgene). SypHer-2 plasmid is available from Addgene.
9. PDGF, 10 μg/mL in Opti-MEM, which is 1000 × stock solution. Should be stored in aliquots at −70 °C. On the day of the experiment, the aliquot should be kept on blue ice (**Example 2**).
10. Mammalian cells (NIH-3T3 fibroblasts, ATCC #CRL-1658).
11. Cell starvation and imaging medium: MEM supplemented with either 2.2 g/L sodium bicarbonate and 20 mM HEPES (if CO_2 supply for thermostating imaging chamber is available) or just 20 mM HEPES (use bicarbonate-free media in case that no CO_2 supply for thermostating box is available in order to prevent alkalinization of the imaging media), 100 U/mL penicillin and 100 μg/mL streptomycin, 2 mM L-glutamine, and 0.5% BSA (*see* **Note 1**).

Table 1
A comparison of HyPer and SypHer probe properties

Name	Excitation maximum (nm)	Emission maximum (nm)	Signal type	pKa in cells (mostly reduced HyPer)	Brightness compared t o HyPer (%)	Sensitivity to H_2O_2 in vitro	Fold change of probe signal	Oxidation rate constant ($M^{-1} \cdot s^{-1}$)	Reduction rate in cells compared to HyPer (%)	Oligomeric state	Addgene number
HyPer	420 500	516	Excitation ratio, fluorescence intensity, FLIM	8.6 ± 0.01	100	20–30 nM	3 ± 0.4[a]	$5 \cdot 10^5$	100	Monomer	Evrogen, (cat. no. FP941)
HyPer-2	420 500	516			250		6 ± 0.3[a]	$1.2 \cdot 10^5$	50 ± 6	Dimer	42211
HyPer-3	420 500	516			100			$2.5 \cdot 10^5$	100 ± 2	Monomer	42131
HyPerRed	575	605	Fluorescence intensity	8.7 ± 0.04	260		2 ± 0.2[b]	$3.0 \cdot 10^5$	70 ± 6	N/A	48249
SypHer	420 500	516	Excitation ratio, fluorescence intensity	8.6 ± 0.01	90	Insensitive	2.2 ± 0.2[a,c]			Monomer	48250
SypHer-2	420 500	516			210		7.6 ± 0.6[a,c]			Dimer	–
SypHer3s	410 505	525		7.8 ± 0.02	260		19 ± 0.8[a,c]			Monomer	108118
HyPerRed-C199S	575	605	Fluorescence intensity	8.7 ± 0.04	240		2.2 ± 0.2[b]			N/A	48252

[a]Fold change in ratio
[b]Fold change in fluorescence
[c]Between pH 6.0 and 8.0 in cytoplasm of HeLa Kyoto cells under nigericin-monensin pH titration

2.2 Equipment

1. Glass-bottom Petri dishes 35 cm^2.
2. Standard equipment for eukaryotic cell culturing—sterile laminar flow cabinet and cultural box to maintain 5% CO_2 and temperature of 37 °C.
3. Confocal microscope or wide-field fluorescent microscope with thermostating chamber.

Lasers providing violet (405 nm) and blue (488 or 496 nm) laser lines are needed for ratiometric imaging of green HyPer versions. A 559 nm laser is required when using HyPerRed. Alternatively, green probes can be imaged in single-wavelength mode with only blue excitation, but it will lead to a decreased signal compared to the ratiometric variant. Alternatively, a wide-field fluorescent microscope adapted for sequential imaging in two channels can be used: (1) excitation in 400–430 nm range, emission in 510–550 nm range, and (2) excitation in 480–500 nm range, emission in 510–550 nm range. For HyPerRed imaging the microscope should be equipped with a green excitation (567 or 547 nm) laser line or a RFP (or TX2) filter cube on the wide-field system.

3 Methods

Below, Example 1 describes the stimulation of H_2O_2 production by DAAO fused with HyPer probe in NIH-3T3 cells. Example 2 describes H_2O_2 detection by HyPer upon PDGF stimulation in NIH-3T3 cells.

3.1 Transfection

1. Plate NIH-3T3 cells on 35 mm glass-bottom dishes in DMEM supplemented with 10% FCS and 20 mM of L-glutamine, 50,000 cells/dish. Allow cells to grow to 60–80% confluence.
2. Change medium for Opti-MEM without phenol red supplemented with 2.2 g/L sodium bicarbonate without serum, 1 mL per dish.
3. Add 100 μL serum-free Opti-MEM to a sterile Eppendorf tube (1.5 mL). Quickly add 3 μL of X-tremeGene 9 transfection reagent (*see* **Note 2**). Vortex rapidly.
4. Example 1: Add to the mix 1 μg of DAAO-HyPer plasmid DNA (*see* **Note 3**). Mix rapidly by vortexing; allow to rest for at least 10 min at room temperature for liposome-DNA complex formation.
5. Example 2: Add to the mix 1 μg of HyPer plasmid DNA (*see* **Note 3**). Mix rapidly by vortexing; allow to rest for at least 10 min at room temperature for liposome-DNA complex formation.

6. Add transfection complex to the cells on dish in a dropwise manner. Shake or swirl gently.
7. Incubate at least for 4–5 h in the cultural box with 5% CO_2 (*see* **Note 4**).
8. Replace cell medium with 2 mL DMEM supplemented with 10% FCS and 20 mM of L-glutamine. This step will increase the reduced/oxidized balance of HyPer at the start of experiment. Twenty-four hours after transfection cells will be ready for imaging (*see* **Note 5**).
9. Additional incubation before imaging: In Example 1 no additional conditions are needed. For Example 2 starve cells in 2 mL cell starvation and imaging medium for at least 2 h.
10. Wash cells gently with 1 mL cell starvation and imaging medium, and replace cell medium with 2 mL fresh cell starvation and imaging medium.

3.2 Microscopy

3.2.1 Primary Visualization of Cells

Sixteen to 24 h after transfection, check the effectiveness of transfection by performing excitation with a mercury lamp. Use the GFP filter set in "visual" mode of the microscope. In fact, it is not necessary to get 100% of the cells transfected, but higher abundance of fluorescent cells in a field of view permits obtaining more information about the synchronicity of the cells' response during the imaging series. Use the GFP or FITC filter set in "visual" or "live" mode of the microscope. For green HyPer versions, irradiation with blue light leads to photoconversion of the deprotonated form (500 nm) of the chromophore to the dark state which will look like rapid photobleaching of HyPer. Use the excitation of the protonated form (420 nm) to avoid this effect (*see* Table 2).

3.2.2 Microscopy Settings

Wide-Field Microscope Leica 6000 for Green HyPer

1. Turn on the lamp; it should be turned on at least 10 min before starting the microscopy.
2. Activate the "FLUO" mode.
3. Set binning (none, 4 × 4). Note that the binning function reduces the number of pixels on the CCD chip and will decrease resolution and increase a signal-to-noise ratio. However, this procedure will dramatically reduce phototoxicity.
4. Keep the acquisition time as low as possible. Then set the channels' filters. Activate filter cube "FRET" equipped with a dichroic mirror (440/520 nm) and external excitation and emission filter wheels (*see* **Note 6**).
5. For FRET cube, set up channel 1 to excitation filter CFP (BP 427/10) to excite the 420 nm peak, and for channel 2 set up excitation filter YFP (BP 504/12) to excite the 500 nm peak. Emission filter for both channels is YFP (BP 542/27).

Table 2
Troubleshooting suggestions

Subheadings	Problem	Possible reason	Solution
Subheading 3.1	Low level of transfection or co-transfection	Non-optimal transfection protocol	Optimize transfection conditions: titrate DNA/transfection reagent according to manufacturer's protocol
		Low quality of plasmid DNA	Transform bacteria and prepare fresh DNA samples
Subheading 3.2	Low fluorescence brightness	Low expression level of HyPer	Incubate cells for another 12–24 h to increase protein expression
		Non-optimal imaging settings	Check spectral setting of light sources, detectors, or filter cubes
Subheading 3.2	Rapid decrease in the fluorescence intensity excited in 480–520 nm range	Photoconversion of the chromophore into the non-fluorescent state	Wait for several minutes, or illuminate the sample with a brief pulse of 400–430 nm light
Subheading 3.2	No H_2O_2 generation after D-amino acid addition	Imaging medium pH is not 7.2–7.4	Change imaging medium. Check that CO_2 valve in the microscope's environmental chamber is "off" if you have cell imaging solution buffered by a **non**-bicarbonate-based system (HEPES, PBS, etc.) or "on" if you have cell imaging solution buffered by sodium bicarbonate
		Low concentration of D-alanine	Increase D-alanine concentrations (up to 15 mM) or prepare a fresh stock of D-alanine
Subheading 3.2	No H_2O_2 production upon growth factor stimulation	No or too short time of serum starvation	Increase the serum starvation time (up to 6 h)
		Inappropriate cell culture medium	Change cell medium, the FCS lot, or manufacturer. Make sure that complete medium was not stored for more than 2 months
		Too high or too low cell passage number	Thaw fresh cells or wait for several passages, respectively

6. Set light intensity (2–5) for the channels and exposure times (30–500 ms). Fluorescence should be clearly visible and have similar intensity in both channels.
7. Set up time delay between frames. This depends on the experimental system and can vary from 300 ms to15 min.

Confocal Microscope for Green HyPer Time-Lapse Imaging

1. Activate 405 nm and 488 (496) nm lasers; let them warm up for 15 min.
2. Set the XYT mode for acquisition of time series for two-dimensional images.
3. Set the scanning resolution to 512 × 512. This is the resolution useful for many models; can be changed further to higher or lower (*see* **Note 7**).
4. Set pinhole size to 2 airy. Set PMT voltage to 630–850 V. Lower PMT voltage values make images less noisy; higher values increase the noise but allow using lower laser intensity, thereby decreasing photodynamic damage of the cells.
5. Set up the beam path for 500 nm excitation (channel 2): excitation by 488/496 nm laser with 500–530 nm emission.
6. Set the laser power: intensity of the 405 nm laser to 4–10% and the intensity of the 488 nm 10 mW Ar laser to 1–15%. Laser intensity may vary depending on the optics, lasers, and system type. Fluorescence should be clearly visible and have rather equal intensity in both channels.
7. Set up the beam path for 420 nm excitation: excitation by 405 nm laser with 500–530 nm emission. Every line should be at first exited by 405 nm laser and only then with 488 nm laser to avoid dark state conversion of HyPer.
8. Set the "Speed Beam frequency" 400–600 Hz.
9. Set up time delay between frames. This depends on the experimental system and can vary from 300 ms to 15 min.

3.2.3 Imaging of H_2O_2 Produced by DAAO or NOX in NIH-3T3 Cells

1. Warm up the microscope thermal box to 37 °C if temperature control will be maintained during the experiment. It takes 1–2 h for the temperature to equilibrate; therefore, it is convenient to turn the heating chamber on overnight.
2. Place the Petri dish with the cells onto the microscope stage.
3. Set up focus in "visual" or "live" mode.
4. Take a test frame.
5. Select a region of interest (ROI) in the frame with an individual transfected cell or several ones, and measure the signal intensity in the ROI.
6. Set the fluorescence intensity in each channel to adjust the ROI intensity in the range between 30 and 80 a.u. (for 8-bit system)

by changing the light intensity or exposure time. Make sure that the focus is stable.

7. Start time series recording. The ratio signal should stay stable over at least 5 min before addition of any agents.
8. Test the effect of the addition of MEM without any additional drugs to the HyPer ratio by adding 100 μL of MEM before stimulation. Add MEM in a drop-by-drop manner during ratio recording, and wait up to 3 min afterward.
9. Example 1, D-alanine addition: add drop-by-drop D-alanine diluted in 200 μL of prewarmed MEM to the Petri dish to a final concentration 0.5–8.0 mM. Continue data acquisition for 15–30 min.

   Example 2, PDGF addition: add drop-by-drop PDGF diluted in 200 μL of prewarmed MEM to the Petri dish to a final concentration 10 ng/mL. Continue data acquisition for 30–60 min.
10. Make a control of the HyPer reduction state by adding H_2O_2 to a final concentration of 0.18 mM.

Do not touch the microscope stage or the Petri dish during the imaging. Figures 2 and 3 show results for Example 1 and Example 2, respectively.

3.2.4 Analysis of the Results

Time series can be analyzed with ImageJ software. For H_2O_2 dynamics plotting stacks corresponding to 420 nm and 500 nm excitations of HyPer are usually converted to 32 bit after background subtraction. A 420 nm stack is thresholded to remove the background values ("Not a number" function of 32-bit thresholding). A 500 nm stack should be divided by the corresponding 420 nm stack frame by frame using the "Image calculator" command of ImageJ. For HyPerRed signal analysis or single-wavelength detection of green HyPer, stacks are calculated in the same manner, but without thresholding and division of channels frames. For confocal microscopy data processing, *see* Subheading "5.5 Image Processing" in [31].

The analysis protocol can be automated using prewritten ImageJ macro FluoQ [32], which enables fast analysis of multiparameter time-lapse fluorescence microscopy data with minimal manual input. FluoQ provides statistical analysis of all standard measured parameters and delivers the results in multiple graphic and numeric displays. FluoQ is a free software and available from its author on www.github.com. Time course of HyPer fluorescence is calculated for regions of interest inside the imaged cell and then plotted in OriginPro, MathLab, or Excel.

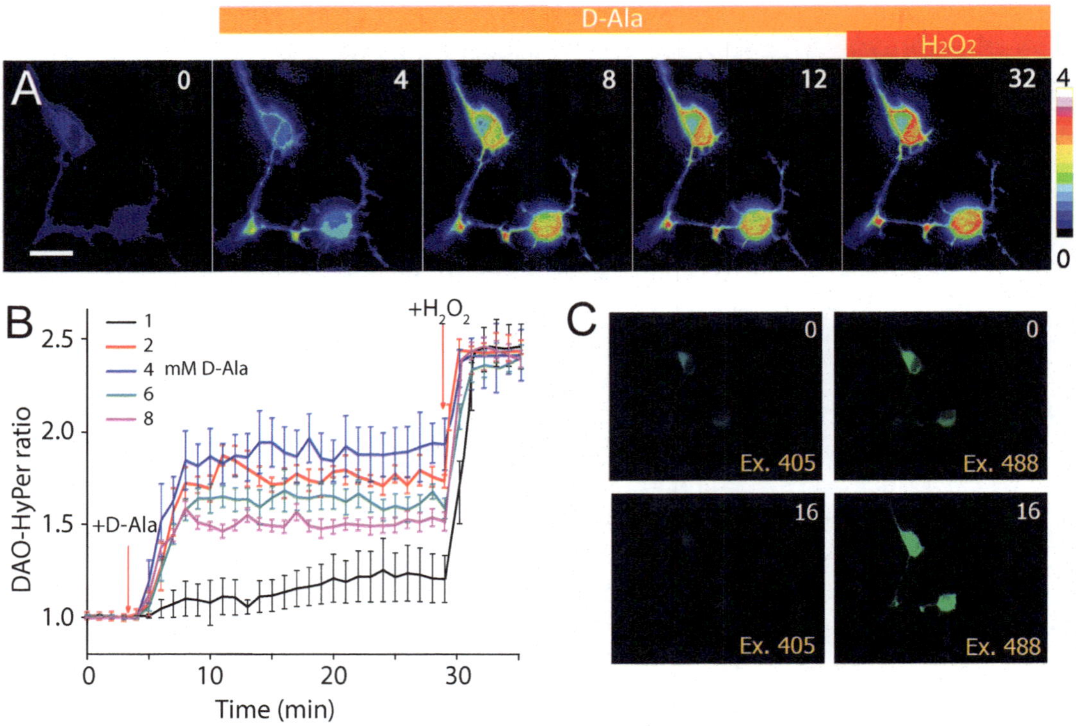

Fig. 2 HyPer-DAAO fusion in NIH-3T3 cells. (**A**) Wide-field ratiometric images of NIH-3T3 cells transfected with a HyPer-DAAO fusion. Numbers indicate time points in minutes. Scale bars 20 μm. (**B**) Time course of ratiometric HyPer-DAAO changes in NIH-3T3 cells upon addition of different concentrations of D-Ala with error bars showing standard error of the mean. (**C**) Images of cells at 405 and 488 nm excitation before (0) and after 16 min (16) of D-Ala addition

4 Notes

1. Cell culture: Avoid HEPES concentration above 25 mM as it is toxic to cells and leads to spontaneous ROS production in the culture medium.
2. Choosing the transfection method: For cell lines known to be easily transfected (HEK293, HeLa, NIH-3T3), we recommend using liposomal transfection with FuGene6, FuGene HD, or X-tremeGene9 reagents. Similar efficiency can be obtained with Lipofectamine LTX, Lipofectamine 2000, or Lipofectamine 3000, depending on the cell type. For hard-to-transfect cell lines such as RAW 264.7 macrophages, we recommend lentiviral transduction or adeno-associated viral transduction. Primary culture of neurons and glial cells from mouse embryos can be transfected with calcium phosphate transfection, liposomal agents (Lipofectamine LTX, Lipofectamine 3000, etc.) or viral transduction. The efficiency of transient transfection of neurons is usually very low (15–25 transfected neurons per 150 000 cells/dish).

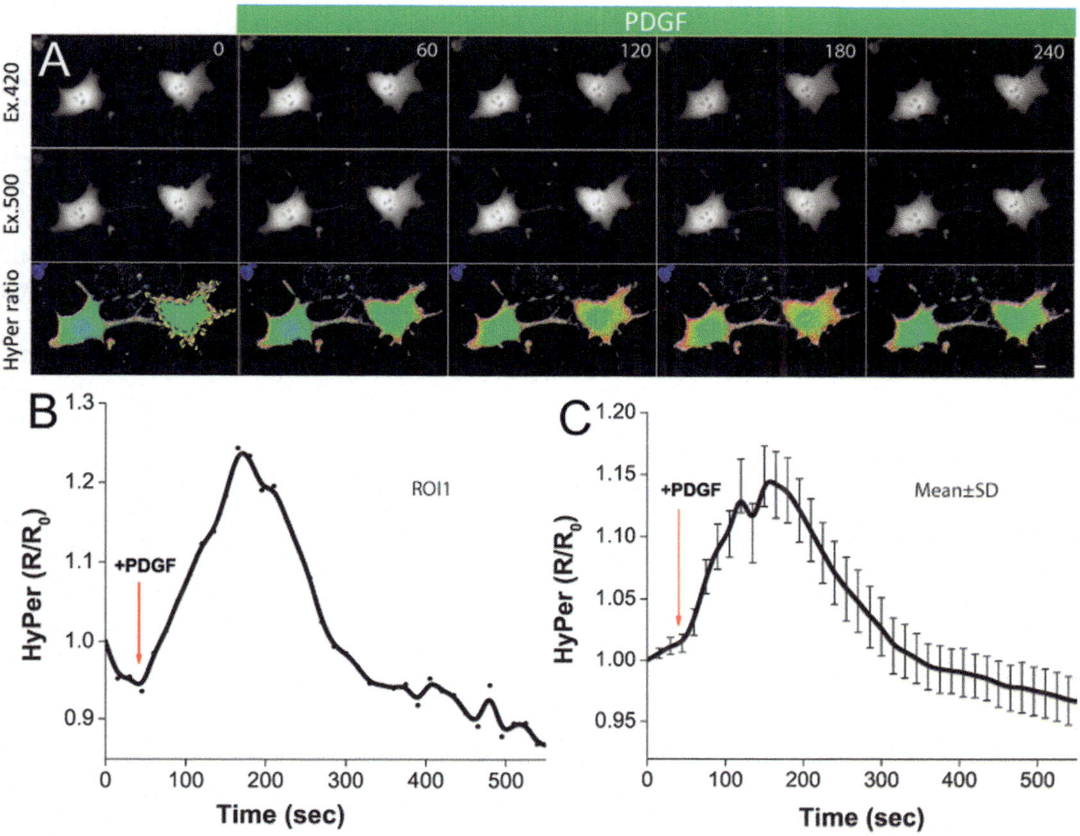

Fig. 3 PDGF stimulation of NIH-3T3 cells. NIH-3T3 cells were transfected with HyPer and stimulated with PDGF. (**A**) Wide-field images of cells in 420 nm, 500 nm, and ratio channels. Numbers indicate time points in seconds. Scale bar 10 μm. (**B** and **C**) Time course of ratiometric HyPer changes in NIH-3T3 cells upon addition of 10 ng/mL PDGF in selected ROI (**A**, white square). Panel (**B**) shows individual cell data; on panel (**C**) the data from seven cells were averaged

3. DNA quality for transfection: The total volume of the plasmid DNA should not exceed 5 μL for efficient transfection (this corresponds to the concentration of plasmid DNA exceeding 0.2 mg/mL). If your DNA sample is less concentrated, precipitate DNA with ethanol, and dissolve in a small volume of MQ water (or TAE buffer). Also note that the preparation of DNA should not be contaminated with proteins and the ratio of the absorption at a wavelength of 260 nm (nucleic acid) to the absorption at 280 nm (total protein) should not be lower than 1.75.

4. Change of the cell media for transfection: Replacement of culture medium to Opti-MEM without antibiotics or FCS increases efficiency of liposomal transfection with reagents such as FuGene6, FuGene HD, or X-tremeGene 9, but it is possible to add the transfection mix directly into the medium (DMEM supplemented with 10% FCS and 20 mM of

L-glutamine), in which the cells were plated. Change the medium 8–24 h after transfection. Usually transfection efficiency decreases then to less than 10% when compared to Opti-MEM.

5. Optimal time after transfection: Optimal time for imaging can vary from 18 to 48 h after transfection and usually depends on the signal/background ratio and the expression level of HyPer. HyPer-2 is two- to threefold brighter in mammalian cells when compared to HyPer and HyPer-3 but is somewhat less sensitive to H_2O_2.

   Imaging medium: Optimal imaging medium depends on the experimental system. If you image for 10–50 min, PBS or other isotonic buffer solutions (Krebs-Ringer, Tyrode, etc.) can be used. If cell culture medium in a CO_2-lacking microscope chamber is used, the medium should be supplemented with 10–25 mM HEPES. In the presence of 5% CO_2, the medium should be supplemented with 2.2 g/L sodium bicarbonate. Check the formulation of the medium before the experiment. For the long-term imaging (1 h and more) experiments, MEM medium is recommended.

6. Alternative filter cubes for Leica: Alternatively a CFP cube (BP 436/20, 455) for registration of the protonated form of the chromophore (not optimal) and a GFP cube (BP 470/40, 500) for registration of the deprotonated form of HyPer can be used. For HyPerRed use a TX2 (BP 560/40, 595) filter cube.

7. Confocal microscope settings: If cells used do not significantly change their shape during the course of imaging, the xt (line scan) mode can be used as it is less photodamaging for the cells. The lower resolution gives less detailed pictures, while the higher resolution slows the laser beam, which is more photodamaging for the cell. In any experiment with real-time dynamics of fluorescence, two applications are possible: to quantify the fluorescent signal or to make "good looking" pictures. For quantification, the use of low resolution is enough and much less photodamaging.

Acknowledgment

This work was supported by the Russian Science Foundation grant 17-14-01086 (experiments with D-amino acid oxidase), Russian Foundation for Basic Research grant 18-54-74003 (image processing and PDGF experiments), and DFG IRTG 1816. YGE received Sergey Shpiz fellowship. Experiments were partially carried out using the equipment provided by the IBCH core facility (CKP IBCH, supported by the Russian Ministry of Education and Science, grant RFMEFI62117X0018).

References

1. Droge W (2002) Free radicals in the physiological control of cell function. Physiol Rev 82 (1):47–95. https://doi.org/10.1152/physrev.00018.2001
2. Winterbourn CC, Hampton MB (2008) Thiol chemistry and specificity in redox signaling. Free Radical Bio Med 45(5):549–561. https://doi.org/10.1016/j.freeradbiomed.2008.05.004
3. D'Autreaux B, Toledano MB (2007) ROS as signalling molecules: mechanisms that generate specificity in ROS homeostasis. Nat Rev Mol Cell Biol 8(10):813–824. https://doi.org/10.1038/nrm2256
4. Bedard K, Krause KH (2007) The NOX family of ROS-generating NADPH oxidases: physiology and pathophysiology. Physiol Rev 87 (1):245–313. https://doi.org/10.1152/physrev.00044.2005
5. Folkes LK, Christlieb M, Madej E, Stratford MR, Wardman P (2007) Oxidative metabolism of combretastatin A-1 produces quinone intermediates with the potential to bind to nucleophiles and to enhance oxidative stress via free radicals. Chem Res Toxicol 20 (12):1885–1894. https://doi.org/10.1021/tx7002195
6. Serrander L, Jaquet V, Bedard K, Plastre O, Hartley O, Arnaudeau S, Demaurex N, Schlegel W, Krause KH (2007) NOX5 is expressed at the plasma membrane and generates superoxide in response to protein kinase C activation. Biochimie 89(9):1159–1167. https://doi.org/10.1016/j.biochi.2007.05.004
7. Stocker S, Maurer M, Ruppert T, Dick TP (2018) A role for 2-Cys peroxiredoxins in facilitating cytosolic protein thiol oxidation. Nat Chem Biol 14(2):148–155. https://doi.org/10.1038/nchembio.2536
8. Mendez I, Vazquez-Martinez O, Hernandez-Munoz R, Valente-Godinez H, Diaz-Munoz M (2016) Redox regulation and pro-oxidant reactions in the physiology of circadian systems. Biochimie 124:178–186. https://doi.org/10.1016/j.biochi.2015.04.014
9. Rhee SG, Kil IS (2017) Multiple functions and regulation of mammalian peroxiredoxins. Annu Rev Biochem 86:749–775. https://doi.org/10.1146/annurev-biochem-060815-014431
10. Sies H (2015) Oxidative stress: a concept in redox biology and medicine. Redox Biol 4:180–183. https://doi.org/10.1016/j.redox.2015.01.002
11. Zhu Y, Sun J, Wang L, Qi B (2015) Endogenic oxidative stress response contributes to glutathione over-accumulation in mutant *Saccharomyces cerevisiae* Y518. Appl Microbiol Biotechnol 99(17):7069–7078. https://doi.org/10.1007/s00253-015-6629-7
12. Gomes A, Fernandes E, Lima JL (2005) Fluorescence probes used for detection of reactive oxygen species. J Biochem Biophys Methods 65(2–3):45–80. https://doi.org/10.1016/j.jbbm.2005.10.003
13. Yang G, de Castro Reis F, Sundukova M, Pimpinella S, Asaro A, Castaldi L, Batti L, Bilbao D, Reymond L, Johnsson K, Heppenstall PA (2015) Genetic targeting of chemical indicators in vivo. Nat Methods 12 (2):137–139. https://doi.org/10.1038/nmeth.3207
14. Belousov VV, Fradkov AF, Lukyanov KA, Staroverov DB, Shakhbazov KS, Terskikh AV, Lukyanov S (2006) Genetically encoded fluorescent indicator for intracellular hydrogen peroxide. Nat Methods 3(4):281–286. https://doi.org/10.1038/nmeth866
15. Markvicheva KN, Bilan DS, Mishina NM, Gorokhovatsky AY, Vinokurov LM, Lukyanov S, Belousov VV (2011) A genetically encoded sensor for H2O2 with expanded dynamic range. Bioorg Med Chem 19(3):1079–1084. https://doi.org/10.1016/j.bmc.2010.07.014
16. Bilan DS, Pase L, Joosen L, Gorokhovatsky AY, Ermakova YG, Gadella TW, Grabher C, Schultz C, Lukyanov S, Belousov VV (2013) HyPer-3: a genetically encoded H(2)O (2) probe with improved performance for ratiometric and fluorescence lifetime imaging. ACS Chem Biol 8(3):535–542. https://doi.org/10.1021/cb300625g
17. Ermakova YG, Bilan DS, Matlashov ME, Mishina NM, Markvicheva KN, Subach OM, Subach FV, Bogeski I, Hoth M, Enikolopov G, Belousov VV (2014) Red fluorescent genetically encoded indicator for intracellular hydrogen peroxide. Nat Comm 5:5222. https://doi.org/10.1038/ncomms6222
18. Zhao Y, Araki S, Wu J, Teramoto T, Chang YF, Nakano M, Abdelfattah AS, Fujiwara M, Ishihara T, Nagai T, Campbell RE (2011) An expanded palette of genetically encoded Ca(2) (+) indicators. Science 333(6051):1888–1891. https://doi.org/10.1126/science.1208592
19. Hanson GT, Aggeler R, Oglesbee D, Cannon M, Capaldi RA, Tsien RY, Remington SJ (2004) Investigating mitochondrial redox potential with redox-sensitive green

fluorescent protein indicators. J Biol Chem 279 (13):13044–13053. https://doi.org/10.1074/jbc.M312846200
20. Berg J, Hung YP, Yellen G (2009) A genetically encoded fluorescent reporter of ATP: ADP ratio. Nat Methods 6(2):161–166. https://doi.org/10.1038/nmeth.1288
21. Poburko D, Santo-Domingo J, Demaurex N (2011) Dynamic regulation of the mitochondrial proton gradient during cytosolic calcium elevations. J Biol Chem 286 (13):11672–11684. https://doi.org/10.1074/jbc.M110.159962
22. Whitaker M (2010) Genetically encoded probes for measurement of intracellular calcium. Methods Cell Biol 99:153–182. https://doi.org/10.1016/B978-0-12-374841-6.00006-2
23. Boselli A, Piubelli L, Molla G, Pilone MS, Pollegioni L, Sacchi S (2007) Investigating the role of active site residues of *Rhodotorula gracilis* D-amino acid oxidase on its substrate specificity. Biochimie 89(3):360–368. https://doi.org/10.1016/j.biochi.2006.10.017
24. Pollegioni L, Diederichs K, Molla G, Umhau S, Welte W, Ghisla S, Pilone MS (2002) Yeast D-amino acid oxidase: structural basis of its catalytic properties. J Mol Biol 324 (3):535–546
25. Matlashov ME, Belousov VV, Enikolopov G (2014) How much H(2)O(2) is produced by recombinant D-amino acid oxidase in mammalian cells? Antioxid Redox Signal 20 (7):1039–1044. https://doi.org/10.1089/ars.2013.5618
26. Pollegioni L, Falbo A, Pilone MS (1992) Specificity and kinetics of *Rhodotorula gracilis* D-amino acid oxidase. Biochim Biophys Acta 1120(1):11–16
27. Bogdanova YA, Schultz C, Belousov VV (2017) Local generation and imaging of hydrogen peroxide in living cells. Curr Protocols Chem Biol 9(2):117–127. https://doi.org/10.1002/cpch.20
28. Chen KC, Zhou Y, Zhang W, Lou MF (2007) Control of PDGF-induced reactive oxygen species (ROS) generation and signal transduction in human lens epithelial cells. Mol Vis 13:374–387
29. Matlashov ME, Bogdanova YA, Ermakova GV, Mishina NM, Ermakova YG, Nikitin ES, Balaban PM, Okabe S, Lukyanov S, Enikolopov G, Zaraisky AG, Belousov VV (2015) Fluorescent ratiometric pH indicator SypHer2: applications in neuroscience and regenerative biology. Biochim Biophys Acta 1850(11):2318–2328. https://doi.org/10.1016/j.bbagen.2015.08.002
30. Ermakova YG, Pak VV, Bogdanova YA, Kotlobay AA, Yampolsky IV, Shokhina AG, Panova AS, Marygin RA, Staroverov DB, Bilan DS, Sies H, Belousov VV (2018) SypHer3s: a genetically encoded fluorescent ratiometric probe with enhanced brightness and an improved dynamic range. Chem Commun (Camb) 54(23):2898–2901. https://doi.org/10.1039/c7cc08740c
31. Mishina NM, Markvicheva KN, Fradkov AF, Zagaynova EV, Schultz C, Lukyanov S, Belousov VV (2013) Imaging H2O2 microdomains in receptor tyrosine kinases signaling. Methods Enzymol 526:175–187. https://doi.org/10.1016/B978-0-12-405883-5.00011-9
32. Stein F, Kress M, Reither S, Piljic A, Schultz C (2013) FluoQ: a tool for rapid analysis of multiparameter fluorescence imaging data applied to oscillatory events. ACS Chem Biol 8 (9):1862–1868. https://doi.org/10.1021/cb4003442

Chapter 16

Imaging Intracellular H_2O_2 with the Genetically Encoded PerFRET and OxyFRET Probes

Balázs Enyedi and Miklós Geiszt

Abstract

Hydrogen peroxide (H_2O_2) is an important signaling intermediate with various regulatory and effector functions. Despite its significance, the subcellular organization of H_2O_2 signals is poorly understood. Introducing novel techniques for the intracellular detection of H_2O_2 would be essential for a more complete understanding of its role in cellular signaling. We previously reported the development of two novel fluorescence resonance energy transfer (FRET)-based protein sensors that showed opposite emission ratio changes upon reaction with H_2O_2. In this chapter, we detail the methods for using OxyFRET and PerFRET for the assessment of changes in subcellular H_2O_2 levels.

Key words Hydrogen peroxide, PerFRET, OxyFRET, Genetically encoded sensor, Fluorescent biosensor imaging

1 Introduction

Reactive oxygen species (ROS) are oxygen-derived compounds that can rapidly interact with and modify target molecules including proteins, lipids, and nucleic acids. Among the several chemically distinct molecules belonging to ROS, H_2O_2 has emerged as a particularly important compound with diverse signaling and effector functions [1, 2]. Regulated production of H_2O_2 by NOX family NADPH oxidases is now considered an integral part of homeostasis, but the exact function of some NOX enzymes remains to be identified [3]. A major obstacle in ROS research is the lack of techniques enabling the precise and specific measurement of intracellular ROS production. Most of the fluorescent and luminescent dyes which are currently in use for this purpose have several shortcomings including the lack of specificity and the inability to provide subcellular resolution of ROS signals [4]. For example, dichlorofluorescein (DCF) is still frequently used in ROS measurement studies despite the detailed description of the possible artifacts—including DCF-induced ROS formation—that are associated with

Ulla G. Knaus and Thomas L. Leto (eds.), *NADPH Oxidases: Methods and Protocols*, Methods in Molecular Biology, vol. 1982, https://doi.org/10.1007/978-1-4939-9424-3_16,

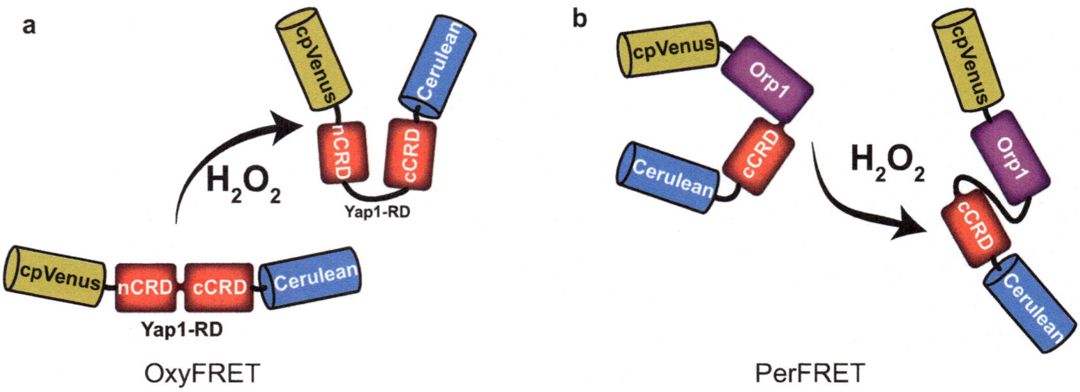

Fig. 1 Schematic representation of the OxyFRET **(a)** and PerFRET **(b)** probe structures. H_2O_2-mediated oxidation of central cysteine-rich domains induces conformational changes that either increase **(a)** or decrease **(b)** the efficiencies of fluorescence resonance energy transfer between Cerulean (donor) and Venus (acceptor) proteins

the application of this dye [4]. Genetically encoded probes offer advantages over traditional ROS measurement methods [5]. For example, using different targeting sequences, it is possible to express the sensors at various intracellular locations [6]. This feature offers the possibility of measuring ROS signals at well-defined subcellular locations. The development of novel genetically encoded redox sensors including the roGFP2-Orp1 fusion protein [7] and Hyper [8] were very important steps in the field, and these probes have been successfully applied in different experimental settings [5, 9].

We developed and characterized two novel H_2O_2-sensors (OxyFRET and PerFRET) that produce FRET signal changes upon oxidation by H_2O_2 [10]. These redox sensors are based on the Orp1-Yap1 redox relay system of *S. cerevisiae*. In both probes, H_2O_2-sensitive regions were inserted between variants of the Cerulean and Venus fluorescent proteins (Fig. 1). OxyFRET contains the N- and C-terminal cysteine-rich domains of Yap1 (nCRD and cCRD), whereas the sensor part of PerFRET is composed of the Orp1 protein and the Yap1 cCRD. In this article, we describe the methodology for measuring H_2O_2 signals in HeLa cells expressing OxyFRET or PerFRET.

2 Materials

2.1 Cell Culture and Transfection Reagents

1. HeLa cells (ATCC® CCL-2™).
2. Dulbecco's modified Eagle's medium (Lonza) supplemented with 10% fetal calf serum (FCS), 50 U/mL penicillin, and 50 μg/mL streptomycin (Lonza).
3. Lipofectamine™ 2000 transfection reagent (Invitrogen).

4. UltraMEM™ reduced serum media (Lonza).
5. Thirty-five-millimeter glass-bottom dishes (*see* **Note 1**).
6. OxyFRET and PerFRET mammalian expression plasmids (available from the authors upon request).
7. EC-1 medium: 133 mM NaCl, 3.1 mM KCl, 1.2 mM $CaCl_2$, 0.5 mM KH_2PO_4, 0.5 mM $MgSO_4$, 2 mM $NaHCO_3$, 5 mM glucose, and 5 mM Na-HEPES, pH 7.4.
8. H_2O_2 (30% stock—Sigma).

2.2 Equipment

Any inverted fluorescent widefield or confocal laser microscope with an excitation wavelength around 435 nm can be used to image the FRET-based CFP/YFP variant Cerulean/Venus OxyFRET and PerFRET probes. The setup should enable rapid collection of emissions from both donor (Cerulean) and FRET (Venus) channels. This requires an external emission filter wheel allowing rapid switching between the two emission channels (within milliseconds). Alternatively, donor and FRET emissions can be simultaneously acquired using a dual-camera setup or a beam splitter on a fluorescent widefield microscope or two detectors on a confocal microscope (*see* **Note 2**). We routinely use the following microscope setup in our experiments:

1. Axio Observer D1 (Zeiss) equipped with a 40 × 1.4 oil immersion objective (Fluar, Zeiss).
2. Excitation wavelength of 435 nm set by a random-access monochromator connected to a xenon-arc lamp (DeltaRAM, Photon Technology International).
3. Cascade II camera (Photometrics) equipped with a dual-view emission splitting system. The latter harbors a dichroic mirror (505dcxr) splitting the Cerulean and FRET (Venus) emissions. Appropriate bandpass filters to image Cerulean and FRET (Venus) emissions are 480/30 and 535/30 (e.g., from Chroma Technology).

3 Methods

3.1 Transfection

1. Seed HeLa cells at a sufficient density 1 day prior to transfection in DMEM supplemented with 10% FCS to reach a confluence of 40–60% on the day of transfection.
2. Various gene transfection methods exist; in **steps 3–6** we describe the process of transfection using cationic lipids, by incubating cells with a mixture of DNA and a transfection reagent over the course of a few days.

3. Add 2 μL of Lipofectamine™ 2000 to 100 μL serum-free UltraMEM™ in a 1.5 mL Eppendorf tube, and mix briefly by vortexing.
4. Add 2 μg of OxyFRET or PerFRET vector to 100 μL serum-free UltraMEM™ in a 1.5 mL Eppendorf tube, and mix briefly by vortexing.
5. Mix the diluted DNA with the diluted Lipofectamine™ 2000 reagent; then incubate for 5–10 min at room temperature.
6. Add the 200 μL of DNA-Lipofectamine™ 2000 complex dropwise to the media covering the cells, and return them into a CO_2 incubator.

3.2 Measurement of Intracellular H_2O_2 in Live Cells by Fluorescent Widefield Microscopy

1. OxyFRET and PerFRET expression become sufficient for imaging by 16–24 h post-transfection (*see* **Note 3**).
2. To carry out experiments at room temperature without a CO_2 controlled chamber, change media over the cells to 1.8 mL of HEPES-containing media (e.g., EC-1) 10–20 min prior to imaging (*see* **Note 4**).
3. Place the 35 mm dish on the microscope, and find a group of cells expressing the OxyFRET or PerFRET sensor (*see* **Note 5**).
4. Acquire images of the cells every 10 s over the course of the experiment with appropriate settings for donor (Cerulean) and FRET (Venus) emission ratio imaging. The following describes appropriate settings on a fluorescent widefield microscope.
   (a) Set microscope to collect the donor (Cerulean) and FRET (Venus) images by exciting at 435 nm and collecting emission at 485 nm and 535 nm, respectively (*see* **Note 6**).
   (b) Acceptor emission, although not necessary, can be collected by exciting at 515 nm and collecting emission at 535 nm (*see* **Note 7**).
   (c) Set the excitation time and illumination intensity according to the expression levels and camera sensitivity to ensure a signal-to-background ratio of at least two- to threefold (*see* **Note 8**).
5. Measure endogenous H_2O_2 production (or the effect of exogenous H_2O_2) by applying the stimulant (or H_2O_2 in bolus) during acquisition (*see* **Note 9**).

3.3 Image Processing and Data Analysis

Image processing can be done in any image analysis software such as Metamorph or open-source programs such as ImageJ, Fiji, or CellProfiler. The following steps should be performed to process the data and calculate FRET ratio values using Fiji [11] (Fig. 2):

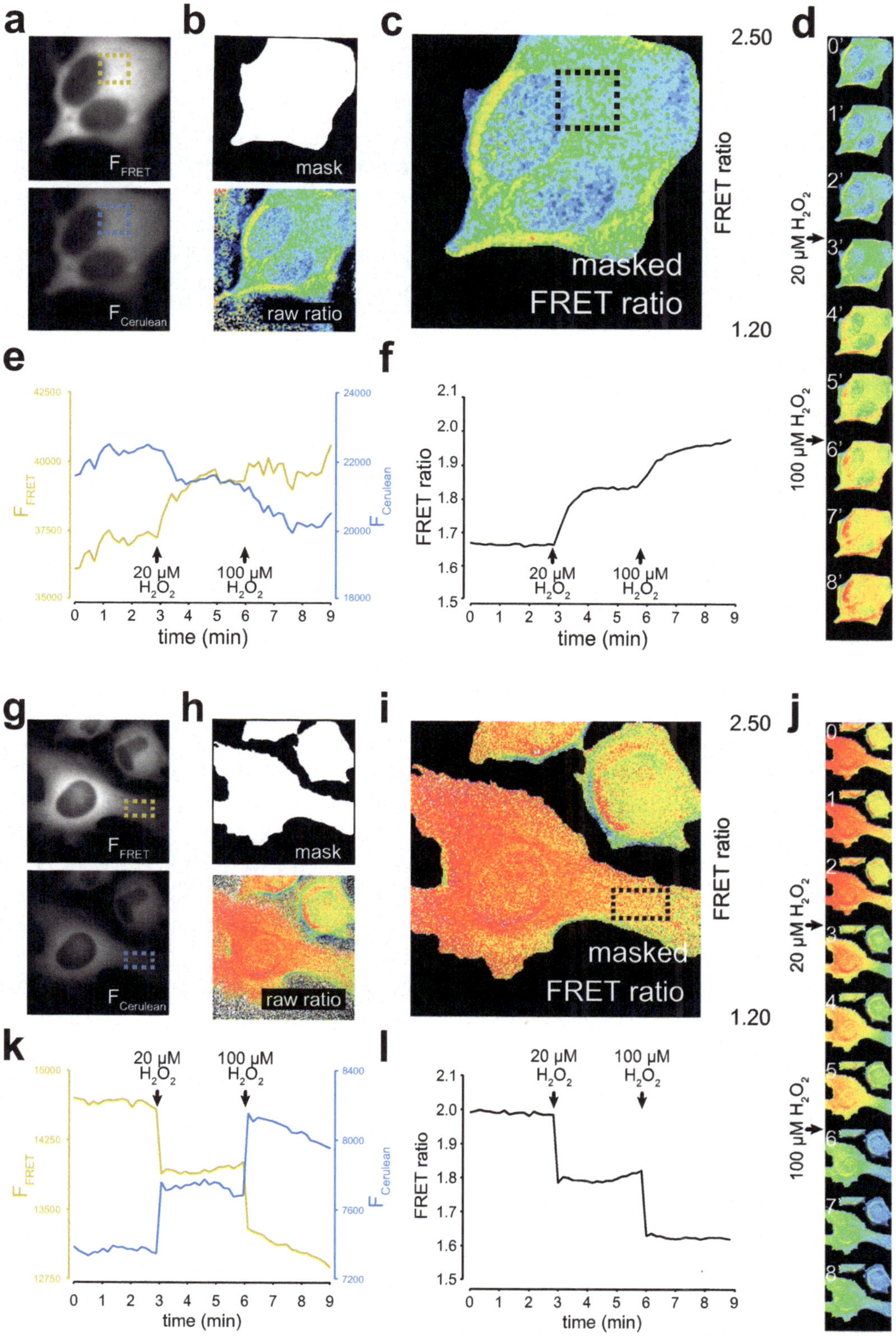

Fig. 2 H_2O_2 stimulation of HeLa cells expressing the OxyFRET or PerFRET probe. Masked FRET ratio images of HeLa cells expressing (**a–f**) OxyFRET or (**g–l**) PerFRET calculated from the (**a**, **g**) background-corrected F_{FRET} and $F_{Cerulean}$ images by multiplying the (**b**, **h** top) mask image series generated from thresholded F_{FRET} and

1. Subtract the background from raw F_{FRET} and $F_{Cerulean}$ images frame-by-frame in Fiji by globally subtracting the average intensity of a region not covered by cells (*see* **Note 10**).
2. Create a background mask from the sum of the F_{FRET} and $F_{Cerulean}$ images (Process → Image calculator → Add (32 bit)) by applying automatic thresholds on the stack (Image → Adjust → Auto Threshold) (*see* **Note 11**) after median-filtering (Process → Filters → Filters (2 pixels)). Divide the 8-bit binary image with a fixed value of 255 to set the mask values to "1" and the background values to "0." Finally, convert the image to 32 bits, and change the background "0" values to NaNs (following thresholding: Process → Math → NaN Background).
3. Calculate the ratio image by dividing the background-corrected F_{FRET} and $F_{Cerulean}$ images (Process → Image calculator → divide (32 bit)), and apply the 16_colors LUT. To remove the background noise, multiply the ratio image with the Image mask.
4. Measure changes in a given region of the FRET ratio image by selecting a region of interest and plotting the average intensity in the region vs. time.

4 Notes

1. The thickness of the glass coverslip should be chosen to match the numerical aperture (NA) of the objective (0.17 mm for a high NA objective).
2. On widefield microscopes a sensitive EMCCD/sCMOS camera with good dynamic range (12 bit or greater) is preferred as a detector. The camera/detector settings must remain constant for the detection of the donor and FRET channels if they are measured sequentially. Also, ensure that no sample movement occurs between the channels.
3. OxyFRET and PerFRET expression levels can be evaluated on an inverted fluorescent microscope using the standard "CFP" or "GFP" filter sets. Imaging should be conducted 24–72 h after DNA transfection. Longer times result in decreased expression, and signals may be too weak if performed less than 16 h post-transfection.

Fig. 2 (continued) $F_{Cerulean}$ images with the (**b**, **h** bottom) raw ratio image series. (**d**, **j**) Montage of masked FRET ratio image series of HeLa cells stimulated with H_2O_2 at the indicated timepoints. (**e**, **k**) Intensity profiles of selected regions in the individual F_{FRET} (yellow) and $F_{Cerulean}$ (cyan) channels and (**f**, **l**) corresponding FRET ratios from the same experiment

4. Phenol red-containing media should be avoided as it increases background fluorescence and interferes with the measurements.
5. Once adequate cells have been found, close the shutter to avoid any unnecessary illumination of the sample.
6. In single-chain FRET biosensors such as OxyFRET and PerFRET, where the stoichiometry of the donor and acceptor is fixed, the ratio of acceptor fluorescence to donor fluorescence (FRET ratio) signal can be used to conveniently measure and quantify changes in FRET efficiency. Similarly to other CFP-YFP based biosensors such as the Ca^{2+}-sensor cameleons [12], numerator and denominator images can be taken using a single excitation wavelength of the donor (CFP or Cerulean). In this scenario, the FRET ratio would be $(F_{FRET})/(F_{Cerulean})$, where F_{FRET} is the fluorescent image from the YFP (Venus) channel using the Cerulean excitation wavelength and $F_{Cerulean}$ is the fluorescent image from the Cerulean channel using the Cerulean excitation wavelength.
7. Several methods exist to calculate normalized FRET efficiency, with sensitized emission (SE) being the simplest method to correct for the spectral cross talk between fluorophores. To this end, images of the donor, FRET, and acceptor channels must be collected when only the donor or acceptor is present by measuring fluorescence in cells that express Cerulean or Venus proteins alone. This is performed prior to measuring the single-chain FRET biosensors to determine the cross talk correction factors which is specific to the fluorescent proteins and the optical setup used. Once the correction factors are determined, they can be used to calculate the normalized donor and FRET intensities and the corrected FRET ratio. However, when using single-chain FRET biosensors, this normalization is routinely not used, as the relative changes in FRET ratio are not affected by the correction.
8. To prevent phototoxicity and bleaching, longer excitation times are preferred over higher illumination intensities.
9. We routinely stimulate with 10× concentration stimulants or exogenous H_2O_2 (e.g., 200 μL 10× H_2O_2 mixed into 1.8 mL media). This can be added carefully during acquisition, or alternatively, a perfusion system can be used to change the media over the cells and stimulate them.
10. Make sure to select a region for background subtraction by inspecting the entire time lapse, not just a single image, in order to prevent false background subtractions from potential fluorescent objects moving into the field of view during the time lapse.

11. The adequate thresholding method has to be determined for every imaging setup; we thus recommend testing various thresholding algorithms before making a choice.

Acknowledgment

This work was supported by grants from the Hungarian Research Fund (OTKA NF72669) and from the National Office for Research and Technology (OMFB-01680/2009). The work was also financed by the Higher Education Institutional Excellence Programme (FIKP) of the Ministry of Human Capacities in Hungary, within the framework of the Molecular Biology thematic programme of the Semmelweis University and by "Lendület" grants from the Hungarian Academy of Sciences to both MG (LP2011-001/2011) and BE (LP2018-13/2018). EB is also grateful for the support of the Premium postdoctoral research program (PPD-016/2017) of the Hungarian Academy of Sciences.

References

1. Murphy MP, Holmgren A, Larsson NG et al (2011) Unraveling the biological roles of reactive oxygen species. Cell Metab 13:361–366
2. Lambeth JD, Neish AS (2014) Nox enzymes and new thinking on reactive oxygen: a double-edged sword revisited. Annu Rev Pathol 9:119–145
3. Sirokmány G, Donkó Á, Geiszt M (2016) Nox/duox family of NADPH oxidases: lessons from knockout mouse models. Trends Pharmacol Sci 37:318–327
4. Kalyanaraman B, Darley-Usmar V, Davies KJ et al (2012) Measuring reactive oxygen and nitrogen species with fluorescent probes: challenges and limitations. Free Radic Biol Med 52:1–6
5. Bilan DS, Belousov VV (2017) New tools for redox biology: from imaging to manipulation. Free Radic Biol Med 109:167–188
6. Enyedi B, Várnai P, Geiszt M (2010) Redox state of the endoplasmic reticulum is controlled by ero1l-alpha and intraluminal calcium. Antioxid Redox Signal 13:721–729
7. Gutscher M, Sobotta MC, Wabnitz GH et al (2009) Proximity-based protein thiol oxidation by H_2O_2-scavenging peroxidases. J Biol Chem 284:31532–31540
8. Belousov VV, Fradkov AF, Lukyanov KA et al (2006) Genetically encoded fluorescent indicator for intracellular hydrogen peroxide. Nat Methods 3:281–286
9. Albrecht SC, Barata AG, Grosshans J et al (2011) In vivo mapping of hydrogen peroxide and oxidized glutathione reveals chemical and regional specificity of redox homeostasis. Cell Metab 14:819–829
10. Enyedi B, Zana M, Donkó A, Geiszt M (2013) Spatial and temporal analysis of NADPH oxidase-generated hydrogen peroxide signals by novel fluorescent reporter proteins. Antioxid Redox Signal 19:523–534
11. Schindelin J, Arganda-Carreras I, Frise E et al (2012) Fiji: an open-source platform for biological-image analysis. Nat Methods 9:676–682. https://doi.org/10.1038/nmeth.2019
12. Whitaker M (2010) Genetically encoded probes for measurement of intracellular calcium. Methods Cell Biol 99:153–182. https://doi.org/10.1016/B978-0-12-374841-6.00006-2

Chapter 17

Quantitative Imaging of Endogenous and Exogenous H_2O_2 Gradients in Live Zebrafish Larvae

Mark Jelcic, Balázs Enyedi, and Philipp Niethammer

Abstract

Quantitative aspects of extracellular H_2O_2 signaling in animals, such as its spatiotemporal dynamics within tissues, remain little understood. Here we detail an optimized, experimental setup for measuring the dynamics and physiological consequences of extracellular H_2O_2 application to live tissues by intravital biosensor imaging in zebrafish larvae.

Key words Hydrogen peroxide, Diffusion, Gradient, Zebrafish, Biosensor, Intravital, Quantitative

1 Introduction

1.1 An Assay to Address Quantitative Questions of H_2O_2 Tissue Signaling

Increasing evidence suggests that extracellular hydrogen peroxide (H_2O_2) acts as a tissue signal, for instance, by regulating inflammation, healing, and regeneration [1–11]. Underlying molecular mechanisms are beginning to be discovered, including redox signaling through redox-sensitive cysteines in signaling proteins [12, 13]. Yet, several important quantitative questions remain unresolved. For example, how far and fast does extracellular H_2O_2 diffuse in intact tissues? What determines its diffusion length, that is, the range on which it can mediate bystander and paracrine effects? Are these length scales consistent with the physiological effects attributed to H_2O_2? Are there distinct zones in tissues, where extracellular H_2O_2 causes signaling without oxidative damage? Quantitatively addressing such questions requires an animal model (1) in which defined concentrations of H_2O_2 can be locally applied, (2) H_2O_2 diffusion through the tissue can be precisely monitored, and (3) physiological responses, for example, morphological changes, or changes in gene expression, can be measured as a function of distance from the local source of H_2O_2.

Ulla G. Knaus and Thomas L. Leto (eds.), *NADPH Oxidases: Methods and Protocols*, Methods in Molecular Biology, vol. 1982, https://doi.org/10.1007/978-1-4939-9424-3_17,

1.2 The Zebrafish Model

Over the last decade, transparent zebrafish larvae have become a popular animal model for investigating wound healing and regeneration and for testing the regulatory role of extracellular H_2O_2 in these processes [10, 14]. The main advantages of the zebrafish model include its amenability for direct intravital imaging and pharmacological approaches and the availability of zebrafish larvae at high numbers. The generation time of zebrafish is ~2–3 months, which allows for rapid implementation of genetic manipulations, for example, through gene editing or fluorescent transgene expression, in a tissue-specific manner if desired. The wound responses of zebrafish are comparable to mammals. Zebrafish and mammals possess largely similar repertoires of immune cells [15, 16]. Unlike the mammalian skin, larval zebrafish epidermis is covered by freshwater. Thus, injury of zebrafish epidermis resembles mucosal injury of mammals, namely, injury of the epithelial linings of the upper digestive tract (esophagus, mouth), which are likewise immersed in a freshwater-like solution, that is, saliva.

1.3 A Modified Zebrafish Tail Fin Injury Assay for Local H_2O_2 Application

In zebrafish larvae, H_2O_2 generated by an epithelial NADPH oxidase at the wound site attracts immune cells and mediates tail fin regeneration through oxidation of Src kinases, probably through generating chemotactic oxidized lipid mediators [1, 3, 17]. Wound-associated H_2O_2 production has been visualized using chemical and genetically encoded fluorescent reporters in live larvae [1]. Studying these endogenous H_2O_2 tissue gradients has provided valuable insights into the physiological functions of reactive oxygen species signaling in live animals.

In typical experiments, the tail fin tips of ~3–4 days old zebrafish larvae are amputated, and wound margin H_2O_2 is measured with appropriate fluorescent sensors as previously described [8]. Endogenous H_2O_2 wound patterns dissipate over approximately 60 min after tail fin tip amputation and have been studied by time-lapse microscopy and genetic/pharmacologic manipulation. Below, we detail a modified experimental strategy that allows for the generation of artificial H_2O_2 gradients at zebrafish tail fin wound margins through H_2O_2 bath application. In our setup, the amplitude and duration of the wound H_2O_2 signal can be experimentally controlled to reveal, for example, sufficiency of H_2O_2 for molecular (e.g., transcriptional activation) or morphological (e.g., wound healing, regeneration, inflammation) tissue responses, as well as dose dependency. In addition, tissue mechanisms of extracellular H_2O_2 uptake, diffusion, and metabolism can be studied by live imaging and reaction-diffusion modeling as previously demonstrated [18].

Hydrophilic compounds, such as H_2O_2, cannot easily diffuse through intact zebrafish epidermis and must be locally supplied through an epidermal opening, such as a wound [19]. Thus, the physiological activity of these compounds is hampered by the extraordinarily fast kinetics of wound closure in zebrafish larvae [19]. This must be kept in mind for hydrophilic agonist/antagonist experiments using this model, including the experiments described below. Fortunately, the recent discovery of osmotic cues as key signals for wound closure and leukocyte recruitment in zebrafish opened up ways to experimentally circumvent this complication [17, 19, 20]. Namely, bathing larvae in an isotonic solution during injury strongly delays wound closure, without being toxic for incubation times <8 h. While the wound is open, it can be efficiently perfused with hydrophilic substances, such as H_2O_2, for example, to probe their tissue metabolism, spatiotemporal dynamics [18], or paracrine signaling functions in live tissue.

In this chapter, we provide an experimental outline that can be modified according to individual research requirements. This protocol was originally developed to quantitatively study extracellular H_2O_2 reaction-diffusion and responses to local H_2O_2 application. We note that it may be adapted to study other hydrophilic agonists of interest, for example, chemokines or growth factors.

2 Materials

2.1 Reagents

1. pCS2+-HyPer: HyPer (Evrogen) subcloned into the pCS2+ expression vector (*see* **Note 1**).
2. pCS2FA-transposase (Tol2kit construct 396) [21].
3. QIAquick Gel Extraction Kit (Qiagen).
4. NotI restriction enzyme.
5. mMESSAGE mMACHINE® SP6 Transcription Kit (Life Technologies).
6. MEGAclear™ Transcription Clean-Up Kit (Life Technologies).
7. E3 embryo medium: 5 mM NaCl, 0.17 mM KCl, 0.33 mM $CaCl_2$, 0.33 mM $MgSO_4$.
8. *N*-phenylthiourea (PTU): 0.2 mM in E3 embryo medium.
9. pME-HyPer: HyPer subcloned into pME-MCS (Tol2kit construct 237) [21] (*see* **Note 2**).
10. p5E-*bactin2* (Tol2kit construct 299) [21].
11. p3E-polyA (Tol2kit construct 302) [21].
12. pDestTol2CG2 (Tol2kit construct 395) [21].
13. TE buffer, pH 8.0.
14. LR Clonase™ II Plus enzyme (Invitrogen).
15. One Shot® TOP10 Chemically Competent *E. coli* (Invitrogen).

16. Luria Broth (LB).
17. Ampicillin solution (1000×): 100 mg mL^{-1} in H_2O.
18. QIAprep Spin Miniprep Kit (Qiagen).
19. KCl solution: 1 M in H_2O.
20. Tricaine solution (20×): 4 mg mL^{-1} Ethyl 3-aminobenzoate methanesulfonate (Sigma), 10 mg mL^{-1} Na_2HPO_4, pH 7.2 (*see* **Note 3**).
21. Isotonic E3 embryo medium: 145 mM NaCl, 0.17 mM KCl, 0.33 mM $CaCl_2$, 0.33 mM $MgSO_4$.
22. Low-melting agarose solution: 1%, dissolved in E3 embryo medium (*see* **Note 4**).
23. Isotonic low-melting agarose solution: 1%, dissolved in isotonic E3 embryo medium (*see* **Note 4**).
24. H_2O_2 solution (Sigma).

2.2 Equipment

1. Hair loop tool: Squeeze a single hair into a 20–200 μL pipette tip and make a short loop protruding out of the tip.
2. Plastic and glass petri dishes.
3. Sterile plastic transfer pipettes.
4. P-2000 Laser-based micropipette puller (Sutter Instrument Co., or a comparable apparatus capable of pulling microinjection capillaries).
5. Nanoject II Auto-Nanoliter Injector (Drummond, or a comparable microinjection apparatus capable of nanoliter volume injections).
6. 3.5 in. long borosilicate glass microinjection capillaries (Drummond).
7. Dissecting microscope.
8. Olympus MVX10 Macroview microscope (or a comparable stereomicroscope equipped for fluorescence imaging).
9. Micro Needle Knife (Fine Science Tools).
10. Nikon Eclipse Ti microscope equipped with a ×20 plan-apochromat NA 0.75 air objective lens, an Andor Clara CCD (charge-coupled device) camera, a light-emitting diode light source (Lumencor), and a motorized stage (or a comparable inverted microscope).
11. LED light source equipped with 438/57 and 475/28 band-pass filters (Lumencor) (or a comparable light source equipped for excitation at 420 and 500 nm).
12. 535/30 emission filter (Chroma Technology) (or a comparable emission filter equipped for acquisition at 520 nm).

13. 35 mm glass bottom microwell dishes (MatTek Corporation).
14. Four-chamber 35 mm glass bottom dish with a 20 mm microwell (Cellvis).

3 Methods

Inconsistencies in tissue thickness and structure necessitate the use of genetically encoded ratiometric H_2O_2 sensors with high dynamic range to obtain consistent and comparable H_2O_2 gradient measurements in live animals. The fluorescent protein-based H_2O_2 sensor HyPer [22] meets these criteria and has been successfully used by multiple groups to measure H_2O_2 gradients in live zebrafish larvae [1, 4, 17–19, 23, 24]. HyPer-expressing zebrafish larvae can be obtained through the injection of in vitro transcribed mRNA into one-cell stage embryos or by the generation of stable transgenic lines that use tissue-specific promoters to drive HyPer expression. The following sections describe detailed methods utilizing HyPer to measure H_2O_2 gradients in live, larval zebrafish tail fins.

3.1 In Vitro Preparation and Zebrafish Embryo Injection of Capped HyPer mRNA

1. Linearize ~3 μg of the pCS2+ HyPer construct using the NotI restriction enzyme for ~2 h at 37 °C.
2. Separate the linearized plasmid from any undigested DNA using agarose gel electrophoresis. Excise and purify the linearized DNA from the gel using the QIAquick Gel Extraction Kit according to the manufacturer's instructions. Elute the linearized fragment in 30 μL of ddH_2O.
3. Prepare capped mRNA with the mMESSAGE mMACHINE® SP6 Transcription Kit using ~1 μg of the linearized plasmid as the template in a 20 μL reaction according to the manufacturer's instructions. Incubate the in vitro transcription reaction for a minimum of 2 h at 37 °C.
4. Add 1 μL of TURBO DNase (provided with mMESSAGE mMACHINE® SP6 Transcription Kit), mix, and incubate for 15 min at 37 °C.
5. Purify the in vitro transcribed mRNA with the MEGAclear™ Transcription Clean-Up Kit according to the manufacturer's instructions (*see* **Note 5**).
6. Dilute the capped HyPer mRNA to 0.5 μg μL^{-1} in ddH_2O, and inject ~2.3 nL into one-cell stage embryos of wild-type AB or *casper* fish.
7. Grow larvae to 3–4 days post fertilization (dpf), changing to fresh E3 medium each day prior to imaging. Use larvae that have ubiquitous expression of HyPer for imaging experiments (*see* **Note 6**).

3.2 Preparation of Transgenic HyPer-Expressing Zebrafish

1. Generate a *bactin2*:HyPer construct by carrying out a multisite Gateway® LR reaction with the p5E-*bactin2*, pME-HyPer, p3E-polyA entry clones, and the pDestTol2CG2 destination vector utilizing the LR Clonase™ II Plus enzyme mix according to the manufacturer's instructions.
2. Transform 3 μL of the reaction into 50 μL of One Shot® TOP10 cells (*see* **Note 7**).
3. Pick single colonies and grow up each in 5 mL LB/ampicillin broth overnight.
4. Purify the construct using the QIAprep Spin Miniprep Kit according to the manufacturer's instructions. Elute in 50 μL of TE buffer or ddH_2O (*see* **Note 8**).
5. Prepare capped Tol2 transposase mRNA in vitro using the pCS2FA-transposase construct as the template (*see* **Note 9**).
6. Prepare a 0.1 M KCl solution containing 10.85 ng μL^{-1} of the *bactin2*:Hyper construct and 10.85 ng μL^{-1} of the Tol2 transposase mRNA. Inject ~2.3 nL of this solution into the cytosol of one-cell stage AB or *casper* embryos.
7. At 2–3 dpf, sort out injected larvae that have mosaic HyPer and cardiac EGFP expression. Raise these animals to sexual maturity.
8. Identify germline transgenic founder (F0) fish by crossing the injected adults with wild-type fish and then screening the 2–3 dpf embryos for HyPer and cardiac EGFP expression. Grow the HyPer-expressing F1 embryos to sexual maturity (*see* **Note 10**).
9. Perform all imaging experiments on the progeny of F1 adults mated to wild-type fish.

3.3 Animal Preparation and Tail Fin Injury to Image Endogenous, Wound-Induced HyPer Signal

Damage to larval zebrafish epidermis triggers H_2O_2 production that is generated on a timescale of minutes and can be accurately captured by wide-field microscopy at low magnification (20×) [1]. We typically mount and capture signals from multiple samples (up to four larvae) during one experiment using a motorized stage (*see* Fig. 1).

1. Anesthetize HyPer-expressing 2–4 dpf zebrafish larvae in anesthetizing E3 medium (E3 containing 0.2 mg mL^{-1} Tricaine) (*see* **Note 11**).
2. Apply ~200 μL of anesthetizing E3 medium to the center of a 35 mm glass bottom microwell dish. Transfer larvae to this droplet with a plastic transfer pipette (*see* **Note 12**).
3. With a hair loop tool, move the larvae to the center of the glass bottom dish. If you are imaging multiple larvae, orient them in order that the anterior and posterior ends of all larvae are

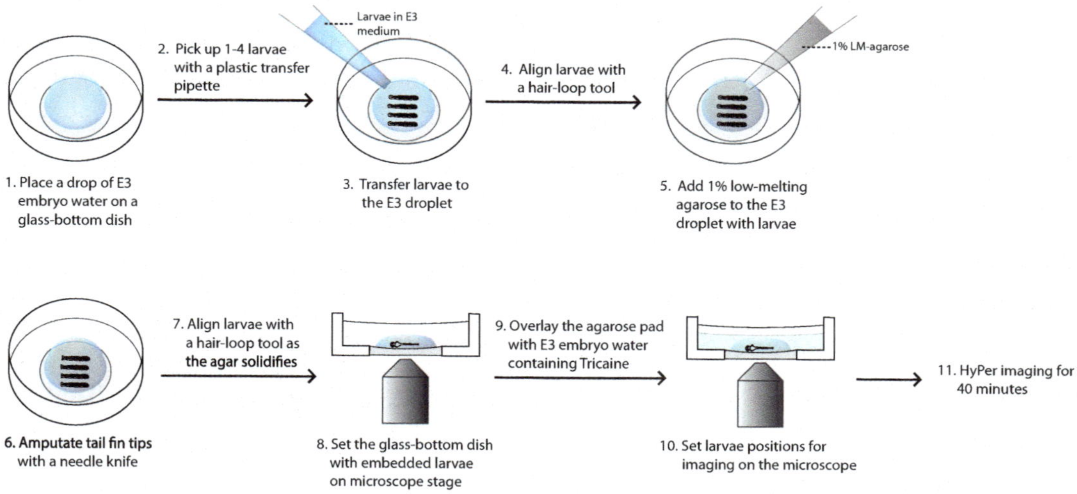

Fig. 1 Experimental schematic for imaging the endogenous, wound-induced HyPer signal in 2–4 dpf zebrafish larvae. 1–4 larvae are embedded in an agarose pad on a glass bottom dish and wounded by tail fin tip amputation (**steps 1–7**). After the agarose solidifies, the dish is set on the microscope stage, and the agarose pad is overlaid with E3 medium (**steps 8** and **9**). Finally, the coordinates of each sample are set, and HyPer imaging is initiated at ~3 min after wounding and proceeds for 40 min (**steps 10** and **11**)

aligned. Next, stack the larvae in column along the dorsal-ventral axes of the animals in the center of the glass bottom dish. Flatten out the tail fins very gently with a hair loop tool.

4. To mount the larvae, gradually mix in ~200 μL of 1% low-melting agarose (dissolved in E3 medium) to the droplet with a plastic transfer pipette. Slowly pipette up and down to create a ~1:1 mixture. After mixing, leave a final volume of ~200 μL (*see* **Note 13**).
5. Using a needle knife, amputate the tail fin tips. Then, quickly realign the larvae, making sure that they are lying with their tail fins flat on the glass bottom dish. We begin imaging the embryos at 3 min after wounding, so note the time of injury (*see* **Note 14**).
6. Fasten the glass bottom dish to the inverted microscope stage. Gently add at least 200 μL of anesthetizing E3 medium to the center of the agarose pad to prevent the sample from drying out during the experiment.

3.4 Imaging the Endogenous, Wound-Induced HyPer Signal

1. Prior to any experiments, mount an unwounded HyPer-expressing larva on a glass bottom microwell dish, and position the sample on the inverted microscope so that the unwounded tail fin is in the field of view.
2. Have the camera set to 8× binning. Using the 20× objective lens, set the illuminating light power and exposure times of the 500 (YFP_{500}) and 420 nm (YFP_{420}) excitation wavelengths so

that the baseline HyPer ratio (YFP_{500}/YFP_{420}) of unwounded tail fin epithelium is ~1. Use unwounded tail fin epithelium at about half the distance from the tip of the tail fin to the tip of the notochord as the reference point (*see* **Note 15**).

3. Set the coordinates of each experimental sample, and choose a focal plane in which the amputated tail fin is clearly in focus under bright-field illumination with a 20× objective lens (*see* **Note 16**).
4. The positions of the experimental samples on the microscope should be set by ~3 min after wounding. At this time, begin imaging. Acquire an image every 1–2 min for at least 40 min (imaging will be completed 43 min after wounding). All images should be acquired at a temperature of 26–28 °C.

3.5 Preparation of Zebrafish Larvae for Imaging Exogenous H_2O_2 Gradients

This method has been optimized for multi-sample imaging in a four-chamber glass bottom dish. This setup permits imaging of multiple samples exposed to four separate conditions (such as four different H_2O_2 concentrations) in a single experiment. We typically mount 1–4 larvae in a single chamber, allowing us to image up to 16 different samples in a single experiment (*see* Fig. 2).

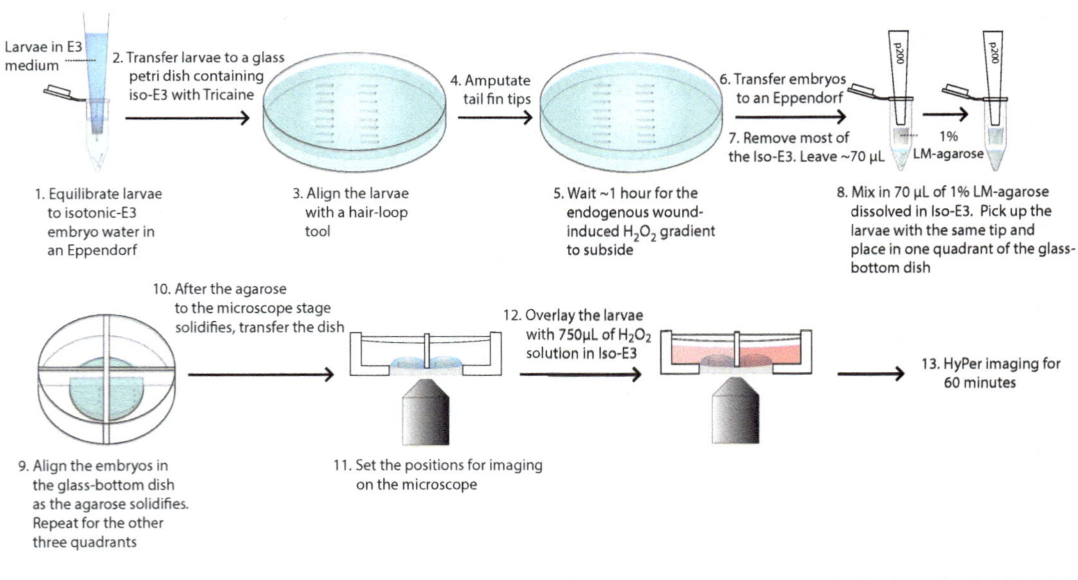

Fig. 2 Experimental schematic for imaging exogenous, H_2O_2-mediated HyPer signals in the larval zebrafish tail fin. Anesthetized 3.5–4 dpf zebrafish larvae are equilibrated to isotonic E3 medium and aligned for tail fin tip amputation in a glass petri dish (**steps 1–3**). Tail fin tips are amputated with a sharpened needle knife, and larvae are incubated in the isotonic E3 bath until the endogenous, wound-induced H_2O_2 gradient has subsided (**steps 4** and **5**). Next, the larvae are transferred to an Eppendorf tube, submerged in a low-melting agarose solution, and transferred to a glass bottom dish (**steps 6–9**). After all the samples are mounted, the glass bottom dish is transferred to the motorized microscope stage, and the coordinates for each sample are set. Finally, the samples are overlaid with H_2O_2 solution, and HyPer imaging is initiated at 60 min after wounding and proceeds for 60 min (**steps 10–13**)

1. Anesthetize HyPer-expressing 3.5–4 dpf zebrafish larvae in anesthetizing E3 medium (*see* **Note 17**).
2. Using a plastic transfer pipette, add 1–4 larvae to an Eppendorf tube containing 1 mL of isotonic E3 medium (iso-E3) supplemented with 0.2 mg mL^{-1} Tricaine. Allow the larvae to settle to the bottom of the tube.
3. Remove the majority of the medium with a transfer pipette, leaving only enough for the larvae to remain completely submerged. Add 1 mL of fresh anesthetizing iso-E3 to the tube. Pipette up and down very slowly with a transfer pipette to suspend the larvae in the medium. Allow them to settle to the bottom of the tube. Repeat this one more time to fully equilibrate the larvae to iso-E3 (*see* **Note 18**).
4. Move the larvae to a glass petri dish filled with anesthetizing iso-E3. Orient the larvae in an optimal position for tail fin tip amputation with a hair loop tool (*see* **Note 19**).
5. Amputate tail fin tips using a freshly sharpened needle knife (*see* **Note 20**).
6. At ~40–45 min after amputation, begin mounting the larvae in the four-chamber glass bottom dish. Transfer larvae that are to be exposed to the same condition in a single chamber to an Eppendorf tube containing 1 mL of anesthetizing iso-E3, and allow them to settle to the bottom of the tube (*see* **Note 21**).
7. Remove the majority of the medium with a transfer pipette, leaving ~70 μL which is enough to completely submerge 4 larvae.
8. Cut off approximately half of a 200 μL plastic pipette tip with a fresh razor blade. With the cut pipette tip and a P200 micropipette, dispense 70 μL of 1% isotonic low-melting agarose to the Eppendorf tube containing the larvae.
9. Gently pipette up the larvae in the same cut pipette tip, and transfer them to a single chamber of the four-chamber glass bottom dish. Quickly align the larvae along their dorsal-ventral axes, and flatten out the tail fins with a hair loop tool before the agarose solidifies.
10. Repeat **steps 6–9** three more times, to mount larvae in the remaining chambers of the glass bottom dish (*see* **Note 22**).
11. Once the agarose solidifies in all the chambers, fasten the glass bottom dish to the inverted microscope stage.

3.6 Imaging Exogenous H_2O_2 Gradients in the Larval Zebrafish Tail Fin

1. Prior to any experiments, mount an unwounded HyPer-expressing larva on a glass bottom microwell dish, and position the sample on the inverted microscope so that the unwounded tail fin is in the field of view.

2. Have the camera set to 8× binning. Using the 20× objective lens, set the illuminating light power and exposure times of the 500 (YFP_{500}) and 420 nm (YFP_{420}) excitation wavelengths in order that the baseline HyPer ratio (YFP_{500}/YFP_{420}) of unwounded tail fin epithelium is ~1. Use unwounded tail fin epithelium at about half the distance from the tip of the tail fin to the tip of the notochord as the reference point (*see* **Note 15**).
3. Freshly prepare H_2O_2 dilutions in iso-E3 supplemented with 0.2 mg mL^{-1} Tricaine (*see* **Note 23**).
4. After fastening the glass bottom dish to the microscope stage, set the coordinates of each experimental sample, and choose a focal plane in which the amputated tail fin is clearly in focus under bright-field illumination with a 20× objective lens (*see* **Note 16**).
5. At 1 h after amputation, add 750 μL of H_2O_2 in anesthetizing iso-E3 to each chamber.
6. Immediately begin image acquisition. Acquire an image every 1–2 min for at least 60 min. All images should be acquired at a temperature of 26–28 °C.

3.7 Image Processing

To calculate HyPer ratio images and spatiotemporal heatmaps of the time-lapse ratio data, we routinely use open-source programs such as Fiji or CellProfiler combined with custom Python scripts based on the Pandas, SciPy, NumPy, and scikit-image libraries. The following steps should be performed to process the data (*see* Fig. 3):

1. Subtract background frame by frame in Fiji by globally subtracting the average intensity of a region outside the larvae from the raw YFP_{500} and YFP_{420} images (*see* **Note 24**).
2. Create a background mask from the YFP_{500} image by applying automatic thresholding on the stack (Image → Adjust → Auto Threshold; *see* **Note 25**) after median filtering (Process → Filters → Filters (2 pixels)). Divide the 8-bit binary image with a fix value of 255 to set the mask values to "1" and the background values to "0." Finally convert the image to 32 bits, and change the background "0" values to NaNs (following thresholding: Process → Math → NaN Background).
3. Calculate the ratio image by dividing the background corrected YFP_{500} and YFP_{420} images (Process → Image calculator → divide (32 bits)), and apply the 16_colors LUT. To remove the background noise, multiply the ratio image with the Image mask.
4. Manually mask out the region of the notochord (following manual selection: Edit → Clear). This is required to exclude ratio data from regions outside the tail fin tissue. Finally, crop a rectangular region of the image series, in which the

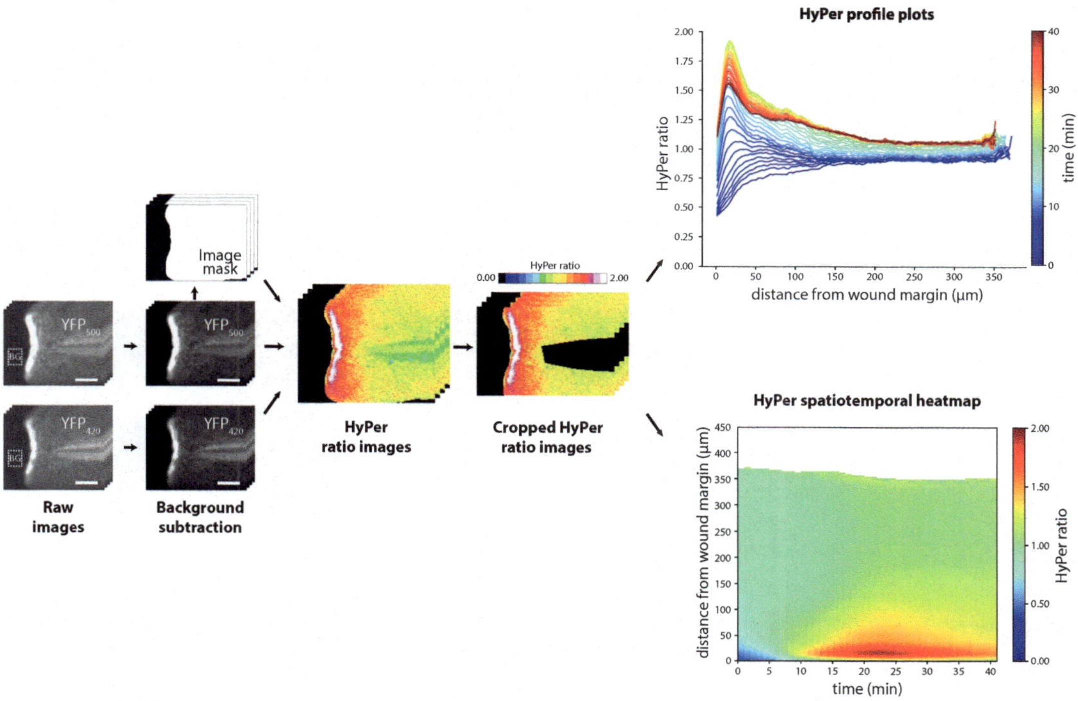

Fig. 3 Workflow of HyPer image processing. HyPer ratio images of wounded TG (bactin2: HyPer) zebrafish are calculated from YFP_{500} and YFP_{420} images after performing background (BG) subtraction and multiplication of the ratio images with a mask derived from the thresholded YFP_{500} image. The notochord region is then manually masked out, and the wounded area of the tail fin is cropped along the vertical axis of the image. The ratio value of every pixel in the time series is then remapped and expressed as a function of binned distance from the wound margin. The data is shown as an average profile plot or spatiotemporal heatmap. Scale bars, 100 μm

contracting/relaxing wound margin fills the vertical axis of the images along the whole time-lapse (*see* **Note 26**).

5. The final step of image processing is to transform the time-lapse image series into averaged profile plots and heatmaps, in which HyPer ratio values of the tail fin are expressed as a function of binned distance from the wound margin. This can be performed in programs such as Python or MATLAB by first applying a distance transform on the cropped HyPer ratio images. The distance transform is an operator which assigns points inside a foreground region (tail fin) the distance closest to the boundary. In our case, the boundary is the wound margin, which is delineated by the background NaN values of the cropped ratio images. HyPer ratio values of each image pixel are thus assigned an exact distance measure from the wound margin. This data is then binned for plotting as individual profile plots for every image timepoint or as heatmaps of the HyPer ratio values as a function of time and distance from the wound margin.

3.8 Possible Downstream Applications and Variations

The amputated larval zebrafish tail fin can be pictured (at least we do it) as a transparent, quasi-1D tissue reaction chamber, a type of "biological microfluidics chamber," in which the spatiotemporal behavior of extracellular H_2O_2 can be microscopically measured with clearly defined temporal (e.g., time of H_2O_2 addition to bath or injury) and spatial (wound margin) reference points. These reference points later allow normalization and integration of quantitative imaging data from different animal experiments into aggregated data matrices that are amenable to downstream analysis by quantitative modeling approaches. For example, by biosensor calibration and implementation of a simplified 1D reaction-diffusion model, we recently inferred extracellular H_2O_2 reaction rates from in vivo data sets generated by the above protocols [18]. Although applied here for studying extracellular H_2O_2, the outlined approach can be, in principle, extended to other hydrophilic substances, such as peptide growth factors or chemokines, given that appropriate imaging reporters are available to monitor their biochemical activity on the cells in the tissue (e.g., Ca^{2+} biosensors for GPCR ligands).

The biosensor imaging assays can be also combined with assays that probe for possible downstream roles of extracellular H_2O_2 gradients, such as leukocyte recruitment, wound closure, regeneration, or spatial regulation of transcriptional regulators (e.g., through in situ hybridization, fluorescence promoter assays). For instance, it could be tested how diffusion length and tissue concentration of extracellular H_2O_2 correlate with functional outputs in vivo. Close spatiotemporal correlations would hereby indicate direct regulatory relationships, and vice versa.

4 Notes

1. Other vectors containing an SP6 RNA polymerase promoter site can be used in place of the pCS2+ vector.
2. Other middle entry vectors, such as the Gateway® pDONR™221 Vector (Life Technologies), can also be used in place of the pME-MCS vector.
3. Prepare anesthetizing E3 medium for larvae imaging fresh each time by diluting Tricaine solution to 1× in E3 embryo medium. Store the 20× Tricaine solution at 4 °C.
4. We find that it is best to prepare these solutions fresh each time. Store the solutions at 42 °C to keep from solidifying. When using pharmacological compounds, add compound to the 1% low-melting agarose immediately before use and mix well.
5. Determine the mRNA concentration by reading the absorbance with a spectrophotometer at 260 nM. Perform agarose

gel electrophoresis on an aliquot of the in vitro transcribed mRNA to test for possible degradation. The 20 μL in vitro transcription reaction generally yields ~20 μg of mRNA. 0.5 μg μL^{-1} is the ideal concentration for mRNA injection into zebrafish embryos. Elute the mRNA from the MEGA-clear™ Kit filter cartridge in 25 μL to increase the final concentration of the mRNA.

6. Pigment formation in AB larvae should be inhibited by maintaining the larvae in E3 medium containing 0.2 mM *N*-phenylthiourea (PTU) starting at 24 h post fertilization (hpf). Change this medium daily up to the day of imaging. *Casper* fish, or other pigmentation mutant lines, can be used as alternative.
7. After heat shock, grow up the cells in 250 μL of SOC for at least 1 h at 37 °C shaking. Plate the entire reaction on a LB/ampicillin plate. A successful reaction usually results in a plate with hundreds of colonies. The plate typically consists of clear and opaque colonies. When selecting colonies for overnight growth, select only the clear colonies. These colonies almost always express the construct of interest while opaque colonies never do.
8. Check for positive clones using restriction enzyme digest. We commonly use the PvuII enzyme to test our LR reaction clones.
9. Prepare the Tol2 transposase mRNA using the exact same method described earlier for preparing capped HyPer mRNA with the change of template DNA being the sole variation.
10. We find that screening ~10 injected adults usually results in the discovery of at least one founder fish. When screening F1 embryos for HyPer expression, be sure to obtain clutches of at least 50 embryos each because the transgene will not be transmitted to all progeny.
11. All E3 media should contain 0.2 mg mL^{-1} Tricaine throughout the entire experiment to keep the larvae anesthetized at all times. In experiments with pharmacological compounds, directly add the compound to the anesthetizing E3 medium, and incubate the larvae for 45–60 min prior to wounding. The pharmacological compounds should also be added to the 1% low-melting agarose used for mounting the samples and to the medium overlaying the agarose that prevents it from drying out during the experiment. The effective drug bathing concentrations for zebrafish larvae are typically higher than those used in cell culture or cell-free experiments. Common working concentrations range between ~1 and 100 μM.
12. Multiple larvae can be imaged during the course of a single experiment if a motorized stage is available. We find that

imaging up to four larvae per experiment provides a sufficient amount of time to set up the samples on the microscope for imaging after tail fin tip amputation. Attempting to image more than four samples is not recommended because this usually requires more than the desired 3 min after wounding to set up the samples on the microscope.

13. When mixing in the 1% low-melting agarose, do your best not to move the larvae out of position. This will save you time from having to realign them while the agarose is solidifying.

14. We find that amputating the tail fins up to the tip of the notochord provides us with consistent and comparable results.

15. Ratiometric imaging of HyPer is optimal with excitation wavelengths of 500 and 420 nm and with emission acquired at 516 nm. Because of potential photooxidation of HyPer at strong illuminations, we recommend using neutral density filters and high-sensitivity CCD cameras to reduce the need for strong illumination. Exposure times of up to 300 ms per wavelength can be used. We recommend using high binning (4–8×) because subcellular resolution is not essential.

16. We only use one focal plane for HyPer imaging; consequently, it is important that the larvae tail fins are mounted flat on the glass bottom dish. A tail fin that is not flat will be impossible to focus in a single plane with bright-field illumination. Imaging of these samples will provide inconsistent and incomparable results. We recommend that you do not image these samples.

17. We recommend using 3.5–4 dpf larvae for this experiment. Larvae of this age had the most consistent penetrance and perfusion of exogenous H_2O_2 into the gaping wound.

    In experiments with pharmacological compounds, directly add the compound to the anesthetizing E3 medium, and incubate the larvae for 45–60 min prior to wounding. The pharmacological compounds should be added to the iso-E3 used for equilibration and wounding, the 1% isotonic low-melting agarose used for mounting the samples, and to the H_2O_2 solution that the samples are immersed in during the experiment.

18. Exogenous H_2O_2 can enter the zebrafish larvae through a breach in the epidermis. Amputating the tail fin tips of larvae in their natural hypotonic environment results in rapid wound closure and epithelial barrier reconstitution. Amputation in an isotonic environment blocks wound closure, leaving the wound agape and permitting consistent perfusion of H_2O_2 into the larvae. Prior to amputation, we wash the larvae three times in 1 mL of anesthetizing iso-E3 to equilibrate the larvae to an isotonic environment.

Larvae can stick to the sides of the plastic Eppendorf tube and transfer pipette. Sticking to these surfaces can cause damage to the tail fin epithelium and therefore should be avoided. Always pipette very gently and transfer fish slowly from one medium to another in order to avoid sticking. If larvae get stuck to the Eppendorf tube or transfer pipette, gently pipette the medium up and down until the larvae are knocked off the surface. It is also very important to avoid exposing the fish to air. Any exposure to air will damage the tail fin epithelium of the larvae and render them useless for experiments.

19. Do not amputate the tail fin tips on a plastic petri dish. We find that H_2O_2 perfusion into the wounded tissue is commonly delayed and inconsistent across samples when amputations are done on a plastic surface. We find that amputating tail fins on a glass surface with a sharpened needle knife typically relieves us of these perfusion delays and inconsistencies.

    We recommend aligning the larvae in a pattern that will allow you to amputate a tail fin from one sample and then quickly move on to amputate the next sample without having to realign with a hair loop tool. Minimizing the time difference between tail fin tip amputations synchronizes the endogenous, wound-induced HyPer signal of each sample which is required to subside prior to imaging. Minimal time difference between amputations also results in more consistent and comparable H_2O_2 perfusion into the tissue of the wounded larvae.

20. Always use a sharpened needle knife for tail fin tip amputations. Amputations with a blunt needle knife produce inconsistent and jagged wounds. These wounds are not ideal and result in delayed and inconsistent H_2O_2 perfusion into the wounded tissue across samples.

21. Samples are imaged at ~1 h after tail fin tip amputation. This waiting time is required for the endogenous, wound-induced HyPer signal to subside. We find that it generally takes ~15–20 min to mount larvae in all four chambers of the glass bottom dish. In order to begin imaging at 1 h after amputation, we begin mounting at 40–45 min after amputation. Use this 40–45 min waiting period to prepare the fresh H_2O_2 dilutions that you plan to use for the experiment.

22. To simplify imaging and image processing, mount the larvae in all chambers so that they are all facing the same direction (anterior and posterior ends are aligned).

23. It is best to freshly prepare the H_2O_2 dilutions of interest before each experiment. We recommend preparing the dilutions during the 40–45 min waiting period that is required prior to mounting the larvae. If pharmacological compounds

are being used in an experiment, the compound should be added to the H_2O_2 dilution.

24. Make sure to select a region for background subtraction based on the whole time-lapse, not just a single image, in order to prevent false background subtractions from potential fluorescent objects moving into the field of view during the time-lapse.
25. The adequate thresholding method has to be determined for every imaging setup; we thus recommend to test various thresholding algorithms before making a choice.
26. Depending on the experimental setup, significant wound margin movements can be observed such as closure and contraction during the initial phase of hypotonic wounding and relaxation 30–60 min post isotonic wounding.

Acknowledgments

Research was supported by the NIH/NIGMS grant R01GM099970, an American Asthma Foundation Scholar award to PN, the MSKCC Functional Genome Initiative, and in part through the NIH/NCI Cancer Center Support grant P30CA008748.

References

1. Niethammer P, Grabher C, Look AT, Mitchison TJ (2009) A tissue-scale gradient of hydrogen peroxide mediates rapid wound detection in zebrafish. Nature 459(7249):996–999
2. Razzell W, Evans IR, Martin P, Wood W (2013) Calcium flashes orchestrate the wound inflammatory response through DUOX activation and hydrogen peroxide release. Curr Biol 23(5):424–429
3. Yoo SK, Starnes TW, Deng Q, Huttenlocher A (2011) Lyn is a redox sensor that mediates leukocyte wound attraction in vivo. Nature 480(7375):109–112
4. Yoo SK, Freisinger CM, LeBert DC, Huttenlocher A (2012) Early redox, Src family kinase, and calcium signaling integrate wound responses and tissue regeneration in zebrafish. J Cell Biol 199(2):225–234
5. Rieger S, Sagasti A (2011) Hydrogen peroxide promotes injury-induced peripheral sensory axon regeneration in the zebrafish skin. PLoS Biol 9(5):e1000621
6. Suzuki N, Mittler R (2012) Reactive oxygen species-dependent wound responses in animals and plants. Free Radic Biol Med 53 (12):2269–2276
7. Moreira S, Stramer B, Evans I, Wood W, Martin P (2010) Prioritization of competing damage and developmental signals by migrating macrophages in the Drosophila embryo. Curr Biol 20(5):464–470
8. Enyedi B, Niethammer P (2013) H2O2: a chemoattractant? Methods Enzymol 528:237–255
9. Enyedi B, Niethammer P (2015) Mechanisms of epithelial wound detection. Trends Cell Biol 25(7):398–407
10. Niethammer P (2016) The early wound signals. Curr Opin Genet Dev 40:17–22
11. Love NR, Chen Y, Ishibashi S, Kritsiligkou P, Lea R, Koh Y et al (2013) Amputation-induced reactive oxygen species are required for successful Xenopus tadpole tail regeneration. Nat Cell Biol 15(2):222–228
12. Poole LB (2015) The basics of thiols and cysteines in redox biology and chemistry. Free Radic Biol Med 80:148–157
13. Stöcker S, Van Laer K, Mijuskovic A, Dick TP (2018) The conundrum of H2O2 signaling

and the emerging role of peroxiredoxins as redox relay hubs. Antioxid Redox Signal 28 (7):558–573

14. Redd MJ, Cooper L, Wood W, Stramer B, Martin P (2004) Wound healing and inflammation: embryos reveal the way to perfect repair. Philos Trans R Soc Lond B Biol Sci 359 (1445):777–784
15. Trede NS, Langenau DM, Traver D, Look AT, Zon LI (2004) The use of zebrafish to understand immunity. Immunity 20(4):367–379
16. Bennett CM, Kanki JP, Rhodes J, Liu TX, Paw BH, Kieran MW et al (2001) Myelopoiesis in the zebrafish, *Danio rerio*. Blood 98 (3):643–651
17. Enyedi B, Kala S, Nikolich-Zugich T, Niethammer P (2013) Tissue damage detection by osmotic surveillance. Nat Cell Biol 15 (9):1123–1130
18. Jelcic M, Enyedi B, Xavier JB, Niethammer P (2017) Image-based measurement of H2O2 reaction-diffusion in wounded zebrafish larvae. Biophys J 112(9):2011–2018
19. Gault WJ, Enyedi B, Niethammer P (2014) Osmotic surveillance mediates rapid wound closure through nucleotide release. J Cell Biol 207(6):767–782
20. Enyedi B, Jelcic M, Niethammer P (2016) The cell nucleus serves as a mechanotransducer of tissue damage-induced inflammation. Cell 165 (5):1160–1170
21. Kwan KM, Fujimoto E, Grabher C, Mangum BD, Hardy ME, Campbell DS, Parant JM, Yost HJ, Kanki JP, Chien CB (2007) The Tol2kit: a multisite gateway-based construction kit for Tol2 transposon transgenesis constructs. Dev Dyn 236(11):3088–3089
22. Belousov VV, Fradkov AF, Lukyanov KA, Staroverov DB, Shakhbazov KS, Terskikh AV, Lukyanov S (2006) Genetically encoded fluorescent indicator for intracellular hydrogen peroxide. Nat Methods 3(4):281–286
23. Pase L, Layton JE, Wittmann C, Ellett F, Nowell CJ, Reyes-Aldasoro CC, Varma S, Rogers KL, Hall CJ, Keightley MC, Crosier PS, Grabher C, Heath JK, Renshaw SA, Lieschke GJ (2012) Neutrophil-delivered myeloperoxidase dampens the hydrogen peroxide burst after tissue wounding in zebrafish. Curr Biol 22(19):1818–1824
24. Gauron C, Meda F, Dupont E, Albadri S, Quenech'Du N, Ipendey E, Volovitch M, Del Bene F, Joliot A, Rampon C, Vriz S (2016) Hydrogen peroxide (H_2O_2) controls axon pathfinding during zebrafish development. Dev Biol 414(2):133–141

Chapter 18

Kinetic Analysis of Phagosomal ROS Generation

Sophie Dupré-Crochet, Marie Erard, and Oliver Nüβe

Abstract

Phagosomal ROS generation is critical for our immune defense against microbial infections. Quantitative assessment of phagosomal ROS production is required to understand the complex relationship between the phagocyte and the microbe, in particular for pathogens that resist phagosomal destruction. ROS detection is difficult due to the transient nature of the reactive species and their multiple interactions with the environment. Direct labeling of phagocytic prey with a ROS-sensitive dye allows to target the dye into the phagosome and to follow the kinetics of phagosomal ROS production on a single phagosome base. Here we describe the basic labeling procedure, the quality assessment, and the imaging technique to achieve this kinetic analysis.

Key words NADPH oxidase, Phagosome, Microbial killing, Yeast, ROS-sensitive dyes, Covalent labeling, DCF, $DCFH_2$-SE, Imaging, Flow cytometry

1 Introduction

The primary role of the phagocyte NADPH oxidase, NOX2, is the production of superoxide anion, which is converted to other reactive oxygen species (ROS), inside phagosomes to kill ingested microbes [1, 2]. To this end, the phagocyte assembles and activates the enzyme specifically at the phagosomal membrane immediately upon formation of the phagosome. The importance of its role for host defense in infection becomes evident when NOX2 is absent or nonfunctional as in chronic granulomatous disease [3]. Some pathogens either resist to ROS in the phagosome or inhibit NOX2 activation [4, 5]. To investigate the mechanism of phagosomal ROS production and the associated pathologies, quantification of ROS inside the phagosome is required. From the wide choice of ROS detection methods, only few are compatible with localized, time-resolved detection. The phagosome is particularly challenging because of the harsh intraphagosomal conditions in terms of changing pH, ROS, proteases, and salt concentrations [6].

Ulla G. Knaus and Thomas L. Leto (eds.), *NADPH Oxidases: Methods and Protocols*, Methods in Molecular Biology, vol. 1982, https://doi.org/10.1007/978-1-4939-9424-3_18,

Covalent labeling of the phagocytic prey with a ROS-sensitive dye assures that the dye is localized inside the phagosome. Dichlorodihydrofluorescein ($DCFH_2$) is a fluorescein derivative that becomes fluorescent upon oxidation. Dichlorofluorescein (DCF) is the main oxidation product [7, 8]. $DCFH_2$ is not specific and should be considered as a general indicator of ROS (*see* **Notes 1** and **2**). We describe a general method to attach $DCFH_2$ to heat-killed yeast as well as other particles containing free amine groups. The procedure involves preparation of the particles, the labeling reaction, and a second reaction to de-protect the dye. Since the dye is sensitive to chemical- and photooxidation, all steps need to be performed at low oxygen tension and as little light exposure as possible. The quality of the preparation is best evaluated by flow cytometry. $DCFH_2$-labeled yeast can be opsonized and used for time-resolved fluorescence microscopy or flow cytometry [9–11]. Double or triple labeling of the particles with several dyes allows the detection of multiple parameters with the same particle [12, 13]. We describe here the case of a reference dye that is rather stable in the phagosomal environment, which allows the independent quantification of phagocytosis. We will also provide a basic protocol for live cell imaging of phagosomal ROS production. The imaging conditions strongly depend on the features of the microscope. A number of controls are required to find the appropriate settings.

2 Materials

1. 100 mM sodium bicarbonate buffer, 0.84 g $NaHCO_3$ dissolved in 100 mL of water and then adjusted with 1 M HCl at pH 8.3 to suspend yeast during the reaction.
2. Clean and dry DMSO.
3. 1.5 M hydroxylamine solution at pH 8.5 prepared the same day.
4. $DCFH_2$-SE (2′,7′-dichlorodihydrofluorescein diacetate, succinimidyl ester, Molecular Probes D-2935) 5 mg, resuspended in 0.5 mL DMSO, sufficient for 2×10^9 yeast (check appropriate number of particles when used with other microorganisms or beads) (*see* **Note 3**).
5. Alexa405-SE (Molecular Probes A30000) 1 mg, resuspended in 0.1 mL DMSO, divide in five aliquots, evaporate DMSO in SpeedVac, store at −20 °C for approximately 6 months, and resuspend each aliquot in 20 μL DMSO just before use.
6. 1 mL syringes, 23G syringe needles.
7. Inert gas (N_2 or argon) from bottle or wall outlet.
8. Dark room with dim red light if possible.

9. Centrifuge for yeast (1100 × *g*).
10. Slowly rotating wheel or shaker for tubes during reaction.
11. Human serum (H4522, Sigma).
12. Chamber for live cell microscopy.
13. Flow cytometry with 488 nm laser.
14. Wide-field inverted fluorescence microscope with appropriate filters for GFP or FITC detection (e.g., excitation at 480/15 nm, 495-nm beam splitter, and emission at 525/50 nm) and a fast, sensitive camera driven by a software for time-lapse acquisition (*see* **Note 4**).

3 Methods

3.1 Particle Labeling with $DCFH_2$

The following protocol as outlined in Fig. 1 has been developed for heat-killed yeast (*Saccharomyces cerevisiae*). We have also used this protocol for zymosan and *Candida glabrata*. In the literature, a similar labeling has been used for living microorganisms and for inert polystyrene beads [12, 14, 15] (*see* **Note 5**):

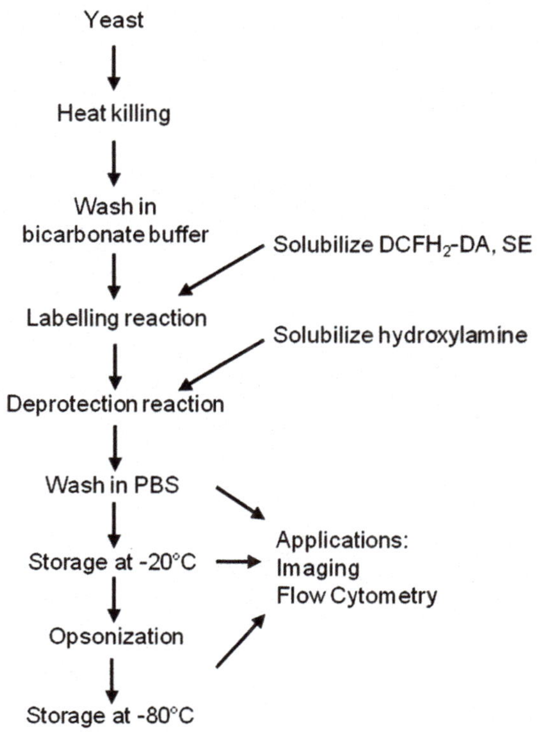

Fig. 1 Workflow of the labeling procedure. Step-by-step representation of the labeling procedure

1. Yeasts, *S. cerevisiae*, are grown in YPD medium at 30 °C at 220 rotations per minute until they reach an exponential growth rate (absorbance at 600 nm between 0.2 and 0.4). They are spun down at 1100 × *g* and washed twice with PBS. They are resuspended at approximately 10^9 cells/mL and heat killed by boiling at >95 °C for 60 min. Clumps are removed by filtration through filter paper of coarse porosity. Heat-killed yeasts are washed in H_2O, adjusted to 10^9 particles/mL, and stored at −20 °C. This preparation is stable for several years (*see* **Note 6**).
2. For labeling, wash yeasts twice in bicarbonate buffer, and centrifuge at 1100 × *g*.
3. Degas 10 mL bicarbonate buffer with nitrogen gas in a closed 15 mL Erlenmeyer flask for 60 min. The flask is closed with a rubber gasket pierced with two syringe needles. One of the needles is connected to the gas outlet and reaches into the buffer solution; the other does not reach into the buffer and serves for the escape of the gas from the flask.
4. Wash yeast once with degassed bicarbonate buffer.
5. Resuspend yeast in degassed bicarbonate buffer at 10^9/mL, and transfer into a black Eppendorf vial sealed with a rubber gasket.
6. Degas Eppendorf vial with syringe needles through gasket.
7. Add $DCFH_2$-SE to yeast (250 μL dye solution for 10^9 yeast) by injecting through the gasket with a small (1 mL) syringe (*see* **Note 7**).
8. Incubate for 1 h at room temperature with constant gentle agitation, and protect from light at all times (*see* **Note 8**).
9. Prepare the solution of hydroxylamine 1.5 M during the incubation time.
10. Prepare 25 mL at 210 mg/mL of hydroxylamine in distilled water, and adjust pH to 8.5 with a 5 M NaOH solution while stirring the solution. The solution will be at 3 M hydroxylamine and shall be diluted 1:2 to obtain the final solution at 1.5 M (*see* **Note 9**).
11. Add hydroxylamine solution directly to the yeast suspension to obtain 100 mM final concentration. Incubate for 45 min at room temperature. The hydroxylamine solution de-protects the ester groups of $DCFH_2$ (Fig. 2) and neutralizes dye molecules that are not bound to the particles.
12. The yeast needs to be washed intensively to remove unbound dye. Centrifuge for 5 min at 1100 × *g*, and wash 3 times with degassed PBS. Recover the first supernatant for later testing (*see* **Note 10**).

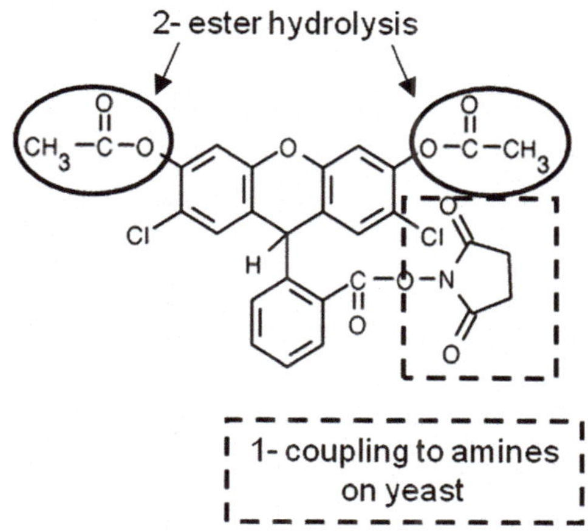

Fig. 2 Formula of 2′,7′-dichlorodihydrofluorescein diacetate, succinimidyl ester (OxyBURST® Green DCFH2-DA, SE) (Adapted from Molecular Probes, Thermo Fisher). The dashed box indicates the site of amine labeling in the first reaction. The full circles indicate the protective ester groups that need to be removed with hydroxylamine in the second reaction

13. Incubate in degassed PBS overnight at room temperature with gentle agitation.
14. Add three more washes with PBS and recover the last supernatant for testing.
15. Store aliquots in light-protected vials under inert gas if possible at −20 °C. Practically, aliquots of 20 μL with 10^9 yeasts/mL in Eppendorf tubes are prepared and then briefly flushed with N_2 and then closed.
16. Preparation for phagocytosis requires opsonization (in most cases). The labeled yeasts are diluted ten times and then opsonized with human serum. An equal volume of yeast and serum should be mixed (e.g., 50 μL of yeast at 10^9 yeasts/mL and 50 μL of human serum). The sample is then incubated for 1 h at 37 °C, washed twice with PBS, and then resuspended in a suitable buffer for phagocytosis (*see* Subheading 3.3). The label does not seem to affect the outcome in terms of phagocytosis. Opsonized yeasts are stored at −80 °C. Even at that temperature, the opsonization is not stable, and opsonized particles should not be stored for more than 1 month.

3.2 Assessment of Labeling Quality

The quality of the labeling is best assessed by flow cytometry. All particles should be labeled, and the fluorescence intensity of the population should be homogeneous (Fig. 3). However, a factor of 10 in fluorescence intensity between the weakest and the strongest particles is common. Treatment of the particles with H_2O_2 (50 mM) and HRP (50 U/mL) for 24 h at 22 °C shifts the

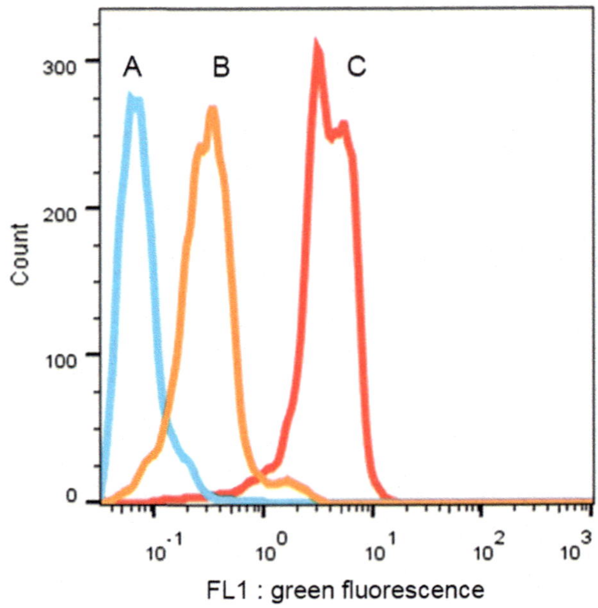

Fig. 3 Analysis of labeled particles by flow cytometry. Comparison of fluorescence intensity (excitation 488 nm, emission 536 $\pm$ 20 nm) of unlabeled yeast particles (A, blue), non-oxidized $DCFH_2$-labeled particles (B, orange), and $DCFH_2$-labeled particles oxidized by treatment with 50 mM H_2O_2, 50 U/mL HRP for 24 h (C, red)

whole population to higher fluorescence [7]. For freshly labeled yeast, DCF fluorescence shifts by a factor of $\approx$10 from untreated to H_2O_2/HRP-treated particles (*see* **Note 11**). After double labeling, the fluorescence for both dyes detected by flow cytometry should be presented on a two-parameter dot-plot. The points should form a cloud around a diagonal from left bottom to upper right [6] indicating that particles that are weakly labeled with one dye tend to be also weakly labeled with the other dye. Strongly labeled particles are highly fluorescent with both dyes.

3.3 Time-Lapse Imaging of Phagocytosis of $DCFH_2$-Labeled Particles

Opsonized labeled yeasts are readily internalized by primary phagocytic cells and by neutrophil-like or macrophage-like cell lines. We have mostly used neutrophil-like, differentiated PLB985 cells (Fig. 4). The actual imaging conditions depend strongly on the microscope setup. We will first describe a standard protocol and then list appropriate control experiments.

1. Use an inverted fluorescent microscope equipped with filters for GFP or FITC detection (e.g., excitation at 480/15 nm, 495-nm beam splitter, and emission at 525/50 nm) and a sensitive rapid camera for fluorescence monitoring (*see* **Note 12**).
2. Freshly prepare phagocytic cells (10^6 cells/mL) in a physiological buffer (e.g., 10 mM HEPES, pH 7.4, 140 mM NaCl, 5 mM

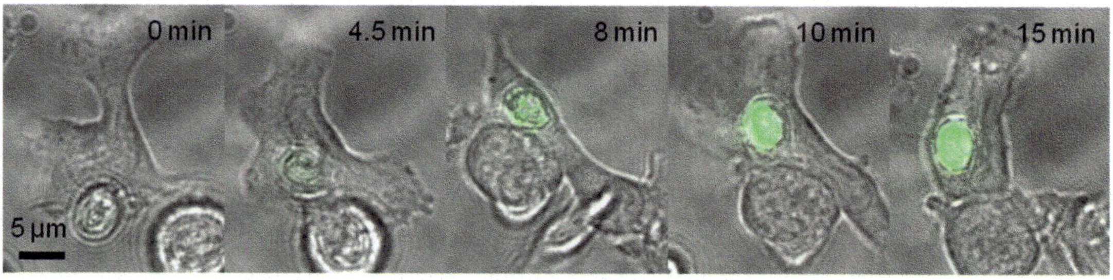

Fig. 4 Images of DCF-labeled yeast in phagosomes. Superimposed images of transmitted light and green fluorescence showing the increase of phagosomal DCF fluorescence over time. Time 0 is near the closure of the phagosome. A neutrophil-like PLB985 cell has internalized a $DCFH_2$-labeled, opsonized yeast particle

KCl) containing divalent ions (1 mM Mg^{2+}, 2 mM Ca^{2+}), 1 mM glucose, and a low concentration of protein (e.g., 1% human serum). Keep cells on ice.

3. Put cells on 25 mm coverslips fixed in a heated chamber (30–37 °C, below 30 °C no phagocytosis occurs) in an inverted microscope in 100 µL buffer, and let them settle on the coverslip. Add 300 µL of buffer in order to avoid dehydration.
4. Add 10 µL $DCFH_2$-labeled yeast (at 10^8yeast/mL) to the coverslip, and let them settle on the phagocytes.
5. Choose a field with phagocytes surrounded by the yeasts (*see* **Note 13**).
6. Start time-lapse acquisition using an established protocol. Typically take images with 10 ms exposure at an interval of 10 s for a period of 30 min.
7. As the phagocyte itself and the phagosome within the phagocyte move in every direction, readjustment of the *z*-focus is often required. Therefore, the images have to be monitored during acquisition and the microscope focus adjusted accordingly.
8. End the recording if (1) the phagosome is lost from focus or from the field of view, (2) significant bleaching occurs, and (3) photooxidation of $DCFH_2$ renders all yeast particles fluorescent, even those that are not internalized.
9. Analyze data by choosing an ROI for each phagosome, adjusting this ROI to the *x*-*y*-movement throughout the recording, and measure mean gray value of the ROI for each image. Automated analysis is possible with the Multi Dimensional Motion Analysis module from MetaMorph software (Molecular Devices) or similar analysis modules.

3.4 Control Steps for Imaging Protocol

1. A number of controls are required to optimize this protocol for each microscopy setup. The appropriate exposure time for each image depends on the intensity of the excitation light, the

quality of the optical path, and the sensitivity of the camera. DCF is sensitive to photobleaching. $DCFH_2$ is also sensitive to photooxidation that renders a fluorescent product. 1. To optimize between these two opposing constraints, it is useful to start with $DCFH_2$-labeled yeast alone to see how much light they support before photooxidation occurs.

2. Then, DCF yeasts should be prepared by deliberate oxidation (50 mM H_2O_2 and 50 U/mL HRP for 24 h) [7]. These oxidized yeasts will indicate the exposure time to obtain a signal with a satisfactory signal-to-noise ratio under the microscope.
3. Repeated exposure of these DCF yeasts will also indicate whether photobleaching occurs under the chosen conditions. Photooxidation is not a linear phenomenon, for example, 10×300 ms exposure does not have the same effect as 5×600 ms exposure.
4. Control experiments either with cells that do not have a functional NADPH oxidase (PLB $gp91^{phox}$ knockout cells) or in the presence of an inhibitor of the NADPH oxidase will confirm that any increase in DCF fluorescence during the course of phagocytosis is indeed due to phagosomal ROS production.
5. The rapid and strong ROS production by most phagocytes may oxidize all available dye molecules on the particle. Thus, the increase in DCF fluorescence may end due to dye saturation instead of the end of ROS production by the phagocyte [6, 7] (*see* **Note 14**).

4 Notes

1. $DCFH_2$ in its various forms is widely used since the 1960s [16]. It is by no means a perfect dye. It has poor specificity, it is photoactivated and photobleached, and it has a moderate pH sensitivity. In the phagosome, $DCFH_2$ most likely detects H_2O_2 and HOCl [7]. Today, however, it is the only ROS-sensitive dye commercially available as a succinimidyl ester. It would be extremely useful to have other dyes available for covalent labeling. Amplex Red would be a good candidate. Hydrocyanines are an interesting option because they exist in several colors and the oxidized variants are frequently used for protein labeling. Their succinimidyl esters are commercially available (Lumiprobe, https://www.lumiprobe.com; Biotium, https://biotium.com). It may be possible to label particles with cyanine dyes and then reduce them to obtain ROS-sensitive particles [17]. We are not aware that this has been performed to date.

2. Genetically encoded biosensors could be an alternative to chemical labeling [18, 19]. Biosensors based on fluorescent proteins such as Hyper or roGFP are frequently used in redox biology. For measuring ROS production in the phagosome, the sensor should be placed on or near the surface of the living microorganism. The synthesis of the bacterial or fungal cell wall is a complicated issue that is only partially understood. Targeting of a biosensor to the cell wall should be possible but has not been reported. The known biosensors are reversible and interact with the cellular antioxidant machinery. It is not clear whether they would be appropriate for ROS detection in the phagosome. The harsh conditions of the phagosome (pH, proteases, ROS, salt) may substantially alter the photophysical properties of these sensors.
3. The package size of $DCFH_2$-DA, SE (5 mg), is larger than required for most phagocytosis applications. Weighing the dye to prepare aliquots most likely causes some oxidation. Taking up the whole package into a solvent (DMSO), preparing aliquots, and evaporating the solvent may be an option. We have not assessed whether this is feasible without a major loss of dye or reactivity.
4. Particles such as yeasts are moving up and down in the *Z*-direction along with the deformations of the phagocytes. In a wide-field microscope, there is no spatial selection along the *Z*-axis, and the fluorescence is integrated over a depth of several microns. On the contrary, with a confocal microscope, the axial sectioning leads to the detection of a single fluorescent plane. In confocal microscopy, the phagosome constantly moves in and out of the observation plane and renders any quantification difficult. If the reader wants to use a confocal microscope, an option is to record a *Z*-stack composed of a few planes at each time-point and to calculate the mean gray value of the yeast fluorescence for each plane. The highest value, corresponding approximately to the equatorial plane, could be taken for further quantification.
5. Labeling of live cells is possible but faces two major problems. The actual labeling reaction in bicarbonate buffer at pH 8–9 is not toxic to many bacteria, yeast, or even *Leishmania* (unpublished observation). The solvent DMSO may be toxic, but its concentration in the reaction medium can be reduced by using highly concentrated reagent solutions. On the contrary, the de-protection reaction with hydroxylamine is toxic to most microorganisms. Some references in the literature do not mention the second reaction. Avoiding the hydroxylamine step may be necessary to keep the cells alive, but the amount of "active" dye on the organisms may be much smaller. The second problem concerns the cost of the experiment. $DCFH_2$-DA, SE

(D2935 Molecular Probes), is sold in packages of 5 mg, a quantity enough to label >10^9 cells. Preparing aliquots of the reactive dye is difficult as light and oxygen should be avoided as much as possible. For dead particles, it is best to use the entire package in one reaction and store the labeled particle. Living bacterial or fungal cells cannot be easily stored. They may either die or proliferate. We do not know whether storage in 25% glycerol at −80 °C is compatible with method. So, the labeling reaction has to be performed immediately before the experiment, which costs an entire package of the reagent and several hours of work (*see also* **Note 2**).

6. Several techniques for "heat killing" of yeast exist. We have used the same technique for two types of yeast (*S. cerevisiae* and *C. glabrata*) (*see* Subheading 3.1, **step 1**) and observed similar labeling results. An advantage of this method is the large number of sites available for the labeling. The presence of a large number of attached dye molecules available for oxidation increases the amount of ROS detected before the saturation of the dye occurs and thus increases the detection window accessible for monitoring the labeled prey. Other heat-killing protocols, namely, lower temperature such as 70 °C for 10 min [20], may change the number of available amines for this reaction. The killing method also will change the surface of the yeast, which may affect the phagocytic process.
7. Labeling with multiple dyes is possible. In particular, it is possible to label the yeast with Alexa 405-SE (Invitrogen A30000). Its fluorescence may serve as a reference in a ratiometric approach [6, 13]. The fluorescence of Alexa 405 is quite stable with respect to pH or oxidation even within the phagosome. A solution of Alexa 405-SE (10 mg/mL in DMSO) can be added to the yeast together with $DCFH_2$-SE. Typically, 10 μL of Alexa 405-SE solution is added to 10^9 yeasts in 1 mL at the same time as $DCFH_2$-SE addition. Other succinimidyl ester dyes react under the same experimental conditions, but whether they are stable inside the phagosome needs to be tested for each dye.
8. Protection of $DCFH_2$ against oxygen and light is tedious but greatly improves the outcome. Any oxidation of the dye during the reactions, washing steps, and storage will increase the background due to DCF being bound on the particle (*see* paragraph on quality assessment).
9. Hydroxylamine is toxic; wear gloves and eye protection, and dispose unused solution as chemical waste. The solution needs to be freshly prepared, at best during the first reaction period. The pH adjustment of this solution takes time. Concentrated, 5 mM NaOH needs to be added dropwise under stirring while

monitoring pH with an electrode. If the solution is not properly stirred, pH may overshoot above 8.5 and then needs to be lowered by adding HCl.

10. At the end of the labeling procedure, yeasts are rose colored, and the supernatant is light yellow due to unbound, oxidized dye. By oxidizing the dye (10 mM H_2O_2 + 10 U/mL HRP for 1 h) and measuring its absorbance (A), the concentration (C) can be determined.

$$C = A/(\varepsilon \times 1).$$

l is the optical path length of the measuring cuvette, usually 1 cm. The extinction coefficient (ε) of unbound DCF at pH 7.4 and 488 nm is 94,000 M^{-1} cm^{-1} [7]. The dye should be in excess to insure optimal labeling; thus, at least some of the dye should remain unbound. Furthermore, at the end of the washing steps, no unbound dye should be detectable to make sure that the remaining dye is firmly attached to the particles.

11. With time, $DCFH_2$ oxidizes even under ideal storage conditions. It is worthwhile to test the basal and H_2O_2/HRP-induced fluorescence by flow cytometry when older particles are used.

12. DCF particles are strongly fluorescent in the middle of the visual spectrum. Co-detection with fluorescent proteins is difficult since the proteins tend to have weaker fluorescence. Furthermore, excitation and emission spectra of DCF as well as fluorescent proteins are overlapping. Visualizing proteins tagged with mCherry in the same cell as fully oxidized DCF particles has been possible in some but not all cases. Co-detection of DCF with the calcium indicator fura-2 is possible due to their clearly distinct excitation wavelengths [9]. The red fluorescent pH indicator pHrodo is suitable for co-detection with DCF, even on the same particle [12].

13. **Steps 5** and 7 require training. An experienced researcher will recognize cells that are likely to take up a particle and he/she will lose less cells due to bad focus adjustment.

14. The amount of label on each particle depends on the number of accessible amine groups. Therefore, the type and size of the particle and its preparation method have a strong influence on the maximal signal that can be obtained. Phagocytes are strong ROS producers and can rapidly saturate the entire label on the particles.

Acknowledgments

This work was supported by grants from the FRM (Foundation for Medical Research, DCM20121225747). We wish to thank our past colleagues who have contributed to this method and Elodie Hudik for her excellent technical assistance.

References

1. Dupre-Crochet S, Erard M, Nusse O (2013) ROS production in phagocytes: why, when, and where? J Leukoc Biol 94:657–670
2. Nauseef WM (2008) Biological roles for the NOX family NADPH oxidases. J Biol Chem 283:16961–16965
3. O'Neill S, Brault J, Stasia MJ, Knaus UG (2015) Genetic disorders coupled to ROS deficiency. Redox Biol 6:135–156
4. Imlay JA (2008) Cellular defenses against superoxide and hydrogen peroxide. Annu Rev Biochem 77:755–776
5. Seider K, Heyken A, Lüttich A, Miramón P, Hube B (2010) Interaction of pathogenic yeasts with phagocytes: survival, persistence and escape. Curr Opin Microbiol 13:392–400
6. Nault L, Bouchab L, Dupré-Crochet S, Nüsse O, Erard M (2016) Environmental effects on ROS-detection - learning from the phagosome. Antioxid Redox Signal 25:564–576
7. Tlili A, Dupré-Crochet S, Erard M, Nüße O (2011) Kinetic analysis of phagosomal production of reactive oxygen species. Free Radic Biol Med 50:438–447
8. Wardman P (2007) Fluorescent and luminescent probes for measurement of oxidative and nitrosative species in cells and tissues: progress, pitfalls, and prospects. Free Radic Biol Med 43:995–1022
9. Dewitt S, Laffafian I, Hallett MB (2003) Phagosomal oxidative activity during beta2 integrin (CR3)-mediated phagocytosis by neutrophils is triggered by a non-restricted Ca2+ signal: Ca2+ controls time not space. J Cell Sci 116:2857–2865
10. Tlili A, Erard M, Faure MC, Baudin X, Piolot T, Dupre-Crochet S, Nusse O (2012) Stable accumulation of p67(phox) at the phagosomal membrane and ROS production within the phagosome. J Leukoc Biol 91:83–95
11. Song ZM, Bouchab L, Hudik E, Le Bars R, Nüsse O, Dupré-Crochet S (2017) Phosphoinositol 3-phosphate acts as a timer for reactive oxygen species production in the phagosome. J Leukoc Biol 101:1155–1168
12. Bernardo J, Long HJ, Simons ER (2010) Initial cytoplasmic and phagosomal consequences of human neutrophil exposure to *Staphylococcus epidermidis*. Cytom Part A 77A:243–252
13. Russell DG, VanderVen BC, Glennie S, Mwandumba H, Heyderman RS (2009) The macrophage marches on its phagosome: dynamic assays of phagosome function. Nat Rev Immunol 9:594–U84
14. Vanderven BC, Yates RM, Russell DG (2009) Intraphagosomal measurement of the magnitude and duration of the oxidative burst. Traffic 10:372–378
15. Kamen LA, Levinsohn J, Cadwallader A, Tridandapani S, Swanson JA (2008) SHIP-A increases early oxidative burst and regulates phagosome maturation in macrophages. J Immunol 180:7497–7505
16. Chen X, Zhong Z, Xu Z, Chen L, Wang Y (2010) 2′,7′-Dichlorodihydrofluorescein as a fluorescent probe for reactive oxygen species measurement: forty years of application and controversy. Free Radic Res 44:587–604
17. Kundu K, Knight SF, Willett N, Lee S, Taylor WR, Murthy N (2009) Hydrocyanines: a class of fluorescent sensors that can image reactive oxygen species in cell culture, tissue, and in vivo. Angew Chemie Int Ed Engl 48:299–303
18. Erard M, Dupre-Crochet S, Nusse O (2018) Biosensors for spatiotemporal detection of reactive oxygen species in cells and tissues. Am J Physiol Integr Comp Physiol 314(5): R667–R683. https://doi.org/10.1152/ajpregu.00140.2017
19. Schwarzlander M, Dick TP, Meyer AJ, Morgan B (2016) Dissecting redox biology using fluorescent protein sensors. Antioxid Redox Signal 24:680–712
20. Seider K, Brunke S, Schild L, Jablonowski N, Wilson D, Majer O, Barz D, Haas A, Kuchler K, Schaller M, Hube B (2011) The facultative intracellular pathogen *Candida glabrata* subverts macrophage cytokine production and phagolysosome maturation. J Immunol 187:3072–3086

Chapter 19

Imaging Intestinal ROS in Homeostatic Conditions Using L-012

Emer Conroy and Gabriella Aviello

Abstract

Reactive oxygen species (ROS) are critical redox regulators of cellular dynamics controlling homeostasis. Although numerous fluorescent probes are currently available to measure ROS in cell-based assays, the short-lived nature of these molecules renders their detection challenging in more complex biological systems, such as the gastrointestinal tract in vivo. However, in the past decade, significant progress has been made in the development of novel imaging technologies and probes, facilitating ROS quantification with high sensitivity, selectivity, and temporal resolution. The IVIS Spectrum (PerkinElmer) is an optical imaging system for small animal imaging allowing precise and noninvasive visualization of fluorescent or bioluminescent signals. Here, we describe a reproducible and comprehensive method for the measurement of physiological intestinal NADPH oxidase-derived ROS by using the chemiluminescent probe L-012. Using transgenic mice deficient in Nox isoforms expressed in the intestinal mucosa, we delineate the contribution of gut epithelial *versus* immune cell NADPH oxidase activity in homeostatic conditions. We also discuss L-012 probe specificity and potential alternatives for in vivo studies.

Key words L-012, IVIS, NADPH oxidase, NOX1, NOX2, p22^{phox}, Intestine, Microbiota, Inflammatory bowel disease

1 Introduction

The intestinal epithelium constantly faces the unique challenge of interacting with nearly 100 trillion commensal bacteria, which symbiotically contribute to the host's health. Disruption of the gut epithelial barrier by noxious stimuli (e.g., diet, pathogens, and chemicals) perturbs immune and stromal mechanisms regulating tolerance and responsiveness against microbes and induces an inflammatory response aimed to eradicate the cause of damage [1]. Reactive oxygen species (ROS) play a key role in maintaining intestinal homeostasis by regulating pathogen killing, host-microbe interactions, and tissue repair mechanisms [2], and therefore, the

Emer Conroy and Gabriella Aviello contributed equally to this work.

Ulla G. Knaus and Thomas L. Leto (eds.), *NADPH Oxidases: Methods and Protocols*, Methods in Molecular Biology, vol. 1982, https://doi.org/10.1007/978-1-4939-9424-3_19, © Springer Science+Business Media, LLC, part of Springer Nature 2019

ability to measure ROS in vivo is crucial for the study of these dynamics. The primary sources of superoxide ($O_2^{\cdot -}$) and hydrogen peroxide (H_2O_2) in the intestine are membrane-bound multimeric complexes, named nicotinamide adenine dinucleotide phosphate (NADPH) oxidases (NOX/DUOX). NOX1 and DUOX2 are expressed in the intestinal epithelium, while NOX2 is expressed in *lamina propria* innate immune cells. Their function and localization along the gastrointestinal tract is well documented, while the expression of other NADPH oxidases, e.g., NOX4 in fibroblasts and smooth muscle cells or DUOX1 in lymphocytes, has also been reported but remains less well characterized [2]. Loss-of-function mutations in genes encoding for proteins forming the phagocytic NOX2 complex are the cause of chronic granulomatous disease (CGD), an inherited immunodeficiency characterized by severe and relapsing life-threatening bacterial and fungal infections [3]. Patients with CGD can also develop comorbid intestinal manifestations, including diarrhea, pancolitis, and perianal disease [4]. The important role of NADPH oxidase-derived ROS in regulating intestinal pathophysiology is demonstrated by the ~40% of CGD patients developing an inflammatory condition mimicking Crohn's disease [5] and also by the recent identification of inactivating missense *NOX1* and *DUOX2* variants in very early onset inflammatory bowel disease (VEO-IBD) patients [6–8].

The lifetime of ROS in biological systems varies from nanoseconds to seconds depending on the cellular environment and the chemical identity of ROS. Since ROS signaling is regulated in a spatiotemporal manner, an ideal probe should be (a) fast, (b) highly sensitive (up to nanomolar concentrations), (c) able to produce a linear response for quantification purposes, (d) devoid of background noise, and (e) cell permeable for intracellular loading and proper subcellular compartmentalization. In the past two decades, fluorescent dyes have been commonly utilized to measure ROS levels primarily but not exclusively in cell-based assays. The majority of these indicators fall into two classes: small-molecule dyes (e.g., DCFH-DA, DHE, mitoSOX) and genetically encoded fluorescent redox probes (e.g., HyPer, rxYFP, roGFP). The first class comprises molecules diverse in structure and degree of sensitivity and selectivity that simply change their fluorescence in response to redox dynamics [9], while the second class includes probes that are introduced into the cell as DNA and will react to ROS as expressed proteins [10]. The major advantage of genetically encoded fluorescent redox probes over ROS dyes is their reversible oxidation and ability to record compartmentalized ROS due to addition of targeting signal sequences, thus allowing real-time dynamic measurements; however genetic manipulation is not always possible and the creation of transgenic mice stably expressing roGFP1-based redox sensors has just recently been realized [11].

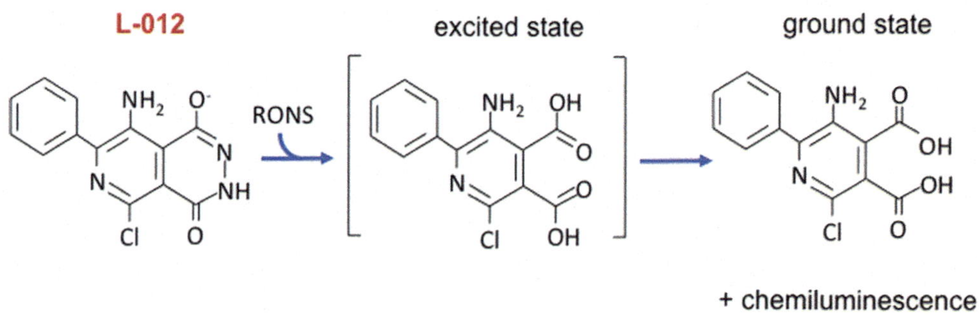

Fig. 1 Chemical structure of L-012 and proposed chemiluminescent emission following reaction with reactive oxygen and nitrogen species (adapted from [15])

Chemiluminescence is a noninvasive, inexpensive, and highly sensitive alternative to fluorescence for the study of ROS generation in biologically complex systems, including living animals. The basic principle of chemiluminescence is the measurement of emitted light, counted in photons, due to the reaction between chemical reagents, such as luminol or lucigenin and ROS. In vitro, luminol measures both extracellular and intracellular $O_2^{\cdot -}$, H_2O_2 and hydroxyl radical, while lucigenin seems to be more specific for extracellular $O_2^{\cdot -}$. Redox cycling of lucigenin is a major limiting factor for its application to detect and quantify superoxide. In case of luminol and analogs, typically a catalyst such as the enzyme horseradish peroxidase (HRP) is used to trigger oxidation; however, HRP does not cross cell membranes, and substrate/HRP-dependent chemiluminescence can only be used to measure ROS in the extracellular milieu. In vivo these two luminescent substrates have been used to distinguish different phases of the inflammatory process [12]. L-012 (8-amino-5-chloro-7-phenyl-pyrido[3,4-d] pyridazine-1,4(2H,3H)dione) (Fig. 1) is a more sensitive chemical analog of luminol, showing higher luminescence yield than luminol and lucigenin upon reaction with ROS. L-012 has been specifically employed to detect $O_2^{\cdot -}$ derived from NOX NADPH oxidases, and it shows a stable signal, well distributed in the mouse under pro-inflammatory conditions, including bacterial infection [13], LPS-induced sepsis [14], and liver disease [15]. It has been reported that L-012 does not react directly with $O_2^{\cdot -}$ in vitro but rather is oxidized by HRP/H_2O_2 to the corresponding radical. The latter reacts then with $O_2^{\cdot -}$ forming an endoperoxide that decomposes to an excited luminescence emitting an intermediate state (Fig. 1) [16]. Thus, one-electron oxidation of L-012 to its radical will serve as "activation" of the probe to react with $O_2^{\cdot -}$ and produce chemiluminescence. How this chemical reaction translates into the in vivo situation is not yet clear. In the presence of peroxynitrite-derived radicals, L-012 can also be used to measure certain reactive nitrogen species, in particular peroxynitrite

($ONOO^-$), and therefore, it is often employed to detect both reactive oxygen and nitrogen species (RONS) (refer to Subheading 3.4 for details). Regardless of the chemical identity of the oxidant, the L-012 probe can detect one-electron oxidants (or peroxidatic activity) in biological systems.

Using transgenic mice deficient in Nox isoforms expressed in the intestinal mucosa, we describe here a sensitive and reproducible method to monitor the physiological production of NADPH oxidase-derived oxidants in vivo and ex vivo, and we delineate the contribution of epithelial *versus* myeloid NADPH oxidase function in homeostatic conditions. Since in the gut, epithelial cells and phagocytes generate $O_2^{\cdot -}$ through Nox1 and Nox2, respectively, partial Nox inactivation will be achieved using *Nox1*$^{-/-}$or *Cybb*$^{-/-}$ (Nox2 knockout) mice, while total Nox inactivation will be obtained using mice deficient in the Nox1–4 partner protein p22^{phox} (*Cyba*$^{-/-}$). The IVIS Spectrum (PerkinElmer) is a very sensitive optical imaging system for small animals in a preclinical setting allowing noninvasive longitudinal studies of disease progression and gene expression patterns. Following injection, the probe L-012 is oxidized in presence of ROS/RONS, and the emitted light after some absorption and scattering by tissue can be detected and quantified by the IVIS Spectrum giving a measure of the in vivo levels of RONS.

2 Materials

1. Mice from any background (e.g., C57BL/6J or BALB/c) can be used. Here we used transgenic mice with impaired ROS production due to the genetic ablation of Nox1 (*Nox1*$^{-/-}$), Nox2 (*Cybb*$^{-/-}$) (Jackson Laboratories), or p22^{phox} (*Cyba*$^{-/-}$) (kindly provided by Ulla G. Knaus) and their wild-type (WT) littermates, all on a C57BL/6 background.
2. Individually ventilated cages (IVC) set in specific pathogen-free conditions for a better control of housing conditions (e.g., microbiota stability).
3. Isoflurane (Vetflurane®, Virbac).
4. IVIS Spectrum (PerkinElmer) integrated with gas anesthesia and oxygen supplies.
5. Living Image® 4.5.2 (64-bit) software for data acquisition and analysis.
6. 8-Amino-5-chloro-2,3-dihydro-7-phenyl-pyrido[3,4-d]pyridazine (L-012) sodium salt (Tocris).
7. Sterile water or saline (0.9% sodium chloride) for injection.
8. Dissection kit, e.g., scissors, tweezers, and black cards.

9. Black tape.
10. Hair shaver or depilation cream.

3 Methods

The primary component of the IVIS Spectrum, the charge-coupled device (CCD) camera, is cooled to −90 °C reducing the dark current and associated noise arising from thermal energy in the CCD detector. Without cooling, the dark current and noise in the CCD detector would increase during the long exposures necessary to detect low-level light reducing the signal to noise ratio. Mice are imaged on a heated platform in a light-excluding chamber with integrated gas anesthesia. The resulting image is a 2-D low-resolution image of the optical luminescent signal from the animal overlaid with a gray scale photograph (a light is switched on momentarily) to facilitate co-localization of optical signal from the probe with the animal. The luminescent image is displayed as a heat map, typically using the rainbow color scale going from blue-low to red-high with maximum and minimum values of the scale being set automatically or manually as desired. For optimum acquisition of a chemiluminescent image, it is advised to conduct an initial pilot study to determine kinetics, time to image postinjection, and system parameters. Kinetics will vary depending on injection route, animal weight and metabolism, animal handling, and anesthesia conditions. These should all be determined during a pilot study and be kept consistent to reduce variation. Time to image postinjection will be determined by kinetics, and system parameters will be set based on the magnitude of the signal. A group number of 3–5 mice may be sufficient to conduct a pilot study.

3.1 Pilot Study

1. Initialize the IVIS Spectrum system and wait for the CCD camera to cool if necessary (*see* **Note 1**).
2. Set exposure time to 1 s initially, but this should be varied depending on quality of image obtained (*see* **Note 2**), F/stop to 1 (*see* **Note 3**), and binning to medium (*see* **Note 4**).
3. Shave or chemically depilate the mouse abdomen under anesthesia before imaging (*see* **Note 5**).
4. Prepare a fresh solution of 2 mg/mL L-012 dissolved in sterile water or saline (*see* **Note 6**), and administer intraperitoneally (i.p.) at the dose of 20 mg/kg by giving 0.1 mL/10 g of body weight (*see* **Note 7**).
5. Anesthetize mice at an induction dose of 4–5% isoflurane and maintenance dose of 2% isoflurane (oxygen flowmeter1000–1500 mL/min) in the induction chamber,

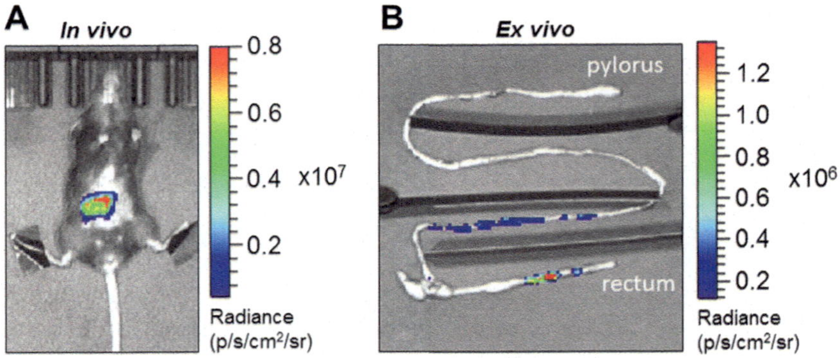

Fig. 2 Anesthetized mouse imaged 3 min post L-012 injection (20 mg/kg, i.p.) (**A**) and whole intestinal tract captured 15 min postinjection (**B**). In in vivo optical imaging, spatial resolution is low, and the precise localization of the optical emission is not possible, while the ex vivo image shows that the L-012-mediated signal is localized mainly in the middle colon and ileum. F/stop is set to 1, binning to medium (8 × 8 pixels), and exposure time is 5 s. Note: images are on a different scale. Panel A and panel B minimum and maximum displayed values are 4.64×10^5 to 8.02×10^6 and 1.20×10^5 to 1.36×10^6 p/s/cm^2/sr, respectively. Luminescence is expressed as radiance (p/s/cm^2/sr)

and when anesthetized place the mouse in a supine position in the nose cone in the chamber on the heated platform, and secure lower limbs with black tape (*see* **Note 8**).

6. Capture images at different time points starting as soon as possible postinjection of L-012 and at a higher frequency when rate of change of the optical signal is expected to be relatively fast, e.g., for i.p. L-012 administration 1, 3, 5, and 10 min postinjection would be ideal. Figure 2A shows the chemiluminescent signal of a WT mouse 3 min after L-012 i. p. injection at an exposure time of 5 s.
7. Review images, plot signal against time, and determine the best time point to image after injection, optimum exposure, and binning. Refer to Subheading 3.3 for instructions on image analysis.
8. Using these settings, image non-injected mice to determine if the autoluminescent background is detectable at these settings (*see* Subheading 3.4). If so, a pre-baseline image before injection is advisable for the main study (*see* **Note 9**).
9. In order to visualize in more detail $O_2^{\cdot -}$ production along the intestinal tract, euthanize mice by cervical dislocation, dissect the whole intestinal tract (from the pylorus to the rectum) immediately, and place the intestine on a black card. Ex vivo imaging of the small and large intestines in Fig. 2B shows the chemiluminescence signal, detected mainly in the middle colon and ileum, 15 min post L-012 i.p. injection.

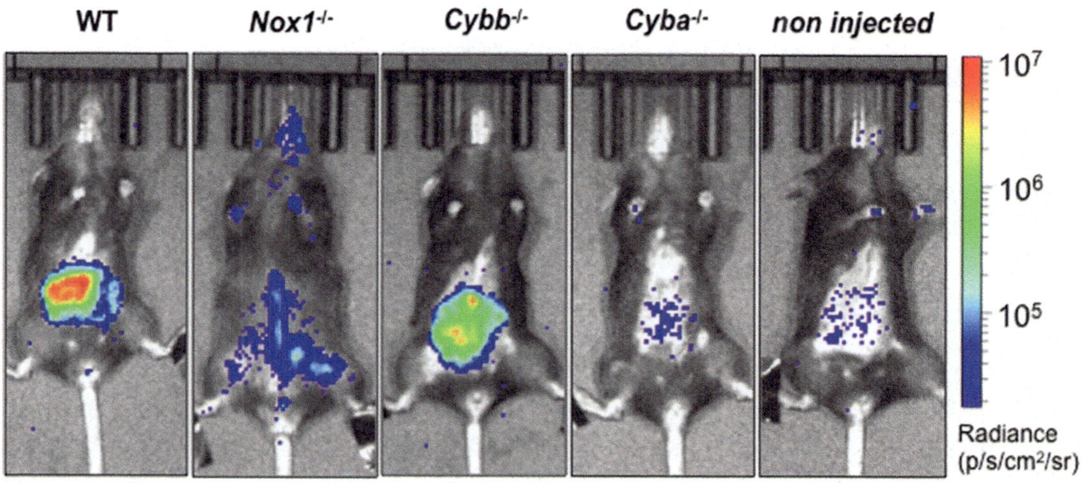

Fig. 3 Superoxide detection in WT, *Nox1*$^{-/-}$, *Cybb*$^{-/-}$ (Nox2 knockout), and *Cyba*$^{-/-}$ (p22^{phox} knockout) mice 3 min post i.p. administration of L-012 (20 mg/kg), compared with a non-injected mouse. Minimum scale is set to show autoluminescence in a non-injected mouse and to minimize background noise. Exposure times varied from 5 to 180 s. Luminescence is expressed as radiance (p/s/cm^2/sr)

3.2 Main Study

1. Follow **steps 1–5** as in pilot study with the following amendments.
2. If as per **step 8** above it has been determined that a pre-baseline image is required, this must be done between **steps 3** and **4** (*see* **Note 10**).
3. Capture images at a set time postinjection and for an exposure time as determined from pilot study. Figure 3 shows ROS imaging by L-012 in WT, *Nox1*$^{-/-}$, *Cybb*$^{-/-}$, and *Cyba*$^{-/-}$ mice 3 min postinjection. Exposure times varied from 5 to 180 s, F/stop was 1, and binning was medium (8 × 8 pixel binning).

3.3 Image Processing and Data Analysis

1. Image analysis using Living Image® 4.5.2 (64-bit).
2. Open image and set units to Radiance (p/s/cm^2/sr) (*see* **Note 11**).
3. Image will initially be scaled automatically (*see* **Note 12**).
4. Set the minimum scale to just above the noise. One way to do this consistently is to set the minimum scale level to twice the standard deviation of the background noise. Pixels exceeding this minimum value (just above the noise floor) will now be displayed (as shown in Figs. 4 and 5).
5. For high signal levels, large region of interests (ROIs) can be drawn around optical signal, which can include some background. For low signal levels, ROI background should not be included as contribution may no longer be negligible (*see* **Note 13**).

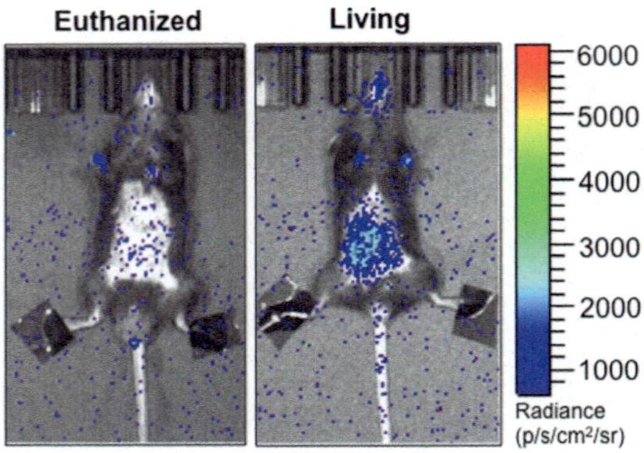

Fig. 4 Bioluminescent images overlaid with photographic images of representative non-injected euthanized *versus* living WT mice. Scale is set to show the autoluminescent signal from the living animal by setting the minimum value displayed to twice the standard deviation of the background noise. F/stop is set to 1, binning to medium (8 × 8 pixels), and the exposure time is 180 s. Luminescence is expressed as radiance (p/s/cm^2/sr)

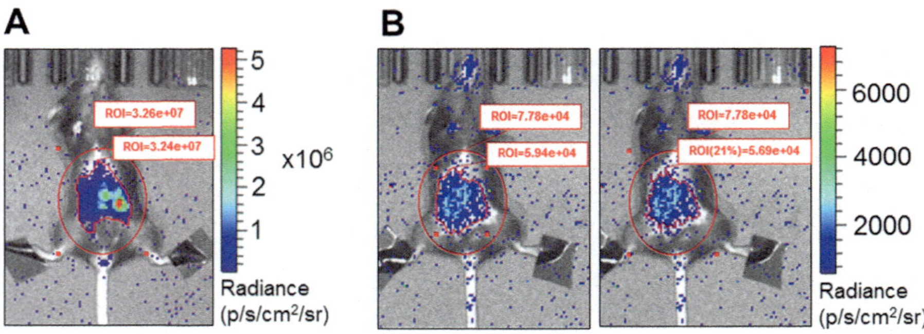

Fig. 5 Representative large region of interest (ROI) *versus* specific ROI for high signal level (**A**) and large ROI *versus* specific ROI (manual and automatic) for low signal level (**B**). Luminescence is expressed as radiance (p/s/cm^2/sr)

6. ROIs can either be drawn manually or automatically (*see* **Note 14**).
7. If assumptions associated with parametric statistical tests are not met, a Kruskal-Wallis nonparametric test followed by Dunn's correction for multiple comparisons may be used to obtain *P* values to evaluate statistical significance between experimental groups.

3.4 Interpretation of Data

When imaging in a very low-level light setting, other sources of background light from the animal not related to the probe must be considered. Low-level chemiluminescence is thought to be associated with electronically excited species created during normal metabolic oxidative reactions, and this can lead to an ultra-weak photon emission from living organisms. This is also referred to as biophoton emission, spontaneous photon emission, or autoluminescence, and it has been well documented in plants, animals, and humans [17–20]. Although this optical emission is very low and often negligible when compared with optical signal of interest, in a low-level light setting, this autoluminescent signal may comprise a significant percentage of detected signal, and thus, it should be measured prior to injection of L-012. A persistent luminescent signal from feed in the gut can also contribute to in vivo [21] and ex vivo (Authors' experience) background.

Knockout mouse models are very useful tools for understanding many biological processes, including the production of ROS by different NADPH oxidase isoforms and cell types in the gut. Figure 3A shows that WT and *Cybb*$^{-/-}$ (Nox2 knockout) mice produce the highest amounts of ROS in vivo in the steady state, while *Nox1*$^{-/-}$ mice have a strongly reduced but still detectable ROS signal when compared to WT mice; notably, ROS generation in *Cyba*$^{-/-}$mice is nearly abolished. This suggests that the contribution of Nox2-derived $O_2^{\cdot -}$ by innate immune cells in the gut *lamina propria* is negligible in the absence of inflammatory stimuli and it indicates that epithelial rather than phagocytic $O_2^{\cdot -}$ (or its dismutation product H_2O_2) plays an important role in regulating barrier dynamics in homeostasis.

Although the *Cyba*$^{-/-}$ model (global Nox1–4 inactivation due to p22^{phox}-deficiency) suggests that, in absence of immune cell activation, L-012-mediated chemiluminescence is a measure of $O_2^{\cdot -}$, several studies have employed the L-012 probe to measure RONS [22, 23]. In the cell, superoxide radical anion and nitric oxide radical combine to form the strong oxidant peroxynitrite ($OONO^-$). $OONO^-$ (or $ONOO^-$-derived radicals: $^{\cdot}NO_2$, $^{\cdot}OH$, $CO_3^{\cdot -}$) may then oxidize L-012. Nitric oxide ($^{\cdot}NO$) is formed by NO synthases (NOS) isoforms (1 or neuronal, 2 or inducible, and 3 or endothelial) using NADPH, molecular oxygen, and the amino acid L-arginine. The most efficient (micromolar concentrations) and long-lasting (up to days) source of $^{\cdot}NO$ is inducible NOS2 (iNOS), which is highly upregulated during intestinal infection and inflammation. Residual oxidation of the L-012 probe could also be explained by $^{\cdot}NO_2$ radical generated from peroxidase/H_2O_2/nitrite reaction rather than from $OONO^-$.

In conclusion, although L-012 represents a robust tool to measure NADPH oxidase-derived oxidant(s) at the intestinal mucosa level in the steady state, care must be taken in the interpretation of results when studying inflammatory models, including colitis [24] or using transgenic mice where upregulation of iNOS will cause positive signals. In these circumstances L-012 should be considered a probe suitable for general RONS evaluation. Possible alternatives to L-012 in vivo are described in **Note 15.**

4 Notes

1. Although the dark current background is significantly reduced due to cooling, it still needs to be subtracted from the images, as does the other main source of machine background, the electronic bias. The system automatically takes a range of reference backgrounds at a variety of different exposure times at a preset time during downtime, and these are automatically subtracted from the acquisition along with electronic bias when radiance units are selected. If the software is closed, the camera is only cooled to take background images and is not actively cooled until system is initialized. The electronic bias is automatically checked directly prior to an acquisition and modified if necessary.
2. Increasing exposure time increases the amount of incident light on the detector and will improve signal to noise ratio.
3. The F/stop is related to the geometry of the system and is inversely proportional to the diameter of the aperture in front of the lens. If the camera is saturated, increasing the f-number will reduce the amount of light incident on the detector, and in a low-level light setting, minimizing f-number will maximize the amount of incident light.
4. Binning pools the signal from a subset of pixels into one super pixel. For example, if a binning of 8 (medium in Living Image® software) is selected, the signal from 8 × 8, i.e., 64 pixels, is added together, increasing the signal to noise ratio at the cost of spatial resolution. In this example the signal will increase by 64, thus improving the signal to noise ratio and facilitating the detection of even lower signal levels. As the spatial resolution in this modality is not limited by the pixel size but rather the light scattering, this is a reasonable way to improve sensitivity.
5. Fur will attenuate the emitted light from the mice. White fur causes an attenuation of more than 50% of the signal, and the darker hair of C57BL/6 mice will have even more of an impact (*see* Technical note from PerkinElmer). It is thus very important to be consistent with the area shaved. If using chemical depilation, check that the product is not phosphorescent and emitting detectable light.
6. Keep the L-012 solution on ice in aluminum foil, and warm up to room temperature before injection.
7. If the sequence is inverted, i.e., mice will be anesthetized first and then i.p. injection takes place, a background reading can be performed before injection, and photon emission can be recorded immediately after injection. Thus, early signals (up to 3 min) can be recorded, and continuous recording can

determine the lag time, slope, and peak. Kielland et al. [14] reported that intraperitoneal and intravenous injection routes result in higher signals but also larger standard deviation, while the subcutaneous route results in a more stable but lower signal with less intragroup variation. In addition, the subcutaneous route of injection might be considered a refinement in animal welfare as it avoids intraperitoneal injection by inexperienced researchers and reduces variability.

8. Securing the animals in exactly the same position each time is very important to reduce attenuation differences due to position of legs, for example. Tape should be black and as matt as possible so that it does not reflect any scattered light emitted by the animal.

9. In shaved C57BL/6 mice, the abdominal autoluminescent total flux was approximately 3.6×10^4 photons/s ($n = 3$) or an average radiance of 840 photons/s/cm^2/sr with a 180 s exposure, F/stop of 1, medium binning (8×8 pixels) measured over an average abdominal area of 3.4 cm^2. Figure 4 compares a euthanized and a living mouse under identical conditions suggesting that the signal is due to the autoluminescence in this case and that there is no contribution from luminescent food in the gut, a possibility that should be considered, as ex vivo imaging of feed (ENVIGO TEKLAD Rodent Global Diet, alfalfa-free) by the Authors confirmed that it could be a source of optical signal.

10. Anesthesia potentially affects the biodistribution of L-012; pre-baseline images should be taken, and animals should be fully recovered before injection of probe (*see* also **Note** 7).

11. The IVIS Spectrum is calibrated annually with a National Standards and Technology (NIST) source allowing the software to convert the number of electrons generated in the CCD detector by the incident light to the number of photons detected *per* second. When the radiance unit is selected, counts are converted to an absolute calibrated number that takes into account any changes in exposure, F/stop, and binning. Total flux is measured in photons/second (system assumes isotropic emission) and is used for quantifying the total optical signal measured in a selected region of interest (ROI). Total flux (p/s) is converted automatically by the software into average radiance (p/s/cm^2/sr) in the ROI by dividing total flux by the area (cm^2) of the ROI and 4π steradians [a steradian (sr) is a measure of a 3-D angle].

12. When the image is scaled automatically, the minimum scale is set to suppress noise, and the maximum scale is set to the maximum value in the image. Pixel values less than the minimum are not displayed, and pixel values above the maximum scale are set to the maximum color according to the scale.

13. Figure 5A shows that the "fitted" ROI and an oval ROI that includes some background do not differ significantly in value for higher signal levels and that the additional contribution from the machine background can be considered negligible. For lower signal levels, the background contribution may no longer be considered negligible as demonstrated in Fig. 5B (right image).
14. For automatically drawing a ROI on a noisy image, set the minimum scale to threshold the signal (suggestion is twice the standard deviation of background noise). Select Auto1 and move marker to signal area. Vary the % threshold level until the ROI aligns with the edges of your signal. In Fig. 5B (right image), the threshold was set to 21%. The % threshold specifies the minimum percent of peak pixel intensity that a pixel must have to be included in a ROI. The boundary of the ROI when using the threshold feature therefore depends on the maximum signal in the image and should just be used as a tool rather than as an absolute value to be kept consistent when measuring total optical signal.
15. Other probes may be considered as alternative to L-012 in vivo. Coelenterazine is a luminescent enzyme substrate for apoaequorin and Renilla luciferases. It is a cell membrane permeable probe able to quantify $O_2^{\cdot -}$ levels produced during oxidative phosphorylation in pathophysiological conditions in living animals [25], and it has been used to study the development of intestinal inflammation [26]. Coelenterazine is able to quantify intracellular $O_2^{\cdot -}$, and it does not exhibit redox cycling. Unfortunately, its mechanism is similar to luminol, and therefore, possible interaction with RNS, autoxidation, and unfavorable signal to background ratio must be taken into account.

    An ex vivo image in Fig. 2B shows the L-012-mediated chemiluminescent signal in the ileum. Gene expression analyses of intestinal NADPH oxidases have shown higher expression of Duox2 and lower expression of Nox1 in the mouse ileum [27]. These results, together with the evidence that Nox1–4 inactivation in *Cyba*$^{-/-}$ mice does not impinge on H_2O_2-producing Duox2 activity in homeostatic conditions [28], raise the question whether L-012 detects H_2O_2 as well. At the ileal pH, $O_2^{\cdot -}$ could possibly be converted into hydroperoxyl radical ($HO_2^{\cdot}$, $pKa = 4.8$), which being a much stronger oxidant ultimately oxidizes L-012 to its radical. Alternatively, in absence of infections or inflammation, H_2O_2 produced by Duox2 could be sensibly inferior to signals resulting from the dismutation of Nox1-derived $O_2^{\cdot -}$. For more direct H_2O_2 measurements, another chemiluminescence-based probe, named HyPerBlu (Lumigen), might be suitable to generate a robust high-intensity signal that can be detected by CCD imaging systems, but in vivo data are not available.

H_2O_2 can be detected by boronate-based fluorescence probes; however, boronate/boronic acids react within milliseconds with $OONO^-$ as opposed to reacting with H_2O_2 where the oxidation is much slower [29]; therefore they can only be used to detect H_2O_2 in absence of $ONOO^-$ and are unsuitable for the use during colitis studies.

For general monitoring of oxidative stress, optical reporter mice harboring the OKD48 (Keap1-dependent Oxidative stress Detector, No-48) transgene based on the dual regulating mechanism in the Keap1-Nrf2 pathway have been created (Trans Genic Inc., Japan). Although potentially usable during disease, the main limitation of these mice is the weak optical signal emitted in physiological conditions [30].

Acknowledgments

The authors thank Prof. Ulla G. Knaus for fruitful discussions and Dr. Jacek Zielonka for critical reading of the manuscript. EC is supported by the Irish Cancer Society Collaborative Cancer Research Centre BREAST-PREDICT. GA is supported by the European Crohn's and Colitis Organisation (ECCO) and the Medical Research Council (MRC).

References

1. Peterson LW, Artis D (2014) Intestinal epithelial cells: regulators of barrier function and immune homeostasis. Nat Rev Immunol 14 (3):141–153. https://doi.org/10.1038/nri3608
2. Aviello G, Knaus UG (2017) ROS in gastrointestinal inflammation: rescue Or Sabotage? Br J Pharmacol 174(12):1704–1718. https://doi.org/10.1111/bph.13428
3. O'Neill S, Brault J, Stasia MJ, Knaus UG (2015) Genetic disorders coupled to ROS deficiency. Redox Biol 6:135–156. https://doi.org/10.1016/j.redox.2015.07.009
4. Marciano BE, Rosenzweig SD, Kleiner DE, Anderson VL, Darnell DN, Anaya-O'Brien S, Hilligoss DM, Malech HL, Gallin JI, Holland SM (2004) Gastrointestinal involvement in chronic granulomatous disease. Pediatrics 114 (2):462–468
5. Marks DJ, Miyagi K, Rahman FZ, Novelli M, Bloom SL, Segal AW (2009) Inflammatory bowel disease in CGD reproduces the clinicopathological features of Crohn's disease. Am J Gastroenterol 104(1):117–124. https://doi.org/10.1038/ajg.2008.72
6. Hayes P, Dhillon S, O'Neill K, Thoeni C, Hui KY, Elkadri A, Guo CH, Kovacic L, Aviello G, Alvarez LA, Griffiths AM, Snapper SB, Brant SR, Doroshow JH, Silverberg MS, Peter I, McGovern DP, Cho J, Brumell JH, Uhlig HH, Bourke B, Muise AA, Knaus UG (2015) Defects in NADPH oxidase genes NOX1 and DUOX2 in very early onset inflammatory bowel disease. Cell Mol Gastroenterol Hepatol 1(5):489–502. https://doi.org/10.1016/j.jcmgh.2015.06.005
7. Parlato M, Charbit-Henrion F, Hayes P, Tiberti A, Aloi M, Cucchiara S, Begue B, Bras M, Pouliet A, Rakotobe S, Ruemmele F, Knaus UG, Cerf-Bensussan N (2017) First identification of biallelic inherited DUOX2 inactivating mutations as a cause of very early onset inflammatory bowel disease. Gastroenterology 153(2):609–611 e603. https://doi.org/10.1053/j.gastro.2016.12.053
8. Schwerd T, Bryant RV, Pandey S, Capitani M, Meran L, Cazier JB, Jung J, Mondal K,

Parkes M, Mathew CG, Fiedler K, McCarthy DJ, WGS500 Consortium; Oxford IBD cohort study investigators; COLORS in IBD group investigators; UK IBD Genetics Consortium, Sullivan PB, Rodrigues A, Travis SPL, Moore C, Sambrook J, Ouwehand WH, Roberts DJ, Danesh J; INTERVAL Study, Russell RK, Wilson DC, Kelsen JR, Cornall R, Denson LA, Kugathasan S, Knaus UG, Serra EG, Anderson CA, Duerr RH, McGovern DP, Cho J, Powrie F, Li VS, Muise AM, Uhlig HH (2018) NOX1 loss-of-function genetic variants in patients with inflammatory bowel disease. Mucosal Immunol 11(2):562–574. https://doi.org/10.1038/mi.2017.74

9. Woolley JF, Stanicka J, Cotter TG (2013) Recent advances in reactive oxygen species measurement in biological systems. Trends Biochem Sci 38(11):556–565. https://doi.org/10.1016/j.tibs.2013.08.009
10. Ren W, Ai HW (2013) Genetically encoded fluorescent redox probes. Sensors 13 (11):15422–15433. https://doi.org/10.3390/s131115422
11. Wagener KC, Kolbrink B, Dietrich K, Kizina KM, Terwitte LS, Kempkes B, Bao G, Muller M (2016) Redox indicator mice stably expressing genetically encoded neuronal roGFP: versatile tools to decipher subcellular redox dynamics in neuropathophysiology. Antioxid Redox Signal 25(1):41–58. https://doi.org/10.1089/ars.2015.6587
12. Tseng JC, Kung AL (2013) In vivo imaging method to distinguish acute and chronic inflammation. J Vis Exp (78). https://doi.org/10.3791/50690
13. Pizzolla A, Hultqvist M, Nilson B, Grimm MJ, Eneljung T, Jonsson IM, Verdrengh M, Kelkka T, Gjertsson I, Segal BH, Holmdahl R (2012) Reactive oxygen species produced by the NADPH oxidase 2 complex in monocytes protect mice from bacterial infections. J Immunol 188(10):5003–5011. https://doi.org/10.4049/jimmunol.1103430
14. Kielland A, Blom T, Nandakumar KS, Holmdahl R, Blomhoff R, Carlsen H (2009) In vivo imaging of reactive oxygen and nitrogen species in inflammation using the luminescent probe L-012. Free Radic Biol Med 47 (6):760–766. https://doi.org/10.1016/j.freeradbiomed.2009.06.013
15. Han W, Li H, Segal BH, Blackwell TS (2012) Bioluminescence imaging of NADPH oxidase activity in different animal models. J Vis Exp (68):3925. https://doi.org/10.3791/3925
16. Zielonka J, Lambeth JD, Kalyanaraman B (2013) On the use of L-012, a luminol-based chemiluminescent probe, for detecting superoxide and identifying inhibitors of NADPH oxidase: a reevaluation. Free Radic Biol Med 65:1310–1314. https://doi.org/10.1016/j.freeradbiomed.2013.09.017
17. Pospisil P, Prasad A, Rac M (2014) Role of reactive oxygen species in ultra-weak photon emission in biological systems. J Photochem Photobiol B 139:11–23. https://doi.org/10.1016/j.jphotobiol.2014.02.008
18. Van Wijk R, Kobayashi M, Van Wijk EP (2006) Anatomic characterization of human ultra-weak photon emission with a moveable photomultiplier and CCD imaging. J Photochem Photobiol B 83(1):69–76. https://doi.org/10.1016/j.jphotobiol.2005.12.005
19. Birtic S, Ksas B, Genty B, Mueller MJ, Triantaphylides C, Havaux M (2011) Using spontaneous photon emission to image lipid oxidation patterns in plant tissues. Plant J 67 (6):1103–1115. https://doi.org/10.1111/j.1365-313X.2011.04646.x
20. Kobayashi M (2014) Highly sensitive imaging for ultra-weak photon emission from living organisms. J Photochem Photobiol B 139:34–38. https://doi.org/10.1016/j.jphotobiol.2013.11.011
21. Troy T, Jekic-McMullen D, Sambucetti L, Rice B (2004) Quantitative comparison of the sensitivity of detection of fluorescent and bioluminescent reporters in animal models. Mol Imaging 3(1):9–23. https://doi.org/10.1162/153535004773861688
22. Van Dyke K, Ghareeb E, Van Dyke M, Van Thiel DH (2007) Ultrasensitive peroxynitrite-based luminescence with L-012 as a screening system for antioxidative/antinitrating substances, e.g. Tylenol (acetaminophen), 4-OH tempol, quercetin and carboxy-PTIO. Luminescence 22(4):267–274. https://doi.org/10.1002/bio.959
23. Matziouridou C, Rocha SDC, Haabeth OA, Rudi K, Carlsen H, Kielland A (2018) iNOS- and NOX1-dependent ROS production maintains bacterial homeostasis in the ileum of mice. Mucosal Immunol 11(3):774–784. https://doi.org/10.1038/mi.2017.106
24. Asghar MN, Emani R, Alam C, Helenius TO, Gronroos TJ, Sareila O, Din MU, Holmdahl R, Hanninen A, Toivola DM (2014) In vivo imaging of reactive oxygen and nitrogen species in murine colitis. Inflamm Bowel Dis 20 (8):1435–1447. https://doi.org/10.1097/MIB.0000000000000118
25. Bronsart LL, Stokes C, Contag CH (2016) Chemiluminescence imaging of superoxide anion detects beta-cell function and mass. PLoS One 11(1):e0146601. https://doi.org/10.1371/journal.pone.0146601

26. Bronsart L, Nguyen L, Habtezion A, Contag C (2016) Reactive oxygen species imaging in a mouse model of inflammatory bowel disease. Mol Imaging Biol 18(4):473–478. https://doi.org/10.1007/s11307-016-0934-0
27. Grasberger H, Gao J, Nagao-Kitamoto H, Kitamoto S, Zhang M, Kamada N, Eaton KA, El-Zaatari M, Shreiner AB, Merchant JL, Owyang C, Kao JY (2015) Increased expression of DUOX2 is an epithelial response to mucosal dysbiosis required for immune homeostasis in mouse intestine. Gastroenterology 149(7):1849–1859. https://doi.org/10.1053/j.gastro.2015.07.062
28. Pircalabioru G, Aviello G, Kubica M, Zhdanov A, Paclet MH, Brennan L, Hertzberger R, Papkovsky D, Bourke B, Knaus UG (2016) Defensive mutualism rescues NADPH oxidase inactivation in gut infection. Cell Host Microbe 19(5):651–663. https://doi.org/10.1016/j.chom.2016.04.007
29. Sedgwick AC, Sun X, Kim G, Yoon J, Bull SD, James TD (2016) Boronate based fluorescence (ESIPT) probe for peroxynitrite. Chem Commun (Camb) 52(83):12350–12352. https://doi.org/10.1039/c6cc06829d
30. Oikawa D, Akai R, Tokuda M, Iwawaki T (2012) A transgenic mouse model for monitoring oxidative stress. Sci Rep 2:229. https://doi.org/10.1038/srep00229

Chapter 20

Hydro-Cy3-Mediated Detection of Reactive Oxygen Species In Vitro and In Vivo

Bejan J. Saeedi, Bindu Chandrasekharan, and Andrew S. Neish

Abstract

Reactive oxygen species (ROS) are potent signaling molecules with critical roles in cellular pathology and homeostasis. They are produced in all cell types via a diverse array of cellular machinery, giving rise to an equally diverse repertoire of molecular effects. These range from cytotoxic killing of microbes to alteration of the cellular transcriptional response to stress. Despite their importance, research into ROS has been difficult given their inherent instability and transient signaling properties. Herein we describe methods for the use of the redox-sensitive probe hydro-Cy3 for the detection and quantification of ROS both in vitro and in vivo.

Key words Reactive oxygen species, ROS, Hydro-Cy3, Oxidative stress, NADPH oxidase, NOX

1 Introduction

Reactive oxygen species (ROS) have potent cytotoxic properties and have long been recognized as critical effectors of immune cells [1, 2]. In these cells, ROS are generated in specialized compartments known as phagosomes by NADPH oxidase 2 (NOX2) [3, 4]. Nonimmune cell types can also produce ROS [5]. In the case of the colonic epithelium, this occurs primarily through the actions of NOX1 and can result in cytosolic superoxide [6–8]. At high levels, superoxide and hydroxyl free radicals can elicit significant molecular damage to cellular components and may be termed redox distress [9, 10]. Low levels of superoxide-derived hydrogen peroxide, however, can instead act as a second messenger to mediate myriad cellular signaling events. This state of physiologic ROS that mediate events critical for cellular homeostasis can be termed redox eustress [10, 11]. H_2O_2 can oxidize reactive thiol groups on a variety of cellular proteins, altering their activity and subsequent downstream signaling events [12, 13]. For example, cytosolic H_2O_2-mediated oxidation of the E3 ubiquitin ligase adaptor KEAP1 can result in the activation of the transcription factor

Ulla G. Knaus and Thomas L. Leto (eds.), *NADPH Oxidases: Methods and Protocols*, Methods in Molecular Biology, vol. 1982, https://doi.org/10.1007/978-1-4939-9424-3_20, © Springer Science+Business Media, LLC, part of Springer Nature 2019

Nrf2, a master regulator of the cellular antioxidant response [14–16]. This can occur at sub-cytotoxic concentrations of H_2O_2, suggesting a mechanism whereby physiological levels of certain ROS can prepare the cell for more potent oxidative stressors [8, 17].

Both forms of redox stress described above play critical roles in pathology and homeostasis. As such, research into the nature of ROS in vitro and in vivo has surged. Because of the inherent instability of ROS, it is difficult to accurately measure these molecules in various cellular contexts. In recent years, development of the hydrocyanine class of redox sensors has allowed for more in-depth research in this area [18].

Hydrocyanines are a class of redox-sensitive probes that can be used in vitro and in vivo for detection of ROS [18, 19]. They can be synthesized in a one-step reaction from fluorescent cyanine molecules using NaBH4 to reduce the iminium cations. This results in chemical and physical properties unique to the hydrocyanines that makes them particularly useful for the detection of ROS (Fig. 1). In their reduced form, hydrocyanines are weakly fluorescent due to disrupted π conjugation. However, oxidation results in restoration of π conjugation and therefore strong fluorescence. In addition, reduced hydrocyanines are membrane permeable. This allows for the uptake of the dye into live cells without permeabilization. Upon exposure to an oxidative environment within a cell, the dye becomes membrane impermeable and is trapped. Fluorescence can then be quantified as a read out of the oxidative environment within cells throughout the course of various treatments. An example of this is illustrated in Fig. 2.

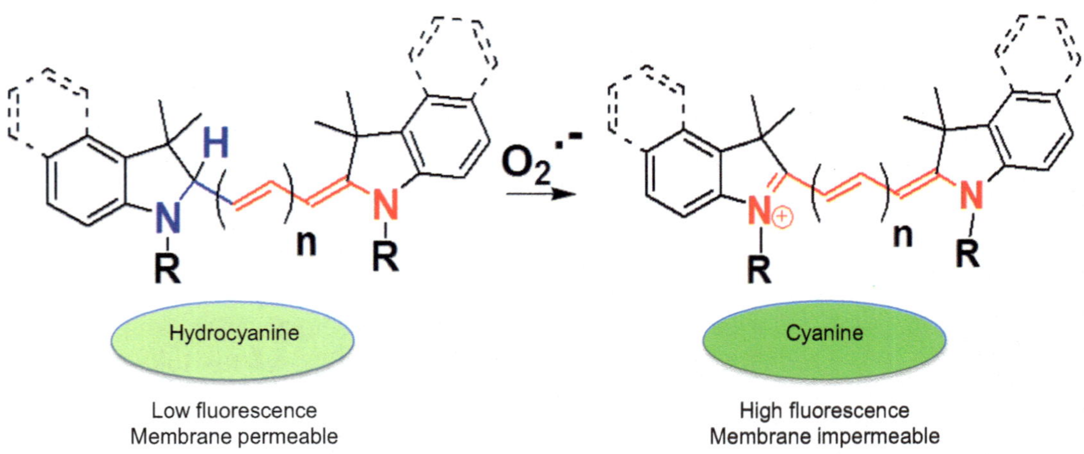

Fig. 1 Mechanism of ROS detection by hydrocyanines. Hydrocyanines are weakly fluorescent, membrane permeable molecules. Oxidation of the hydrocyanine molecule upon exposure to superoxide or hydroxyl radical results in the formation of a strongly fluorescent cyanine that is membrane impermeable and therefore trapped in cells experiencing oxidative stress

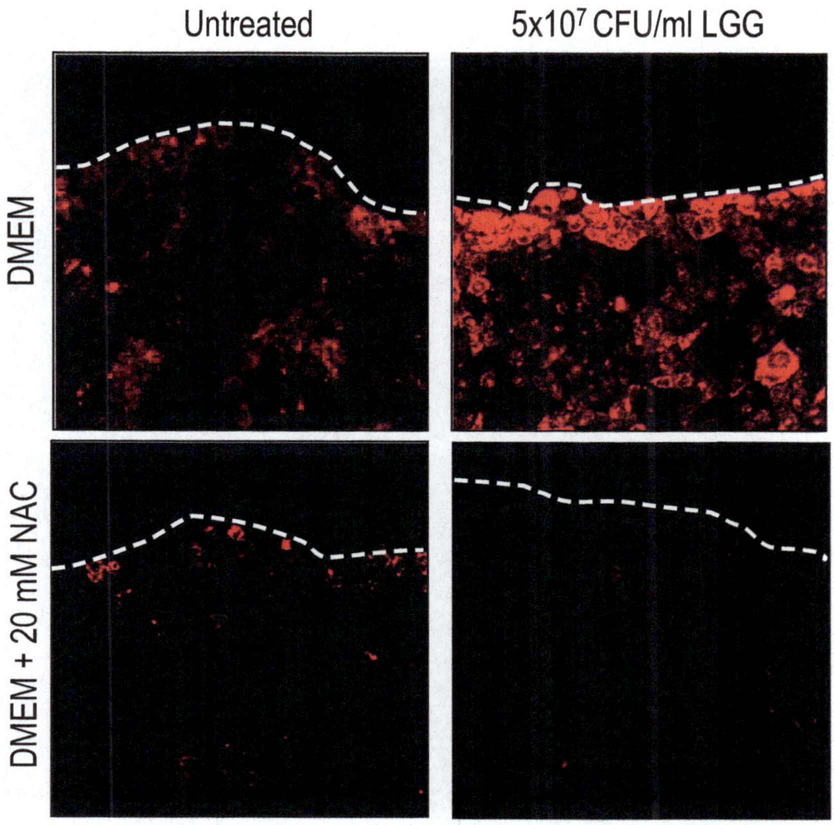

Fig. 2 Detection of ROS generation in vitro. Fluorescent images show scratch-wounded epithelium in culture. Dashed line denotes leading edge of scratch wound. Untreated cells exhibit low hydro-Cy3 fluorescence (upper left). Addition of 5×10^7 colony-forming units (CFU)/mL of the bacterium *Lactobacillus rhamnosus* GG results in a robust increase in fluorescence, i.e., ROS generation 5 min after treatment. These effects are abrogated upon treatment with the antioxidant *N*-acetylcysteine (*NAC* lower right)

The low toxicity of hydrocyanines can also be used to probe ROS signaling in vivo [8, 20, 21]. In this context, the hydrocyanine dyes can be administered by intraperitoneal injection into mice. The dye will circulate and be trapped in tissues undergoing oxidative stress. A tissue of interest can subsequently be dissected and analyzed by confocal fluorescence microscopy as shown in Fig. 3. The stability of the fluorescent signal permits evaluation of tissues up to several hours after administration.

A variety of different hydrocyanine dyes exist, each with unique properties and different specificities to detect various oxidant species. In this review, we focus on hydro-Cy3, a hydrocyanine molecule selectively reactive to superoxide and hydroxyl radical species, with limited reactivity to reactive nitrogen species, H_2O_2, or Fe^{2+} [19]. Hydro-Cy3 has high sensitivity, detects nanomolar levels of ROS, and shows a linear relationship ($r^2 = 0.99$) between fluorescence intensity and concentration of radical oxidants (1–30 nm

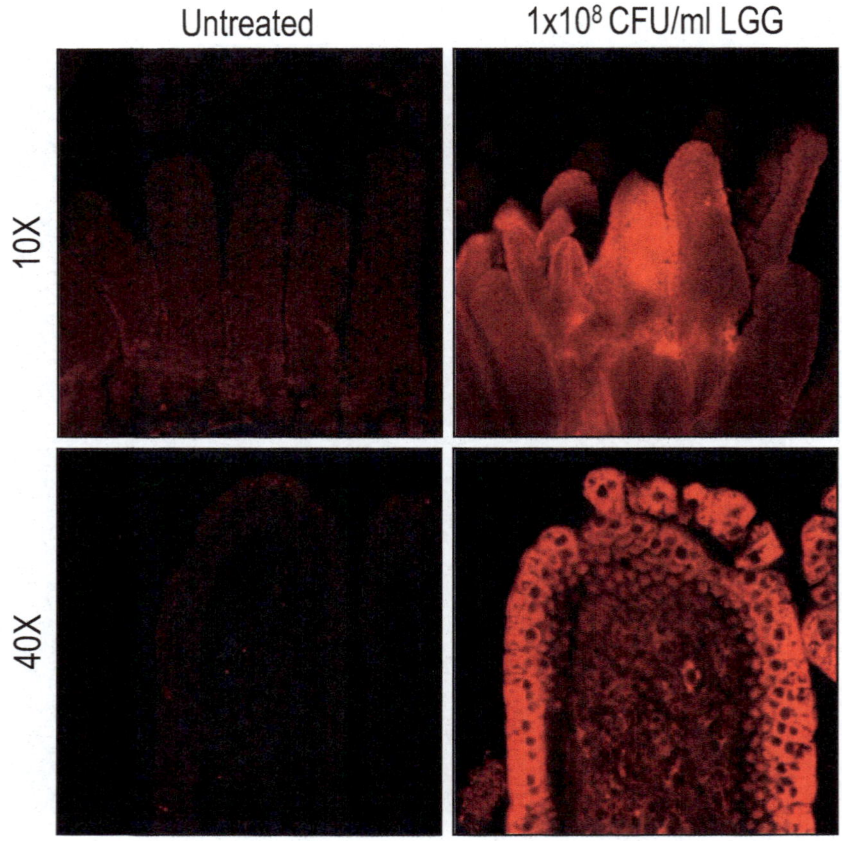

Fig. 3 Detection of ROS generation in vivo. Images of hydro-Cy3 fluorescence in whole-mount preparation of small intestine from untreated mice (left) or mice gavaged 1×10^8 colony-forming units of *Lactobacillus rhamnosus* GG (right). Images taken 1 h after administration of LGG at 10× (upper row) and 40× (lower row) magnifications

range) [19]. These features make it a useful tool for the investigation of oxidative stress and redox signaling. Here, we describe methods to utilize hydro-Cy3 to detect ROS generation both in vitro and in vivo.

2 Materials

1. Caco2 cells (ATCC HTB-37).
2. T75 flask.
3. Black-walled 96-well cell culture plates.
4. Cell culture medium: Dulbecco's Modified Eagle's Medium with and without phenol red (*see* **Note 1**) supplemented with 10% fetal bovine serum, 2 mM L-glutamine, 100 units/mL penicillin, and 100 μg/mL streptomycin.

5. Hanks balanced salt solution (HBSS), with calcium and magnesium, without phenol red (*see* **Note 1**).
6. Hydro-Cyanine3 (ROSstar™ 550, Li-Cor) available as a 2.5 mM stock solution in DMSO. Stock solution can be stored at −20 °C for up to 3 months. Hydro-Cy3 should be diluted to a working concentration of 100 μM in DMEM without phenol red for in vitro experiments or in HBSS without phenol red for in vivo experiments. These working stocks should be used immediately to avoid autoxidation (*see* **Note 2**).
7. Fluorescent microplate reader.
8. Multichamber slides.
9. 18 × 18 mm coverslips, glass microscope slides.
10. 1 cc syringes.
11. 27 gauge needless.
12. Isoflurane.
13. Anesthetic vaporizer with induction chamber and nosecone.
14. Flexible rubber feeding tube (3.5 French).
15. Scalpel.

3 Methods

Perform all steps at room temperature unless otherwise described.

3.1 *In Vitro* Determination of Reactive Oxygen Species

1. Grow Caco2 cells to confluence in a T75 flask.
2. Aspirate media, wash with sterile PBS, and add 1 mL trypsin/EDTA solution. Incubate for 5 min at 37 °C, 5% CO_2. Resuspend cells in DMEM with phenol red, and adjust to a concentration of 5×10^4 cells/mL.
3. Plate 200 μL of Caco2 cells on a 96-well black-walled cell culture plate, and allow to grow to confluence at 37 °C, 5% CO_2.
4. Load cells with hydro-Cy3: Dilute hydro-Cy3 in DMEM without phenol red to a concentration of 100 μM. Remove media from Caco2 cells, and replace with hydro-Cy3 containing media. Incubate in the dark at 37 °C, 5% CO_2 for 1 h.
5. Remove hydro-Cy3 containing media, and wash monolayers with Hanks balanced salt solution (HBSS).
6. Add treatment of interest in HBSS, and incubate in the dark at 37 °C, 5% CO_2. To ensure that fluorescence is due to the production of ROS and not due to autofluorescence or alterations in mitochondrial membrane potential (*see* **Note 3**), an antioxidant such as *N*-acetyl-cysteine can be added as a control.

This sample should show low fluorescence as demonstrated in Fig. 2.

7. At desired time point, aspirate HBSS and wash cells with fresh HBSS.
8. Read plate immediately using a fluorescent microplate reader with an excitation wavelength of 546 nm and an emission wavelength of 561 nm (*see* **Note 4**).
9. Plot data as relative fluorescence units as compared to control (*see* **Note 5**).

3.2 Confocal Microscopy for ROS Generation In Vitro

1. Prepare Caco-2 cells as described in Subheading 3.1, **steps 1** and **2**.
2. Plate Caco-2 cells on multichamber slides, and grow to confluence at 37 °C, 5% CO_2.
3. Replace cell culture media with fresh DMEM without phenol red containing 100 μM hydro-Cy3, and incubate in the dark at 37 °C, 5% CO_2 for 1 h.
4. Remove media and wash cells with HBSS.
5. Add treatment of interest in DMEM without phenol red, and incubate in the dark at 37 °C, 5% CO_2 for desired period of time.
6. At desired time point, remove media and again wash with HBSS.
7. If desired, counterstain with cell-permeant dye (*see* **Note 6**).
8. Apply coverslip and image at 40× by laser scanning confocal microscopy with a helium-neon excitation laser at 543 nm and a 505–530 nm band pass filter set. Imaging should be performed immediately and if possible without fixation (*see* **Notes** 7 and **8**).

3.3 In Vivo Determination of ROS Generation

Carry out the following procedure using 6–8-week-old mice.

1. Fast mice for 16 h prior to injection of hydro-Cy3.
2. Make working solution of hydro-Cy3 by diluting in HBSS without phenol red to a final concentration of 100 μM.
3. Load a 1 cc syringe with hydro-Cy3 working solution and administer 200 μL by intraperitoneal injection. If colonic epithelial ROS generation is of primary interest, hydro-Cy3 can be administered intrarectally to anesthetized mice. Induce anesthesia using an anesthetic vaporizer with 3–5% isoflurane in an induction chamber at a flow rate 0.5–1 L/min. After onset of anesthesia, remove mouse and place in nosecone with 1–3% isoflurane. Using a small gauge (3.5 French) rubber feeding tube, administer 100 μL of 7.5 μM hydro-Cy3 intrarectally.

4. Wait 15 min to allow for hydro-Cy3 uptake into tissues.
5. Administer treatment of interest. For example, ROS-generating bacterial species can be administered by oral gavage or by rectal administration as described above. Figure 3 depicts resulting fluorescence in whole-mount intestinal samples 1 h after gavage with 1 × 10^8 CFU of *Lactobacillus rhamnosus* GG.
6. Dissect tissue of interest (i.e., colon). Using a scalpel, cut tissue into 2–5 mm pieces and whole mount on glass slides. Best results are obtained with rapid preparation and analysis (*see* **Note 5**).
7. Image using laser confocal microscopy at 40× with a helium-neon excitation laser at 543 nm and a 505–530 nm band pass filter set.
8. Quantification of ROS can be performed by imaging at random positions along each slide and quantifying fluorescence intensity using the ImageJ software package from the National Institutes of Health.

4 Notes

1. Phenol red has an absorption spectrum close to that of hydrocyanine around physiological pH and hence may cause artifacts in the measurement of fluorescence intensity [22]. Furthermore, phenol red anions may also react with hydrocyanine [23].
2. Stability and autofluorescence: Compared to other ROS dyes such as DHE, hydro-Cy3 exhibits less autoxidation (as measured in PBS, pH 7.4) due to lower resonance stabilization of oxidation intermediates, resulting in an enhanced half-life of approximately 3 days [19]. Hydro-Cy3 is also the dye of choice as no cellular toxicity has been reported up to concentrations of 1 mM, as determined by the trypan blue cell viability assay [19].
3. Pitfalls: Recently it has been shown that oxidized hydro-Cy3 mainly accumulates in the mitochondrial matrix, but fluorescence intensity declines once the mitochondrial membrane potential ($\Delta\Psi$m) is compromised [24]. In a series of experiments using the mitochondrial inhibitor oligomycin, uncoupler FCCP, and the common pesticide paraquat (ROS inducer), it has been demonstrated that hydro-Cy3 fluorescence rapidly declines upon mitochondrial depolarization, irrespective of cellular ROS. Thus, it is advisable that results of hydro-Cy 3 fluorescence should be interpreted with caution in

experiments that involve loss or compromise mitochondrial membrane integrity.

4. Sensitivity: Hydro-Cy3 can detect nanomolar concentrations of ROS (1–30 nm) in vitro and exhibits a linear relationship ($r^2 = 0.99$) between hydrogen peroxide concentration in Fenton's reagent and relative fluorescence intensity/units (RFU) measured at 560 nm [19].
5. All cells and tissues will show stress-induced ROS during the preparation as media/blood supply is disestablished. Biologically valid ROS-induced fluorescence is best discerned as signal above and distinct from that seen in control cells/tissues. Best results are achieved with rapid preparation and analysis.
6. Counterstains: Hydro-Cy3 can be used efficiently with counter stains like Hoechst 33342 (nuclear marker). As the cells are not fixed, a cell-permeant counterstain must be used for this step.
7. Fixation: Hydro-Cy3 enables imaging of ROS in live cells and in vivo mouse models. Fixation of tissues cannot guarantee preservation of the fluorescence intensity. Hence, it is best to image the cells/tissue without fixation as the fluorescence intensity remains stable for up to approximately 3 days.
8. Quantification: Relative fluorescence between sample of interest and control is the preferred mode of data presentation. Hydro-Cy3 cannot be used for absolute quantification of ROS. ROS by their nature are unstable and especially in vivo are rapidly dismutated; thus these approaches are not appropriate for the quantification of a given species (e.g., superoxide vs. H_2O_2).

Acknowledgments

The authors acknowledge support from the NIH AI64462 and CA179424.

References

1. Yang Y, Bazhin AV, Werner J et al (2013) Reactive oxygen species in the immune system. Int Rev Immunol 32:249–270
2. Bogdan C, Röllinghoff M, Diefenbach A (2000) Reactive oxygen and reactive nitrogen intermediates in innate and specific immunity. Curr Opin Immunol 12:64–76
3. Dahlgren C, Karlsson A (1999) Respiratory burst in human neutrophils. J Immunol Methods 232:3–14
4. Singel KL, Segal BH (2016) NOX2-dependent regulation of inflammation. Clin Sci (Lond) 130:479–490
5. Lambeth JD (2004) NOX enzymes and the biology of reactive oxygen. Nat Rev Immunol 4:181–189
6. Kikuchi H, Hikage M, Miyashita H et al (2000) NADPH oxidase subunit, gp91(phox) homologue, preferentially expressed in human colon epithelial cells. Gene 254:237–243
7. Leoni G, Alam A, Neumann P-A et al (2013) Annexin A1, formyl peptide receptor, and NOX1 orchestrate epithelial repair. J Clin Invest 123:443–454
8. Jones RM, Luo L, Ardita CS et al (2013) Symbiotic lactobacilli stimulate gut epithelial

proliferation via Nox-mediated generation of reactive oxygen species. EMBO J 32:3017–3028
9. Lushchak VI (2014) Free radicals, reactive oxygen species, oxidative stress and its classification. Chem Biol Interact 224:164–175
10. Go Y-M, Jones DP (2017) Redox theory of aging: implications for health and disease. Clin Sci (Lond) 131:1669–1688
11. Sies H, Berndt C, Jones DP (2017) Oxidative stress. Annu Rev Biochem 86:715–748
12. Winterbourn CC, Hampton MB (2008) Thiol chemistry and specificity in redox signaling. Free Radic Biol Med 45:549–561
13. Schieber M, Chandel NS (2014) ROS function in redox signaling and oxidative stress. Curr Biol 24:R453–R462
14. Moi P, Chan K, Asunis I et al (1994) Isolation of NF-E2-related factor 2 (Nrf2), a NF-E2-like basic leucine zipper transcriptional activator that binds to the tandem NF-E2/AP1 repeat of the beta-globin locus control region. Proc Natl Acad Sci U S A 91:9926–9930
15. McMahon M, Thomas N, Itoh K et al (2006) Dimerization of substrate adaptors can facilitate cullin-mediated ubiquitylation of proteins by a "tethering" mechanism: a two-site interaction model for the Nrf2-Keap1 complex. J Biol Chem 281:24756–24768
16. Dinkova-Kostova AT, Holtzclaw WD, Cole RN et al (2002) Direct evidence that sulfhydryl groups of Keap1 are the sensors regulating induction of phase 2 enzymes that protect against carcinogens and oxidants. Proc Natl Acad Sci U S A 99:11908–11913
17. Jones RM, Desai C, Darby TM et al (2015) Lactobacilli modulate epithelial cytoprotection through the Nrf2 pathway. Cell Rep 12:1217–1225
18. Sadlowski CM, Maity S, Kundu K et al (2017) Hydrocyanines: a versatile family of probes for imaging radical oxidants in vitro and in vivo. Mol Sys Des Eng 2:191–200
19. Kundu K, Knight SF, Willett N et al (2009) Hydrocyanines: a class of fluorescent sensors that can image reactive oxygen species in cell culture, tissue, and in vivo. Angew Chem Int Ed Engl 48:299–303
20. Ardita CS, Mercante JW, Kwon YM et al (2014) Epithelial adhesion mediated by pilin SpaC is required for Lactobacillus rhamnosus GG-induced cellular responses. Appl Environ Microbiol 80:5068–5077
21. Alam A, Leoni G, Quiros M et al (2016) The microenvironment of injured murine gut elicits a local pro-restitutive microbiota. Nat Microbiol 1:15021
22. Pizarro G, Ríos E (2004) How source content determines intracellular Ca2+ release kinetics. Simultaneous measurement of [Ca2+] transients and [H+] displacement in skeletal muscle. J Gen Physiol 124:239–258
23. Shapovalov SA, Kiseleva YS (2008) The association of cyanine dye cations with phenol red anions in aqueous solutions. Russ J Phys Chem 82:972–977
24. Zhdanov AV, Aviello G, Knaus UG et al (2017) Cellular ROS imaging with hydro-Cy3 dye is strongly influenced by mitochondrial membrane potential. Biochim Biophys Acta 1861:198–204

Part IV

Regulation of NADPH Oxidase

Chapter 21

Phosphorylation of gp91^{phox}/NOX2 in Human Neutrophils

Houssam Raad, Riad Arabi Derkawi, Asma Tlili, Sahra A. Belambri, Pham My-Chan Dang, and Jamel El-Benna

Abstract

The phagocyte NADPH oxidase NOX2 was the first NOX family member to be discovered. It is responsible for the production of reactive oxygen species that are required for bacterial killing and host defense. Activated NOX2 is an enzymatic complex composed of two membrane proteins, p22^{phox} and gp91^{phox} (renamed NOX2), which form the cytochrome b_{558}, and four cytosolic proteins, p47^{phox}, p67^{phox}, p40^{phox}, and the small GTPase Rac2. Except for Rac2, all proteins from the complex become phosphorylated during neutrophil activation, suggesting the importance of this process in NOX2 regulation. The phosphorylation of the cytosolic components, and in particular p47phox, has been extensively studied; however, the phosphorylation of the membrane proteins was less studied, in part due to the lack of good antibodies and accurate membrane solubilization techniques. In this chapter, we describe a method we have used to study NOX2 phosphorylation, which is based on the labeling of the intracellular ATP pool with ^{32}P prior to applying a stimulus inducing protein phosphorylation. We also describe the solubilization of membrane-bound gp91^{phox}/NOX2 and analysis by immunoprecipitation, polyacrylamide gel electrophoresis, electrophoretic transfer, phosphoamino acid analysis, and autoradiography. This protocol can also be used to study the possible phosphorylation of other NOX family members.

Key words NOX2, gp91phox, NADPH oxidase, Neutrophils, Protein phosphorylation, Respiratory burst

1 Introduction

Phagocytic cells such as neutrophils, monocytes, and macrophages play a central role in defending the host against pathogens by producing high quantities of reactive oxygen species (ROS) [1, 2]. The enzyme responsible for this function is called the NADPH oxidase [3, 4]. This is a multicomponent enzyme complex, which is dormant in unstimulated cells and can be activated by various stimuli, such as phorbol myristate acetate (PMA), *N*-formyl-methionyl-leucyl-phenylalanine (fMLF), and opsonized

Houssam Raad and Riad Arabi Derkawi contributed equally to this work.

Ulla G. Knaus and Thomas L. Leto (eds.), *NADPH Oxidases: Methods and Protocols*, Methods in Molecular Biology, vol. 1982, https://doi.org/10.1007/978-1-4939-9424-3_21,

zymosan or bacteria [5, 6]. In its activated form, the NADPH oxidase complex mediates the transfer of one electron from cytosolic NADPH to O_2 to produce superoxide anion ($O_2^{\cdot-}$), the precursor of other toxic ROS, such as hydrogen peroxide (H_2O_2), hydroxyl radical ($OH^{\cdot}$), and hypochlorous acid (HOCl) [4, 7].

The NADPH oxidase consists of two membrane-bound proteins, $p22^{phox}$ and $gp91^{phox}$, (aka NOX2) which form the flavocytochrome b_{558} and four cytosolic proteins, $p47^{phox}$, $p67^{phox}$, $p40^{phox}$, and Rac1/2 [8, 9]. The $gp91^{phox}$ subunit is the central catalytic core of the oxidase and has a hydrophobic N-terminal domain and six putative transmembrane helices that coordinate two heme groups [9]. The more hydrophilic C-terminal domain is located in the cytosol and contains a flavoprotein domain, which is homologous to known flavoprotein dehydrogenase (FAD) binding sequences and has a consensus sequence for a putative NADPH-binding site [9]. Several homologues of $gp91^{phox}$ have been described, now members of the NADPH oxidase family or NOX, including $gp91^{phox}$ as NOX2, NOX1 to NOX5, and DUOX1 and DUOX2 [10]. The latter NOXs are expressed in several tissues (the lung, kidney, and colon) and in various cell types (epithelial cells, endothelial cells, vascular smooth muscle cells) [10].

NADPH oxidase activation in phagocytes is accompanied by the phosphorylation of all phox proteins, i.e., $p47^{phox}$, $p67^{phox}$, $p40^{phox}$, $p22^{phox}$, and $gp91^{phox}$ [11–15]. The phosphorylation of the cytosolic components has been known for years and that of p47phox has been extensively studied [16]. In contrast, the phosphorylation of the membrane $gp91^{phox}$ has been less investigated.

This chapter describes the protocol we use to study NOX2 phosphorylation in human neutrophils. Briefly, intracellular ATP is first labeled with ^{32}P prior to stimulating the cells to induce changes in protein phosphorylation. NOX2 is then solubilized from neutrophils (Fig. 1), immunoprecipitated using a specific antibody, resolved on sodium dodecyl sulfate-polyacrylamide gel

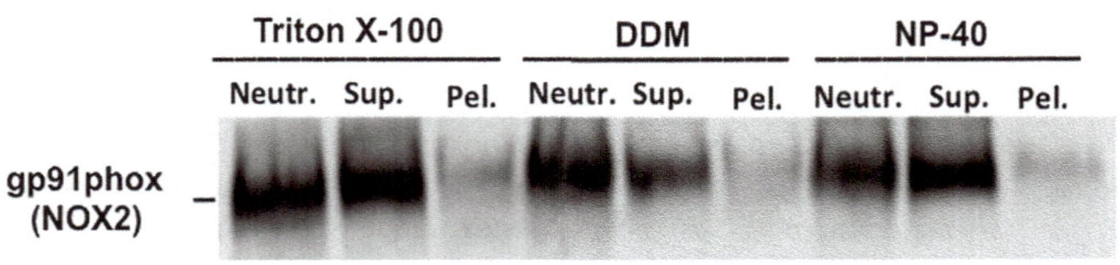

Fig. 1 Solubilization of $gp91^{phox}$/NOX2 by different detergents. Human neutrophils (40×10^6/mL) were isolated and lyzed in lysis buffer containing either 2% Triton X-100, 2% DDM, or 2% NP-40 and centrifuged at $100{,}000 \times g$. Proteins from the supernatants and pellets (1.5×10^6 cell eq) were analyzed by SDS-PAGE, transferred to nitrocellulose, and analyzed by Western blotting with a specific anti-NOX2 antibody. Neutr., total neutrophil lysate; Sup., supernatant containing solubilized proteins; Pel., pellet

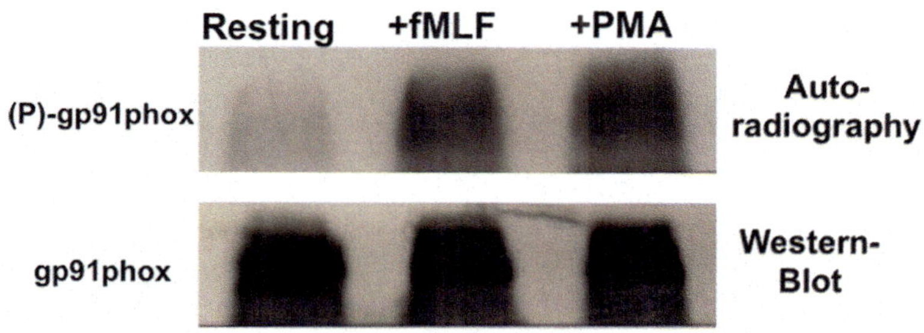

Fig. 2 Representative autoradiogram showing gp91^{phox}/NOX2 phosphorylation in human neutrophils. Human neutrophils were isolated and labeled with (^{32}P) H_3PO_4 and then stimulated with fMLP (10^{-6} M for 2 min) or PMA (200 ng/mL for 6 min). Resting and fMLP- or PMA-activated neutrophils were lyzed, gp91^{phox} was immunoprecipitated, and the immunoprecipitates were analyzed by SDS-PAGE. After transfer, phosphorylated gp91^{phox} was detected by autoradiography, and the whole protein was detected by Western blotting with a specific antibody

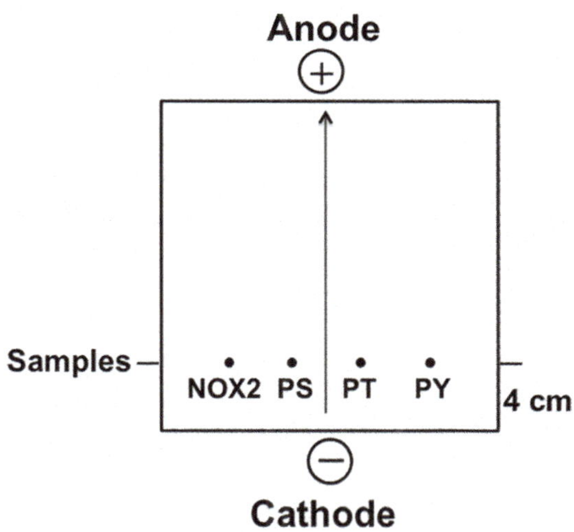

Fig. 3 Diagram of a cellulose plate showing marks used for phosphoamino acids analysis. The NOX2 sample is spotted 4 cm from the left side (cathode side) and 4 cm from the bottom of the plate. Phosphoserine (PS), phosphothreonine (PT), and phosphotyrosine (PY) were used as control markers and were spotted every 4 cm. The central arrow represents the migration direction

electrophoresis (SDS-PAGE), electro-transferred to membranes, and visualized by means of autoradiography and Western blotting (Fig. 2). The protocol for identifying the type of phosphorylated amino acids is also described (Figs. 3 and 4).

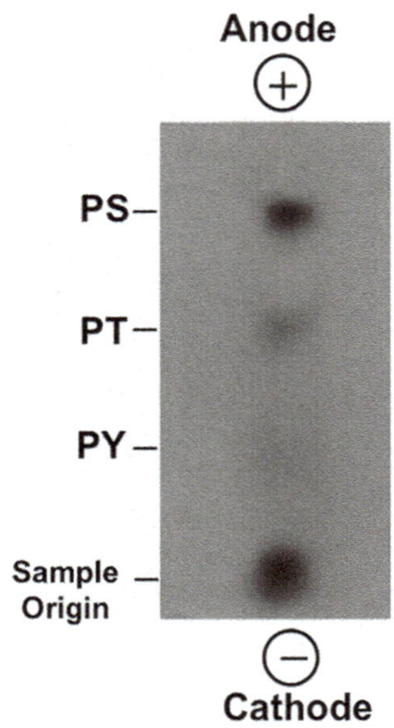

Fig. 4 A representative autoradiogram showing phosphoamino acid analysis of gp91[phox]/NOX2. ^{32}P-labeled neutrophils were stimulated with PMA (200 ng/mL for 6 min) and lyzed, gp91[phox]/NOX2 was immunoprecipitated, and the immunoprecipitates were analyzed by SDS-PAGE and transferred to PVDF membranes. The sample on the membrane was hydrolyzed before being subjected to phosphoamino acid analysis and autoradiography. Phosphoserine (PS), phosphothreonine (PT), and phosphotyrosine (PY) were used as control markers and stained by ninhydrin. The migration of the markers is indicated to the left of the figure

2 Materials

2.1 Buffers and Solutions

1. Dextran T500: 2% solution of 10 gr of dextran T500 (Pharmacosmos, Denmark) dissolved in 500 mL of 0.9% NaCl and filter-sterilized. Store at 4 °C for up to 4 weeks (*see* **Note 1**).
2. Phosphate-free loading buffer: 20 mM Hepes (pH 7.4), 150 mM NaCl, 5.4 mM KCl, 5.6 mM D-glucose, 0.8 mM $MgCl_2$, and 0.025% bovine serum albumin. Filter-sterilize and store for up to 3 weeks at 4 °C (*see* **Note 1**).
3. Lysis buffer: 10 mM HEPES pH 7.4, 8 mM NaCl, 80 mM KCl, 0.8 mM EDTA, 25 mM NaF, 5.4 mM Na_3VO_4, 2 mM β-glycerophosphate, 1 mM levamisole, 1 mM p-nitrophenyl phosphate (p-NPP), P8340 protease inhibitor cocktail (1:1000 dilution; Sigma), 8 μg/mL chymostatin, 0.08 mM DTT, 2% Triton X-100 (*see* **Note 2**). Final lysis buffer: 0.5 mM PMSF and 0.1 mM diisopropyl fluorophosphate (DFP) added to the lysis buffer above just before use (*see* **Note 3**).

4. Immunoprecipitation buffer: Lysis buffer containing 1% Triton X-100.
5. Laemmli sample buffer (2×): 125 mM Tris–HCl (pH 6.8), 6% SDS, 8% beta-mercaptoethanol, 18% glycerol, 5 mM EDTA, 5 mM EGTA, 10 μg/mL leupeptin, 10 μg/mL pepstatin, 10 μg/mL aprotinin, 10 mM NaF, 5 mM $NaVO_3$, and 2 mM p-NPP. Store at −20 °C for up to 12 months.
6. Transfer buffer: 50 mM Tris base, 95 mM glycine, 0.08% SDS, and 20% methanol.
7. TBS-Tween (TBST) buffer: 25 mM Tris–HCl pH 7.5, 150 mM NaCl, 0.05% Tween 20.
8. Nonfat dried milk (NFDM) solutions: 5% and 1% solutions prepared in TBST buffer.
9. ^{32}P-Orthophosphoric acid ($H_3{}^{32}PO_4$) in water, HCl-free, carrier-free: 10 mCi/mL (Perkin-Elmer-NEN).
10. Thin-layer electrophoresis (TLE) buffer: 2 L prepared with 1890 mL water +100 mL acetic acid +10 mL pyridine, pH 3.5.
11. Ninhydrin stain solution: 0.25% in acetone.
12. Phorbol myristate acetate (PMA) stock: 1 mg/mL in sterile DMSO is stored at −70 °C in aliquots.
13. *N*-formyl-methionyl-leucyl-phenylalanine (fMLF) stock: 10 mM in sterile DMSO is aliquoted for storage at −20 °C.
14. Protease inhibitors: Leupeptin and aprotinin are dissolved at 10 mg/mL in PBS; pepstatin is dissolved at 10 mg/mL in DMSO and stored at −20 °C in aliquots.
15. Phenylmethylsulfonyl fluoride (PMSF): 0.5 M in DMSO, prepared fresh before use.
16. The anti-human gp91^{phox} (clone 54.1) is from Santa Cruz Biotechnology.
17. ECL reagents are from GE Healthcare.

3 Methods

3.1 Isolation of Human Neutrophils

Blood is collected from healthy adult volunteers into citrate dextrose as the anticoagulant. Neutrophils are isolated by a classical technique using dextran sedimentation and Ficoll density gradient centrifugation as follows:

1. Mix 25 mL of whole blood and 25 mL of the 2% dextran T500 solution (1% final) in 50 mL tubes. Gently mix by inverting the tubes several times, and allow to sediment at room temperature for 20–40 min (*see* **Note 4**).

2. Gently collect the upper layer containing the leukocytes into centrifuge tubes, and discard the pellet containing red cells.
3. Centrifuge the collected upper layer at 400 × *g* for 8 min at room temperature (*see* **Note 5**).
4. After centrifugation, the pellets contain leukocytes and some contaminating erythrocytes, while the supernatants contain plasma, dextran, and platelets. Discard the supernatants by gently inverting the tubes, resuspend each pellet in 5 mL of loading buffer, and pool several pellets from the same donor (*see* **Note 6**).
5. Prepare a gradient by gently layering the cells on top of the Ficoll solution (1–3 volume of cell suspension/1 volume Ficoll) avoiding mixing the two layers, and centrifuge the tubes at 400 × *g* for 30 min at 22 °C without the centrifuge brake (*see* **Note 7**).
6. After centrifugation, discard the upper layers and mononuclear cells, which are at the interface of the buffer and Ficoll using suction.
7. Gently disperse the pellet containing neutrophils and contaminating erythrocytes (*see* **Note 8**).
8. Remove contaminating erythrocytes by hypotonic lysis: add 15 mL of ice-cold sterile 0.2% NaCl for 40 s, mix gently, and restore isotonicity by adding 15 mL of 1.6% NaCl (*see* **Note 9**).
9. Centrifuge the tubes at 400 × *g* for 8 min at 4 °C. Aspirate the red supernatant with gentle suction, and resuspend the neutrophil pellet in 5–10 mL of phosphate-free loading buffer.
10. Dilute the cells in trypan blue (1/100), and count them using a hemocytometer.

3.2 DFP Treatment (See Note 3)

1. Adjust the cell concentration to 50–100 × 10^6/mL in phosphate-free loading buffer.
2. Add the serine protease inhibitor DFP (2.7 mM) by diluting the stock solution (5.4 M) 1:2000, and incubate for 20 min on ice at 4 °C while gently mixing every 5 min (*see* **Note 3**).
3. Add 10 vol. of phosphate-free loading buffer, centrifuge at 400 × *g* at 4 °C, discard the supernatant in 1 N NaOH, and resuspend the neutrophil pellet in phosphate-free loading buffer.

3.3 ^{32}P-Labeling and Stimulation of Neutrophils

As large amounts of [^{32}P]-orthophosphoric acid are used during cell labeling, good shielding with 1.5-cm-thick Plexiglas is necessary at all times. Waste containers and bags must be kept in 1.5-cm-thick Plexiglas boxes. The investigator should wear protective eyewear, a laboratory coat, and two pairs of gloves. To avoid laboratory contamination, it is important to check the gloves with a Geiger counter after each manipulation of ^{32}P and to discard them in the shielded boxes, if contaminated.

1. Dilute the cells to 100 × 10^6/mL in phosphate-free loading buffer, and transfer to a 50-mL conical tube.
2. Add 0.5 mCi ^{32}P-orthophosphoric acid per 10^8 cells/mL (^{32}P-orthophosphoric acid ($H_3{}^{32}PO_4$ in water, HCl-free (10 mCi/mL)), and incubate for 60 min at 30 °C in a 30 °C water bath with gentle mixing every 10–15 min.
3. Centrifuge the cells at 400 × *g* for 8 min, discard the supernatant in an appropriate waste container, and resuspend the cells in phosphate-free loading buffer +1 mM $CaCl_2$ + 0.7 mM $MgCl_2$ (*see* **Note 10**).
4. To stimulate the cells, dilute them to 20 × 10^6 cells/mL, and preincubate the suspension at 37 °C for 10 min prior to adding PMA (200 ng/mL) for 8 min or 1 μM fMLP for 1 min.
5. The reaction is stopped by adding 10 volumes of ice-cold loading buffer, mixing gently, and centrifuging at 400 × *g* for 8 min at 4 °C. Discard the supernatants, and eliminate residual buffer with a pipette and suction.

3.4 Neutrophil Lysis

We describe here a lysis technique that we have found useful to preserve phosphorylated proteins in neutrophil lysates:

1. Keep the tubes on ice, and resuspend the cell pellets (to 50 × 10^6 cells/mL) with a pipette in ice-cold lysis buffer containing 2% Triton X-100, 0.5 mM PMSF, and 0.1 mM DFP (*see* **Note 11**). As for DFP treatment of the cells, all manipulations with DFP must be done under the hood with the same safety rules (*see* **Note 3**).
2. Sonicate the cells (3 × 10 s) at 30% power with a microtip sonicator under the hood and behind shields.
3. Incubate the cell lysate on ice for 15 min to complete solubilization, and centrifuge at 100,000 × *g* for 30 min at 4 °C in a TL100 ultracentrifuge (Beckman Inc.). Recover the cleared cell lysate for immunoprecipitation.

3.5 gp91^{phox}/NOX2 Immunoprecipitation

1. Wash protein A-Sepharose beads by adding 10 vol. of PBS, mix by inverting, and centrifuge at 900 rpm (150 × *g*) for 3 min. To minimize contamination, screw-cap tubes are used for immunoprecipitation.
2. Discard the supernatant and add 10 vol. of immunoprecipitation buffer containing 1 mg/mL BSA to the beads, place on a rotating wheel for at least 30 min, centrifuge at 900 rpm for (150 × *g*) for 3 min, and discard the supernatant.

3. Dilute the cleared cell lysate 1:2 (from Subheading 3.4, **step 3**) in 2× lysis buffer without Triton X-100 (final concentration 1%), then add the diluted lysates to the tubes containing each 50 μL of the washed protein A beads, and add the anti-gp91^{phox} antibody (dilution 1:200).
4. Incubate overnight at 4 °C with end-over-end rotation on a rotating wheel.
5. The next day, centrifuge the tubes and transfer the supernatant to new tubes; store at −70 °C in a shielded box until further use.
6. Wash the beads four times with 1 mL of immunoprecipitation buffer by gently inverting the tubes and centrifuging at 900 rpm (150 × *g*) for 3 min to recover the bead pellet. The tube caps are replaced by new ones at every centrifugation.
7. Elute bound gp91^{phox} by adding 2 × Laemmli sample buffer (100 μL), vortex, and boil for 5 min.
8. Centrifuge the samples at 15,000 × *g* for 30 s, and use the supernatant for electrophoresis and Western blotting.

3.6 Electrophoresis and Western Blotting

The samples are then subjected to 10% SDS-PAGE using standard techniques [17]. The separated proteins are electro-transferred to nitrocellulose or PVDF membranes [18] using the transfer buffer described above. After transfer, one membrane is wrapped in plastic film and placed in a cassette with two screens for autoradiography at −70 °C for 24–72 h (Fig. 2-Autoradiography). The second membrane is used to detect gp91^{phox} by Western blotting with a specific antibody in order to check the amount of gp91^{phox} immunoprecipitated from each sample:

1. Incubate the membrane in 5% NFDM in TBST for 30 min at room temperature on a shaker to block any non-specific protein-binding sites.
2. Add the anti-gp91^{phox} antibody (1:5000 dilution) in TBST and 1% milk, and incubate for 1 h at room temperature.
3. Wash three times with TBS-Tween buffer for 10 min each wash.
4. Add a goat anti-mouse antibody (1:5000 dilution, Santa Cruz Biotechnology) labeled with horseradish peroxidase (HRP) or alkaline phosphatase (AP), and incubate for 1 h at room temperature.
5. Wash three times with blotto for 10 min each wash.
6. Develop with ECL (HRP) or BCIP/NBT (AP): to prepare the latter, add 100 μL of NBT solution and 100 μL of BCIP solution to 10 mL of carbonate buffer. Incubate the membrane for 15–30 min at room temperature. Stop the reaction by washing with H_2O (Fig. 2-Western Blot).

3.7 Phosphoamino Acid Analysis of gp91^{phox}/NOX2

This technique is adapted from Boyle et al. [19]. The phosphorylated proteins are first hydrolyzed in hot HCl, and the amino acids are then separated by thin-layer electrophoresis (TLE) using standard phosphorylated amino acids as migration controls. Phosphorylated amino acids are detected by autoradiography or ninhydrin staining for the standards.

3.7.1 Hydrolysis of gp91^{phox} in HCl

1. After immunoprecipitation and SDS-PAGE, transfer the protein to PVDF membranes, and perform autoradiography, as described in Subheading 3.6 (*see* **Note 12**).
2. Cut the PVDF area containing the band of interest (^{32}P-labeled gp91phox) into pieces about 2 × 2 mm, and place in an Eppendorf tube (*see* **Note 13**).
3. Add 100 μL of 6 N HCl, and incubate for 90 min at 110 °C.
4. Stop the reaction by cooling down the tube and centrifugation at 16,000 × *g* for 2.5 min.
5. Transfer the supernatant containing the released amino acids to a new tube, and dry in a SpeedVac™ concentrator.
6. Wash the peptides by adding 100 μL H_2O, vortex, and dry in the SpeedVac™ concentrator.

3.7.2 Thin-Layer Electrophoresis (TLE)

Although some groups use the HTLE-7000 thin-layer electrophoresis apparatus, we routinely use a horizontal gel apparatus (Multiphor II electrophoresis system, Pharmacia) cooled with a circulating water bath at 4 °C.

1. Dissolve the amino acids in 10 μL of TLE buffer pH 3.5 using a pipette tip and vortexing. Centrifuge the sample briefly to ensure that no residual particulate material will be applied to the cellulose plate.
2. Transfer the phosphoamino acids to a new tube, and add 2 μg of standard markers (phosphoserine, phosphothreonine, phosphotyrosine).
3. Choose a cellulose plate (20 × 20 cm, VWR-Merck) with no irregularities, and mark the location of the sample on the back of the plate with a permanent marker, i.e., 4 cm from the left side and 4 cm from the bottom of the plate (Fig. 3). The sample is then spotted at that location.
4. Fill the buffer tanks with TLE buffer, and place a double layer of 3 mm Whatman electrophoresis wicks in the tank.
5. Take a new double layer of 3 mm Whatman paper with the same dimensions as the cellulose plate (20 × 20 cm), and make holes corresponding to the positions of the samples. Wet the blotter by soaking it in the TLE buffer, and let the excess buffer drain off.

6. Wet the plate by carefully placing the Whatman blotter on it with the holes aligned over the samples. Press gently on the paper until the plate is evenly wetted, and then remove the blotter. This step allows the samples to concentrate into one spot.
7. Place the wet plate on the horizontal TLE apparatus, apply the wetted wicks to the plate by gently pressing such that they cover 1.5 cm of the plate, place the cover on the apparatus, connect the electrodes to the power supply, and start the run toward the cathode at 1100 V at 6 °C for 45 min.
8. Stop the electrophoresis, disconnect the apparatus, and air-dry the plate under the hood.
9. To visualize the phosphoamino acid standards, spray the plate with 0.25% ninhydrin, and heat the plate at 65 °C for at least 10 min. Purple spots will appear that correspond to the position of p-Ser, p-Thr, and p-Tyr.
10. Place the plate in a cassette with one screen for autoradiography at −70 °C for 24–72 h (Fig. 4).

4 Notes

1. All solutions used for cell preparation and cell manipulation, such as dextran solution and phosphate-free loading buffer, are prepared with sterile water or sterile 0.9% NaCl (injection grade) from VWR or any other company.
2. Solutions used for cell lysis and protein analysis are prepared in very pure water (resistivity:18.2 MΩ-cm).
3. Other serine protease inhibitors have been used, but using DFP yields the best results. DFP protects proteins from degradation by neutrophil serine proteases but is extremely toxic. Appropriate safety precaution must be used, like handling only under a fume hood and wearing gloves and laboratory coats. All tips and other materials contaminated with DFP must be discarded in a 1 N NaOH-containing waste container placed in the hood and disposed appropriately.
4. Room temperature must be around 22 °C; if it is higher, incubate at 4 °C for longer time.
5. This centrifugation step is used to concentrate the cells and thus to use less Ficoll. If only a small volume of blood is used, the supernatant can be layered directly onto the Ficoll.
6. It is important to use phosphate-free buffer for all procedures prior to ^{32}P labeling in order to starve the cells.
7. When layering the cells on Ficoll, take care to avoid mixing the two media.

8. Never vortex the cells at this point as it can activate and damage them.
9. Never lyse the erythrocytes without resuspending the pellet first, as the cells will tend to aggregate. The tube can be gently tapped on the bench to dislodge the pellet prior to lysis.
10. To avoid radioactive contamination, radioactive supernatants are best discarded by pipetting with plastic disposable transfer pipettes rather than by decanting and placing the liquid in proper waste containers.
11. As $gp91^{phox}$ is a membrane protein, we first compared different detergents to extract it. Results presented in Fig. 1 show that Triton X-100 gave the best results. Therefore, Triton X-100 has been used throughout the protocol.
12. For phosphoamino acid analysis, PVDF membranes were used instead of nitrocellulose because it is resistant to boiling in 6 N HCl. PVDF must be soaked in 100% methanol before use in transfer buffer.
13. To locate the phosphorylated band of interest, before autoradiography, the PVDF membrane is taped to a used film, and the new unexposed film is taped to the old one.

Acknowledgments

This work was supported by grants from the Institut National de la Santé et de la Recherche Médicale (INSERM), the Centre National de la Recherche Scientifique (CNRS), Université Paris Diderot, Labex Inflamex, and Association Vaincre la Mucoviscidose (VLM). The authors also thank Dr. Martine Torres for her critical review and editing of the manuscript.
Houssam Raad, Riad Arabi Derkawi, contributed equally to this work.

References

1. Roos D, van Bruggen R, Meischl C (2003) Oxidative killing of microbes by neutrophils. Microbes Infect 5:1307–1315
2. Nauseef WM (2007) How human neutrophils kill and degrade microbes: an integrated view. Immunol Rev 219:88–102
3. Babior BM (1999) NADPH Oxidase: an update. Blood 93:1464–1476
4. El-Benna J, Dang PM, Gougerot-Pocidalo MA et al (2005) Phagocyte NADPH oxidase: a multicomponent enzyme essential for host defenses. Arch Immunol Ther Exp (Warsz) 3:199–206
5. Bokoch GM (1995) Chemoattractant signaling and leukocyte activation. Blood 86:1649–1660
6. El-Benna J, Hurtado-Nedelec M, Marzaioli V et al (2016) Priming of the neutrophil respiratory burst: role in host defense and inflammation. Immunol Rev 273:180–193
7. Hampton MB, Kettle AJ, Winterbourn CC (1998) Inside the neutrophil phagosome: oxidants, myeloperoxidase, and bacterial killing. Blood 12:3007–3017

8. Groemping Y, Rittinger K (2005) Activation and assembly of the NADPH oxidase: a structural perspective. Biochem J 386:401–416
9. Vignais PV (2002) The superoxide-generating NADPH oxidase: structural aspects and activation mechanism. Cell Mol Life Sci 59:1428–1459
10. Bedard K, Krause KH (2007) The NOX family of ROS-generating NADPH oxidases: physiology and pathophysiology. Physiol Rev 87:245–313
11. El-Benna J, Faust LP, Babior BM (1994) The phosphorylation of the respiratory burst oxidase component p47phox during neutrophil activation. Phosphorylation of sites recognized by protein kinase C and by proline-directed kinases. J Biol Chem 269:23431–23436
12. Dang PM, Morel F, Gougerot-Pocidalo MA et al (2003) Phosphorylation of the NADPH oxidase component p67(PHOX) by ERK2 and P38MAPK: selectivity of phosphorylated sites and existence of an intramolecular regulatory domain in the tetratricopeptide-rich region. Biochemistry 42:4520–4526
13. Bouin AP, Grandvaux N, Vignais PV et al (1998) p40(phox) is phosphorylated on threonine 154 and serine 315 during activation of the phagocyte NADPH oxidase. Implication of a protein kinase C-type kinase in the phosphorylation process. J Biol Chem 273:30097–30103
14. Regier DS, Waite KA, Wallin R et al (1999) A phosphatidic acid-activated protein kinase and conventional protein kinase C isoforms phosphorylate p22(phox), an NADPH oxidase component. J Biol Chem 274:36601–36608
15. Raad H, Paclet MH, Boussetta T et al (2009) Regulation of the phagocyte NADPH oxidase activity: phosphorylation of gp91phox/NOX2 by protein kinase C enhances its diaphorase activity and binding to Rac2, p67phox, and p47phox. FASEB J 23:1011–1022
16. El-Benna J, Dang PM, Gougerot-Pocidalo MA et al (2009) p47phox, the phagocyte NADPH oxidase/NOX2 organizer: structure, phosphorylation and implication in diseases. Exp Mol Med 41:217–225
17. Laemmli UK (1970) Cleavage of structural proteins during the assembly of the head of bacteriophage T4. Nature 227:680–685
18. Towbin H, Staehlin T, Gordon J (1979) Electrophoretic transfer of proteins from polyacrylamide gels to nitrocellulose sheets: procedure and some applications. Proc Natl Acad Sci U S A 76:4350–4354
19. Boyle WJ, Van Der Geer P, Hunter T (1991) In: Hunter T, Sefton BM (eds) Phosphopeptide mapping and phosphoamino acid analysis by two-dimensional separation on thin-layer cellulose plates. Academic Press, Inc, San Diego, CA, pp 110–149

Chapter 22

The Molecular Regulation and Functional Roles of NOX5

David J. R. Fulton

Abstract

NOX (*N*ADPH *ox*idases) are a family of NADPH-dependent transmembrane enzymes that synthesize superoxide and other reactive oxygen species. There are seven isoforms (NOX1–5 and DUOX1–2) which derive from a common ancestral NOX. NOX enzymes are distinguished by different modes of activation, the types of ROS that are produced, the cell types where they are expressed, and distinct functional roles. NOX5 was one of the earliest eukaryotic Nox enzymes to evolve and ironically the last isoform to be discovered in humans. In the time since its discovery, our knowledge of the regulation of NOX5 has expanded tremendously, and we now have a more comprehensive understanding of the molecular mechanisms underlying NOX5-dependent ROS production. In contrast, the cell types where NOX5 is robustly expressed and its functional significance in health and disease remain an underdeveloped area. The goal of this chapter is to provide an up-to-date overview of the mechanisms regulating NOX5 function and its importance in human physiology and pathophysiology.

Key words NOX5, NADPH oxidase 5, Reactive oxygen species, Posttranslational regulation, Function

1 Introduction

NOX5 was first reported by two independent groups in 2001 based on a blast search using the C-terminus of gp91^{phox} as bait to identify novel transcripts [1] or Nox3 [2]. The deduced mRNA encoded a new isoform with a novel extended N-terminal EF-hand calcium binding domain and conserved six transmembrane-spanning and C-terminal NADPH binding reductase domains. While NOX5 shares the same basic structural organization as other members of the NOX family, it is distinguished by its unique N-terminal EF-hand motif that confers calcium-dependent regulation of enzyme activity. In cells transfected with NOX5 cDNA, low levels of superoxide can be detected under steady-state conditions. The addition of calcium-mobilizing stimuli triggers a rapid and robust increase in superoxide that returns to baseline over the time course of minutes. Unique to NOX5 is the ability to generate ROS without an absolute requirement for specialized accessory

Ulla G. Knaus and Thomas L. Leto (eds.), *NADPH Oxidases: Methods and Protocols*, Methods in Molecular Biology, vol. 1982, https://doi.org/10.1007/978-1-4939-9424-3_22,

proteins which bind to and regulate the activity of NOXes1–4. NOX5 mRNA expression was detected in the testis, spleen, and lymph nodes and also in vascular cells [1, 2]. While initially the regulation of NOX5 was simple, we now appreciate that there are multiple splice variants of NOX5, multiple intronic and exonic SNPs, the molecular mechanisms regulating enzymatic activity and ROS production extend beyond calcium, NOX5 actually binds to numerous cytosolic proteins that are important for its activity, NOX5 is expressed in a wide variety of cell types and activity and expression levels are modified in various disease states, and while NOX5 has been lost from the genomes of rats and mice, novel NOX5 transgenic mice have been created that provide unique perspectives into its biological roles. In this chapter, we shall summarize these aspects of NOX5 biology as well as advances in the specific tools that have been developed to identify the biological functions of NOX5.

2 Molecular Regulation of NOX5 Activity

2.1 Topology

The basic structural organization of NOX5 is conserved with the other Nox isoforms and consists of six transmembrane α-helices that are connected by five loops (referred to as the A–E loops) [3]. The topology of the transmembrane domain places loops A, C, and E on the outside of the cell (plasma membrane-localized NOX5), and conversely loops B and D are located in the cytoplasm. In organelle-resident NOX5, loops A, C, and E would lie within the lumen, and loops B and D would be cytosolic. Regardless of whether localized to membranes of the plasma membrane or organelles, NADPH binding occurs in the cytosol and oxygen reduction occurs on the opposite side of the membrane, in the extracellular or luminal space. The six transmembrane helices are preceded by an additional N-terminal α-helix and polybasic region that lies parallel to the inner side of the membrane. Based on the crystal structure of a related NOX5 isoform, the transmembrane region exhibits a triangular shape with the base toward the inner membrane and a narrower apex in the direction of the outer membrane [4]. NOX5 has been proposed to assemble into a transmembrane homotetramer. This is mediated through C-terminus motifs, and the ability of truncated C-terminal NOX5 constructs to impair enzyme activity suggests that tetramerization is critical for function and/or enzyme stability [5] although the mechanisms involved are incompletely understood.

2.2 Oxygen Binding and Electron Reduction

Histidine residues located in the third and fifth transmembrane (H286, H300, H374, and H387 using the sequence for NOX5αV1 and based on homology to a recently published crystal structure [4]) coordinate the binding of two type-b heme groups within a cavity formed by the first five transmembrane helices. The hemes are proposed to reside in an ordered configuration, with one closer to the inner membrane and the other closer to the outer membrane. Close to the other heme is a cavity that is suggested to facilitate oxygen binding through the coordinated actions of K243 and H304 [4]. The crystal structure of the transmembrane region does not support the direct binding of molecular oxygen to the heme but rather a noncovalent interaction that involves hydrophilic side chains [4]. Thus, electrons flow from NADPH bound to the C-terminal through FAD which binds directly to the lower heme, and electrons move from the lower heme to the upper heme via V380 (or W381) where, in coordination with histidine and lysine residues, they promote reduction of resident molecular oxygen. Further insights gained from the proposed three-dimensional structure of NOX5 are the interactions between the loops and other functional domains. The cytosolic loops (B and D) have been shown to be important in the regulation of other NOX enzymes [6], and in NOX5, the B-loop has also been shown to interact with L558 in the C-terminal dehydrogenase domain. In the D-loop, R362, R363 may be in direct contact with the C-terminal F737. The C-terminal F residue is highly conserved among NOX enzymes and proposed to function as a regulatory toggle switch that enables NADPH binding [4].

2.3 Calcium-Dependent Regulation

The N-terminal extension of NOX5 [1] is a structural feature unique among the NOX isoforms and is analogous to that found in the DUOXes [7]. NOX5 activity has an absolute requirement for calcium, and ROS are not generated in a calcium-free buffer or from truncated NOX5 enzymes lacking the N-terminal region [1, 8, 9]. NOX5 contains four EF-hand motifs that operate as two pairs which are distinguished by a difference in the affinity for calcium, with the proximal N-terminal pair binding calcium with a lower affinity than the distal pair [8, 10]. The proposed model for calcium-dependent ROS production is that elevation of calcium promotes occupation of both pairs of EF hands which triggers a conformational change in the N-terminus that enables it to bind to a C-terminal region to initiate electron flow from NADPH to oxygen. An auto-inhibitory domain that restricts enzyme activity has been identified in the C-terminus of NOX5 and designated the regulatory EF-hand binding domain (REFBD, aa 656–679). The calcium-bound N-terminus of NOX5 binds to the REFBD and displaces it, removing the auto-inhibition and facilitating enzyme activation [8, 11]. EF-hand binding promotes conformational changes that enable the B-loop to interact with a C-terminal region (540–555 NOX5V1) and Lys363 of the D-loop to interact with the C-terminal F737 to induce a NADPH binding conformation that promotes electron transfer [4]. These numbers are based on the crystal model, and more rigorous analysis of the enzymatic properties of these residues is warranted. Despite this robust working model, caveats remain on the regulation of NOX5 activity including the amount of calcium required to fully activate NOX5 is surprisingly high and secondly the ability of NOX5 to produce ROS under conditions where resting levels of intracellular calcium do not change appreciably. These observations suggest that additional mechanisms may influence NOX5 activity.

2.4 Phosphorylation

In the characterization of calcium-dependent activity in cells, NOX5 was found to be surprisingly insensitive to cytoplasmic calcium levels with a concentration of approximately 1 μM required to elicit 50% of maximal activity [8]. This makes the activation of NOX5 in most cells unlikely unless there are supraphysiological increases in intracellular calcium. In 2007, additional pathways were identified that enable robust activation of NOX5 without changes in intracellular calcium. PMA was found to activate NOX5 in a slow and sustained manner that contrasted the acute and transient changes elicited by calcium-mobilizing agents. PMA was able to activate NOX5 without elevating intracellular calcium and could synergize with low or submaximal concentrations of calcium-mobilizing agonists to elicit significantly greater levels of ROS production [9, 12]. PMA also stimulated the direct phosphorylation of NOX5 within a C-terminal region that contains a cluster of putative PKC phosphorylation sites (T512 and S516, NOX5V1). Mutation of these residues to the phospho-null residue, alanine (which is structurally similar but cannot be phosphorylated), attenuated the ability of PMA to activate NOX5. In contrast, mutation of these residues to negatively charged amino acids such as glutamic acid, which can mimic the negative charge imparted by phosphorylation, resulted in an enhanced ability to generate ROS [9]. The PKC family member that mediates NOX5 phosphorylation was found to be PKCα as genetic silencing of PKCα reduced the ability of PMA to activate NOX5 and a constitutively active form of PKCα was sufficient to robustly increase NOX5 activity [13]. Other PKC isoforms may also play a role, although the evidence is less clear. The silencing of PKCε also attenuated the ability of PMA to stimulate ROS production from NOX5; however, the expression of two different constitutively active constructs of PKCε strongly reduced NOX5 activity. How PKCε can have such seemingly divergent effects on NOX5 is not yet clear. It remains possible that the active forms of PKCε traffic to different subcellular locations, that endogenous PKCε participates in cross talk between PKC isoforms, and that there are differences in substrate specificity between endogenous and overexpressed enzymes. Other PKC isoforms can negatively regulate NOX5 activity as suggested by the increased PMA-stimulated NOX5 activity observed in cells lacking PKCδ [13]. PKC can also activate other kinases, and the ability of PMA to stimulate NOX5 may, at least in part, involve the kinase ERK [14]. The MEK1 inhibitors (PD98059 and U0126), silencing of MEK1, or expression of a dominant negative MEK1 reduces the ability of PMA to stimulate NOX5. ERK can stimulate the phosphorylation of S516 and to a lesser degree by T512. However, the expression of a constitutively active form of MEK1 failed to increase NOX5 activity suggesting that the MEK/ERK1/2 pathway is necessary but not sufficient to regulate the PMA-dependent activation of NOX5 [14] and that PMA activation requires other signaling molecules such as PKCα.

NOX5 can also be activated by a number of other kinases. The inhibition of calcium-calmodulin-dependent pathways was shown to attenuate NOX5 activity [15] in endothelial cells, but the pathways involved were not identified. Calcium-mobilizing agonists are known to trigger the activation of calcium-dependent kinases, and inhibitors of calcium-calmodulin-dependent kinases have been shown to reduce NOX5 activity [16, 17]. CAMKII was shown to phosphorylate NOX5 on S493 (NOX5V1) which increases NOX5 activity [16]. The tyrosine kinase c-ABL has been shown to mediate activation of NOX5 in response to hydrogen peroxide, and while the kinase activity of ABL is necessary, it is not yet known if this involves the direct tyrosine phosphorylation of NOX5 or accessory proteins (see below) [18]. ABL has been shown to mediate NOX5 activation in sperm and cancer cells [19, 20]. Given the presence of numerous other motifs in NOX5, it is likely that several other kinases can regulate NOX5 activity, but these await further identification. In summary, while increased calcium is a major factor stimulating ROS production from NOX5, posttranslational phosphorylation can elicit robust increases in NOX5 activity at baseline levels of intracellular calcium, and these pathways can synergize to produce much greater NOX5 activity at lower levels of intracellular calcium. The cooperative interaction between NOX5 phosphorylation and calcium-dependent occupation of the EF hands provides for additional levels of control and diversifies the ability to produce an appropriate level of ROS in response to a stimulus (summarized in Table 1).

Table 1
Kinases that regulate NOX5 activity

Kinase	Site of phosphorylation on NOX5α (V1)	Impact on activity	Refs.
PKCα	S508, T512, S516	↑	[13]
PKCε	ND[a]	↓	[13]
PKCδ	ND[a]	↓	[13]
CAM kinase II	S493	↑	[16]
ERK	S516	↑	[14]
Abl	ND (predicted Y105)	↑	[18]

[a] *ND* not determined

2.5 Accessory Proteins

Unlike the other NOX isoforms, the initial characterization of NOX5 revealed that it generates ROS in a calcium-dependent manner when expressed as a single gene product [1]. Nox1–4 require the co-expression of several specialized accessory proteins (P22PHOX (CYBA), P47PHOX (NCF1), P67PHOX (NCF2), NOXO1, NOXA1, 21). A prior study has shown that NOX5 can bind p22phox, but others have shown quite convincingly that it is not needed for NOX5 activity [22, 23]. Although NOX5 does not bind to established NOX regulatory subunits, it may bind to other proteins that are less specialized and more ubiquitously expressed to regulate its activity. In time, the latter concept has been validated, and a number of proteins have been shown to bind NOX5 and regulate its activity (summarized in Table 2).

2.6 Calmodulin

Calmodulin was the first protein-binding partner demonstrated to associate with NOX5 and regulate its activity [24]. A calmodulin binding site was identified in close proximity to the C-terminal, NADPH binding site (aa689–707 of NOX5V1), and calmodulin was shown to sensitize enzyme activity and ROS production to ambient calcium levels. Although equivalent calmodulin binding sites are well conserved on other NOX enzymes and present in NOX enzymes from organisms that don't express calmodulin [4], the binding of calcium-bound calmodulin to NOX5, but not to other NOX isoforms, was shown to increase its sensitivity to calcium. Whether other proteins bind to this region and perform a similar role for other NOX enzymes or in prokaryote cells that lack

Table 2
NOX5 protein-binding partners and effect on NOX5 activity

Protein	Binding site on NOX5	NOX5 activity	Refs.
Calmodulin	C-terminal aa689–707	↑	[24]
c-Abl	ND[a]	↑	[18]
PKCα	ND[a]	↑	[13]
Cav-1	ND[a]	↓	[26]
Hsp90	C-terminal (aa490–550)	↑	[35, 37]
Hsp70	C-terminal	↓	[36]
CHIP	ND[a]	↓	[36]
Hsp40	C-terminal	↑	[38]
Hop	C-terminal	↑	[38]
p23	C-terminal	↑	[38]

[a] *ND* not determined

calmodulin remains to be determined. The C-terminal location of the REFBD and the calmodulin binding motif suggest that calmodulin binding also promotes electron transfer, as with other calcium-dependent NADPH-dependent proteins such as eNOS. Calmodulin binding to NOX5 provides another pathway for calcium sensitization in addition to phosphorylation.

2.7 Caveolin-1

The transmembrane protein caveolin-1 (Cav-1) is the major structural protein of the plasma membrane organelle, caveolae. Cav-1 is essential for caveolae formation, and loss of Cav-1 expression contributes to numerous hyper-proliferative pathologies including pulmonary hypertension and cancer [25, 26]. Cav-1 has a complex biology, and in addition to its architectural role in the biogenesis of caveolae, it binds to and regulates the activity and subcellular location of numerous signaling molecules. Cav-1 has been proposed to facilitate the activation of NOX2 by concentrating receptors, signaling molecules, and accessory proteins within close proximity in microdomains of the plasma membrane [27, 28]. This is mediated in part through the formation of caveolae which possess a unique lipid environment and also through the direct interaction been Cav-1 and NOX [26]. In an apparent contradiction to this model, Cav-1 has been also shown to bind directly to NOX5 (and NOX2) and inhibits NOX1–3 and NOX5 enzyme activity. The ability of Cav-1 to inhibit NOX5 is mediated by its scaffolding domain (aa 82 and 101) and can be mimicked by a scaffolding domain peptide fused to a cell-penetrating sequence. This relationship seems paradoxical at first where caveolae are important for NOX activity, yet Cav-1 inhibits NOX enzyme activity, but is actually well documented for other enzymes that are regulated by Cav-1 including eNOS. Many studies have reported that eNOS activity is directly inhibited by caveolin-1, and yet eNOS must reside at plasma membrane caveolae for maximal activity [29, 30]. In this paradigm, the influx of calcium displaces Cav-1 from eNOS and activates the enzyme. Consistent with this, calcium-bound calmodulin has also been shown to robustly displace Cav-1 from NOX2 [26]. Whether this relationship exists for NOX5 remains to be determined, but is a likely eventuation. Thus Cav-1 expression and the structural integrity of caveolae are important for optimal NOX activation via compartmentalized signaling and calcium influx, but Cav-1 is also an important repressor of ROS production through direct enzyme binding as well as inhibition of pro-inflammatory signaling [26]. Accordingly, in disease states where there is a loss of Cav-1 expression, this may result in increased expression and activity of NOX enzymes and greater production of ROS as has been shown by numerous groups [26, 31–33]. NOX-derived ROS may also impact cellular signaling pathways which increase the phosphorylation of Cav-1 [34]. Cav-1 phosphorylation on Y14 has been shown to be important for the internalization of caveolae and also the degradation of Cav-1. Thus, the ability of ROS to stimulate Cav-1 phosphorylation may subsequently alter NOX activity.

2.8 Protein Kinases

Several kinases have been reported to bind NOX5 including ABL1 and PKCα, and the interaction between these molecules is regulated in a stimulus-dependent manner with increased binding seen with active forms of kinases [13, 18]. It is not yet clear whether other kinases capable of activating NOX5, including CAMKII, stably associate with NOX5.

2.9 Molecular Chaperones

NOX5 is a client protein of Hsp90 which is ubiquitously expressed in most cells. Hsp90 inhibition robustly compromises the ability of NOX5 to produce superoxide, and chronic Hsp90 inhibition triggers NOX5 protein degradation [35–38]. Direct binding of NOX5 and Hsp90 is documented by co-immunoprecipitation, proximity ligation, and bioluminescence resonance energy transfer studies [35]. Hsp90 inhibitors stimulate the dissociation of Hsp90 from NOX5 and reciprocal binding of Hsp70, which serves to recruit CHIP (carboxyl terminus of Hsp70-interacting protein), an Hsp70-regulated E3 ubiquitin ligase [36]. Increased Hsp70 binding promotes NOX5 ubiquitination and subsequent degradation by the proteasome. How Hsp90 binding impacts NOX5 enzyme activity at the molecular level remains incompletely understood. Using other Hsp90 client proteins as examples, Hsp90 might serve to regulate the tertiary structure of NOX5 and integrity of substrate or cofactor binding pockets, the coordination of loops, the insertion of functional heme moieties, its subcellular localization, or the regulation of kinase activity. A particularly interesting observation is that acute inhibition of Hsp90 selectively inhibits superoxide production but not hydrogen peroxide [35]. While revealing about the mechanisms by which these ROS are produced, how Hsp90 orchestrates these changes is not yet known. In studies of the hydrogen peroxide-producing enzymes, NOX4 and DUOX, the extracellular loops have been shown to dismutate superoxide to hydrogen peroxide [39, 40]. Hsp90 may therefore have an important role as a chaperone that helps coordinate the stability or position of the extracellular loops of NOX5. A major confounder to this mechanism is the intracellular location of Hsp90 and how it would impact the extracellular (or intra-organelle) loops of NOX5 [9, 22]. Hsp90 has also been shown to regulate the formation of NOX5 oligomers. The C-terminus of NOX5 has been demonstrated to mediate the formation of tetrameric NOX5 complexes [5] which is also where Hsp90 binds. Inhibition of Hsp90 results in the loss of higher molecular weight complexes of NOX5 and decreased interaction between monomers suggesting that Hsp90 facilitates the interaction between NOX5 monomers which may help to facilitate enzyme stability and superoxide production

[38]. Increasing Hsp70 expression can displace Hsp90 and alter the balance of enzyme stability and ROS production in favor of reduced superoxide and decreased enzyme stability. This mechanism provides an opportunity for drugs (GGA and BGP-15) that can acutely increase Hsp70 expression to promote the degradation of NOX5 (and NOX1–3) and reduce ROS production. Alternatively, a shift in the balance of Hsp90/Hsp70 levels, which has been reported in several disease states, may function to alter NOX activity and ROS production. Examples of this include the mouse $ApoE^{-/-}$ model of atherosclerosis where increased Hsp90 levels are seen in vascular lesions and Hsp90 inhibitors effectively reduced the size and complexity of lesions along with reduced expression of NOX1 and NOX2 and reduced ROS [37]. In a model of type II diabetes, Hsp90 inhibitors potently reduce ROS production and NOX expression in isolated leukocytes, lung tissue, human blood vessels, and aorta from db/db mice [36]. Hsp90 also has established roles in ROS-associated diseases such as cancer [41] and inflammation [42] and, in addition to NOX5, has been shown to also be important for the activity of NOX1–3 [35].

2.10 Subcellular Localization

In human prostate cancer cells, human endothelial cells, and transfected cells, NOX5 has been found predominantly on endomembranes with a distribution consistent with the endoplasmic reticulum (ER) [9, 12, 22, 43, 44]. ER expression has been observed with both WT NOX5 using antibody-based approaches in fixed cells and GFP-NOX5 fusion proteins, and these results are also consistent with studies documenting the ER expression of other NOX isoforms [45–48]. NOX5 has also been detected at the plasma membrane [12, 43] where presumably it is ideally positioned to efficiently release superoxide into the extracellular space. NOX5 has been shown to translocate from intracellular membranes to the plasma membrane in a process that is mediated by a polybasic domain within the N-terminus of NOX5 (PBR-N) that is juxtaposed to the transmembrane region, which is able to bind to phosphatidylinositol 4,5-bisphosphate, a plasma membrane-enriched phospholipid. This region of NOX5 is highly conserved across species and may underlie the variable expression pattern of NOX5 at both the ER and plasma membrane. NOX5 has an additional polybasic domain in the C-terminal region (PBR-C) which does not affect its intracellular location but does impact the regulation of enzyme activity. How plasma membrane NOX5 generates extracellular superoxide is well appreciated, but whether the endomembrane NOX5 actually produces extracellular superoxide remains poorly understood. Recent studies show that plasma membrane NOX5 is required to generate extracellular superoxide [49]. How NOX5 traffics to the plasma membrane is poorly understood. For NOX2, transport to the plasma membrane occurs via the conventional secretory pathway and is mediated by Sar1 and Stx5.

NOX5 reaches the plasma membrane via another yet to be identified pathway. Glycosylation may be an important distinguishing factor as the other NOX isoforms are glycosylated, whereas NOX5 is not, and the introduction of artificial glycosylation sites to NOX5 alters the mechanisms by which it traffics to the plasma membrane [49].

2.11 Protein S-Nitrosylation

An intimate relationship exists between superoxide and nitric oxide (NO). Superoxide was shown over 30 years ago to react avidly with NO and inactivate NO-dependent signaling [50]. More recently, NO has been shown to inhibit NOX activity and superoxide production [51] including NOX5 with a rank-order potency of NOX1 $\geq$ NOX3 > NOX5 > NOX2, whereas NOX4-dependent production of hydrogen peroxide was refractory to the effects of NO [52]. How NO suppresses NOX activity was shown to be due to a direct posttranslational modification of the enzyme but not due to tyrosine nitration, glutathionylation, or phosphorylation. Perhaps not surprisingly, the major modification detected was S-nitrosylation. Four major sites of S-nitrosylation, C107, C246, C519, and C694, were identified by mass spectrometry. Of these sites, S-nitrosylation of C694 was the most significant in reducing NOX5 activity. Inhibition of NOX5-derived superoxide was seen with both NO donors and endogenously produced NO in both endothelial cells and LPS-challenged smooth muscle. Increased S-nitrosylation was reversed by increases in calcium and also the denitrosylating enzymes, thioredoxin 1 and GSNO reductase [52]. Functionally, these results advance the concept that NO curtails the activity of NOX5 and other NOX isoforms and that loss of NO, as occurs with many disease states, may reciprocally increase ROS production.

2.12 SUMOylation

Protein SUMOylation (SUMO, small ubiquitin-related modifier) provides protection from cellular stress, including excessive levels of oxidative stress [53]. Increased SUMOylation was found to significantly reduce NOX5 (and other NOX isoform) activity in a variety of cell types, but the mechanisms by which SUMOylation decreased ROS production remain unresolved. Direct enzyme activity is decreased in isolated (cell disrupted) assays with excess calcium and NADPH, but the level of enzyme expression and phosphorylation of NOX5 remain unchanged. The SUMO story is weakened by the lack of evidence supporting the posttranslational attachment of SUMO to NOX5 [54]. Furthermore, the ability of SUMO to reduce the activity of all of the NOX enzymes (NOX1–5) suggests a more general mechanism that may involve actions on universal regulators such as accessory proteins, heme insertion, or the lipid environment.

3 Genetics of NOX5

The gene for human NOX5 is located on chromosome 15 and is capable of encoding six distinct isoforms (Fig. 1). Multiple isoforms of NOX5 arise from alternate promoter use and different translational start sites as well as the splicing of exon6. Diversity is found only at the N-terminus, and there is complete conservation of the enzyme through the transmembrane region and C-terminus. Additions and subtractions to the N-terminus can have profound effects on the function of the NOX5 isoforms (Table 3). The α (V1) and β (V2) isoforms are approximately equal in activity, generating equivalent amounts of superoxide and hydrogen peroxide, regardless of the type of stimulation (calcium- or PMA-initiated). The NOX5α (V1) contains a small N-terminal extension compared to NOX5β (V2), and it remains to be determined whether this confers any unique properties in the regulation of NOX5 activity. In contrast, the γ (V3), δ (V4), and ε (V5) isoforms are all catalytically inactive in transfected cells and fail to generate detectable levels of superoxide or hydrogen peroxide at low or high levels of expression [55]. For the NOX5 V3 and V4 isoforms, an N-terminal insertion between the third and fourth EF hand could possibly disrupt calcium binding or the ability of this region of the N-terminus to interact with the C-terminus that is required to trigger electron flow. The majority of the calcium-dependent regulation appears to be mediated by the distal EF-hand pair as mutation within the first pair does not appreciably alter NOX5 function [8]. The "short"

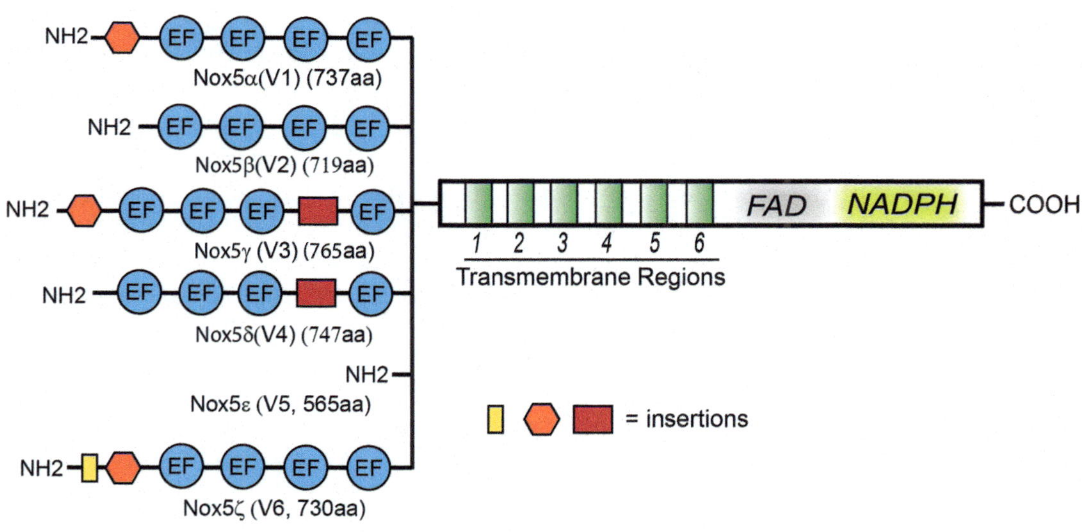

Fig. 1 Isoforms of NOX5. Differential splicing and alternative promoter use lead to the expression of six different isoforms of NOX5 that are distinguished by unique insertions and deletions within the extreme N-terminus. (*EF* EF-hand calcium binding motif, *FAD* flavin adenine dinucleotide, *NADPH* nicotinamide adenine dinucleotide phosphate)

Table 3
Splice isoforms of NOX5 and respective enzyme activity

Isoform	Active	Refs.
NOX5α, NOX5V1	Yes	[1, 9, 55]
NOX5β, NOX5V2	Yes	[1, 55]
NOX5γ, NOX5V3	No	[55]
NOX5δ, NOX5V4	No	[55]
NOX5ε, NOX5V5	No	[55]
NOX5ζ, NOX5V6	?/Likely	[21]

NOX5ε (V5) lacks all of the N-terminal EF hands and is completely inactive in transfected cells. It is not yet known whether NOX5ζ (V6) is active, but based on homology to the other isoforms with functional EF hands, it is likely to be active. The relative expression levels of NOX5 isoforms vary with cell type, and V3–5 can function as dominant negatives when co-expressed with the active isoforms (V1 and V2) as they have intact C-terminal regions which are critical for oligomerization and N-terminal insertions that disrupt enzyme activity [5, 55].

3.1 Polymorphisms

In comparison to other members of the NOX family, the gene for NOX5 is relatively large (~132 kb versus ~31 kb for Nox1 and ~33 kb Nox2). Within the gene for NOX5, several single nucleotide polymorphisms have been identified in intragenic, intronic, and exonic regions that may influence NOX5 expression and activity. The ORF for NOX5 V1 has greater than 108 nonsense, missense, and synonymous SNPs (NCBI) with frequencies varying from a minor allele frequency (MAF) of 0.0005–0.31. Several missense SNPs have been found in NOX5 within regions of the enzyme that may impact enzyme function including N-terminal EF hands, transmembrane regions, and the C-terminus reductase. One SNP that is validated (rs34406284) and two non-validated SNPs (rs369517329, rs370082662) are missense mutations that code for stop codons. Individuals harboring these SNPs in NOX5 would express truncated enzymes that are likely to be inactive. The enzymatic consequences of the more frequent SNPs with an impact in critical areas of the enzyme have been assessed [56]. Mutations, representing 15 of these SNPs, were generated in NOX5 and activity assessed. Of the 15 SNPs, 7 individual mutations resulted in the expression of full-length NOX5 that was catalytically inactive (M95K, S254R, T271M, R437Q, R548H, G560R, V707A, NOX5V1). A few SNPs had relatively high frequencies in the population studied such as the R548H (10.62%). All of the SNPs resulted in reduced or no change in activity, and there were no gain-of-function mutants. The demographics of SNPs in the NOX5 gene are varied, for example, the SNP encoding R548H was more common among Asians and Africans versus Europeans with South Americans in between (0.115, African Americans; 0.208, Kenyans; 0.119, Nigerians; and 0 for Western and Northern Europeans; 0 from Great Britain, Italy, and Spain; 0.086 for Mexicans; 0.046 for Puerto Ricans; 0.067 Columbians). SNPs that encoded a truncated and inactive form of NOX5 (W272Ter) were found predominantly in Africans (0.074 in African Americans, 0.0625 in Kenyans, and 0.108 in Nigerians) and not detected in other populations. The SNP encoding the M95K mutant (M77K in NOX5V2) was partially active. M95 lies between the two pairs of EF hands and in close proximity to a cluster of hydrophobic amino acids. Mutation of the hydrophobic M77 to a charged K residue (M95K, rs112069106) yielded an enzyme with greatly impaired calcium-dependent activity. Interestingly, a more conservative SNP encoding the mutation of M95 to the hydrophobic V (rs34097994) abrogated the loss of calcium-dependent activity seen in the M95K mutant NOX5. The changes in NOX5 activity encoded by these SNPs provide important mechanistic insights. Calmodulin binds to target proteins via critical methionine (M) residues. Given that M95 is the only methionine residue in this region between EF-hand pairs in NOX5, it may well be the critical moiety that mediates the interaction with the C-terminal REFBD and triggers enzyme activation. Alternatively, SNPs in intragenic regions may influence the expression of NOX5 and NOX5 isoforms by altering promoter activity and, in introns, by altering RNA splicing. The high frequency of SNPs that truncate NOX5 or render it inactive raises the distinct possibility that there are humans devoid of functioning NOX5. Like many genes, including the other NOXes, NOX5 may not be essential for reproduction and vitality, and some species such as mice and rats thrive without a gene for NOX5. Whether NOX5 predisposes or protects from disease is a more likely proposition and beyond a subgenome-wide association identifying NOX5 as a gene linked to changes in Lp-PLA2 activity in subjects taking rosuvastatin [57], little is known of the role of NOX5 in disease. NOX5 SNPs were not associated with salt-dependent blood pressure [58].

4 Functional Significance of NOX5

A growing body of literature has discovered numerous physiological and pathophysiological roles of NOX5 that are detailed below. In comparison to NOXes1–4, identification of the functional significance of NOX5 has been delayed by its later discovery, the lack of appropriate tools, and its absence from the genomes of mice and rats.

4.1 Cell-Specific Expression of NOX5

A promoter region of NOX5 was recently cloned and incorporates binding sites for NF-kB, AP-1, and STAT1/STAT3 [59]. However, it is doubtful this region contains all of the regulatory elements as well as those that enable cell-specific expression, and more intensive characterization is warranted.

The highest levels of NOX5 expression have been reported in the spleen and testis, and lower levels are seen in the vasculature, the gastrointestinal tract, fetal organs, and various cancers ([1, 2, 55], Protein Atlas, [60, 61]). Cell types expressing NOX5 include vascular smooth muscle [1, 55, 59, 60, 62, 63], endothelial cells [15, 22, 60], fibroblasts [64], podocytes [17], proximal tubule [65], monocytes [66, 67], oligodendrocytes [68], prostate cancer cells [69], T-cells [70], epithelial cells [71], platelets [72], anaplastic large cell lymphoma [73], esophageal adenocarcinoma cells [74], melanoma [75], breast cancer [76], and lung cancer [19]. Increased NOX5 expression can be triggered by a variety of mechanisms including acid treatment of esophageal adenocarcinoma cells, thrombin treatment of endothelial cells [22, 77], and treatment of smooth muscle cells with angiotensin II, endothelin-1 [15], PDGF [62], TNFα [55], leptin [71], and IFNγ [59]. Signaling molecules such as STAT5 [76] and RhoA kinase [74] have been shown to increase NOX5 expression. On the other hand, AngI-7 has been shown to downregulate NOX5 [78]. Epigenetics may also influence NOX5 expression, and NOX5 promoter methylation is associated with decreased transcription [79]. The expression level of NOX5 splice variants also varies across tissues with NOX5α (V1) the predominant isoform in blood vessels [55], and NOX5β and NOX5δ (V2 and V3) are highest in endothelial cells [22], NOX5α-γ (V1–V4) in smooth muscle cells [22], NOX5β (V2) in podocytes [17], and NOX5ε (V6) in Barrett's esophageal adenocarcinoma cells [77].

4.2 Cardiovascular System

NOX5 was first detected in human vascular smooth muscle cells and, to a lesser extent, in endothelial cells [1]. Over time, NOX5 has been increasingly detected in wide variety of human and primate blood vessels [22, 55, 60, 63, 80], and expression is highest in the intima (endothelial cells) and media (smooth muscle). Elevated levels of NOX5 have been reported in human atherosclerosis [60], human myocardial infarction [80], and human hypertension [15] but not early-stage vascular lesions in primates [63]. The primary isoforms expressed in human blood vessels are the functional isoforms NOX5α (V1) and NOX5β (V2) [55], whereas endothelial cells in culture express V1, V4 (inactive), and V5 (inactive), and vascular smooth muscle cells express NOX5 V1–V4 [22]. Why the inactive variants of NOX5 are expressed is poorly understood and may reflect a downregulation or moderation of NOX5 activity through their actions to inhibit the activity of the function NOX5 V1/V2 variants. Whether NOX5 contributes to cardiovascular disease in humans remains uncertain, and there is no genetic information to suggest protection or susceptibility in humans lacking functional variants. In isolated cells, functional roles of NOX5 have been easier to identify as detailed below. Recent reports in a novel NOX5 transgenic mouse show that smooth muscle NOX5 promotes endothelial dysfunction and aberrant vascular remodeling without a significant change in blood pressure [81].

4.3 Endothelium

Human endothelial cells express three isoforms NOX5α, NOX5δ, and NOX5ε (V1, V4, and V5) that contribute to ROS in response to thrombin and ionomycin [22] with presumably the V4 and V5 buffering the actions of the V1 isoform. In transduced mouse endothelial cells in situ, NOX5 generates superoxide that scavenges nitric oxide and impairs endothelium-derived relaxation [82]. NOX5 expression in smooth muscle cells in a transgenic mouse also impairs endothelial function [81]. In human endothelial cells, NOX5 alters signaling through multiple kinase pathways (increased phosphorylation of Erk, Jnk, P38, and JAK2); stimulates proliferation [22], migration [83], and angiogenesis [83]; and, at high concentrations, promotes apoptosis [55].

4.4 Smooth Muscle

Vascular smooth muscle cells from human coronary arteries, aorta, and blood vessels of the spleen and lung [1, 15, 55, 60, 62] express NOX5. In cultured vascular smooth muscle, multiple splice variants have been measured (NOX5 V1–V4). NOX5-depedendent ROS production in vascular smooth muscle cells stimulates Erk phosphorylation and increases proliferation [55, 62], migration [84], and increased expression of calcium-activated potassium channels [84].

4.5 Myocardium

NOX5 is expressed in the human heart, both within cardiomyocytes and intramyocardial blood vessels. Acute myocardial infarction increased NOX5 expression which was observed in both the endothelium and smooth muscle of myocardial vessels as well as in myocytes [80]. Both necrotic and non-necrotic and hypertrophied areas expressed NOX5 which makes interpretation of the functional significance ambiguous.

4.6 Kidney

NOX5 expression is increased in the kidneys of humans with diabetes, and the majority of expression is localized to podocytes. Exposure of isolated human podocytes to angiotensin II resulted in a robust upregulation of NOX5 expression, whereas high glucose had no effect. Only the V2 isoform of NOX5 was induced, and the other isoforms were not detected. The functional significance of NOX5 was assessed in the kidneys using a novel transgenic mouse that enables cell-specific TET-inducible NOX5 V2 expression in mice which do not harbor an endogenous NOX5 gene. Mice expressing NOX5 in podocytes had cytoskeletal changes, loss of the glomerular filtration barrier, and increased blood pressure [17]. Others have confirmed that diabetes increases the expression of NOX5 in human diabetic kidneys but have identified mesangial cells as the primary site of expression [85]. Silencing NOX5 reduced the ability of high glucose to elevate ROS levels in human mesangial cell in vitro. In transgenic mice that overexpress NOX5 in mesangial and smooth muscle cells in a TET-inducible manner, there was evidence for exacerbation of STZ-induced diabetic renal injury. Individuals with hypertension also have increased expression of NOX5 in the proximal tubules [65]. These findings add to the concept that increased levels or activity of angiotensin II which is seen in both diabetes and hypertension may underlie increased NOX5 expression in the kidney.

4.7 The Immune System

The initial discovery of NOX5 found that it was highly expressed in the spleen and lymph nodes but not in circulating lymphocytes [1]. However, the functional role of NOX5 in immune cells is poorly understood. NOX5 expression has been documented in monocyte-derived dendritic cells and is important for differentiation into dendritic cells along with p22phox [67]. While NOX5 was shown to interact with p22phox, the functional implications remain unclear, and several other groups have convincingly shown that it is not required for NOX5 activity. NOX5 expression has also been shown in THP-1 cells where it mediates calcium-dependent increases in ROS and in macrophages in atherosclerotic lesions [66]. T-cells can also express NOX5 which is downregulated by IL-15 along with the upregulation of antioxidant genes [86], and NOX5 V1 upregulation is important for T-cell transformation by human T-cell leukemia virus type I [70].

4.8 The Reproductive System

High levels of NOX5 are seen in the testis and in particular the seminiferous tubules and spermatozoa [18, 87]. NOX5 has been shown to mediate the calcium-dependent production of superoxide in sperm which increases motility [20]. The prostate expresses a low level of NOX5 which is increased in malignant cells [88]. In female reproductive organs, NOX5 can be detected in the uterus, ovaries, and placenta [1, 2]. In the absence of clear genetic data, functional roles of NOX5 remain uncertain in human reproduction. However, in model system such as drosophila, there is an important role. NOX5 facilitates the muscular contractions important for egg laying [89] by increasing intracellular calcium. Increased calcium stimulates dNOX which then produces ROS that in turn promotes greater release of calcium in a positive feedback loop that stimulates rigorous contractions.

4.9 Cancer

Increased levels of ROS have long been associated with cancer, at the initiation stage, as well as support increased rates of proliferation, migration, and invasiveness. The inability of broad-spectrum antioxidants to prevent or treat cancer has diminished the appreciation of ROS as a driver of neoplasia. However, in cells in vitro, increased expression of NOX promotes cellular transformation [90]. NOX5 expression has been detected in a number of cancers [91] or cancer cell lines including epithelial cells [71], breast cancer [76], melanoma [75], prostate cancer [69], lung cancer [19], lymphoma [73], Barrett's carcinoma [92], pancreatic cancer [93], endometrial cancer, liver cancer, thyroid cancer, hairy cell leukemia [94], and esophageal cancer [77]. In DU145 prostate cancer cells, NOX5 mediates calcium-dependent ROS production which contributes to cellular proliferation. In another prostate cancer cell line, PC-3, the silencing of NOX5 decreased proliferation and stimulated apoptosis [69]. A number of studies have linked NOX5 as proliferative mechanism linking excess acid in Barrett's esophagus (BE) [77, 95, 96]. Acid stimulates increased NOX5 expression in esophageal adenocarcinoma cells by activating CREB [77] and STAT5 [97]. In esophageal cells, the isoform induced by acid is NOX5 V5 which lacks EF hands and is inactive [1, 55]. In addition, it is unlikely that calcium could stimulate the activity of a NOX5 variant lacking EF hands, and thus other mechanisms are likely at play. NOX5 protein expression can be detected in human cancers with varying frequencies including prostate (81% were positive for NOX5), ovarian (70%), melanoma (64%), breast (61%), glioblastoma (61%), colon (60%), lung (56%), and lymphoma (44%) [91].

5 Tools for Working with NOX5

NOX5 plasmids are most easily obtained from Addgene (NOX5V2, Plasmid #69354). Epitope-tagged versions and isoforms have been published [9, 49, 55]. Antibodies are available from many vendors but not all show an appropriate sized band which may vary depending on the most highly expressed isoform (NOX5V1 ~ 84 kDa, NOX5V2 ~ 82 kDa, NOX5V3 ~ 85 kDa, NOX5V4 ~ 86 kDa, NOX5V5 ~ 64 kDa). NOX5 is not glycosylated and epitope-tagged versions reveal a single band that is close to the predicted molecular weight [55]. NOX5 is also detergent insoluble in triton and NP40 and requires SDS to efficiently extract it for Western blots.

6 Summary

With the enzymatic characterization and a recent crystal structure of a NOX5 ortholog from cyanobacteria, our understanding of the complex molecular regulation of NOX5 is now quite mature. While there may be a few new revelations ahead, the multitude of pathways that lead to NOX5 activation and ROS production are well understood. Gaps in our knowledge include the mechanisms of subcellular targeting and whether superoxide is released from intracellular sites, why a tetrameric structure is important for stability and activity, whether other kinases can alter activity, the biological roles of inactive NOX5 variants, clarification of the NOX5 promoter, and the role of SNPs in defining the importance of NOX5 in human disease. We look forward to these discoveries over the next several years.

Acknowledgments

This work was supported by NIH grants R01HL124773, R01HL125926.

References

1. Banfi B, Molnar G, Maturana A, Steger K, Hegedus B, Demaurex N, Krause KH (2001) A Ca(2+)-activated NADPH oxidase in testis, spleen, and lymph nodes. J Biol Chem 276 (40):37594–37601
2. Cheng G, Cao Z, Xu X, van Meir EG, Lambeth JD (2001) Homologs of gp91phox: cloning and tissue expression of Nox3, Nox4, and Nox5. Gene 269(1–2):131–140
3. Jackson HM, Kawahara T, Nisimoto Y, Smith SM, Lambeth JD (2010) Nox4 B-loop creates an interface between the transmembrane and dehydrogenase domains. J Biol Chem 285 (14):10281–10290
4. Magnani F, Nenci S, Millana Fananas E, Ceccon M, Romero E, Fraaije MW, Mattevi A (2017) Crystal structures and atomic model of NADPH oxidase. Proc Natl Acad Sci U S A 114(26):6764–6769
5. Kawahara T, Jackson HM, Smith SM, Simpson PD, Lambeth JD (2011) Nox5 forms a functional oligomer mediated by self-association of its dehydrogenase domain. Biochemistry 50 (12):2013–2025
6. von Lohneysen K, Noack D, Wood MR, Friedman JS, Knaus UG (2010) Structural insights into Nox4 and Nox2: motifs involved in function and cellular localization. Mol Cell Biol 30 (4):961–975
7. Rigutto S, Hoste C, Grasberger H, Milenkovic M, Communi D, Dumont JE, Corvilain B, Miot F, De Deken X (2009) Activation of dual oxidases Duoxl and Duox2: differential regulation mediated by camp-dependent protein kinase and protein kinase C-dependent phosphorylation. J Biol Chem 284(11):6725–6734
8. Banfi B, Tirone F, Durussel I, Knisz J, Moskwa P, Molnar GZ, Krause KH, Cox JA (2004) Mechanism of Ca2+ activation of the NADPH oxidase 5 (NOX5). J Biol Chem 279 (18):18583–18591
9. Jagnandan D, Church JE, Banfi B, Stuehr DJ, Marrero MB, Fulton DJ (2007) Novel mechanism of activation of NADPH oxidase 5. calcium sensitization via phosphorylation. J Biol Chem 282(9):6494–6507
10. Wei CC, Reynolds N, Palka C, Wetherell K, Boyle T, Yang YP, Wang ZQ, Stuehr DJ (2012) Characterization of the 1st and 2nd EF-hands of NADPH oxidase 5 by fluorescence, isothermal titration calorimetry, and circular dichroism. Chem Cent J 6(1):29
11. Tirone F, Radu L, Craescu CT, Cox JA (2010) Identification of the binding site for the regulatory calcium-binding domain in the catalytic domain of NOX5. Biochemistry 49 (4):761–771
12. Serrander L, Jaquet V, Bedard K, Plastre O, Hartley O, Arnaudeau S, Demaurex N, Schlegel W, Krause KH (2007) NOX5 is expressed at the plasma membrane and generates superoxide in response to protein kinase C activation. Biochimie 89(9):1159–1167
13. Chen F, Yu Y, Haigh S, Johnson J, Lucas R, Stepp DW, Fulton DJ (2014) Regulation of NADPH oxidase 5 by protein kinase C isoforms. PLoS One 9(2):e88405
14. Pandey D, Fulton DJ (2011) Molecular regulation of NADPH oxidase 5 via the MAPK pathway. Am J Physiol Heart Circ Physiol 300 (4):H1336–H1344
15. Montezano AC, Burger D, Paravicini TM, Chignalia AZ, Yusuf H, Almasri M, He Y, Call-

era GE, He G, Krause KH, Lambeth D, Quinn MT, Touyz RM (2010) Nicotinamide adenine dinucleotide phosphate reduced oxidase 5 (Nox5) regulation by angiotensin II and endothelin-1 is mediated via calcium/calmodulin-dependent, rac-1-independent pathways in human endothelial cells. Circ Res 106 (8):1363–1373

16. Pandey D, Gratton JP, Rafikov R, Black SM, Fulton DJ (2011) Calcium/calmodulin-dependent kinase II mediates the phosphorylation and activation of NADPH oxidase 5. Mol Pharmacol 80(3):407–415
17. Holterman CE, Thibodeau JF, Towaij C, Gutsol A, Montezano AC, Parks RJ, Cooper ME, Touyz RM, Kennedy CR (2014) Nephropathy and elevated BP in mice with podocyte-specific NADPH oxidase 5 expression. J Am Soc Nephrol 25(4):784–797
18. El Jamali A, Valente AJ, Lechleiter JD, Gamez MJ, Pearson DW, Nauseef WM, Clark RA (2008) Novel redox-dependent regulation of NOX5 by the tyrosine kinase c-Abl. Free Radic Biol Med 44(5):868–881
19. Dho SH, Kim JY, Kwon ES, Lim JC, Park SS, Kwon KS (2015) NOX5-L can stimulate proliferation and apoptosis depending on its levels and cellular context, determining cancer cell susceptibility to cisplatin. Oncotarget 6 (36):39235–39246
20. Musset B, Clark RA, DeCoursey TE, Petheo GL, Geiszt M, Chen Y, Cornell JE, Eddy CA, Brzyski RG, El Jamali A (2012) NOX5 in human spermatozoa: expression, function, and regulation. J Biol Chem 287 (12):9376–9388
21. Bedard K, Krause KH (2007) The NOX family of ROS-generating NADPH oxidases: physiology and pathophysiology. Physiol Rev 87 (1):245–313
22. BelAiba RS, Djordjevic T, Petry A, Diemer K, Bonello S, Banfi B, Hess J, Pogrebniak A, Bickel C, Gorlach A (2007) NOX5 variants are functionally active in endothelial cells. Free Radic Biol Med 42(4):446–459
23. Prior KK, Leisegang MS, Josipovic I, Lowe O, Shah AM, Weissmann N, Schroder K, Brandes RP (2016) CRISPR/Cas9-mediated knockout of p22phox leads to loss of Nox1 and Nox4, but not Nox5 activity. Redox Biol 9:287–295
24. Tirone F, Cox JA (2007) NADPH oxidase 5 (NOX5) interacts with and is regulated by calmodulin. FEBS Lett 581(6):1202–1208
25. Fridolfsson HN, Roth DM, Insel PA, Patel HH (2014) Regulation of intracellular signaling and function by caveolin. FASEB J 28 (9):3823–3831
26. Chen F, Barman S, Yu Y, Haigh S, Wang Y, Dou H, Bagi Z, Han W, Su Y, Fulton DJ (2014) Caveolin-1 is a negative regulator of NADPH oxidase-derived reactive oxygen species. Free Radic Biol Med 73:201–213
27. Hu G, Ye RD, Dinauer MC, Malik AB, Minshall RD (2008) Neutrophil caveolin-1 expression contributes to mechanism of lung inflammation and injury. Am J Physiol Lung Cell Mol Physiol 294(2):L178–L186
28. Lobysheva I, Rath G, Sekkali B, Bouzin C, Feron O, Gallez B, Dessy C, Balligand JL (2011) Moderate caveolin-1 downregulation prevents NADPH oxidase-dependent endothelial nitric oxide synthase uncoupling by angiotensin II in endothelial cells. Arterioscler Thromb Vasc Biol 31(9):2098–2105
29. Garcia-Cardena G, Oh P, Liu J, Schnitzer JE, Sessa WC (1996) Targeting of nitric oxide synthase to endothelial cell caveolae via palmitoylation: implications for nitric oxide signaling. Proc Natl Acad Sci U S A 93 (13):6448–6453
30. Fulton D, Babbitt R, Zoellner S, Fontana J, Acevedo L, McCabe TJ, Iwakiri Y, Sessa WC (2004) Targeting of endothelial nitric-oxide synthase to the cytoplasmic face of the Golgi complex or plasma membrane regulates Akt-versus calcium-dependent mechanisms for nitric oxide release. J Biol Chem 279 (29):30349–30357
31. Pavlides S, Tsirigos A, Vera I, Flomenberg N, Frank PG, Casimiro MC, Wang C, Fortina P, Addya S, Pestell RG, Martinez-Outschoorn UE, Sotgia F, Lisanti MP (2010) Loss of stromal caveolin-1 leads to oxidative stress, mimics hypoxia and drives inflammation in the tumor microenvironment, conferring the "reverse Warburg effect": a transcriptional informatics analysis with validation. Cell Cycle 9 (11):2201–2219
32. Karuppiah K, Druhan LJ, Chen CA, Smith T, Zweier JL, Sessa WC, Cardounel AJ (2011) Suppression of eNOS-derived superoxide by caveolin-1: a biopterin-dependent mechanism. Am J Physiol Heart Circ Physiol 301(3): H903–H911
33. Zhao YY, Zhao YD, Mirza MK, Huang JH, Potula HH, Vogel SM, Brovkovych V, Yuan JX, Wharton J, Malik AB (2009) Persistent eNOS activation secondary to caveolin-1 deficiency induces pulmonary hypertension in mice and humans through PKG nitration. J Clin Invest 119(7):2009–2018
34. Basset O, Deffert C, Foti M, Bedard K, Jaquet V, Ogier-Denis E, Krause KH (2009) NADPH oxidase 1 deficiency alters caveolin phosphorylation and angiotensin II-receptor

localization in vascular smooth muscle. Antioxid Redox Signal 11(10):2371–2384

35. Chen F, Pandey D, Chadli A, Catravas JD, Chen T, Fulton DJ (2011) Hsp90 regulates NADPH oxidase activity and is necessary for superoxide but not hydrogen peroxide production. Antioxid Redox Signal 14 (11):2107–2119
36. Chen F, Yu Y, Qian J, Wang Y, Cheng B, Dimitropoulou C, Patel V, Chadli A, Rudic RD, Stepp DW, Catravas JD, Fulton DJ (2012) Opposing actions of heat shock protein 90 and 70 regulate nicotinamide adenine dinucleotide phosphate oxidase stability and reactive oxygen species production. Arterioscler Thromb Vasc Biol 32(12):2989–2999
37. Madrigal-Matute J, Fernandez-Garcia CE, Gomez-Guerrero C, Lopez-Franco O, Munoz-Garcia B, Egido J, Blanco-Colio LM, Martin-Ventura JL (2012) HSP90 inhibition by 17-DMAG attenuates oxidative stress in experimental atherosclerosis. Cardiovasc Res 95(1):116–123
38. Chen F, Haigh S, Yu Y, Benson T, Wang Y, Li X, Dou H, Bagi Z, Verin AD, Stepp DW, Csanyi G, Chadli A, Weintraub NL, Smith SM, Fulton DJ (2015) Nox5 stability and superoxide production is regulated by C-terminal binding of Hsp90 and CO-chaperones. Free Radic Biol Med 89:793–805
39. Takac I, Schroder K, Zhang L, Lardy B, Anilkumar N, Lambeth JD, Shah AM, Morel F, Brandes RP (2011) The E-loop is involved in hydrogen peroxide formation by the NADPH oxidase Nox4. J Biol Chem 286 (15):13304–13313
40. Ueyama T, Sakuma M, Ninoyu Y, Hamada T, Dupuy C, Geiszt M, Leto TL, Saito N (2015) The extracellular A-loop of dual oxidases affects the specificity of reactive oxygen species release. J Biol Chem 290(10):6495–6506
41. Calderwood SK, Neckers L (2016) Hsp90 in cancer: transcriptional roles in the nucleus. Adv Cancer Res 129:89–106
42. Thangjam GS, Birmpas C, Barabutis N, Gregory BW, Clemens MA, Newton JR, Fulton D, Catravas JD (2016) Hsp90 inhibition suppresses NF-kappaB transcriptional activation via Sirt-2 in human lung microvascular endothelial cells. Am J Physiol Lung Cell Mol Physiol 310(10):L964–L974
43. Kawahara T, Lambeth JD (2008) Phosphatidylinositol (4,5)-bisphosphate modulates Nox5 localization via an N-terminal polybasic region. Mol Biol Cell 19(10):4020–4031
44. Brar SS, Corbin Z, Kennedy TP, Hemendinger R, Thornton L, Bommarius B, Arnold RS, Whorton AR, Sturrock AB, Hueckstead TP, Quinn MT, Krenitsky K, Ardie KG, Lambeth JD, Hoidal JR (2003) NOX5 NAD(P)H oxidase regulates growth and apoptosis in DU 145 prostate cancer cells. Am J Physiol Cell Physiol 285(2):C353–C369
45. Ambasta RK, Kumar P, Griendling KK, Schmidt H, Busse R, Brandes RP (2004) Direct interaction of the novel nox proteins with p22phox is required for the formation of a functionally active NADPH oxidase. J Biol Chem 279(44):45935–45941
46. Bayraktutan U, Blayney L, Shah AM (2000) Molecular characterization and localization of the NAD(P)H oxidase components gp91-phox and p22-phox in endothelial cells. Arterioscler Thromb Vasc Biol 20(8):1903–1911
47. Van Buul JD, Fernandez-Borja M, Anthony EC, Hordijk PL (2005) Expression and localization of NOX2 and NOX4 in primary human endothelial cells. Antioxid Redox Signal 7 (3–4):308–317
48. Chen K, Kirber MT, Xiao H, Yang Y, Keaney JF Jr (2008) Regulation of ROS signal transduction by NADPH oxidase 4 localization. J Cell Biol 181(7):1129–1139
49. Kiyohara T, Miyano K, Kamakura S, Hayase J, Chishiki K, Kohda A, Sumimoto H (2018) Differential cell surface recruitment of the superoxide-producing NADPH oxidases Nox1, Nox2 and Nox5: the role of the small GTPase Sar1. Genes Cells 23(6):480–493
50. Gryglewski RJ, Palmer RM, Moncada S (1986) Superoxide anion is involved in the breakdown of endothelium-derived vascular relaxing factor. Nature 320(6061):454–456
51. Selemidis S, Dusting GJ, Peshavariya H, Kemp-Harper BK, Drummond GR (2007) Nitric oxide suppresses NADPH oxidase-dependent superoxide production by S-nitrosylation in human endothelial cells. Cardiovasc Res 75 (2):349–358
52. Qian J, Chen F, Kovalenkov Y, Pandey D, Moseley MA, Foster MW, Black SM, Venema RC, Stepp DW, Fulton DJ (2012) Nitric oxide reduces NADPH oxidase 5 (Nox5) activity by reversible S-nitrosylation. Free Radic Biol Med 52(9):1806–1819
53. Bossis G, Melchior F (2006) SUMO: regulating the regulator. Cell Div 1:13
54. Pandey D, Chen F, Patel A, Wang CY, Dimitropoulou C, Patel VS, Rudic RD, Stepp DW, Fulton DJ (2011) SUMO1 negatively regulates reactive oxygen species production from NADPH oxidases. Arterioscler Thromb Vasc Biol 31(7):1634–1642

55. Pandey D, Patel A, Patel V, Chen F, Qian J, Wang Y, Barman SA, Venema RC, Stepp DW, Rudic RD, Fulton DJ (2012) Expression and functional significance of NADPH oxidase 5 (Nox5) and its splice variants in human blood vessels. Am J Physiol Heart Circ Physiol 302(10):H1919–H1928
56. Wang Y, Chen F, Le B, Stepp DW, Fulton DJ (2014) Impact of Nox5 polymorphisms on basal and stimulus-dependent ROS generation. PLoS One 9(7):e100102
57. Chu AY, Guilianini F, Grallert H, Dupuis J, Ballantyne CM, Barratt BJ, Nyberg F, Chasman DI, Ridker PM (2012) Genome-wide association study evaluating lipoprotein-associated phospholipase A2 mass and activity at baseline and after rosuvastatin therapy. Circ Cardiovasc Genet 5(6):676–685
58. Han X, Hu Z, Chen J, Huang J, Huang C, Liu F, Gu C, Yang X, Hixson JE, Lu X, Wang L, Liu DP, He J, Chen S, Gu D (2017) Associations between genetic variants of NADPH oxidase-related genes and blood pressure responses to dietary sodium intervention: the GenSalt study. Am J Hypertens 30 (4):427–434
59. Manea A, Manea SA, Florea IC, Luca CM, Raicu M (2012) Positive regulation of NADPH oxidase 5 by proinflammatory-related mechanisms in human aortic smooth muscle cells. Free Radic Biol Med 52(9):1497–1507
60. Guzik TJ, Chen W, Gongora MC, Guzik B, Lob HE, Mangalat D, Hoch N, Dikalov S, Rudzinski P, Kapelak B, Sadowski J, Harrison DG (2008) Calcium-dependent NOX5 nicotinamide adenine dinucleotide phosphate oxidase contributes to vascular oxidative stress in human coronary artery disease. J Am Coll Cardiol 52(22):1803–1809
61. Lambeth JD (2007) Nox enzymes, ROS, and chronic disease: an example of antagonistic pleiotropy. Free Radic Biol Med 43 (3):332–347
62. Jay DB, Papaharalambus CA, Seidel-Rogol B, Dikalova AE, Lassegue B, Griendling KK (2008) Nox5 mediates PDGF-induced proliferation in human aortic smooth muscle cells. Free Radic Biol Med 45(3):329–335
63. Stanic B, Pandey D, Fulton DJ, Miller FJ Jr (2012) Increased epidermal growth factor-like ligands are associated with elevated vascular nicotinamide adenine dinucleotide phosphate oxidase in a primate model of atherosclerosis. Arterioscler Thromb Vasc Biol 32 (10):2452–2460
64. Weyemi U, Redon CE, Aziz T, Choudhuri R, Maeda D, Parekh PR, Bonner MY, Arbiser JL, Bonner WM (2015) Inactivation of NADPH oxidases NOX4 and NOX5 protects human primary fibroblasts from ionizing radiation-induced DNA damage. Radiat Res 183 (3):262–270
65. Yu P, Han W, Villar VA, Yang Y, Lu Q, Lee H, Li F, Quinn MT, Gildea JJ, Felder RA, Jose PA (2014) Unique role of NADPH oxidase 5 in oxidative stress in human renal proximal tubule cells. Redox Biol 2:570–579
66. Manea A, Manea SA, Gan AM, Constantin A, Fenyo IM, Raicu M, Muresian H, Simionescu M (2015) Human monocytes and macrophages express NADPH oxidase 5; a potential source of reactive oxygen species in atherosclerosis. Biochem Biophys Res Commun 461 (1):172–179
67. Marzaioli V, Hurtado-Nedelec M, Pintard C, Tlili A, Marie JC, Monteiro RC, Gougerot-Pocidalo MA, Dang PM, El-Benna J (2017) NOX5 and p22phox are 2 novel regulators of human monocytic differentiation into dendritic cells. Blood 130(15):1734–1745
68. Accetta R, Damiano S, Morano A, Mondola P, Paterno R, Avvedimento EV, Santillo M (2016) Reactive oxygen species derived from NOX3 and NOX5 drive differentiation of human oligodendrocytes. Front Cell Neurosci 10:146
69. Holl M, Koziel R, Schafer G, Pircher H, Pauck A, Hermann M, Klocker H, Jansen-Durr P, Sampson N (2016) ROS signaling by NADPH oxidase 5 modulates the proliferation and survival of prostate carcinoma cells. Mol Carcinog 55(1):27–39
70. Shigemura T, Shiohara M, Kato M, Furuta S, Kaneda K, Morishita K, Hasegawa H, Fujii M, Gorlach A, Koike K, Kamata T (2015) Superoxide-generating Nox5alpha Is functionally required for the human T-cell leukemia virus type 1-induced cell transformation phenotype. J Virol 89(17):9080–9089
71. Mahbouli S, Der Vartanian A, Ortega S, Rouge S, Vasson MP, Rossary A (2017) Leptin induces ROS via NOX5 in healthy and neoplastic mammary epithelial cells. Oncol Rep 38 (5):3254–3264
72. Bartimoccia S, Carnevale R, Sanguigni V, De Falco E, Frati G, Loffredo L, Plebani A, Soresina A, Pignatelli P, Violi F (2016) NOX 5 is expressed in platelets from patients with chronic granulomatous disease. Thromb Haemost 116(1):198–200
73. Carnesecchi S, Rougemont AL, Doroshow JH, Nagy M, Mouche S, Gumy-Pause F, Szanto I (2015) The NADPH oxidase NOX5 protects against apoptosis in ALK-positive anaplastic large-cell lymphoma cell lines. Free Radic Biol Med 84:22–29

74. Hong J, Li D, Cao W (2016) Rho kinase ROCK2 mediates acid-induced NADPH oxidase NOX5-S expression in human esophageal adenocarcinoma cells. PLoS One 11(2): e0149735
75. Antony S, Jiang G, Wu Y, Meitzler JL, Makhlouf HR, Haines DC, Butcher D, Hoon DS, Ji J, Zhang Y, Juhasz A, Lu J, Liu H, Dahan I, Konate M, Roy KK, Doroshow JH (2017) NADPH oxidase 5 (NOX5)-induced reactive oxygen signaling modulates normoxic HIF-1alpha and p27(Kip1) expression in malignant melanoma and other human tumors. Mol Carcinog 56(12):2643–2662
76. Dho SH, Kim JY, Lee KP, Kwon ES, Lim JC, Kim CJ, Jeong D, Kwon KS (2017) STAT5A-mediated NOX5-L expression promotes the proliferation and metastasis of breast cancer cells. Exp Cell Res 351(1):51–58
77. Fu X, Beer DG, Behar J, Wands J, Lambeth D, Cao W (2006) cAMP-response element-binding protein mediates acid-induced NADPH oxidase NOX5-S expression in Barrett esophageal adenocarcinoma cells. J Biol Chem 281 (29):20368–20382
78. Pai WY, Lo WY, Hsu T, Peng CT, Wang HJ (2017) Angiotensin-(1-7) inhibits thrombin-induced endothelial phenotypic changes and reactive oxygen species production via NADPH oxidase 5 downregulation. Front Physiol 8:994
79. Zhu C, Yu ZB, Chen XH, Ji CB, Qian LM, Han SP (2011) DNA hypermethylation of the NOX5 gene in fetal ventricular septal defect. Exp Ther Med 2(5):1011–1015
80. Hahn NE, Meischl C, Kawahara T, Musters RJ, Verhoef VM, van der Velden J, Vonk AB, Paulus WJ, van Rossum AC, Niessen HW, Krijnen PA (2012) NOX5 expression is increased in intramyocardial blood vessels and cardiomyocytes after acute myocardial infarction in humans. Am J Pathol 180(6):2222–2229
81. Montezano A, Harvey A, Rios F, Beatie W, McPherson L, Thomson J, Holterman CE, Kennedy C, Touyz RM (2017) 151 Nox5 induces vascular dysfunction and arterial remodelling independently of blood pressure elevation in ang ii-infused nox5-expressing mice. Heart 103(Suppl 5):A111–A111
82. Zhang Q, Malik P, Pandey D, Gupta S, Jagnandan D, Belin de Chantemele E, Banfi B, Marrero MB, Rudic RD, Stepp DW, Fulton DJ (2008) Paradoxical activation of endothelial nitric oxide synthase by NADPH oxidase. Arterioscler Thromb Vasc Biol 28 (9):1627–1633
83. Pi X, Xie L, Portbury AL, Kumar S, Lockyer P, Li X, Patterson C (2014) NADPH oxidase-generated reactive oxygen species are required for stromal cell-derived factor-1alpha-stimulated angiogenesis. Arterioscler Thromb Vasc Biol 34(9):2023–2032
84. Gole HK, Tharp DL, Bowles DK (2014) Upregulation of intermediate-conductance Ca2+-activated K+ channels (KCNN4) in porcine coronary smooth muscle requires NADPH oxidase 5 (NOX5). PLoS One 9(8):e105337
85. Jha JC, Banal C, Okabe J, Gray SP, Hettige T, Chow BSM, Thallas-Bonke V, De Vos L, Holterman CE, Coughlan MT, Power DA, Skene A, Ekinci EI, Cooper ME, Touyz RM, Kennedy CR, Jandeleit-Dahm K (2017) NADPH oxidase Nox5 accelerates renal injury in diabetic nephropathy. Diabetes 66 (10):2691–2703
86. Kaur N, Naga OS, Norell H, Al-Khami AA, Scheffel MJ, Chakraborty NG, Voelkel-Johnson C, Mukherji B, Mehrotra S (2011) T cells expanded in presence of IL-15 exhibit increased antioxidant capacity and innate effector molecules. Cytokine 55(2):307–317
87. Sabeur K, Ball BA (2007) Characterization of NADPH oxidase 5 in equine testis and spermatozoa. Reproduction 134(2):263–270
88. Juhasz A, Ge Y, Markel S, Chiu A, Matsumoto L, van Balgooy J, Roy K, Doroshow JH (2009) Expression of NADPH oxidase homologues and accessory genes in human cancer cell lines, tumours and adjacent normal tissues. Free Radic Res 43(6):523–532
89. Ritsick DR, Edens WA, McCoy JW, Lambeth JD (2004) The use of model systems to study biological functions of Nox/Duox enzymes. Biochem Soc Symp (71):85–96
90. Suh YA, Arnold RS, Lassegue B, Shi J, Xu X, Sorescu D, Chung AB, Griendling KK, Lambeth JD (1999) Cell transformation by the superoxide-generating oxidase Mox1. Nature 401(6748):79–82
91. Antony S, Wu Y, Hewitt SM, Anver MR, Butcher D, Jiang G, Meitzler JL, Liu H, Juhasz A, Lu J, Roy KK, Doroshow JH (2013) Characterization of NADPH oxidase 5 expression in human tumors and tumor cell lines with a novel mouse monoclonal antibody. Free Radic Biol Med 65:497–508
92. Hong J, Li D, Wands J, Souza R, Cao W (2013) Role of NADPH oxidase NOX5-S, NF-kappaB, and DNMT1 in acid-induced p16 hypermethylation in Barrett's cells. Am J Physiol Cell Physiol 305(10):C1069–C1079
93. Mochizuki T, Furuta S, Mitsushita J, Shang WH, Ito M, Yokoo Y, Yamaura M, Ishizone S, Nakayama J, Konagai A, Hirose K, Kiyosawa K, Kamata T (2006) Inhibition of

NADPH oxidase 4 activates apoptosis via the AKT/apoptosis signal-regulating kinase 1 pathway in pancreatic cancer PANC-1 cells. Oncogene 25(26):3699–3707

94. Kamiguti AS, Serrander L, Lin K, Harris RJ, Cawley JC, Allsup DJ, Slupsky JR, Krause KH, Zuzel M (2005) Expression and activity of NOX5 in the circulating malignant B cells of hairy cell leukemia. J Immunol 175 (12):8424–8430
95. Zhou X, Li D, Resnick MB, Wands J, Cao W (2013) NADPH oxidase NOX5-S and nuclear factor kappaB1 mediate acid-induced microsomal prostaglandin E synthase-1 expression in Barrett's esophageal adenocarcinoma cells. Mol Pharmacol 83(5):978–990
96. Li D, Cao W (2014) Role of intracellular calcium and NADPH oxidase NOX5-S in acid-induced DNA damage in Barrett's cells and Barrett's esophageal adenocarcinoma cells. Am J Physiol Gastrointest Liver Physiol 306 (10):G863–G872
97. Si J, Behar J, Wands J, Beer DG, Lambeth D, Chin YE, Cao W (2008) STAT5 mediates PAF-induced NADPH oxidase NOX5-S expression in Barrett's esophageal adenocarcinoma cells. Am J Physiol Gastrointest Liver Physiol 294(1):G174–G183

Chapter 23

Using Synthetic Peptides for Exploring Protein-Protein Interactions in the Assembly of the NADPH Oxidase Complex

Edgar Pick

Abstract

The NADPH oxidase complex, responsible for reactive oxygen species (ROS) generation by phagocytes, consists of a membrane-associated flavocytochrome b_{558} (a heterodimer of NOX2 and p22^{phox}) and the cytosolic components p47^{phox}, p67^{phox}, Rac(1 or 2), and p40^{phox}. NOX2 carries all redox stations through which electrons flow from NADPH to molecular oxygen, to generate the primary ROS, superoxide. For the electron flow to start, a conformational change in NOX2 is required. The dominant hypothesis is that this change is the result of the interaction of NOX2 with one or more of the cytosolic components (NADPH oxidase assembly). At the most basic level, assembly is the sum of several protein-protein interactions among oxidase components. This chapter describes a reductionist approach to the identification of regions in oxidase components involved in assembly. This approach consists of "transforming" one component in an array of overlapping synthetic peptides and assessing binding to the peptides of another component, represented by a recombinant protein. The peptides are tagged with biotin, at the N- or C-terminus, and immobilized on streptavidin-coated 96-well plates. The protein partners are expressed with a 6His tag and added to the plates in the fluid phase. Binding of the protein to the peptides is quantified by a kinetic ELISA, using a peroxidase-conjugated anti-polyhistidine antibody. Protein-peptide binding assays were applied successfully to (a) identifying the binding site on one component (represented by peptides) for another component (proteins), (b) precisely defining the "binding sequence," (c) acquiring information on the binding site in the partner protein, (d) investigating the effect of conformational changes in proteins on binding to peptides, (e) determining the effect of physicochemical modification of peptides on binding of proteins, and (f) identifying epitopes recognized by anti-oxidase component antibodies by binding of antibody to peptide arrays derived from the component.

Key words NADPH oxidase, Superoxide, Cytochrome b_{558}, NOX2, p67^{phox}, Synthetic peptides, Peptide walking, Protein-protein interaction, Protein-peptide binding, Kinetic ELISA

1 Introduction

The love of complexity without reductionism makes art; the love of complexity with reductionism makes science

—Edward Osborne Wilson

Ulla G. Knaus and Thomas L. Leto (eds.), *NADPH Oxidases: Methods and Protocols*, Methods in Molecular Biology, vol. 1982, https://doi.org/10.1007/978-1-4939-9424-3_23, © Springer Science+Business Media, LLC, part of Springer Nature 2019

1.1 NADPH Oxidase Assembly

The superoxide ($O_2^{\cdot-}$)-generating NADPH oxidase (briefly, "oxidase") complex of phagocytes consists of a membrane-localized protein, flavocytochrome b_{558}, and five cytosolic proteins: p47*phox*, p67*phox*, p40*phox*, the small GTPase Rac(1 or 2), and Rho GDP dissociation inhibitor (RhoGDI) (reviewed in Refs. [1, 2]). Flavocytochrome b_{558} is a heterodimer of a 91 kDa glycosylated flavoprotein, known as NOX2, which is associated with a second protein of 22 kDa (p22*phox*). The only catalytically active component, responsible for $O_2^{\cdot-}$ generation, is NOX2; all the other proteins have exclusively regulatory functions.

NOX2 contains 570 residues and comprises six transmembrane α-helices linked by three outside-facing loops and two cytosol-facing loops, and a cytosolic segment, which extends from residue 288 to 570, also known as the dehydrogenase region (DHR). NOX2 contains three redox stations, carrying the electrons from NADPH to O_2. These are an NADPH binding site and non-covalently bound FAD, present in the DHR, and two non-identical hemes, linked to the third and fifth transmembrane helices. From an evolutionary perspective, the DHR is homologous to the prokaryotic protein ferredoxin-NADP$^+$ reductase, which also possesses NADP- and FAD-binding domains [3].

$O_2^{\cdot-}$ production by the oxidase does not occur in resting phagocytes, being initiated by the interaction of membrane receptors with a variety of stimulatory ligands, such as phagocytosed microorganisms. The electron flow through the redox centers in NOX2 is initiated by a conformational change in NOX2, consequent to its interaction with the cytosolic components which translocate to the membrane. Thus, one can look upon oxidase activation as the functional expression of the assembly of the NOX2-p22*phox* dimer with the cytosolic components, generating a functional NADPH-consuming, $O_2^{\cdot-}$-generating membrane-anchored complex (reviewed in Refs. [4, 5]). Under physiological conditions, p47*phox*, p67*phox*, and Rac are all required for the induction of $O_2^{\cdot-}$ production, but whether direct interaction of all components with the NOX2 DHR is required is a yet unsettled question. A hypothesis attributing to p67*phox* a predominant or even exclusive role in the induction of a conformational modification in NOX2 is gaining popularity [6–8]. In this hypothesis, the formation of the assembled oxidase complex is based on a "propagated wave" model [9], in which protein-protein interactions are linked to the sequential occurrence of conformational changes, with one protein serving as the inducing agent and the second, as target [10]. Thus, conversion of Rac from the GDP- to the GTP-bound form results in a change in Rac (reviewed in Ref. [11]); next, modified Rac-GTP interacts with p67*phox* [12] and alters its conformation [8], this being followed by the binding of

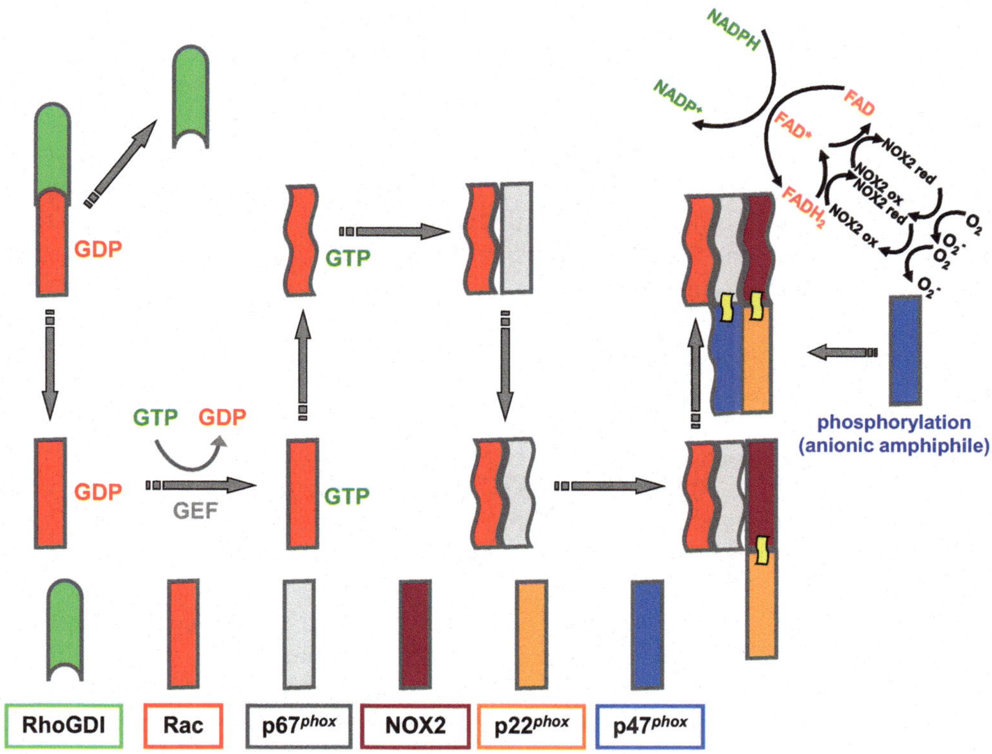

Fig. 1 The "propagated wave" model of NADPH oxidase assembly. In this model, the assembly and consequent activation of the NADPH oxidase complex is seen as a series of linked conformational changes. Rac is modified by conversion from the GDP-bound to the GTP-bound form by a guanine nucleotide exchange factor (GEF), following dissociation from RhoGDI. It next interacts with p67*phox*, causing a conformational change in p67*phox*, which enables p67*phox* to bind to NOX2. Binding of the conformationally changed p67*phox* to NOX2 induces a conformational modification in NOX2. p47*phox* binds to p22*phox* as a consequence of a conformational change in p47*phox*, caused by phosphorylation of specific serines (in vivo) or by anionic amphiphiles (in vitro). Binding of p47*phox* to p22*phox* is likely to promote the direct interaction of p47*phox* with NOX2 and, by virtue of the existence of a p47*phox*-p67*phox* complex in the resting cell, assist in bringing p67*phox* to the proximity of NOX2. Plain rectangles represent the NADPH oxidase components in their native form; rectangles in wavy shape represent components following a conformational change

modified p67*phox* to NOX2 and the induction of a restructuring of NOX2, which culminates in the activation of the electron flow from NADPH to oxygen (Fig. 1). A second path contributing to the p67*phox*-NOX2 interaction comprises the removal of auto-inhibition in p47*phox* by phosphorylation of critical serines, which enables its interaction with p22*phox* [13, 14]. By virtue of the existence of p47*phox*-p67*phox* complexes, the binding of p47*phox* to p22*phox* makes it a carrier of p67*phox* to the membrane environment of cytochrome b_{558}, with direct contacts between p47*phox* and NOX2 assisting in the establishment of p67*phox*-NOX2 bonds. There is accumulating evidence for the conformational change in p67*phox* being the expression of the relief of an intramolecular auto-inhibitory bond. Thus, C-terminal truncation of p67^{phox} at residues 243, 210, and 199 led to a progressive increase in binding to flavocytochrome b_{558}, with p67^{phox}(1–199) exhibiting maximal binding [15].

The take-home message of these data is that activation of NOX2 is based on a series of protein-protein interactions, likely occurring sequentially, resulting in an assembled complex comprising a now catalytically active NOX2. These interactions are predominantly intermolecular and involve the canonical interaction modules: src homology 3 (SH3), proline-rich region (PRR), tetratricopeptide repeat (TPR), phox homology (PX), and phox and Bem1 (PB1). Some of the intermolecular interactions require previous disengagements of existing intramolecular bonds by phosphorylation of specific residues or the effect of unsaturated fatty acid (as is the case in p47^{phox} and, possibly, in p67^{phox}?) or as the result of a preliminary protein-protein interaction (RacGTP-p67^{phox}).

Finally, assembly of the NADPH oxidase also involves protein-phospholipid interactions. Both p47^{phox} and p40^{phox} contain PX domains at their N-terminus which attach to specific phosphoinositides in the cytosolic aspect of the membrane, p47^{phox} showing a preference for phosphatidyl-4,5-bisphosphate whereas p40^{phox}, for phosphatidylinositol-3 phosphate [16]. These interactions involve basic and hydrophobic amino acids "attracted" by the negatively charged membrane phosphoinositides. Protein-phospholipid interactions are also critical for the binding of Rac to the membrane and, thus, for carrying p67^{phox} to the proximity of NOX2. These latter interactions are either electrostatic, between the C-terminal polybasic region and the positively charged insert region in Rac and anionic membrane phospholipids in the membrane, or hydrophobic, between the isoprenyl group at the C-terminus of Rac and membrane phospholipids (reviewed in Ref. [11]).

1.2 Protein-Protein Interactions: Not Intending to Be a Survey

Tell me who your friends are and I will tell you who you are.
Popular proverb, also attributed to Vladimir Ilyich Ulyanov (Lenin)

Protein-protein interactions are essential for so many biological processes as to make any survey incompatible with the purpose of this chapter. The reader is directed to a comprehensive description of the essentials of protein-protein recognition [17] and to a recent overview of the use of structural bioinformatics for the prediction of protein-protein complexes and their inhibition [18]. A very wide array of methodologies for the analysis of protein-protein interactions in vitro and in vivo are available, and their number and sophistication are increasing all the time. Comprehensive descriptions are available in two books [19, 20] and two recent reviews [21, 22], but these represent only the tip of a constantly growing iceberg. Valuable advice on the proper design of binding assays and a warning about the common flaws encountered in their execution, such as the failure to know the affinity of the reactants, appear in Ref. [23]. In another publication, the authors warn against the reliance on a single assay, especially when experiments involve the use of whole cell extracts and overexpressed proteins [24].

1.3 Protein-Peptide Interactions: Gulliver Meets the Lilliputians

Protein-peptide binding assays represent a methodological approach to the study of protein-protein interaction in which the binding of one protein to synthetic peptides, representing selected sequences of another protein, is assessed. The basic assumption for the legitimacy of such assays is that protein-peptide interactions mimic with reasonable fidelity protein-protein interactions, with the peptide representing an interacting sequence present in one protein binding to a "receptor-like" structure in the partner protein. An excellent review of the structural basis and practical applications of protein-peptide binding is highly recommended for those who would like to use this strategy in exploring the assembly of the oxidase complex [25]. Similar to interaction between proteins, binding is mediated by "hot spot" residues, and it was found that peptide "hot spot" residues exhibit a high frequency of Phe, Leu, Trp, Tyr, and Ile (*see* Subheading 3.4.5, for an example derived from our work). Although most peptides might lack a defined structure in the free form, they acquire a conformation upon binding, adapting to the structure of the bound protein partner, which does not change substantially. A recent review focusing on the methodological aspects of protein-peptide binding assays is the most appropriate introduction to the subject at the conceptual and "benchtop" levels [26]. The issues discussed will be covered in detail as applied to the situations particular to the study of interactions between oxidase components (*see* Subheading 3.3).

Protein-peptide binding assays can be designed in two "directions"; in one, binding of protein A to peptides derived from protein B is measured; in the second, binding of protein B to peptides derived from protein A is measured. Protein-peptide binding assays serve a double purpose: to identify a liaison between two proteins and to define the binding site(s) on one protein for the other, ideally in both partners of the interaction. Unlike canonical protein-protein binding assays, the peptide-based methods require screening a relatively large number of peptides, covering the full length or, at least, segments of one protein. For initial screening, arrays of overlapping peptides are used (10–15 residues-long), with an offset of 1–4 residues. Initial experiments are done with unpurified peptides, the goal being to find clusters of vicinal peptides binding the target protein. Confirmatory work is performed next with purified peptides, and if the preliminary results are corroborated, the likely limits of the binding sequence are defined by a series of procedures to be described in Subheading 3.4.

Protein-peptide binding assays became feasible and easy to perform by the introduction of the "multipin" method of peptide synthesis by the Mimotopes company (Clayton, Victoria, Australia) in the early 1980s of the past century [27]. We first started using such peptide libraries made by Mimotopes, known as PepSets, cleaved and supplied in powder form, for identifying domains in p47*phox* mediating binding of p67*phox* [28]. Numerous variations of this methodology are available today, predominantly in the form of peptide spots arranged as arrays on membrane sheets [26, 29, 30], and several companies are providing peptide SPOT arrays (*see* **Note 1**).

1.4 From the Method to Results: Our Experience with Applying Protein-Peptide Binding Assays to the Study of Oxidase Assembly

We first applied this method by assessing the binding of p67*phox* protein to 47*phox* peptides [28], based on the awareness that a p47*phox*-p67*phox* complex, also containing p40*phox*, is present in the cytosol of resting phagocytes [31]. Overlapping synthetic 15-mer peptides, with a four-residue offset, covering the full length of p47*phox* (residues 1–390), with a biotin tag at the N-terminus attached by the intermediary of a Ser-Gly-Ser-Gly (SGSG) spacer, were attached to streptavidin-coated 96-well plates and incubated with recombinant full-length p67*phox*(1–526). After removal of unbound p67*phox*, the peptide-attached protein was quantified by ELISA using a polyclonal goat anti-human p67*phox* antibody followed by a peroxidase-conjugated rabbit anti-goat IgG antibody and the quantification of bound peroxidase by 3,3′,5.5-′-tetramethylbenzidine (TMB) and H_2O_2. The reaction was stopped by H_2SO_4 and absorbance at 450 nm read against a reagent blank in a microplate reader. These succinct methodological details, typical for "endpoint" reading, are provided with the purpose to be compared to the presently used markedly modified technique (*see* Subheading 3.3). Using this assay, the binding site for p67*phox* in p47*phox* was found to comprise residues 357–371, corresponding to the canonical polyproline type II structure, revealed by other methodological approaches. Preincubation of p67*phox* with p47*phox* protein prevented binding to p47*phox* peptides. To the best of my knowledge, this is the first published successful use of a protein-peptide binding assay to demonstrate a protein-protein interaction in the oxidase and to identify the binding site on one of the partners.

We next attempted to use this method for the study of the interaction between a membrane-attached oxidase component and cytosolic components. In this study, overlapping peptides, with a two-residue offset, covering the full length of p22*phox*, with a biotin tag at the N-terminus attached by the intermediary of a SGSG spacer, were attached to streptavidin-coated 96-well plates and incubated with recombinant full-length p47*phox*(1–390) or p67*phox*(1–526) [32]. These experiments showed binding of p47*phox* to a PRR in p22*phox*, corresponding to residues 151–160, as expected from earlier work [13, 14]. However, some of the results raised questions about the ability of protein-peptide assays to faithfully reproduce events occurring in vivo. Thus, binding of p67*phox* to the PRR was also found, and binding of p47*phox* to the PRR in p22*phox* was not dependent on the "opening" of an auto-inhibitory loop in p47*phox*, consequent to phosphorylation in vivo or the effect of anionic amphiphiles, in vitro. These results might be explained by easier access of p47*phox* (and p67*phox*?) to a 15-mer peptide anchored to a surface than to the C-terminus of the p22*phox* protein in its native configuration. An important methodological issue did become apparent in this early work, namely, the importance of the orientation of the surface-bound peptide. When the 151–160 sequence was at the C-terminus of the peptide, it bound preferentially p67*phox*; when it was at the N-terminus, it bound preferentially p47*phox*.

Protein-peptide binding assays are especially valuable when one of the proteins is difficult to obtain in a purified form or cannot be expressed as a recombinant protein. This is the case with most NOX proteins, as best exemplified by NOX2. We, thus, used the protein-peptide binding assay with the purpose of identifying the binding site for p67*phox* in the DHR of NOX2 [33]. The region extending from residue 288 to 570 was represented by 91 overlapping 15-mer peptides with an offset of three residues. The peptides were biotinylated and surface-attached to streptavidin-coated 96-well plates and exposed to recombinant p67*phox*, with recombinant p47*phox* and maltose binding periplasmic protein from *E. coli* serving as control proteins. In this study, several key modifications in the methodology of the assay were introduced, representing the technique presently used in our laboratory and described in detail in Subheading 3. These were the following:

(a) We use commercially available streptavidin-coated 96-well plates instead of plates coated manually by us (*see* Subheading 2.2).
(b) All recombinant proteins were expressed with a 6His tag at the N-terminus.
(c) This made possible the use of a single anti-polyhistidine peroxidase-conjugated antibody for detecting binding of all recombinant proteins and did not necessitate a secondary antibody.
(d) In order to accommodate low-affinity interactions, the protein to peptide binding stage was performed at 4 °C for 18 h.
(e) Also, all plate washing procedures were done using a programmable plate washer, fitted with a "cell wash head," originally intended for washing cultured cell monolayers (*see* Subheading 2.5).
(f) Peroxidase-dependent color development was quantified by the linear kinetics of the increase in absorbance (kinetic ELISA), as described in Refs. [34, 35], assuring a more

quantitative assessment of the amount of protein bound to the fixed amount of peptide than the customary "endpoint" reading.

(g) When indicated, peptides were bound to the wells in two orientations, exposing either the C-terminus or the N-terminus to the fluid phase.

(h) In order to eliminate the possibility that the protein-peptide interaction was based on purely electrostatic interaction or on the nonspecific participation of the protein tag, such as polyhistidine, binding reactions were performed at various ionic strengths or in the presence of imidazole (which competes with histidine).

(i) In order to check for sequence specificity, peptides were subjected to scrambling and/or replacement of individual residues.

(j) With the purpose of obtaining information on the region in the protein interacting with the peptide, the protein partners were subjected to mutations, deletions, and truncations.

Examples of such approaches are provided by replacing cysteines in peptides, suspected of being critical for binding, by arginines or serines [33], truncating p67^{phox} at several points [33], or replacing cysteines in p67^{phox}, by serines [36, 37]. In a recent study, a novel binding mechanism was described, based on the establishment of disulfide bonds between NOX2 peptides containing a Cys-Gly-Cys (CGC) triad, in which the near vicinal cysteines are engaged in an intramolecular disulfide bond, and cysteine(s) in p67^{phox} [37]. The formation of such bonds was prevented by exposing peptides to reducing agents or treating p67^{phox} with thiol alkylating agents, such as *N*-ethylmaleimide [37].

An illustration of the use of the canonical technique is the detection of a cluster of five peptides, encompassing NOX2 residues 357–383, binding p67^{phox} [33] (Fig. 2; also *see* insert). All five peptides shared a CGC triad, and binding of p67^{phox} was maximal to peptides 24 (residues 357–371) and 28 (residues 369–383), in which the CGC triad was at the C-terminus (peptide 24) or N-terminus (peptide 28) of the peptides (Fig. 3).

1.5 Peptides Binding Oxidase Components Should Not "A Priori" Be Expected to Inhibit Oxidase Activation

Based on the concept that assembly of the oxidase complex is an obligatory precondition for activation, one could expect synthetic peptides derived from one oxidase component, found to bind another component, to interfere with oxidase activation. The search for synthetic peptides, derived from oxidase components, as potential inhibitors of oxidase activation is an active area of research in the hope that these might serve as therapeutic agents in diseases associated with excessive production of ROS (reviewed in Ref. [38]). Doubts about equaling regions participating in binding with regions involved in activity were already raised by the

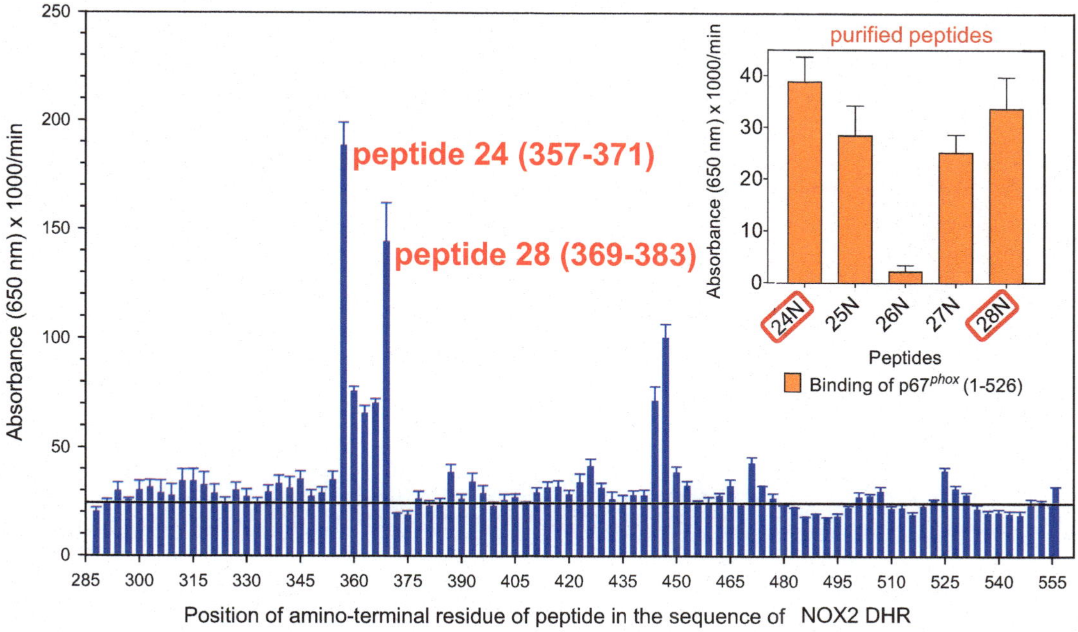

Fig. 2 Binding of p67^{phox} (1–526) to an array of overlapping peptides covering the NOX2 DHR (residues 288–570). Peptides consisted of 15 residues and a SGSG spacer sequence linking biotin to the N-terminus. They overlapped by 12 residues, had an offset of 3 residues, and were not purified. The horizontal line above the *x*-axis represents the binding value of p67^{phox} to wells in the absence of peptides. The insert illustrates the binding to purified peptides 24–28. Results represent means ± SE of three experiments. This figure is a modified version of Fig. 3, panel **A**, in Ref. [33]

finding that activation of the oxidase by p67^{phox} was dependent on an "activation domain," represented by residues 199–210 [39], or an "extended activation domain," represented by residues 190–208 [40], but binding of p67^{phox} to NOX2 required only residues N-terminal to residue 199 [15]. Striking examples of binding/activity dichotomy are also offered by the following findings:

(a) Peptides corresponding to the PRR in p22^{phox} (residues 151–160), which bind p47^{phox} (an essential step in oxidase activation, *see* Refs. [13, 14]), did not interfere with oxidase activation in a cell-free system [32].

(b) NOX2 peptides 24 and 28, corresponding to the sequences 357–371 and 369–383, found to bind p67^{phox} [33], did not interfere significantly with oxidase activation in a cell-free system, and when a low level of inhibition was found, this was not sequence specific, as shown by its persistence when cysteine 369 was mutated to arginine [41].

(c) The introduction of an intramolecular disulfide bond between cysteines 369 and 371, which markedly enhanced binding of p67^{phox} to NOX 2 peptide 24 (357–371) [37], had only a minor augmenting effect on its weak activation inhibitory action (I. Dahan and E. Pick, unpublished results).

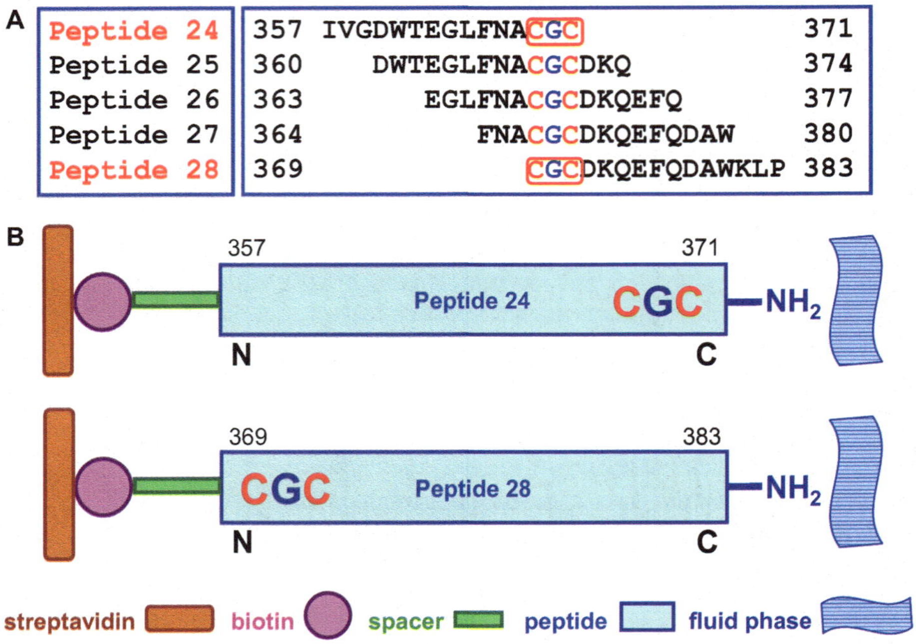

Fig. 3 Sequence analysis of the cluster of NOX2 peptides 24–28 found to bind p67^{phox}(1–526). (**A**) All five peptides share the ^{369}CGC^{371} triad. (**B**) Location of the CGC triad in peptides 24 and 28, which were found to be the most active in binding p67^{phox}(1–526) (Fig. 2). In peptide 24, the CGC triad is at the C-terminus; in peptide 28 the CGC triad is at the N-terminus. Peptides 25, 26, and 27, in which the CGC triad is located closer to the center of the sequence, exhibit lower protein binding abilities. Panel **a** in this figure is a modified version of Fig. 5, panel **A**, in Ref. [33]

1.6 A Side-Alley: Protein-Peptide Binding Assays for Identifying Epitopes Recognized by Antibodies

We have also used protein-peptide binding assays to identify or confirm sequences in oxidase components recognized by antibodies. One example was confirming the p22^{phox} sequence ^{182}GPQV^{185} as the epitope recognized by monoclonal anti-p22^{phox} antibody 449 [42]. The assay was also successfully used for confirming residues ^{381}KLPKIAVDG^{389} in NOX2 as the epitope recognized by mouse monoclonal anti-NOX2 antibody 54.1 [43] (Fig. 4) and identifying NOX2 residues ^{357}IVG^{359} as the sequence recognized by polyclonal anti-protein disulfide isomerase A3 (PDIA3) antibody H-220 (Ref. [36] and E. Bechor and E. Pick, unpublished results).

1.7 Preferences, Variations, and Caveats on the Theme of Protein-Peptide Binding Assays

Before engaging in the description of the methodological details of protein-peptide binding assays, a few general principles governing the proper execution of these assays and its many variations are listed.

1. Protein-peptide binding assays are bidirectional in the sense that the investigator can choose which protein in an assay, meant to mimic protein-protein interaction, will be represented by peptides and which, by the protein. Whereas the

A

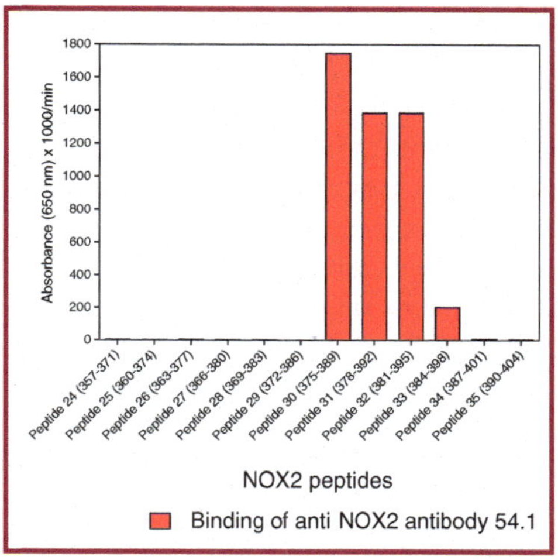

B

NOX2 Peptide No.	Sequence
24	357 IVGDWTEGLFNACGC 371
25	360 DWTEGLFNACGCDKQ 374
26	363 EGLFNACGCDKQEFQ 377
27	366 FNACGCDKQEFQDAW 380
28	369 CGCDKQEFQDAWKLP 383
29	372 DKQEFQDAWKLPKIA 386
30	375 EFQDAWKLPKIAVDG 389
31	378 DAWKLPKIAVDGPFG 392
32	381 KLPKIAVDGPFGTAS 395
33	384 KIAVDGPFGTASEDV 398
34	387 VDGPFGTASEDVFSY 401
35	390 PFGTASEDVFSYEVV 404

Fig. 4 Use of protein-peptide binding assay for identifying the epitope in the NOX2 DHR recognized by mouse monoclonal anti-NOX2 antibody 54.1 (IgG_1). In this assay, we measured binding of anti-NOX2 antibody 54.1 to an array of NOX2 peptides attached to streptavidin-coated wells via N-terminal biotin. Binding of mouse anti-NOX2 IgG was measured by peroxidase-conjugated anti-mouse IgG (H + L). (**A**) The antibody was bound by three peptides (30, 31, and 32) and to a much lesser degree to a fourth peptide (33). (**B**) Based on the sequence shared by peptides 30–32, we concluded that the epitope comprises NOX2 residues 381–389. This sequence closely resembles the published epitope (residues 383–390; Ref. [43]). Results of one characteristic experiment are shown

option of a peptide array is always at hand, the availability of a preferably recombinant protein partner is more limited.

2. A limitation of the assays is that it will work best if the search for binding site sequence is small enough to be contained in a 15-mer peptide or smaller. Also, it has a lesser chance of working if the binding site is composed of multiple domains that are nonadjacent in the linear representation of the protein, unless one domain is dominant (= sufficient for binding).

3. A further limitation of the assay is that peptides, even if representing a binding region, are unlikely to possess a conformation identical to that assumed by the region represented by the peptide, in the intact protein.

4. For preliminary assays, unpurified peptides can be used, but following the identification of one or more clusters of binding peptides, the use of purified peptides is essential. From our experience, a purity level of ≥70% is sufficient.

5. The use of purified recombinant proteins with a 6His tag is recommended. The proteins are first purified by metal affinity chromatography on Nickel Sepharose and further purified by preparative fast protein liquid chromatography (FPLC) gel filtration. Gel filtration is advantageous not only as an additional purification step but also by providing information on

the M_r of the native nondenatured protein and for freeing the protein solution of unwanted buffer components, derived from the metal affinity purification step, such as imidazole and high concentrations of NaCl. This leads to the ideal combination of purified peptides binding purified proteins.

6. The assay is based on the assumptions that both partners in the binding reaction are present in aggregate-free and monomeric form. To assure that this is the case, the peptides (usually obtained in powder form) have to be dissolved in the proper solvent and at a concentration adequate for complete solubilization. The same applies to the proteins, these exhibiting the advantage of M_r information being provided by the gel filtration step and a high probability of being soluble in aqueous buffers. Further information on these issues appears in Subheading 2.
7. A seemingly technical issue is in fact of theoretical significance. This is the fact that in all binding assays described in this chapter, peptides are surface-bound and proteins are in the fluid phase. Attempts to devise an assay in which the situation was reversed did not yield satisfactory results principally because binding of proteins to the wells was much less reproducible than that of biotinylated peptides to streptavidin-coated plates and detection of bound peptides required a fundamental change of methodology. However, surface attachment of biotinylated peptides also generated problems, such as the degree of accessibility of surface-anchored 15-residue molecules to the markedly larger proteins "free floating" in the fluid phase. We solved part of this problem by the introduction of a spacer between biotin and the peptide and, most significantly, by assessing binding of the proteins to peptides in which biotin was attached to either the N- or C-terminus, with appropriate spacers. This resulted in either the C- or the N-terminus of the peptide being more accessible to the protein.
8. The only modification of the surface-bound peptides technique is in experiments in which the effect of an agent on the peptides is tested. As an example, in recent work, we examined the effect of air-derived oxygen or of H_2O_2 on the ability of NOX2 peptides to bind p67^{phox} [37]. In these experiments, peptides were in solution in test tubes and air was bubbled via tubing immersed in the peptide solution or H_2O_2 was added. Next, the treated peptides were transferred to streptavidin-coated 96-well plates, and the procedure continued as described for untreated peptides attached to the wells. This comprised washing steps that removed the agents used to treat the peptides before the addition of the protein the binding of which was measured.

9. As already alluded to in Subheading 1.4, the conditions of performing the binding assay are critical and should always be taken into consideration when results are interpreted or when attempting to reproduce the findings. Variations should be considered for the following parameters: (a) amount of peptide added per well (can only be decreased because any increase is limited by the fixed amount of streptavidin coating the wells), (b) length of time of binding the peptides to the wells (no evidence for significant effect due to the very high affinity of biotin for streptavidin), (c) composition of the "blocking" buffer, (d) composition of the buffer containing the protein, (e) amount of protein added per well (this is, probably, the moist significant variable), (f) details of washing the wells free of unbound peptide and protein, and (g) length of time allowed for binding of protein to the peptide, temperature, and motion of plate during binding.
10. In the procedures to be described in this chapter, the bound protein is detected by a peroxidase-conjugated anti-polyhistidine antibody, having the advantage of being appropriate for the detection of all recombinant 6His-tagged proteins. In the past we used antibodies recognizing specific oxidase components, but this method was abandoned for a number of reasons: the requirement for a second conjugated antibody step, occasional lack of access to the epitope due to its participation in the binding to the peptide (steric hindrance), and unexpected cross-reactions with peptides unrelated to the protein against which the antibody was raised.
11. We use exclusively a peroxidase-conjugated antibody and the colorimetric method for measuring peroxidase activity. Among the many strategies for the detection of bound protein, antibody-based detection provided the best signal-to-noise ratio [44]. We found no need for signal-amplifying procedures, such as chemiluminescence.
12. The quantitative assessment of the binding of protein to the peptides is a critical parameter. As mentioned in Subheading 1.4, we use a kinetic ELISA instead of the routine "endpoint" reading. This choice is motivated by the need to avoid misleading equal "endpoint" readings for binding of different amounts of protein. Such a result is the consequence of absorbance plateaus being reached, at different time intervals, by reactions with different V_{max} values, due to consumption of detection reagent(s) or to reaching the upper limit of the absorbance range of the microplate reader.
13. Even when using the kinetic ELISA, the assay does not allow the determination of kinetic parameters of the binding reaction, such as "association rate constant" (k_+) and "dissociation

rate constant" (k_-), as well as equilibrium constants, as described in detail in Ref. [23].

14. Finally, a whole range of means are available for refining the rough results obtained by protein-peptide binding assays. This refers both to the peptides and the bound proteins. In order to define more precisely the binding epitope, maximally binding peptide(s) can be subjected to scrambling, truncation, alanine scan, positional scanning, and more (*see* **Note 2**). Finding the binding epitope on the bound protein is more difficult, but truncations, deletions, and mutations might be helpful in achieving this goal.
15. On rarer occasions, the effect of one protein on the binding of another protein to peptides is measured. To distinguish between the protein to be bound and the one affecting binding, only the first is tagged with 6His, whereas the latter is left untagged. A variation on this theme is assessing the binding to peptides of chimeric proteins consisting of a fusion of two proteins or of segments of two proteins, in comparison to the binding of the chimera moieties, represented by individual proteins.

2 Materials

2.1 Chemicals and Reagents

2.1.1 Peptides

Two categories of synthetic peptides are used. The first comprises overlapping 15-residue peptides spanning the full-length or selected segments of membrane or cytosolic oxidase components, with an overlap of 11–14 residues and an offset of 4 to 1 residues. A typical example of five 15-mer overlapping peptides with an offset of three residues, corresponding to residues 288–314 of NOX2, is shown below:

```
1  288 FWRSQQKVVITKVVT 302
2     291 SQQKVVITKVVTHPF 305
3        294 KVVITKVVTHPFKTI 308
4           297 ITKVVTHPFKTIELQ 311
5              300 VVTHPFKTIELQMKK 314
```

The peptides have a biotin tag at the N-terminus, attached by a SGSG spacer, and a C-terminal amide (*see* **Note 3**). The purity of the peptides ranges from 60% to 70% (based on random sampling by the manufacturer) but is not known at the level of individual peptides. The peptides are provided in powder form in individual tubes and should be stored at −20 °C under vacuum in desiccator jars containing silica gel as desiccant till dissolved (*see* **Note 4**).

The second category consists of individual peptides of a purity of ≥70%, selected on the basis of the preliminary screening using unpurified peptides. These are also 15-residues-long, with a biotin tag at the N-terminus (attached by a SGSG spacer) and a C-terminal amide. Some peptides are synthesized with a biotin tag at the C-terminus (attached by either a Gly-Ser-Gly-Lys or an aminoethoxy ethoxy acetic acid-Lys spacer) and an acetyl at the N-terminus (*see* **Note 5**). Peptide purity is checked by reversed phase chromatography and the molecular mass confirmed by MALDI-mass spectroscopy, both performed by the peptide manufacturers. It is essential that these data are made available by the manufacturer on delivery of the peptides.

2.1.2 Chemicals and Reagents Required for the Expression of Recombinant Oxidase Cytosolic Components

All recombinant proteins used in protein-peptide binding experiments have an N-terminal 6His tag.

1. *E. coli* competent cells (Rosetta 2(DE3)pLysS; Novagen).
2. pET-30a expression vector (Novagen).
3. LB Broth, Lennox (Accumedia, Neogen).
4. Isopropyl β-D-1-thiogalactpyranoside (IPTG, Sigma-Aldrich).
5. Triton X-100, BioXtra (Sigma-Aldrich). Prepare a 10% (v/v) solution in H_2O.
6. Protease inhibitor mixture cOmplete, EDTA-free (Sigma-Aldrich).

2.1.3 Chemicals Required for the Purification of Recombinant Oxidase Cytosolic Components

1. Imidazole (Sigma-Aldrich). Prepare a 2 M solution in H_2O, bringing it to pH 7.4 with 3 M HCl.
2. Ni Sepharose 6 Fast Flow (GE Healthcare).
3. *E. coli* lysis buffer also serving as binding buffer for metal affinity chromatography on Ni Sepharose (20 mM sodium phosphate buffer, pH 7.4, 0.5 M NaCl, and 20 mM imidazole).
4. Washing buffer for metal affinity chromatography on Ni Sepharose (20 mM sodium phosphate buffer, pH 7.4, 0.5 M NaCl, and 40 mM imidazole).
5. Elution buffer for metal affinity chromatography on Ni Sepharose (20 mM sodium phosphate buffer, pH 7.4, 0.5 M NaCl, and 300–500 mM imidazole).
6. Buffer used in FPLC gel filtration experiments (phosphate-buffered saline, PBS) consists of 137 mM NaCl, 2.7 mM KCl, 4.3 Na_2HPO_4, 1.4 mM KH_2PO_4, and 2 mM NaN_3, pH 7.3.
7. Gel filtration markers kit for protein molecular weights 12,000–200,000 Da (Sigma-Aldrich).
8. "Protein assay dye reagent concentrate" for measuring protein concentration by the Bradford assay (Bio-Rad).

9. Bovine gamma globulin standard (2 mg/mL) for Bradford assay (Pierce, Thermo Scientific).
10. NuPage 12% Bis-Tris electrophoresis gels, 1 mm gel thickness (Invitrogen, Life Technologies).
11. NuPage MOPS SDS running buffer (Invitrogen, Life Technologies).
12. NuPage LDS sample buffer (4×) and NuPage reducing agent (10×) (Invitrogen, Life Technologies).
13. Precision Plus SDS-PAGE protein standards, unstained (10–250 kDa) (Bio-Rad).
14. "Instant Blue" protein gel stain (Expedeon).

2.1.4 Chemicals and Reagents Required for Performing Protein-Peptide Binding Experiments

1. Tween 20 (polyethylene glycol sorbitan monooleate; Sigma-Aldrich). A 20% v/v solution is made in H_2O_2.
2. Casein sodium salt from bovine milk (Sigma-Aldrich).
3. 1-Methyl-2-pyrrolidinone (NMP; biotech grade, Sigma-Aldrich).
4. Phosphate-buffered saline (PBS) supplemented with 0.1% v/v Tween 20 (PBST), made by mixing 5 mL Tween 20% with 995 mL PBS. Filter through 0.45 μm pore size filter unit and store at 4 °C.
5. PBS supplemented with 1% w/v casein sodium salt (PBScas1). Stir magnetically till fully dissolved and filter through 0.45 μm pore size filter unit. Store at 4 °C.
6. PBS supplemented with 0.1% v/v Tween 20 and 1% w/v casein sodium salt (PBSTcas1). Stir magnetically till fully dissolved and filter through 0.45 μm pore size filter unit. Store at 4 °C.
7. PBS supplemented with 5% w/v casein sodium salt (PBScas5). Stir magnetically till fully dissolved, and centrifuge at 38,000 × *g* for 3 h at 4 °C in a fixed-angle rotor (SS34) in a RC5C or RC5C Plus centrifuge (Sorvall) (the reason for opting for centrifugation instead of filtering is the difficult passage of concentrated casein solutions through 0.45 μm pore size filters). Store in aliquots frozen at −20 °C.
8. Monoclonal peroxidase-conjugated anti-polyhistidine antibody (mouse IgG2a isotype; Sigma-Aldrich). The antibody is obtained in lyophilized form and is reconstituted by the addition of 1 mL/vial of a mixture (v/v) of 50% H_2O and 50% glycerol, corresponding to a concentration of 2.5–5.5 mg/mL IgG. The solution is aliquoted in volumes of 50 μL and stored at −20 °C.

9. 3,3′,5,5′-Tetramethylbenzidine (TMB) + Substrate-Chromogen (Dako), for use in the colorimetric peroxidase-based assay, to read binding of peroxidase-conjugated anti-polyhistidine antibody to 6His tagged proteins.

2.2 Disposable Plasticware

1. Ninety-six-well microplates, coated with streptavidin, serve as the basic tool for the performance of all assays described in this chapter. A large variety of such plates are available commercially. In most of the experiments performed with the core methodology, we used "BioBind Assembly Streptavidin Coated Strip 1x8 plates," Cat. No. 95029263 (Thermo Scientific). The wells have a coated area corresponding to a volume of 200 μL and a biotin binding capacity of >12 pmol biotin/well (*see* **Note 6**). This type of plate was used in all experiments described in Refs. [33, 36, 37]. Recently, the manufacture of this plate was discontinued. Upon searching for an alternative, we found "Pierce Streptavidin Coated Plate (clear, 8-well strips)," Cat. No. 15121 (Pierce Biotechnology, Thermo Scientific) as being very similar to the plates used before. The wells have a coated area corresponding to a volume of 200 μL and a biotin binding capacity of ~10 pmol biotin/well (*see* **Note 7**).
2. Adhesive sealing tape for sealing filled 96-well plates (Thermo Scientific).
3. "Protein LoBind" 1.5 mL conical tubes (Eppendorf), or "Non-Stick Surface" 1.5 mL conical tubes (Labcon), for the preparation of dilutions and the storage of dissolved peptides and recombinant proteins. These tubes, made of treated polypropylene, are recommended in order to reduce binding of peptides and proteins to the tube wall. Glass and polystyrene tubes should not be used.
4. Disposable centrifuge columns (polypropylene, 10 mL capacity), with polyethylene bottom filter (30 μm pore size) (Pierce, Thermo Scientific), were found very useful for batch metal affinity purification of 6His-tagged recombinant proteins (both binding to gel and elution from gel).
5. Centrifugal concentrators, 10,000-molecular-weight-cut-off, 4- and 15-mL (Amicon Ultra, Millipore, Merck KGaA), for the concentration of all recombinant cytosolic components.
6. Disposable Rapid-Flow Filter Unit, 0.45 μm pore size aPES membrane, 250 or 500 mL capacity (Nalgene-Thermo Scientific), for filtration of all solutions used in protein-peptide binding experiments.

2.3 Small- and Medium-Sized Equipment for General Use and for Preparation of Recombinant Oxidase Components

1. Electronic single-channel pipettors (range from 0.5 to 1000 μL) (Eppendorf). These have a "dispensing" mode, very useful for adding small equal amounts of reagents to 96-well plates.
2. Multipette Plus (manual) or Multipette Stream (electronic) pipettors (Eppendorf) and various Combitips (Eppendorf), for distributing equal amounts of reagents in 96-well plates.
3. Finnpipette digital 12-channel pipette (50–300 μL range) (Thermo Scientific).
4. Rotating tube mixer Rotamix RM1 (ELMI).
5. XCell SureLock Mini-Cell for SDS-PAGE of mini-gels (Invitrogen, Life Technologies).
6. Electrophoresis power supply Power Pac 300 (Bio-Rad).
7. Thermomixer Comfort, rotary mixer and heater/cooler (Eppendorf).
8. Table top microcentrifuge (Minispin, Eppendorf).
9. FPLC gel filtration columns HiLoad 10/60 Superdex 75 prep grade (fractionation range: 3–70 kDa), for purification of $p47^{phox}$, $p67^{phox}$ (1–212) and Rac, and HiLoad 10/60 Superdex 200 prep grade (fractionation range: 10–600 kDa), for purification of $p67^{phox}$(1–526) (GE Healthcare). Columns are fitted with a coolant jacket.

2.4 Large Equipment of General Use and for Preparation of Recombinant Oxidase Components

1. "Warm room" set at 37 °C, containing shaking platform (such as Innova 2100 platform shaker, New Brunswick Scientific), for culturing transformed *E. coli* cultures before induction by IPTG.
2. Innova 4230 or C24KC refrigerated incubator shaker (New Brunswick Scientific). These incubators are suitable for growth of *E. coli* at 18 °C, for expression of recombinant oxidase components by IPTG induction.
3. Refrigerated low-speed centrifuge (up to 7000 × *g*), with a swing-out rotor (e.g., Sorvall RC-3B or RC-3C and H-6000A rotor).
4. Refrigerated high-speed centrifuge (up to 48,000 × *g*), with fixed-angle rotor (e.g., Sorvall RC-5 or RC-5 Plus and SS-34 rotor).
5. High-intensity ultrasonic processor (500-W) fitted with a 13 mm tip probe (Sonics and Materials). This is used for disruption of *E. coli* expressing recombinant oxidase components.
6. Spectrophotometer (double-beam) UV/visible (Uvikon 943, Kontron Instruments).
7. Akta Basic 10 chromatography system, to be used for FPLC gel filtration of recombinant oxidase components (GE Healthcare).

8. Coolant circulator for gel filtration columns (Fryka).
9. Compressed helium gas tank for bubbling through FPLC buffer.

2.5 Large Equipment Specifically Intended for Performance of Protein-Peptide Binding Assays

1. High-intensity ultrasonic processor (400-W) fitted with cup horn (Sonics and Materials). This is required for assisting in the solubilization of peptides.
2. Tunable microplate readers. For reading of protein-peptide binding assays, as described in Subheading 3, we used Spectramax 340, fitted with software SoftMax Pro, Version 5.2, and, more recently, Versamax, fitted with software SoftMax Pro, Version 6.5.1 (both instruments are manufactured by Molecular Devices).
3. Vibrating platform shaker for 96-well plates (Mini-orbital shaker, 0–1900 RPM, 1.6 mm orbit, Bellco Glass, or Titramax 100, 150–1350 RPM, 1.5 mm orbit, Heidolph).
4. Plate washer for 96-well plates (Wellwash Versa, software version 1.02.5; Thermo Scientific). The washer was fitted with a 2 × 8 cell wash head, instead of the standard wash head (*see* **Note 8**).

3 Methods

3.1 Dissolving Peptides and Storing Peptide Solutions

The procedure used by us is described in Refs. [33, 36, 37, 41].

1. Dissolve peptides in a mixture of 75 parts NMP and 25 parts water (v/v), in polypropylene tubes, to a concentration of 1.5 mM, to serve as stock solutions for further dilution (*see* **Notes 9** and **10**).
2. Subject the peptide solutions to sonic disruption, using five 10 s pulses, at 90% amplitude, in a 400-W ultrasonic processor equipped with a cup horn filled with ice-water mixture, in which the tubes containing the peptide solution are immersed. If the peptides are properly solubilized, the solutions should not contain particles or be opalescent.
3. Divide the peptide stock solutions into 100 μL aliquots in low protein binding polypropylene tubes, and keep frozen at −75 °C.
4. Prepare working solutions freshly by a two-stage dilution procedure on the day of the experiment. First, dilute the 1.5 mM peptide stock solutions in 75% NMP/25% water to a concentration of 100 μM in PBS, lowering the concentration of NMP to 5%. Dilute these intermediary stocks further to a concentration of 1 μM in PBSCas1, reducing the concentration of NMP to 0.05% (*see* **Note 10**).

3.2 Preparing Recombinant NADPH Oxidase Components

All recombinant cytosolic components are expressed in *E. coli* as N-terminal 6His-tagged proteins. The methodology is described in Ref. [45]. The key steps are briefly summarized below:

3.2.1 Expression of Recombinant p47^{phox}, p67^{phox}, and Rac1

1. Transform *E. coli* competent cells (Rosetta 2(DE3)pLysS, Novagen) with the expression vector pET-30a-6His Kan^R (Novagen), carrying cDNA encoding each of the three cytosolic components, following a standard protocol, as described in pET System Manual, 11th edition, Novagen.
2. Induce bacteria with 0.4 mM IPTG, and grow at 18 ° C for 18 h in a refrigerated incubator shaker. Induction at 18 °C is important for maximizing the recovery of the recombinant proteins in the soluble fraction after disruption of the bacteria.
3. Test by SDS-PAGE for successful induction of the desired protein by comparing bacterial cell extracts before and after induction by IPTG.
4. Sediment bacteria by centrifugation at 3500 × *g*, and resuspend in lysis buffer supplemented with protease inhibitor mixture cOmplete, EDTA-free (one tablet per 50 mL buffer). For 1 L of original bacterial culture, use 40–50 mL lysis buffer.
5. Freeze the bacterial suspension at −75 °C. Frozen suspensions can be kept frozen till processed further. This freezing step is essential for successful recovery of soluble recombinant protein.
6. Allow suspension to thaw slowly, and add Triton X-100 to a final concentration of 1% v/v.
7. Subject the suspension (~50 mL) to sonic disruption in a glass beaker immersed in an ice/water mixture using a 500-W ultrasonic processor with 13 mm tip probe, for two to three 5 min intervals, at an amplitude of 20%, and alternating cycles of sonication for 2 s and 2 s of rest.
8. Centrifuge at 48,000 × *g* for 30 min, and decant supernatant containing the soluble protein.
9. Measure protein concentration by the Bradford assay [46], modified for use with 96-well microplates (*see* Bio-Rad Technical Bulletin 1177 EG and **Note 11**).

3.2.2 Two-Stage Purification of Recombinant p47^{phox}, p67^{phox}, and Rac1

Purification by Metal Affinity Chromatography on Ni Sepharose

1. Mix 3 mL washed packed Ni Sepharose beads with soluble fraction derived from sonic disruption of bacteria from 1 L culture (~40 mL). Incubate for 1 h at room temperature with top/bottom rotation using a rotating tube mixer at 10 RPM.
2. Transfer content into a centrifuge column with bottom filter, and allow the fluid to flow by gravity into collecting tube.
3. Wash beads twice with 15 mL volumes of binding buffer and twice with 15 mL volumes of washing buffer.

4. Add 10 mL/column elution buffer, seal bottom and top apertures, and incubate for 30 min at room temperature with top/bottom rotation using a rotating tube mixer at 10 RPM.
5. Allow the eluate to run by gravity into a collecting tube, and repeat procedure from one to four times (labelled as eluates 1, 2, 3, and 4).
6. Measure protein concentrations in all eluates, and analyze by SDS-PAGE, for purity. The purity requirements for the performance of protein-peptide binding assays are usually not achieved by purification on Ni Sepharose only (*see* **Note 12**).

Purification by FPLC Gel Filtration

1. Proceed to purification by gel filtration on HiLoad 16/60 Superdex 200 column, for p67^{phox}(1–526), and on HiLoad 16/60 Superdex 75 column, for p67^{phox}(1–212), p47^{phox}(1–390), and Rac. Pool eluates with significant protein concentration from Ni Sepharose and concentrate to volumes of 2.5 mL per one column purification (= 2% of column volume), using Amicon Ultra 15 mL (10 kDa molecular-weight-cut-off) centrifugal concentrators. Centrifuge at 12,000 × *g* for 30 min in a microcentrifuge at 4 °C, and use supernatant for gel filtration.
2. Inject material into a column refrigerated by a coolant circulator, set at 4 °C, and perform gel filtration using an Akta Basic 10 chromatography system with PBS, refrigerated by being kept in an ice-filled bucket, degassed by bubbling of helium, as the running buffer, at a flow rate at 1 mL/min. Record absorbance at 280 nm and collect 2 mL fractions. Analyze fractions by SDS-PAGE and pool those of highest purity.
3. Supplement purified recombinant proteins with 30% v/v glycerol, divide in small aliquots in polypropylene tubes with low protein binding quality, and store at −75 °C. Avoid repeated thawing/freezing. In this state, they are active for an unlimited time period (*see* **Note 13**).

3.3 Protein-Peptide Binding Assays Designed for the Study of Protein-Protein Interactions in the Assembly of the NADPH Oxidase Complex

3.3.1 A Few Caveats

1. This chapter is based on our experience in performing protein-peptide binding assays, as applied to oxidase components, and is not intended to provide the model for a universal methodology. Variations of the method have been discussed in recent reviews [25, 26].
2. The method is not intended to be used in the ever-expanding field of high-throughput screening, for which the many variations of the SPOT technology were developed [29, 30].
3. It represents a compendium of what we have learned over a period of more than two decades by working on the interaction of proteins with peptides within the realm of the NOX2 oxidase complex. Our approach, which we named "peptide walking," was initiated by measuring the ability of overlapping peptides,

derived from the sequences of Rac [47, 48], p47^{phox} [28], p22^{phox} [32], and NOX2 [41], to inhibit oxidase activation in vitro (reviewed in Ref. [38]). This was followed by measuring binding of one oxidase component to peptides derived from another component [28, 32, 33, 36, 37].

4. Our more recent work departs from merely screening for binding sequences and focuses on the in-depth analysis of domains identified by screening, by subjecting selected peptides to truncations, residue replacements, introduction of disulfide bonds, dimerization, and changes in orientation vs. the well surface [33, 36, 37].
5. We also subject the recombinant proteins under study to modifications which might affect binding to peptides. These comprise truncations, deletions, and single or multiple mutations, as well as exposure to chemical agents affecting specific residues.

3.3.2 The Canonical Protein-Peptide Binding Assay

1. To the wells of a streptavidin-coated 96-well plate, add 300 μL/well of blocking solution (PBSTcas1), and keep for 1 h at room temperature without shaking. Although most presently manufactured plates are supposed to be pre-blocked, we prefer to perform an additional blocking procedure, using a casein-based solution (*see* **Notes 6** and **7**).
2. Wash the blocked wells four times with 300 μL PBST/well, using a plate washer fitted with a cell wash head (*see* **Note 8**). Although the plate washer is programmed to leave the wells empty of liquid, remove rests of liquid by inverting the plate on absorbent filter paper.
3. Add 200 μL/well of biotinylated peptides, as a 1 μM solution in PBScas1, and shake plate for 1 h at room temperature on a vibrating platform shaker at 700 RPM. This corresponds to 200 pmol peptide per well. Since the wells have a binding capacity of ~10 pmol biotin, the peptides are in a 20-fold excess.
4. Prepare at least triplicate identical wells for each peptide-protein combination, and a higher number of replicate wells are an advantage.
5. Make sure to always run "blank wells," lacking peptides. Such wells are subject to blocking with PBSTcas1, and, following washing, wells are filled with 200 μL/well of PBScas1. From this point on, the wells are treated as those containing peptides.
6. Wash plate four times with 300 μL PBST/well.
7. Exchange the protein solution in PBS (which served as the buffer in the gel filtration purification step) to PBScas1 by mixing 4 parts protein, brought to a concentration of

1.875 μM, with 1 part PBScas5. This results in a final protein concentration of 1.5 μM and a final concentration of casein of 1% w/v.

8. Add 200 μL/well of protein solution. This corresponds to 300 pmol protein per well. Assuming that the streptavidin-binding sites were fully saturated by binding ~10 pmol of peptide, this represents a 30-fold excess of protein over peptide (*see* **Note 14**). Cover the plate with adhesive sealing tape, and place on a vibrating platform shaker, located in a cold room, and shake plates for 18 h at 4 °C, at 700 RPM (*see* **Note 15**). Having wells to which no protein was added was not found necessary; we never observed "false positives" due to binding of anti-polyhistidine antibody to peptides.
9. Wash plate four times with 300 μL PBST/well.
10. Dilute the monoclonal peroxidase-conjugated anti-polyhistidine antibody 1/3000 v/v (= 3 μg/mL) with PBSTcas1, and add 200 μL/well. Shake plate for 1 h at room temperature on a vibrating platform shaker at 700 RPM (*see* **Note 16**).
11. Wash plate four times with 300 μL PBST/well.
12. Wash plate two times with 300 μL PBS/well.
13. Add 200 μL/well of TMB + substrate-chromogen.
14. After adding the TMB reagent, place plate immediately in the plate carrier of the tunable microplate reader, and read the change in absorbance at 650 nm of the selected wells in the kinetic mode, at room temperature. A key step in the correct expression of the results is to select the linear segment of the increase in absorbance curve. This is easily done using the software associated with the plate reader (Figs. 5 and 6) (*see* **Note 17**). Results are expressed as V_{max} (change in absorbance at 650 nm × 1000/min). Looking carefully at the "rough" data (shape of the curves, time and absorbance values when a plateau is reached, kinetics of "blank" wells, variability within replicate wells), before applying any software-mediated manipulation, is essential for a correct interpretation of the results.
15. For data analysis, plotting and presenting results of protein-peptide binding assays, we use "Prism Version 7" software (GraphPad Software).
16. On some occasions, binding values of protein to wells lacking peptides ("blank wells") are deducted from values of protein bound to peptide, and results are presented as "peptide-bound *minus* blank." This makes graphical presentation simpler but masks valuable information and, if applied to data to be published, should be accompanied by listing of the "blank" values in the figure legend.

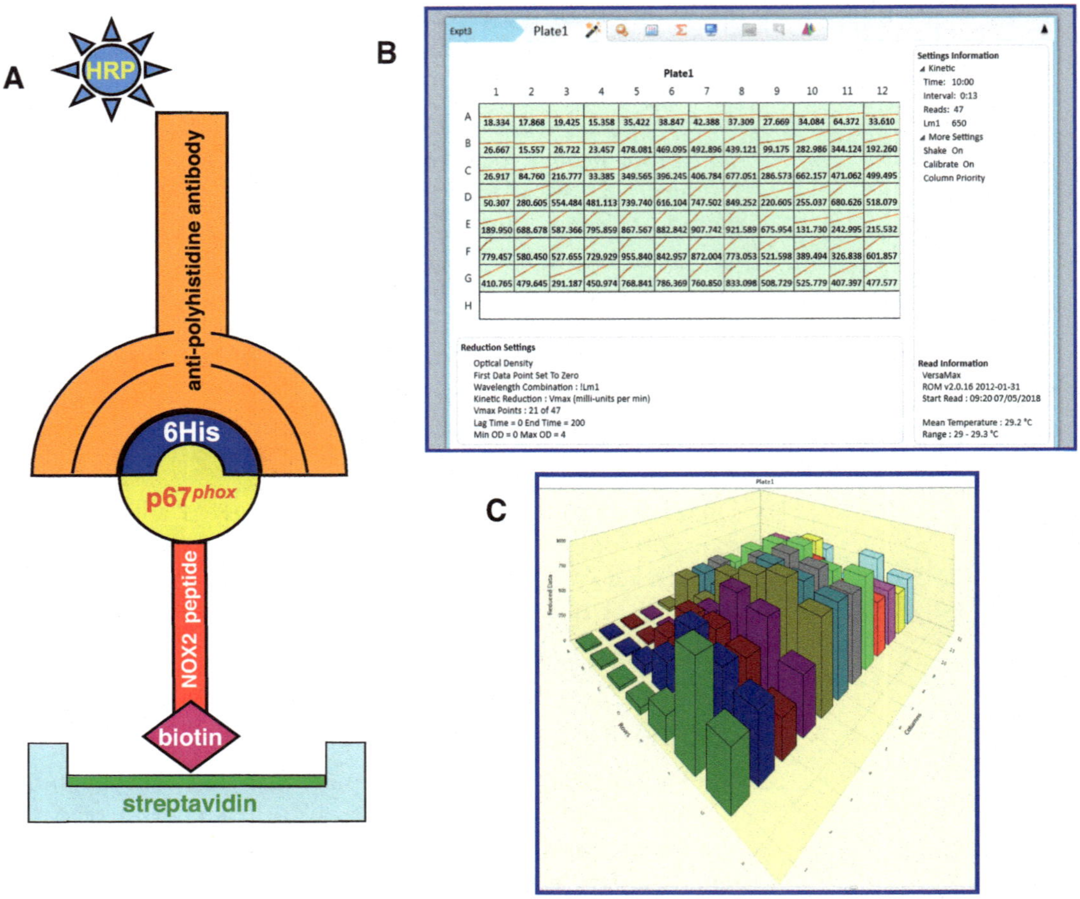

Fig. 5 An example of a protein-peptide binding assay. (**A**) Schematic representation of the method. (**B**) The typical "look" of the read out of a kinetic ELISA. The orange curves are the original increase in absorbance records (end time = 600 s); the black curves are the chosen linear segments (end time = 200 s). The numbers in the wells represent the calculated V_{max} values (milliabsorbance/min) based on the linear segments of the curves. The maximal absorbance value was four, and the first absorbance reading value was set to zero. (**C**) Three-dimensional representation of V_{max} values shown in panel **B**. Panel **A** in this figure is a modified version of Fig. 2, panel **A**, in Ref. [33]

3.3.3 The "Two Antibodies" Variation of the Protein-Peptide Binding Assay

In this variation of the assay, the procedure described in Subheading 3.3.2 is followed up to point 10. From this point on, proceed as follows:

1. Dilute the first unconjugated antibody in PBSTcas1 to a concentration found optimal in preliminary experiments, and add 200 μL/well. Shake plate for 1 h at room temperature on a vibrating platform shaker at 700 RPM (*see* **Note 18**).
2. Wash plate four times with 300 μL PBST/well.
3. Dilute the peroxidase-conjugated second antibody, normally an anti-mouse, anti-rabbit, or anti-goat IgG (H + L), in

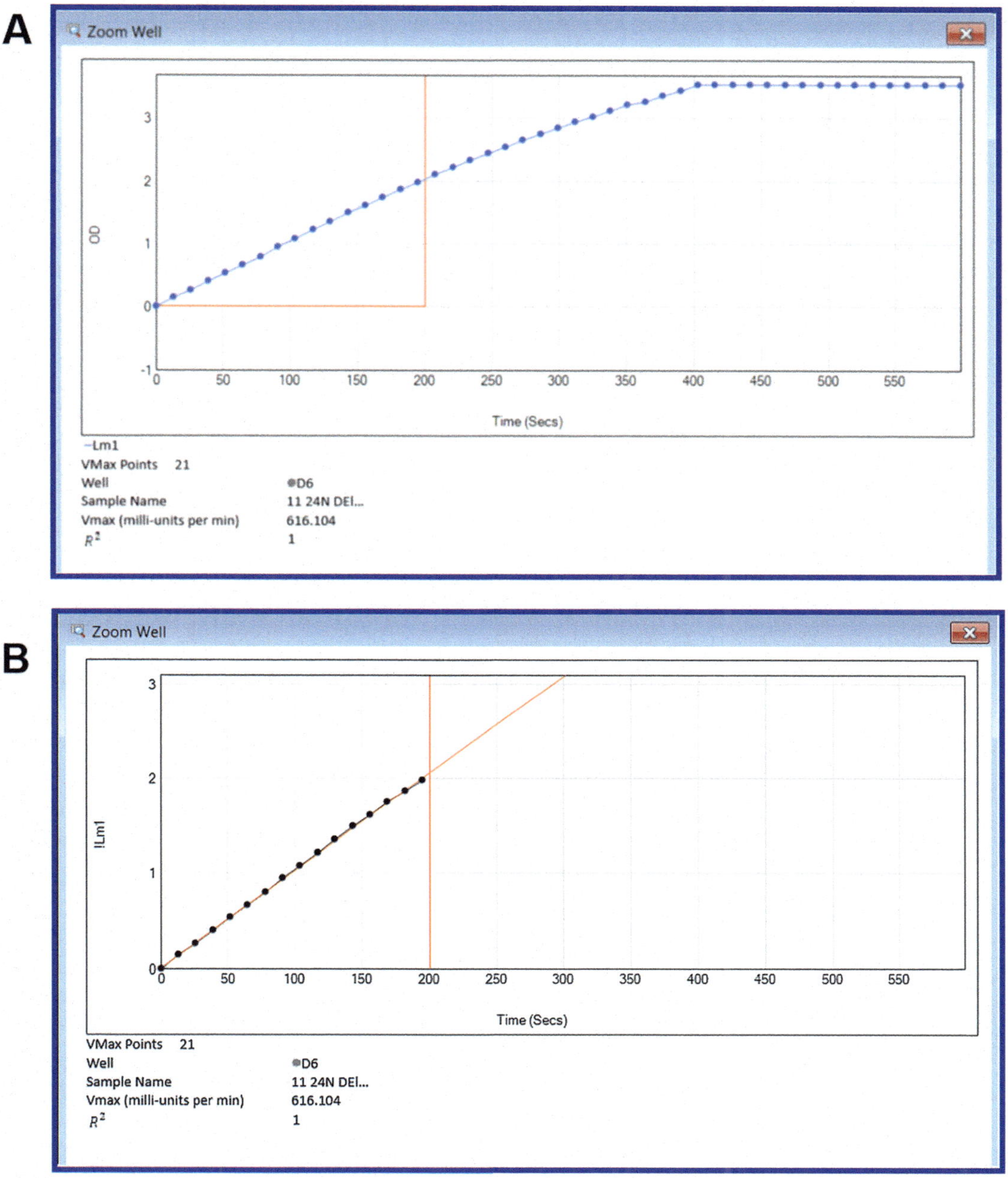

Fig. 6 Detailed look of an individual well in the plate illustrated in Fig. 5. (**A**) Raw data of well D6 of plate shown in Fig. 5, panel **B** (blue dots and blue curve). One can see that the curve reaches a plateau at 400 s and the curve tends to become nonlinear starting at 200 s. (**B**) The end time was reduced to 200 s and a linear plot generated (black dots and red curve). The V_{max} value is 616.104 and the R^2 equals 1

PBSTcas1 to the concentration found optimal, and add 200 μL/well. Shake plate for 1 h at room temperature on a vibrating platform shaker at 700 RPM (*see* **Note 19**).

4. Wash plate four times with 300 μL PBST/well.

5. Wash plate two times with 300 μL PBS/well.
6. From this point on, follow procedure as described at **steps 13** and **14** of Subheading 3.3.2.

3.3.4 Subjecting Peptides to Modifying Procedures Ahead of Attachment to the Wells of Microplates

On some occasions, the effect of modifying agents on the peptides is studied. In one approach, the agents are applied in solution to peptides immobilized to the wells. A disadvantage of this is that contact of the agents with the peptides depends on the way of attachment of the peptide to the surface, with different segments of the peptide being exposed to the fluid phase. For this reason, we prefer to expose the peptides to the agents when both are in free form in solution. In this case, the following methodology is used:

1. Prepare stock solution of peptides as described in Subheading 3.1, at a concentration of 1.5 mM, in 75% NMP/25% H_2O_2 v/v.
2. Dilute the stock solution to a concentration of 100 μM in PBS; the peptide will be further diluted to 1 μM in PBS v/v when mixed with the agent to be studied (*see* **Note 20**). These dilutions are to be made in low protein binding polypropylene tubes.
3. Prepare the agent to be used at a concentration 100-fold higher than the desired final concentration. As an example, when the effect of H_2O_2 was studied at a final concentration of 10 mM, a stock solution of 1 M was prepared in H_2O (*see* Ref. [37]).
4. Mix in a low protein binding 1.5 mL tube 980 μL PBS, 10 μL peptide (100 μM), and 10 μL H_2O_2 (1 M). This will result in final concentrations of 1 μM peptide and 10 mM H_2O_2.
5. Seal the tubes with the attached stoppers, and place in a Thermomixer Comfort rotary mixer and incubate for 2 h at 24 °C at 700 RPM.
6. Block plates as described in Subheading 3.3.2, **steps 1** and **2**.
7. Add to wells 200 μL of the (peptide + agent) mixtures, and shake plate for 1 h at room temperature on a vibrating platform shaker at 700 RPM.
8. Continue as described in Subheading 3.3.2, **steps 5–14**.

3.4 The Assay in Action: Examples of Using the Protein-Peptide Binding Method

This section comprises examples of applying protein-peptide binding assays to specific issues.

3.4.1 Detection of a Region in NOX2 Binding a Modified Form of p67*phox*

In the course of studies on the role of the N-terminal src 3 homology (SH3) domain of p67*phox* in the activation of the NOX2 oxidase, we deleted residues 259–279, located at the center of the domain, and compared the binding of wild-type p67^{phox}(1–526) and p67^{phox}(1–526)Δ(259–279) to an array of 91 overlapping 15-mer peptides, with an offset of three residues, covering the whole length of the NOX2 DHR (residues 288–570). As seen in Fig. 7, panel B, p67^{phox}(1–526)Δ(259–279) was bound most prominently to peptide 28 (residues 369–383), with an affinity markedly exceeding the moderate binding of wild-type p67^{phox}(1–526) (Fig. 7, panel A), described before (Ref. [33] and

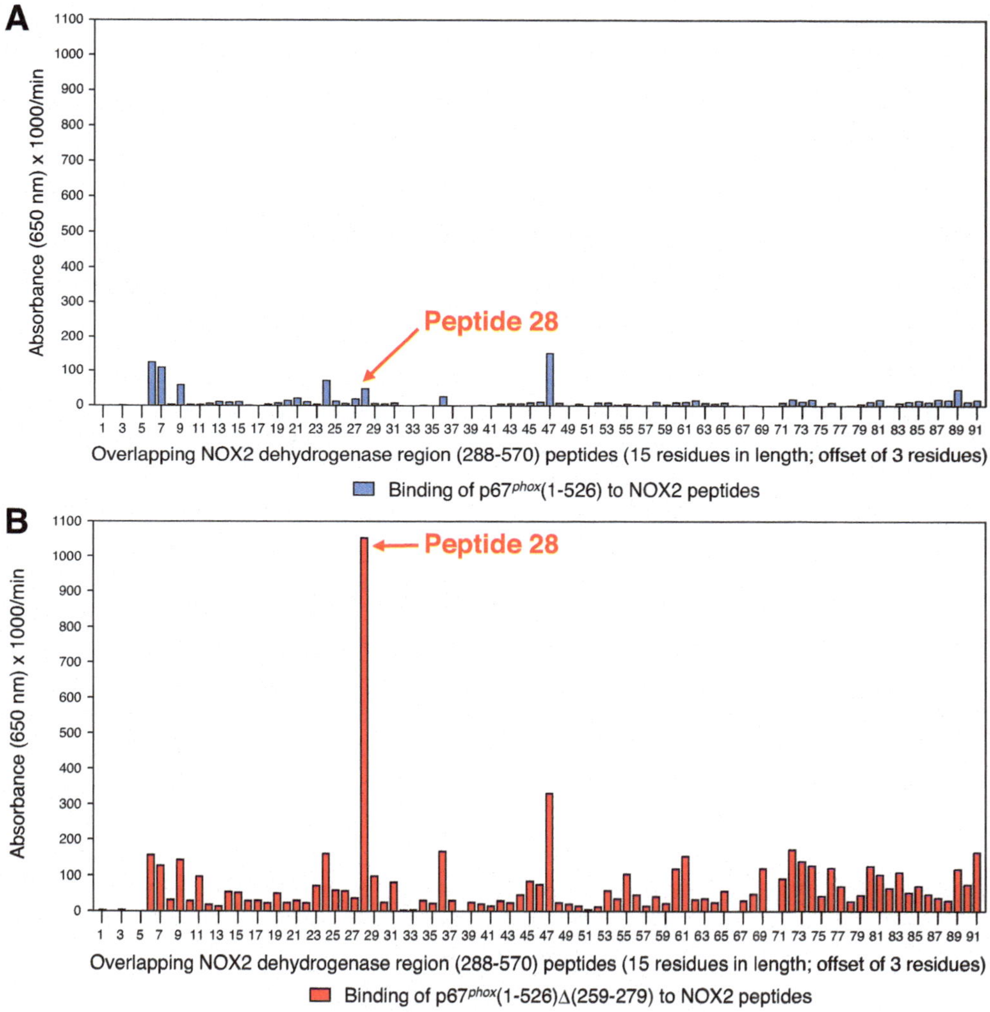

Fig. 7 "Peptide walking" through NOX2 DHR peptides with conformationally modified p67^{phox}. This experiment is an example of the use of protein-peptide binding method for detecting changes in the structure of an NADPH oxidase component. p67^{phox}(1–526) was subjected to deletion of residues 259–279, from within the N-terminal SH3 domain, and the binding of the deletion mutant to NOX2 DHR peptides (residues 288–570) was compared to that of the wild-type protein. As seen in panel **B**, the deletion mutant exhibited pronounced binding to peptide 28 (residues 369–383), whereas the wild-type protein lacked this ability (panel **A**). This type of experiment is performed with arrays of unpurified 15-mer peptides with N-terminal biotin and an offset of three residues. Results of one characteristic experiment are shown

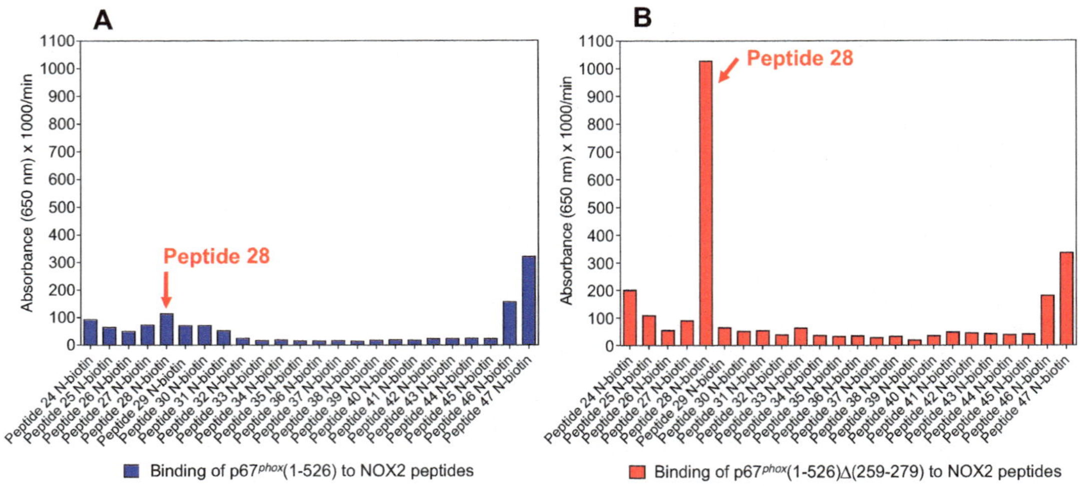

Fig. 8 Refining preliminary results obtained by protein-peptide binding assay with unpurified peptides by using purified peptides. The experiment illustrated in Fig. 7 was repeated by using peptides with a degree of purity of ≥70% and by restricting the range of residues covered by the peptides. We measured the binding of wild-type p67^{phox}(1–526) and p67^{phox}(1–526)Δ(259–279) to purified NOX2 DHR peptides (15-mer, three-residue offset, N-terminal biotin), covering the range of residues 357–440. Maximal binding of p67^{phox}(1–526)Δ(259–279) to peptide 28 (369–383) was confirmed (panel **B**), as well as the much lower binding of wild-type p67^{phox} (panel **A**). The minor binding of both wild-type and deleted proteins to peptide 47 (426–440), found in the experiment illustrated in Fig. 7, was also confirmed. Results of one characteristic experiment are shown

Fig. 2). Such screens are performed with arrays of unpurified peptides, and the results are confirmed by focusing on regions of positive binding, by using purified peptides (Fig. 8).

3.4.2 Characterizing the "Binding Sequence" in a Peptide

We next proceeded to characterize the region, within NOX2 peptide 28, participating in binding of p67^{phox}(1–526)Δ(259–279). A ^{369}CGC^{371} triad was shown to be involved in the binding of wild-type p67^{phox}(1–526) to NOX2 peptides 24 and 28 [33, 37]. As seen in Fig. 9, replacing Cys369 or both cysteines 369 and 371 by arginine(s) in peptide 28, attached by an N-terminal biotin (28 N-biotin), did not affect the binding of p67^{phox}(1–526)Δ(259–279). However, scrambling of the peptide or replacing the NOX2 sequence by the aligned NOX4 sequence abolished binding (*see* **Note 21**). Also, when peptide 28 was attached by a C-terminal biotin (28 C-biotin), exposing the N-terminus to the fluid phase, the enhanced ability to bind p67^{phox}(1–526)Δ(259–279) was lost. These data suggest that the "binding sequence" is likely to be located in the C-terminal half of the peptide and does not involve the ^{369}CGC^{371} triad.

3.4.3 Identifying the "Binding Sequence" by Truncations of the Peptide

More precise definition of the sequence engaged in binding of p67^{phox}(1–526)Δ(259–279) can be obtained by truncation of the peptide from the N- and C-termini. An example of such an

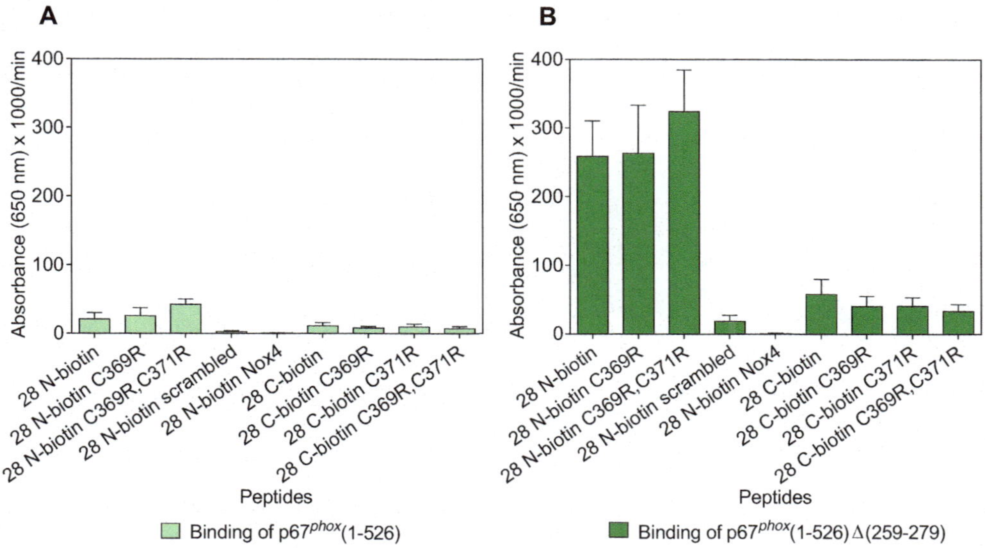

Fig. 9 On the path to identify the region, in NOX2 peptide 28, responsible for binding p67^{phox}(1–526)Δ (259–279). Wild-type p67^{phox} (panel **A**) and deleted p67^{phox} (panel **B**) were tested for binding to NOX2 peptide 28 with biotin attached to the N- or C-terminus and subject to scrambling, replacement of Cys369 or both Cys369 and Cys371 by arginine(s), and replacement of the NOX2 369–383 sequence by the corresponding sequence in NOX4. The results indicate that the region binding p67^{phox}(1–526)Δ(259–279) is located in the C-terminal half of peptide 28 and that cysteines in the CGC triad do not participate in binding. Results represent means ± SE of three experiments

approach is shown in Fig. 10. NOX2 peptide 28 (residues 369–383), found to bind p67^{phox}(1–526)Δ(259–279) preferentially, when compared to that of p67^{phox}(1–526) (Figs. 7 and 8), was subjected to serial N-terminal three-residue truncations and several C-terminal truncations (Fig. 10, panel C). p67^{phox}(1–526)Δ (259–279) was bound to the full-length peptide 28 and to two peptides, truncated at residues 372 and 375, at levels similar to that to the full-length peptide (Fig. 10, panel B). This suggests that the "binding sequence" consists principally of residues 375–383.

3.4.4 Free Peptide in Solution Competes with Surface-Bound Peptide for Binding of the Interacting Protein

Whenever a positive result is obtained in protein-peptide binding assays, it is recommended to look for the ability of free peptide to prevent binding of the protein to the surface-attached peptide. For this purpose, the peptide under investigation has to be synthesized in two forms: (a) N-terminal biotin-spacer-peptide-amide or C-terminal biotin-spacer-peptide-acetyl, to serve as surface-attached peptides, and (b) N-terminal acetyl-peptide-amide, to serve as competing peptide. The protein under study is incubated, as a 1.5 μM solution in PBScas1, with the non-biotinylated peptide, added to the protein at concentrations of 15 or 30 μM, for 1 h at room temperature. As a control, an identical volume of PBScas1, containing 0.75 or 1.5% v/v NMP, is added to a sample of the protein (75% NMP is present in the solvent used to prepare the

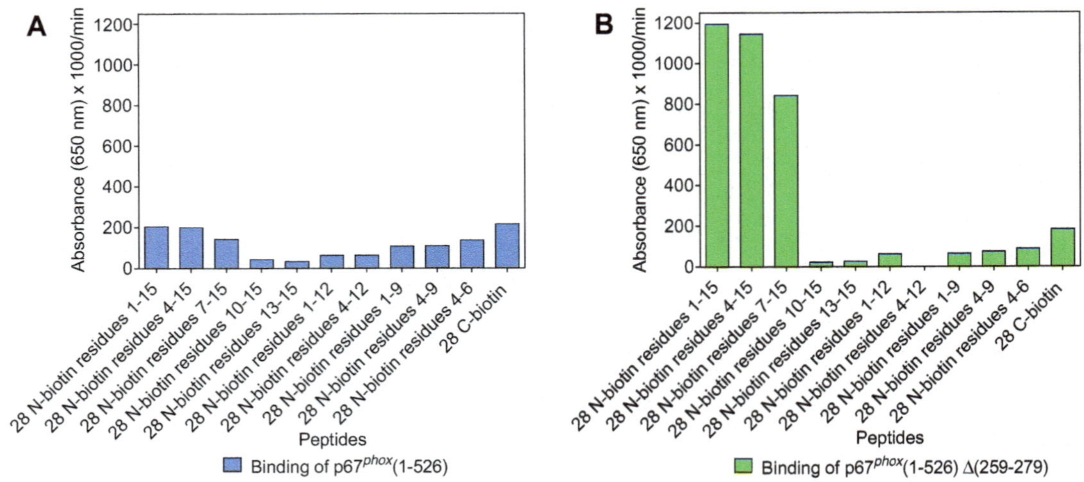

C

Nox2 peptide 28 truncations

Peptide 28	1-15	369	CGCDKQEFQDAWKLP	383
Peptide 28	4-15	372	DKQEFQDAWKLP	383
Peptide 28	7-15	375	EFQDAWKLP	383
Peptide 28	10-15	378	DAWKLP	383
Peptide 28	13-15	381	KLP	383
Peptide 28	1-12	369	CGCDKQEFQDAW	380
Peptide 28	4-12	372	DKQEFQDAW	380
Peptide 28	1-9	369	CGCDKQEFQ	377
Peptide 28	4-9	372	DKQEFQ	377
Peptide 28	4-6	372	DKQ	374

Fig. 10 Identification of the "binding sequence" in NOX2 peptide 28. NOX2 peptide 28, with an N-terminal biotin tag, was subjected to N-terminal truncations down to three residues, and several additional truncated peptides were synthesized. As seen in panel **B**, p67^{phox}(1–526)Δ(259–279) was bound, in addition to the full-length peptide 369–383, also by peptides truncated at residues 372 and 375. This suggests that the binding site for the deleted form of p67^{phox} comprises residues 375–383 (panel **C**). There was no significant binding of wild-type p67^{phox} to any of the peptide 28 truncations (panel **A**). Results of one characteristic experiment are shown

1.5 mM peptide stock solutions). An additional control consists of incubating the protein with a non-biotinylated peptide, other than the one to be competed for. In an example of such an experiment, we assessed the binding of p67^{phox}(1–526) and p67^{phox}(1–526)Δ(259–279) to NOX2 peptides 24 and 28, in the absence of free non-biotinylated peptide and following pre-exposure to either free peptide 24 or 28. As shown in Fig. 11, binding of p67^{phox}(1–526)Δ(259–279) to surface-attached peptide 28 was inhibited by free peptide 28 but not by free peptide 24.

3.4.5 Mapping the Sequence in a Protein the Deletion of Which Enhances Binding of the Protein to a Peptide

In this example, we explored the minimal region in p67^{phox}(1–526), the deletion of which leads to enhanced binding to NOX2 peptide 28. The original sequence deleted from p67^{phox}(1–526) (residues 259–279) was truncated from the N- or C-terminus, resulting in deletion mutants in which the deleted segments varied in size from

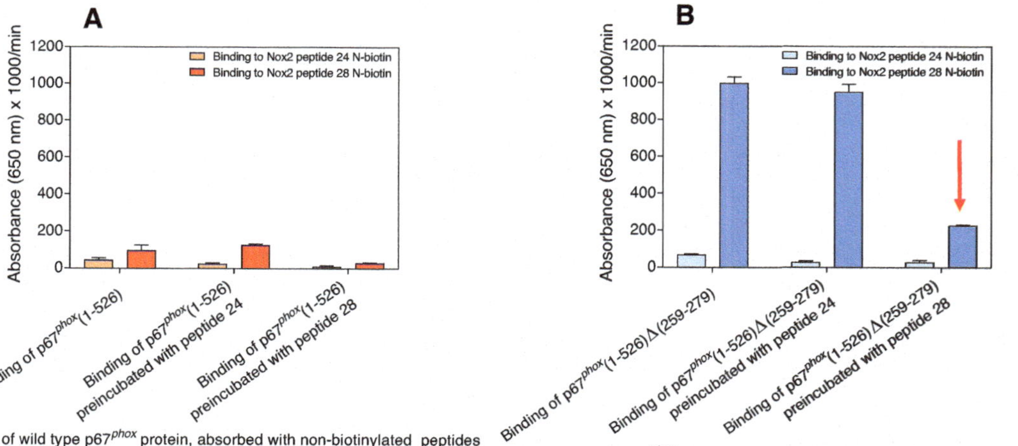

Binding of wild type p67^{phox} protein, absorbed with non-biotinylated peptides 24 or 28, to surface-attached N-biotinylated peptides 24 and 28

Binding of deleted p67^{phox} protein, absorbed with non-biotinylated peptides 24 or 28, to surface-attached N-biotinylated peptides 24 and 28

Fig. 11 Free NOX2 peptide 28 in solution prevents binding of p67^{phox}(1–526)Δ(259–279) to surface-attached peptide 28. Preincubation of p67^{phox}(1–526)Δ(259–279) with a 10- to 20-fold excess, in molar terms, of non-biotinylated peptide 28 in solution interferes with binding of deleted p67^{phox} to biotinylated peptide 28 attached to streptavidin-coated wells (panel **B**; 77% inhibition of binding compared to binding of deleted p67^{phox} pre-incubated with buffer; see red arrow). Preincubation with non-biotinylated peptide 24 had no effect on binding. There is only minor binding of p67^{phox} (1-526) to surface-attached peptide 28, which is also reduced by non-biotinylated peptide 28 in solution (panel **A**). Results represent means ± SE of three experiments

15 to 3 residues (*see* insert in Fig. 12). The binding of the proteins, with truncated deletions, to NOX2 peptide 28 was measured, and, for comparison, their binding to NOX2 peptide 24 was also assessed. As seen in Fig. 12, deletion of three residues located at the center of the original (259–279) deletion (residues 265–267 or 268–270) was sufficient for causing enhanced binding of p67^{phox}(1–526) to NOX2 peptide 28. It is of interest that the two groups of three residues contain amino acids encountered frequently at interaction "hot spots" (*see* Ref. [25]), and one possible interpretation is that this region participates in an intramolecular interaction, the disengagement of which is responsible for enhanced binding to NOX2 peptide 28.

3.4.6 More Examples of the Use of the Protein-Peptide Binding Assay

The assay was used for clarifying the role of Cys369 in NOX2. A ^{369}Cys to Arg (C369R) mutation in NOX2 in humans causes chronic granulomatous disease (CGD) of the X91^{+} form, with normal expression of a nonfunctional NOX2 [49]. Cys369 was replaced by Arg in NOX2 peptide 24 and assayed for binding of p67^{phox}. As reported in Ref. [33], such a "mutation" in the peptide resulted in lack of binding of p67^{phox}.

Yet another example for the use of the assay is its application to the study of the effect of the introduction of an intra-peptide disulfide bond between cysteines 369 and 371 in NOX2 peptide 24. As described in Ref. [37], the generation of such a bond by

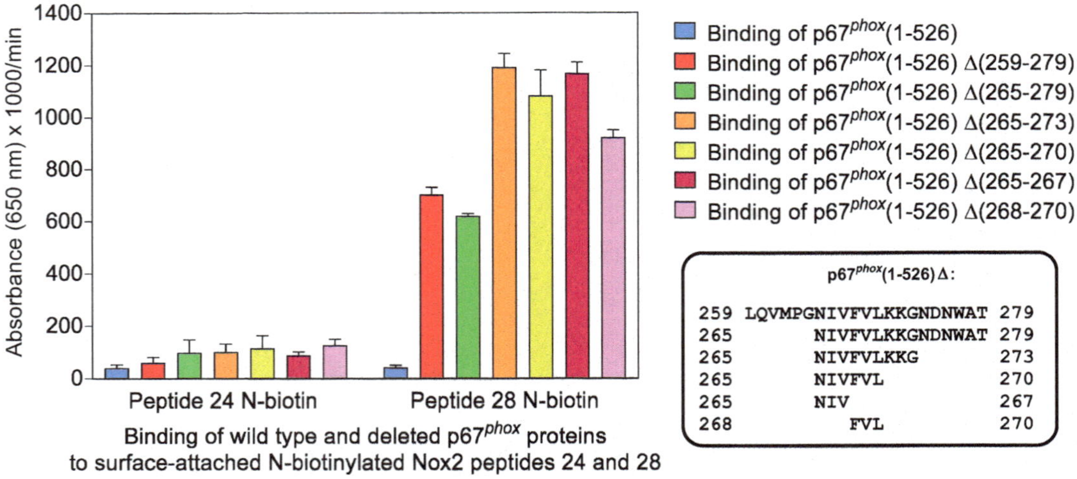

Fig. 12 Identification of the minimal region in p67^{phox}(1–526)Δ(259–279), the deletion of which leads to enhanced binding to NOX2 peptide 28. The original sequence deleted from p67^{phox}(1–526) (residues 259–279) was truncated from the N- and C-terminus, resulting in deletion mutants in which 15, 9, 6, and 2 variations of 3 residues were deleted (*see* insert). The various deletion mutants were tested for binding to peptide 28 with N-terminal biotin and, as a control, to peptide 24. Deletion of residues 265–267 or 268–270 was sufficient for causing enhanced binding of p67^{phox}(1–526) to NOX2 peptide 28, exceeding the level of binding of the "mother" mutant, p67^{phox}(1–526)Δ(259–279). Results represent means ± SE of three experiments

oxidation by H_2O_2 or purposeful synthesis led to a marked increase in the binding of p67^{phox}, a finding that suggested that disulfide bonds are established between cysteines 369 and/or 371 in NOX2 and cysteines in p67^{phox}, as a stabilizing mechanism in oxidase assembly in vivo.

The assay was also used to gather information on the reciprocal binding site in p67^{phox} for NOX2 peptides 24 and 28. As described in Ref. [33], the binding site to peptide 24 is located within residues 1–186, whereas binding to peptide 28 requires the full length of p67^{phox} (residues 1–526).

Finally, in two recent reports, use was made of the assay to study the role of cysteines in p67^{phox} in binding to NOX2 peptide 24. Mutating the four cysteines present in p67^{phox}(1–212) or all nine cysteines present in p67^{phox}(1–526) to serines resulted in the abolishment of binding of both truncated and full-length p67^{phox} mutants to peptide 24 [36, 37]. In yet another example of the need for care in the interpretation of protein-peptide binding assay results, the lack of binding of p67^{phox} Cys → Ser mutants to peptide 24 was associated with only a partial decline in the ability of mutated p67^{phox} to support oxidase activation in vitro.

4 Notes

1. Companies manufacturing peptide SPOT arrays are JPT Peptide Technologies, GenScript, and ProImmune-thinkpeptides, but this list is far from complete and is not based on personal experience.
2. Several companies offer design tools for performing these and other applications (*see* https://www.genscript.com/ and http://www.mimotopes.com/, as examples).
3. PepSets™ are synthetic peptide libraries made by Mimotopes (Clayton, Victoria, Australia). These are sets of chemically synthesized peptides, made on a small scale using parallel array synthesis techniques. They are made to yield at least 1 micromole of each peptide or, at a larger scale, at least 3 micromoles. PepSets can comprise hundreds, or even thousands, of peptides. They are most commonly used "as synthesized" (unpurified) for primary screens.
4. No special measures were applied to storage of peptides containing cysteine and methionine residues. If oxidation of cysteines was suspected or intentionally induced, the amount of free thiol was determined by the binding of the thiol probe monobromobimane [50] followed by FPLC gel filtration on a Superdex Peptide column, to assess the ratio of free to oxidized thiols, as described in Ref. [37].
5. The purified peptides used in the experiments described in this chapter were manufactured by Mimotopes (Clayton, Victoria, Australia) or Bachem (Bubendorf, Switzerland), but this in no way represents a recommendation for using peptides synthesized by these companies.
6. These plates, consisting of 12 detachable 8-well strips, allowing the use of only parts of the plate, were supposed to be pre-blocked, but no details were available concerning the nature of the blocker used. As a precautionary procedure, we always applied our casein-based blocking solution (PBSTcas1) before adding the biotinylated peptides.
7. These plates, also consisting of 12 detachable 8-well strips, are supplied blocked with Blocker™ BSA. This is a 1% w/v solution of bovine serum albumin in PBS. The precise conditions of blocking by the manufacturer are not known. We chose to block these plates, too, with our blocking solution (PBSTcas1) in spite of the fact that the manufacturer recommends using the plates as supplied.
8. The cell wash head is originally intended to be used with plates harboring adherent cell cultures and is designed for gentle washing and lack of turbulence. This is achieved by the ability to plan accurately the conditions of liquid aspiration and

dispensing from and to the wells. This comprises the possibility to regulate location of aspiration and dispensing on the well surface, the depth reached by the washing head in the course of aspiration and dispensing, and the aspiration speed and time. The use of a cell wash head was found to be ideal for detecting low-affinity binding.

9. The characteristics of the synthetic peptides are provided by the manufacturer (molecular weight, purity, and quantity). Further information can be obtained via "peptide calculators" accessible via the Internet (http://www.bachem.com/service-support/peptide-calculator/ and https://www.genscript.com/peptide_technical_resources.html?src=emailelq201708182c).

10. If the peptide solutions contain particles or are opalescent, this is a sign of incomplete solubilization. If the peptide has a low hydropathy index [51] (hydrophilic), chances are that further dilution in aqueous buffers (PBS and PBScas1) to concentrations of 100 μM and 1 μM will result in solubilization. Peptides with a very high hydropathy index (hydrophobic) might not allow full solubilization, a situation that may prevent their use in the assay. Procedures for dealing with hydrophobic peptides are available at websites, such as http://www.mimotopes.com/files/editor_upload/File/PeptidesAndAntibodies/PU3007-1-Handling-Cleaved-PepSets.PDF or https://www.bachem.com/fileadmin/user_upload/pdf/Technical_Notes/Solubilization_of_Peptides.pdf. The 1.5 mM stock solutions in NMP/H_2O_2 are stable for long time intervals when stored at −75 °C (we used peptides dissolved 5 years ago). No special precautions are applied to peptides containing cysteine and methionine residues, except avoiding too vigorous mixing, such as by using a "vortex"-type apparatus. Unlike dimethyl sulfoxide, which has oxidizing properties, NMP does not, and this represents an additional reason for its use as a solvent. Also, proper aliquoting of the peptide solutions should assure that repeated freeze/thaw cycles are avoided. We routinely prepare the 100 μM and 1 μM dilutions of peptides on the day of performing the protein-peptide binding experiments. 1 μM solutions are kept in ice/water mixtures during the experiment and discarded at the end of the experiment. The 100 μM solutions can be refrozen to −75° C and reused in the future, provided that the peptides do not have a high hydropathy index and that repeated cycles of freezing/thawing are avoided.

11. It is recommended to use the same protein standard throughout in order to compare protein concentrations over long time periods. No external protein standard reflects the true protein concentration of the recombinant proteins, but what is important is to maintain the same level of "error." We use bovine gamma globulin, as the standard.

12. The concentration of the protein added to the peptide array is the most significant variable in the optimization of protein-peptide binding assays (our experience and Ref. [26]). Because of this, we prefer to further purify proteins, eluted from Ni Sepharose gel, by preparative gel filtration. In addition to the benefit derived from working with proteins of high purity, gel filtration provides important information on the native state of the protein, such as molecular weight, shape, and the presence of polymers.
13. Highly purified cytosolic components have the tendency to self-aggregate, even in the presence of 30% glycerol. Repeated thawing/freezing is likely to promote aggregation. Aggregated protein is inappropriate for protein-peptide binding assays, and its presence distorts the results of protein concentration measurements. It is, thus, highly recommended to subject frozen proteins, after thawing, to centrifugation at 12,000 × *g* at 4 °C, for 30 min, in a tabletop microcentrifuge accommodating 1.5 mL conical tubes. Measure protein concentration in the supernatants, and keep the supernatants at 4° C during the performance of the assay. The material can be refrozen at −75° C for further storage, but re-centrifugation and rechecking protein concentration are recommended when used in a fresh assay.
14. Because the amount of peptide bound to the well cannot exceed 10 pmol, the concentration of the protein is the major variable available for the optimization of the system. We found a concentration of 1.5 μM protein as being optimal for the experimental conditions described in this chapter, but when these are changed, it is advisable to adjust the concentration of protein added.
15. Whereas the binding of peptides to the streptavidin-coated wells is of very high affinity and there was no need for optimization of the conditions of binding, the conditions of binding of the proteins to peptides were in need of optimization. In preliminary experiments, we examined the effect of protein concentration, temperature, and length of incubation on binding and, in a limited number of experiments, the effect of ionic strength and detergents. In general, incubation of proteins with peptide at room temperature and at 37° C promoted binding but at the cost of higher nonspecific binding values. Prolonging the time of incubation had the same double effect. We found that lowering the temperature to 4° C and extending the time of incubation to 18 h resulted in a preference for specific binding, reproducible results, and low background values.
16. In preliminary stages of setting up protein-peptide binding assays, it is recommended to try using specific antibodies

against the bound protein, as a confirmatory measure. Such antibodies might not always be satisfactory because of unexpected cross-reactions with the peptides or inability to detect the protein because of steric hindrance.

17. The settings are shown for the Versamax microplate reader, fitted with software SoftMax Pro, Version 6.5.1. These are absorbance reading at 650 nm; minimum absorbance, 0; maximum absorbance, 4; kinetic mode; time, 10 min; lag time, 0; reading interval, 0.13 min; number of reads, 47; kinetic reduction to linear segment chosen by investigator; units, V_{max} = (milliabsorbance/min); first data point set to zero; 10-s shake before first read; and 3-s shake between reads.
18. A large variety of antibodies are available against components of the oxidase complex, and it is beyond the scope of this chapter to list these. Based on our experience, it is wise to test a variety of antibodies recognizing different epitopes in the protein because some epitopes might become less accessible when the protein is bound to the surface-attached peptide.
19. The peroxidase-conjugated secondary antibodies, used in our work, were supplied by Jackson ImmunoResearch Laboratories.
20. When peptides are meant to be attached to the wells of streptavidin-coated plates, the 1 μM solutions are prepared in PBScas1, to prevent nonspecific binding to the plates. When intended for interacting with a reagent in solution, casein present at a concentration of 1% w/v (= 10 mg/mL) might compete with the peptide present at a concentration of 1 μM (equaling ~2 μg/mL, for 15-mer peptides), for interaction with the agent.
21. The activity of NOX 4 is not regulated by cytosolic components [52], and, thus, conformationally modified $p67^{phox}$ is not expected to bind to NOX4 peptide 28.

Acknowledgments

This work was supported by grants No. 49/09, 300/13, and 144/17 from the Israel Science Foundation, by the Roberts Fund, by the Joseph and Shulamit Salomon Fund, and by the Blavatnik Center for Drug Discovery of Tel Aviv University. Edgar Pick would like to thank his collaborators and students, without whom this work would not have been possible, at both the practical and conceptual levels. These comprise Drs. Gili Joseph, Giora Morozov, Iris Dahan, Irina Issaeva, Ofra Lotan, Natalia Sigal, Yevgeny Berdichevsky, Anat Zahavi, Rive Sarfstein, Ariel Mizrahi, Maya Amichai, and Edna Bechor and Ms. Tania Fradin, Mrs. Yara Gorzalczany, Mr. Shahar Molshanski-Mor, and Mrs. Meirav Rafalowski.

References

1. Quinn MT, Gauss KA (2004) Structure and regulation of the neutrophil respiratory burst oxidase: comparison with nonphagocyte oxidases. J Leukoc Biol 76:760–781
2. Sumimoto H (2008) Structure, regulation and evolution of Nox-family NADPH oxidases that produce reactive oxygen species. FEBS J 275:3249–3277
3. Karplus PA, Daniels MJ, Herriott JR (1991) Atomic structure of ferredoxin-NADP$^+$ reductase: prototype for a structurally novel flavoenzyme family. Science 251:60–66
4. Nauseef WM (2004) Assembly of the phagocyte NADPH oxidase. Histochem Cell Biol 122:277–291
5. Groemping Y, Rittinger K (2005) Activation and assembly of the NADPH oxidase: a structural perspective. Biochem J 386:401–416
6. Kreck ML, Freeman JL, Abo A, Lambeth JD (1996) Membrane association of Rac is required for high activity of the respiratory burst oxidase. Biochemistry 35:15863–15682
7. Gorzalczany Y, Sigal N, Itan M, Lotan O, Pick E (2000) Targeting of Rac1 to the phagocyte membrane is sufficient for the induction of NADPH oxidase assembly. J Biol Chem 275:40073–40081
8. Sarfstein R, Gorzalczany Y, Mizrahi A, Berdichevsky Y, Molshanski-Mor S, Weinbaum C, Hirshberg M, Dagher M-C, Pick E (2004) Dual role of Rac in the assembly of NADPH oxidase: tethering to the membrane and activation of p67^{phox}. A study based on mutagenesis of p67^{phox}-Rac1 chimeras. J Biol Chem 279:16007–16016
9. Dale H (1936) Some recent extensions of the chemical transmission of the effects of nerve impulses. Nobel Lecture, December 12, 1936. From Nobel Lectures Physiology or Medicine, 1922–1941, Elsevier Publishing Company, Amsterdam, 1965. https://www.nobelprize.org/nobel_prizes/medicine/laureates/1936/dale-bio.html
10. Mizrahi A, Berdichevsky Y, Ugolev Y, Molshanski-Mor S, Nakash Y, Dahan I, Alloul N, Gorzalczany Y, Sarfstein R, Hirshberg M, Pick E (2006) Assembly of the phagocyte NADPH oxidase complex: chimeric constructs derived from the cytosolic components as tools for exploring structure-function relationships. J Leukoc Biol 79:881–895
11. Pick E (2014) Role of the Rho GTPase Rac in the activation of the phagocyte NADPH oxidase: outsourcing a key task. Small GTPases 5 (1):e27952
12. Diekmann D, Abo A, Segal AW, Hall A (1994) Interaction of Rac with p67phox and regulation of phagocytic NADPH oxidase activity. Science 265:531–533
13. Leto TL, Adams AG, de Mendez I (1994) Assembly of the phagocyte NADPH oxidase: binding of Src homology 3 domains to proline-rich targets. Proc Natl Acad Sci U S A 91:10650–10654
14. Sumimoto H, Kage Y, Nunoi H, Sasaki H, Nose T, Fukumaki Y, Ohno M, Minakami S, Takeshige K (1994) Role of Src homology 3 domains in assembly and activation of the phagocyte NADPH oxidase. Proc Natl Acad Sci U S A 91:5345–5349
15. Dang PM-C, Cross AR, Quinn MT, Babior BM (2002) Assembly of the neutrophil respiratory burst oxidase: a direct interaction between p67^{phox} and cytochrome b_{558} II. Proc Natl Acad Sci U S A 99:4262–4265
16. Kanai F, Liu H, Field SJ, Akbary H, Matsuo T, Bown GE, Cantley LC, Yaffe MB (2001) The PX domains of p47phox and p40phox bind to lipid products of PI(3)K. Nat Cell Biol 3:675–678
17. Kleanthous C (ed) (2000) Protein – protein recognition. Oxford University Press, Oxford
18. Sudha G, Nussinov R, Srinivasan N (2014) An overview of recent advances in structural bioinformatics of protein-protein interactions and a guide to their principles. Prog Biophys Mol Biol 116:141–150
19. Fu H (ed) (2004) Protein-protein interactions. Methods and Applications. Humana Press, Totowa, NJ
20. Golemis ES, Adams PD (eds) (2005) Protein-protein interactions, 2nd edn. Cold Spring Harbor Laboratory Press, Cold Spring Harbor, New York
21. Rao VS, Srinivas K, Sujini GN, Kumar GNS (2014) Protein-protein interaction detection: methods and analysis. Int J Proteomics 2014:147648
22. Hayes S, Malacrida B, Kiely M, Kiely PA (2016) Studying protein-protein interactions: progress, pitfalls and solutions. Biochem Soc Trans 44:994–1004
23. Pollard TD (2010) A guide to simple and informative binding assays. Mol Biol Cell 21:4061–4067
24. Mackay JP, Sunde M, Lowry JA, Crossley M, Matthews JM (2007) Protein interactions: is seeing believing? Trends Biochem Sci 32:530–531

25. London N, Movshovitz-Attias D, Schueler-Furman O (2010) The structural basis of peptide-protein binding strategies. Structure 18:188–199
26. Katz C, Levy-Beladev L, Rotem-Bamberger S, Rito T, Rüdiger SGD, Friedler A (2011) Studying protein-protein interactions using peptide arrays. Chem Soc Rev 40:2131–2145
27. Rodda SJ, Maeji NJ, Tribbick G (1996) Epitope mapping using multipin peptide synthesis. In: Morris GE (ed) Epitope mapping protocols. Humana Press, Totowa, NJ
28. Morozov I, Lotan O, Joseph G, Gorzalczany Y, Pick E (1998) Mapping of functional domains in $p47^{phox}$ involved in the activation of NADPH oxidase by "peptide walking". J Biol Chem 273:15435–15444
29. Reinecke U, Volkmer-Engert R, Schneider-Mergener J (2000) Applications of peptide arrays prepared by the SPOT-technology. Curr Opin Biotechnol 12:59–64
30. Cushman I (2008) Utilizing peptide SPOT arrays to identify protein interactions. In: Taylor GP (ed) Current protocols in protein science. Wiley Interscience, Hoboken, NJ, pp 18.10.1–18.10.9
31. Park JW, Ma M, Ruedi JM, Smith RM, Babior BM (1992) The cytosolic components of the respiratory burst oxidase exist as a M_r approximately 240,000 complex that acquires a membrane-binding site during activation of the oxidase in a cell-free system. J Biol Chem 267:17327–17332
32. Dahan I, Issaeva I, Gorzalczany Y, Sigal N, Hirschberg M, Pick E (2002) Mapping of functional domains in the $p22^{phox}$ subunit of flavocytochrome b_{559} participating in the assembly of the NADPH oxidase complex by "peptide walking". J Biol Chem 277:8421–8432
33. Dahan I, Smith SME, Pick E (2015) A Cys-Gly-Cys triad in the dehydrogenase region of Nox2 plays a key role in the interaction with $p67^{phox}$. J Leukoc Biol 98:859–874
34. Tsang VCW, Wilson BC, Madison SE (1980) Kinetic studies of a quantitative single-tube enzyme-linked immunosorbent assay. Clin Chem 26:1255–1260
35. Tsang VCW, Wilson BC, Peralta JM (1983) Quantitative single tube, kinetic dependent enzyme-linked immunosorbent assay (K-ELISA). Methods Enzymol 92:391–403
36. Bechor E, Dahan I, Fradin T, Berdichevsky Y, Zahavi A, Federman Gross A, Rafalowski M, Pick E (2015) The dehydrogenase region of the NADPH oxidase component Nox2 acts as a protein disulfide isomerase (PDI) resembling PDIA3 with a role in the binding of the activator protein $p67^{phox}$. Front Chem 3:3
37. Fradin T, Bechor E, Berdichevsky Y, Dahan I, Pick E (2018) Binding of $p67^{phox}$ to Nox2 is stabilized by disulfide bonds between cysteines in the ^{369}Cys-Gly-Cys^{371} triad in Nox2 and in $p67^{phox}$. J Leukoc Biol 104:1023–1039
38. Dahan I, Pick E (2012) Strategies for identifying synthetic peptides to act as inhibitors of NADPH oxidases, or "all that you did and did not want to know about Nox inhibitory peptides". Cell Mol Life Sci 69:2283–2305
39. Han C-H, Freeman JLR, Lee TH, Motalebi SA, Lambeth JD (1998) Regulation of the neutrophil respiratory burst oxidase – identification of an activation domain in $p67^{phox}$. J Biol Chem 273:16663–16668
40. Maehara Y, Miyano K, Yuzawa S, Akimoto R, Takeya R, Sumimoto H (2010) A conserved region between the TPR and activation domains of $p67^{phox}$ participates in activation of the phagocyte NADPH oxidase. J Biol Chem 285:31435–31445
41. Dahan I, Molshanski-Mor S, Pick E (2012) Inhibition of NADPH oxidase activation by peptides mapping within the dehydrogenase region of Nox2 - a "peptide walking" study. J Leukoc Biol 91:501–515
42. Burritt JB, Fritel GN, Dahan I, Pick E, Roos D, Jesaitis AJ (2000) Epitope identification for human neutrophil flavocytochrome b monoclonals 48 and 449. Eur J Hematol 65:407–413
43. Baniulis D, Nakano Y, Nauseef WM, Banfi B, Cheng G, Lambeth DJ, Burritt JB, Taylor RM, Jesaitis AJ (2005) Evaluation of two anti-$gp91^{phox}$ antibodies as immunoprobes for Nox family proteins: mAb 54.1 recognizes recombinant full-length Nox2, Nox3 and the C-terminal domains of Nox1-4 and reacts with GRP 58. Biochim Biophys Acta 1752:186–196
44. Hurst R, Hook B, Slater MR, Hartnett J, Storts DR, Nath N (2009) Protein-protein interaction studies on protein arrays: effect of detection strategies on signal-to-background ratios. Anal Biochem 392:45–53
45. Mizrahi A, Berdichevsky Y, Casey PJ, Pick E (2010) A prenylated $p47^{phox}$-$p67^{phox}$-Rac1 chimera is a quintessential NADPH oxidase activator. Membrane association and functional capacity. J Biol Chem 285:25485–25499
46. Bradford MM (1976) A rapid and sensitive method for the quantitation of microgram quantities of protein utilizing the principle of protein-dye binding. Anal Biochem 72:248–254

47. Joseph G, Gorzalczany Y, Koshkin V, Pick E (1994) Inhibition of NADPH oxidase activation by synthetic peptides mapping within the carboxy-terminal domain of small GTP-binding proteins. Lack of amino acid sequence specificity and importance of the polybasic motif. J Biol Chem 269:29024–29031
48. Joseph G, Pick E (1995) "Peptide walking" is a novel method of mapping functional domains in proteins. Its application to the Rac1-dependent activation of NADPH oxidase. J Biol Chem 270:29079–29082
49. Leusen JHW, Meischl C, Eppink MHM, Hilarius PM, de Boer M, Weening RS, Ahlin A, Sanders L, Goldblatt D, Skopczynska H, Bernatowska E, Palmblad J, Verhoeven AJ, van Berkel WJH, Roos D (2000) Four novel mutations in the gene encoding gp91^{phox} of human NADPH oxidase: consequences for oxidase activity. Blood 95:666–673
50. Kosower EM, Kosower NS (1995) Bromobimane probes for thiols. Methods Enzymol 251:133–148
51. Kyte J, Doolittle RF (1982) A simple method for displaying the hydropathic character of a protein. J Mol Biol 157:105–132
52. Bedard K, Krause K-J (2007) The Nox family of ROS-generating NADPH oxidases: physiology and pathophysiology. Physiol Rev 87:245–313

Chapter 24

Rational Design and Delivery of NOX-Inhibitory Peptides

Eugenia Cifuentes-Pagano and Patrick J. Pagano

Abstract

A growing appreciation of NADPH oxidases (NOXs) as mediators of fundamental physiological processes and as important players in myriad diseases has led many laboratories on a search for specific inhibitors to help dissect the role in a given pathway or pathological condition. To date, there are only a few available inhibitors with a demonstrated specificity for a given isozyme. Among those, peptidic inhibitors have the advantage of being designed to target very specific protein-protein interactions that are essential for NOX activity. Herein, we provide the techniques to deliver these inhibitors both in cell culture as well as in vivo.

Key words NADPH oxidase, Isoform-specific inhibitors, Peptidic inhibitors, Nox2ds-tat, NoxA1ds

1 Introduction

NADPH oxidases (NOXs), major sources of reactive oxygen species (ROS) in the cell, have emerged as key players both in physiological signal transduction and in disease. Dysregulation of this family of enzymes causing an excess production of reactive oxygen species has been demonstrated to promote the development of pathophysiological conditions including inflammation, cardiovascular diseases, neurodegenerative diseases, and cancer.

To ameliorate the negative impact that dysregulation of NOXs have in disease, and to understand the role that specific isozymes of the enzyme play in a given pathway, intense efforts have been made to develop NOX-specific inhibitors [1]. The main goal is targeting the source of ROS rather than the ROS themselves as had been attempted for a long time with a variety of chemical and enzyme antioxidants. Given the complexity of subunit interactions involved in the assembly of active enzyme and similarities among isozymes in this family, the task of obtaining isoform-specific inhibitors has proven to be extremely difficult. Although the limitation of reduced oral bioavailability is a common refrain, evolving technologies and routes of delivery methods are being employed to avoid this fate (see below for one example) as is the case with other biologics. That

Ulla G. Knaus and Thomas L. Leto (eds.), *NADPH Oxidases: Methods and Protocols*, Methods in Molecular Biology, vol. 1982, https://doi.org/10.1007/978-1-4939-9424-3_24, © Springer Science+Business Media, LLC, part of Springer Nature 2019

notwithstanding, one major advantage that makes this class of inhibitors uniquely valuable is that unlike small molecule inhibitors, they can be designed to competitively block primary protein-protein interactions by precisely replicating a unique peptide-binding domain. This rational approach is intended to minimize off-target effects.

Our laboratory has developed two isoform-specific NOX inhibitors that have proven useful to dissect the role of these isozymes in a variety of processes. One of these inhibitors was directed toward the canonical NOX2 isozyme, Nox2ds-tat (aka gp91ds-tat) [2, 3], and the other was directed against the canonical NOX1 isozyme, NoxA1ds [4]. Below we discuss both of these peptides to exemplify the different strategies used in their design.

Nox2ds-*tat* (*see* Fig. 1) is the most widely used NOX2-specific inhibitor. It contains a sequence that mimics a portion of the second intracellular loop of NOX2 (loop B) known to be a site of interaction between NOX2 and p47^{phox} subunits [5, 6], and it is linked to a short 9-amino acid peptide span of HIV-*tat* viral coat protein (HIV-*tat*), which has been shown to deliver "cargo" proteins into cells [7]. Rodent and human versions of the peptide differ only in one amino acid; the isoleucine in the rodent version is replaced by a valine in the human version. Using human heterologous systems expressing only one of the NOX isozyme systems at a time, we demonstrated that human Nox2ds-*tat* is specific for the canonical NOX2 [3]. In in vivo experiments, rodent Nox2ds-*tat* attenuated angiotensin II-induced increases in blood pressure [2] and suppressed angioplasty-induced superoxide production and neointimal hyperplasia of the carotid artery [8]. Moreover, Quesada et al. demonstrated that treatment with rodent Nox2ds-*tat* is able to reverse atherosclerotic plaque formation in aorta from ApoE$^{-/-}$ mice fed a high fat diet [9]. The studies described above are a small sample of a wide array of studies that have employed

Rodent Nox2ds-*tat*: [H]-R-K-K-R-R-Q-R-R-R-C-S-T-R-**I**-R-R-Q-L-[NH$_2$]
Human Nox2ds-*tat*: [H]-R-K-K-R-R-Q-R-R-R-C-S-T-R-**V**-R-R-Q-L-[NH$_2$]

Scrambled controls:

R. Scmb Nox2ds-tat: [H]-R-K-K-R-R-Q-R-R-R-C-L-R-V-T-R-Q-S-R-[NH$_2$]
H. Scmb Nox2ds-tat: [H]-R-K-K-R-R-Q-R-R-R-C-L-R-I-T-R-Q-S-R-[NH$_2$]

Fig. 1 Sequence of Nox2ds-tat containing the tat sequence followed by a sequence of 9 amino acids corresponding to loop B of NOX2 (highlighted in gray). Sequence from rodent and human differ in the isoleucine to valine substitution shown in bold. In scramble control peptides, the loop B sequence is scrambled, but the tat sequence is kept intact

p67^{phox}:	190 -V-A-Q-L-A-K-K-D-Y-L-G-K-A-T-V-VA-S-V-V- 210
Rodent NoxA1ds:	[H]-E-P-M-D-**A**-L-G-K-A-K-V-[NH$_2$]
Human NoxA1ds:	[H]-E-P-V-D-**A**-L-G-K-A-K-V-[NH$_2$]
Scrambled controls:	
R. Scmb NoxA1ds:	[H]-L-M-K-G-P-D-A-E-K-V-A-[NH$_2$]
H. Scmb NoxA1ds:	[H]- L-V-K-G-P-D-A-E-K-V-A -[NH$_2$]

Fig. 2 Sequence of NoxA1ds corresponding to amino acids from NoxA1 containing the activation domain (highlighted in gray) and 5 amino acids upstream of it which do not share homology to the equivalent region on p67phox. Shown in bold is the Ala used to substitute Phe-199. Shown in red is the methionine from the mouse NoxA1 that replaces the valine in the human version of the protein. Scrambled control peptides for each version are also shown

both human and rodent versions of Nox2ds-*tat* as a tool to define the physiological role of NOX2 and to study the therapeutic potential of NOX2 inhibition.

NoxA1ds (*see* Fig. 2) was designed to target the interaction between NOXA1 and NOX1 of the canonical NOX1 oxidase isozyme by replicating the amino acid region on NOXA1 that is equivalent to the "activation domain" of p67^{phox} [10]. The amino acid sequence of this peptide includes residues between NOXA1 and p67^{phox} that are homologous (residues 200–205) and nonhomologous (residues 195–198) to confer effective blockade and specificity toward the NOX1 system, per se. It also comprises a substitution of Phe-199 for Ala, a modification previously shown to reduce NOX1 activity [4, 11].

This peptide was originally used to delineate the role of NOX1in hypoxia-induced human endothelial cell ROS production and VEGF-stimulated migration. Subsequently, it has been used to test the role of NOX1 in uniaxial stretch-induced phenotypic transitioning of rat vascular smooth muscle cells from a contractile to synthetic phenotype [12], and more recently human NoxA1ds was used as a modality to target senescence-associated endothelial dysfunction [13].

In this chapter, protocols for the delivery of peptidic inhibitors both in cell culture and in vivo are detailed.

2 Materials

1. Peptidic inhibitors: Peptides for use in our studies have generally been synthesized at the Tufts University Core (http://www.tucf.org/peptidesynthesis-f.html) with acetylation of the N-terminus and amidation of the C-terminus. Purity of the

peptide stock is usually near 90% (*see* **Note 1**). Any peptide synthesis core that meets the same quality standards may be contracted. Please refer to and reference Rey et al. 2001 [2], Csanyi et al. 2011 [3], and Ranayhossaini et al. 2013 [4] for more detail.

2. Acetic acid-saline solution: 0.01 N acetic acid in 0.9% NaCl (28 μL glacial acetic acid into 50 mL 0.15 M NaCl).
3. Culture media (recommended formulation is dependent on cell type). For example, human pulmonary artery endothelial cells (HPAEC) purchased from Lonza (Basel, Switzerland) are maintained in the growth medium (EBM-2) suggested by Lonza, while rat aortic smooth muscle cells (RASMC) (Lonza, Walkersville, MD, USA) were grown in DMEM (Cell-gro) with 4.5 g/L glucose, L-glutamine, and sodium pyruvate containing 10% heat-inactivated fetal bovine serum (FBS) and 100 U/mL penicillin and 100 μg/mL streptomycin (Invitrogen).
4. Opti-MEM™ (Gibco).
5. Lipofectamine™ LTX with PLUS™ reagent (Invitrogen).
6. Phorbol 12-myristate 13-acetate (PMA): 20 mM stock dissolved in DMSO (store at −20 °C).
7. Amplex Red reagent (Life Technologies/Invitrogen): 20 mM dissolved in DMSO (store at -20° C in 100 μL aliquots).
8. *AR assay buffer (1× stock):* 25 mM HEPES, 0.12 M NaCl, 3 mM KCl, 1 mM $MgCl_2$. Adjust the pH to 7.4 with HCl.
9. *2× Amplex Red/ HRP solution:* 55 μL of 20 mM Amplex Red, 3.5 μL of 1 U/μL HRP stock, 5.5 mL AR assay buffer (prepared fresh at the time of assay).
10. NADPH (12.5× stock): 0.45 M NADPH in 1× AR assay buffer (prepared fresh at the time of assay).
11. Alzet™ osmotic minipumps (volumes, rate of infusion, and length of treatment should be chosen according to animal model and outcome of interest). Insertion of pumps can be for delivery of peptides intravenously (i.v.), subcutaneously (s.c.), or intraperitoneally (i.p.). Delivery s.c. or i.p. will be described here in detail. For all three routes of administration and particularly i.v. delivery, see Alzet's protocols (http://www.alzet.com/products/guide_to_use/implantation_and_explantation.html).
12. Surgical supplies.
13. CompAir XLT nebulizer (Omron) for aerosolized delivery to lungs and airways.
14. Fiberglass nebulization chambers for aerosolized delivery.

3 Methods

3.1 Design of Peptidic Inhibitors

1. The design of specific and efficient peptidic inhibitors is based on the knowledge of protein-protein interactions essential for the catalytic activity of the enzyme (*see* Fig. 3). Some of the strategies employed for identification of regions of interaction among the various subunits of NOX (mainly NOX2 isozyme) include the use of peptide phage display libraries [6] and "peptide walking" [14]. Many of these interactions have been published and described in various reviews [15–17]. Dahan and Pick [18] have proposed a host of potential strategies for inhibition. Nox2ds-*tat* and NoxA1ds are good examples of inhibitors based on available information and rational design; in the case of Nox2ds-*tat*, the strategy involved the use of the tat sequence to overcome the difficulty of penetration [2, 3], while in the case of NoxA1ds, the strategy involved the inclusion of amino acids outside the activation domain to afford specificity [4].
2. Once the sequence of the region of interest is determined, use searching engines like BLAST™ (basic local alignment search tool, https://blast.ncbi.nlm.nih.gov) as described in Rey et al. 2001 [2] to assess conservation or lack thereof of this sequence in other proteins. The objective is to choose a peptide with no or very minimal homology to other proteins.
3. Based on the selected peptide sequence, generate its scrambled peptide control by randomly changing the order of the amino acid sequence and checking for the minimum number of hits in the search engine. In the case of a chimeric peptide containing the *tat* sequence, only the selected peptide sequence is scrambled; the *tat* is kept intact.

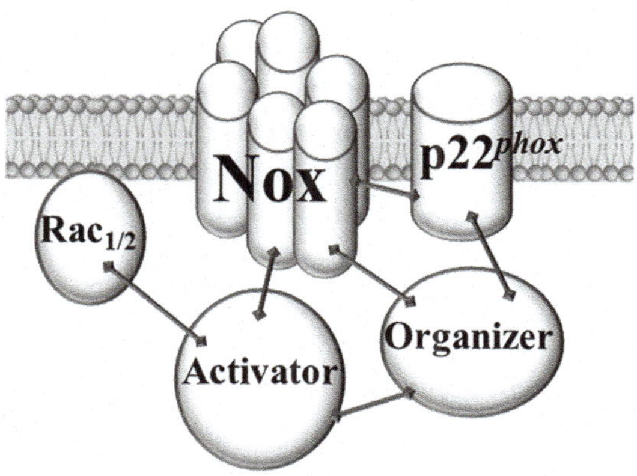

Fig. 3 Protein-protein interactions important for the assembly of a fully active prototype NOX

4. Analysis of the hydrophobicity of the selected peptide will determine if the inclusion of a tat sequence is necessary (*see* **Note 2**).
5. Testing of the designed peptide should include effect on NOX activity, scavenging properties, specificity, cell penetration, toxicity, etc.
6. Although rational design is expected to yield very selective inhibitors, success is dependent not just on the sequence chosen but also on how tight the interactions being targeted are, as was implicated in the failure of multiple attempts to target NOX4 intramolecular or NOX4-p22^{phox} interactions [19].

3.2 Test of Peptide Inhibitors in Cells In Vitro

In order to determine specificity and potency of inhibitors, it is ideal to have a heterologous cell system expressing all the subunits needed for the NOX isoform of interest. A stably transfected cell line commonly used for this purpose is COS-phox, developed by Dr. Dinauer [20], which expresses all the components necessary for NOX2 activity. For other NOX isoforms, transiently transfected HEK-293 or COS7 cells can be used.

3.2.1 Cell Transfection

1. Grow HEK-293 or COS7 cells in 100 mm plates to 70% confluency.
2. Change media to Opti-MEM (Gibco) (13.5 mL).
3. In a 15 mL conical tube containing 2.7 mL of Opti-MEM, add 20 μg total plasmid DNA (if plasmids for four subunits are going to be transfected, then add 5 μg of each). Mix.
4. Add 6.75 μL PLUS reagent, mix, and let stand for 5 min at room temperature.
5. Add 30 μL Lipofectamine LTX, mix, and let stand for 30 min.
6. Add DNA-Lipofectamine mix dropwise to cells.
7. Incubate overnight at 37 °C.

3.2.2 Cell Treatments

1. Grow cells of choice in appropriate media to 70–80% confluency (*see* **Note 3**).
2. Serum starve the cells to promote cell cycle synchronization by changing the growth medium for a medium containing no more than 10% of the original media's growth factors/serum concentrations (*see* **Note 3**), and incubate overnight.
3. Prepare peptide solution by weighing the appropriate amount of peptide to have at least 100× working solution and resuspending it in acetic acid in saline solution (*see* **Note 4**). Peptide solution is prepared freshly the day of the experiment. The 1:100 dilution or greater ensures that the acetic acid-saline solution has no effect on the pH of the medium.

4. Pretreat the cells with the peptide, scrambled peptide, or vehicle for 30 min before addition of stimuli.
5. Stimulate the cells with phorbol 12-myristate 13-acetate, PMA (1–5 μM) for NOX2 activity, or 10 μM ionomycin plus 1 μM PMA for NOX5. For NOX1 and NOX4, activity is compared to that of untransfected control cells.

3.2.3 ROS Assay

Assess ROS production in the presence and absence of the inhibitors using whole cells or homogenates of stimulated cells. In general, the ROS assay of choice in this case is Amplex Red as it is more sensitive than the gold standard assay, cytochrome *C*, as described by Csanyi 2011 et al. [3]. Assay can be done with whole cells or homogenates of stimulated cells.

1. Disrupt cells/tissues in ice-cold disruption buffer (HBSS containing protease inhibitor cocktail) by glass-on-glass homogenizer (tissues) or five freeze/thaw cycles (cells), and pass through a 30-gauge needle five times to further lyse the cells. Centrifuge the cell lysate at 1000 × *g* for 10 min at 4 °C to remove unbroken cells, nuclei, and debris. Throughout all these procedures, extreme care should be taken to maintain the lysate at a temperature close to 0 °C.
2. Measure protein concentrations.
3. Prepare working solutions (Subheading 2, **step 8–10**) and cell or cell homogenates for ROS assays.

   If whole cells are going to be tested in a 384-well plate, seed ~40,000 cells per well in Opti-MEM. If cell homogenates are going to be tested, amount of protein needs to be optimized for each tissue/cell type. For 384-well plate, test in a range between 0.1 and 1 μg/well. Dilute appropriate amount of cell homogenate in AR assay buffer.
4. Prepare H_2O_2 standards by serial dilution of stock H_20_2 in AR assay buffer starting at 10 μM concentration. Plate 25 μL/well of each dilution, and then add 25 μL/well of 2× AR/HRP solution.
5. For a 384-well plate, add cell homogenate and then 2× Amplex Red/ HRP solution to wells. Read four points for baseline fluorescence. Add NADPH and read for 1 h.
   - **Reaction** (in microliters)

     Cells or cell homogenate: 21
2× AR/HRP: 25
0.45 mM NADPH: 4
6. Fluorescence measurements are made using a microplate reader with a 530 nm excitation and a 590 nm emission filter.
7. Data are calculated as the rate of RFU/min.

8. Although depending on the cell type and endogenous SOD levels, superoxide dismutation happens spontaneously. To measure superoxide production, SOD can be added to the reaction mixture to ensure full conversion of superoxide into H_2O_2.
9. Repeat the test using cells expressing each of the other isoforms to determine isoform specificity.

3.3 Use of Peptide Inhibitors in Cells In Vitro

1. Treat cells as described above (*see* Subheading 3.2.2, **step 1–3**).
2. Pretreat the cells of interest with the peptide, scrambled peptide, or vehicle for 30 min–4 h before addition of stimuli.
3. Stimulate the cells with the agent to be tested (i.e., angiotensin II, TNFα, etc.) for the length of time necessary for the response of interest (e.g., migration, proliferation, protein expression). If treatment time with stimulant is longer than 6 h, at 4 h intervals after the stimulus administration, add peptide again to the same final concentration as the initial administration (*see* **Note 5**).
4. Assess biochemical and physiological readouts of interest.

3.4 Application of Peptide Inhibitors In Vivo

The application of peptides in animals can be performed by simple injections intraperitoneally or procedures that are less stressful for the animal such as via minipump or aerosolization.

3.4.1 Preparation of Osmotic Minipumps for In Vivo Delivery

1. Prepare Alzet osmotic minipumps containing either vehicle (0.01 N acetic acid in saline solution), stimulus solution (e.g., angiotensin II delivered at a rate of 0.75 mg/kg/day), or combined stimulus + peptide (10 mg/kg/day) or combined stimulus + scrambled peptide (10 mg/kg/day).
2. Calculate the amounts of peptide needed:

   Amount of peptide per day = Dosing rate (in mg/kg/day) × Weight of animal (in gm)/1000.

   Volume of infusion per day (μL/day) = 24 × mean pumping rate ascribed by the manufacturer (in μL/h).

   Concentration of peptide solution (mg/μL) = Amount of peptide per day/Volume of infusion per day.

   Amount (weight) of peptide per pump = Concentration of peptide solution × volume of pump.

   This amount of peptide is multiplied per the number of animals per group and resuspended in acetic acid-saline solution in the appropriate volume required by each pump (for an example of calculations, *see* **Note 6**), calculating ~10–20% in excess for handling. It is important also to take into account the solubility of the peptide for the dosage chosen **in case treatments require** more than one pump.

3. Sterilize the peptide and vehicle solutions by loading them into a syringe and passing the contents through a 0.22 μm sterile disc filter.
4. While wearing gloves, and in a sterile culture hood, fill pumps using insertion "needle" provided by the pump manufacturer, being careful not to introduce bubbles into the solutions. Then, immerse pumps into a sterile 0.9% NaCl saline solution in 50 mL conical tubes, and cover the tubes to maintain sterility.
5. Place tubes at 37 °C for 3 to 4 h in an incubator to allow pumping to commence (priming of the pump).

3.4.2 Surgical Implantation of Osmotic Minipumps

1. Anesthetize the animal as required, shave abdomen or nape of the neck, and place animal on a heating pad.
2. For i.p. delivery, under aseptic conditions, make a midline incision of 1/3 in. on the skin; separate skin from peritoneal wall using forceps. For s.c. delivery make a similar incision in the scapular region in the nape (backside) of the neck.
3. Cut into the peritoneal wall or subdermal scapula, with care not to damage internal organs, to make an incision just large enough to insert pump. Using a hemostat spread the tissue to create a pocket for the pump.
4. Insert mini-osmotic pump holding it from the bottom and sliding it into the pocket, delivery portal side first.
5. Close the wound opening and skin using silk sutures or wound clips.
6. Apply betadine solution to the wound, and keep the animal on the heating pad until it recovers from the anesthesia.
7. When the animal is awake and moving, return it to cage and monitor for postoperative pain/distress (examples include failure to eat/drink, hovering in the corner, shallow breathing, and failure to groom). To alleviate pain and distress, analgesic (buprenex –0.5 m g/kg s.c.) can be administered once during recovery post minipump implantation, and repeat every 6–12 h for 24 h post-surgery. Monitor animals daily.
8. Depending of the experiment, animals are maintained under normal husbandry conditions, following appropriate IACUC regulations, for the appropriate length of time.
9. Take appropriate measurements according to outcome of interest.

3.4.3 Aerosolization

In order to deliver a therapeutic dose of the inhibitor in the form of an aerosol of respirable particles, a nebulizer is used. The mist to be inhaled is created by a flow of air going through the narrow opening in the nebulizer cup.

1. Prepare peptide solution. Depending on the number of animals that can be accommodated in an enclosed chamber, calculate the amount of peptide necessary for the desired dose. For example, for a four-animal chamber, to have a dose of 5.1 mg/kg for a rat of 275 g (=1.4 mg/rat), weigh 5.6 mg of peptide.
2. Dilute appropriate amount of peptide in a minimum of 5 mL (*see* **Note 7**).
3. Place the animals in the chamber.
4. Set nebulizer compressor to an air flow rate of 6 L/min (equivalent to a nebulizer rate of 0.2 mL/min).
5. Aerosolize for 30 min per treatment.
6. Repeat treatment as appropriate (*see* **Note 8**).

4 Notes

1. Commonly used Nox2ds-tat has a molecular weight of 2452.98 gr/mol for the rodent form RKKRRQRRRCSTR**I**RRQL and 2438.96 gr/mol for the human form that has the isoleucine substituted by a valine. For NoxA1ds the molecular weight is 1157.4 gr/mol for the rodent form, EP**M**DALGKAKV, and 1167.38 gr/mol for the human version that has the methionine substituted by a valine.
2. Hydrophobicity can be calculated using the calculator in: https://www.peptide2.com/N_peptide_hydrophobicity_hydro philicity.php. For example, the values obtained for Nox2ds are as follows: hydrophobic, 22.22%; acidic, 0%; basic, 33.33%; and neutral, 44.44%. These values suggest that hydrophobicity of Nox2ds sequence is low so it required a carrier, such as the *tat* sequence, to allow for membrane translocation. In contrast to the values calculated for NoxA1ds, hydrophobic is 54.55%, acidic is 18.18%, basic is 18.18%, and neutral is 9.09% that suggested its ability to permeate the plasma membrane.
3. Depending on the cells used, the level of confluency may negatively affect NOX activity; thus preliminary assays are required to determine the optimal confluency to achieve an optimal NOX activity.
4. The sensitivity of the cells to serum starvation should also be determined beforehand, i.e., some cells such as endothelial cells do not tolerate lower than 10% of the original growth factors/ serum concentrations in the growing media.
5. Dilution of peptides in acetic acid-saline solution is done to minimize peptide binding to the plastic or glass vial.

6. Until the stability of the peptide inhibitor in the preparation or animal can be determined, repeated doses of peptide may ensure that the desired concentration is maintained for the duration of the experiment.
7. For a 7-day infusion of peptide at 10 mg/kg/day in a 30-gram mouse using an Alzet 2001 pump (specifications for a 7-day pump: 1 μL/h mean pumping rate, 200 μL reservoir volume):

   Amount of peptide per day = 10 (mg/kg/day) × 30 (gr)/1000 = 0.3 mg /mouse/day

   Volume of infusion per day = 24 × 1 μL/h = 24 μL/day

   Concentration of peptide solution (mg/μL) = 0.3 mg/24 μL = 0.0125 mg/μL (=12.5 mg/mL)

   Amount of peptide per pump: 0.0125 mg /μL × 200 μL = 2.5 mg

   For four mice in each group (2.5 mg × 4 = 10 mg), 12 mg of peptide should be resuspended in a 960 μL acetic acid-saline solution (a 20% excess allows for small losses during pump loading).
8. For protracted treatment protocols, for example, for hypoxia regimen lasting 3 weeks, NoxA1ds was aerosolized to rodents, as described, every 3 days for the duration of the treatment. The profound right ventricular hypertrophy and increased Fulton index induced by hypoxia were significantly ameliorated by the aerosolization with NoxA1ds (unpublished data).

Acknowledgments

Research in the Pagano Lab is supported by the National Institutes of Health Grants R01HL079207 and P01HL103455-01 and receives support from the Institute for Transfusion Medicine and the Hemophilia Center of Western Pennsylvania.

References

1. Cifuentes-Pagano ME, Meijles DN, Pagano PJ (2015) Nox inhibitors & therapies: rational design of peptidic and small molecule inhibitors. Curr Pharm Des 21(41):6023–6035
2. Rey FE, Cifuentes ME, Kiarash A, Quinn MT, Pagano PJ (2001) Novel competitive inhibitor of NAD(P)H oxidase assembly attenuates vascular O 2 - and systolic blood pressure in mice. CircRes 89:408–414
3. Csanyi G et al (2011) Nox2 B-loop peptide, Nox2ds, specifically inhibits the NADPH oxidase Nox2. Free Radic Biol Med 51 (6):1116–1125
4. Ranayhossaini DJ et al (2013) Selective recapitulation of conserved and nonconserved regions of putative NOXA1 protein activation domain confers isoform-specific inhibition of Nox1 oxidase and attenuation of endothelial cell migration. J Biol Chem 288 (51):36437–36450
5. DeLeo FR, Quinn MT (1996) Assembly of the phagocyte NADPH oxidase: molecular

interaction of oxidase proteins. J Leukoc Biol 60(6):677–691

6. DeLeo FR et al (1995) Mapping sites of interaction of p47-phox and flavocytochrome b with random-sequence peptide phage display libraries. Proc Natl Acad Sci U S A 92 (15):7110–7114
7. Fawell S et al (1994) Tat-mediated delivery of heterologous proteins into cells. Proc Natl Acad Sci U S A 91(2):664–668
8. Jacobson GM et al (2003) Novel NAD(P)H oxidase inhibitor suppresses angioplasty-induced superoxide and neointimal hyperplasia of rat carotid artery. Circ Res 92(6):637–643
9. Quesada IM et al (2015) Selective inactivation of NADPH oxidase 2 causes regression of vascularization and the size and stability of atherosclerotic plaques. Atherosclerosis 242 (2):469–475
10. Han CH, Freeman JL, Lee T, Motalebi SA, Lambeth JD (1998) Regulation of the neutrophil respiratory burst oxidase. Identification of an activation domain in p67(phox). J Biol Chem 273(27):16663–16668
11. Maehara Y et al (2010) A conserved region between the TPR and activation domains of p67phox participates in activation of the phagocyte NADPH oxidase. J Biol Chem 285 (41):31435–31445
12. Rodriguez AI et al (2015) MEF2B-Nox1 signaling is critical for stretch-induced phenotypic modulation of vascular smooth muscle cells. Arterioscler Thromb Vasc Biol 35(2):430–438
13. Meijles DN et al (2017) The matricellular protein TSP1 promotes human and mouse endothelial cell senescence through CD47 and Nox1. Sci Signal 10(501):eaaj1784
14. Joseph G, Pick E (1995) "Peptide walking" is a novel method for mapping functional domains in proteins. Its application to the Rac1-dependent activation of NADPH oxidase. J Biol Chem 270(49):29079–29082
15. Brandes RP, Kreuzer J (2005) Vascular NADPH oxidases: molecular mechanisms of activation. Cardiovasc Res 65(1):16–27
16. Brandes RP, Weissmann N, Schroder K (2014) Nox family NADPH oxidases: molecular mechanisms of activation. Free Radic Biol Med 76C:208–226
17. Sumimoto H (2008) Structure, regulation and evolution of Nox-family NADPH oxidases that produce reactive oxygen species. FEBS J 275 (13):3249–3277
18. Dahan I, Pick E (2012) Strategies for identifying synthetic peptides to act as inhibitors of NADPH oxidases, or "all that you did and did not want to know about Nox inhibitory peptides". Cell Mol Life Sci 69 (14):2283–2305
19. Csanyi G, Pagano PJ (2013) Strategies aimed at Nox4 oxidase inhibition employing peptides from Nox4 B-LOOP and C-terminus and p22 (phox) N-terminus: an elusive target. Int J Hypertens 2013:842827
20. Price MO et al (2002) Creation of a genetic system for analysis of the phagocyte respiratory burst: high-level reconstitution of the NADPH oxidase in a nonhematopoietic system. Blood 99(8):2653–2661

Chapter 25

High-Throughput Screening of NOX Inhibitors

Jacek Zielonka, Monika Zielonka, Gang Cheng, Micael Hardy, and Balaraman Kalyanaraman

Abstract

Development of new, selective inhibitors of nicotinamide adenine dinucleotide phosphate oxidase (NOX) isoforms is important both for basic studies on the role of these enzymes in cellular redox signaling, cell physiology, and proliferation and for development of new drugs for diseases carrying a component of increased NOX activity, such as several types of cancer and cardiovascular and neurodegenerative diseases. High-throughput screening (HTS) of large libraries of compounds remains the major approach for development of new NOX inhibitors. Here, we describe the protocol for the HTS campaign for NOX inhibitors using rigorous assays for superoxide radical anion and hydrogen peroxide, based on oxidation of hydropropidine, coumarin boronic acid, and Amplex Red. We propose using these three probes to screen for and identify new inhibitors, by selecting positive hits that show inhibitory effects in all three assays. Protocols for the synthesis of hydropropidine and for confirmatory assays, including oxygen consumption measurements, electron paramagnetic resonance spin trapping of superoxide, and simultaneous monitoring of superoxide and hydrogen peroxide, are also provided.

Key words High-throughput screening, Superoxide radical anion, Hydrogen peroxide, Hydropropidine, Coumarin boronic acid, Amplex Red, Fluorescence, EPR spin trapping, Seahorse extracellular flux analyzer

1 Introduction

The members of the family of nicotinamide adenine dinucleotide phosphate (NADPH) oxidase enzymes (NOX1-5, DUOX1-2) are specialized multicomponent enzymes, activatable or constitutively active, for which a sole function is to transfer electrons from the NADPH coenzyme to molecular oxygen, producing superoxide radical anion ($O_2^{\bullet-}$) and hydrogen peroxide (H_2O_2), a product of dismutation of $O_2^{\bullet-}$ (Fig. 1) [1, 2]. Different isoforms of NADPH oxidases have been implicated in cellular redox signaling, immune response to pathogens, and pathological conditions, including cancer and cardiovascular and neurodegenerative diseases [3–5]. The availability of specific pharmacological inhibitors, therefore, provides both basic scientific and translational potential. However, the

Ulla G. Knaus and Thomas L. Leto (eds.), *NADPH Oxidases: Methods and Protocols*, Methods in Molecular Biology, vol. 1982, https://doi.org/10.1007/978-1-4939-9424-3_25,

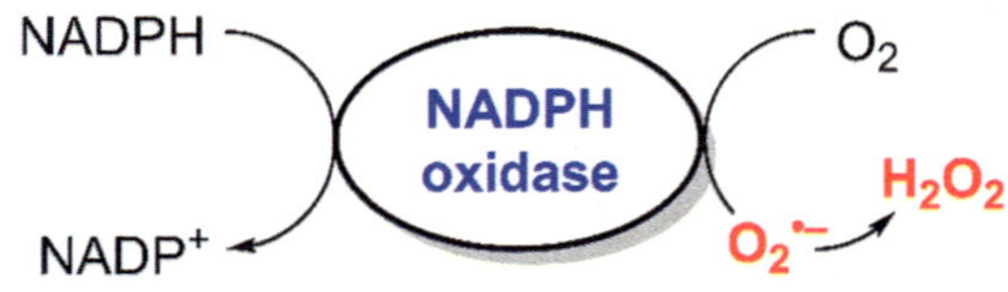

Fig. 1 Scheme showing the enzymatic activity of NADPH oxidases (NOX)

development of specific NOX enzyme inhibitors is not trivial because no crystal structure of any member of the NOX family of enzymes is available, limiting the ability to use computational tools or rational design of new inhibitors. Thus, new approaches to discovering inhibitors of NADPH oxidases are mostly based on "untargeted" screening of the libraries of chemical compounds, including large libraries available in the specialized screening centers [6–15]. Typically, high-throughput screening (HTS) assays are based on the detection of oxidants generated by NOX enzymes, using chemiluminescent or fluorescent probes [16–18]. Many of those probes, however, do not react directly with $O_2^{\bullet-}$ or H_2O_2 and require "activation" via electron transfer mechanisms or a catalyst. Thus, many compounds inhibiting the luminescent (chemiluminescence, fluorescence) signal may act by blocking the activation step or competing for the catalyst with the probe rather than by blocking the formation of $O_2^{\bullet-}$ or H_2O_2. This leads to a high rate of false-positive hits, necessitating cost- and time-consuming confirmatory assays. In fact, a recent report on myeloperoxidase inhibitors was based on the initial screening campaign for NOX inhibitors [19]. The results of confirmatory assays were critical to accurately assign the mode of action of the identified positive hits. In several cases, the claims of NOX-inhibitory activity have not been confirmed using more rigorous assays and/or monitoring NOX activity in cell-free systems in independent laboratories [11, 13]. Recent development in fluorescent probes for $O_2^{\bullet-}$ and H_2O_2 (Fig. 2) enables the use of probes reacting directly with $O_2^{\bullet-}$ (hydropropidine, HPr^+) or H_2O_2 (coumarin boronic acid, CBA), avoiding many limitations of the previously used probes [14, 20–22]. Upon the reaction between HPr^+ and $O_2^{\bullet-}$, a highly specific product, 2-hydroxypropidium (2-OH-Pr^{++}), is formed. The fluorescence yield of this product is further increased upon binding to DNA [20]. In the presence of H_2O_2, CBA is oxidized to 7-hydroxycoumarin (COH, also known as umbelliferone). In addition, Amplex Red (10-acetyl-3,7-dihydroxyphenoxazine) undergoes oxidation by H_2O_2 to resorufin in a peroxidase-catalyzed reaction. Understanding the chemistry behind the detection of $O_2^{\bullet-}$ and H_2O_2 using these probes opened the possibility of rigorous monitoring of NOX activity in a high throughput manner [17, 23–27]. Correlation of the results obtained with three probes with different probing mechanisms will enable identification of

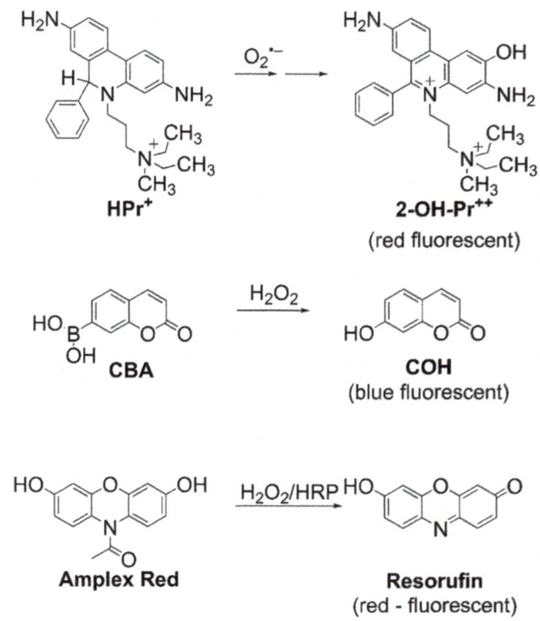

Fig. 2 Chemical structures of probes and their oxidation products used for HTS of NOX inhibitors

NOX inhibitors with higher confidence, as compared with previously used probes.

Here, we describe the protocol for HTS of NOX inhibitors in intact cells using three probes of different ROS sensing mechanisms, including HPr^+ for monitoring $O_2^{\bullet-}$ generation and CBA for monitoring H_2O_2 production. The third probe, Amplex Red, is used to detect H_2O_2 in the presence of horseradish peroxidase (HRP). We also provide protocols for cytotoxicity measurement, electron paramagnetic resonance (EPR) spin trapping, oximetry, and simultaneous monitoring of $O_2^{\bullet-}$ and H_2O_2 for confirmatory purposes for potential NOX inhibitors identified in the HTS campaign.

2 Materials

2.1 Components for the Synthesis of Hydropropidine

1. 0.1 g (0.15 mmol) of propidium iodide.
2. 20 mL of methanol (MeOH).
3. 6 mg (0.16 mmol) of sodium borohydride.
4. 20 mL of dichloromethane (DCM).
5. Argon gas.
6. 1 g of anhydrous sodium sulfate.

2.2 Cell Incubation and Differentiation Components

1. Cell growth medium: Roswell Park Memorial Institute (RPMI) 1640 medium containing 10% fetal bovine serum (FBS), 100 units/mL penicillin, and 100 μg/mL streptomycin (*see* **Note 1**).
2. 0.1 M all-trans retinoic acid (ATRA) in ethanol. The solution may be stored at −80 °C for at least 1 year.

2.3 HTS Assay Components

1. HBSS containing 25 mM 4-(2-hydroxyethyl)-1-piperazineethanesulfonic acid (HEPES), pH 7.4, and 0.1 mM diethylenetriaminepentaacetic acid (DTPA). Adjust pH to 7.4 using concentrated HCl or NaOH if necessary.
2. 0.25 mM solution of hydropropidine (HPr^+) in HBSS containing 0.5 mg/mL DNA, 25 mM HEPES, pH 7.4, 0.1 mM DTPA, and 0.1 mg/mL bovine serum albumin (BSA) (*see* **Note 2**).
3. 0.5 mM solution of coumarin boronic acid (CBA) in HBSS containing 25 mM HEPES, pH 7.4, 0.1 mM DTPA, and 0.1 mg/mL BSA (*see* **Note 2**).
4. 0.25 mM solution of Amplex Red in HBSS containing 0.5 U/mL HRP, 25 mM HEPES, pH 7.4, 0.1 mM DTPA, and 0.1 mg/mL BSA (*see* **Note 2**).
5. 5 μM solution of phorbol 12-myristate 13-acetate (PMA) in HBSS containing 25 mM HEPES, pH 7.4, 0.1 mM DTPA, and 0.1 mg/mL BSA (*see* **Note 2**).

2.4 ATP Measurements (See Note 3)

1. RPMI medium without phenol red and bicarbonate.
2. Adenosine triphosphate (ATP) releasing reagent (lysis buffer, component of a kit for bioluminescence-based ATP measurements).
3. ATP bioluminescence reagent (mix of luciferase and cofactors, component of a kit for bioluminescence-based ATP measurements).

2.5 Seahorse XF96-Based Oximetry

1. RPMI medium without phenol red and bicarbonate.
2. 10 μM PMA solution in RPMI medium without phenol red and bicarbonate, containing 0.1 mg/mL BSA.

2.6 EPR Spin Trapping of $O_2^{\bullet-}$

1. HBSS containing 25 mM HEPES, pH 7.4, and 0.1 mM DTPA.
2. 1 M 5-(Diethoxyphosphoryl)-5-methyl-1-pyrroline-*N*-oxide (DEPMPO) in water (*see* **Note 2**).
3. 20 μM PMA in HBSS containing 25 mM HEPES, pH 7.4, 0.1 mM DTPA, 0.1 mg/mL BSA (*see* **Note 2**).

2.7 Simultaneous Measurements of $O_2^{\bullet -}$ and H_2O_2

1. HBSS containing 25 mM HEPES, pH 7.4, and 0.1 mM DTPA.
2. Mixture of hydroethidine (HE, 6.7 mM), CBA (33 mM), and PMA (0.33 mM) in dimethyl sulfoxide (DMSO) (*see* **Note 2**).
3. A mixture of superoxide dismutase (SOD, 50×) and catalase (CAT, 50×): prepare a solution of SOD (5 mg/mL) and CAT (5 kU/mL) in water (*see* **Note 4**).

3 Methods

3.1 Preparation of HPr^+

1. Prepare HPr^+ (**steps 2–12**) inside a well-ventilated hood.
2. Deoxygenate 20 mL of MeOH by slowly passing argon gas through it for 10 min.
3. Deoxygenate 20 mL of DCM by slowly passing argon gas through it for 10 min.
4. Dissolve sodium borohydride (6 mg) in 1 mL of deoxygenated MeOH.
5. Dissolve propidium iodide (0.1 g) in 5 mL of deoxygenated MeOH.
6. Add sodium borohydride solution dropwise to the solution of propidium iodide, stirring constantly.
7. After 10 min of stirring under Ar atmosphere, add 1 mL of deoxygenated water and 5 mL of deoxygenated DCM, and vigorously shake the mixture for 1 min.
8. Allow the two phases to separate and transfer the DCM phase (lower) into an empty 20 mL vial.
9. Repeat **steps 7–8** five times and combine the DCM extracts.
10. Add 1 g of anhydrous sodium sulfate to the DCM extract, and stir the suspension for 10 min.
11. Filter the suspension and place the filtrate in an empty 20 mL glass vial.
12. Remove the solvent under vacuum (using a rotary evaporator) or by flushing the solution with argon or nitrogen gas.
13. Store the solid at −20 °C or lower.
14. Dissolve in deoxygenated aqueous solution of 1 mM HCl to a final concentration of 1 mM, freeze in liquid nitrogen, and store at −80 °C (*see* **Note 5**).

3.2 Preparation of Differentiated HL60 Cells Expressing NOX2 Isoform (See Note 1)

1. Grow HL60 cells in RPMI 1640 medium supplemented with FBS and antibiotics.
2. Count cells in suspension and prepare a suspension of 10^5 cells/mL.

3. Add ATRA to cells to a final concentration of 1 μM.
4. Incubate cells for 4 days.
5. Inspect cell morphology to confirm differentiation, as described elsewhere [28, 29]. Collect and save some cells to test for NOX2 expression (*see* **Note 6**).

3.3 HTS of NOX Inhibitors

1. Place the suspension of differentiated HL60 (*d*HL60) cells in 50 mL centrifuge tube(s), and spin down the cells (*see* **Note 7**).
2. Remove the supernatant and resuspend the cells in pre-warmed HBSS containing 25 mM HEPES, pH 7.4, and 0.1 mM DTPA.
3. Determine the cell density by cell counting, and dilute the cells to the density of 10^5 cells/mL.
4. Transfer the cell suspension into black 384-well plates; use 30 μL/well (*see* **Note 8**).
5. Add ~100 nL of stock solutions of the compounds to be tested (the chemical library) (*see* **Notes 9** and **10**).
6. Incubate the cells with compounds for 30 min at 37 °C in a carbon dioxide (CO_2)-free incubator under ambient oxygen (21% O_2).
7. After incubation, add 10 μL/well of freshly prepared solution of the probe (5× solution) in HBSS containing 25 mM HEPES, pH 7.4, 0.1 mM DTPA, and 0.1 mg/mL BSA (*see* **Note 8**).
8. Add 10 μL/well of 5 μM solution of PMA in HBSS containing 25 mM HEPES, pH 7.4, 0.1 mM DTPA, and 0.1 mg/mL BSA (*see* **Note 8**).
9. Incubate the cells with reactive oxygen species (ROS) probes for 60 min at 37° C in a CO_2-free incubator (*see* **Note 11**).
10. Measure fluorescence intensity using a plate reader. The parameters for excitation and emission filters for each probe are shown in Table 1 (*see* **Note 12**).
11. Repeat **steps 1–10** for each ROS probe.

3.4 HTS Data Analysis

1. For each plate, determine the *Z*′ value using control positive (PMA, no inhibitor, "control+") and control negative (PMA + known inhibitor, "control−") samples in columns 1, 2, 23, and 24 (*see* **Note 9**). Use the following formula:

$$Z' = 1 - \frac{3\mathrm{SD}_{\mathrm{control+}} + 3\mathrm{SD}_{\mathrm{control-}}}{|\mathrm{mean}_{\mathrm{control+}} - \mathrm{mean}_{\mathrm{control-}}|}$$

2. Normalize the fluorescence intensity data for each sample (I_x), so that the mean fluorescence intensity of sample control+ (I_{c+})

Table 1
Fluorescence parameters used for monitoring the oxidation products of HPr^+, CBA, and Amplex Red probes

Probe	Excitation wavelength (nm)	Emission wavelength (nm)
HPr^+	485	574
CBA	355	460
Amplex Red	535	595

is set to 0% and the mean fluorescence intensity of sample control− (I_{c-}) is set to −100%, using the following formula:

$$NI_x = \frac{I_x - I_{C+}}{I_{C+} - I_{C-}} \times 100\%$$

3. Identify the positive hits, exhibiting normalized fluorescence (NI_x) values (Fig. 3) below a set threshold (*see* **Note 13**).
4. Perform **steps 1–3** for each probe used.
5. Identify the compounds that are positive hits in all three assays (Fig. 4). These compounds are most likely to inhibit NADPH oxidase activity in the cellular model tested.
6. Evaluate the positive hits for the presence of pan assay interference compounds (PAINS) [30, 31]. Remove those compounds from the list of positive hits (*see* **Note 14**).
7. The remaining positive hits are potential inhibitors of NADPH oxidases and should be further evaluated using an array of confirmatory assays. *See* Fig. 5 for the complete workflow (*see* **Note 15**).

3.5 Cytotoxicity Measurements by Monitoring Cellular ATP

1. Load an empty 96-deep well plate (2 mL well volume) with potential NOX inhibitors (1.0 μL of 1000× solution in DMSO or other solvent), identified during the HTS campaign.
2. Place the suspension of *d*HL60 cells in 50 mL centrifuge tube (s), and spin down the cells (*see* **Note 1**).
3. Remove the supernatant, and resuspend the cells in pre-warmed bicarbonate- and phenol red-free RPMI.
4. Determine the cell density by cell counting, and dilute the cells in RMPI medium to the density of 1×10^5 cells/mL.
5. Transfer the cell suspension into the 96-deep well plate pre-loaded with potential inhibitors (1.0 mL/well). Mix the cell suspension with inhibitors by three aspirating/dispensing cycles.
6. Place the plate in CO_2-free incubator (37 °C) for 120 min.

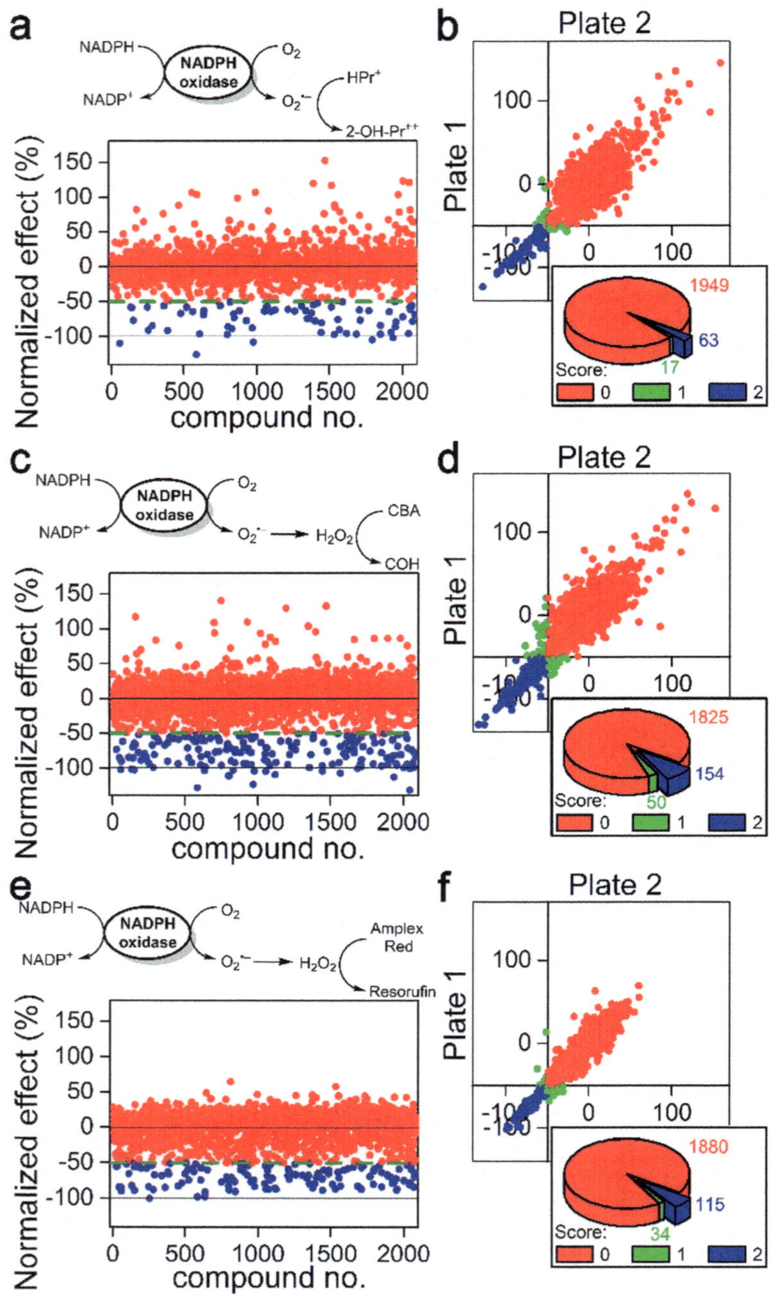

Fig. 3 Results of screening of a library of bioactive compounds (~2000) using three probes: HPr^+ (50 μM) in the presence of DNA (0.1 mg/mL) as a probe for O_2Superscript>/Superscript> (**a** and **b**) and CBA (100 μM, **c** and **d**) or Amplex Red (50 μM) in the presence of HRP (0.1 U/mL, **e** and **f**) as probes for H_2O_2. *d*HL60 cells were stimulated with PMA (1 μM) to induce NOX2 activity (**a**, **c**, and **e**). Schemes of oxidation of the probes and the results of screening after normalization (**b**, **d**, and **f**). Plate-to-plate reproducibility data and number of negative/inconclusive/positive hits for each assay. Score "0" (red color) corresponds to negative, "1" (green color) to inconclusive, and "2" (blue color) to

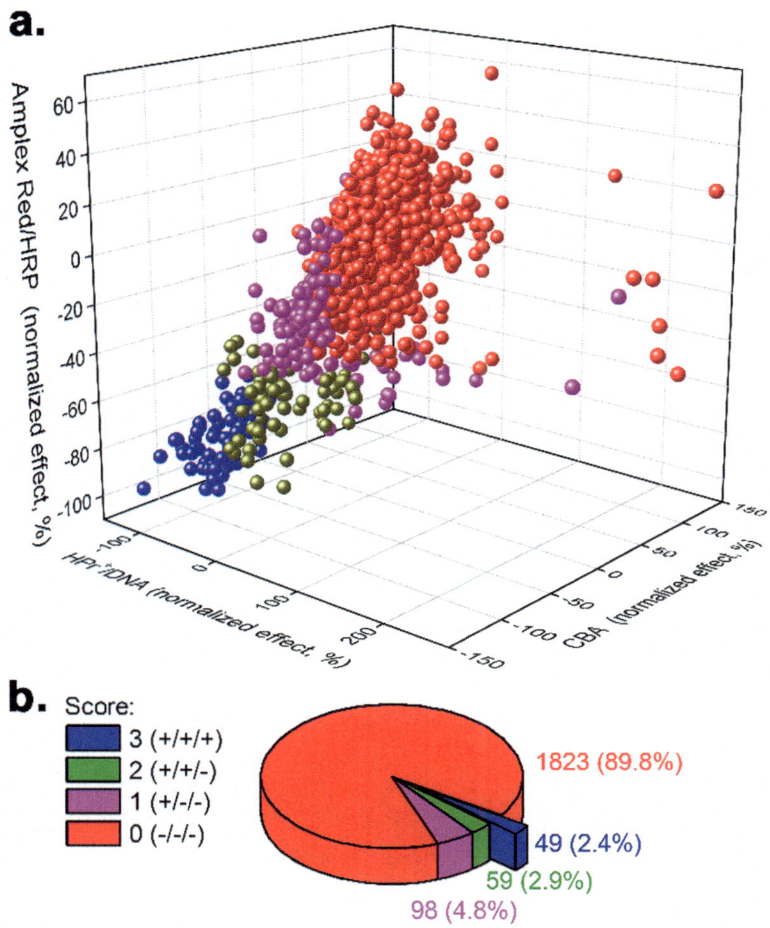

Fig. 4 Results of the screening of a library of bioactive compounds (~2000) using three probes: HPr^+ in the presence of DNA as a probe for O_2Superscript>/ Superscript> and CBA or Amplex Red in the presence of HRP as probes for H_2O_2. (**a**) Correlation of the results of the three assays for NOX2 activity. (**b**) Results of screening as a percentage of positive hits in one, two, or all three assays. (This research was originally published in *Journal of Biological Chemistry*. Zielonka, J., Zielonka, M., VerPlank, L., Cheng, G., Hardy, M., Ouari, O., Ayhan, M. M., Podsiadly, R., Sikora, A., Lambeth, J. D., & Kalyanaraman, B. Mitigation of NADPH oxidase 2 activity as a strategy to inhibit peroxynitrite formation. *J. Biol. Chem.* 2016; 291:7029-7044. © the American Society for Biochemistry and Molecular Biology)

Fig. 3 (continued) positive hits. (This research was originally published in *Journal of Biological Chemistry*. Zielonka, J., Zielonka, M., VerPlank, L., Cheng, G., Hardy, M., Ouari, O., Ayhan, M. M., Podsiadly, R., Sikora, A., Lambeth, J. D., & Kalyanaraman, B. Mitigation of NADPH oxidase 2 activity as a strategy to inhibit peroxynitrite formation. *J. Biol. Chem.* 2016; 291:7029-7044. © the American Society for Biochemistry and Molecular Biology)

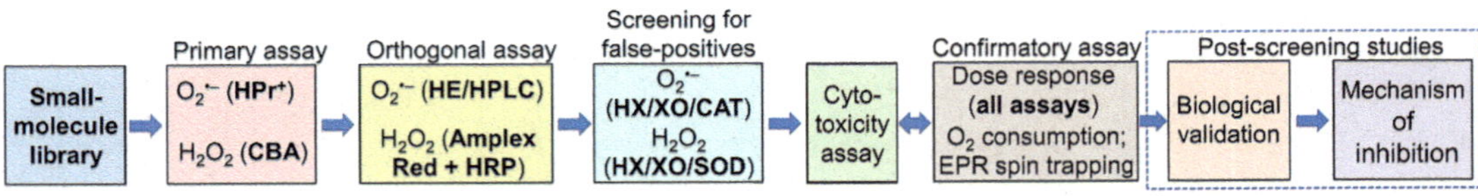

Fig. 5 The workflow scheme for the screening of NOX inhibitors. (This research was originally published in *Journal of Biological Chemistry*. Zielonka, J., Cheng, G., Zielonka, M., Ganesh, T., Sun, A., Joseph, J., Michalski, R., O'Brien, W. J., Lambeth, J. D., & Kalyanaraman, B. High-throughput assays for superoxide and hydrogen peroxide: design of a screening workflow to identify inhibitors of NADPH oxidases. *J. Biol. Chem.* 2014; 289:16176-16189. © the American Society for Biochemistry and Molecular Biology)

7. Spin down the cells by centrifuging the plate for 1 min at 300 × *g*.
8. Remove supernatant using a 12-channel pipettor.
9. Add ATP-releasing reagent to cell pellets (40 μL/well) using a 12-channel pipettor. Shake the plate for 3 min.
10. Transfer 20 μL of the lysate to 96-well solid bottom white plate (for luminescence measurements) using a 12-channel pipettor.
11. Add the ATP reagent containing luciferase and cofactors using a 12-channel pipettor, and immediately measure luminescence intensity (*see* **Note 16**).

3.6 Oxygen Consumption Measurements Using the Seahorse XF96 Extracellular Flux Analyzer

1. Hydrate the Seahorse XF96 cartridge overnight according to the manufacturer's instructions.
2. Place the suspension of *d*HL60 cells in 50 mL centrifuge tube (s), and spin down the cells (*see* **Note 1**).
3. Remove the supernatant, and resuspend the cells in pre-warmed bicarbonate- and phenol red-free RPMI (assay medium).
4. Determine the cell density by cell counting, and dilute the cells in the assay medium to the density of 2.5×10^5 cells/mL.
5. Transfer the cell suspension into the Seahorse XF96 well plate, 80 μL/well, to a total of 2×10^4 cells/well (*see* **Note 17**).
6. Spin down the cells by centrifuging the plate for 3 min at 100 × *g*.
7. Gently add 100 μL of the assay medium, without disturbing the cells at the bottom. Keep the plate at 37 °C in a CO_2-free incubator.
8. Load the Seahorse XF96 cartridge port A with solutions of potential NOX inhibitors (20 μL of 10× solution), identified during the HTS campaign. Use RPMI containing 0.1% BSA as the solvent. Additional wells containing the same amounts of the organic solvent (typically DMSO) as introduced with the inhibitors also should be included. In case of limited solubility of the inhibitors in the medium, preincubation of cells with the

inhibitors (1× concentration in the medium) before starting the measurements of oxygen consumption is also possible (*see* **Note 17**).

9. Load port B with 22.2 μL of 10 μM PMA solution in RPMI containing 0.1% BSA.
10. Place the cartridge in the Seahorse XF96 instrument, and start the calibration process.
11. Set up the oxygen consumption rate (OCR) measurements over 240 min including basal OCR monitoring, followed by injection of potential NOX inhibitors (port A) at the 60 min time point and injection of PMA solution (port B) at the 120 min time point (*see* **Note 18**).
12. Insert the plate with cells when prompted and start measurements.

3.7 EPR Spin Trapping of $O_2^{\bullet -}$

1. Place the suspension of *d*HL60 cells in 50 mL centrifuge tube (s), and spin down the cells (*see* **Note 1**).
2. Remove the supernatant, and resuspend the cells in pre-warmed HBSS containing 25 mM HEPES, pH 7.4, 0.1 mM DTPA.
3. Determine the cell density by cell counting, and dilute the cells in RMPI medium to the density of 1.25×10^5 cells/mL.
4. To a 1.5 mL microcentrifuge tube, add 10 μL of potential NOX inhibitor (10 μL of 10× solution), identified during the HTS campaign.
5. Add 80 μL of cell suspension (1.25×10^5 cells/mL), and mix by three aspirating/dispensing cycles.
6. Incubate cells with inhibitor for 30 min.
7. Add 5 μL of 1 M DEPMPO spin trap in water.
8. Add 5 μL of 20 μM PMA in HBSS containing 25 mM HEPES, pH 7.4, 0.1 mM DTPA, and 0.1 mg/mL BSA, and mix by three aspirating/dispensing cycles.
9. Incubate the mixture for 1 h.
10. Transfer the solution into the EPR capillary (50 or 75 μL), seal the capillary, and place in the EPR instrument.
11. Measure EPR spectra. Average multiple scans if necessary to obtain an acceptable signal-to-noise ratio.
12. Repeat **steps 4–11** for each inhibitor to be tested, and compare to a control sample without inhibitors. All samples should be run using the same number of averaged spectra.
13. Quantify the spin adduct by double integration of the EPR peaks, and compare the integrated peak areas of the PMA-stimulated superoxide spin adduct between control and inhibitor-treated cells.

3.8 Rapid HPLC-Based Simultaneous Measurements of $O_2^{\bullet -}$ and H_2O_2

1. Load the empty 96-deep well plate (2 mL well volume) with potential NOX inhibitors (1.2 μL of 1000× solution in DMSO or other solvent), identified during the HTS campaign.
2. Place the suspension of *d*HL60 cells in 50 mL centrifuge tube (s), and spin down the cells (*see* **Note 1**).
3. Remove the supernatant, and resuspend the cells in pre-warmed HBSS containing 25 mM HEPES, pH 7.4, 0.1 mM DTPA.
4. Determine the cell density by cell counting, and dilute the cells in HBSS containing 25 mM HEPES, pH 7.4, and 0.1 mM DTPA to the density of 1×10^5 cells/mL.
5. Transfer the cell suspension into the 96-deep well plate pre-loaded with potential inhibitors (1.2 mL/well). Mix the cell suspension with inhibitors by three aspirating/dispensing cycles.
6. Incubate the cells with inhibitors for 30 min.
7. Preload a second 96-deep well plate with a mix of HE (6.7 mM), CBA (33 mM), and PMA (0.33 mM) in DMSO (3 μL/well).
8. Transfer cells and inhibitors suspensions (1 mL/well) to a plate preloaded with HE, CBA, and PMA. Mix the cell suspension with inhibitors by three aspirating/dispensing cycles.
9. Incubate the cells for 1 h at 37 °C, protected from light.
10. Add SOD + CAT solution (50×, 20 μL/well) to stop oxidation of the probes (HE, CBA).
11. Spin down the cells by centrifuging the plate for 1 min at 300 × *g*.
12. Transfer the supernatants into high-performance liquid chromatography (HPLC) vials (400 μL/vial), a 96-well plate (250 μL/well), or a 384-well plate (80 μL/well), depending on the HPLC autosampler compatibility.
13. Run rapid HPLC analyses of HE, 2-OH-E^+, CBA, and COH, as described elsewhere [14, 15] (*see* **Note 19**). The levels of 2-OH-E^+ and COH reflect the production of $O_2^{\bullet -}$ and H_2O_2, respectively.

4 Notes

1. The cell culture medium depends on the cell type and should be based on the requirements for normal cell growth, unless intended otherwise. In the protocol described here, the cell culture and differentiation conditions are provided for HL60 promyelocytic leukemia cells, differentiated into neutrophil-

like cells to study the NOX2 isoform. These cells are grown and assayed in suspension. Some modifications of the protocols may be needed to study cellular models of NOX isoforms using adherent cells.

2. The solutions should be prepared immediately before being added to the cells and should be shielded from light. Any leftover solutions should be discarded and not reused.
3. For ATP measurements, any commercially available bioluminescence kit based on ATP-dependent luciferase-catalyzed oxidation of luciferin should be appropriate. Final solutions should be prepared immediately before measurements and shielded from light.
4. From our experience, different preparations of commercial catalase exhibit peroxidatic activity to different extents. To minimize the peroxidatic activity, we recommend using catalase from *Corynebacterium glutamicum* (Sigma, cat. no. 02071).
5. The purity of the prepared HPr^+ should be confirmed by HPLC to ensure the amount of the oxidation product, propidium dication, does not exceed 1%. If necessary, HPLC-based purification of the probe can be carried out, as described elsewhere [20]. HPLC purification using mobile phase containing trifluoroacetic acid (TFA) will result in anion exchange into TFA, which should be considered when calculating the molecular weight of the product. The concentration of the stock solution of HPr^+ should be determined by UV-Vis spectroscopy at 350 nm, using the extinction coefficient of $1.15 \times 10^4\ M^{-1}\ cm^{-1}$ [20]. No absorption band above 400 nm, indicative of the presence of HPr^+ oxidation product (s), should be observed.
6. The expression of the NOX isoform of interest should be confirmed by immunoblotting and/or other technique. This also applies to cytosolic components, when appropriate.
7. As mentioned in **Note *1***, the protocol described here is for cells grown in suspension. Adherent cells could be grown directly in a 384-well plate and assayed while attached to the plate bottom. Alternatively, cells could be grown in a cell culture flask, harvested, and assayed in suspension, similar to the protocol for *d*HL60 cells. The choice of experimental protocol should be based on the optimal conditions to measure the activity of NADPH oxidase, as determined by the calculated *Z′* value.
8. When screening hundreds or thousands of compounds, an automatic dispensing/pipetting instrument should be used for fast and accurate pipetting of the assay components.
9. For each assay plate, columns 1 and 23 are reserved for DMSO (or other solvent used in the solutions of inhibitors), and these

wells will be used as control+ samples. Columns 2 and 24 are reserved for known inhibitors of NADPH oxidase (diphenyleneiodonium or phenylarsine oxide for the NOX2 isoform), and these wells will be used as control- samples. From these samples, the Z' parameter should be determined for each plate, as should its reproducibility between plates and the days of measurements for each probe/experimental condition.

10. Typically, the chemical libraries of compounds contain DMSO solutions, with concentrations in the millimolar range (5–10 mM). To minimize the final concentration of DMSO, small volumes of the solutions should be added to the cell medium to keep the total DMSO concentration below 0.5% (v/v). The maximum DMSO content, which does not significantly affect the activity of NADPH oxidase in the studied cellular model, should be experimentally determined. Low volumes are typically added using specialized instruments (e.g., equipped with pin tools) available in HTS centers. For most labs, more cost-effective solutions, including disposable pin tools, delivering 100–130 nL/well may be used.
11. The incubation time as well as optimal density should be determined experimentally, by determining Z' values and choosing the conditions yielding the maximal Z' value. The plate should contain control+ and control− samples (*see* **Note 9**) in at least triplicates for each experimental condition. For the HTS campaign, the Z' value should be equal to or higher than 0.5. The best approach is to run the initial experiment, using different cell densities and varying other conditions (concentrations of probes, activators, etc.), in the real-time (kinetic) mode and calculate Z' values for all time points before selecting the optimal time point. It may be best to choose the time at which the signal is close to reaching plateau, so that no large changes are expected due to small shifts in the total time of incubation (from starting the incubation with the probes through signal measurement) between all samples within and between the plates.
12. The fluorescence parameters used in our lab are listed in Table 1. However, small shifts in the wavelengths used may be acceptable. The excitation/emission maxima for each probe are as follows: 2-OH-Pr^{++}/DNA, 508 nm/575 nm [20]; COH, 332 nm/450 nm [21]; and resorufin, 570 nm/583 nm [25].
13. Typically, the threshold is set at −50% of the normalized intensity (NI_x). Minimum inhibitory activity for the threshold should be equal to three standard deviations of the positive control ($SD_{control+}$). The choice of threshold will affect the number of positive hits to be evaluated in the confirmatory

assays, possibly with higher rates of false-positive hits. However, this will also enable identification of weaker, but potentially more selective, inhibitors, the potency of which could be later improved using medicinal chemistry approaches.

14. PAINS are compounds that are frequent hitters in most screening campaigns [30] and are deemed unsuitable for probe/drug development due to their promiscuity. PAINS could be eliminated from the identified positive hits using an online filter: http://www.cbligand.org/PAINS/search_struct.php.

15. The positive hits identified during the HTS campaign may work by direct interaction with NOX isoform or show the activity due to its cytotoxicity or the interference with the upstream events affecting NOX activity. Thus, additional (confirmatory) assays need to be performed to understand the mechanism of the inhibitory effects observed, as shown in Fig. 5. These include cytotoxicity screening, oxygen consumption measurements, EPR spin trapping of $O_2^{\bullet-}$, and HPLC-based monitoring of the products from HPr^+ and CBA oxidation. The dose response should be also tested. *See* Fig. 6 for an example of the confirmatory assays used for an identified inhibitor of NOX2 activity in *d*HL60 cells [14]. Further studies establishing the mechanisms of inhibition using cell-free assays and NOX isoform selectivity are also required for complete characterization of the mechanism of action and selectivity of the identified inhibitor [11, 13, 32].

16. A decrease in the total ATP level may indicate the cytotoxic activity of the positive hits. It is also possible that the compound exhibits cytotoxicity only in the presence of a NOX activator (e.g., PMA), and therefore this assay should be performed in duplicate plates, one with non-stimulated and one with stimulated cells, if appropriate.

17. Alternatively, cells (at the density of 2.5×10^5 cells/mL) may be preincubated with inhibitors before being loaded onto the plate. In such case, only the PMA solution needs to be loaded into the cartridge. The advantage of this approach is that preparation of 10× solutions of the inhibitors in RPMI could be avoided, which may be necessary in the case of limited solubility of the potential inhibitors. The advantage of direct injection of 10× solutions during OCR measurements is that each sample is its own control, when determining the effect of the potential inhibitor on basal OCR (mitochondrial respiration).

18. The basal OCR value is, in most cases, due largely to mitochondrial respiration, as it is inhibitable by injection of rotenone (complex I inhibitor) and antimycin A (complex III inhibitor) [14]. Injection of PMA induces NOX-dependent

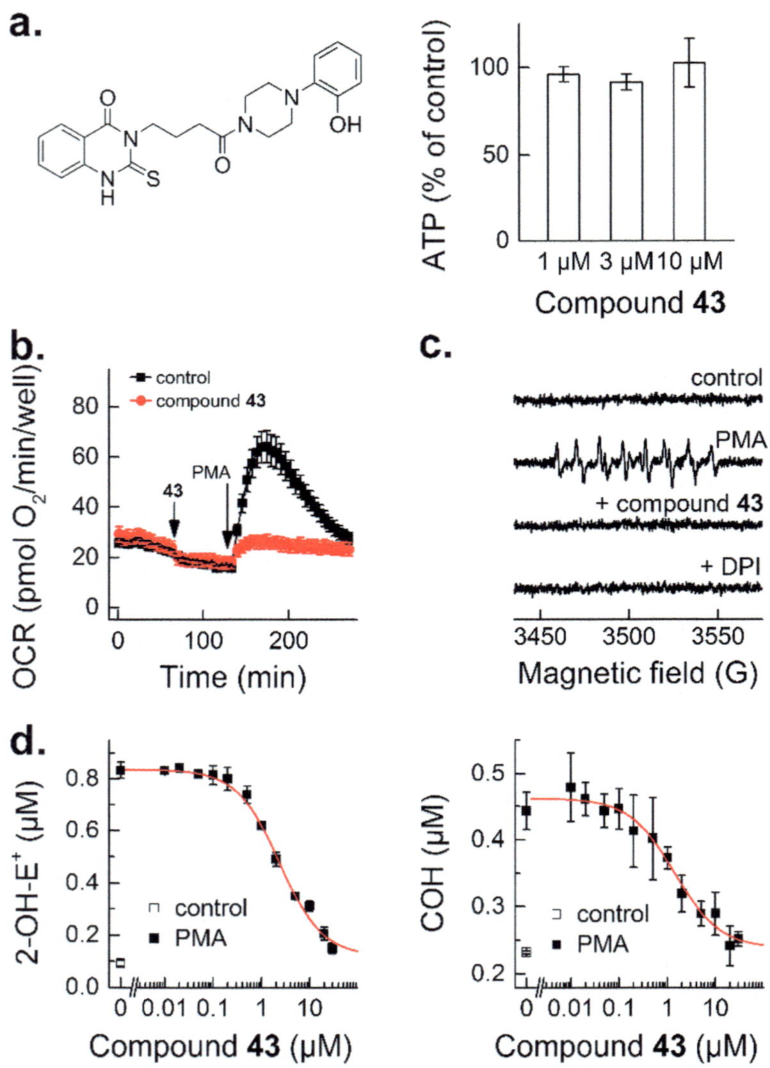

Fig. 6 Examples of confirmatory assays used for characterization of the positive hits from the HTS campaign. (**a**) Structure of the identified hit (compound 43 from ref. 14) and its effect on total cellular ATP level. (**b** and **c**) Effect of the identified hit on the PMA-stimulated oxygen consumption rates (**b**) and formation of DEPMPO superoxide spin adduct (**c**). (**d**) Concentration dependence of the effect of compound 43 on PMA-stimulated probe oxidation by *d*HL60 cells in the HPLC-based assays for simultaneous monitoring of O_2Superscript>/Superscript> and H_2O_2. (This research was originally published in *Journal of Biological Chemistry*. Zielonka, J., Cheng, G., Zielonka, M., Ganesh, T., Sun, A., Joseph, J., Michalski, R., O'Brien, W. J., Lambeth, J. D., & Kalyanaraman, B. High-throughput assays for superoxide and hydrogen peroxide: design of a screening workflow to identify inhibitors of NADPH oxidases. *J. Biol. Chem.* 2014; 289:16176-16189. © the American Society for Biochemistry and Molecular Biology)

oxygen consumption and reflects NOX activity. Injection of the compound may affect both mitochondrial respiration (basal OCR) and NOX-dependent oxygen consumption (PMA-stimulated OCR), as shown for the VAS2870 compound [14]. Notably, inhibition of mitochondrial respiration by rotenone does not block the PMA-induced oxidative burst, indicating that both processes are independent in the cellular model tested.

19. Rapid HPLC analyses of HE, CBA, 2-OH-E^+, and COH are performed using HPLC equipped with UV-Vis absorption, fluorescence detectors, and a column Ascentis Express Phenyl-Hexyl 50 mm × 4.6 mm, 2.7 μm (Supelco), as described elsewhere [14, 15]. The HE and CBA levels are monitored using a UV-Vis absorption detector set at 370 and 290 nm, respectively, and the 2-OH-E^+ and COH levels are monitored using a fluorescence detector with excitation set at 370 nm and emission set at 565 nm. Retention times are 19 and 36 s for HE and 2-OH-E^+, and 21 and 25 s for CBA and COH, respectively.

Acknowledgment

This work was supported by NIH grants NCI U01 CA178960 and R01 AA022986 to B.K.

References

1. Lambeth JD (2004) NOX enzymes and the biology of reactive oxygen. Nat Rev Immunol 4:181–189
2. Bedard K, Krause KH (2007) The NOX family of ROS-generating NADPH oxidases: physiology and pathophysiology. Physiol Rev 87:245–313
3. Cifuentes-Pagano E, Meijles DN, Pagano PJ (2014) The quest for selective nox inhibitors and therapeutics: challenges, triumphs and pitfalls. Antioxid Redox Signal 20:2741–2754
4. Bedard K, Whitehouse S, Jaquet V (2015) Challenges, Progresses, and Promises for Developing Future NADPH Oxidase Therapeutics. Antioxid Redox Signal 23:355–357
5. Diebold BA, Smith SM, Li Y, Lambeth JD (2015) NOX2 as a target for drug development: indications, possible complications, and progress. Antioxid Redox Signal 23:375–405
6. Cifuentes-Pagano E, Csanyi G, Pagano PJ (2012) NADPH oxidase inhibitors: a decade of discovery from Nox2ds to HTS. Cell Mol Life Sci 69:2315–2325
7. Smith SM et al (2012) Ebselen and congeners inhibit NADPH oxidase 2-dependent superoxide generation by interrupting the binding of regulatory subunits. Chem Biol 19:752–763
8. Borbely G et al (2010) Small-molecule inhibitors of NADPH oxidase 4. J Med Chem 53:6758–6762
9. Seitz PM et al (2010) Development of a high-throughput cell-based assay for superoxide production in HL-60 cells. J Biomol Screen 15:388–397
10. Cifuentes-Pagano E et al (2013) Bridged tetrahydroisoquinolines as selective NADPH oxidase 2 (Nox2) inhibitors. Medchemcomm 4:1085–1092
11. Seredenina T et al (2015) A subset of N-substituted phenothiazines inhibits NADPH oxidases. Free Radic Biol Med 86:239–249
12. Gianni D et al (2010) A novel and specific NADPH oxidase-1 (Nox1) small-molecule inhibitor blocks the formation of functional

invadopodia in human colon cancer cells. ACS Chem Biol 5:981–993

13. Hirano K et al (2015) Discovery of GSK2795039, a novel small molecule NADPH oxidase 2 inhibitor. Antioxid Redox Signal 23:358–374
14. Zielonka J et al (2014) High-throughput assays for superoxide and hydrogen peroxide design of a screening workflow to identify inhibitors of NADPH oxidases. J Biol Chem 289:16176–16189
15. Zielonka J et al (2016) Mitigation of NADPH oxidase 2 activity as a strategy to inhibit peroxynitrite formation. J Biol Chem 291:7029–7044
16. Maghzal GJ, Krause KH, Stocker R, Jaquet V (2012) Detection of reactive oxygen species derived from the family of NOX NADPH oxidases. Free Radic Biol Med 53:1903–1918
17. Zielonka J et al (2017) Recent developments in the probes and assays for measurement of the activity of NADPH oxidases. Cell Biochem Biophys 75:335–349
18. Wardman P (2007) Fluorescent and luminescent probes for measurement of oxidative and nitrosative species in cells and tissues: progress, pitfalls, and prospects. Free Radic Biol Med 43:995–1022
19. Li Y et al (2015) Thioxo-dihydroquinazolinone compounds as novel inhibitors of myeloperoxidase. ACS Med Chem Lett 6:1047–1052
20. Michalski R, Zielonka J, Hardy M, Joseph J, Kalyanaraman B (2013) Hydropropidine: a novel, cell-impermeant fluorogenic probe for detecting extracellular superoxide. Free Radic Biol Med 54:135–147
21. Zielonka J, Sikora A, Joseph J, Kalyanaraman B (2010) Peroxynitrite is the major species formed from different flux ratios of co-generated nitric oxide and superoxide: direct reaction with boronate-based fluorescent probe. J Biol Chem 285:14210–14216
22. Kalyanaraman B, Hardy M, Zielonka J (2016) A critical review of methodologies to detect reactive oxygen and nitrogen species stimulated by NADPH oxidase enzymes: implications in pesticide toxicity. Curr Pharmacol Rep 2:193–201
23. Zielonka J, Kalyanaraman B (2010) Hydroethidine- and MitoSOX-derived red fluorescence is not a reliable indicator of intracellular superoxide formation: another inconvenient truth. Free Radic Biol Med 48:983–1001
24. Zielonka J, Sikora A, Hardy M, Joseph J, Dranka BP, Kalyanaraman B (2012) Boronate probes as diagnostic tools for real time monitoring of peroxynitrite and hydroperoxides. Chem Res Toxicol 25:1793–1799
25. Zielonka J et al (2012) Global profiling of reactive oxygen and nitrogen species in biological systems: high-throughput real-time analyses. J Biol Chem 287:2984–2995
26. Michalski R, Michalowski B, Sikora A, Zielonka J, Kalyanaraman B (2014) On the use of fluorescence lifetime imaging and dihydroethidium to detect superoxide in intact animals and ex vivo tissues: a reassessment. Free Radic Biol Med 67:278–284
27. Debski D et al (2016) Mechanism of oxidative conversion of Amplex(R) Red to resorufin: pulse radiolysis and enzymatic studies. Free Radic Biol Med 95:323–332
28. Martin SJ, Bradley JG, Cotter TG (1990) HL-60 cells induced to differentiate towards neutrophils subsequently die via apoptosis. Clin Exp Immunol 79:448–453
29. Dufer J, Biakou D, Joly P, Benoist H, Carpentier Y, Desplaces A (1989) Quantitative morphological aspects of granulocytic differentiation induced in HL-60 cells by dimethylsulfoxide and retinoic acid. Leuk Res 13:621–627
30. Baell JB, Holloway GA (2010) New substructure filters for removal of pan assay interference compounds (PAINS) from screening libraries and for their exclusion in bioassays. J Med Chem 53:2719–2740
31. Baell J, Walters MA (2014) Chemistry: chemical con artists foil drug discovery. Nature 513:481–483
32. Pick E (2014) Cell-free NADPH oxidase activation assays: "in vitro veritas". Methods Mol Biol 1124:339–403

Chapter 26

Protein–Protein Interaction Assay to Analyze NOX4/p22^{phox} Heterodimerization

Sharon O'Neill and Ulla G. Knaus

Abstract

The stabilization and activation of NOX4 through its binding with p22^{phox} are well documented; however little is known of the precise manner by which these two proteins interact. In recent years, the field of proteomics has undergone tremendous development with the introduction of many novel methods for the identification and characterization of protein–protein interactions (PPIs). To enhance our understanding of structural determinants leading to the association between NOX4 and p22^{phox}, we developed a binary luciferase reporter assay (NanoBiT®) to quantitatively assess NOX4-p22^{phox} heterodimerization. The complementation reporter quantitatively determines the accurate, reduced, or failed complex assembly, which can be confirmed and further interrogated by analyzing NOX4 catalytic activity (H_2O_2 release), protein expression, and dimer localization. This association-based PPI technique represents both a much-needed expansion of the NOX4 lead discovery tool box and a versatile method to probe the architecture of NOX and DUOX complexes in the future.

Key words NADPH oxidase, NOX4, p22^{phox}, Protein–protein interaction (PPI), Heterodimerization, NanoBiT®, Bioluminescence

1 Introduction

NADPH oxidases (NOX/DUOX) are multimeric, reactive oxygen species (ROS)-generating enzymes that require dimerization of NOX or DUOX isoforms with a partner protein (p22^{phox} and DUOXA1/2, respectively), which then serves as a membrane-integrated platform for assembly with cytosolic proteins to form the active oxidase complex. Exceptions are NOX5, which seems not to utilize a dimerization partner, and the NOX4-p22^{phox} dimer that appears to generate H_2O_2 constitutively without the need for additional components. Heterodimer formation of NOX2 and NOX4 with p22^{phox} (and by extension likely for NOX1 and NOX3) occurs after incorporation of two bis-histidine coordinated low-potential hemes [1–3], followed by trafficking (with or without Golgi processing) to the plasma membrane or intracellular

Ulla G. Knaus and Thomas L. Leto (eds.), *NADPH Oxidases: Methods and Protocols*, Methods in Molecular Biology, vol. 1982, https://doi.org/10.1007/978-1-4939-9424-3_26, © Springer Science+Business Media, LLC, part of Springer Nature 2019

membrane compartments. Protein–protein interactions (PPIs) between oxidase components are not only required for heterodimerization but also for activation-induced assembly of NOX. The assembly process is induced by post-translational modifications and GDP to GTP exchange on RAC1 or RAC2 (NOX1-3), translocation of the cytosolic proteins to the membrane-bound heterodimer, and association with the dimer and with each other via specific interaction domains [4, 5]. Any disruption of these associations or changes in binding affinity, as observed in innate immune cells of chronic granulomatous disease (CGD) patients with loss-of-function NOX2 complex variants, will result in reduced or absent ROS production. CGD mutations revealed that the stability of both proteins, NOX2 and $p22^{phox}$, is dependent on efficient heterodimerization in neutrophils. Furthermore, changes in the dimerization partner (DUOXA1 versus DUOXA2) affected localization of DUOX enzymes [6], while imperfect heterodimer assembly due to mutations or deletions can alter oxidase localization and/or result in changes in the reactive species being produced [7–9]. Although NOX1-4 proteins share the same dimerization partner, the binding surfaces differ, as evidenced when mutation of an amino acid in the last transmembrane domain of $p22^{phox}$ abrogated NOX1-3 binding without affecting NOX4 [10]. These unique contact sites in NOX1-4 and in DUOX1-2 heterodimers could be exploited for selective oxidase isoform targeting by chemical compounds if sensitive and quantitative assay systems can be established for high-throughput screening.

Detecting protein–protein interactions is important for discerning molecular mechanisms controlling signaling and degradation pathways or regulating transcription networks; however these interaction domains are also prime drug targets since defined hot spots or allosteric sites can act as binding regions for drugs [11]. Different approaches exist to analyze PPIs in vivo [12], although for determining PPIs in NADPH oxidases under basal and/or activated conditions co-immunoprecipitation has been the main tool. Co-immunoprecipitation from detergent-containing extracts is not quantitative, often not efficient, and dependent on selective antibodies with high affinity. More recently, proximity ligation assays and FRET-based analyses were employed to test PPIs between NOX components and NOX-associated proteins [13–15]; however suitable antibodies are scarce, and attaching fluorescent proteins to the C-terminus of NOX abrogates catalytic activity [3].

Bioluminescence has long been used as a sensitive readout for gene expression, with luciferases being fused to proteins or split into fragments to monitor protein interactions [16]. Luciferases are also part of the bioluminescence resonance energy transfer (BRET) method, where energy from a donor luciferase enzyme is transferred to an acceptor fluorescent protein, primarily using Renilla luciferase (Rluc) as donor and yellow fluorescent protein (YFP) as acceptor. Recently a brighter, more stable, luciferase, NanoLuc (Nluc), together with an improved coelenterazine analogue termed furimazine was engineered [17] and has been successfully used as a BRET donor coupled with the fluorescent protein Venus [18]. Historically, a major drawback for using luciferase or BRET fusions has been the size of the fusion tag (Fluc 61 kDa, Rluc 36 kDa, Nluc 19 kDa, YFP/Venus 26 kDa), which often perturbs interaction dynamics or induces conformational changes in complexes. In 2015, Dixon and coworkers reported a split NanoLuc system, termed NanoBiT®, which was created by splitting NanoLuc into two protein epitope tags, SmBiT (SB) with 11 amino acids and LgBiT (LB) with 157 amino acids. The fragments were optimized by mutation to exhibit low intrinsic affinity, allowing the interaction of higher affinity target proteins ($K_D < 10\ \mu M$) to bring the split luciferase together [19]. Reconstituted NanoBiT® luciferase activity is linear in cells, quantifiable, and stable for up to 2 hours, and the SB-LB association is reversible. To date, three reports have used NanoBiT® luminescence for PPI analysis [3, 20, 21]. The minimal SB tag fused to NOX4 proved to be highly enabling for monitoring NOX4/p22^{phox} heterodimerization, catalytic activity, and intracellular localization of the oxidase without perturbing naïve conformation and provides the first assay system to quantitatively determine interaction surfaces and hot spot residues of a membrane-bound NADPH oxidase heterodimer (Fig. 1) [3]. This NanoBiT® assay can serve as a high-throughput screening platform to identify selective inhibitors of NOX4, a therapeutic target in several pathologies.

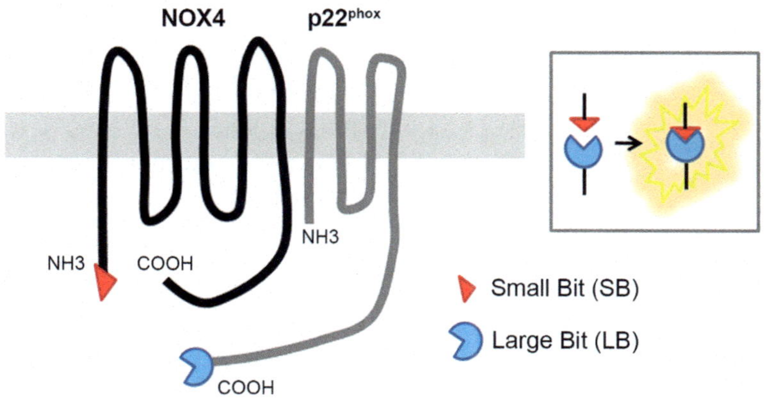

Fig. 1 Schematic of the NanoBiT® protein–protein interaction reporter system using a functional interacting SB-NOX4/ p22-LB pair. Interaction between SB and LB (modified Nluc) is induced by heterodimerization of the NOX4/p22^{phox} complex, producing a glow-type luminescence signal in the presence of furimazine (Nano-Glo®). In the absence of heterodimerization or when placement of the SB-LB tags hinders NanoLuc enzyme formation (incorrect distance, opposing sides of complex), luminescence will not be generated

2 Materials

1. Mammalian cells deficient in endogenous NOX and $p22^{phox}$ proteins, for example, NCI-H661 cancer cells or other cell types, such as HEK293 cells modified with CRISPR/Cas 9 to generate NOX4 and $p22^{phox}$ deletions.
2. RPMI medium supplemented with 10% fetal bovine serum (FBS).
3. Opti-MEM medium (reduced serum medium) with and without phenol red.
4. Trypsin/EDTA.
5. 10× HVA solution: 1 mM homovanillic acid (HVA), 40 U/mL horseradish peroxidase (HRP, Type VI-A), 1× phosphate buffered saline with Ca^{2+} and Mg^{2+} ($PBS^{+/+}$).
6. H_2O_2 standards prepared using H_2O_2 (30% w/v solution) and 1× phosphate buffered saline with Ca^{2+} and Mg^{2+} ($PBS^{+/+}$).
7. HVA stop solution: 100 mM glycine, 100 mM $NaHCO_3$, 25 mM ethylenediaminetetraacetic acid (EDTA), pH 12.0, adjust with 1 M NaOH.
8. Human *NOX4* (NM_016931.3; amino acids 1-578, wild type and various mutations) and *CYBA* ($p22^{phox}$) (NM_000101.3; amino acids 1-195 or C-terminal deletions, and various mutations) DNA.
9. *NOX4* and *CYBA* cDNAs fused with small bit (SB) or large bit (LB) additions/insertions and linkers of various sizes and compositions cloned into pcDNA3.1 (CMV promoter) [3] or into pBiT (HSV-TK promoter, includes N- or C-terminal SB or LB with predesigned linkers, Promega).
10. Lipid-based transfection reagent for transient transfection such as Dreamfect Gold or Lipofectamine 3000.
11. 1× RIPA lysis buffer: 50 mM Tris–HCl, pH 7.5, 150 mM NaCl, 1% (w/v) sodium deoxycholate, 1% (w/v) NP-40, 0.1% (w/v) sodium dodecyl sulfate (SDS).
12. Bicinchoninic acid (BCA) assay.
13. Nano-Glo® Live Cell Assay System: Nano-Glo® Live Cell Substrate and Nano-Glo® LCS Dilution Buffer (furimazine substrate, Promega).
14. Luminometer.
15. Fluorimeter.

3 Methods

3.1 Cell Culture

1. Culture cells (e.g., NCI-H661) in RPMI medium supplemented with 10% FBS and incubate at 37 °C and 5% CO_2.
2. Passage cells every 3–4 days by washing with 2 mL of Trypsin/EDTA, followed by detaching with 1 mL Trypsin/EDTA at room temperature and splitting at a 1:3 to 1:5 ratio.

3.2 Transfection of NCI-H661 Cells for Measurement of H_2O_2 Production

1. Seed NCI-H661 cells at a confluency of 5×10^4 cells per well in RPMI medium containing 10% FBS in a sterile 12-well plate and allow to attach for 24 h at 37 °C and 5% CO_2.
2. Remove culture medium and replace with 1 mL of fresh RPMI containing 10% FBS or with Opti-MEM, depending on the transfection reagent used (*see* **Note 1**).
3. Transfect cells with NOX4 SB fusion cDNA and p22^{phox} LB fusion cDNA. Use 2:1 ratio of NOX4 to p22^{phox} cDNA (e.g., 250 ng NOX4 and 125 ng p22^{phox} per well for a 12-well plate) (optimal pairs for NOX4 listed in [3]). Use sterile H_2O to dilute the DNA to 100 ng/μL. Control transfections include mock (empty vector), p22^{phox} WT alone, NOX4 WT + p22^{phox} WT, and if available a negative SB-NOX4 or p22-LB control such as catalytically inactive NOX4 or complex disrupting mutations (*see* **Note 2**).
4. Perform the transient transfections as per the specific instructions for the transfection reagent used. Use Opti-MEM medium to dilute the DNA/lipid reagent mix. Mix carefully and add to medium above cells (*see* **Note 3**).
5. Incubate the transfected cells at 37 °C and 5% CO_2 for 48 h. Replace medium after 24 h with fresh medium.

3.3 HVA Assay for Determination of Extracellular H_2O_2 Production

The HVA assay is based on the oxidation of homovanillic acid (HVA) to its fluorescent dimer in the presence of H_2O_2 and horseradish peroxidase. The formation of pseudo-stable fluorescent dimers is proportional to the total H_2O_2 produced [22, 23].

1. Following transfection, remove the culture medium and wash the cells with pre-warmed 1× PBS$^{+/+}$.
2. Prepare the 10× HVA solution (protect from light) and add to the cells together with 1× PBS$^{+/+}$ to reach a final concentration of 1× HVA solution (final volume depends on well size, usually 500–1000 μL).
3. Cover the plates with aluminum foil to protect the reaction from light, and incubate at 37 °C for 1 h (*see* **Note 4**).
4. Prepare the H_2O_2 standards (0–10 μM, see below) by adding 1 mM H_2O_2 to 1× PBS$^{+/+}$ and 10× HVA solution in a 1.5 mL

Table 1
Pipetting scheme for HVA standard curve

μM H_2O_2	Volume of 1 mM H_2O_2 (μL) to add	Volume of 1× PBS$^{+/+}$ (μL) to add	Volume of 10× HVA solution (μL) to add
0	0	450	50
1	0.5	450	50
2	1	450	50
4	2	450	50
6	3	450	50
8	4	450	50
10	5	450	50

Eppendorf tube, cover with aluminum foil to protect the reaction from light, and incubate at 37 °C for 1 h.

5. Following 1 h of incubation, add 75 μL of HVA stop solution to 500 μL of the PBS/HVA supernatant recovered from each well and mix by inversion (*see* **Note 5**). Similarly, add 75 μL of HVA stop solution to each tube prepared for the standard curve and mix by inversion (Table 1). The stopped reaction is stable for 2 h in the dark.
6. Remove 100 μL from each tube (samples and standards) and immediately dispense into a 96-well flat-bottomed white plate.
7. Read sample fluorescence which correlates with extracellular H_2O_2 production on a fluorimeter (320 nm excitation, 420 nm emission wavelength, slit width = 3).
8. Use the H_2O_2 standard curve to convert fluorescence readings into nmol H_2O_2. The nmol H_2O_2 calculations should be normalized to the amount of protein in every sample tested. Final concentrations are shown as nmol H_2O_2/h/mg protein (calculate mg of protein by obtaining the μg/μL concentration of protein in each well of the plate and multiplying this by the volume of RIPA buffer used to detach cells, e.g., 100 μL). Alternatively trypsinize and count cells after the H_2O_2 measurements and standardize by cell number (*see* **Note 6**).

3.4 Transfection of NCI-H661 Cells for Measurement of Protein–Protein Interaction

1. Seed NCI-H661 cells at a confluency of 7.5×10^3 cells per well in RPMI medium containing 10% FBS in a sterile 96-well white flat-bottomed plate, and allow to attach for 24 h at 37 °C and 5% CO_2 (*see* **Note 7**).
2. Remove the culture medium and replace with 200 μL of fresh RPMI containing 10% FBS.

3. Use sterile H_2O to dilute the cDNA to 1 ng/μL. Transfect cells with 10 ng p22^{phox} SB or LB fusion cDNA, 10 ng NOX4 SB or LB fusion cDNA, and 147 ng empty vector (e.g., pBluescript cDNA) per well. Control transfections include mock (empty vector), a negative control, and a positive control as in Subheading 3.2, **step 3**, or [3] or commercially available (NanoBiT™ PPI starter system) (*see* **Notes 7–9**).
4. Perform the transient transfections as per the specific instructions for the transfection reagent used. Use Opti-MEM medium to dilute the DNA/lipid reagent mix. Depending on the transfection reagent used, the DNA/lipid mix can be added to cells in complete medium or in Opti-MEM. Replace medium after 4–6 h (Opti-MEM) or after 16–24 h (complete medium) into fresh complete medium without wash step.
5. Incubate the transfected cells at 37°C and 5% CO_2 for 48 h.

3.5 Nano-Glo® Live Cell Assay to Quantify NOX4/p22^{phox} Heterodimerization

The Nano-Glo® Live Cell Assay uses a single-addition non-lytic detection reagent to measure the luminescence from living cells that express either NanoBiT® or NanoLuc®.

1. Following the transfection period, remove the culture medium, and replace with 100 μL of Opti-MEM medium without phenol red per well. Incubate the cells for 1 h at 37 °C and 5% CO_2.
2. Prepare the Nano-Glo® Live Cell Reagent by adding the Nano-Glo® Live Cell Substrate to the Nano-Glo® LCS Dilution Buffer to give a 5× stock (*see* **Note 10**).
3. Set the inside temperature of the luminometer to 37 °C.
4. Add 25 μL of the 5× Nano-Glo® Live Cell Reagent to each well, and gently mix by hand or using an orbital shaker (*see* **Note 11**).
5. Measure the luminescence immediately following addition of the Nano-Glo® Live Cell Reagent (peak emission 460 nm) (*see* **Notes 12–15**). For comparison of SB/LB pairs, take the luminescence reading at a predetermined time point of maximal amplitude (e.g., 15 min) for plotting as bar graphs (in RLU or % of control, taking the optimal pairing or a solvent control as 100% value) (Fig. 2).

4 Notes

1. When adding fresh medium or reagents to cells, add gently to the side of the plate/well so as to avoid detaching any cells from the surface of the plate or well.

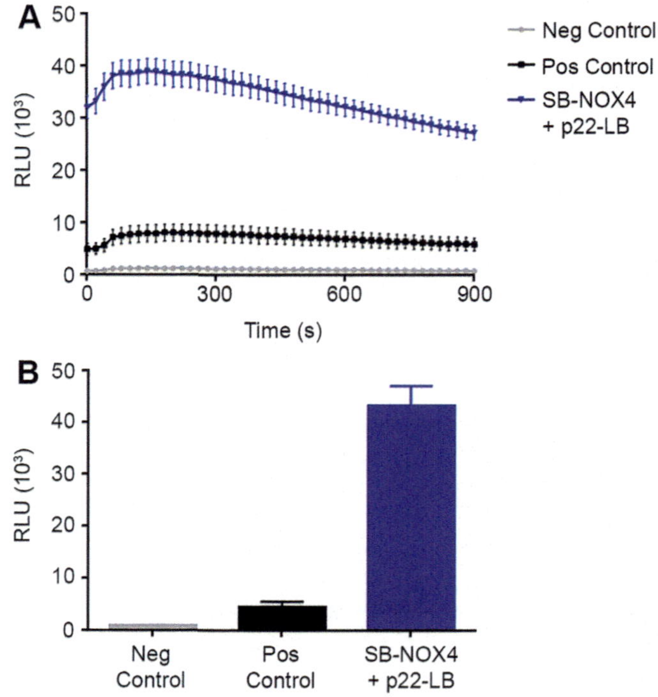

Fig. 2 Representative luminescence curves over time for the SB-NOX4/p22-LB interaction. (**A**) Luminescence signal obtained by the interaction of SB-NOX4 and p22-LB (blue) or control pairs (gray, black) monitored over a period of 15 min. Note that the used positive control pair (PRKACA/PRKAR2A) was much less effective in generating luminescence than the NOX4/p22^{phox} complex. Negative control was HaloTag®-SB. (**B**) Bar graph representing the peak luminescence signal at 2–3 min

2. Transfection of cells should be performed (at least) in triplicate for each combination of NOX4 and p22^{phox} cDNA used. A master mix containing a triplicate amount of DNA, lipid transfection agent, and Opti-MEM medium should be prepared for each transfection, and 100 μL of the mix should be added to each well of a 12-well plate.
3. A method of transfection that allows consistently high levels of transfection without cell toxicity should be used. For NCI-H661 cells, lipid-based transfection reagents are optimal but need to be tested for their efficiency. For cell types that detach easily, polylysine coating and seeding 36–48 h before transfection is recommended.
4. Both the HVA solution and horseradish peroxidase (HRP) used in the HVA assay are light-sensitive. All reagents should be prepared in the dark and stored in a light-protected container. Similarly, during the 1h incubation of cells with the HVA/HRP reagent mix, plates should be wrapped in foil before removing from the hood and during the incubation

period. Once the HVA stop solution is added to the cells, the assay is no longer light-sensitive.

5. Check in the microscope that cells have not lifted up, in particular when using pharmacological compounds such as NOX inhibitors. Minor quantities of floating cells can be removed from the supernatant by a short spin (1500 rpm) in an Eppendorff tube.
6. Analysis should include immunoblots for protein expression and if feasible localization studies of the SB/LB pairs. To perform immunoblots, wash the cells after the HVA assay has been performed in 1× PBS, and detach from the plates by scraping the cells into ice-cold RIPA buffer containing protease and phosphatase inhibitors. Protein samples should be prepared without β-mercaptoethanol. Rather, Bond-Breaker TCEP solution can be added to the protein samples as a replacement for β-mercaptoethanol. Samples should not be boiled, but instead used immediately or stored overnight at 4 °C, and brought to room temperature before loading onto the SDS gel.
7. The amount of transfected cDNAs for the NanoBiT® assay was adjusted to monitor the concentration-dependent 1:1 interaction of dimerization partners. The expression levels of NOX4 and p22^{phox} are therefore too low for determining differences in H_2O_2 production and protein expression, in particular for mutant analysis, necessitating an independent second transfection followed by HVA assay. For better reproducibility and workflow, it may be beneficial to generate inducible cell lines or to introduce the SB and LB fusion tags by CRISPR/Cas 9 technology into endogenous NOX4 and p22^{phox}.
8. The bioluminescence of firefly luciferase (FLuc) expressed in mammalian cells is sensitive to extracellular H_2O_2 due to an unidentified mechanism [24]. While Renilla luciferase (RLuc8) seemed less affected, H_2O_2 concentrations were not measured, impeding comparisons between luciferases. The NOX4-p22^{phox} complex overexpressed in NCI-H661 cells generated up to 60 nmol H_2O_2/h/mg protein. H_2O_2 levels in this range did not affect NanoBiT® luminescence [3].
9. For 96-well transfection formats, the DNA concentrations are very low. Prepare a mastermix for each transfection condition (minimum triplicate wells) and add a small volume (e.g., 50 μL) of Opti-MEM to each mastermix tube. Due to the low quantity of DNA being transfected (10 ng SB and 10 ng LB), empty vector DNA (e.g., pBluescript-SK$^+$) should be added to each mastermix to a give final quantity of DNA of 0.5 μg per triplicate. This allows for the accurate addition of lipid transfection reagent as per reagent protocols. Divide the volume of

DNA/lipid/Opti-MEM transfection mix equally among the number of wells to be transfected.

10. Positive and negative controls are provided by Promega; however, the luminescence obtained by the functional SB-NOX4/p22-LB pair far exceeded that obtained by the Promega's constitutive positive control (PRKACA/PRKAR2A) in the same cell type. An experimental KRAS-LB/cRAF G12C-SB pair showed a high luminescence signal comparable to the SB-NOX4/p22-LB signal [3].
11. The Nano-Glo® Live Cell Reagent should be made fresh for each experiment. During the 1 h incubation period, gently equilibrate the Nano-Glo® LCS Dilution Buffer to room temperature (≤25 °C). When removing the Nano-Glo® Live Cell Substrate from storage at −20 °C, mix well and keep on ice. Add the Nano-Glo® Live Cell Substrate to the Nano-Glo® LCS Dilution Buffer immediately before starting the assay. The reconstituted reagent will lose 10% activity in 8 h RT or in 2 days at 4 °C.
12. For the Nano-Glo® Live Cell Assay, minimize the edge effect in a 96-well format by filling empty wells adjacent to the sample wells and the inter-well space with 200 μL 1× PBS following transfection of the cells and incubation at 37 °C and 5% CO_2.
13. The luminescence can be recorded at a single time point or continuously for up to 2 h. In the experiments described here, luminescence was recorded every 20 s for 15 min.
14. For compound screening or inhibitor experiments, assess unspecific interactions of solvents or compounds using a cell-free NanoLuc® control (10 nM 6×His-NanoLuc enzyme (903479, Promega) diluted in PBS). DMSO as solvent was well tolerated up to 2.7–3% (v/v) final concentration [3]; other solvents (ethanol, methanol) depressed the luminescence signal by 10–15% at a final concentration of 3% (v/v) (Technical Manual Nano-Glo® Luciferase Assay System, Promega).
15. Purified NanoLuc enzyme was screened against a LOPAC^{1280} library (10 μM) [17], showing >20% enzyme inhibition by 0.5% of the library compounds. Higher false-positive hit rates have been observed in the SB-NOX4/p22-LB NanoBiT® assay (unpublished observations), emphasizing the need for robust counter-screens.

Acknowledgments

This work was supported by Science Foundation Ireland (UGK) and the MolCellBiol Programme (Programme for Research in Third-Level Institutions, co-funded under the EU Regional

Development Fund) (SON). We thank M. Mathis and R. Bouhelal of the Novartis Institutes for Biomedical Research, Switzerland, and Promega Corporation, Wisconsin, USA, for the collaborative effort in establishing this assay and for providing reagents.

References

1. DeLeo FR, Burritt JB, Yu L, Jesaitis AJ, Dinauer MC, Nauseef WM (2000) Processing and maturation of flavocytochrome b558 include incorporation of heme as a prerequisite for heterodimer assembly. J Biol Chem 275 (18):13,986–13,993
2. Biberstine-Kinkade KJ, DeLeo FR, Epstein RI, LeRoy BA, Nauseef WM, Dinauer MC (2001) Heme-ligating histidines in flavocytochrome b (558): identification of specific histidines in gp91(phox). J Biol Chem 276 (33):31,105–31,112
3. O'Neill S, Mathis M, Kovacic L, Zhang S, Reinhardt J, Scholz D, Schopfer U, Bouhelal R, Knaus UG (2018) Quantitative interaction analysis permits molecular insights into functional NOX4 NADPH oxidase heterodimer assembly. J Biol Chem 293(23):8750–8760. https://doi.org/10.1074/jbc.RA117.001045
4. Groemping Y, Rittinger K (2005) Activation and assembly of the NADPH oxidase: a structural perspective. Biochem J 386. (Pt 3:401–416. https://doi.org/10.1042/BJ20041835
5. Sumimoto H (2008) Structure, regulation and evolution of Nox-family NADPH oxidases that produce reactive oxygen species. FEBS J 275 (13):3249–3277. https://doi.org/10.1111/j.1742-4658.2008.06488.x
6. Luxen S, Noack D, Frausto M, Davanture S, Torbett BE, Knaus UG (2009) Heterodimerization controls localization of Duox-DuoxA NADPH oxidases in airway cells. J Cell Sci 122. (Pt 8:1238–1247. https://doi.org/10.1242/jcs.044123
7. Ameziane-El-Hassani R, Morand S, Boucher JL, Frapart YM, Apostolou D, Agnandji D, Gnidehou S, Ohayon R, Noel-Hudson MS, Francon J, Lalaoui K, Virion A, Dupuy C (2005) Dual oxidase-2 has an intrinsic Ca2+-dependent H2O2-generating activity. J Biol Chem 280(34):30,046–30,054
8. Ueyama T, Sakuma M, Ninoyu Y, Hamada T, Dupuy C, Geiszt M, Leto TL, Saito N (2015) The extracellular A-loop of dual oxidases affects the specificity of reactive oxygen species release. J Biol Chem 290(10):6495–6506. https://doi.org/10.1074/jbc.M114.592717
9. Takac I, Schroder K, Zhang L, Lardy B, Anilkumar N, Lambeth JD, Shah AM, Morel F, Brandes RP (2011) The E-loop is involved in hydrogen peroxide formation by the NADPH oxidase Nox4. J Biol Chem 286 (15):13,304–13,313. https://doi.org/10.1074/jbc.M110.192138
10. von Lohneysen K, Noack D, Jesaitis AJ, Dinauer MC, Knaus UG (2008) Mutational analysis reveals distinct features of the Nox4-p22 phox complex. J Biol Chem 283 (50):35,273–35,282. https://doi.org/10.1074/jbc.M804200200
11. Stynen B, Tournu H, Tavernier J, Van Dijck P (2012) Diversity in genetic in vivo methods for protein-protein interaction studies: from the yeast two-hybrid system to the mammalian split-luciferase system. Microbiol Mol Biol Rev 76(2):331–382. https://doi.org/10.1128/MMBR.05021-11
12. Xing S, Wallmeroth N, Berendzen KW, Grefen C (2016) Techniques for the analysis of protein-protein interactions in vivo. Plant Physiol 171(2):727–758. https://doi.org/10.1104/pp.16.00470
13. Ambasta RK, Kumar P, Griendling KK, Schmidt HH, Busse R, Brandes RP (2004) Direct interaction of the novel Nox proteins with p22phox is required for the formation of a functionally active NADPH oxidase. J Biol Chem 279(44):45,935–45,941
14. Prior KK, Wittig I, Leisegang MS, Groenendyk J, Weissmann N, Michalak M, Jansen-Durr P, Shah AM, Brandes RP (2016) The endoplasmic reticulum chaperone calnexin is a NADPH oxidase NOX4 interacting protein. J Biol Chem 291(13):7045–7059. https://doi.org/10.1074/jbc.M115.710772
15. Al Ghouleh I, Meijles DN, Mutchler S, Zhang Q, Sahoo S, Gorelova A, Henrich Amaral J, Rodriguez AI, Mamonova T, Song GJ, Bisello A, Friedman PA, Cifuentes-Pagano ME, Pagano PJ (2016) Binding of EBP50 to Nox organizing subunit p47phox is pivotal to cellular reactive species generation and altered vascular phenotype. Proc Natl Acad Sci U S A 113(36):E5308–E5317. https://doi.org/10.1073/pnas.1514161113

16. Remy I, Michnick SW (2006) A highly sensitive protein-protein interaction assay based on Gaussia luciferase. Nat Methods 3 (12):977–979. https://doi.org/10.1038/nmeth979
17. Hall MP, Unch J, Binkowski BF, Valley MP, Butler BL, Wood MG, Otto P, Zimmerman K, Vidugiris G, Machleidt T, Robers MB, Benink HA, Eggers CT, Slater MR, Meisenheimer PL, Klaubert DH, Fan F, Encell LP, Wood KV (2012) Engineered luciferase reporter from a deep sea shrimp utilizing a novel imidazopyrazinone substrate. ACS Chem Biol 7(11):1848–1857. https://doi.org/10.1021/cb3002478
18. Mo XL, Fu H (2016) BRET: nanoluc-based bioluminescence resonance energy transfer platform to monitor protein-protein interactions in live cells. Methods Mol Biol 1439:263–271. https://doi.org/10.1007/978-1-4939-3673-1_17
19. Dixon AS, Schwinn MK, Hall MP, Zimmerman K, Otto P, Lubben TH, Butler BL, Binkowski BF, Machleidt T, Kirkland TA, Wood MG, Eggers CT, Encell LP, Wood KV (2016) NanoLuc complementation reporter optimized for accurate measurement of protein interactions in cells. ACS Chem Biol 11 (2):400–408. https://doi.org/10.1021/acschembio.5b00753
20. Oh-Hashi K, Hirata Y, Kiuchi K (2016) SOD1 dimerization monitoring using a novel split NanoLuc, NanoBit. Cell Biochem Funct 34 (7):497–504. https://doi.org/10.1002/cbf.3222
21. Duellman SJ, Machleidt T, Cali JJ, Vidugiriene J (2017) Cell-based, bioluminescent assay for monitoring the interaction between PCSK9 and the LDL receptor. J Lipid Res 58 (8):1722–1729. https://doi.org/10.1194/jlr.D074658
22. Ruch W, Cooper PH, Baggiolini M (1983) Assay of H2O2 production by macrophages and neutrophils with homovanillic acid and horse-radish peroxidase. J Immunol Methods 63(3):347–357
23. Martyn KD, Frederick LM, von Loehneysen K, Dinauer MC, Knaus UG (2006) Functional analysis of Nox4 reveals unique characteristics compared to other NADPH oxidases. Cell Signal 18(1):69–82. https://doi.org/10.1016/j.cellsig.2005.03.023
24. Czupryna J, Tsourkas A (2011) Firefly luciferase and RLuc8 exhibit differential sensitivity to oxidative stress in apoptotic cells. PLoS One 6 (5):e20073. https://doi.org/10.1371/journal.pone.0020073

Part V

Function of NADPH Oxidase

Chapter 27

Isolation of Redox-Active Endosomes (Redoxosomes) and Assessment of NOX Activity

Weam S. Shahin and John F. Engelhardt

Abstract

Reactive oxygen species (ROS) convey signals essential for proliferation, maintenance, and senescence of a growing list of cell types. Compartmentalization of these signals is integral to cell viability as well as the signaling pathways ROS direct. Redox-active endosomes (redoxosomes) are formed downstream of several ligand-activated receptors. NADPH oxidase (NOX) is a main component of redoxosomes, which recruits multiple proteins (Rac1, NOX2, $p67^{phox}$, SOD1). Isolation of redoxosomes and evaluation of how superoxide ($O_2^{\cdot -}$) production directs receptor signaling at the level of the endosome have enabled a better understanding of biologic processes controlled by ROS. In this chapter, we will first review the major signaling pathways that utilize redoxosomes and components that control its redox-dependent functions. We will then outline biochemical and biophysical methods for the isolation and characterization of redoxosome properties.

Key words ROS, Redoxosomes, Iodixanol, Immuno-affinity isolation, Lucigenin, EPR, NOX, Rac1, TRAF

1 Introduction

Endosomal NOX2-dependent ROS perform a critical role in signal transmission following pro-inflammatory stimuli such as cytokines (TNFα or IL-1ß) [1] or growth factors (lysophosphatidic acid, LPA) [2] binding to their receptors. This binding is followed by endocytosis and activation of NADPH oxidase (NOX2 and/or NOX1) in the endosome [3, 4]. Activation of NOX2 is required for signal transmission and recruitment of downstream effectors. In this context, ROS generated by redoxosomes act as second messengers to facilitate the redox-dependent activation of the IL-1β receptor (IL-1R) and TNFα receptor (TNFR1).

Redoxosome formation after IL-1β or TNFα stimulation starts with endocytosis of the ligand-activated receptor complex. Endocytosis is required for endosomal superoxide production by NOX and at least in some cell lines is facilitated by lipid rafts that

Ulla G. Knaus and Thomas L. Leto (eds.), *NADPH Oxidases: Methods and Protocols*, Methods in Molecular Biology, vol. 1982, https://doi.org/10.1007/978-1-4939-9424-3_27, © Springer Science+Business Media, LLC, part of Springer Nature 2019

sequester inactive NOX complexes at the plasma membrane [1]. Rac1 is required for recruitment of NOX2 from the cell membrane to the nascent endosome [3], and Vav1, a Rac1 guanine exchange factor and activator of NOX2, appears to be integral to this process [5]. Different cell types can differentially utilize NOX1 or NOX2 for endosomal ROS production after IL-1β or TNFα stimulation [3, 6, 7], suggesting that there is some level of redundancy in these pathways. Activation of NOX1 and NOX2 requires the recruitment of cytosolic coactivator subunits of the NOX complex prior to superoxide production. NOXO1, NOXA1, and Rac bind to the NOX1/$p22^{phox}$ heterodimer to activate superoxide production [8]. The cytosolic proteins Rac, $p67^{phox}$, $p47^{phox}$, and $p40^{phox}$ are recruited to activate superoxide production by the NOX2/$p22^{phox}$ complex [9]. Superoxide ($O_2^{\cdot -}$) produced by endosomal-activated NOX leads to the production of H_2O_2—through either spontaneous dismutation or superoxide dismutase 1 (SOD1)-facilitated dismutation—which acts as an intermediate signal to drive local redox events at the surface of the endosome required for effector docking of subunits in the receptor and downstream NFκB activation [3, 4].

Superoxide is produced within the lumen of the redoxosome. This is supported by the fact that superoxide dismutase 1 (SOD1) quenches superoxide when loaded into endosomes at the time of IL-1β and TNFα stimulation and by inhibition of effector recruitment to redoxosomal IL-1R and TNFR1 and downstream of NOX-dependent activation of NFκB [3, 4, 10]. NADPH-dependent $O_2^{\cdot -}$ can be rapidly transported across the redoxosomal membrane through a DIDS-sensitive chloride channel [10]. Importantly, $O_2^{\cdot -}$ is not permeable to membranes, but if spontaneous dismutation occurs within the redoxosomal lumen, the product H_2O_2 can freely diffuse across the membrane. This feature of redoxosomes allows for localized high concentrations of H_2O_2 to facilitate redox-signaling events without significantly changing the overall redox status of the cell.

Recruitment of cytoplasmic SOD1 to IL-1β- and TNF-α-stimulated redoxosomes support the notion that dismutation of superoxide on the endosomal surface likely contributes to the redox signal event through the production of local H_2O_2 gradients [3, 4]. SOD1 is also essential for IL-1β-mediated activation of NFκB [10]. NOX activity generates voltage, osmotic, and pH gradients across redoxosomal membranes. Since $O_2^{\cdot -}$ is relatively impermeable to cellular membranes [11, 12], the presence of transmembrane channel(s) to control these gradients could be integral to redoxosome functions [10]. SOD1 regulates redoxosomal ROS production through direct association with Rac1 [13]. In this context, SOD1 binding to Rac1 inhibits GTP hydrolysis and promotes activation of NOX. But at higher concentrations of H_2O_2, Rac1 is oxidized and SOD1 is dissociated, leading to

Rac1-GTP hydrolysis and inactivation of NOX. Thus, SOD1/Rac1 interactions directly with the NOX complex act as a redox sensor that regulates redoxosomal NOX activity [13].

The coordinated recruitment of NOX only into endosomes with ligand-activated receptors spatially restricts $O_2^{\cdot -}$ and H_2O_2 production to the sites where redox signaling is required. In this context H_2O_2 production by redoxosomes mediates the recruitment of TRAFs to IL-1R and TNFR1 and is required for downstream signal activation of NFκB. TRAF2 and TRAF6 recruitment to TNFR1 and IL-1R1, respectively, is essential for the formation of an active IKK kinase (IKKK) complex on these receptors. Active IKKK facilitates phosphorylation and activation of the IKK complex. Activation of the IKK complex then leads to phosphorylation of IκB (inhibitor of NFκB) and translocation of NFκB to the nucleus, where it activates transcription of pro-inflammatory pathways [1].

Several methods have been used to dissect the biology of redoxosomes and redox-dependent events required for signaling. Iodixanol density gradient isolation is based on the size and density of different organelles and protein complexes. Larger size and heavier vesicles will travel faster through the gradient and band at a higher density. Using this approach at equilibrium, groups of vesicles with varying densities can be isolated and separate from other major subcellular compartments. This technique is successful at separating the redoxosome from the lower-density plasma membrane and the higher-density peroxisomes and mitochondrial compartments [3]. We have also combined density gradients with immuno-affinity isolation to further refine redoxosome isolation. Given that Rac1, a coactivator of NOX1 and NOX2, is recruited to receptor-activated redoxosomes [3], infection of cells with recombinant adenovirus expressing HA-tagged Rac1 has allowed for immuno-isolation of redoxosomes using anti-HA-tag antibody-coated beads following TNFα stimulation.

Chemiluminescent probes such as lucigenin [3, 4, 7, 10] and luminol [10] have been successfully used to investigate superoxide production in isolated redoxosomes. Lucigenin, a membrane-permeable compound, is widely used for detection of superoxide [14]. Detection with lucigenin is based upon its ability to emit photons upon reaction with superoxide. NADPH is supplied to the exterior of the endosomal preparations to initiate the reaction and will be utilized by the NOX complex that resides on the cytoplasmic side of the redoxosome. The lucigenin reaction is a three-step process. The third step results in an emission of a photon with a wavelength of ~470 nm. This release can be detected using a luminometer [14].

Since there is some controversy about the specificity of lucigenin, and more so luminol, for superoxide, NADPH-dependent $O_2^{\cdot -}$ production in isolated endosomes from cytokine-stimulated cells has been unambiguously shown using electron paramagnetic resonance (EPR) [3]. EPR is based on the paramagnetic nature of radicals (i.e., all radicals possess unpaired electron(s) that are attracted to magnetic fields). Upon application of a strong magnetic field to radicals, a small percentage of unpaired electrons will align themselves in the magnetic field. Application of microwave range electromagnetic energy can induce shifts of the unpaired electrons into special spin states. Distinctive molecules can be identified by the unique absorption signatures from the aligned electrons.

At physiologic pH, the half-life of a superoxide radical is too short to directly detect its signature by EPR. To avoid this problem, spin traps are used. Spin traps are specific molecules designed to swiftly react with transient radicals to form a new radical adduct that is more stable. Transient radicals can be identified by using spin traps with unique paramagnetic signatures [15]. Therefore, EPR has the advantage of being very specific for various types of free radicals. 5,5-Dimethylpyrroline-N-oxide (DMPO) is a commonly utilized spin trap in the detection of superoxide radicals.

2 Materials

2.1 Cell Culture and Homogenization Buffer

1. MCF-7 cells, a human mammary epithelial tumor cell line obtained from ATCC.
2. Cell culture media: MCF-7 cells were grown in minimum essential medium (MEM) with Eagle's salts and L-glutamine, 1% minimum essential medium nonessential amino acids, 10% fetal bovine serum (FBS), 1% penicillin/streptomycin, and 30 nM sodium selenite. Cells were maintained at 37 °C in 5% CO_2/95%O_2.
3. Tissue culture incubator.
4. Serum-free medium: minimum essential medium with Eagle's salts and L-glutamine, 1% minimum essential medium nonessential amino acids, 1% penicillin/streptomycin, and 30 nM sodium selenite.
5. Phosphate-buffered saline (PBS): 8.1 mM Na_2HPO_4, 1.47 mM KH_2PO_4, 138 mM NaCl, 2.67 mM KCl, pH 7.4.
6. Tumor necrosis factor alpha (TNFα).
7. Interleukin-1ß (IL-1ß).
8. Homogenization buffer: 0.25 M sucrose, 10 mM triethanolamine, 1 mM EDTA, 1 mM phenylmethylsulfonyl fluoride, and 100 mg/mL aprotinin.
9. Duall tissue grinder.

2.2 Viral Vectors

Recombinant adenovirus carrying HA-tagged Rac1 (Ad.HA-Rac1) or control vector GFP (Ad.CMV-GFP).

2.3 Density Gradient Preparation

Stock solution of iodixanol 60 % (OptiPrep, Axis-Shield). Prepare 24 and 20% iodixanol in homogenization buffer.

2.4 Immuno-affinity Isolation Reagents

1. Dynabeads M-500 subcellular (Invitrogen).
2. Dynal MPC Magnet (Invitrogen).
3. Secondary anti-rat antibody.
4. Primary rat anti-HA-tag antibody.
5. 0.1 M borate buffer, pH 9.5: dissolve 618.3 mg boric acid in 90 mL distilled water, adjust pH to 9.5 with NaOH, and then adjust volume to 100 mL.
6. 0.1% bovine serum albumin (BSA)-PBS (wt/vol): dissolve 100 mg BSA in 100 mL PBS.
7. 0.2 M Tris–HCl, pH 8.5—BSA: dissolve 2.4228 g of Tris base in 90 mL of distilled water, adjust pH to 8.5 using HCl, and then adjust volume to 100 mL. Then, add 100 mg of BSA.

2.5 Lucigenin Assay

1. Luminometer: Berthold LB9505 Multi-Channel Biolumat (AutoLumat).
2. Freshly prepared 0.5 mM lucigenin (bis-N-methylacridinium nitrate) solution in water, keep on ice in dark.
3. Freshly prepared 2 mM NADPH in distilled water under nitrogen, keep on ice.

2.6 Electron Paramagnetic Resonance (EPR) Spectrometry

1. Bruker electron paramagnetic resonance (EPR) spectrometer.
2. Freshly prepared 2 mM NADPH in distilled water under nitrogen, keep on ice.
3. Freshly prepared 100 mM 5,5-dimethyl-1-pyrroline *N*-oxide (DMPO) and 200 μM diethylenetriaminepentaacetic acid (DTPA) in PBS under nitrogen.

3 Methods

3.1 Cell Culture, Infection, and Stimulation

1. Grow MCF-7 cells to 80% confluence in cell culture medium 150 mm culture plates.
2. Replace medium with half volume of serum-free medium containing adenovirus carrying HA-tagged Rac1 (Ad.HA-Rac1) or Ad.CMV-GFP at multiplicity of infection (MOI) 500 particles/cell.
3. Incubate for 2 hours at 37 °C then add an equal volume of fresh medium containing 20% FBS.

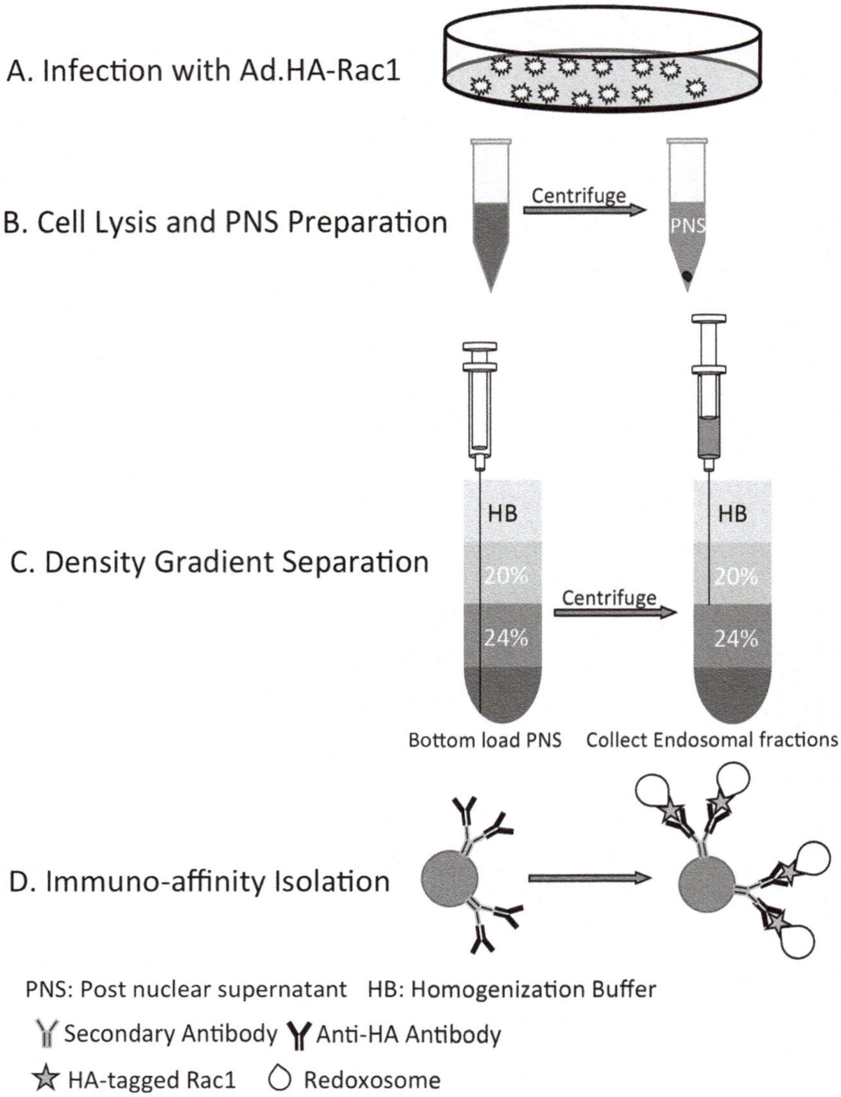

Fig. 1 Isolation of Rac1-containing redoxosomes from endosomal preparations. (**A**) MCF-7 cells are infected with Ad.HA-Rac1 or Ad.CMV-GFP. (**B**) Cells are lysed to liberate endosomes and enriched with the post-nuclear supernatant (PNS). (**C**) The PNS is loaded at the bottom of the iodixanol gradient and centrifuged to separate endosomes. (**D**) Collected endosomal fractions are immediately used for immuno-affinity isolation of redoxosomes using rat anti-HA antibody-coated beads

4. Incubate for another 22 h for a total of 24 h at 37 °C (Fig. 1a).
5. Replace medium with fresh growth medium.
6. 48 h later, stimulate Ad.HA-Rac1-infected MCF-7 cells with TNFα (0.5 ng/mL) or IL-1ß (1 ng/mL) for 20 min, and then isolate vesicular fractions (*see* **Note 1**).

3.2 Cell Lysis and Isolation of Endosomal Fractions

After cell lysis, an iodixanol gradient was used to separate the endosomes from the post-nuclear supernatant [4].

1. Wash cells three times with ice-cold phosphate-buffered saline (PBS). Use about 10 mL/wash (*see* **Note 2**).
2. Add 1 mL of ice-cold PBS to cells after the last wash.
3. Scrap cells into 1.5 mL Eppendorf tubes using a disposable plastic scrapper.
4. Centrifuge cells at 3000 × *g* for 5 min at 4 °C and aspirate the supernatant.
5. Resuspend the cell pellets in 0.5 mL of homogenization buffer.
6. Homogenize cell pellets in a Duall tissue grinder, and then centrifuge at 2000 × *g* at 4 °C for 10 min. Discard the pellet and collect the post-nuclear supernatant (PNS) (Fig. 1b) (*see* **Note 3**).
7. Prepare a two-step gradient of iodixanol in a SW55Ti centrifuge tube by adding 2 mL of 24% iodixanol in homogenization buffer to the bottom of a SW55Ti centrifuge tube. Then slowly add 2 mL of 20% iodixanol in homogenization buffer on top of the previous layer.
8. Combine the PNS with a 60% iodixanol solution to obtain a final concentration of 32% iodixanol and then bottom load under the two-step gradient of 24 and 20% iodixanol in homogenization buffer using a syringe.
9. Fill the tube with homogenization buffer.
10. Centrifuge samples at 88,195 × *g* for 2 h at 4 °C.
11. After centrifugation, a milky band should be visible at every interface between iodixanol gradient steps (Fig. 1c).
12. Collect fractions from the top to the bottom of the centrifuge tube at 4 °C (300 μL per fraction), and use immediately for NADPH oxidase activity and immuno-isolation assays (*see* **Note 4**).

3.3 Immuno-affinity Isolation of Redoxosomes

MCF-7 cells expressing HA-tagged Rac1 were used to isolate redoxosomes, since Rac1 is recruited to the NOX2 complex only in ligand-activated endosomes. Redoxosomes were enriched using HA-tag antibody-coated beads [4].

3.3.1 Coating Dynabeads M-500 Subcellular with the Anti-HA Antibody

1. Make a homogeneous suspension of Dynabeads M-500 in PBS by vortexing for 2 min.
2. Conjugate the secondary antibody (anti-rat antibody) 10 μg per 10^7 beads to Dynabeads M-500. Make sure the final concentration of beads in the coating reaction is 4×10^8 beads/mL in 0.1 M borate buffer, pH 9.5. Mix thoroughly by pipetting up and down (*see* **Note 5**).

3. Incubate the mix for 24 h at 25 °C with slow rocking.
4. Place the beads into the magnet for 5 min and discard supernatant.
5. Wash the beads two times by resuspending the beads in 0.1% BSA-PBS for 5 min at 4 °C, then placed into the magnet for 5 min, and remove the supernatant.
6. Wash the beads once in 0.2 M Tris–HCl, pH 8.5 with 0.1% BSA for 24 h at 4 °C (*see* **Note 6**).
7. Wash the beads once in 0.1% BSA-PBS for 5 min at 4 °C, then placed into the magnet for 5 min, and remove the supernatant.
8. Conjugate the linker coated beads to 4 μg of primary anti-HA antibody per 10^7 beads and mix thoroughly. Make sure to reach a final concentration of 4×10^8 beads/mL in the conjugation buffer (*see* **Note 7**).
9. Incubate overnight at 4 °C with slow rotation.
10. Place into the magnet for 2–3 min and remove the supernatant.
11. Wash four times by resuspending the beads in 0.1% BSA-PBS for 5 min at 4 °C, and then place into the magnet for 2–3 min and remove the supernatant.
12. Resuspend the coated beads in PBS containing 2 mM EDTA, 5% BSA, and protease inhibitors at a concentration of 1.4×10^7 to 1.4×10^8 beads/700 μL and use for isolation of redoxosomes (*see* **Note 8**).

3.3.2 Isolation of Redoxosomes (Fig. 1d)

1. Mix 50 μg of vesicular fractions in 300 μL PBS from Subheading 3.2, **step 12**, with 700 μL of coated beads from Subheading 3.3.1, **step 12** (*see* **Note 8**).
2. Incubate the mixture for 6 h at 4 °C with slow rocking.
3. Place the tube into the magnet and pipette the supernatant into a clean labeled tube.
4. Wash the beads in the same tube three times (15 min each) in PBS containing 2 mM EDTA, 5% BSA, and protease inhibitors.
5. Resuspend beads with HA-enriched endosomes in PBS, and use immediately to measure $O_2^{\cdot -}$ levels.
6. Save wash supernatants for analysis.

3.4 Measuring the Rate of $O_2^{\cdot -}$ Generation Using a Lucigenin-Based System

Superoxide production is quantified using lucigenin. The assay is based upon the fact that lucigenin is a membrane-permeable compound that emits photons upon contact with superoxide. Thus, relative rates of superoxide production can be quantified with or without cytokine stimulation in the presence or absence of NADPH. The level of NADPH-dependent superoxide production in isolated redoxosomes is an indicator of NOX2 activity [3, 4, 10].

1. Turn the luminometer on, set the wavelength at 470 nm, and set up the measurement for five readings/second over the course of 3 min.
2. Add 10 μL of 0.5 mM lucigenin to 5 μg of vesicular proteins in 940 μL PBS in a chemiluminescence (CL) tube (*see* **Note 9**).
3. Incubate the mix in darkness at room temperature for 10 min.
4. Measure the background changes in luminescence without addition of NADPH (*see* **Note 10**).
5. To initiate the reaction, add 50 μL of 2 mM NADPH for a final concentration of 100 μM of NADPH and 5 μM lucigenin (*see* **Note 9**).
6. Transfer the CL tube immediately to the luminometer, and measure the changes in luminescence over the course of 3 min (five readings).
7. Use the slope of the luminescence curve (relative light units [RLU] per minute) ($r > 0.95$) to calculate the rate of $O_2^{\cdot-}$ formation as an index of NADPH oxidase activity (RLU/min μg protein).

3.5 Estimates of $O_2^{\cdot-}$ Producing Capacity by Redox-Active Endosomes Using EPR

A spin trap (DMPO) is used to bind to the transient superoxide radical to generate a more stable adduct. Upon exposure to high magnetic field, a small fraction of the unpaired electrons will align to the magnetic field. This produces a unique absorption signature that can be used to detect specific molecules. EPR is very specific for different types of free radicals. As outlined here, the rate of superoxide production is measured in the presence or absence of NADPH to quantify NOX2 activity in isolated redoxosomes from cytokine-stimulated MCF-7 cells [3].

1. Turn the EPR spectrometer on and let the instrument warm up for 30 min.
2. Set up an EPR acquisition with the following parameters: frequency, 9.78 GHz; center field, 3480 G; sweep rate, 80 G/21 s; power, 20.3 mW; receiver gain, 5.02×10^4; modulation frequency, 100 kHz; time constant, 81.92 ms; modulation amplitude, 1.0 G; sweep time, 20.972 s; number of scans per spectrum, 15; and resolution, 1024 points.
3. Measure an EPR spectrum of an empty EPR tube to ensure that there are no background signals from either the EPR tube or the instrument resonator. In anaerobic conditions (under nitrogen), mix 225 μL of HA-enriched endosomes from control (unstimulated) or TNFα-stimulated MCF-7 cells with 250 μL of PBS containing 100 mM of the spin trap DMPO (the spin trap) and 200 μM DTPA in a total volume of 475 μL.

4. Initiate the reaction by adding 25 μL of 2 mM NADPH to a final concentration of 100 μM NADPH, 50 mM DMPO, and 100 μM DTPA (*see* **Note 9**).
5. Incubate at 37 °C for 10 min.
6. Immediately load each sample to EPR tube, and scan with EPR spectrometer for ~5 min at room temperature.
7. Using DMPO incubated with xanthine and xanthine oxidase (X/XO) and identical EPR parameters, generate standard curve of peak height vs. nanomolar DMPO-OH (*see* **Note 11**).
8. Using the standard curve, quantify EPR spectra by peak-to-peak height of the second (low field) DMPO-OH line.

4 Notes

1. Control cells including non-infected and non-stimulated MCF-7 cells should be included to assure the specificity and set the background levels for the downstream assays.
2. Cell lysis, separation, and collection of endosomal fractions are temperature-sensitive steps that need to be carried out at 4 °C.
3. During the lysis step, care should be taken to break open the cells, but not fragment the cellular membrane. Fragmenting the cellular membrane by lysing of cells too vigorously will lead to plasma membrane vesiculation (i.e., formation of vesicles from the membrane compartment). This will contaminate plasma membrane NOX complexes within the endosome fraction. The optimal lysis procedure should be adjusted for a given cell type by visually inspecting cells during the lysis procedure under the microscope.
4. Collected endosomal fractions should be used immediately to measure NOX activity or for immune-affinity isolation of Rac1-enriched endosomes. Freezing will destroy the enzymatic activity, likely by dissociating the NOX complex components from the membrane of endosomes.
5. The surface of the beads, when activated, will covalently bind to amino and sulfhydryl groups in the antibody. Given that the crosslinking is random, not all antibodies will be active for binding antigen. Thus, it is advised that only the secondary anti-IgG antibody be cross-linked to beads, thus ensuring that the primary antibody (in this case anti-HA) will be properly oriented with its Fab antigen-binding site facing the target organelle (Fig. 1d).
6. The beads are washed in Tris buffer after coating with secondary antibody to block the bead surface. This step prevents nonspecific binding of primary antibodies or the organelles to

the bead surface, making sure that the primary antibody is binding only to the secondary antibody and that subsequently the HA-enriched endosomes are binding only to the primary antibody.

7. The amount of secondary or primary antibody conjugated to the beads can be estimated by measuring the amount of proteins remaining in the supernatant after each coating reaction.
8. Determining the appropriate number of beads and the volume of vesicular fractions for immune-affinity isolation depends on cell type, the efficiency of infection, and the abundance of target redoxosomes in the vesicular fractions. Thus, optimization is required.
9. NADPH, lucigenin, and spin trap solutions are rapidly degrading, light sensitive, and need to be freshly prepared under nitrogen and protected from light.
10. Measurement of background levels of lucigenin-dependent luminescence, in the absence of NADPH, should always be 1000-fold lower than maximally induced values in the presence of NADPH. If this is not the case, there is likely contamination in the endosomal faction (i.e., plasma membrane or mitochondria).
11. Short EPR scan times for experimental samples might result in lower signal-to-noise ratios than previously observed for isolated redoxosomes produced by longer scan times that average a greater number of spectra. The shorter scan times, on the other hand, minimize potential loss of the DMPO-OH adduct due to unknown enzymatic activity present in the samples.

Acknowledgments

This work was supported by NIH grant R24 DK096518 (to J.F. E.).

References

1. Oakley FD, Abbott D, Li Q, Engelhardt JF (2009) Signaling components of redox active endosomes: the redoxosomes. Antioxid Redox Signal 11(6):1313–1333. https://doi.org/10.1089/ARS.2008.2363
2. Klomsiri C, Rogers LC, Soito L, McCauley AK, King SB, Nelson KJ, Poole LB, Daniel LW (2014) Endosomal H2O2 production leads to localized cysteine sulfenic acid formation on proteins during lysophosphatidic acid-mediated cell signaling. Free Radic Biol Med 71:49–60. https://doi.org/10.1016/j.freeradbiomed.2014.03.017
3. Li Q, Harraz MM, Zhou W, Zhang LN, Ding W, Zhang Y, Eggleston T, Yeaman C, Banfi B, Engelhardt JF (2006) Nox2 and Rac1 regulate H2O2-dependent recruitment of TRAF6 to endosomal interleukin-1 receptor complexes. Mol Cell Biol 26(1):140–154. https://doi.org/10.1128/MCB.26.1.140-154.2006

4. Li Q, Spencer NY, Oakley FD, Buettner GR, Engelhardt JF (2009) Endosomal Nox2 facilitates redox-dependent induction of NF-kappaB by TNF-alpha. Antioxid Redox Signal 11(6):1249–1263. https://doi.org/10.1089/ARS.2008.2407
5. Oakley FD, Smith RL, Engelhardt JF (2009) Lipid rafts and caveolin-1 coordinate interleukin-1beta (IL-1beta)-dependent activation of NFkappaB by controlling endocytosis of Nox2 and IL-1beta receptor 1 from the plasma membrane. J Biol Chem 284 (48):33,255–33,264. https://doi.org/10.1074/jbc.M109.042127
6. Miller FJ Jr, Filali M, Huss GJ, Stanic B, Chamseddine A, Barna TJ, Lamb FS (2007) Cytokine activation of nuclear factor kappa B in vascular smooth muscle cells requires signaling endosomes containing Nox1 and ClC-3. Circ Res 101(7):663–671. https://doi.org/10.1161/CIRCRESAHA.107.151076
7. Li Q, Zhang Y, Marden JJ, Banfi B, Engelhardt JF (2008) Endosomal NADPH oxidase regulates c-Src activation following hypoxia/reoxygenation injury. Biochem J 411(3):531–541. https://doi.org/10.1042/BJ20071534
8. Banfi B, Clark RA, Steger K, Krause KH (2003) Two novel proteins activate superoxide generation by the NADPH oxidase NOX1. J Biol Chem 278(6):3510–3513. https://doi.org/10.1074/jbc.C200613200
9. Lambeth JD (2004) NOX enzymes and the biology of reactive oxygen. Nat Rev Immunol 4(3):181–189. https://doi.org/10.1038/nri1312
10. Mumbengegwi DR, Li Q, Li C, Bear CE, Engelhardt JF (2008) Evidence for a superoxide permeability pathway in endosomal membranes. Mol Cell Biol 28(11):3700–3712. https://doi.org/10.1128/MCB.02038-07
11. Lynch RE, Fridovich I (1978) Permeation of the erythrocyte stroma by superoxide radical. J Biol Chem 253(13):4697–4699
12. Salvador A, Sousa J, Pinto RE (2001) Hydroperoxyl, superoxide and pH gradients in the mitochondrial matrix: a theoretical assessment. Free Radic Biol Med 31(10):1208–1215
13. Harraz MM, Marden JJ, Zhou W, Zhang Y, Williams A, Sharov VS, Nelson K, Luo M, Paulson H, Schoneich C, Engelhardt JF (2008) SOD1 mutations disrupt redox-sensitive Rac regulation of NADPH oxidase in a familial ALS model. J Clin Invest 118 (2):659–670. https://doi.org/10.1172/JCI34060
14. Mizuno T, Kaibuchi K, Ando S, Musha T, Hiraoka K, Takaishi K, Asada M, Nunoi H, Matsuda I, Takai Y (1992) Regulation of the superoxide-generating NADPH oxidase by a small GTP-binding protein and its stimulatory and inhibitory GDP/GTP exchange proteins. J Biol Chem 267(15):10,215–10,218
15. Kopani M, Celec P, Danisovic L, Michalka P, Biro C (2006) Oxidative stress and electron spin resonance. Clin Chim Acta 364 (1-2):61–66. https://doi.org/10.1016/j.cca.2005.05.016

Chapter 28

Model Systems to Investigate NOX-Dependent Cell Migration and Invasiveness

Howard E. Boudreau and Thomas L. Leto

Abstract

There is mounting evidence indicating that reactive oxygen species (ROS) play a crucial role in cell migration and invasion. Our previous studies have demonstrated the NADPH oxidase (NOX) family of enzymes are a source of ROS in different cell types undergoing migration. Several NOX enzymes are induced or activated in processes including wound repair and maintenance of epithelial barriers, as well as in promoting metastatic cell migration and invasiveness. This chapter outlines three different in vitro assays used to examine how NOX enzymes are involved in cell motility: scratch-wound repair, Matrigel invasion, and migration from confluent cell monolayer boundaries created by cell culture inserts. The three methods provide a range of experimental approaches for delineating roles of NOX enzymes in cell migration through manipulation of the expression or activities of the endogenous or overexpressed oxidases.

Key words Cell migration, Cell invasion, Wound healing, NADPH oxidase, NOX1, NOX4, p53, TGF-β

1 Introduction

Cell migration is important in normal tissue morphogenesis, homeostasis, and wound repair, as well as in the dissemination of cancer cells during the metastatic process. NADPH oxidase (NOX/DUOX)-derived reactive oxygen species (ROS) are known to act as signaling molecules regulating various cellular functions including cell migration and invasion [1]. In respiratory epithelium, DUOX1 mediates cell migration in scratch-wound repair assays performed in vitro [2] or in response to chemical wounding of the epithelial layer of mouse airways [3, 4]. NOX1 is thought to support scratch-wound repair of the colon epithelium in mice in a process promoted by gut commensal microbes signaling through formyl peptide receptors [5]. NOX1 is also thought to promote tumor cell invasiveness in partnership with the Src tyrosine kinase substrate, Tsk5 [6]. We previously reported that wild-type and different mutant forms of p53 differentially regulate ROS

Ulla G. Knaus and Thomas L. Leto (eds.), *NADPH Oxidases: Methods and Protocols*, Methods in Molecular Biology, vol. 1982, https://doi.org/10.1007/978-1-4939-9424-3_28, © Springer Science+Business Media, LLC, part of Springer Nature 2019

generation by NOX4 in breast, pancreatic, and lung epithelial tumor lines. We demonstrated tumor-associated mutant p53 proteins enhance TGFβ/SMAD3-dependent NOX4 expression, cell migration, and invasion, whereas wild-type p53 suppresses TGF-β-induced NOX4, ROS, and cell migration [7–9]. Here, we found that epigenetic regulation of the NOX4 promoter is directly involved in TGFβ/SMAD3/mutant p53-mediated cell migration and suggested NOX4 inhibitors could exert anti-metastatic effects in tumors bearing p53 "hotspot" mutations [8]. The pathways by which NOX-derived ROS act downstream to promote cell migration are poorly understood, although some common signaling intermediates have been noted in NOX1- and NOX4-dependent cell migration, including phospho-ERK and phospho-FAK [3, 8].

In this chapter, we describe three different in vitro assays we have used to demonstrate the involvement of NOX-derived ROS in epithelial cell motility. (1) The scratch-wound repair assay is a quick and simple method that allows researchers to investigate the rate at which different cells migrate and provides a system where specific genes or drugs affecting cell migration can be studied. The procedure involves creating a "wound" in a cell monolayer, capturing images at intervals during or at some endpoint following migration from the wound edge as cells repopulate the denuded area. (2) The second assay is a Matrigel® invasion assay that allows for the in vitro assessment of the invasive properties of a cell population. The Matrigel® matrix is a gelatinous mixture of proteins similar in composition to the basement membrane extracellular matrix found in several tissues. In order for most tumor cells to migrate and metastasize to a secondary site, they produce matrix metalloproteinases (MMPs) that allow cells to degrade and cross an extracellular matrix barrier. The Matrigel® invasion assay provides a way to investigate mediators of matrix remodeling or drugs targeted to inhibit their function and prevent tumor metastasis. This assay is also well suited to examine cell migratory responses to chemotactic gradients. (3) The third migration assay makes use of removable cell culture inserts that form a barrier between cells seeded in two separate wells 500 μm apart. Unlike the "scratch-wound" repair assay, removing this barrier creates a clean, cell-free area between confluent cell areas where cells can migrate while not damaging or "wounding" any cells along the migrating cell boundary. Each of these systems offers distinct advantages and disadvantages to study roles of NOX enzymes and the ROS they generate in cell mobility, depending on the question under investigation. Methods 1 and 2 have been successfully employed to study roles of endogenous NOX4 in migration of normal or metastatic epithelial cells in responses to TGF-β or manipulations of p53 and other downstream regulators of NOX4 expression [8]. Method 3 has been used in HCT-116 colon epithelial cells, where the expression of NOX1 and other NOX1-supportive cofactors has been restored by gene transfection [10]. With all the methods described here, the migration time course is performed within 24 h or less, such that most of the repair or repopulation of the "wounded area" is attributed to cell migration, not cell proliferation.

2 Materials

1. Cultured cell lines (all available from ATCC): MDA-MB-231 metastatic breast epithelial cells (endogenous mutant p53-R280K), MCF-10A immortalized human breast epithelial cells (endogenous wild-type p53), H1299 lung epithelial cells p53-*null*, and HCT-116 colon epithelial cells (endogenous wild-type p53) (*see* **Note 1**). All lines are grown in 5% CO_2 atmosphere incubator at 37 °C.
2. MDA-MB-231 cell complete culture medium: DMEM, 10% heat-inactivated fetal bovine serum (HI-FBS) and 100 μg/mL penicillin-streptomycin.
3. MCF-10A cell complete culture medium: DMEM/F12, 5% horse serum (HI), 500 ng/mL hydrocortisone, 20 ng/mL EGF, 100 ng/mL cholera toxin, 10 μg/mL insulin, and 100 μg/mL penicillin-streptomycin.
4. H1299 cell complete culture medium: RPMI 1640, 10% HI-FBS, and 100 μg/mL penicillin-streptomycin.
5. HCT-116 cell complete culture medium: McCoy's 5A medium (modified) containing 10% HI-FBS and 100 μg/mL penicillin-streptomycin.
6. OPTI-MEM transfection low serum medium.
7. 1× phosphate-buffered saline (PBS).
8. Trypsin-EDTA (0.05%).
9. Transforming growth factor-β-1.
10. 6-well tissue culture plate.
11. 96-well tissue culture plate.
12. NOX4 dominant-negative plasmid (NOX4-DN) (*see* **Note 2**).
13. Multichannel pipettor.
14. Control and NOX4 Mission® shRNA plasmids (Sigma) or other shRNAs targeting genes of interest.
15. Lipofectamine® 3000 (ThermoFisher) (*see* **Note 3**).
16. IncuCyte® WoundMaker™ tool (Essen Bioscience).
17. 24-well BD BioCoat™ Matrigel® Invasion Chamber (BD Bioscience)
18. Ibidi® 2-well cell culture silicone inserts.

19. Diff Stain Kit.
20. Phase contrast microscope or other cell imaging system (*see* **Note 4**).

3 Methods

3.1 Scratch-Wound Repair Assay

For this assay, a wound is created across a monolayer of cells by scratching. The "healing" of the wound by cell migration and growth toward the center of the gap is monitored over time. Analysis of cell migration in vitro is a useful assay to investigate alterations in cell migratory capacity in response to different experimental conditions. The scratch-wound assay is a simple, versatile, and cost-effective way to study cell migration and wound healing. We have used this assay to demonstrate involvement of NOX4 in TGF-β-mediated migration of breast and lung epithelial cells. As shown in Fig. 1, we performed the scratch-wound repair assay on MDA-MB-231 metastatic and MCF-10A normal breast epithelial cells and examined the effects of NOX4 dominant-negative (NOX4-DN) or NOX4-shRNA on TGF-β-mediated migration. We have also performed the scratch-wound assay on p53-null

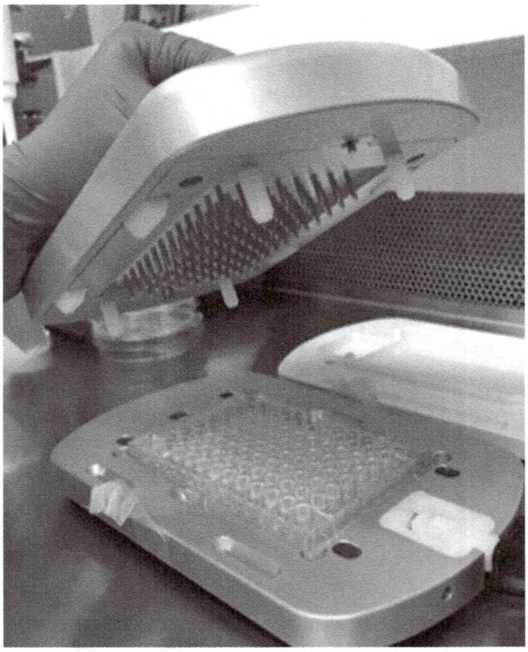

Fig. 1 Image of The WoundMaker™ used for scratch wound assays. This device is a 96-pin tool that generates uniform and reproducible 800 μm scratch wounds on cell monolayers seeded within 96-well culture plates with the push of a lever arm

H1299 lung epithelial cells overexpressing different p53 mutants that are co-transfected along with NOX4-DN [8, 9].

3.1.1 Cell Seeding and Transfection for Scratch-Wound Assay

1. Twenty-four hours before transfecting, seed 4×10^5 MDA-MB-231 cells/well or 5×10^5 MCF-10A cells/well in 2 mL of complete culture media in 6-well tissue culture plates.
2. Transfect the cells when they are approximately 80–90% confluent with a total of 2 μg of plasmid DNA/well. The cells should be transfected with 2 μg of either vector control, NOX4-DN, or NOX4-shRNA plasmid (*see* **Note 2**).
3. Use a 1:1 (v:w) ratio of Lipofectamine® 3000 to plasmid DNA per well (2 μL of Lipofectamine® 3000 to 2 μg of plasmid DNA). Add 2 μL of transfection reagent to 200 μL of OPTI-MEM in microfuge tube, and allow it to incubate at room temperature for 5 min. Next, add the appropriate DNA plasmids to their correct tube, mix thoroughly by pipetting, and allow the tubes to incubate at room temperature for 15 min (*see* **Note 3**). Add the transfection mix dropwise to the medium of cells in the 6-well plate, dispersing it thoroughly with a gentle swirling motion.

3.1.2 Cell Reseeding for Scratch-Wound Assay

1. After the cells have been transfected for 24 h, trypsinize and reseed them into a 96-well tissue culture plate at 2×10^4 cells per well in 100 μL of complete medium.
2. Allow the cells to settle and form a confluent monolayer overnight (*see* **Note 5**).

3.1.3 Wounding and Treatment

1. Make scratch wounds using the IncuCyte® WoundMaker™ tool according to the manufacturer's protocol. The WoundMaker™ is a 96-pin tool that will create reproducible uniform cell-free "scratch" zones 800 μm in width (Fig. 2). For proper instruction on creating wounds on a 96-well plate, watch the video from Essen Bioscience. All wells should contain a medium (*see* **Note 6**).
2. After wounding, use a multichannel pipet to gently wash the cells twice with 100 μL × PBS followed by one wash with 100 μL of serum-free medium to remove any floating cells dislodged during scratch wounding. Allow the cells to incubate in 100 μL of complete medium with or without TGF-β (5 ng/mL).
3. Allow the cells to migrate for approximately 18–24 h before fixing and staining. In general, it is best to fix migrating cells before the wound closure exceeds 75%, beyond which migration rates tend to decrease (*see* **Note 7**).

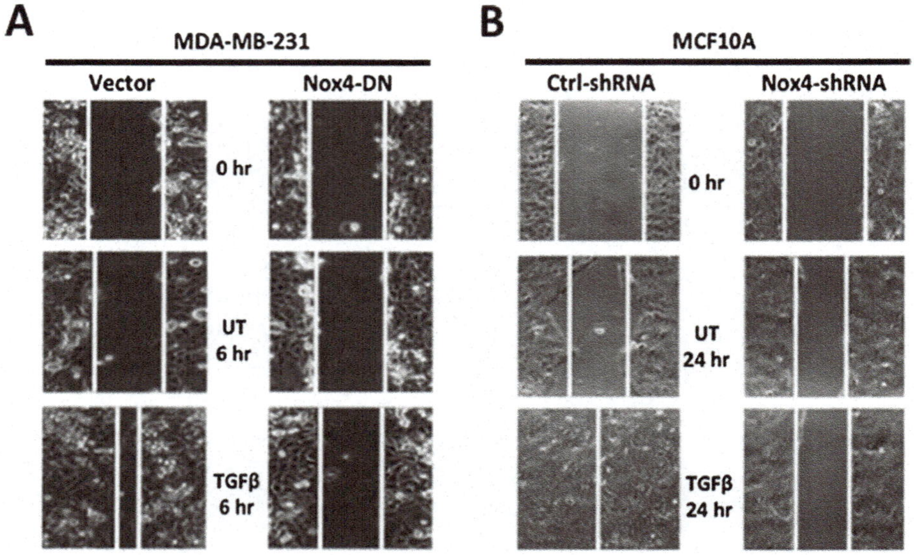

Fig. 2 Scratch-wound repair assays demonstrating the role of NOX4 in TGF-β-stimulated breast epithelial cell migration (6). Highly metastatic MDA-MB-231 cells **(A)** and immortalized normal MCF10A cells **(B)** were transfected with either NOX4 dominant negative (NOX4-DN), NOX4-specific shRNA, or control plasmids and grown to confluency. Scratch wounds were created across the cell monolayer and complete medium was replaced with medium with TGF-β (5 ng/mL) or left untreated (UT) without TGF-β. Phase contrast images were recorded at 0, 6, or 24h post-wounding

3.1.4 Fixing and Staining Scratch-Wound Assay

1. Fix and stain the cells with a Diff staining kit: To fix the cells, aspirate the medium from each well and add 100 μL of Solution A containing methanol for fixation. Leave the cells in Solution A at room temperature for 5 min before removing by aspiration.
2. Stain cells with 100 μL of Solution B (Eosin Y dye). Incubate at room temperature for 5 min and remove.
3. Add 100 μL of Solution C (Azure B dye). Incubate at room temperature for 5 min and remove.
4. Rinse the wells with deionized water once and allow to air-dry.
5. Collect microscopic images for use in quantifying wound gap closure.
6. Calculate the wound gap closure index (GCI) at endpoint time *T*, as follows:

$$\mathrm{GCI} = 1 - \frac{G_T}{G_0}$$

where G_T is the gap size at endpoint time T and G_0 is the starting point gap size [10]. Thus, the *GCI* values range from 0 to 1 for no gap closure to complete gap closure, respectively. Migrating cell boundaries can be hand-traced if they remain

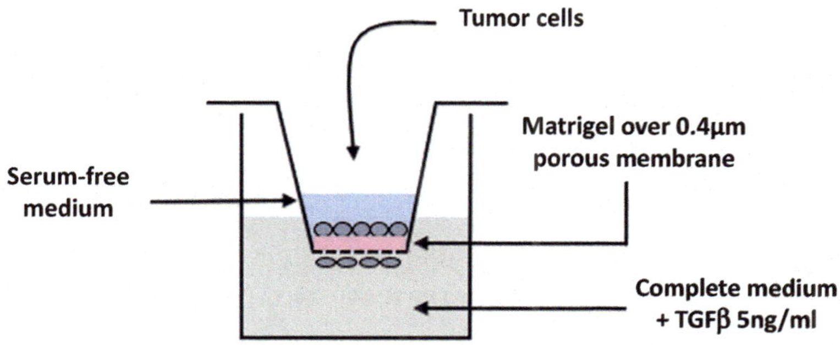

Fig. 3 Schematic representation of the two-chambered Matrigel® trans-well system used for invasion assays. Tumor cells are seeded in serum-free medium in the upper chamber on a porous membrane coated with a layer of Matrigel®. The lower chamber medium contains TGF-β used as the chemoattractant. Tumor cells invade the Matrigel® layer and penetrate the porous membrane toward the TGF-β gradient. At the end of the migration time course, cells remaining in the upper chamber are removed, while those that transmigrated through the membrane are fixed, stained, and counted

relatively straight, but the preferred methodology for accurate migrating edge and wound area detection requires automated image analysis software (*see* **Note 8**).

3.2 Matrigel® Invasion Assay

The Matrigel® invasion assay consists of an upper chamber with a semipermeable membrane coated with Matrigel® that is suspended over a larger well containing a medium or chemoattractant. Cells are placed inside the upper chamber on top of the matrix protein gel and allowed to migrate through the matrix and membrane pores, to the other side of the membrane (Fig. 3). After a suitable migration time course, cells remaining in the upper chamber and matrix are scrubbed out. Migratory cells remaining on the lower membrane surface are then fixed, stained, and counted. To demonstrate the involvement of NOX4 in TGF-β-mediated cell migration, we utilized the Matrigel® migration assay on MDA-MB-231 cells [7]. We have also used this assay to show NOX4 involvement in mutant p53-mediated cell invasion in H1299 lung epithelial cells [8, 9].

3.2.1 Cell Transfection and Seeding

1. Prepare cells for transfection with plasmid DNA in 6-well plates according to the procedures described in Subheading 3.1.1.
2. Take out the Matrigel® invasion chamber (24-well) plate from −20 °C and allow to reach room temperature.
3. Once at room temperature, hydrate the trans-well chamber with 500 µL of serum-free cell culture medium added both to the upper trans-well and lower well chambers and incubate for 2 h at 37 °C and 5% CO_2

4. Resuspend 24-h transfected cells at 5×10^4 cells/mL in serum-free medium.
5. Remove the medium from the upper and lower chambers of the trans-wells.
6. Add 500 μL of the cell suspension (2.5×10^4) to the upper chamber of the Matrigel-coated trans-well. Add 500 μL of complete medium containing TGF-β (5 ng/mL), or without, to the lower chamber.
7. Allow cells to migrate for 24 h at 37 °C and 5% CO_2.

3.2.2 Fixing and Staining Matrigel® Invasion Assay

1. Aspirate the medium from the inside of the Matrigel-coated chamber and from the bottom chamber of the 24-well plate.
2. Use a wet cotton-tipped swab (dipped in water) to gently scrub away the cells from the matrix layer on the upper membrane surface. Repeat with another cotton-tipped swab.
3. Wash the chamber twice with distilled water.
4. Add 500 μL each Diff Stain solution into three rows of wells within a 24-well plate.
5. Sequentially transfer the inserts through each fixative or stain solution allowing 5 min incubation in each solution.
6. Allow the inserts to air dry.
7. Count invading cells accumulated on the membrane lower surface under a microscope.

3.3 Ibidi® Cell Culture-Insert Migration Assay

This migration assay involves use of 2-well adhesive silicone inserts placed into 6- or 12-well culture plates that are used to create two confluent patches of cells separated by a barrier that creates a well-defined, cell-free zone 500 μm in width (Fig. 4) (*see* **Note 9**). Upon reaching confluence, the insert is removed and cells migrate to fill the gap. The method is ideal for cell monolayers that are not easily "scratched" to produce a straight-edged wounded boundary; since no scratching is involved, the cells on the migration boundary are not damaged. We have used this method with HCT-116 human colon epithelial cells to examine the role of NOX1 in cell migration [10]. This tumor line is dedifferentiated and produces undetectable levels of superoxide until transfected with NOX1 and other NOX1-supportive components (NOXO1, NOXA1, and PRDX6) that together restore NADPH oxidase activity and promote cell migration.

3.3.1 Cell Seeding and Transfections

1. Passage HCT-116 cells in complete McCoy's 5A medium. Seed the cells into 6-well plates (2×10^5 cells/mL) in preparation for transfection 24 h later once they reach 75% confluency.
2. Transfect cells with 2 μg total plasmid using Lipofectamine® 3000: For optimum NOX1 reconstitution, use 0.5 μg each of

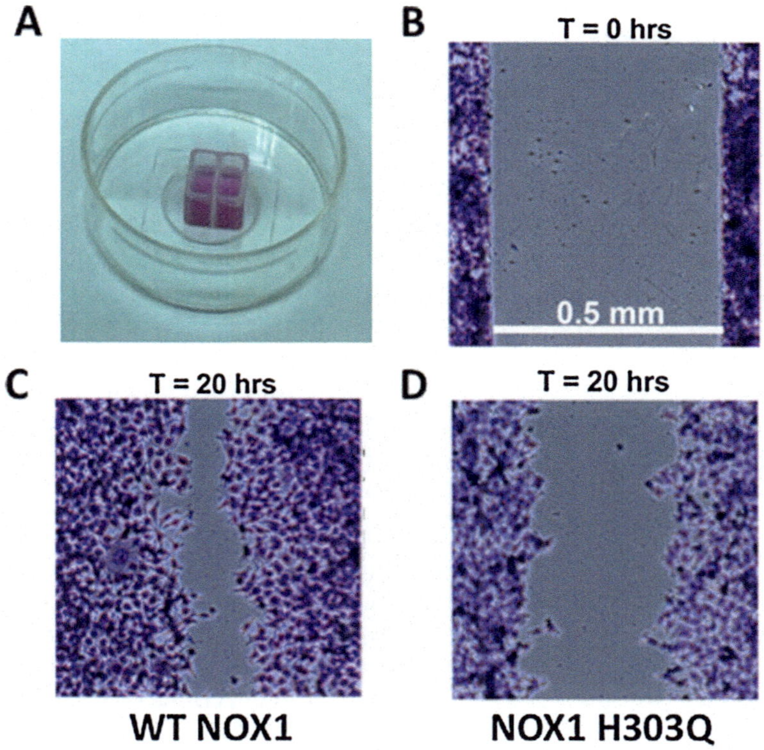

Fig. 4 Cell migration assays demonstrating the role of NOX1 in HCT-116 colon epithelial cell migration between confluent cell monolayer boundaries formed within an Ibidi® 2-well culture insert. (**A**) Image of the Ibidi® culture insert mounted in a glass-bottom dish, at the initial stage of cell seeding (70 μL per well). The barrier between the two wells creates a 500 μM cell-free zone where cells migrate upon removal of the culture insert. Stained images of confluent HCT-116 cells fixed at (**B**) initiation ($T = 0$) and (**C**, **D**) endpoint ($T = 20$ h) of migration assays. Cells expressing WT NOX1 show maximum cell migration (**C**), whereas cells expressing the NOX1 H303Q loss-of-function mutant migrate significantly slower (**D**). Data adapted from previously reported findings [10]

pcDNA3.1 vectors encoding NOX1, NOXO1, NOXA1, and PRDX6 (*see* **Note 10**). Mix all plasmid DNAs together and dilute to 100 μL in OPTI-MEM. Separately dilute Lipofectamine® 3000 with 100 μL in OPTI-MEM and incubate at room temperature for 5 min, then combine the diluted DNAs and lipofection reagents, and incubate for another 20 min at room temperature.

3. Meanwhile, change the medium in the overnight 6-well plate and maintain at 37 °C. Then add the 200 μL of transfection mixes dropwise to each well while swirling gently to disperse. Incubate at 37 °C for 5 h.

4. Prepare Ibidi® 2-well chambers by mounting one per well in 12-well collagen1-coated cell culture plates using sterile

forceps. Inspect under a microscope to confirm the insert bottoms are fully seated with the adhesive side down against the plate surface. There should be no bubbles or gaps, particularly along the barrier separating the two wells (*see* **Note 11**).

5. Trypsinize cells, count them, and adjust to 1×10^6 cells/mL of complete McCoy's 5A medium. Dispense 70 μL of cell suspension per Ibidi® well, being careful to avoid any bubbles at the bottom of the well. Allow cells to settle and adhere in the incubator overnight. To avoid evaporation of the low volumes in these chambers, place these plates in a larger sealed plastic box containing D/W moistened blotting paper. Any remaining cells can be returned to the 6-well plate and incubated for another 2 days, and then harvest for other assays performed in parallel with the migration experiment (e.g., NOX1-dependent superoxide release or Western blotting).
6. Continue growing until the adherent cell layer reaches ~90% confluence, at which time the Ibidi® frames should be removed to create a cell-free zone between the two wells. To do this, grip the top corner of the Ibidi® silicone frames with forceps and remove in one vertical motion without disturbing the adherent cell patch edges in contact with the wall of the inserts (*see* **Note 12**). Gently flush the well with 1 mL fresh McCoy's 5A complete medium to remove any floating cells, and then add another mL of fresh medium. Return to the incubator overnight.
7. Periodically check the progress of cell migration under an inverted phase microscope the next day, being careful not to disturb the cell layers in transit from the incubator.
8. Stop, fix, and stain the cells according to Methods described above in the steps described under Subheading 3.1.4, **steps 1–5**. The maximum migration endpoint reached before fixation should be <75% complete gap closure, which usually occurs within 20–24 h.
9. Calculate the gap closure index (GCI) as described under Methods in Subheading 3.4.1, **step 6**.

4 Notes

1. The human breast epithelial cell lines MDA-MB-231 (highly metastatic) and MCF-10A (normal immortalized) are commonly used cell models for scratch-wound and Matrigel® invasion assays [7]. The number of cells required for each assay will vary depending on the cell transfection and other treatment conditions. The p53-*null* human lung epithelial cell line H1299 is another cell line with a high transfection efficiency

our laboratory routinely uses for both scratch-wound and invasion assays, particularly for p53-related studies [8, 9].

2. The NOX4 dominant-negative plasmid contains cDNA encoding amino acids 1-320 with a C-terminal V5 epitope tag sub-cloned into pcDNA3.1. The resulting NOX4 protein lacks amino acids 321-578 containing the C-terminal FAD and NADPH domains [11]. We have also shown that expression of an inactive mutant form of NOX4 (P437H) attenuated TGF-β and mutant p53-mediated cell migration [9].
3. Depending on cell type, some transfection reagents have a higher transfection efficiency than others or may be less toxic. Our lab has found that Lipofectamine® 3000 works well with both MCF-10A and MDA-MB-231.
4. We collect visible light images of stained cells with a 4× objective lens on the Evos FL Cell Imaging System (Life Technologies).
5. The confluency of the cells at the time of wounding is an important factor. We have found that seeding the cells the evening before wounding (approximately 15–18 h prior) is an adequate amount of time for the cells to settle, attach, and spread into a thin, nearly confluent monolayer. If the cells are too confluent, wounding will cause the cells to tear resulting in a jagged edge rather than a straight-line wound boundary. If scratch wounds observed with other cell lines tested are consistently jagged, consider the alternative wound-healing migration assay described in Subheading 3.3.
6. Essen Bioscience has provided a helpful short video that demonstrates how to properly handle and operate the WoundMaker™: https://www.youtube.com/watch?v=x7pMzJ1VIdA. The WoundMaker™ is a very fragile instrument and must be handled properly to maintain the ability to make reproducible scratch wounds each time. A few important things to remember: (i) never place the WoundMaker™ 96 pin block directly on to the bench top as this will damage the plastic-tipped pins, (ii) clean the pin block after each use with the detergents provided, (iii) each well in the 96-well plate must contain 100 μL media or PBS before making a scratch, and (iv) never press down on the pin block when generating wounds.
7. This is the simple migration endpoint assay, which upon fixation will permit image analysis later. Alternatively, some investigators monitor migration in real time using phase microscopy equipment capable of reproducibly scanning and recording images of multiple well positions over time while maintaining conditions compatible with cell growth.

8. We use ImageJ to process wound closure images, which is available at http://rsbweb.nih.gov/ij/docs/guide/146-29.html. Detailed considerations on the use of ImageJ or MATLAB automated image processing software for migration assays have been described by Treloar and Simpson [12].
9. Ibidi® has provided an online video that provides a useful overview of the methodology (https://ibidi.com/content/145-mv-16). Their product manual can be accessed at the following website: https://ibidi.com/img/cms/products/labware/open_removable/E_8XXXX_CI_2Well/IN_8XXXX_CI_2Well.pdf
10. Substitute or scale back these plasmid amounts depending on the experimental design needed for questions under investigation. Perform a parallel control transfection with at least 0.5 μg EGFP-expressing plasmid to confirm >60% transfection efficiency at 48 h.
11. We routinely scan cell migration images at low power magnification, which can be performed with standard 6- or 12-well plastic cell culture dishes. For higher magnification imaging, seed cells in Ibidi® culture inserts mounted in collagen1-coated, thin-layer glass bottom dishes from MatTek or μ-dishes from Ibidi®.
12. The Ibidi® culture inserts can be used more than once as long as they remain adherent and are free of defects and debris. Wash inserts in sterile distilled water, then 70% ethanol. Leave to dry and store with the adhesive side down in sterile petri dishes.

Acknowledgments

This work was supported by the Intramural Research Program of the National Institute of Allergy and Infectious Diseases, NIH.

References

1. Schroder K (2014) NADPH oxidases in redox regulation of cell adhesion and migration. Antioxid Redox Signal 20:2043–2058
2. Luxen S, Belinsky SA, Knaus UG (2008) Silencing of DUOX NADPH oxidase by promoter hypermethylation in lung cancer. Cancer Res 68:1037–1045
3. Wesley UV, Bove PF, Hristova M, McCarthy S, van der Vliet A (2007) Airway epithelial cell migration and wound repair by ATP-mediated activation of dual oxidase 1. J Biol Chem 282:3213–3220
4. Gorissen SH, Hristova M, Habibovic A, Sipsey LM, Spiess PC, Janssen-Heininger YM, van der Vliet A (2013) Duox oxidase-1 is required for airway epithelial cell migration and bronchial reepitheilialization after injury. Am J Respir Cell Mol Biol 48:337–345
5. Alam A, Leoni G, Wentworth CC, Kwal JM, Wu H, Ardita CS, Swanson PA, Lambeth JD, Jones RM, Nusrat A, Neish AS (2014) Redox signaling regulates commensal-mediated mucosal homeostasis and restitution and requires formyl peptide receptor 1. Mucosal Immunol 7:645–655
6. Gianni D, Diaz B, Taulet N, Fowler B, Courtneidge SA, Bokoch GM (2009) Novel $p47^{phox}$-

related organizers regulate localized NADPH oxidase (Nox1) activity. Sci Signal 2(88):ra54

7. Boudreau HE, Casterline BW, Rada B, Korzeniowska A, Leto TL (2012) Nox4 involvement in TGF-beta and SMAD3-driven induction of the epithelial-to-mesenchymal transition and migration of breast epithelial cells. Free Radic Biol Med 53:1489–1499
8. Boudreau HE, Casterline BW, Burke DJ, Leto TL (2014) Wild-type and mutant p53 differentially regulate NADPH oxidase 4 in TGF-β-mediated migration of human lung and breast epithelial cells. Br J Cancer 110:2569–2582
9. Boudreau HE, Ma WF, Korzeniowska A, Park JJ, Bhagwat MA, Leto TL (2017) Histone modifications affect differential regulation of TGFβ- induced NADPH oxidase 4 (NOX4) by wild-type and mutant p53. Oncotarget 8:44379–44397
10. Kwon J, Wang A, Burke DJ, Boudreau HE, Lekstrom KJ, Korzeniowska A, Sugamata R, Kim YS, Yi L, Ersoy I, Jaeger S, Palaniappan K, Ambruso DR, Jackson SH, Leto TL (2016) Peroxiredoxin 6 (Prdx6) supports NADPH oxidase1 (Nox1)-based superoxide generation and cell migration. Free Radic Biol Med 96:99–115
11. Boudreau HE, Emerson SU, Korzeniowska A, Jendrysik MA, Leto TL (2009) Hepatitis C virus (HCV) proteins induce NADPH oxidase 4 expression in a transforming growth factor beta-dependent manner: a new contributor to HCV-induced oxidative stress. J Virol 83:12,934–12,946
12. Treloar KK, Simpson MJ (2013) Sensitivity of edge detection methods for quantifying cell migration assays. PLoS One 8:e67389

Chapter 29

NADPH Oxidases and Aging Models of Lung Fibrosis

Karen Bernard and Victor J. Thannickal

Abstract

There is a growing recognition that aging is a risk factor for fibrosis that affects a number of organ systems, including the lung. Despite this understanding, most studies of experimental fibrosis have been conducted in young mice that typically resolve injury-induced lung fibrosis over the course of several months. Our studies demonstrate that aged mouse models may recapitulate human disease by generating a more persistent fibrotic response to injury. This is, in part, due to an imbalance in the expression and activity of NADPH oxidase (NOX) enzymes, in particular the NOX4 isoform, and a related deficiency in antioxidant responses in pathogenic myofibroblasts. These pathogenic myofibroblasts acquire features of cellular senescence and become resistant to apoptosis. In this chapter, we present methods and procedures to apply the aging model of lung fibrosis in mice that will allow interrogation of myofibroblast functions and the expression and activity of NOX4 in cells. We provide recommendations for best laboratory practices to assess the severity and resolution of fibrosis in murine models of aging.

Key words NADPH oxidase, Fibrosis, Myofibroblasts, NOX4, Bleomycin injury, Aging, Oxidative stress

1 Introduction

1.1 Evidence for NOX Enzymes in Fibrosis

NADPH oxidases are transmembrane reactive oxygen species (ROS)-generating enzymes that have been implicated in host defense, cellular differentiation, and fate. The NADPH oxidase family comprises seven members (NOX1–5 and DUOX1–2). gp91^{phox} or NOX2 was the first identified member of the NADPH oxidase family [1–3]. NOX2 localizes in the plasma membrane of phagocytes where it releases superoxide ($O_2^{\cdot-}$). NOX2-produced $O_2^{\cdot-}$ and H_2O_2, in the presence of myeloperoxidase (MPO), is converted to the microbicidal HOCl that is critical for host defense [4, 5]. NOX1–4 are anchored to biological membranes via six transmembrane domains. The cytosolic C-terminus of these enzymes contains FAD (flavin adenine dinucleotide) and NADPH-binding domains. The NOX5 catalytic core is similar to that of NOX1–4 with the addition of calcium (Ca^{2+})-binding domains at the N-terminus of the enzyme. The structures of

Ulla G. Knaus and Thomas L. Leto (eds.), *NADPH Oxidases: Methods and Protocols*, Methods in Molecular Biology, vol. 1982, https://doi.org/10.1007/978-1-4939-9424-3_29, © Springer Science+Business Media, LLC, part of Springer Nature 2019

DUOX1 and DUOX2 contain a Ca^{2+}-binding domain and an additional transmembrane helix followed by a peroxidase-like domain that utilizes H_2O_2 to oxidize other substrates [6].

NOX4 has been reported to mostly generate H_2O_2, in a constitutive manner [7]. Our studies have identified NOX4 as one of the most inducible genes in human lung fibroblasts stimulated by the pro-fibrotic cytokine, transforming growth factor-beta 1 (TGF-β1), which is necessary to induce fibrosis in murine models of lung injury and fibrosis [8]. Furthermore, induction of NOX4 with an impaired antioxidant response has been implicated in the aging-associated susceptibility to lung fibrosis [9].

Lung fibrosis is the prototypical reaction to an unremitting wound healing response associated with the accumulation of scar-forming myofibroblasts in lung tissues. Idiopathic pulmonary fibrosis (IPF) is a progressive, life-threatening disease in humans with a median survival of 2–3 years after diagnosis [10]. Myofibroblasts are critical cellular effectors of pulmonary fibrosis. They secrete, deposit, and remodel the extracellular matrix (ECM) that constitutes scar formation. Fibroblasts differentiate into myofibroblasts under the action of TGF-β1 and mechanical tension [11, 12].

It has been estimated that 45% of all deaths result from various degrees of fibrosis affecting multiple organ systems such as the cardiovascular system, the liver, kidneys, and lungs [13]. The role of NOX enzymes in fibrosis extends beyond the lung as exemplified in liver fibrosis where NOX1, 2, and 4 play critical roles in initiating fibrosis through activation of hepatic stellate cells [14–16]. In this context, methods allowing for the determination of NADPH oxidase expression and function in experimental models of fibrosis are critical to elucidate their roles in fibrosis development and progression.

1.2 Aging as a Risk Factor for Lung Fibrosis

IPF is a disease of aging [17–19]. The incidence and prevalence of IPF increase almost exponentially with each decade of life [17]; in this study, two-thirds of IPF patients were older than 60 years at time of presentation, and the mean age of diagnosis for IPF was 66 years. Despite advances in our understanding of the biology of aging and of its role in age-related diseases, mechanistic links between aging and lung fibrosis remain to be established [20]. Genomic instability, telomere attrition, epigenetic alterations, loss of proteostasis, deregulated nutrient sensing, mitochondrial dysfunction, cellular senescence, stem cell exhaustion, and altered intercellular communication are hallmarks of aging that contribute to the predisposition to IPF [21–23]. Our studies have linked the persistence of lung fibrosis in aged mice to the accumulation of senescent and apoptosis-resistant myofibroblasts [9]. In this aging model of murine lung fibrosis, we have shown that a redox imbalance characterized by an increase in NOX4 and a deficiency in Nrf2-mediated antioxidant cellular responses promotes the accumulation of senescent and apoptosis-resistant myofibroblasts [9]. In these studies, e demonstrated that pharmacological and genetic targeting of NOX4 in aged mice led to the reversal of persistent fibrosis by inducing myofibroblast apoptosis. Additionally, we have highlighted a key role of mitochondrial bioenergetics and cellular metabolism in regulating myofibroblast activation [24, 25]. Further studies are required to more precisely define the nature of mitochondrial dysfunction associated with aging-associated persistence of myofibroblasts in fibrotic tissues.

1.3 Models of Lung Fibrosis

Lung fibrosis is reproducibly induced in rodents by inducing injury to the airway epithelium and is widely used as an experimental and preclinical model. Although several agents have been used to induce airway epithelial injury, the most commonly used agent is bleomycin by direct delivery to the lung [26]. While one model cannot fully recapitulate all aspects of human IPF, recommendations have been formulated by a panel of experts regarding the optimal use of animal models in discovery research and drug testing [26]. The nonsurgical, oropharyngeal instillation of bleomycin is recommended for preclinical testing of anti-fibrotic drugs, although other techniques include surgical intratracheal instillation, aerosolization, intraperitoneal administration, and subcutaneous delivery by an implantable pump. Since IPF has a male predilection, the panel recommended carrying out initial studies in male animals and to validate key findings in females [26]. Our group reported that older mice (18 months of age) have a diminished capacity to resolve bleomycin-induced lung fibrosis compared to their younger counterparts (2 months of age) [9]. While many studies have elucidated mechanisms of lung fibrosis in young mice, few have investigated these mechanisms in older mice. Furthermore, few studies have examined the mechanisms of fibrosis resolution, which is impaired in older mice. Old mice are generally considered to be 18–24 months of age, corresponding to an age ranging from 56 to 69 in humans (Table 1). Based on the premise that aging is a risk factor to IPF, we believe that aging models of lung fibrosis may more closely recapitulate features of the human disease, including mechanisms of disease progression.

Despite the large number of compounds that have shown efficacy in ameliorating lung fibrosis in mice, few have translated to the clinic [27]. The reasons for this may include (1) failure to identify the targeted molecule and its associated signaling pathways in IPF tissues, (2) failure to demonstrate that the targeted molecule is involved in disease-promoting pathways in both animal models and cells isolated from IPF patients, (3) failure to consider aging as an important physiological variable, and (4) flaw in the experimental design of preclinical studies, in particular related to the timing of drug administration; we recommend that the anti-fibrotic agent be

Table 1
Estimated age equivalencies for adult mice relative to human subjects

	Mice (months)	Human (years)
Mature adult	3–6	20–30
Middle-aged	10–14	38–47
Old	18–24	56–69

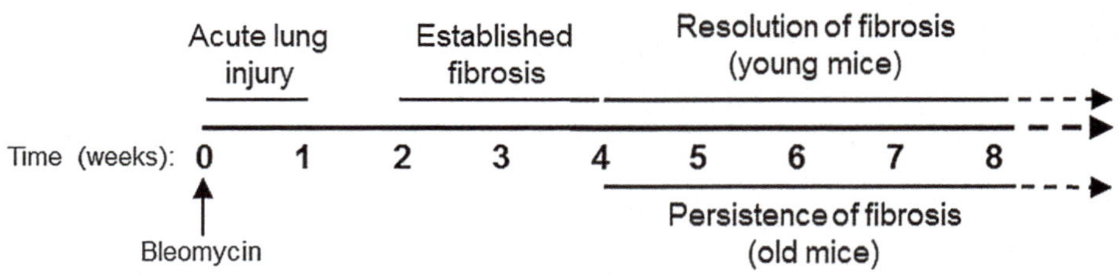

Fig. 1 Timeline of lung fibrosis development and progression in the bleomycin injury model of lung fibrosis in mice

administered during the fibrogenic phase of the disease, which should be no sooner than 8–10 days post lung injury (Fig. 1).

2 Materials

1. C57BL/6 mice, 2 or 18 months of age. Young mice (2 months) and old mice (18 months) are purchased from Jackson Laboratories; old mice may also be obtained from the National Institute of Aging.
2. Bleomycin (30 units, Besse Medical).
3. Phosphate-buffered saline 1× without Ca^{2+} and Mg^{2+}.
4. Isoflurane (Fluriso, VetOne).
5. Vaporizer for isoflurane anesthetic (Ohio Medical Products).
6. Hank's Balanced Salt Solution (HBSS) with Ca^{2+} and Mg^{2+}.
7. HBSS without Ca^{2+} and Mg^{2+}.
8. Collagenase type 4 (Worthington) suspended at 10 mg/mL in sterile HBSS without Ca^{2+} and Mg^{2+}.
9. 100 μm cell strainer.
10. Recombinant porcine TGF-β1 (R&D systems): prepare a stock solution at 1 μg/mL in 4 mM HCl and 0.1% bovine serum albumin, store at 4 °C for a month or aliquot, and keep at −80 °C.

11. Homovanillic acid (HVA) (Sigma-Aldrich): prepare a stock solution of HVA (10 mM) in HBSS without Ca^{2+}, Mg^{2+}, and phenol red, aliquot, and store at −80 °C.
12. Complete serum-containing DMEM cell culture media: DMEM supplemented with 10% fetal bovine serum (FBS), 1% penicillin-streptomycin, 1% L-glutamine, and 0.5% fungizone.
13. Horseradish peroxidase (HRP), type VI (Sigma-Aldrich): prepare a stock solution of HRP (5000 U/mL) in HBSS without Ca^{2+}, Mg^{2+}, and phenol red, aliquot, and store at −80 °C.
14. H_2O_2, 30% solution (Sigma-Aldrich), dilute in dH_2O to make a stock solution of 1 mM.
15. Stop solution: 12 M NaOH, 0.1 M glycine, 25 mM EDTA in dH_2O.
16. Assay media (prepare on the day of the assay): HVA 100 μM, HRP 5 U/mL in HBSS without phenol red but containing Ca^{2+} and Mg^{2+}.
17. QuickZyme Total Collagen Assay (QuickZyme) Kit.
18. Lysis buffer: 50 mM Tris–HCl, pH 8.0, with 150 mM NaCl, 1.0% IGEPAL CA-630 (NP-40), 0.5% sodium deoxycholate, and 0.1% SDS supplemented with protease (cOmplete mini, Roche) and phosphatase (PhosSTOP, Roche) inhibitors.
19. Bicinchoninic acid assay: BCA Protein Assay Kit (Thermo Fisher Scientific).
20. RNeasy® Mini Kit (Qiagen).
21. RNA*later* (Qiagen).
22. iScript reverse transcription supermix for RT-qPCR (Bio Rad).
23. SYBR®Select Master Mix (Applied Biosystems).
24. NOX4-specific primer pairs: forward primer 5′-3′, AGATGTTGGGGCTAGGATTG, and reverse primer 5′-3′, TCTCCTGCTTGGAACCTTCT.
25. StepOnePlus Real-Time PCR System (Applied Biosystems, Foster City, CA).

3 Methods

All procedures involving animals should be approved by the local Institutional Animal Care and Use Committee. Mice are anesthetized with vaporized isoflurane (3%) and placed on a vertical surgery platform; loss of pain sensation is confirmed by toe pinch. Saline or bleomycin (1.25 U/Kg) is then administered via the oropharyngeal route (in a 75 μL volume per instillation). Mice are maintained in the vertical position until no liquid is observed in the oral cavity.

During the procedure the tongue of the mouse is held with forceps, and the nose of the mouse is pinched to facilitate airway delivery.

The weight of the mice is monitored weekly after saline or bleomycin injury. Lung tissues are prepared for analysis of protein, RNA isolation, and histology. Prior to harvesting the lungs, mice are euthanized by CO_2 inhalation followed by thoracotomy. Blood is flushed out of the animal by a perfusion of PBS 1× through the heart, prior to dissecting out the lungs. Specific cell populations may also be extracted for further studies. Lung fibrosis is assayed by measuring the total lung content of hydroxyproline as a measure of collagen deposition.

3.1 Determination of Total Lung Hydroxyproline Content

Hydroxyproline is a major component of fibrillar collagen and generated by the enzyme prolyl 4-hydroxylase as a posttranslational modification of proline [28]. The following method is used to determine total lung hydroxyproline levels at varying times post-bleomycin injury. To study the role of NOX4 expression/activity on lung collagen accumulation, studies may involve NOX4 knock-out mouse model [29]; alternatively, mice may be treated with small-interfering RNAs or pharmacologic inhibitors of NOX4 [30].

1. Dry lungs in an oven at 70 °C for 2 days in 2 mL polypropylene tubes with O-ring, making sure that lids from sample tubes have been removed.
2. Add 1.7 mL of 6 N HCl to each tube and hydrolyze at 95 °C in a block heater for 2 days with occasional vortexing.
3. After acid hydrolysis, hydroxyproline level is determined using the QuickZyme Total Collagen Assay (QuickZyme) Kit according to manufacturer's instructions.
4. We recommend expressing collagen content as μg hydroxyproline per lung, as opposed to μg hydroxyproline per dry lung weight.

Additional endpoints to assess the severity of lung fibrosis in response to bleomycin injury may be considered, including measurement of pulmonary physiology (resistance and compliance) using the flexiVent technology (SCIREQ Inc.) [31], Masson's trichrome staining of collagen fibers on fixed lung tissue slice [32], and microCT imaging of the lungs [32, 33].

3.2 Fibroblast Isolation and Measurement of NOX4 Activity

The following method describes lung fibroblast isolation and measurement of extracellular H_2O_2 released by isolated fibroblasts, a reliable biomarker of NOX4 activity in myofibroblasts [8, 34]. The rate of extracellular H_2O_2 release can be measured using a fluorimetric method based on the dimerization of a substituted phenol (homovanillic acid; HVA) in the presence of a heme peroxidase (horseradish peroxidase, HRP).

1. Prepare a collagenase solution (1 mg/ml) by diluting the collagenase stock (10 mg/mL) in sterile HBSS plus Ca^{2+} and Mg^{2+}; keep on ice until further use.
2. Prepare complete serum-containing DMEM cell culture media.
3. Finely mince lungs with scissors, and incubate with the collagenase solution prepared in **step 1** for 30 min at 37 °C (10 ml collagenase per lung).
4. Filter the cell suspension obtained after **step 3** with a 100 μm cell strainer, and neutralize collagenase by adding 15 mL of complete serum-containing culture media.
5. Centrifuge the cell solution [5 min, 400 × *g* at room temperature (RT)].
6. Discard the supernatant and suspend the cell pellet in culture media plus serum; dispense passage zero cells (P0) into a 10 cm culture dish for expansion.
7. Change cell culture media 3 days post-seeding.
8. Split P0 cells at 80% confluency and seed for H_2O_2 measurement assay in 6-well plate (2.10^5 cells/well; lung fibroblasts are now at P1).
9. At 80% confluency serum-starve fibroblasts with 1% FBS-containing DMEM without phenol red for 16 h.
10. As positive controls for NOX4-derived H_2O_2, growth-arrest fibroblasts with low serum-containing media (*see* **step 9**), and treat with TGF-β1 (5 ng/mL) for 16 h.
11. Rinse fibroblasts once with HBSS (with Ca^{2+} and Mg^{2+}) and add assay media (1 mL/well).
12. Incubate fibroblasts with assay media for 2 h at 37 °C.
13. At the end of **step 12**, collect 1 mL of assay media and add 2.2 μL of stop solution to each sample, vortex, and dispense into a 96-well plate (200 μL/well).
14. Set up a standard curve of known H_2O_2 concentrations (0–10 μM in assay media) using H_2O_2 stock solution (1 mM). For each concentration, let the reaction proceed for 5 min at RT prior to adding the stop solution (2.2 μL/mL).
15. Read fluorescence using a plate reader at an excitation and emission wavelength of 321 and 421 nm, respectively.
16. Count fibroblasts for each experimental condition, and calculate the rate of H_2O_2 release in pmole $H_2O_2/min/10^6$ cells.

Fibroblast-derived H_2O_2 may arise from different sources such as other NADPH oxidases and mitochondrial respiration. To determine extracellular H_2O_2 attributable specifically to NOX4, positive and negative controls of NOX4 activity should be included (*see* **Note 1**).

3.3 Isolation of Total Lung Proteins and Analysis of NOX4 Protein Expression

1. Mince lungs and transfer to ice-cold microcentrifuge tube.
2. Add lysis buffer (1 mL per 100 mg tissue), and break down tissue directly into the microcentrifuge tube using a disposable pestle; keep sample on ice for 15 min with occasional vortexing.
3. Centrifuge tissue lysate for 15 min at 4 °C (16,000 × *g*), and transfer supernatant to a new microcentrifuge tube.
4. Determine total protein concentration using bicinchoninic acid assay (BCA Protein Assay Kit, Thermo Fisher Scientific).
5. Analyze NOX4 protein expression in tissue lysate by SDS-PAGE followed by immunoblotting using anti-NOX4-specific antibody (*see* **Note 2**).

3.4 Isolation of Total RNA and Quantification of NOX4 mRNA Expression

1. Total RNA from mouse lung tissue is isolated using the RNeasy® Mini Kit (Qiagen).This procedure can be carried out using freshly isolated lung tissue, flash frozen lung tissue (in liquid N_2), or lung tissue stabilized in RNA*later* (Qiagen). For efficient isolation of total lung RNA, manufacturer's instructions recommend that the starting material equals or weigh less than 30 mg.
2. Total RNA is reverse transcribed using iScript reverse transcription supermix for RT-qPCR (Bio Rad) as per manufacturer's instructions. We performed real-time PCR reactions for each cDNA sample in duplicate using SYBR®Select Master Mix (Applied Biosystems) and NOX4-specific primer pairs (forward primer 5′-3′: AGATGTTGGGGCTAGGATTG; reverse primer 5′-3' TCTCCTGCTTGGAACCTTCT). Real-time PCR reactions are carried out for 40 cycles (95 °C for 15 sec, 60 °C for 1 min) in a StepOnePlus Real-Time PCR System (Applied Biosystems, Foster City, CA). Real-time PCR data for the target gene are normalized to endogenous β-actin and compared using the $2^{-\Delta\Delta Ct}$ method.

4 Notes

1. TGF-β1 treatment for 16–24 h stimulates NOX4-derived H_2O_2 release and is recommended as a positive control for NOX4 activity. Silencing of NOX4 with small-interfering RNAs (siRNAs) [35] or the use of NOX4-deficient fibroblasts is recommended as a negative control for NOX4 activity.
2. Commercially available NOX4 antibodies have been difficult to validate across different tissues and cells. There may also be variation in the specificity of these antibodies from lot to lot in recognizing NOX4 in lung fibroblast lysates or total lung homogenates; therefore, we strongly recommend testing each lot of antibodies against positive (total cell lysates isolated from

lung fibroblasts treated with TGF-β1) and negative controls (total cell lysates isolated from NOX4-deficient lung fibroblasts).

Acknowledgments

This work was supported by grants from the National Institute of Health: P01 HL114470 and R01 AG046210 (to VJT).

References

1. Holmes B, Page AR, Good RA (1967) Studies of the metabolic activity of leukocytes from patients with a genetic abnormality of phagocytic function. J Clin Invest 46(9):1422–1432
2. Dinauer MC (1993) The respiratory burst oxidase and the molecular genetics of chronic granulomatous disease. Crit Rev Clin Lab Sci 30(4):329–369
3. Pollock JD, Williams DA, Gifford MA, Li LL, Du X, Fisherman J, Orkin SH, Doerschuk CM, Dinauer MC (1995) Mouse model of X-linked chronic granulomatous disease, an inherited defect in phagocyte superoxide production. Nat Genet 9(2):202–209
4. Klebanoff SJ (1968) Myeloperoxidase-halide-hydrogen peroxide antibacterial system. J Bacteriol 95(6):2131–2138
5. Davies MJ, Hawkins CL, Pattison DI, Rees MD (2008) Mammalian heme peroxidases: from molecular mechanisms to health implications. Antioxid Redox Signal 10 (7):1199–1234
6. Bedard K, Lardy B, Krause KH (2007) NOX family NADPH oxidases: not just in mammals. Biochimie 89(9):1107–1112
7. Nisimoto Y, Diebold BA, Cosentino-Gomes D, Lambeth JD (2014) Nox4: a hydrogen peroxide-generating oxygen sensor. Biochemistry 53(31):5111–5120
8. Hecker L, Vittal R, Jones T, Jagirdar R, Luckhardt TR, Horowitz JC, Pennathur S, Martinez FJ, Thannickal VJ (2009) NADPH oxidase-4 mediates myofibroblast activation and fibrogenic responses to lung injury. Nat Med 15(9):1077–1081
9. Hecker L, Logsdon NJ, Kurundkar D, Kurundkar A, Bernard K, Hock T, Meldrum E, Sanders YY, Thannickal VJ (2014) Reversal of persistent fibrosis in aging by targeting Nox4-Nrf2 redox imbalance. Sci Transl Med 6(231):231ra247
10. Ley B, Collard HR, King TE Jr (2011) Clinical course and prediction of survival in idiopathic pulmonary fibrosis. Am J Respir Crit Care Med 183(4):431–440
11. Thannickal VJ, Lee DY, White ES, Cui Z, Larios JM, Chacon R, Horowitz JC, Day RM, Thomas PE (2003) Myofibroblast differentiation by transforming growth factor-beta1 is dependent on cell adhesion and integrin signaling via focal adhesion kinase. J Biol Chem 278 (14):12384–12389
12. Huang X, Yang N, Fiore VF, Barker TH, Sun Y, Morris SW, Ding Q, Thannickal VJ, Zhou Y (2012) Matrix stiffness-induced myofibroblast differentiation is mediated by intrinsic mechanotransduction. Am J Respir Cell Mol Biol 47(3):340–348
13. Thannickal VJ, Zhou Y, Gaggar A, Duncan SR (2014) Fibrosis: ultimate and proximate causes. J Clin Invest 124(11):4673–4677
14. Paik YH, Iwaisako K, Seki E, Inokuchi S, Schnabl B, Osterreicher CH, Kisseleva T, Brenner DA (2011) The nicotinamide adenine dinucleotide phosphate oxidase (NOX) homologues NOX1 and NOX2/gp91(phox) mediate hepatic fibrosis in mice. Hepatology 53 (5):1730–1741
15. Jiang JX, Chen X, Serizawa N, Szyndralewiez C, Page P, Schroder K, Brandes RP, Devaraj S, Torok NJ (2012) Liver fibrosis and hepatocyte apoptosis are attenuated by GKT137831, a novel NOX4/NOX1 inhibitor in vivo. Free Radic Biol Med 53(2):289–296
16. Sancho P, Mainez J, Crosas-Molist E, Roncero C, Fernandez-Rodriguez CM, Pinedo F, Huber H, Eferl R, Mikulits W, Fabregat I (2012) NADPH oxidase NOX4 mediates stellate cell activation and hepatocyte cell death during liver fibrosis development. PLoS One 7(9):e45285

17. Raghu G, Weycker D, Edelsberg J, Bradford WZ, Oster G (2006) Incidence and prevalence of idiopathic pulmonary fibrosis. Am J Respir Crit Care Med 174(7):810–816
18. Fell CD, Martinez FJ, Liu LX, Murray S, Han MK, Kazerooni EA, Gross BH, Myers J, Travis WD, Colby TV, Toews GB, Flaherty KR (2010) Clinical predictors of a diagnosis of idiopathic pulmonary fibrosis. Am J Respir Crit Care Med 181(8):832–837
19. Collard HR (2010) The age of idiopathic pulmonary fibrosis. Am J Respir Crit Care Med 181(8):771–772
20. Thannickal VJ (2013) Mechanistic links between aging and lung fibrosis. Biogerontology 14(6):609–615
21. Lopez-Otin C, Blasco MA, Partridge L, Serrano M, Kroemer G (2013) The hallmarks of aging. Cell 153(6):1194–1217
22. Mora AL, Bueno M, Rojas M (2017) Mitochondria in the spotlight of aging and idiopathic pulmonary fibrosis. J Clin Invest 127 (2):405–414
23. Schafer MJ, Haak AJ, Tschumperlin DJ, LeBrasseur NK (2018) Targeting senescent cells in fibrosis: pathology, paradox, and practical considerations. Curr Rheumatol Rep 20(1):3
24. Bernard K, Logsdon NJ, Ravi S, Xie N, Persons BP, Rangarajan S, Zmijewski JW, Mitra K, Liu G, Darley-Usmar VM, Thannickal VJ (2015) Metabolic reprogramming is required for myofibroblast contractility and differentiation. J Biol Chem 290(42):25,427–25,438
25. Bernard K, Logsdon NJ, Benavides GA, Sanders Y, Zhang J, Darley-Usmar VM, Thannickal VJ (2018) Glutaminolysis is required for transforming growth factor-beta1-induced myofibroblast differentiation and activation. J Biol Chem 293(4):1218–1228
26. Jenkins RG, Moore BB, Chambers RC, Eickelberg O, Konigshoff M, Kolb M, Laurent GJ, Nanthakumar CB, Olman MA, Pardo A, Selman M, Sheppard D, Sime PJ, Tager AM, Tatler AL, Thannickal VJ, White ES, ATS Assembly on Respiratory Cell and Molecular Biology (2017) An official American thoracic society workshop report: use of animal models for the preclinical assessment of potential therapies for pulmonary fibrosis. Am J Respir Cell Mol Biol 56(5):667–679
27. Thannickal VJ, Roman J (2010) Challenges in translating preclinical studies to effective drug therapies in idiopathic pulmonary fibrosis. Am J Respir Crit Care Med 181(6):532–533
28. Srivastava AK, Khare P, Nagar HK, Raghuwanshi N, Srivastava R (2016) Hydroxyproline: a potential biochemical marker and its role in the pathogenesis of different diseases. Curr Protein Pept Sci 17(6):596–602
29. Carnesecchi S, Deffert C, Donati Y, Basset O, Hinz B, Preynat-Seauve O, Guichard C, Arbiser JL, Banfi B, Pache JC, Barazzone-Argiroffo C, Krause KH (2011) A key role for NOX4 in epithelial cell death during development of lung fibrosis. Antioxid Redox Signal 15(3):607–619
30. Gorin Y, Cavaglieri RC, Khazim K, Lee DY, Bruno F, Thakur S, Fanti P, Szyndralewiez C, Barnes JL, Block K, Abboud HE (2015) Targeting NADPH oxidase with a novel dual Nox1/Nox4 inhibitor attenuates renal pathology in type 1 diabetes. Am J Physiol Renal Physiol 308(11):F1276–F1287
31. McGovern TK, Robichaud A, Fereydoonzad L, Schuessler TF, Martin JG (2013) Evaluation of respiratory system mechanics in mice using the forced oscillation technique. J Vis Exp (75):e50172
32. Gilhodes JC, Jule Y, Kreuz S, Stierstorfer B, Stiller D, Wollin L (2017) Quantification of pulmonary fibrosis in a bleomycin mouse model using automated histological image analysis. PLoS One 12(1):e0170561
33. Ruscitti F, Ravanetti F, Essers J, Ridwan Y, Belenkov S, Vos W, Ferreira F, KleinJan A, van Heijningen P, Van Holsbeke C, Cacchioli A, Villetti G, Stellari FF (2017) Longitudinal assessment of bleomycin-induced lung fibrosis by Micro-CT correlates with histological evaluation in mice. Multidiscip Respir Med 12:8
34. Waghray M, Cui Z, Horowitz JC, Subramanian IM, Martinez FJ, Toews GB, Thannickal VJ (2005) Hydrogen peroxide is a diffusible paracrine signal for the induction of epithelial cell death by activated myofibroblasts. FASEB J 19 (7):854–856
35. Bernard K, Logsdon NJ, Miguel V, Benavides GA, Zhang J, Carter AB, Darley-Usmar VM, Thannickal VJ (2017) NADPH oxidase 4 (Nox4) suppresses mitochondrial biogenesis and bioenergetics in lung fibroblasts via a nuclear factor erythroid-derived 2-like 2 (Nrf2)-dependent pathway. J Biol Chem 292(7):3029–3038

Chapter 30

Proteomic Methods to Evaluate NOX-Mediated Redox Signaling

Christopher M. Dustin, Milena Hristova, Caspar Schiffers, and Albert van der Vliet

Abstract

The NADPH oxidase (NOX) family of proteins is involved in regulating many diverse cellular processes, which is largely mediated by NOX-mediated reversible oxidation of target proteins in a process known as redox signaling. Protein cysteine residues are the most prominent targets in redox signaling, and to understand the mechanisms by which NOX affect cellular pathways, specific methodology is required to detect specific oxidative cysteine modifications and to identify targeted proteins. Among the many potential redox modifications involving cysteine residues, reversible modifications most relevant to NOX are sulfenylation (P-SOH) and *S*-glutathionylation (P-SSG), as both can induce structural or functional alterations. Various experimental approaches have been developed to detect these specific modifications, and this chapter will detail state-of-the-art methodology to selectively evaluate these modifications in specific target proteins in relation to NOX activation. We also discuss some of the limitations of these procedures and potential complementary approaches.

Key words NADPH oxidases, DUOX, H_2O_2, Redox signaling, Sulfenylation, *S*-glutathionylation, Dimedone

1 Introduction

The NADPH oxidases (NOX) were originally recognized as enzymes predominantly involved in host defense but are now also widely recognized to have more diverse roles in many aspects of cell biology and cellular signaling [1]. Generally, this relates to regulated production of reactive oxygen species (ROS) by activated NOX/DUOX enzymes, such as superoxide (O_2^-) or hydrogen peroxide (H_2O_2), which in turn act as second messenger molecules that can reversibly modify proteins at susceptible redox-sensitive amino acids, analogous to reversible phosphorylation/dephosphorylation, a process known as redox signaling [2–4]. The concept of redox signaling applies to oxidation of redox-active metal ions in metalloproteins or oxidant-sensitive amino acid side chains

Ulla G. Knaus and Thomas L. Leto (eds.), *NADPH Oxidases: Methods and Protocols*, Methods in Molecular Biology, vol. 1982, https://doi.org/10.1007/978-1-4939-9424-3_30, © Springer Science+Business Media, LLC, part of Springer Nature 2019

such as cysteine and methionine. Since cysteine thiols (P-SH, where P denotes the rest of the protein) are typically the most redox-sensitive among the amino acids, particularly when localized in areas in which interactions with neighboring amino acid side chains can lower the cysteine pK_a and increase its nucleophilicity [5], most attention has been paid to analysis of protein cysteine oxidation as a central mechanism of redox signaling. In addition to direct oxidation of protein cysteines by NOX-derived ROS, such redox modifications can also occur in a "relay" fashion, in which initial oxidation of highly susceptible cysteines (e.g., peroxiredoxins) is propagated by transferring oxidizing equivalents to target proteins through thiol-disulfide exchange mechanisms [6].

The most widely recognized example of redox-based signaling involves the reversible inactivation of protein tyrosine phosphatases, which possess a redox-sensitive active site Cys residue. Oxidation of this Cys inactivates phosphatase function, thereby allowing for extended phosphorylation of target proteins and their prolonged (in)activation [7, 8]. X-ray crystallography data indicate that oxidatively inhibited phosphatase PTP1B actually contains a sulfenyl amide (Fig. 1), likely formed by reaction of initially generated P-SOH with an adjacent amide bond, which may be responsible for its inactivation [9]. Alternatively, others argue that PTP1B is also inactivated by *S*-glutathionylation [10]. Recent studies by us and others have indicated that tyrosine kinases themselves can also be regulated in a more direct fashion by oxidative mechanisms mediated by NOX/DUOX activation. For example, activation of the epidermal growth factor receptor (EGFR) by its cognate ligand epidermal growth factor (EGF) was found to result in NOX2-mediated oxidation of a conserved cysteine within the ATP-binding region (C797) to P-SOH, which is thought to alter electrostatic interactions within the active site and enhance kinase activity [11, 12]. Similarly, the non-receptor tyrosine kinase Src was found to be susceptible to oxidative events induced by, e.g., NOX activation [13–15]. Our recent studies demonstrate that activation of both EGFR and Src in airway epithelial cells in response to protease allergens or other injurious triggers occurs in coordination with cysteine oxidation within these proteins and is mediated by activation of the epithelial NOX isoform DUOX1 [16–19]. Importantly, while NOX/DUOX activation can induce formation of both P-SOH and P-SSG within these proteins in a sequential mechanism, evidence suggests that it is P-SOH rather than P-SSG that enhances their kinase activity [12, 15]. In other cases, *S*-glutathionylation may be the primary mechanism involved in oxidant-mediated enzyme inactivation, as was demonstrated in the case of IKK-β in the context pro-inflammatory signaling [20] or in oxidative processing of ER-localized pools of cellular receptors such as Fas [21].

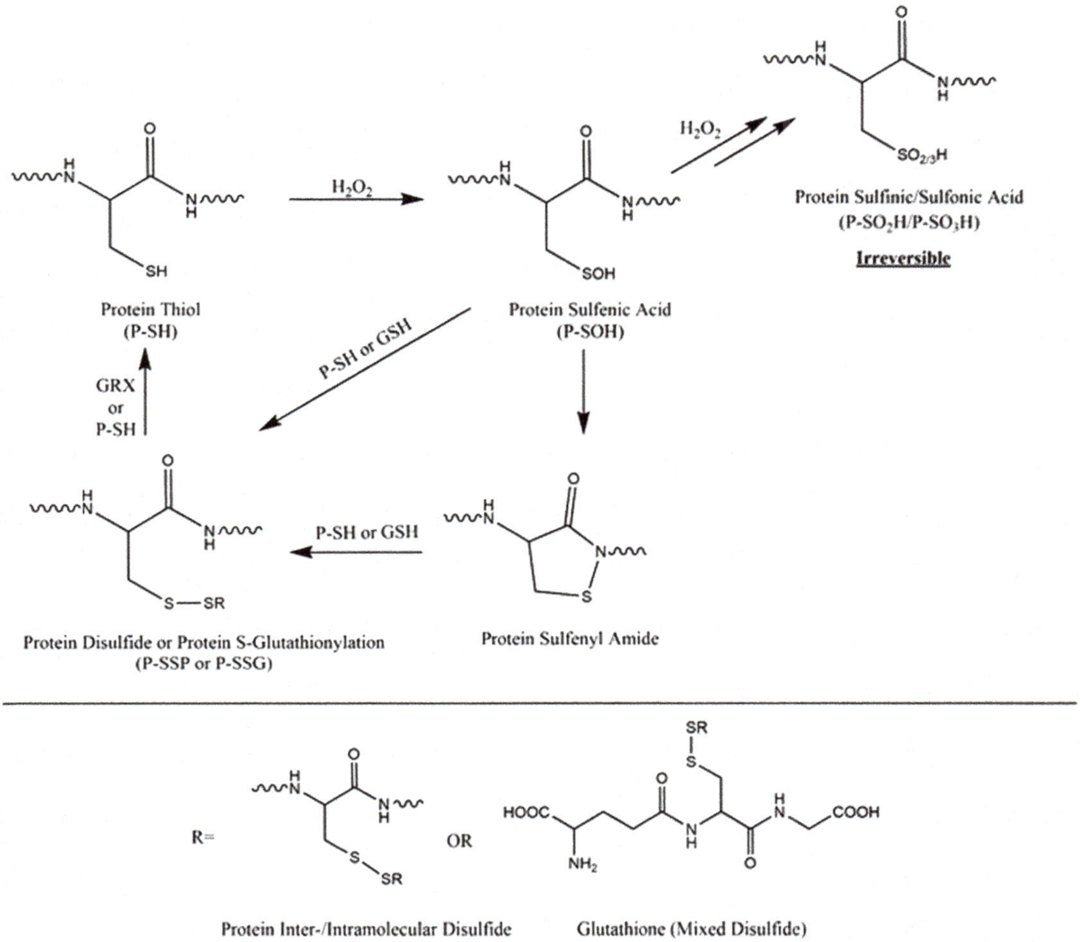

Fig. 1 Schematic representation of potential oxidative intermediates involved in ROS-dependent signaling

A main complicating factor in the analysis of protein cysteine oxidation is its diverse nature (Fig. 1). In the context of NOX-dependent signaling, oxidation of a cysteine (typically by H_2O_2) initially generates a sulfenic acid (P-SOH). Such formation of P-SOH impairs the function of catalytic cysteines and can also alter protein function by other electrostatic mechanisms [12, 15]. However, this product is typically not stable within proteins and readily reacts with other cysteine thiols to form a disulfide bond (P-S-S-R, where R is another protein or small molecule), either by reaction with another cysteine within the same protein (to form an intramolecular disulfide), reaction with P-SH in different proteins (to form an intermolecular disulfide), or reaction with low-molecular-weight thiols such as free cysteine or glutathione (GSH) (resulting in *S*-cysteinylation (P-S-S-Cys) or *S*-glutathionylation (P-SSG)). These various disulfide intermediates can be reduced back to the original reduced thiol (P-SH) by oxidoreductases such as thioredoxins and glutaredoxins [22]. P-SOH is also subject to further oxidation to sulfinic and sulfonic acids (P-$SO_{2/3}H$; Fig. 1), but this is mostly relevant to conditions of severe oxidative stress and is likely not of major significance in the context of physiological cell signaling. It is often unclear which oxidative cysteine modification is most critical for regulating protein function. Formation of P-SOH may be considered a gateway intermediate to facilitate disulfide formation, which often is thought to be the most relevant modification that regulates protein function, but in some cases P-SOH rather than P-SSG may be primarily responsible for the redox signal, and conversion of P-SOH to P-SSG instead serves to prevent overoxidation and allow regeneration of the initial cysteine thiol [12, 15]. It is important to note that the biochemistry of thiol-based redox signaling is more complex and is not restricted to the actions of NOX-derived ROS but can also involve reversible modifications related to production of nitric oxide (NO), a process known as *S*-nitrosylation [23], alkylation by various cellular electrophiles [24], and formation of polysulfides (P-S-$S_{(n)}$-H) due to cellular formation of, e.g., hydrogen sulfide (H_2S) or other polysulfide species [25, 26]. Although these modifications are typically not thought to be directly regulated by NOX/DUOX activation, they could be affected by NOX/DUOX in more indirect ways. Indeed, recent studies indicate that NOX activation can also promote oxidation of polysulfide species resulting in so-called perthiosulfenic acids (P-S-S(n)OH) [27].

Based on these various considerations above, there is a strong need for reliable methodology that allows for identification and quantification of specific protein targets that are oxidized in response to NOX/DUOX activation, as well as the specific oxidative modification(s) (e.g., P-SOH vs P-SSG) within these targets. Much of our current understanding in this regard is based on indirect approaches to assess protein cysteine oxidation, such as isotope-coded affinity tag (ICAT) proteomics approaches [28] or so-called switch techniques in which oxidized proteins are subject to chemical reduction followed by labeling with thiol-reactive detection reagents (e.g., biotin). While some of these biotin switch techniques are based on putatively selective chemistry to reduce specific forms of oxidized cysteine, e.g., ascorbate for *S*-nitrosothiols [29] or glutaredoxin-based reversal of P-SSG [30], such approaches are not always specific and may detect alternative modifications such as P-SOH. Detection of P-SOH has over the years mostly relied on the use of cell-permeable 5,5-dimethyl-1,3-cyclohexanedione (commonly known as dimedone) or related reagents, which are thought to preferentially react with the P-SOH and are unreactive with P-SH, in contrast with previously used reagents such as 4-chloro-7-nitrobenzofurazan [31]. Identification of dimedone adducts in specific proteins can then be confirmed by mass spectrometry or by immunochemical detection with specific antibodies raised against dimedone-adducted Cys residues [32]. To expand analytical approaches, dimedone has also been synthesized as a biotin-conjugated derivative, which allows for facile purification [15, 33] or other utilities taking advantage of biotin-streptavidin interactions, such as fluorophore labeling for overall quantification or subcellular localization of sulfenylated proteins. Many of these reagents are now commercially available (e.g., from Kerafast or Cayman Chemical). While many research groups, including our own [15, 17, 18], have successfully utilized such biotin-conjugated dimedone probes, an intrinsic limitation is their limited cell permeability, which means that they are best used during cell lysis or tissue homogenization, requiring careful steps to avoid artificial oxidation and labeling during such lysis/homogenization procedures. To circumvent this, dimedone has been conjugated to alkyne or azide moieties, which improves cell permeability and allows for their use in intact cells, to probe oxidation in a more in situ manner. Protein adducts with dimedone-alkyne or -azide can then be evaluated using click chemistry reactions after cell lysis [11, 34].

With respect to analysis of protein *S*-glutathionylation [35], many studies have relied on use of an α-GSH antibody (e.g., ViroGen α-GSH) to detect *S*-glutathionylation in specific proteins or to perform immunoprecipitation. While this has allowed for successful detection of *S*-glutathionylated proteins in some cases [36, 37], it is unclear whether this approach works equally well for all *S*-glutathionylated proteins [38], and affinity may in some cases be limited. Another approach is to use a biotin switch method in which reduced cysteines in protein mixtures are first alkylated, after which P-SSG is reduced to the corresponding thiol using an enzymatic glutaredoxin-based reduction step and subsequently recovered by thiol-specific detection methods [39]. While this method can be selective for P-SSG, it is cumbersome and critical controls are needed, since the GSH that is included in the glutaredoxin-based step could also reduce other oxidized forms of cysteine. An alternative approach to assess formation of P-SSG in situ is by preloading cells with, e.g., biotin-GSH conjugates, after which incorporation of biotin into proteins after, e.g., cell stimulation reflects increased *S*-glutathionylation. The use of esterified forms of biotin-GSH (BioGEE) enhances cell permeability and allows for facile loading into cells prior to experimentation [40, 41]. Another advantage of this labeling approach is that subsequent purification of biotin-tagged proteins is facilitated by the fact that the Biotin-G-SS-protein adduct is DTT-cleavable, to allow for easy elution of proteins and minimize contamination with non-specifically bound proteins.

In the following sections, we will describe specific methodology for analysis of P-SOH or P-SSG in specific cell proteins in the context of NOX/DUOX signaling (Fig. 2). Approaches to detect P-SOH will utilize two specific dimedone-based probes: (1) the

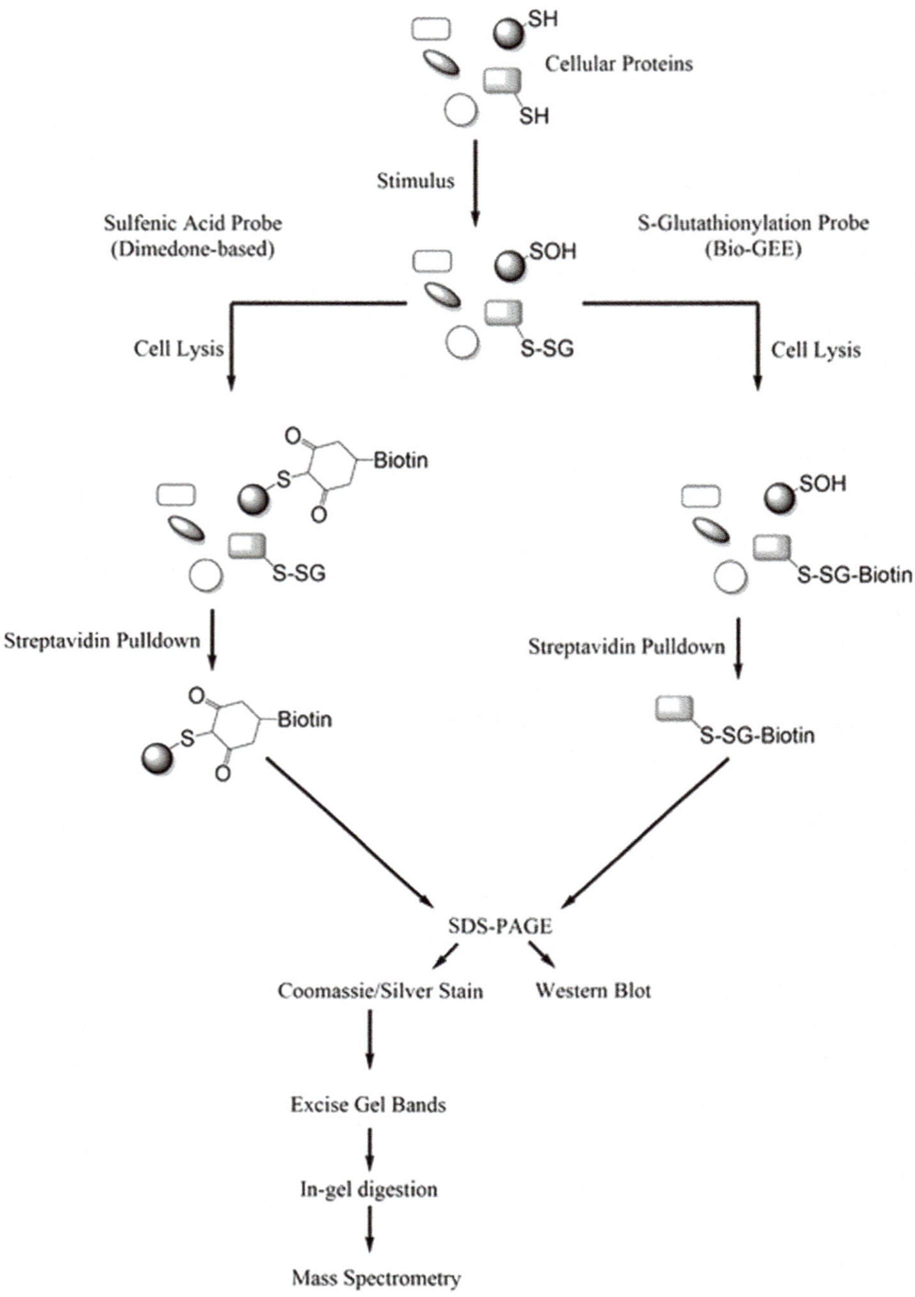

Fig. 2 Flowchart illustrating the different derivatization steps to detect P-SOH or P-SSG by either MS approaches or Western blot of candidate proteins

biotin-conjugated DCP-Bio1 probe and (2) the alkyne-containing DYn-2 probe (which can be linked to biotin azide via click reactions). Biotin tagging of P-SOH is then followed by avidin-based purification for analysis by MS (for unbiased protein screening) or by Western blot analysis of suspected target proteins. The methods described below are largely based on protocols developed by Nelson and co-workers [42] and the Carroll group [2, 12], with some modifications. Analysis of P-SSG, modeled after the method developed by Sullivan and co-workers [40], will be based on cell preloading with BioGEE, after which adduction of cellular biotin-tagged GSH to proteins will be determined as a reflection similar adduction of endogenous GSH (*S*-glutathionylation). We will present some examples in the context of NOX-mediated signaling and discuss some limitations and pitfalls in the "Notes" section. It is important to recognize that NOX enzymes could directly oxidize a target protein but may also work by more oxidative mechanisms (e.g., by activating of other NOXes or mitochondria), and therefore it is important to also consider complementary approaches that address spatial considerations, using, e.g., fluorescent imaging strategies to localize specific oxidative events in proximity to a specific NOX enzyme [43].

2 Materials

Prepare all aqueous solutions in Milli-Q H_2O or water of comparable purity. All reagents should be of the highest available purity and should be stored per manufacturer's recommendation.

2.1 Materials for Identification of Targets of Sulfenylation

1. Cultured cells of interest (*see* **Note 1**).
2. Cell stimulus of choice (e.g., 10 μg/mL EGF in H_2O or 10 mM ATP in H_2O).
3. Phosphate Buffered Saline pH 7.4.
4. Dimethyl sulfoxide (DMSO).
5. 100 mM DCP-Biol (Kerafast) or 100 mM–1.0 M DYn-2 (Kerafast) in DMSO (*see* **Notes 2** and **3**).
6. Catalase (~40,000 Units/mL; Worthington).
7. 1 M *N*-ethylmaleimide in ethanol.
8. Western solubilization buffer (WSB) (50 mM HEPES, 250 mM NaCl, 1.5 mM $MgCl_2$, 1% Triton X-100, 10% glycerol, 1 mM EDTA, 1 mM phenylmethylsulfonyl fluoride (PMSF), 2 mM Na_3VO_4, 10 mg/mL aprotinin, and 10 mg/mL leupeptin (pH 7.4)) (*see* **Note 4**).
9. DYn-2 lysis buffer (50 mM HEPES pH 7.4, 150 mM NaCl, 1% NP-40, 0.1% SDS, 200 μ/mL catalase, EDTA-free protease, and phosphatase inhibitor cocktails (e.g., Thermo Fisher Scientific Halt protease and phosphatase inhibitor cocktail)) (*see* **Note 5**).
10. WSB wash buffer (50 mM HEPES, 250 mM NaCl, 1% Triton X-100, 10% Glycerol, pH 7.4) (*see* **Note 6**).

11. NeutrAvidin high-capacity beads (Thermo Fisher Scientific) or comparable streptavidin agarose beads.
12. Click reaction mix (1 mM Tris[(1-benzyl-1*H*-1,2,3-triazol-4-yl)methyl]amine (TBTA) in 9:1 acetonitrile/DMSO, 5 mM sodium ascorbate in PBS, 500 μM $CuSO_4$ in PBS, and 500 μM biotin azide in DMSO (Kerafast)). Add 10 μL 100 mM TBTA, 10 μL 500 mM sodium ascorbate, 10 μL 50 mM $CuSO_4$, and 900 μL water. Check the pH of the solution with a pH strip. If the pH is 7.0, add 50 μL water and 50 μL 10 mM biotin azide.
13. 1% Sodium dodecyl sulfate (SDS) in H_2O.
14. 6 M Urea in PBS.
15. 1 M NaCl in H_2O.
16. 100 μM Ammonium bicarbonate + 10 mM DTT in H_2O.
17. 100 μM Ammonium bicarbonate in water.
18. Elution buffer (50 mM Tris, 2% SDS, 1 mM EDTA, pH 7.4) in H_2O.
19. Laemmli buffer (the concentration prepared depends on preferred sample volume. The final, diluted concentration that we use is 0.063 M Tris–Cl, 2% SDS, 10% v/v glycerol, 0.02% β-mercaptoethanol, 0.01% bromophenol blue, and pH 6.8).

2.2 Materials for Identification of S-Glutathionylated Proteins

1. Cultured cells of interest (*see* previous section and **Note 1**).
2. Stimulus of choice (*see* previous section).
3. Phosphate-buffered saline, pH 7.4.
4. RIPA lysis buffer (20 mM Tris–HCl (pH 7.5), 150 mM NaCl, 1 mM Na_2EDTA, 1 mM EGTA, 1% NP-40, 1% sodium deoxycholate, 2.5 mM sodium pyrophosphate, 1 mM β-glycerophosphate, 1 mM Na_3VO_4, 1 μg/mL leupeptin, and 1 mM PMSF).
5. 1 M *N*-ethylmaleimide in ethanol.
6. NeutrAvidin high-capacity resin (Thermo Fisher Scientific) or comparable streptavidin agarose resin.
7. 500 mM $NaHCO_3$ in H_2O.
8. 110 mM Glutathione ethyl ester (GEE) in $NaHCO_3$ (*see* **Note 7**).
9. 100 mM EZ link Sulfo-NHS-Biotin.
10. 500 mM NH_4HCO_3.
11. PD MidiTrap G-25 spin columns.
12. 10 mM DTT in H_2O.
13. PBS + 0.1% sodium dodecyl sulfate.
14. Amicon spin filters (0.5 mL, 3 kDa cutoff).
15. Laemmli buffer. (The concentration prepared depends on preferred sample volume. *See* **Note 8**).

3 Methods

The methods described below are broadly applicable to cell-based studies of cell signaling pathways induced by, e.g., cytokines or growth factors, and the role of NOX enzymes can be assessed by, e.g., siRNA-mediated deletion, CRISPR/Cas9 gene editing approaches, or pharmacological inhibitors (although NOX isoform-specific inhibitors are still largely lacking [44]) or by using cells isolated from different genetic mouse models of NOX deficiency. These methods to identify P-SOH or P-SSG have been used successfully in identifying targets of protein oxidation [40, 42, 45–47] and also in various studies by our group that address redox signaling by DUOX1 [15, 17, 18, 41]. Approaches have also been developed for quantitative analysis of, e.g., P-SOH through mass spectrometry utilizing isotope-labeled dimedone approaches, either with or without biotin purification [46, 48], but these are particularly challenging for large proteins with multiple cysteines (requiring diverse protein digestion procedures) or low-abundance proteins that are often involved in cell signaling pathways. The importance of specific cysteines could alternatively be addressed by mutating these cysteines in suspected or confirmed target proteins.

3.1 Cell Culture and Pretreatments

1. Culture cells of interest to desired confluence. Typically, 100,000–200,000 cells per treatment group are sufficient for successful detection of target proteins by Western blotting, but considerably more cells will be needed for MS analysis (*see* **Note 9**).
2. If applicable, perform cell transfections to manipulate, e.g., NOX/DUOX enzymes according to established protocols. Alternatively, pretreat with small molecule NOX/DUOX inhibitors as appropriate.
3. Prior to cell treatments, replace culture medium with serum-free medium.
4. For analysis of P-SOH using DYn-2, preincubate cells with 5 mM DYn-2 reagent in serum-free medium for 30 min (*see* **Note 10**).
5. For analysis of P-SSG, preincubate cells with BioGEE, by changing to serum-free media containing 250 μM BioGEE and incubating for 1 h at 37 °C.
   (a) BioGEE can be purchased commercially (Thermo Fisher Scientific) but can also be easily prepared as follows: Mix 250 μL 110 mM GEE with 250 μL 100 mM Sulfo-NHS-biotin (*see* Subheading 2), and react at RT for 1 h with mixing. Quench the reaction by the addition of 2 mL 500 mM NH_4HCO_3 (*see* **Note 11**). The final concentration of BioGEE is ~10 mM (*see* **Note 12**).

6. Treat cells with appropriate stimulus to activate NOX/DUOX-dependent redox signaling (e.g., ATP to stimulate DUOX1 or EGF to activate NOX2 [15]).
7. At appropriate time points, place cells on ice, remove medium and wash with PBS, and add appropriate lysis buffer (see next section).
   (a) In the case of cells preloaded with BioGEE, wash cells with PBS containing 50 mM NEM prior to cell lysis. This NEM step removes residual unreacted BioGEE and prevents its reaction with proteins during cell lysis.

3.2 Cell Lysis and Avidin Purification of DCP-Bio1- or DYn-2-Tagged Proteins

1. Lyse cells on ice by adding 100 μL WSB (DCP-Bio1) or DYn-2 lysis buffer (*see* Subheading 2) supplemented with 10 mM NEM (10 μL of 1 M Stock/per mL buffer) and 200 units/mL catalase (5 μL 40,000 units/mL stock per mL buffer). (*Optional*) If using DCP-Bio1, add 10 μL of a 100 mM stock in DMSO to the WSB buffer for a final concentration of 1 mM. Incubate on ice for 1 h with gentile rocking. If adding protease inhibitors (e.g., for DYn-2 buffer), do so before adding to cells.
2. Scrape cells, collect lysate into an Eppendorf tube, and sonicate briefly. Centrifuge at 21,000 × *g* for 5 min to remove cell debris. Lysates can be stored at −20 °C until use.
3. Determine protein concentration using a BCA assay, and collect equal amounts of protein from each sample for purification. This amount can range from ~100 μg to ~10 mg (*see* **Note 13**). Ensure that 30–50 μL of lysate is saved for analysis of input controls and/or total streptavidin analysis.
4. Wash NeutrAvidin beads (25–50 μL bead slurry/sample) with 20 mM Tris pH 7.4 buffer 3× using a rotisserie mixer. Collect beads between washes via centrifugation at 1200 × *g* for 5 min. After the final wash, resuspend beads in ~2 volumes of Tris buffer to generate fresh NeutrAvidin bead slurry for use in subsequent steps.
5. If using DYn-2, preclear lysates with around 30 μL fresh NeutrAvidin slurry to remove endogenously biotinylated protein. Centrifuge samples at 1200 × *g*, and add supernatant to new tubes while discarding the beads. Following this preclearing step, perform the click reaction with a 1:1 solution of lysate to click reaction mix (*see* Subheading 2). React for 1 h at RT with mixing. Quench the reaction with 1 mM EDTA.
6. Add protein sample to 0.5 mL Amicon centrifuge filters (3 kDa molecular weight cutoff) and perform buffer exchange with WSB wash three times by centrifugation at 13,500 × *g* for 20–30 min. Discard the flow-through from each spin. This step removes excess unconjugated biotin reagents.

7. Collect retained liquid containing the protein sample (typically ~50 μL), rinse the top of the filter with ~0.2 mL WSB wash, and combine with protein sample.
8. Distribute washed beads evenly between new tubes. To the beads, add the buffer-exchanged protein sample from **steps 6** and 7, and dilute sample/slurry mix to a total volume of 0.5–1 mL with Tris buffer.
9. Rotate beads overnight at 4 °C on a rotisserie mixer.
10. The following day, spin beads at 1200 × *g* for 5 min, and collect supernatant (*see* **Note 14**). Wash beads with 1 mL 0.1% SDS for 30 min with rotisserie mixing. Spin beads and wash with new 1% SDS again for 5 additional minutes. Spin beads and remove 1% SDS.
11. Wash beads with 4 M urea for 30 min with rotisserie mixing. Spin beads, discard supernatant, and wash for an additional 5 min with 4 M urea. Spin beads and discard supernatant.
12. Wash beads with 1 M NaCl for 30 min with rotisserie mixing. Spin beads, and discard supernatant. Wash for an additional 5 min with new 1 M NaCl. Spin beads and discard supernatant.
13. Wash beads with 100 μM ammonium bicarbonate +10 mM DTT for 5 min and rotisserie mixing. Spin beads and discard supernatant (*see* **Note 15**).
14. Wash beads with 100 μM Ammonium bicarbonate for 5 min and rotisserie mixing. Spin beads and discard supernatant.
15. Wash beads with water for 5 min and rotisserie mixing. Spin beads and discard supernatant.
    (a) Wash **steps 10–14** serve to remove proteins that are bound non-specifically to the beads or to biotin-tagged proteins (and would represent false-positives; *see* **Note 16**).
16. Add 100 μL elution buffer. Boil samples at 100 °C for 10 min. Briefly vortex samples and spin at 21,000 × *g* for 5 min. Remove and keep supernatant.
17. Add Laemmli buffer to supernatant and mix briefly (*see* **Note 8**). Samples can be stored at −20 °C until analysis.

3.3 Cell Lysis and Avidin Purification of BioGEE-Tagged Proteins

1. Lyse cells with RIPA buffer supplemented with 50 mM NEM (to avoid thiol exchange and loss of P-SSG during lysis) on ice for 30–60 min with gentle rocking.
2. Scrape cells, collect lysate, and sonicate. Centrifuge to remove cell debris. Lysate can be stored at −20 °C until use.
3. Determine protein content in cell lysate (e.g., using BCA assay), and save 30–50 μL lysate for input and total streptavidin analysis.

4. Dilute equal amounts of each sample (*see* **Note 13**) to 1 mL with RIPA buffer +50 mM NEM.
5. Condition G25 columns with three washes of RIPA buffer via centrifugation at 1000 × *g* for 2 min.
6. Load sample on to column with centrifugation at 1000 × *g* for 2 min. This step is needed to remove unreacted Bio-GSH, as proteins in the sample will be in the flow through.
7. In a separate tube, add 150 μL high-capacity NeutrAvidin bead slurry per sample, and wash twice with 1 mL PBS.
8. Add 50 μL bead slurry to each sample and rotisserie mix for 30 min at 4 °C.
9. Spin sample at 1200 × *g* for 5 min. Discard the beads, save the supernatant, and add it to a new vial.
10. To the supernatant, add an additional 100 μL bead slurry and rotisserie mix overnight at 4 °C.
11. The next day, spin the sample at 1200 × *g* and save the supernatant (*see* **Note 14**).
12. Wash the beads five times for 10–20 min with 1 mL cold RIPA buffer with rotisserie mixing. Spin and discard the supernatant after each wash (*see* **Note 17**).
13. Wash the beads with 1 mL PBS + 0.1% SDS with rotisserie mixing. Spin and discard the supernatant.
14. Resuspend the beads in 500 μL PBS + 0.1% SDS, and mix for 30 min at room temperature.
15. Spin the beads down, and save the supernatant for analysis of non-glutathionylated proteins ("-DTT"; *see* **Note 18**).
16. To elute *S*-glutathionylated proteins, add 500 μL PBS with 0.1% SDS containing 10 mM DTT. Mix for 30 min at room temperature.
17. Spin the sample again and save the supernatant. To ensure all proteins are removed successfully, add another 500 μL PBS with 0.1% SDS and wash. After spinning, add this to the DTT eluted samples. This sample ("+DTT") contains the *S*-glutathionylated proteins (*see* **Note 19**).
18. Concentrate samples using 3 kDa Amicon filters (0.5 mL capacity), and spin at 21,000 × *g* for 30 min until all sample is ~100 μL in volume. If the sample is below 100 μL, dilute with buffer.
19. Add Laemmli buffer to samples and heat at 95 °C for 5 min (*see* **Note 8**). Samples can be stored at −20 °C until analysis.

3.4 Analysis of Avidin-Purified Proteins

Avidin-purified proteins representing either P-SOH or P-SSG (based on Subheadings 3.3 or 3.4) can be analyzed by LC-MS/MS (for unbiased analysis) or by SDS-PAGE and Western blot analysis of candidate proteins of interest. For LC-MS/MS identification, separate proteins by SDS-PAGE, stain gel with Coomassie brilliant blue or silver stain, excise bands, and work up according to standard in-gel digestion protocols. SDS-PAGE allows for identification of major protein bands and facile sample desalting prior to tryptic digestion, but is not strictly required, and standard solution-based digestion methods are also suitable as long as peptides are desalted prior to LC-MS/MS, using, e.g., C18 ZipTips (Millipore, according to manufacturer's instructions).

For analysis of candidate proteins by Western blot, separate proteins by standard SDS-PAGE, transfer separated proteins to nitrocellulose membranes, and probe membranes with either Streptavidin-HRP (to visualize all biotin-tagged proteins) or antibodies against specific proteins of interest, according to established protocols. Importantly, cell lysates (*see* **step 3** in Subheading 3.2 or 3.3) should be analyzed similarly as input loading controls. Figure 3 shows results from a representative analysis of P-SOH in NCI-H292 cells in response to stimulation with either 100 μM ATP or 100 ng/mL EGF, using DYn-2. Figure 4 shows an example of P-SSG analysis in EGFR or Src in murine tracheal epithelial cells in response to stimulation with EGF.

4 Notes

1. In principle, this can be applied to any cultured type, either grown as adherent cultures or in suspension culture, although methodology presented here is based on studies with adherent cultures.
2. While dimedone is not known to react with reduced thiols or most other oxidized forms, it is also reactive with sulfenyl amides, which are typically formed after initial P-SOH formation, but potentially via other mechanisms as well [6, 49, 50]. Kinetic studies suggest that reaction of dimedone with P-SOH is rather slow and may be limited by other cellular reactions of P-SOH, e.g., with other thiols, which may be faster [51]. To overcome this, high concentrations of dimedone are needed to outcompete such reactions, which is only feasible in cases where dimedone-based traps are used in the context of cell lysis buffer (such as DCP-Bio1). Cell preloading with high concentrations of cell-permeable dimedone-based probes (e.g., DYn-2) may impact cellular pathways and even viability or could disrupt functions of, e.g., peroxidase enzymes that form intermediate P-SOH [6].

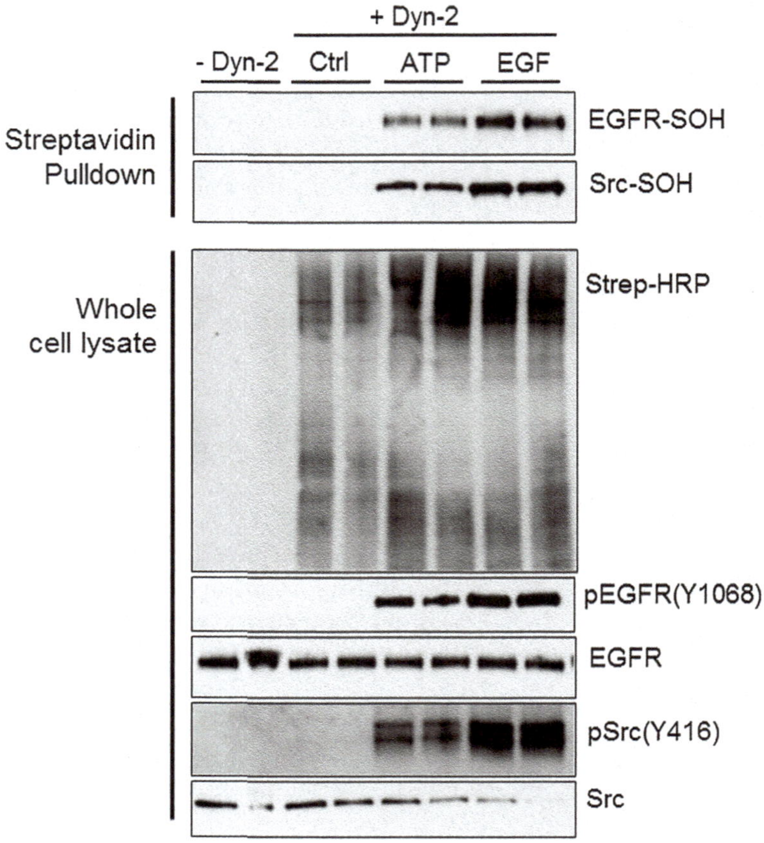

Fig. 3 Increased sulfenylation of EGFR and Src in response to NOX/DUOX stimulation by ATP or EGF. NCI-H292 cells were preloaded with DYn-2, stimulated with 100 μM ATP or 100 ng/mL EGF, conjugated with biotin post-lysis, and streptavidin purified. Avidin-purified samples were subjected to SDS-PAGE and Western blot against Src or EGFR (top panel), and cell lysates were probed with streptavidin-HRP to assess overall protein sulfenylation or with phospho-specific antibodies (pY1068 EGFR, pY416 Src, cell signaling; bottom panel). As a control, DYn-2 is omitted to assess background sulfenylation in unstimulated control conditions

3. Reactivity of P-SOH may be highly variable, and factors other than intrinsic P-SOH reactivity (e.g., steric factors) may influence labeling. Indeed, recent studies using several diverse molecules to trap P-SOH have shown that their ability to trap P-SOH in cells does not always overlap [52], suggesting that different available probes may be needed to cover the full P-SOH proteome.
4. This lysis buffer is chosen because it optimizes solubilization of transmembrane proteins such as EGFR, but other standard lysis buffers can be used as well. For best results, ensure that the pH is adjusted prior to the addition of protease inhibitors.

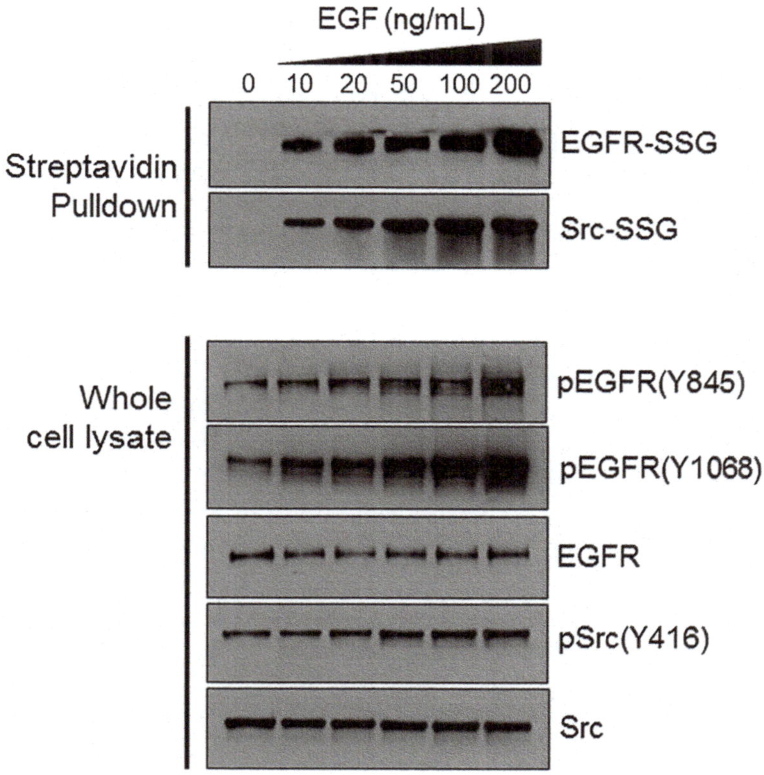

Fig. 4 *S*-glutathionylation of EGFR and Src in response to EGF stimulation. Primary mouse tracheal epithelial cells were preloaded with BioGEE and stimulated with EGF (0–200 ng/mL) for 10 min. Biotin-tagged proteins were purified and analyzed by Western blot for Src and EGFR (top panel). Additionally, whole cell lysates were probed for kinase activation (pY1068 EGFR, pY416 Src) compared with unphosphorylated forms using (phosphor)-specific antibodies (cell signaling, bottom panel)

For application to tissue homogenates, it is recommended that homogenization buffers similarly contain NEM and catalase to minimize artefactual protein sulfenylation during homogenization.

5. Comparable lysis buffer can be used here as well, as long as they don't contain metal chelators or other compounds that may interfere with the click reaction (e.g., EDTA).
6. This buffer should be similar in composition to the lysis buffer used, but additional components (e.g., protease or phosphatase inhibitors) are not necessary at this stage.
7. These reagents are used to synthesize BioGEE, as is described in Subheading 3.1, but BioGEE can also be purchased commercially (Thermo Fisher Scientific).
8. Laemmli buffer is either used 2× or 6× (to avoid unnecessary sample dilution), to yield final 1× concentrations of 0.060 M

Tris–Cl, 2% SDS, 10% v/v Glycerol, 0.02% β-mercaptoethanol, and 0.01% bromophenol blue (pH 6.8).

9. Adherent cells in this case can be substituted for any protein mixture desired, with different considerations, depending on the desired outcome. For example, when analyzing lung homogenates with DCP-Bio1, we will homogenize utilizing the lysis buffer containing whichever probe is desired. Ultimately, this depends on the experimental question and limitations regarding the specific sample type.
10. It is important to restrict the amount of added DMSO to <0.1% of total media volume. If needed, make intermediate dilutions with PBS or media. Additionally, if desired, a control mixture of the DYn-2 vehicle (DMSO) can be used to assess nonspecific protein detection.
11. This should be in at least a fivefold molar excess, ensuring that the amine reactive NHS is quenched by the NH_4HCO_3.
12. BioGEE concentration may need to be optimized, but 250 μM seems to work best in epithelial cells. Loading cells with high concentrations of BioGEE might alter cellular processes and thus induce artifacts. Also, biotin-tagged GSH may not be equally capable of reaching its protein target compared to endogenous GSH.
13. Lysate aliquots containing 100–500 μg of total protein are typically sufficient for detection of biotin-tagged proteins of interest by Western blot analysis. For proteomic identification, larger sample sizes (>1 mg protein) will be required.
14. The proteins of interest should be bound to the beads at this point; however it is beneficial to save the supernatant at this step as a means of assessing any potential issues that may have arisen upon final protein analysis.
15. During the final washes, beads may get stuck in the bottom of the tube. If this happens, simply mix the samples by vortexing before putting them on the rotisserie mixer.
16. Since avidin pulldowns could still contain proteins that coprecipitated and were not directly biotin-tagged, it is recommended to confirm identification of biotin-tagged proteins by first immunoprecipitating proteins of interest (using specific antibodies) and then performing analysis by SDS-PAGE and streptavidin-HRP blotting.
17. As in **Note 6**, a simplified RIPA buffer without protease and phosphatase inhibitors can be used.
18. Supernatant samples eluted in the absence of DTT can be analyzed similarly as "negative controls" to confirm specificity of the DTT elution step in **steps 16** and **17**.

19. Application of BioGEE for identification of specific cysteine targets within proteins/peptides by MS will be limited if they contain multiple cysteines, since the tag will be removed by the reducing conditions used during purification or MS analysis.

Acknowledgments

The authors gratefully acknowledge research support from NHLBI and NIA (grants R01 HL085646, R01 HL138708 and R21 AG055325), as well as Fellowship support from NIH (T32 HL076122 and F31 HL142221).

References

1. Bedard K, Krause K-H (2007) The NOX family of ROS-generating NADPH oxidases: physiology and pathophysiology. Physiol Rev 87 (1):245–313
2. Paulsen CE, Carroll KS (2010) Orchestrating redox signaling networks through regulatory cysteine switches. ACS Chem Biol 5(1):47–62
3. Holmstrom KM, Finkel T (2014) Cellular mechanisms and physiological consequences of redox-dependent signalling. Nat Rev Mol Cell Biol 15(6):411–421
4. Rhee SG (1999) Redox signaling: hydrogen peroxide as intracellular messenger. Exp Mol Med 31:53
5. Poole LB (2015) The basics of thiols and cysteines in redox biology and chemistry. Free Radic Biol Med 80:148–157
6. Stocker S, Van Laer K, Mijuskovic A, Dick TP (2018) The conundrum of hydrogen peroxide signaling and the emerging role of peroxiredoxins as redox relay hubs. Antioxid Redox Signal 28(7):558–573
7. Meng T-C, Fukada T, Tonks NK (2002) Reversible oxidation and inactivation of protein tyrosine phosphatases in vivo. Mol Cell 9 (2):387–399
8. Tonks NK (2005) Redox redux: revisiting PTPs and the control of cell signaling. Cell 121(5):667–670
9. Salmeen A et al (2003) Redox regulation of protein tyrosine phosphatase 1B involves a sulphenyl-amide intermediate. Nature 423 (6941):769–773
10. Barrett WC et al (1999) Regulation of PTP1B via glutathionylation of the active site cysteine 215. Biochemistry 38(20):6699–6705
11. Paulsen CE et al (2011) Peroxide-dependent sulfenylation of the EGFR catalytic site enhances kinase activity. Nat Chem Biol 8 (1):57–64
12. Truong TH et al (2016) Molecular basis for redox activation of epidermal growth factor receptor kinase. Cell Chem Biol 23 (7):837–848
13. Giannoni E, Buricchi F, Raugei G, Ramponi G, Chiarugi P (2005) Intracellular reactive oxygen species activate Src tyrosine kinase during cell adhesion and anchorage-dependent cell growth. Mol Cell Biol 25(15):6391–6403
14. Krasnowska EK et al (2008) N-acetyl-l-cysteine fosters inactivation and transfer to endolysosomes of c-Src. Free Radic Biol Med 45 (11):1566–1572
15. Heppner DE et al (2016) The NADPH oxidases DUOX1 and NOX2 play distinct roles in redox regulation of epidermal growth factor receptor signaling. J Biol Chem 291 (44):23,282–23,293
16. Sham D, Wesley UV, Hristova M, van der Vliet A (2013) ATP-mediated transactivation of the epidermal growth factor receptor in airway epithelial cells involves DUOX1-dependent oxidation of Src and ADAM17. PLoS One 8(1): e54391
17. Hristova M et al (2016) Airway epithelial dual oxidase 1 mediates allergen-induced IL-33 secretion and activation of type 2 immune responses. J Allergy Clin Immunol 137 (5):1545–1556. e1511
18. Habibovic A et al (2016) DUOX1 mediates persistent epithelial EGFR activation, mucous cell metaplasia, and airway remodeling during allergic asthma. JCI Insight 1(18):e88811
19. Gorissen SH et al (2013) Dual oxidase-1 is required for airway epithelial cell migration and bronchiolar reepithelialization after injury. Am J Respir Cell Mol Biol 48(3):337–345

20. Reynaert NL et al (2006) Dynamic redox control of NF-κB through glutaredoxin-regulated S-glutathionylation of inhibitory κB kinase β. Proc Natl Acad Sci 103(35):13,086–13,091
21. Anathy V et al (2012) Oxidative processing of latent Fas in the endoplasmic reticulum controls the strength of apoptosis. Mol Cell Biol 32(17):3464–3478
22. Hanschmann E-M, Godoy JR, Berndt C, Hudemann C, Lillig CH (2013) Thioredoxins, glutaredoxins, and peroxiredoxins—molecular mechanisms and health significance: from cofactors to antioxidants to redox signaling. Antioxid Redox Signal 19(13):1539–1605
23. Hess DT, Matsumoto A, Kim SO, Marshall HE, Stamler JS (2005) Protein S-nitrosylation: purview and parameters. Nat Rev Mol Cell Biol 6(2):150–166
24. Wall SB et al (2014) Detection of electrophile-sensitive proteins. Biochim Biophys Acta 1840 (2):913–922
25. Ida T et al (2014) Reactive cysteine persulfides and S-polythiolation regulate oxidative stress and redox signaling. Proc Natl Acad Sci U S A 111(21):7606–7611
26. Akaike T et al (2017) Cysteinyl-tRNA synthetase governs cysteine polysulfidation and mitochondrial bioenergetics. Nat Commun 8 (1):1177
27. Heppner DE et al (2018) Cysteine perthiosulfenic acid (Cys-SSOH): a novel intermediate in thiol-based redox signaling? Redox Biol 14:379–385
28. Sethuraman M et al (2004) Isotope-coded affinity tag (ICAT) approach to redox proteomics: identification and quantitation of oxidant-sensitive cysteine thiols in complex protein mixtures. J Proteome Res 3 (6):1228–1233
29. Jaffrey SR, Snyder SH (2001) The biotin switch method for the detection of S-nitrosylated proteins. Sci STKE 2001(86): pl1
30. Aesif SW, Janssen-Heininger YMW, Reynaert NL (2010) Protocols for the detection of S-glutathionylated and S-nitrosylated proteins in situ. Methods Enzymol 474:289–296
31. Poole LB (2008) Measurement of protein sulfenic acid content. Curr Protoc Toxicol. 0 17: Unit17.12–Unit17.12. Editorial board, Mahin D. Maines (editor-in-chief) et al.
32. Maller C, Schroder E, Eaton P (2011) Glyceraldehyde 3-phosphate dehydrogenase is unlikely to mediate hydrogen peroxide signaling: studies with a novel anti-dimedone sulfenic acid antibody. Antioxid Redox Signal 14 (1):49–60
33. Klomsiri C et al (2010) Use of dimedone-based chemical probes for sulfenic acid detection: evaluation of conditions affecting probe incorporation into redox-sensitive proteins. Methods Enzymol 473:77–94
34. Yang J et al (2015) Global, in situ, site-specific analysis of protein S-sulfenylation. Nat Protoc 10(7):1022–1037
35. Dalle-Donne I, Rossi R, Giustarini D, Colombo R, Milzani A (2007) S-glutathionylation in protein redox regulation. Free Radic Biol Med 43(6):883–898
36. Markovic J et al (2007) Glutathione is recruited into the nucleus in early phases of cell proliferation. J Biol Chem 282 (28):20,416–20,424
37. Townsend DM et al (2006) A glutathione S-transferase pi-activated prodrug causes kinase activation concurrent with S-glutathionylation of proteins. Mol Pharmacol 69(2):501–508
38. Brennan JP et al (2006) The utility of N, N-biotinyl glutathione disulfide in the study of protein S-glutathiolation. Mol Cell Proteomics 5(2):215–225
39. Lind C et al (2002) Identification of S-glutathionylated cellular proteins during oxidative stress and constitutive metabolism by affinity purification and proteomic analysis. Arch Biochem Biophys 406(2):229–240
40. Sullivan DM, Wehr NB, Fergusson MM, Levine RL, Finkel T (2000) Identification of oxidant-sensitive proteins: TNF-alpha induces protein glutathiolation. Biochemistry 39 (36):11,121–11,128
41. Hristova M et al (2014) Identification of DUOX1-dependent redox signaling through protein S-glutathionylation in airway epithelial cells. Redox Biol 2:436–446
42. Nelson KJ et al (2010) Use of dimedone-based chemical probes for sulfenic acid detection: methods to visualize and identify labeled proteins. Methods Enzymol 473:95–115. https://doi.org/10.1016/S0076-6879(10)73004-4
43. Tsutsumi R et al (2017) Assay to visualize specific protein oxidation reveals spatio-temporal regulation of SHP2. Nat Commun 8(1):466
44. Altenhofer S, Radermacher KA, Kleikers PW, Wingler K, Schmidt HH (2015) Evolution of NADPH oxidase inhibitors: selectivity and mechanisms for target engagement. Antioxid Redox Signal 23(5):406–427
45. Charles RL et al (2007) Protein sulfenation as a redox sensor: proteomics studies using a novel biotinylated dimedone analogue. Mol Cell Proteomics 6(9):1473–1484
46. Yang J, Gupta V, Carroll KS, Liebler DC (2014) Site-specific mapping and

quantification of protein S-sulphenylation in cells. Nat Commun 5:4776

47. Checconi P et al (2015) Redox proteomics of the inflammatory secretome identifies a common set of redoxins and other glutathionylated proteins released in inflammation, influenza virus infection and oxidative stress. PLoS One 10(5):e0127086
48. Seo YH, Carroll KS (2011) Quantification of protein sulfenic acid modifications using isotope-coded dimedone and iododimedone. Angew Chem Int Ed 50(6):1342–1345
49. Forman HJ et al (2017) Protein cysteine oxidation in redox signaling: caveats on sulfenic acid detection and quantification. Arch Biochem Biophys 617:26–37
50. Heppner DE, Janssen-Heininger YMW, van der Vliet A (2017) The role of sulfenic acids in cellular redox signaling: reconciling chemical kinetics and molecular detection strategies. Arch Biochem Biophys 616:40–46
51. Gupta V, Carroll KS (2016) Profiling the reactivity of cyclic C-nucleophiles towards electrophilic sulfur in cysteine sulfenic acid. Chem Sci 7(1):400–415. https://doi.org/10.1039/c5sc02569a
52. Gupta V, Yang J, Liebler DC, Carroll KS (2017) Diverse redoxome reactivity profiles of carbon nucleophiles. J Am Chem Soc 139 (15):5588–5595

Chapter 31

Neutrophil Extracellular Traps

Balázs Rada

Abstract

Neutrophil extracellular traps (NETs) are made of a network of extracellular strings of DNA that bind pathogenic microbes. Histones and several neutrophil granule proteins associated with the DNA framework damage entrapped microorganisms. Reactive oxygen species generated by the neutrophil NADPH oxidase have been shown to be essential to mediate NET release by several stimuli including numerous pathogenic bacteria. Although several methods have been used in the literature to detect NETs in vitro and in vivo, a consensus is urgently needed on the field to standardize the best NET-specific assays. In this chapter, two methods are described in details that can be used to detect NETs and to distinguish them from other mechanisms of neutrophil cell death. While NET-specific, these assays are also relatively simple and straightforward enabling their potential use by a wide audience.

Key words NETs, Neutrophil, NADPH oxidase, MPO-DNA complex, NE-DNA complex, Citrullinated histone, Peptidylarginine deiminase 4, Nuclear decondensation

1 Introduction

Neutrophil extracellular traps (NETs) represent an antimicrobial mechanism of neutrophilic granulocytes (Fig. 1) [1]. The core structure of NETs is extracellular DNA associated with antimicrobial proteins originating from neutrophil granules and nucleus [1]. The major form of NET formation, called suicidal NETosis, leads to the death of neutrophils and is characterized by subsequent morphological changes: disintegration of nuclear membrane, chromatin decondensation, disappearance of plasma membrane, and finally the spill of DNA-based NETs into the extracellular space [2]. On the contrary to this mechanism leading to neutrophil death, vital NETosis has also been described during which process neutrophils remain live and release only parts of their nuclear or mitochondrial DNA [3–6]. NETs have been shown to entrap a wide variety of microbes and provide a crucial innate immune mechanism. Excessive NET release has, however, been associated with numerous diseases [7, 8]. Although NETs have been discovered 14 years ago, specific signaling events leading to NET release

Ulla G. Knaus and Thomas L. Leto (eds.), *NADPH Oxidases: Methods and Protocols*, Methods in Molecular Biology, vol. 1982, https://doi.org/10.1007/978-1-4939-9424-3_31,

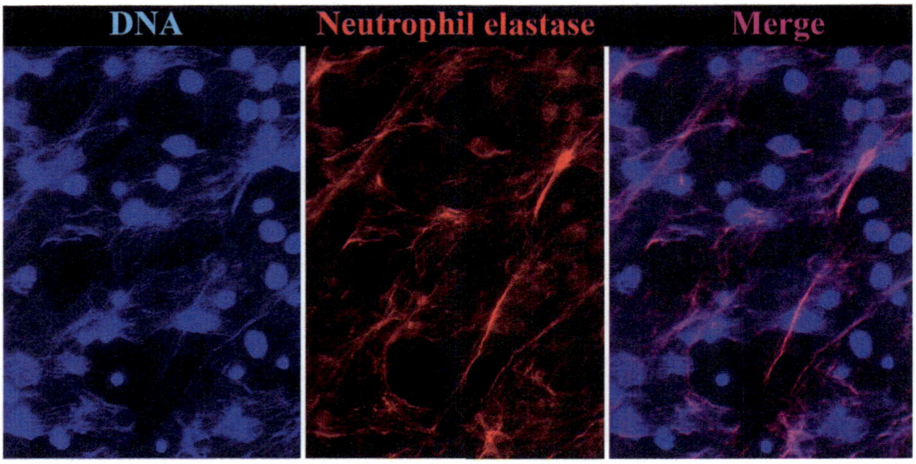

Fig. 1 Human neutrophils were stimulated with 100 nM PMA for 5 h, and immunofluorescence of formed NETs is shown by co-localization of neutrophil elastase (red) with extracellular DNA (blue) (merged image is purple)

are still largely unclear. The phagocytic NADPH oxidase has been early implicated in the process [2]. Pretreatment of neutrophils stimulated with phorbol myristate acetate (PMA) or *Staphylococcus aureus* with a NADPH oxidase inhibitor, diphenylene iodonium (DPI), prevented NET release [2]. Addition of an extracellular source of reactive oxygen species (ROS), glucose oxidase and glucose, circumvented the need for NADPH oxidase-produced ROS and triggered DPI-independent NET formation in human neutrophils [2]. The best evidence for the need of the NADPH oxidase for NET extrusion was provided by experiments done on human neutrophils derived from chronic granulomatous disease (CGD) patients [2]. CGD patients have mutations in one of the subunits of the NADPH oxidase enzyme complex leading to absent or significantly diminished respiratory burst of neutrophils [9]. CGD neutrophils do not expel NETs in response to PMA or *S. aureus*, but do release NETs when an NADPH oxidase-independent ROS source is used [2]. Restoration of NET formation in an X-linked CGD patient by gp91^{phox}-based gene therapy leads to improved clearance of *Aspergillus nidulans*, emphasizing the human clinical importance of NADPH oxidase-mediated NETs in antifungal defense [10]. Although originally the NADPH oxidase had been considered indispensable for the induction of NETs, accumulating evidence later showed that NETs can be triggered in an NADPH oxidase-independent fashion, as well. Broadly speaking, suicidal NET release stimulated by certain bacteria and fungi requires the NADPH oxidase, while extrusion of NETs induced by other bacteria, parasites, and most inflammatory microcrystals does not involve ROS generated by the oxidase [1, 11–14]. Since the NADPH oxidase is also required for intracellular, phagocytic killing of several microbes by neutrophils, the NADPH oxidase cannot be considered a NET-specific protein.

Citrullination of histones by the enzyme peptidylarginine deaminase 4 (PAD4) has been shown to be involved in NET release [15, 16]. PAD4 is highly expressed in neutrophils, translocates to the nucleus upon rise in cytosolic calcium concentrations, and mediates the conversion of arginine residues to citrulline in target proteins, primarily in histones, that was proposed to lead to chromatin decondensation [15, 16]. Although PAD4 had been originally suggested as a specific and universal mediator of NET formation, recent findings suggest that NETs can be released in a PAD4-independent manner, as well. Irrespective of the molecular mechanism of NET formation, the following features are specific for NETs and provide the bases for NET-specific detection methods.

The characteristic morphological changes in the shape of neutrophil nucleus during NET formation can be followed, quantified, and distinguished from other neutrophil death mechanisms by microplate-based fluorescence microscopy [17–19]. Although such methods are NET-specific and can scan large amounts of neutrophils in vitro, they require sophisticated instrumentation and analytical methods that are more suitable for laboratories with neutrophil expertise than with interest in other fields of immunology. Although nuclear swelling at the early stage of NET formation can also be detected by image-based flow cytometry, this method is not capable of recording later stages of NET formation when cellular morphology is not intact anymore [20, 21]. Citrullinated histones can be detected by several methods including immunofluorescence, Western blotting and ELISA. Although citrullinated histones have originally been used as NET markers, they are only specific to PAD4-mediated NET formation and do not inform about PAD4-independent NET extrusion. The most specific NET markers are complexes of extracellular DNA and neutrophil granule proteins which can be detected by their co-localization in the weblike structures of NETs using immunofluorescence (Fig. 1). The same protein-DNA complexes can be also detected as NET remnants in biological biospecimen using hybrid ELISA assays that use a capture antibody specific for the protein component and a dsDNA detection antibody linked to horseradish peroxidase (Fig. 2) [7, 22, 23]. Two primary granule proteins have been mainly used in these assays: myeloperoxidase (MPO) and neutrophil elastase (NE) [7, 22].

In this chapter, these two latter assays (NET detection by immunofluorescence and ELISA) are described in detail because they provide NET specificity, are relatively easy, and can be performed in most laboratories.

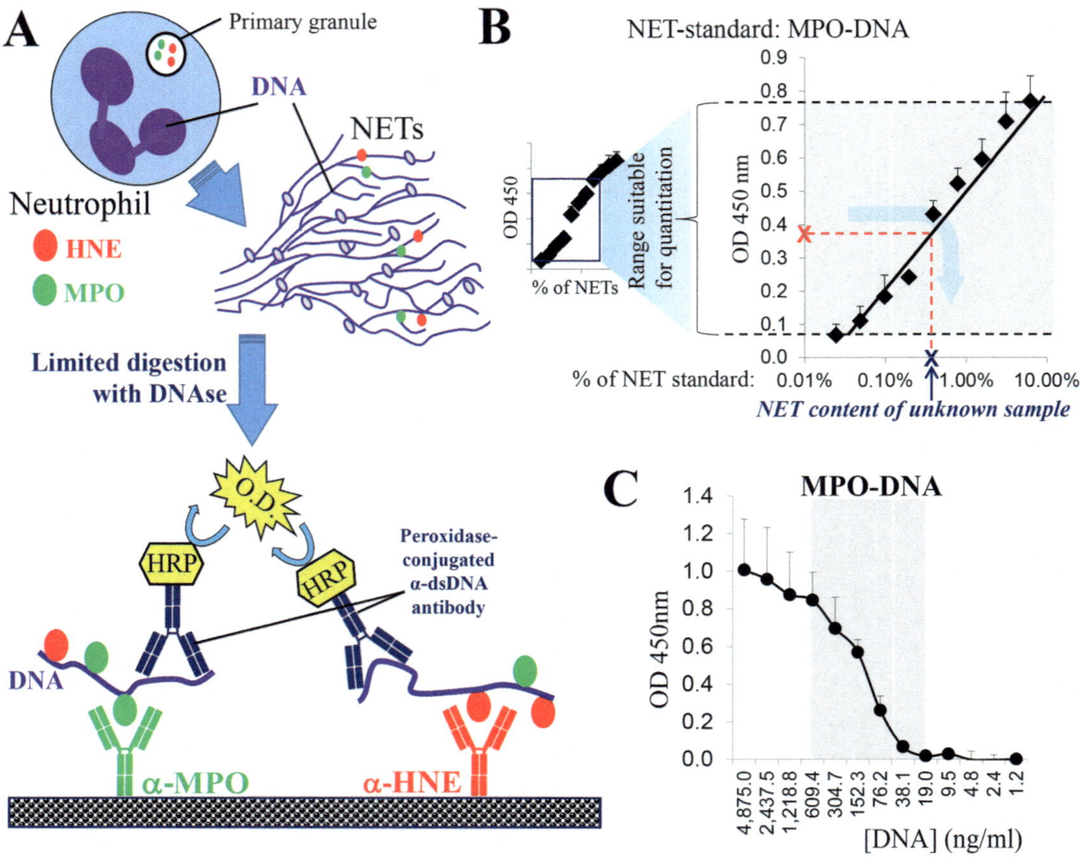

Fig. 2 NET-specific ELISA assays: principles and semiquantitative standard. (**A**) Scheme explaining how the MPO-DNA and HNE-DNA ELISA assays work. (**B**) Serial dilutions (twofold) of the NET standard were prepared and subjected to MPO-DNA ELISA. Measured OD values are plotted against % of NET-standard content. Red "X" indicates a measured OD value of an unknown sample. Gray arrow indicates how the "NET concentration" (blue "X") of the unknown sample is to be determined using the central range of the fitted standard curve. (**C**) DNA concentrations in serially diluted NET-standard samples were determined and plotted against the measured OD values of the MPO-DNA or HNE-DNA ELISA assays. Gray areas indicate the dynamic ranges of the assays

2 Materials

2.1 NET Detection by Immunofluorescence

1. Coverslips.
2. Hank's balanced salt solution (HBSS, 1×).
3. 5 mM D-glucose in HBSS.
4. Assay medium (HBSS + 5 mM glucose + 10 mM HEPES).
5. 4% paraformaldehyde (PFA, PBS).
6. 24-well tissue culture plates.
7. Phosphate-buffered saline (PBS, 1×).

8. Blocking solution (5% bovine serum albumin +5% normal donkey serum + 0.05% Triton X-100).
9. Tween 20.
10. FITC-conjugated antihuman myeloperoxidase (MPO) antibody (mouse, 1:200, Dako, Clone MPO-7, in PBS containing 1% BSA + 1% normal donkey serum + 0.05% Triton X-100).
11. Anti-histone H4 (citrulline 3) antibody (rabbit, 1:500, Millipore in PBS containing 1% BSA + 1% normal donkey serum + 0.05% Triton X-100).
12. ALEXA fluorochrome, anti-rabbit, 594 nm (1:2000 dilution, Invitrogen, donkey anti-rabbit IgG (H + L), in PBS containing 1% BSA + 1% normal donkey serum + 0.05% Triton X-100).
13. Antihuman neutrophil elastase antibody (rabbit, Calbiochem, 1:2000 dilution, in PBS containing 1% BSA + 1% normal donkey serum + 0.05% Triton X-100).
14. 4′,6-Diamidino-2-phenylindole (DAPI) is a fluorescent DNA stain. Prepare a stock solution of 5 mg/mL DAPI dissolved in DMSO, and use it in a 20,000-fold diluted concentration in PBS to stain neutrophil nuclear DNA and NET-DNA.
15. ProLong Antifade Kit (Molecular Probes, Grand Island, NY).
16. Fluorescence microscope (Zeiss AxioCam HRM microscope with Axioplan2 imaging software).

2.2 NET Detection by ELISA

1. Antihuman neutrophil elastase antibody (rabbit, Calbiochem, 1:2000 dilution, PBS).
2. Anti-myeloperoxidase antibody (Rabbit, Millipore, 1:2000 dilution, PBS).
3. 1 μg/mL DNase-1 (Roche, final concentration used for digestion).
4. 20 mM EGTA/PBS.
5. 0.05% (v/v) PBS-Tween 20.
6. Horseradish peroxidase-labelled anti-dsDNA antibody in the cell death detection kit (Roche, 1:500 dilution, PBS).
7. Eon microplate spectrophotometer (BioTek, Gen5 software) or equivalent.
8. 0.2% SYTOX Orange membrane-impermeable DNA-binding dye (Life Technologies, cat#: S11368, 0.2% final concentration/volume).
9. 10 mM HEPES and 5 mM glucose in HBSS (Corning).
10. Varioskan Flash Ver.2.4.3 combined microplate fluorimeter (Thermo Fisher Scientific) or equivalent.
11. 600 nM PMA (Sigma, 6× stock) in HBSS with HEPES and glucose.

12. 1 μM ionomycin.
13. High-binding capacity ELISA plates (Greiner Bio-One).

3 Methods

3.1 NET Detection by Immunofluorescence

1. Place microscope coverslips in wells of a 24-well plate (*see* **Note 1**). Neutrophils will be seeded, stimulated, and exposed to NET-inducing stimuli on the coverslips in the 24-well plate. Once NETs formed and are fixed, coverslips will be taken out from the 24-well plate for further processing.
2. Seed neutrophils at a concentration of 4,000,000 cells/mL in 250 μL assay medium on coverslips. The coverslips must be covered entirely by the medium; otherwise neutrophil density will not be equal throughout the coverslip (*see* **Note 1**).
3. Incubate plate with coverslips for 15 min in cell culture incubator at 37 °C to allow neutrophil adherence. Double check in light microscope that at the end of the incubation time, neutrophils are indeed attached to coverslips. If not, incubate them for another 15 min.
4. Add NET-inducing stimulants to neutrophils in a volume of 50 μL containing stimulants or their solvents at concentrations six times higher than the final one. Fifty microliters of the stimulant's volume will be diluted to a final volume of 300 μL per wells, hence the need to apply higher doses (300/50 = 6).
5. Use 100 nM PMA or 1 μM ionomycin, as positive controls of NET release. PMA is a better choice for studies on NADPH oxidase-mediated NET formation, while ionomycin is more appropriate for works focusing on NET release fueled by calcium and PAD4 activation.
6. For neutrophils without stimulation, add 50 μL assay medium or the solvent of the stimulators per well. Make sure all samples have a final volume of 300 μL.
7. Incubate neutrophils for 4–6 h at 37 °C in cell culture incubator to allow NET formation to take place. When the incubation time is over, handle samples as gently as possible, since formed NETs are very fragile and require special care. Coverslips will next undergo a series of sequential washing steps (*see* **Note 2**).
8. Fix NETs with 4% PFA and incubate for 10 min at room temperature. PFA is the best solution to fix NETs as methanol or ethanol can alter NET structures and even enhance NET release.
9. Wash samples twice with 300 μL sterile PBS for 10 min each time at room temperature.

10. Permeabilize and block samples using 300 μL blocking solution (*see* **Note 3**). Incubate for 30 min at room temperature.
11. Add primary (antihuman MPO-FITC or antihuman neutrophil elastase) antibodies into samples and incubate overnight at 4 °C in dark. To detect PAD4-dependent NET release, anti-citrullinated histone H3 antibody can also be used.
12. While the MPO staining does not require a secondary antibody due to the direct labeling of the primary MPO antibody by FITC, the neutrophil elastase or citrullinated histone H3 staining does need a secondary antibody. Wash coverslips three times in PBS for 5 min at room temperature, and incubate samples with fluorochrome-labelled secondary antibody in PBS for 1 h at room temperature in the dark.
13. Aspirate antibody solutions, and incubate cells in 20,0000-fold diluted DAPI (PBS) for 2 min (*see* **Note 4**).
14. Wash samples two more times in PBS for 5 min each time at room temperature to remove unbound secondary antibodies and DNA stain.
15. Let the coverslips dry on a paper towel, and drop 10 μL mounting medium supplemented with anti-fade solution per coverslip in the middle of the sample (*see* **Notes 5** and **6**). Carefully drop the coverslip on clean, degreased microscope slide with the sample and mounting medium facing down, toward the slide. Wait until the viscous mounting medium slowly spreads and covers the entire sample. Again, NETs are very fragile so be gentle (*see* **Note** 7).
16. Wait for 30 min to let the excess fluid evaporate. Seal the edges of the coverslip with clean nail polish or commercially available sealing liquid to prevent further damage to the sample. The immunofluorescence staining is ready. Store samples at 4 °C in dark to preserve them for later analysis.

3.2 NET Detection by ELISA

1. Apply capture antibodies (anti-MPO or anti-NE, 1:2000 dilution in sterile PBS) to 96-well high-binding capacity ELISA microplates.
2. Incubate ELISA plates overnight at 4 °C to allow binding of capture antibodies (*see* **Note 8**).
3. Apply three repeated washes with PBS-Tween 20 the next day.
4. Block ELISA plates with 5% BSA (200 μL/well).
5. Incubate plates for 2 h at room temperature.
6. Apply again repeated washes with PBS-Tween 20 (200 μL/well).

7. The ELISA plates are now ready to accept the experimental samples. Apply NETs collected from neutrophils in vitro or diluted biological fluids to ELISA plates.
8. Incubate ELISA plates with samples overnight at 4 °C to allow the protein component of NETs to bind to capture antibodies.
9. Wash plates again three times with PBS-Tween20 (200 μL/well) to remove unbound NETs and other components of the samples.
10. Apply detection antibody to ELISA plates (HRP-labelled anti-dsDNA antibody, 1:500, mouse, 50 μL/well), and incubate for 1 h in dark at room temperature.
11. Repeat washes four times with PBS-Tween 20 (200 μL/well).
12. Add TMB peroxidase substrate to each well for 30 min (50 μL/well).
13. Stop reaction by adding 1 M HCl to each well (50 μL/well).
14. Read absorbance at 450 nm using a microplate photometer.
15. Analyze and present results either as absorbance values (relative quantitation) or as percentage of the NET standard (semi-absolute quantitation; *see* **Note 9**).
16. The NET standard consists of pooled supernatants of DNase-digested NETs released from PMA-stimulated neutrophils.
17. To prepare DNAse-digested NETs for the NET standard from each neutrophil donor, stimulate human neutrophils in vitro with 100 nM PMA for 5–6 h, apply 1 U/mL DNAse I to the supernatants for 15 min, stop the action of DNAse by adding 2.5 mM PBS-EGTA, centrifuge the samples (10,000 × *g*. 4 min, 4 °C), and collect supernatants as DNAseI-digested NETs (*see* **Notes 10–12**).

4 Notes

1. Before each step (seeding neutrophils or start of the incubation time to form NETs), make sure that coverslips are at the bottom of the wells on the 24-well microplate as they can easily come up to the surface that will result in unanalyzable samples. If coverslips float on the surface of the 300 μL assay medium, use clean forceps to carefully push the coverslips back to the bottom of the well. Pay attention not to scrape off attached neutrophils as the forceps might remove cells from the coverslip or introduce artifacts in your immunostaining images.
2. Although washing the coverslips following fixation can be done in the wells of the 24-well plate, it requires the researcher to be extra careful as NETs can be easily washed off of the coverslips

by too strong pipetting. Better results can be achieved when coverslips are processed on top of 300 μL PBS droplets using forceps. This way of processing the samples has been described by *Brinkmann* et al. [24] Briefly, stretch a parafilm sheet on top of a 15 mL conical tube rack, and create indentations by pressing your thumb into the parafilm where holes are located [24]. Pipette 300 μL volumes of the solutions the coverslips will be exposed to into the parafilm dents according to the protocol (PBS, PBS with 4% PFA, PBS + Triton X-100). Coverslips will be washed by carefully transferring them between these droplets and dipping them into the solutions in an upside-down position (NETs facing down, into the liquid) [24].

3. Make sure to vortex the normal donkey serum well before applying it to the samples. Well-mixed serum will result in nice and uniform blocking of the NETs.
4. Be accurate with the time of DAPI staining as too long exposure will overstain DNA in your samples prohibiting recording of sharp, publishable images.
5. When drying NET-containing samples on slide before addition of the mounting medium, make sure not to let them dry completely. The best way to do this is one coverslip at a time. You might want to add the mounting medium to the slide before you take the coverslip off of the PBS. Take each coverslip off of PBS, carefully drag its edge on a paper towel (to get rid of excess liquid), and flip coverslips. Make sure you flip them, or you'll contaminate your slides by facing the neutrophil/NET side down on the paper towel.
6. Thaw both of the mounting medium and the anti-fade solutions way ahead of the time of their use to allow enough time for their warm-up. Keep in mind that the mounting medium is extremely viscous.
7. Before turning the coverslips onto the mounting medium, wet a KimTech wipe with ethanol, and gently wipe off the opposite side of the coverslip (facing down on paper towel). This will help you getting better pictures, since cells or debris adhered to the wrong side of the coverslip will appear blurry (in a different plane) on the microscope and give you lesser-quality images. If you want to take pictures that same day, then you won't be able to wipe with ethanol once the coverslips are mounted as they will not be fully dried. However, if you take pictures the next day, then you can wipe the tops of the mounted coverslips with ethanol before taking pictures of the slides.
8. For each incubation step during the ELISA, make sure to use a plate cover to prevent evaporation of the samples.

9. To analyze the data obtained using the NET ELISA assays, calculate the amount of MPO-DNA or HNE-DNA complexes present compared to the "NET standard" (Fig. 2). First, subtract the background absorbance value of the assay medium alone from all your unknown and calibration samples. Correlate the optical density values of the standard samples with their relative concentrations NETs, and establish a linear trend line using the unsaturated range of the standard. The received equation provides the means to convert absorption results into quantitative amounts of NETs. Calculate mean of replicates, and show final results as "MPO-DNA or HNE-DNA complexes (% of standard)."
10. When performing DNAse I digestion of NETs to prepare them for the ELISA assay, make sure that the content of each well is mixed thoroughly to provide an easy access of the DNAse I to expelled NETs.
11. Similarly, mix the contents of each well thoroughly when stopping DNAse I by adding EGTA.
12. NETs generated in vitro in human neutrophils have to be diluted in sterile PBS several-fold to be in the measurement range of the ELISA assays. The degree of dilutions has to be optimized, but diluting the DNAse I-digested NETs 100-fold is a good start in our experience.

Acknowledgments

This work was supported by the NIH grant 1 R01 HL136707-01A1 awarded to B. Rada. I thank previous graduate students Dae-goon Yoo, Madison Floyd, and Payel Sil, for taking the immunofluorescence images presented here and for optimizing and recording detailed protocols that helped writing the "Notes" section of this book chapter.

References

1. Brinkmann V, Reichard U, Goosmann C, Fauler B, Uhlemann Y, Weiss DS, Weinrauch Y, Zychlinsky A (2004) Neutrophil extracellular traps kill bacteria. Science 303 (5663):1532–1535. https://doi.org/10.1126/science.1092385
2. Fuchs TA, Abed U, Goosmann C, Hurwitz R, Schulze I, Wahn V, Weinrauch Y, Brinkmann V, Zychlinsky A (2007) Novel cell death program leads to neutrophil extracellular traps. J Cell Biol 176(2):231–241. https://doi.org/10.1083/jcb.200606027
3. Clark SR, Ma AC, Tavener SA, McDonald B, Goodarzi Z, Kelly MM, Patel KD, Chakrabarti S, McAvoy E, Sinclair GD, Keys EM, Allen-Vercoe E, Devinney R, Doig CJ, Green FH, Kubes P (2007) Platelet TLR4 activates neutrophil extracellular traps to ensnare bacteria in septic blood. Nat Med 13 (4):463–469. https://doi.org/10.1038/nm1565
4. Pilsczek FH, Salina D, Poon KK, Fahey C, Yipp BG, Sibley CD, Robbins SM, Green FH, Surette MG, Sugai M, Bowden MG, Hussain M, Zhang K, Kubes P (2010) A novel mechanism

of rapid nuclear neutrophil extracellular trap formation in response to *Staphylococcus aureus*. J Immunol 185(12):7413–7425. https://doi.org/10.4049/jimmunol.1000675

5. Yipp BG, Petri B, Salina D, Jenne CN, Scott BN, Zbytnuik LD, Pittman K, Asaduzzaman M, Wu K, Meijndert HC, Malawista SE, de Boisfleury Chevance A, Zhang K, Conly J, Kubes P (2012) Infection-induced NETosis is a dynamic process involving neutrophil multitasking in vivo. Nat Med 18 (9):1386–1393. https://doi.org/10.1038/nm.2847
6. Yousefi S, Mihalache C, Kozlowski E, Schmid I, Simon HU (2009) Viable neutrophils release mitochondrial DNA to form neutrophil extracellular traps. Cell Death Differ 16 (11):1438–1444. https://doi.org/10.1038/cdd.2009.96
7. Kessenbrock K, Krumbholz M, Schonermarck U, Back W, Gross WL, Werb Z, Grone HJ, Brinkmann V, Jenne DE (2009) Netting neutrophils in autoimmune small-vessel vasculitis. Nat Med 15 (6):623–625. https://doi.org/10.1038/nm.1959
8. Leffler J, Gullstrand B, Jonsen A, Nilsson JA, Martin M, Blom AM, Bengtsson AA (2013) Degradation of neutrophil extracellular traps co-varies with disease activity in patients with systemic lupus erythematosus. Arthritis Res Ther 15(4):R84. https://doi.org/10.1186/ar4264
9. Kuhns DB, Alvord WG, Heller T, Feld JJ, Pike KM, Marciano BE, Uzel G, DeRavin SS, Priel DA, Soule BP, Zarember KA, Malech HL, Holland SM, Gallin JI (2010) Residual NADPH oxidase and survival in chronic granulomatous disease. N Engl J Med 363 (27):2600–2610. https://doi.org/10.1056/NEJMoa1007097
10. Bianchi M, Hakkim A, Brinkmann V, Siler U, Seger RA, Zychlinsky A, Reichenbach J (2009) Restoration of NET formation by gene therapy in CGD controls aspergillosis. Blood 114 (13):2619–2622. https://doi.org/10.1182/blood-2009-05-221606
11. Almyroudis NG, Grimm MJ, Davidson BA, Rohm M, Urban CF, Segal BH (2013) NETosis and NADPH oxidase: at the intersection of host defense, inflammation, and injury. Front Immunol 4:45. https://doi.org/10.3389/fimmu.2013.00045
12. Rada B (2017) Neutrophil extracellular traps and microcrystals. J Immunol Res 2017:2896380. https://doi.org/10.1155/2017/2896380
13. Yoo DG, Winn M, Pang L, Moskowitz SM, Malech HL, Leto TL, Rada B (2014) Release of cystic fibrosis airway inflammatory markers from *Pseudomonas aeruginosa*-stimulated human neutrophils involves NADPH oxidase-dependent extracellular DNA trap formation. J Immunol 192(10):4728–4738. https://doi.org/10.4049/jimmunol.1301589
14. Rada B, Leto TL (2008) Oxidative innate immune defenses by Nox/Duox family NADPH oxidases. Contrib Microbiol 15:164–187. https://doi.org/10.1159/000136357
15. Wang Y, Li M, Stadler S, Correll S, Li P, Wang D, Hayama R, Leonelli L, Han H, Grigoryev SA, Allis CD, Coonrod SA (2009) Histone hypercitrullination mediates chromatin decondensation and neutrophil extracellular trap formation. J Cell Biol 184(2):205–213. https://doi.org/10.1083/jcb.200806072
16. Li P, Li M, Lindberg MR, Kennett MJ, Xiong N, Wang Y (2010) PAD4 is essential for antibacterial innate immunity mediated by neutrophil extracellular traps. J Exp Med 207 (9):1853–1862. https://doi.org/10.1084/jem.20100239
17. Gupta S, Chan DW, Zaal KJ, Kaplan MJ (2018) A high-throughput real-time imaging technique to quantify NETosis and distinguish mechanisms of cell death in human neutrophils. J Immunol 200(2):869–879. https://doi.org/10.4049/jimmunol.1700905
18. Kraaij T, Tengstrom FC, Kamerling SW, Pusey CD, Scherer HU, Toes RE, Rabelink TJ, van Kooten C, Teng YK (2016) A novel method for high-throughput detection and quantification of neutrophil extracellular traps reveals ROS-independent NET release with immune complexes. Autoimmun Rev 15(6):577–584. https://doi.org/10.1016/j.autrev.2016.02.018
19. Brinkmann V, Goosmann C, Kuhn LI, Zychlinsky A (2012) Automatic quantification of in vitro NET formation. Front Immunol 3:413. https://doi.org/10.3389/fimmu.2012.00413
20. Masuda S, Shimizu S, Matsuo J, Nishibata Y, Kusunoki Y, Hattanda F, Shida H, Nakazawa D, Tomaru U, Atsumi T, Ishizu A (2017) Measurement of NET formation in vitro and in vivo by flow cytometry. Cytometry A 91(8):822–829. https://doi.org/10.1002/cyto.a.23169
21. Zhao W, Fogg DK, Kaplan MJ (2015) A novel image-based quantitative method for the characterization of NETosis. J Immunol Methods 423:104–110. https://doi.org/10.1016/j.jim.2015.04.027

22. Yoo DG, Floyd M, Winn M, Moskowitz SM, Rada B (2014) NET formation induced by *Pseudomonas aeruginosa* cystic fibrosis isolates measured as release of myeloperoxidase-DNA and neutrophil elastase-DNA complexes. Immunol Lett 160(2):186–194. https://doi.org/10.1016/j.imlet.2014.03.003
23. Sil P, Yoo DG, Floyd M, Gingerich A, Rada B (2016) High throughput measurement of extracellular DNA Release and quantitative NET formation in human neutrophils in vitro. J Vis Exp (112). https://doi.org/10.3791/52779
24. Brinkmann V, Laube B, Abu Abed U, Goosmann C, Zychlinsky A (2010) Neutrophil extracellular traps: how to generate and visualize them. J Vis Exp (36). https://doi.org/10.3791/1724

Part VI

Genetic Disorders

Chapter 32

Chronic Granulomatous Disease

Dirk Roos

Abstract

Chronic granulomatous disease is a clinical condition that stems from inactivating mutations in NOX2 and its auxiliary proteins. Together, these proteins form the phagocyte NADPH oxidase enzyme that generates superoxide. Superoxide ($O_2^{\bullet-}$) and its reduced product hydrogen peroxide (H_2O_2) give rise to several additional reactive oxygen species (ROS), which together are necessary for adequate killing of pathogens. Thus, CGD patients, with a phagocyte NADPH oxidase that is not properly functioning, suffer from recurrent, life-threatening infections with certain bacteria, fungi, and yeasts. Here, I give a short survey of the genetic mutations that underlie CGD, the effect of these mutations on the activity of the leukocyte NADPH oxidase, the clinical symptoms of CGD patients, and the treatment options for these patients.

Key words Chronic granulomatous disease, Phagocytes, NADPH oxidase, NOX2, Mutations, *CYBB*, *CYBA*, *NCF1*, *NCF2*, *NCF4*

1 Introduction

The phagocyte NADPH oxidase consists of five structural components. NOX2, also called gp91^{phox} (phox from *ph*agocyte *ox*idase), is the central, enzymatic component. It is located in the plasma membrane and in the membrane of intracellular vesicles of phagocytic leukocytes, i.e., neutrophilic granulocytes, eosinophilic granulocytes, monocytes, and macrophages [1]. NOX2 is a flavocytochrome b_{558}, with an NADPH binding site on the cytosolic side and FAD and two hemes as prosthetic groups. In its active state, it can accept electrons from NADPH and transmit these via FAD and the heme groups to molecular oxygen on the luminal side of the membrane. In this way superoxide is generated on the outside of the phagocytes and inside phagocytic vacuoles, as shown in Fig. 1. For stability, NOX2 needs p22^{phox}, with which it forms a heterodimer. If either NOX2 or p22^{phox} is missing, the other protein is unstable and will be quickly degraded.

In resting phagocytes, the NADPH binding site is inaccessible for NADPH, and as a result, the NADPH oxidase will not be active.

Ulla G. Knaus and Thomas L. Leto (eds.), *NADPH Oxidases: Methods and Protocols*, Methods in Molecular Biology, vol. 1982, https://doi.org/10.1007/978-1-4939-9424-3_32,

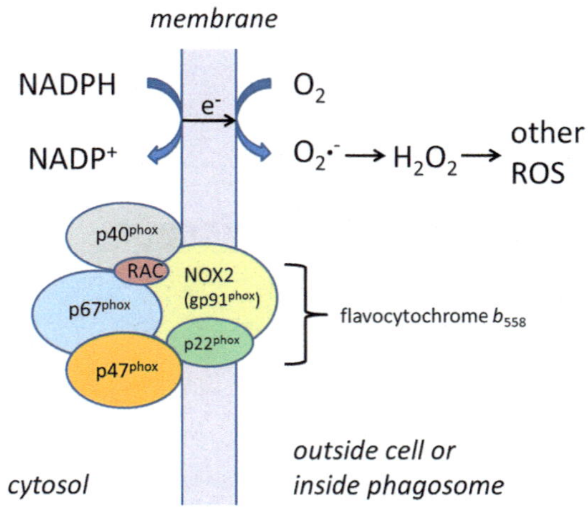

Fig. 1 The subunits of the phagocyte NADPH oxidase and the generation of ROS. The assembled phagocyte NADPH oxidase complex consists of NOX2 ($gp91^{phox}$) and $p22^{phox}$ (together called flavocytochrome b_{558}) in the plasma membrane or the phagosome membrane together with the heterotrimer $p40^{phox}$, $p67^{phox}$, and $p47^{phox}$ that has moved from the cytosol to the flavocytochrome. For this movement and attachment, the active, GTP-bound form of RAC (RAC1 in mononuclear phagocytes, RAC2 in neutrophils) is also required. In this assembled NADPH oxidase complex, NOX2 is able to bind NADPH and transmit its electrons via FAD and two hemes to molecular oxygen at the luminal side of the membrane. This process is schematically illustrated in the upper part of the figure. The resulting superoxide ($O_2^{\bullet-}$) is converted into hydrogen peroxide (H_2O_2) and other reactive oxygen species (ROS), such as singlet oxygen, hydroxyl radicals, hypochlorous acid, and N-chloramines, all of which help to kill ingested and non-ingested microorganisms. All proteins of the phagocyte NADPH oxidase complex are essential for correct functioning of the oxidase. Mutations in the genes that encode these proteins (Table 1) may lead to decreased expression or decreased functioning of these proteins, and thus to impaired killing of microorganisms, which is the cause of chronic granulomatous disease

However, during phagocytosis or at sites of inflammation, a complex of three cytosolic proteins moves to the NOX2/$p22^{phox}$ dimer and changes the 3D structure of NOX2, rendering the NADPH binding site accessible for NADPH and thus starting the enzymatic activity of the oxidase [1, 2]. These cytosolic proteins are $p47^{phox}$, $p67^{phox}$, and $p40^{phox}$. Each of these proteins has a specific role in this activity regulation. Phagocytic or inflammatory stimuli induce phosphorylation of $p47^{phox}$, which then changes conformation and its interaction with the other two cytosolic components. This induces binding of the cytosolic complex to $p22^{phox}$ and to phospholipids in the membrane surrounding NOX2. $p67^{phox}$ will then bind with its activation domain to NOX2, which is essential for the electron flow within NOX2, and with its TPR regions with Rac-GTP. This latter protein is a small GTPase that associates through its insert domain with NOX2 and with its lipid tail in the membrane. Finally, $p40^{phox}$ binds with its PX region to phospholipids in the membrane, a process that enhances superoxide production on phagosome membranes. Altogether, the five components of the phagocyte NADPH oxidase now form a tight complex with high enzymatic activity [1, 2, 4].

All of the abovementioned components of the phagocyte NADPH oxidase are essential for its proper functioning. That is proven by the fact that mutations in any of the genes encoding these components may lead to loss of function and thus to CGD. The best known form of CGD is the X-linked form, with mutations in *CYBB*, located on the X chromosome and encoding NOX2. About two-thirds of all CGD patients suffer from X-CGD, most of them males, of course (Table 1). The genes encoding $p22^{phox}$ (*CYBA*), $p47^{phox}$ (*NCF1*), $p67^{phox}$ (*NCF2*), and $p40^{phox}$ (*NCF4*) are all autosomal genes. Mutations in *NCF1* are found in about 20% of CGD patients, in *CYBA* and *NCF2* each in about 5% of CGD patients, while those in *NCF4* are rare. These numbers vary with social habits, with autosomal recessive CGD being more common in countries with consanguineous marriages. Autosomal CGD patients are found in equal numbers among males and females. The activating protein RAC2 is also essential, because a few patients with CGD-like symptoms have been found with mutations in the *RAC2* gene [4, 5]. Finally, in case of severe glucose-6-phosphate dehydrogenase deficiency, the phagocytes cannot provide enough NADPH for full NADPH oxidase activity, again with CGD-like symptoms as a result [4].

2 Phagocyte NADPH Oxidase Activity

The activity of the phagocyte NADPH oxidase is usually measured in purified blood neutrophils activated by serum-opsonized zymosan (yeast membranes) or serum-opsonized *Escherichia coli* or by fluid stimuli like phorbol myristate acetate (PMA) or formyl-methyl-leucyl-phenylalanine (fMLP or fMLF) after addition of platelet-activating factor (PAF). Methods to detect the NADPH oxidase activity include superoxide assays, hydrogen peroxide assays, or even oxygen consumption assays. Usually, an NADPH oxidase activity assay of CGD neutrophils is followed by an assay to measure the expression of the phagocyte NADPH oxidase protein components and sometimes by measuring the neutrophils' capacity to kill certain bacteria, yeasts, and/or fungi [6]. Details can be found in the chapter by Kuhns in this book.

Is it important to diagnose CGD patients in this way? Would not the clinical symptoms and the causative mutation suffice? The answer is no. CGD can only be diagnosed with certainty by proving the dysfunctional phagocyte NADPH oxidase. Clinical symptoms

Table 1
Properties of the phagocyte NADPH oxidase components and mutations in CGD patients

Component	*NOX2 (gp91^{phox})*	*p22^{phox}*	*p47^{phox}*	*p67^{phox}*	*p40^{phox}*
Amino acids	570	195	390	526	339
Gene locus	*CYBB* Xp21.1	*CYBA* 16q24	*NCF1* 7q11.23	*NCF2* 1q25	*NCF4* 22q13.1
OMIM	*300481	*608508	*608512	*608515	*601488
Exons/ span	13/30 kb	6/8.5 kb	16/40 kb	11/15 kb	10/18 kb
Mutations in CGD	Large deletions (Xb^0), small deletions (Xb^0, Xb^-), insertions (Xb^0), nonsense mutations (Xb^0), missense mutations (Xb^0, Xb^-, Xb^+) splice site mutations (Xb^0, Xb^-), promoter mutations (Xb^-)	Large deletions, small deletions, insertions, nonsense mutations, missense mutations, splice site mutations (all leading to $A22^0$, except one missense mutation $A22^-$)	Unequal crossover with pseudogenes, large deletions, small deletions, nonsense mutations, missense mutations, splice site mutations (all leading to $A47^0$)	Large deletions ($A67^0$), small deletions ($A67^0$, $A67^-$), insertions ($A67^0$), nonsense mutations ($A67^0$), missense mutations ($A67^0$, $A67^-$), splice site mutations ($A67^0$, $A67^-$)	One patient with 10-bp insertion and missense mutation ($A40^-$). Several other patients recently reported [3]
Disease OMIM	#306400	#233690	#233700	#233710	#613960

Abbreviations used: *CYBA* cytochrome-*b* alpha, *CYBB* cytochrome-*b* beta, *NCF* neutrophil cytosol factor, *Xb^0* X-linked (mutation in *CYBB*) without gp91^{phox} protein expression and without NADPH oxidase activity, *Xb^-* X-linked (mutation in *CYBB*) with diminished gp91^{phox} protein expression and diminished NADPH oxidase activity, *Xb^+* X-linked (mutation in *CYBB*) with normal gp91^{phox} protein expression but without NADPH oxidase activity. *A22* autosomal with mutations in p22^{phox} gene *CYBA* (for superscript see Xb). *A47* autosomal with mutations in p47^{phox} gene *NCF1* (for superscript see Xb). *A67* autosomal with mutations in p67^{phox} gene *NCF2* (for superscript see Xb). *A40* autosomal with mutations in p40^{phox} gene *NCF4* (for superscript see Xb) are never specific enough for this conclusion. If a variation is found in one of the "CGD genes," one needs to correlate this finding with strongly decreased phagocyte NADPH oxidase activity. Moreover, if the patient shows residual oxidase activity, this predicts less severe clinical symptoms and better survival chances [7–9] and is therefore an important finding.

Most CGD patients show a total lack of phagocyte NADPH oxidase activity. However, some residual activity can be found in a few patients, depending on the type of mutation and the gene affected. Mutations in *CYBB* can result in residual oxidase activity if they concern amino acid substitutions, small in-frame deletions, or splice site mutations that leave some normal mRNA splicing intact. Usually, such *CYBB* mutations lead to a similar decrease in NOX2 expression and phagocyte NADPH oxidase activity (Xb^{-} phenotype). A few hypomorphic mutations in *CYBA* and *NCF2* are known that also result in decreased expression of $p22^{phox}$ and $p67^{phox}$, respectively, with decreased NADPH oxidase-supporting activity [10, 11]. The big exception in this respect is $p47^{phox}$, because even without its presence, some residual hydrogen peroxide generation, but not superoxide production, by neutrophils can be measured. The reason is probably that $p47^{phox}$ is involved in facilitating electron transfer within NOX2 from FAD to the hemes, but not from NADPH to FAD [12]. Thus, in the absence of $p47^{phox}$, FAD is reduced, unable to donate its electrons to the hemes and oxygen (which would have resulted in superoxide generation) but instead—in an inefficient way—able to directly donate electrons to molecular oxygen, thus generating some hydrogen peroxide. As a result, CGD patients that lack $p47^{phox}$ show a milder clinical course than other CGD patients [7–9, 13, 14].

3 Mutations in CGD

CGD occurs in about 1:200,000 live births [13]. This estimation varies among countries, probably influenced by the awareness of clinicians for the possibility of a CGD diagnosis and the facilities to properly detect it. Final confirmation of this diagnosis requires genetic analysis. This can be provided by Sanger sequencing or next-generation sequencing with SNP arrays, such as Ion Torrent [15]. For genetic counseling, family studies, and finding donors for bone marrow transplantation or gene therapy, the disease-causing mutation needs to be identified.

3.1 Mutations in *CYBB*

Mutations in *CYBB* range from single nucleotide variations to deletions of several megabases [16]. In the latter cases, deletion of proximal genes to *CYBB* may occur, such as those for Duchenne muscular dystrophy, Kell blood group antigen, ornithine transcarbamylase, and dynein light chain Tctex-type 3, leading to additional clinical complaints [4]. Missense mutations can result in Xb^0 phenotype (no NOX2 expression, no oxidase activity), Xb^- phenotype (partial NOX2 expression, partial oxidase activity, see above), or Xb^+ phenotype (normal NOX2 expression, no oxidase activity). Nonsense mutations, either direct or by frameshift inducing a stop codon further downstream, lead to premature termination of protein synthesis. Such shortened proteins—if expressed at all—are usually unstable and do not support NADPH oxidase activity. In most cases, the mRNA of such proteins is quickly degraded by the nonsense-mediated mRNA decay (NMD) process. Insertions in *CYBB* are relatively rare. Most of these are due to smaller or larger duplications [16], but some concern pieces of cDNA that have been inserted by so-called transposable elements of the long interspersed element-1 (LINE-1) family [17].

Missplicing of mRNA can be caused directly by mutations in donor or acceptor splice sites but can also be caused by a single nucleotide variation that activates a latent splice site within an exon or an intron, which is then preferred over the original splice site. This can lead to complete or partial exon skipping or to inclusion of (part of) an intron in the mature mRNA. Another possibility is missplicing caused by destroying an exonic splicing enhancer (ESE) site, which again can lead to exon skipping [18]. ESEs are binding sites for serine- and arginine-rich (SR) proteins that act as splicing enhancers of pre-mRNA. Usually, such altered mRNA is quickly destroyed in the NMD pathway, leading to absence of protein expression.

Mutations in *CYBB* mostly affect males, because males have only one X chromosome. However, due to skewed lyonization, females can have a majority of leukocytes in which either the paternal or the maternal X chromosome is inactivated. If the active X chromosome is carrying a mutation in *CYBB*, then the majority of their leukocytes will lack NADPH oxidase activity and cause a CGD phenotype. This situation can be detected with a nitro blue tetrazolium (NBT) slide test, dihydrorhodamine-1,2,3 (DHR) flow cytometry, or NOX2 expression flow cytometry. In all of these assays, the NADPH oxidase activity or the NOX2 expression is measured in individual phagocytes. If a large majority of these cells shows lack of NADPH oxidase activity or lack of NOX2 expression, the female under investigation is regarded as a CGD patient if she also shows the clinical symptoms.

About 12–15% of the mutations in *CYBB* result from de novo mutations, i.e., without detectable mutations in the mother. Yet, one must be careful, because during the mother's embryonic development, somatic mutations may occur, also in her germ line cells. These mutations are not detectable in her leukocyte DNA, or only in part of that DNA, but can be transmitted to the offspring [9].

3.2 Mutations in Other CGD Genes

NCF1 is a gene on chromosome 7, on each side surrounded by a pseudo-*NCF1* gene. These pseudogenes closely resemble *NCF1*, but the most important difference is a GT deletion at the start of exon 2. Therefore, these pseudogenes do not encode an active $p47^{phox}$ protein. However, unequal crossing over between *NCF1* and one of these pseudogenes during meiosis, leading to loss of (part of) *NCF1*, is a frequently encountered mutational mechanism leading to CGD [19, 20]. Therefore, this so-called delta-GT is the mutation found in about 90% of all $p47^{phox}$-deficient CGD patients. In fact, this is not a GT deletion within *NCF1*, but exchange of *NCF1* for one of its pseudogenes that contains a GT deletion. Usually, these patients carry this mutation on both alleles, but it can also occur on one allele in combination with another mutation on the other allele [21]. These other *NCF1* mutations are rarely found, partly perhaps because their detection is difficult due to interference of the pseudogene sequences. A special example of this last problem is our recent identification of individuals within Ashkenazi Jews with two *NCF1* genes on one chromosome, one of which carried a nonsense mutation and the other was functional. These chromosomes contained only one pseudogene [22].

As with the (non-GT deletion) mutations in *NCF1*, those in *CYBA* and *NCF2* encompass the whole scale of deletions, insertions, missense, nonsense, and splice site variations [21]. Most mutations result in total lack of protein expression, but a few are known to result in expression of proteins with diminished function (hypomorphic mutations, see above). Until now, only one CGD patient has been described with dysfunctional $p40^{phox}$ due to mutations in *NCF4* [23]. The phagocyte NADPH oxidase activity of this patient was disturbed when the phagocytes were activated with particular stimuli, but not when activated with fluid stimuli, leading to the idea that $p40^{phox}$ is especially involved in enhancing the NADPH oxidase activity in phagocytic membranes. Recently, several additional CGD patients with mutations in *NCF4* have been recognized [3]. If phagocyte NADPH oxidase activity induced by PMA is tested, a standard procedure in many laboratories, such patients will be missed, but with oxidase activity induced by, e.g., serum-opsonized *E. coli*, the abnormally low activity will reveal these patients. In sharp contrast to classic CGD, $p40^{phox}$-deficient neutrophils kill fungi and yeast like normal neutrophils do; however, $p40^{phox}$-deficient neutrophils fail to kill *S. aureus* bacteria, like classic CGD neutrophils do. The $p40^{phox}$-deficient CGD patients suffer mainly from hyperinflammation, consisting mostly of granulomatous lesions of the gastrointestinal tract and skin, as well as peripheral infections, but not from any of the invasive bacterial and fungal infections seen in classic CGD. In general, $p40^{phox}$-deficient CGD seems to be clinically milder than classic CGD.

Mutations in patients with CGD-like symptoms have been found both in the *RAC2* gene and in the *G6PD* gene [4, 5]. Since RAC2 is involved not only in activating NOX2 but also other neutrophil functions, the patients suffer from additional clinical problems [4, 5]. Glucose-6-phosphate dehydrogenase (G6PD) is an enzyme of the hexose-monophosphate shunt that reduces $NADP^+$ to NADPH. This is an essential process for the cellular defense against oxidative stress. Total G6PD deficiency is therefore incompatible with life. Usually, G6PD deficiencies manifest themselves only in erythrocytes, because these cells cannot synthesize new proteins. Sometimes, however, the effect of G6PD deficiency is also detectable in leukocytes, e.g., when a mutation leads to a very unstable protein. In that case, the supply of NADPH is too low to provide phagocytes with sufficient substrate for the NADPH oxidase, thus leading to a defect in adequate antimicrobial defense [24]. Prenatal detection of CGD in pregnancies at risk is now a standard procedure. It has been proven to be reliable for X-CGD as well as AR-CGD [25, 26].

4 Clinical Symptoms of CGD

4.1 Infections

Patients with CGD generally suffer from bacterial and fungal infections from infancy or early childhood [1, 2, 4, 8, 9, 13, 14]. Bacterial species that are frequently encountered are *Staphylococcus aureus*, *Burkholderia cepacia* species, *Nocardia* species, and certain gram-negative enteric bacilli, including *Serratia marcescens* and *Salmonella* species. Mycobacterial species pose also a threat for CGD patients, and these patients can develop severe local or systemic disease with BCG, an attenuated strain of *Mycobacterium bovis*, after being vaccinated with BCG. Very dangerous are fungal infections with molds and yeasts, with *Aspergillus fumigatus* and *Aspergillus nidulans* as the most common pathogens.

Common sites of infection are the lungs, the lymph nodes, the skin and soft tissues, the gastrointestinal tract, the liver, and the bones. Pneumonia and sepsis due to *Aspergillus* or *B. cepacia* are the most frequent cause of death. Cutaneous abscesses and lymphadenitis represent the next most common types of infection in CGD and are typically caused by *S. aureus*, followed by various gram-negative organisms, including *B. cepacia* complex and *Serratia marcescens*. Hepatic (and perihepatic) abscesses are also quite common in CGD and are usually caused by *S. aureus*.

4.2 Inflammatory Complications

Inflammatory manifestations are common in CGD [1, 4, 27]. In case infections cannot be cleared, chronic inflammatory cell reactions may develop, consisting of lymphocytes and histiocytes, which can then organize to form granulomas. This is one of the hallmarks of CGD and causes various clinical symptoms of obstruction. A dysregulated inflammatory response as a result of NOX2 inactivation may develop due to impaired ingestion of pathogens, increased pro-inflammatory cytokine production as a result of changes in redox-regulated signaling, and diminished antigen presentation.

An important type of chronic inflammation in CGD is a form of inflammatory bowel disease that closely resembles Crohn's disease and affects a substantial fraction of CGD patients [28, 29]. The colon is typically involved, but the ileum and other parts of the gastrointestinal tract can also be affected. The symptoms can range from mild diarrhea to a debilitating syndrome of bloody diarrhea and malabsorption that can even necessitate colectomy. Other types of chronic inflammation include noninfectious arthritis, gingivitis, chorioretinitis or uveitis, glomerulonephritis, and—rarely—discoid or even systemic lupus erythematosus.

5 Treatment of CGD

Patients suspected of CGD and presenting with acute infections should be treated as soon as possible with antibiotics or antifungal drugs, even before culture results are available [4]. In these cases, the antibiotics should be chosen to provide strong coverage for *S. aureus* and gram-negative bacteria, including *B. cepacia* complex (e.g., a combination of nafcillin and ceftazidime or a carbapenem; most aminoglycosides are typically ineffective against *Burkholderia*). If the infection fails to respond within 24–48 h, then more aggressive diagnostic procedures should be instituted to identify the responsible microorganism. Additional empirical changes in antibiotic coverage may be warranted, such as adding high-dose intravenous trimethoprim/sulfamethoxazole to cover ceftazidime-resistant *B. cepacia* and *Nocardia*. If fungus is identified or strongly suspected, antifungal therapy should be started even before the diagnosis is confirmed. Aspergillus infections of the lungs and bones are the most common fungal infections in CGD and often require prolonged treatment. In general, the newer triazole antifungals are better tolerated and have better activity than amphotericin derivatives.

Surgical drainage (and sometimes excision) of infected lymph nodes and abscesses involving the liver, skin, rectum, kidney, and brain is often necessary for healing, particularly for the visceral abscesses. In infection-free periods, most CGD patients require prophylactic antibacterial treatment with trimethoprim/sulfamethoxazole (dicloxacillin or trimethoprim in sulfa-allergic patients) and prophylactic antifungal treatment with itraconazole (or voriconazole). Prophylactic treatment with interferon-γ is also recommended by most [1, 4, 30], but not all [31] clinicians. Corticosteroids are used to treat clinically significant granulomatous or other inflammatory complications but should be used with caution, given the underlying microbial killing defect.

At present, the only curative treatment for CGD patients is allogeneic hematopoietic stem cell transplantation (HSCT). Reduced intensity conditioning of the patients' bone marrow is accepted as the preferred approach, with less acute and long-term side effects. Overall survival is now about 90%, with acute graft-versus-host disease (GvHD) 4–8% and chronic GvHD 7–12% [32, 33]. In a retrospective study, the clinical outcome of CGD patients after HSCT was compared with CGD patients treated conventionally. Survival was similar, but transplanted patients had less serious infections, less hospital admissions, and less episodes of surgery [34].

For patients without an HLA-compatible donor, gene therapy would be a good alternative to HSCT. Conventional gene therapy trials for CGD have been based on transfer of a therapeutic transgene into autologous hematopoietic stem and progenitor cells (HSPCs) via retroviral vectors that semi-randomly integrate in the genome. Although this initially led to clinical improvement in the patients, methylation of the viral promoter has resulted in transgene silencing over time and therefore loss of therapeutic benefit [35, 36]. Furthermore, insertional activation of several proto-oncogenes led to clonal dominance of corrected cells, eventually leading to myelodysplasia [36]. Current efforts are now directed at self-inactivating gamma retroviral vectors or lentiviral vectors that lack vector-enhancing sequences.

New attempts at gene therapy have been directed at gene correction through introducing site-specific double-strand DNA breaks by engineered homing endonucleases, such as zinc finger nucleases (ZFN) [37–39] or transcription activator-like effector nucleases (TALENs) [40], followed by homologous recombination of the mutated gene region with a similar exogenous wild-type cDNA source through endogenous DNA repair mechanisms. The introduction of site-specific double-strand DNA breaks greatly enhances correction efficiency. With the ZFN technique, Merling et al. [41] succeeded in converting an *NCF1* pseudogene into an active *NCF1* gene in iPS cells derived from four $p47^{phox}$-deficient CGD patients. iPS cells are epigenetically reprogrammed somatic cells with a high reproduction capacity that can be induced to differentiate into all possible somatic cells, including myeloid cells. The neutrophil-like cells derived from the corrected iPS cells of the patients showed restored $p47^{phox}$ expression and NADPH oxidase activity.

Yet another method of gene correction uses the clustered regularly interspaced short palindromic repeat (CRISPR)-Cas9 nuclease system, which does not work with a nuclease (Cas9) that is guided by proteins binding to a specific site in the DNA, such as ZFNs or TALENs, but by a short RNA stretch that fits to the gene at the site to be corrected [42]. Here, iPS clones were obtained from an X-CGD patient with a splicing defect in *CYBB* (intron 1 −11T>G). After correction of *CYBB*, the clones were successfully differentiated into monocytes and macrophages with restored $gp91^{phox}$ mRNA and ROS production. Since it is much simpler to construct site-specific RNA than protein, the CRISPR-Cas9 technique is a promising tool for patient-specific gene correction. At present, it is not yet possible to apply this technique to HPSCs for permanent correction of sufficient phagocyte precursors in the bone marrow. Limitations are efficiency of gene editing in HSPCs and maintaining the bone marrow-reconstituting potential of these fragile cells throughout the whole procedure of transfection, expansion, and selection.

References

1. Dinauer M, Newburger P, Borregaard N (2015) Phagocyte system and disorders of granulopoiesis and granulocyte function. In: Nathan DG, Orkin SH (eds) Hematology of infancy and childhood. Elsevier/Saunders, Philadelphia, pp 773–847
2. Nunes P, Demaurex N, Dinauer MC (2013) Regulation of the NADPH oxidase and associated ion fluxes during phagocytosis. Traffic 14:1118–1131
3. Van de Geer et al (2018) J Clin Invest 128:3957–3975
4. Roos D, Holland SM, Kuijpers TW (2014) Chronic granulomatous disease. In: Ochs HD, Smith CIE, Puck JM (eds) Primary immunodeficiency diseases, a molecular and genetic approach, 3rd edn. Oxford University Press, New York, pp 689–722
5. Alkhairy OK, Rezaei N, Graham RR et al (2015) *RAC2* loss-of-function mutation in two siblings with characteristics of common variable immunodeficiency. J Allergy Clin Immunol 135:1380–1384
6. Roos D (2016) Chronic chranulomatous disease. Br Med Bull 118:53–66
7. Kuhns DB, Alvord WG, Heller T et al (2010) Residual NADPH oxidase and survival in chronic granulomatous disease. N Engl J Med 363:2600–2610
8. Köker MY, Camcıoğlu Y, van Leeuwen K et al (2013) Clinical, functional, and genetic characterization of chronic granulomatous disease in 89 Turkish patients. J Allergy Clin Immunol 132:1156–1163
9. Wolach B, Gavrieli R, de Boer M et al (2017) Chronic granulomatous disease: Clinical, functional, molecular, and genetic studies. The Israeli experience with 84 patients. Am J Hematol 92:28–36
10. Leusen JH, Bolscher BG, Hilarius PM et al (1994) 156Pro-->Gln substitution in the light chain of cytochrome b558 of the human NADPH oxidase (p22-phox) leads to defective translocation of the cytosolic proteins p47-phox and p67-phox. J Exp Med 180:2329–2334
11. Roos D, van Buul JD, Tool AT et al (2014) Two CGD families with a hypomorphic mutation in the activation domain of p67phox. J Clin Cell Immunol 5:pii:1000231
12. Cross AR, Curnutte JT (1995) The cytosolic activating factors p47-phox and p67-phox have distinct roles in the regulation of electron flow in NADPH oxidase. J Biol Chem 270:6543–6548
13. Winkelstein JA, Marino MC, Johnston RB Jr et al (2000) Chronic granulomatous disease. Report on a national registry of 368 patients. Medicine (Baltimore) 79:155–169
14. Van den Berg JM, van Koppen E, Ahlin A et al (2009) Chronic granulomatous disease: the European experience. PLoS One 4:e5234
15. Roos D, Tool AT, van Leeuwen K et al (2017) Biochemical and genetic diagnosis of chronic granulomatous disease. In: Seger R, Roos D, Segal B, Kuijpers TW (eds) Chronic granulomatous disease: genetics, biology and clinical management. Nova Publishers, New York, pp 231–300
16. Roos D, Kuhns DB, Maddalena A et al (2010) Hematologically important mutations: X-linked chronic granulomatous disease (third update). Blood Cells Mol Dis 45:246–265
17. de Boer M, van Leeuwen K, Geissler J et al (2014) Primary immunodeficiency caused by an exonized retroposed gene copy inserted in the *CYBB* gene. Hum Mutat 35:486–496
18. De Boer M, van Leeuwen K, Geissler J et al (2017) Mutation in an exonic splicing enhancer site causing chronic granulomatous disease. Blood Cells Mol Dis 66:50–57
19. Görlach A, Lee P, Roesler J et al (1997) A p47-phox pseudogene carries the most common mutation causing p47-phox deficient chronic granulomatous disease. J Clin Invest 100:1907–1918
20. Hayrapetyan A, Dencher PC, van Leeuwen K et al (2013) Different unequal cross-over

events between *NCF1* and its pseudogenes in autosomal p47(phox)-deficient chronic granulomatous disease. Biochim Biophys Acta 1832:1662–1672

21. Roos D, Kuhns DB, Maddalena A et al (2010) Hematologically important mutations: The autosomal recessive forms of chronic granulomatous disease (second update). Blood Cells Mol Dis 44:291–299
22. De Boer M, Gavrieli R, van Leeuwen K et al (2018) A false-carrier state for the c.579G>A mutation in the NCF1 gene in Ashkenazi Jews. J Med Genet 55:166–172
23. Matute JD, Arias AA, Wright NA et al (2009) A new genetic subgroup of chronic granulomatous disease with autosomal recessive mutations in p40 phox and selective defects in neutrophil NADPH oxidase activity. Blood 114:3309–3315
24. Van Bruggen R, Bautista JM, Petropoulou T et al (2002) Deletion of leucine-61 in glucose-6-phosphate dehydrogenase leads to chronic nonspherocytic anemia, granulocyte dysfunction, and increased susceptibility to infections. Blood 100:1026–1030
25. De Boer M, Bolscher BGJM, Sijmons RH et al (1992) Prenatal diagnosis in a family with X-linked chronic granulomatous disease with the use of the polymerase chain reaction. Prenat Diagn 12:773–777
26. De Boer M, Singh V, Dekker J et al (2002) Prenatal diagnosis in two families with autosomal, p47-*phox*-deficient chronic granulomatous disease due to a novel point mutation in *NCF1*. Prenat Diagn 22:235–240
27. Rieber N, Hector A, Kuijpers TW et al (2012) Currents concepts of hyperinflammation in chronic granulomatous disease. Clin Dev Immunol 2012:252460
28. Marciano BE, Rosenzweig SD, Kleiner DE et al (2004) Gastrointestinal involvement in chronic granulomatous disease. Pediatrics 114:462–468
29. Magnani A, Brosselin P, Beauté J et al (2014) Inflammatory manifestations in a single-center cohort of patients with chronic granulomatous disease. J Allergy Clin Immunol 134:655–662
30. Marciano BE, Wesley R, De Carlo ES et al (2004) Long-term interferon-gamma therapy for patients with chronic granulomatous disease. Clin Infect Dis 39:692–699
31. Martire B, Rondelli R, Soresina A et al (2008) Clinical features, long-term follow-up and outcome of a large cohort of patients with chronic granulomatous disease: an Italian multicenter study. Clin Immunol 126:155–164
32. Gűngőr T, Teira P, Slatter M et al (2014) Reduced intensity conditioning and HLA-matched haematopoietic stem-cell transplantation in patients with chronic granulomatous disease ; a prospective multicentre study. Lancet 383:436–448
33. Morillo-Gutierrez B, Beier R, Rao K et al (2016) Treosulfan based conditioning for allogeneic HSCT in children with chronic granulomatous disease: a multicentre experience. Blood 128:440–448
34. Cole T, Pearce MS, Cant AJ et al (2013) Clinical outcome in children with chronic granulomatous disease managed conservatively or with hematopoietic stem cell transplantation. J Allergy Clin Immunol 132:1150–1155
35. Ott MG, Seger R, Stein S et al (2007) Advances in the treatment of chronic granulomatous disease by gene therapy. Curr Gene Ther 7:155–161
36. Stein S, Ott MG, Schultze-Strasser S et al (2010) Genomic instability and myelodysplasia with monosomy 7 consequent to EVI1 activation after gene therapy for chronic granulomatous disease. Nat Med 16:198–204
37. Nakayama M (2010) Homologous recombination in human iPS and ES cells for use in gene correction therapy. Drug Discov Today 15:198–202
38. Urnov FD, Rebar EJ, Holmes MC et al (2010) Genome editing with engineered zinc-finger nucleases. Nat Rev Genet 11:636–646
39. Zou J, Sweeney CL, Chou BK et al (2011) Oxidase-deficient neutrophils from X-linked chronic granulomatous disease iPS cells: functional correction by zinc finger nuclease-mediated safe harbor targeting. Blood 117:5561–5572
40. Mussolino C, Cathomen T (2012) TALE nucleases: tailored genome engineering made easy. Curr Opin Biotechnol 23:644–650
41. Merling RK, Kuhns DB, Sweeney CL et al (2017) Gene-edited pseudogene resurrection corrects $p47^{phox}$-deficient chronic granulomatous disease. Blood Adv 4:270–278
42. Flynn R, Grundmann A, Renz P et al (2015) CRISPR-mediated genotypic and phenotypic correction of a chronic granulomatous disease mutation in human iPS cells. Exp Hematol 43:838–848

Chapter 33

Diagnostic Testing for Chronic Granulomatous Disease

Douglas B. Kuhns

Abstract

Chronic granulomatous disease (CGD) is a rare genetic immunodeficiency associated with recurrent bacterial infections, granulomas, and increased mortality. It is characterized by the inability of phagocytes (neutrophils, monocytes, etc.) to generate reactive oxygen species (ROS), a major component of the microbicidal repertoire of phagocytes. Diagnosis of patients with CGD is commonly based on the assessment of ROS production by neutrophils. Multiple assays to assess ROS production are described—a flow cytometric dihydrorhodamine assay and a histochemical nitroblue tetrazolium assay, both of which can be used to visualize ROS production in individual cells, and two quantitative assays—$O_2^{\cdot-}$ reduction of ferricytochrome c and a ROS-dependent, luminol-enhanced chemiluminescence assay that will quantitate the response of a population of cells. In addition, two approaches to identify the defective phox protein defect are described—standard immunoblotting and flow cytometry of neutrophils stained with phox-specific antibodies. When determining the status of a patient, several assays should be used to assess ROS production and identify the protein defect. The results of these assays should agree and can be used to develop a comprehensive package, which includes confirmation of a diagnosis of CGD, identification of the specific protein target for genetic sequencing, and an indication of the prognosis for the patient.

Key words Chronic granulomatous disease, NOX2, ROS production, Phox proteins

1 Introduction

One of the primary function of neutrophils is the ingestion (phagocytosis) and subsequent killing of microorganisms. This process requires the assembly of NOX2, a multicomponent phagocyte NADPH oxidase (phox) enzyme complex consisting of at least three cytosolic components, $p47^{phox}$, $p67^{phox}$ [1, 2], and $p40^{phox}$ [3] and two membrane components, $p22^{phox}$ [4] and $gp91^{phox}$ [5] that constitute a membranous flavocytochrome $b_{558.}$ When these subunits associate during phagocytosis, the enzyme complex catalyzes a single electron reduction of molecular O_2 to superoxide anion radical ($O_2^{\cdot-}$) using the reducing potential of NADPH. NADPH is generated by the oxidation of glucose through the pentose phosphate pathway. $O_2^{\cdot-}$ either spontaneously or enzymatically converts to $H_2O_{2.}$ When neutrophil granules fuse with the

Ulla G. Knaus and Thomas L. Leto (eds.), *NADPH Oxidases: Methods and Protocols*, Methods in Molecular Biology, vol. 1982, https://doi.org/10.1007/978-1-4939-9424-3_33,

phagosome, myeloperoxidase, localized in the azurophilic granules, can then catalyze the formation of hypochlorous acid from H_2O_2 and Cl^-. The effects of these reactive O_2 species (ROS), along with neutrophil granular antimicrobial peptides and lysosomal hydrolases, result in the killing and eventual digestion of the ingested microorganism.

CGD, first described in the 1950s [6], is a rare genetic disease (approximately 1 case in 200,000 people) without ethnic preference and with a risk of death of 1–5% per year. In patients with CGD, the production of $O_2^{\cdot-}$ and other ROS by neutrophils, monocytes, macrophages, and eosinophils is impaired, leading to recurrent infections, granulomatous complications, and premature death. CGD is caused by defects in any one of five subunits of phagocyte NADPH oxidase, including gp91^{phox} (gene name, *CYBB*, ~70% of patients), p22^{phox} (*CYBA*, <5%), p47^{phox} (*NCF1*, ~25%), p67^{phox} (*NCF2*, <5%) [7], p40^{phox} (*NCF4*, 25 patients) [8, 9], and Rac2 [10, 11]; in rare cases, severe glucose-6-phosphate dehydrogenase deficiency can result in CGD [12].

Because of their abundance, neutrophils are generally used for assessment of ROS production. However, neutrophils have a relatively short lifetime in vitro; therefore, analysis of neutrophil function should address several limitations—the number of cells required for the assay, the type of cell preparation needed (isolated neutrophils versus whole blood), the overall incubation time for the assay, the complexity of the assay, and the rapidity of data collection. Since neutrophils cannot be stored or frozen and maintain viability, neutrophils from a normal subject are generally assayed in parallel to validate the results. Additionally, isolation of neutrophils can take 1–2 h, limiting the time available for functional assays. The use of multi-well microplates and microplate readers has reduced the number of cells required and facilitated the collection of data. Experience in the handling of neutrophils and the time constraints of the assays can limit the availability of this testing to laboratories that specialize in assessment of neutrophil function.

Assays used in the diagnosis and characterization of patients with CGD are described below. Because of the limitations of each assay, often multiple assays are needed to obtain a comprehensive laboratory evaluation. Quantitation of $O_2^{\cdot-}$ can be used to predict risk in CGD patients but cannot differentiate a CGD patient from a highly lyonized X-linked CGD carrier or identify the specific defective phox subunit. A DHR assay can differentiate a CGD patient from an X-linked CGD carrier but cannot identify the specific defective phox subunit. Immunoblotting can often identify the specific defective phox subunit for genetic sequencing but cannot predict the consequence, i.e., residual functionality of the mutant protein.

2 Materials

2.1 Equipment

1. Clinical, low-speed centrifuge.
2. 37 °C Water bath.
3. Automated blood cell counter.
4. Flow cytometer.
5. Light microscope with photographic accessories.
6. 37 °C Incubator with rotator.
7. 8-Chamber slide culture chamber.
8. Analytical balance.
9. Visible light spectrophotometer.
10. Plastic cuvettes.
11. Microcentrifuge.
12. Polypropylene tubes (because some tubes may induce spontaneous neutrophil activation, tubes should be pretested).
13. Luminometer.
14. Microtip probe sonicator.
15. Electrophoresis chamber and power supply.
16. Immunoblot transfer apparatus.
17. Rotary shaker.

2.2 Reagents

2.2.1 Reagents for Isolation of Neutrophils (Subheading 3.1)

1. Ficoll-Paque Premium.
2. Phosphate-buffered saline (PBS) 1× and 10×.
3. 3% Dextran: 55 mL 10× PBS, 500 mL sterile distilled H_2O, 16.65 g dextran (200,000–300,000 MW, clinical grade). Dissolve dextran in PBS and sterilize with a 0.2 μm filter. Store at 4 °C and pre-warm to room temperature prior to use.
4. Hypotonic lysis buffer: 15.0 mL 10× PBS, 600 mL sterile distilled H_2O.
5. Hypertonic lysis buffer: 110 mL of 10× PBS, 500 mL sterile distilled H_2O.
6. Hanks' balanced salt solution without divalent cations, no phenol red (HBSS[−]).

2.2.2 Reagents for Dihydrorhodamine 123 ROS Assay (Subheading 3.2.1)

1. Lysis buffer: 100 mL Sterile distilled water, 0.83 g ammonium chloride, 0.2 mL 0.5 M EDTA, pH 8.0, 0.2 g potassium bicarbonate. Mix and sterilize through a 0.2 μm filter.
2. Flow buffer: 500 mL HBSS[−], 0.5 g albumin (from human serum, fraction V), 1 mL 0.5 M EDTA, pH 8.0. Mix reagents well and sterilize through a 0.2 μm filter.

3. Catalase (10 mg/mL): 10 mL Phosphate-buffered saline (PBS), 100 mg catalase. Dissolve catalase in PBS. Store at −85 °C in 250 μL aliquots.
4. Dihydrorhodamine 123 (DHR, 25 mM): 1.156 mL Dimethyl sulfoxide (DMSO), 10 mg DHR. Dissolve DHR in DMSO and store at −85 °C in 25 μL aliquots wrapped with foil to protect from light.
5. Phorbol 12-myristate 13-acetate (PMA, 2 μg/mL}: 2.5 mL DMSO, 5 mg PMA: Prepare stock solution of PMA (2 mg/mL) by dissolving PMA in DMSO. Store at −85 °C in 100 μL aliquots. Dilute stock PMA solution to a 2 μg/mL working concentration in flow buffer.

2.2.3 Reagents for NBT Assay (Subheading 3.2.2)

1. HBSS with Ca^{2+} and Mg^{2+} (HBSS[+]) with 10 mM HEPES (HBSS/HEPES), pH 7.35–7.40: 500 mL HBSS[+], 5.0 mL 1.0 M HEPES. Prepare daily and adjust the pH within the range of 7.35–7.40.
2. HBSS/HEPES with 1% bovine serum albumin (BSA): 100 mL HBSS/HEPES, 1.0 g bovine serum albumin. Mix BSA with HBSS/HEPES and sterilize through a 0.2 μm filter. Store at 4 °C.
3. Nitroblue tetrazolium (NBT, 0.2%): 5.0 mL HBSS/HEPES, 1.0 mg NBT. Prepare daily by mixing NBT with HBSS/HEPES and sterilize through a 0.2 μm filter.
4. PMA (20 μg/mL): 4.95 mL 50% DMSO/sterile water(v/v), 50 μL PMA (2 mg/mL). Mix and store as 100 μL aliquots at −80 °C.
5. Safranin O (0.6%).
6. Diphenyleneiodonium chloride (DPI, 20 mM in 10 mM DMSO): 1.562 mL DMSO, 0.438 mL H_2O, 10 mg DPI. Mix well and store in 7 μL aliquots at −80 °C.

2.2.4 Reagents for Assay of $O_2^{\cdot -}$ Production by Ferricytochrome c Reduction (Subheading 3.2.3)

1. HBSS[+] with 10 mM HEPES (HBSS/HEPES), pH 7.35–7.40.
2. Cytochrome c (200 μM): 10 mL HBSS/HEPES, 26 mg cytochrome c.
3. Catalase (10 mg/mL).
4. Superoxide dismutase (SOD, 10 mg/mL): 2.0 mL PBS, 20 mg SOD. Dissolve SOD in PBS and store at −85 °C in 100 μL aliquots.
5. Phorbol 12-myristate 13-acetate (PMA, 20 μg/mL).

2.2.5 Reagents for Luminol-Enhanced Chemiluminescence ROS Assay (Subheading 3.2.4)

1. HBSS[−].
2. HBSS[+].
3. PBS.
4. Luminol (50 mM): 4 mL DMSO, 35.44 mg luminol. Mix well and store in single-use amber tubes in 50 μL aliquots at −80 °C, shielded from light.

2.2.6 Reagents for Immunoblotting Analysis of NADPH Oxidase Subunits (Subheading 3.3)

1. Running buffer: 1900 mL distilled H_2O, 100 mL NuPAGE™ MOPS SDS Running Buffer (20×).
2. Transfer buffer: 2547 mL H_2O, 300 mL methanol, 3 mL NuPAGE™ Antioxidant, 150 mL NuPAGE™ Transfer Buffer (20×).
3. Sample buffer: 650 μL H_2O, 250 μL NuPAGE™ LDS Sample Buffer (4×), 100 μL 10× protease inhibitors such as complete™ Protease Inhibitor Cocktail.
4. NuPAGE™ Antioxidant.
5. NuPAGE™ Sample Reducing Agent (10×).
6. NuPAGE Bis-Tris 10% and 4–12% gradient mini-gels.
7. Tris-buffered saline (TBS): 3920 mL H_2O, 80 mL 1 M Tris, pH 7.4, 116.8 g NaCl.
8. TBS/0.5% Tween: 3900 mL H_2O, 80 mL 1 M Tris, pH 7.4, 116.8 g NaCl, 20 mL Tween 20.
9. Sodium azide (1 M): 10 mL H_2O, 0.65 g sodium azide.
10. Pre-stained protein standards.
11. Immobilon-P transfer membrane.
12. Membrane-blocking solution (5% nonfat milk): 200 mL TBS, 10 g dry milk, 100 μL Tween 20, 200 μL 1 M sodium azide (to prevent bacterial growth).
13. Goat-anti-p22^{phox} (#1535), anti p47^{phox} (#1588), anti-p67^{phox} (#1589), and anti-gp91^{phox} (#1622) antisera [obtained from Drs. Harry Malech and Tom Leto, NIAID, NIH]: 50 mL TBS, 50 μL antibody, 2.5 g nonfat dry milk, 25 μL Tween 20, 50 μL 1 M sodium azide. Prepare fresh every 6 months.
14. Anti-NCF4 Abcam ab76158.
15. Anti-actin: EMD Millipore MAB1501.
16. Anti-goat IgG-horseradish peroxidase-labeled conjugate (1:10,000 dilution): 192 mL TBS, 19.2 μL anti-goat IgG, 9.6 g nonfat milk, 96 μL Tween 20.
17. Peroxidase substrate (for color development) or enhanced chemiluminescence western blotting substrate.

2.2.7 Reagents for Flow Cytometric 7D5/Flavocytochrome b_{558} Staining of Neutrophils (Subheading 3.4)

1. FITC-labeled anti-flavocytochrome b_{558} monoclonal antibody (7D5).
2. FITC-labeled IgG_1 isotype antibody.
3. OptiLyse® C, No-Wash Lysing Solutions.
4. Flow buffer (2% BSA in PBS): 100 mL PBS, 2.0 g BSA. Mix reagents and filter through 0.2 μm filter.

2.2.8 Reagents for Flow Cytometry Assay of $p47^{phox}$ (Subheading 3.5)

1. PBS.
2. IntraPrep permeabilization kit.
3. Goat anti-$p47^{phox}$ antibody (Abcam, ab179457).
4. Goat IgG_1 isotype antibody (Abcam, ab1727301).
5. AlexaFluor488-labeled goat anti-mouse IgG.

3 Methods

3.1 Isolation of Neutrophils

Human neutrophils are the most abundant nucleated cell in circulating blood ($2–8 \times 10^6$ per mL). Assays that avoid neutrophils isolation are preferred because of the spontaneous priming that occurs during isolation [13]. However, most assays require isolated neutrophils to eliminate any contribution of other leukocytes and blood components.

1. Collect patient blood in sodium heparin (5–10 U/mL final concentration) or in acid citrate dextrose (ACD-A) blood collection tubes (*see* **Notes 1–3**).
2. Dispense 15 mL HBSS[−] to 50 mL conical centrifuge tube.
3. Transfer 15–20 mL anticoagulant-treated blood to the 50 mL tube.
4. Carefully underlay blood mixture with 12 mL Ficoll-Paque (*see* **Note 4**).
5. Spin for 30 min (20 °C) at $500 \times g$ with the brake off.
6. Aspirate mononuclear cells, remaining plasma, and Ficoll-Paque from the tube, leaving the neutrophil/erythrocyte-rich pellet.
7. Dilute the pellet with HBSS[−] for a final volume of 20 mL.
8. Add 20 mL of 3% dextran. Invert the tube several times to mix. Allow erythrocytes to sediment at $1 \times g$ for 20 min at room temperature (*see* **Note 5**).
9. Transfer the neutrophil-rich supernatant fluid to a new tube and dilute with HBSS.
10. Spin at $300 \times g$ for 10 min. Aspirate the supernatant fluid.

11. Lyse the remaining erythrocytes with 10 mL hypotonic lysis buffer, followed by the quick addition (within 30 s) of 10 mL hypertonic lysis buffer to restore the isotonicity. Dilute further with 20 mL HBSS[−] (*see* **Note 6**).
12. Spin at 300 × *g* for 10 min, aspirate the supernatant fluid, wash, and repeat process twice. Repeat lysis if there is significant erythrocyte contamination.
13. Resuspend neutrophils in HBSS[−] and count the cells using a hemocytometer or automated cell counter (*see* **Notes 7–10**).

3.2 Generation of Reactive Oxygen Species

Neutrophils isolated from patients with CGD have a defect in the NADPH oxidase and are unable to generate ROS, resulting in an O_2-dependent bactericidal defect. Four methods are described for clinical and research laboratory studies of the respiratory burst: (1) dihydrorhodamine (DHR) assay, (2) NBT assay, (3) cytochrome c reduction, and (4) luminol-enhanced chemiluminescence. The production of ROS has become an important tool to perform risk assessment in patients with CGD. Patients whose neutrophils exhibit the lowest ROS generation (<1% of normal generation) have reduced survival compared with the survival in patients with higher ROS generation (3–10% of normal). Moreover, survival in CGD is a continuous function of ROS production, suggesting that therapeutic interventions that result in an increase in ROS generation should provide a survival benefit to patients with CGD [7].

3.2.1 Dihydrorhodamine Assay

NOX2 activity of individual neutrophils can be measured by flow cytometry using dihydrorhodamine 123 (DHR) [14], the most commonly used assay to establish a diagnosis of CGD. DHR is only a weak fluorescence and can be passively loaded into neutrophils. Upon activation, $O_2^{\cdot -}$ produced by NOX2 is transformed by superoxide dismutase into H_2O_2. Endogenous cellular peroxidases (e.g., myeloperoxidase in neutrophils) use H_2O_2 to oxidize DHR into the strongly fluorescent molecule, rhodamine 123, easily detectable using a flow cytometer (*see* **Note 11**). To minimize cell-to-cell diffusion of H_2O_2, exogenous catalase is added to scavenge extracellular H_2O_2. Since DHR is localized to the cytoplasm, and catalase is present in the extracellular fluid, the DHR assay primarily detects the intracellular production of ROS. Neutrophils can be easily distinguished from other leukocyte populations by flow cytometry using forward light scattering (FS) and right-angle light scattering (SS).

1. Collect patient blood sample in K_2EDTA blood collection tube. A blood sample from a normal subject is also needed to serve as a control to validate the assay.
2. Add 12 mL lysis buffer (pre-warmed to 37 °C) to a conical 15 mL polypropylene centrifuge tube.

3. Add 1.2 mL blood to the 12 mL lysis buffer.
4. Cap the tube and invert to mix.
5. Incubate at 37 °C for 5 min.
6. Spin at 600 × *g* for 5 min and aspirate supernatant. Wash cells with 12 mL of flow buffer and repeat spin. If there are red blood cells remaining, repeat lysis.
7. Resuspend the cells in flow buffer to 2×10^6 cells/mL and transfer 400 μL of cells in each of the two polypropylene tubes labeled "Basal" and "PMA."
8. Add 2 μL of 25 mM DHR and 2.5 μL of 10 mg/mL catalase per tube and vortex gently.
9. Incubate at 37 °C in a shaking water bath for 5 min.
10. Add 100 μL PMA or flow buffer to appropriate tubes and incubate for 15 min at 37 °C in shaking water bath.
11. Analyze the samples on a flow cytometer within 25 min of adding DHR. Use forward and side scatter to gate on the neutrophil population and collect at least 5000 events in the neutrophil gate. If (e.g., in CGD patients) the expected number of DHR(+) cells is ≤1% then count 200,000 events in the neutrophil gate for a more accurate measurement (*see* **Notes 12–14**).

3.2.2 NBT Assay

The nitroblue tetrazolium (NBT) test is a qualitative histochemical assay used to detect the production of ROS by oxidation-reduction reactions in phagocytes. NBT, yellow in solution, is an electron acceptor and is reduced in the presence of $O_2^{\cdot-}$ to formazan, an insoluble blue-black precipitate. Either whole blood or isolated neutrophils are added to the NBT solution in a chamber slide and stimulated with PMA for 15–30 min at 37 °C. During the incubation, the neutrophils adhere to the slide. The slide is air-dried, counterstained with 0.6% safranin O, and examined using a 100× objective of a microscope. The NBT test yields a visual record of ROS generation by individual cells.

1. Label conical tubes with the sample identification as well as "Basal," "Basal +DPI," "PMA," and "PMA+DPI."
2. Transfer 0.1 mL isolated neutrophils (1×10^7 cells/mL) into each appropriately labeled tube.
3. Add 0.4 mL of 1% BSA in HBSS/HEPES.
4. Add 0.5 mL of 0.2% NBT solution.
5. Add 2.5 μL of 20 mM DPI to the tubes labeled "Basal+DPI" and "PMA+DPI." Mix well and incubate all tubes at 37 °C for 15 min.

6. After incubation, add 5 μL of 20 μg/mL PMA to appropriate tubes and mix well.
7. Transfer 500 μL (in duplicate) to a well of a chamber slide.
8. Incubate the slide for 15 min at 37 °C.
9. Wash the slide with HBSS/HEPES.
10. Fix the slide with methanol.
11. Counterstain with 0.6% safranin for 5 min.
12. Allow the slide to dry before viewing the cells on the microscope.
13. Enumerate the %NBT(+) cells by counting 200 neutrophils, differentiating between NBT(+) and NBT(−) (*see* **Notes 15** and **16**).

3.2.3 Assay of $O_2^{\cdot -}$ Production by Ferricytochrome c Reduction

The production of $O_2^{\cdot -}$ can be detected spectrophotometrically by the reduction of ferricytochrome c. Because $O_2^{\cdot -}$ causes a stoichiometric reduction of ferricytochrome c to ferrocytochrome c, the resultant increase in the absorption spectrum at 549.5 nm can be used to quantitate the production of $O_2^{\cdot -}$. The addition of superoxide dismutase (SOD) acts as a negative control to validate the specificity of $O_2^{\cdot -}$-dependent ferricytochrome c reduction. Spectrophotometric determination of reduced cytochrome c at $O.D._{549.5\ nm}$ enables the calculation of the amount of extracellular $O_2^{\cdot -}$ generated using the micromolar extinction coefficient of 0.0211 [15]. Superoxide dismutase is added to an identical tube to control for the nonspecific reduction of cytochrome c (*see* **Note 17**).

1. Add 200 μL catalase to 10 mL cytochrome c solution.
2. Resuspend neutrophils at 2×10^6 cells/mL in HBSS/HEPES.
3. Label 6×1 mL polypropylene microtubes (#1–6).
4. Add 500 μL of cytochrome c/catalase solution to each tube.
5. Add 10 μL of SOD to tube #3 and tube #6.
6. Add 5 μL PMA to tubes #4–6.
7. Add 500 μL neutrophil suspension to each tube and immediately place tubes on an end-over-end rotator (12 rpm) in a 37 °C incubator and incubate for either 10 min or 60 min (note: in a 60-min incubation, neutrophils from normal subjects, or carriers of CGD with DHR(+) neutrophils >20%, neutrophils should be diluted 1:5 in HBSS/HEPES (0.4×10^6/mL) for tubes #4–6).
8. At the end of the incubation, immediately place tubes on ice and spin at $500 \times g$ for 15 min at 4 °C.
9. Transfer supernatants to cuvettes and read on spectrophotometer, using an analytical wavelength of 549.5 nm and

background isosbestic wavelengths of 541 and 556 nm. The background wavelengths represent the intersections of the spectral scans of oxidized and reduced (determined after the addition of a few grains of dithionite) cytochrome c.

10. The net absorbance of each sample is determined using the following formula:

$$O.D._{549.5\ nm} - (((O.D._{560\ nm} - O.D._{541.0\ nm}) * (8.5/15)) + O.D._{541.0\ nm})$$

11. The basal production of $O_2^{\cdot -}$ generated (nmoles/10^6 cells) is calculated by subtracting the net absorbance of the SOD-containing sample (tube # 3) from the mean of net absorbance of tubes #1 and 2, and dividing the difference by 0.0211, the micromolar extinction coefficient for reduced cytochrome c. Repeat the calculations for tubes #4–6. For samples incubated for 60 min with reduced cell numbers (20%), multiply by 5 to normalize to 10^6 cells (*see* **Notes 18–20**).

3.2.4 Luminol-Enhanced Chemiluminescence ROS Assay

An alternative assay to measure ROS production is luminol-enhanced chemiluminescence. This versatile assay, in addition to its quick and easy setup, offers the capability to test several different subjects and stimuli on the same plate using reduced numbers of cells while providing high sensitivity. The luminescence readings are taken every 1–5 min for up to 2 h and the data are expressed as relative light units (RLU). Different stimuli exhibit different kinetics, e.g., fMLF induces a rapid respiratory burst that decays quickly, whereas PMA (20–100 ng/mL) induces a peak of luminescence within 5–15 min that slowly decays by 2 h. Results from normal subjects and patients can be assessed simultaneously and monitored kinetically or compared by calculating the area under the curve (AUC).

1. Dilute freshly prepared neutrophils to a concentration of 1×10^6 cells/mL in HBSS/HEPES and transfer 100 μL (100,000 cells) to 96-well white polypropylene, flat-bottom plate.
2. Add 50 μL luminol (400 μM) to each well for a final concentration of 100 μM. Wrap plate with foil.
3. Preincubate plate with cells and luminol for at least 10 min at 37 °C in CO_2 incubator.
4. While plate is incubating, prepare stimuli at 4× solutions. PMA is a convenient and potent activator of the respiratory burst and can be used at final concentrations up to 1 μg/mL. fMLF can be used at concentrations from 10^{-9} to 10^{-6} M. Bacteria can be used at ratios of between 1 and 1000 microbes per

neutrophil. However, if bacteria are to be used as a stimulus, human AB serum $(10\%)_{final}$ should be added as a source of complement. Ideally a pool of AB serum should be prepared from freshly isolated blood and frozen in aliquots. Transfer 75 μL of 4× stimuli to corresponding wells in a separate clear 96-well plate.

5. After preincubation is complete, transfer 50 μL of stimuli to the corresponding wells of the preincubated cell plate. Use a multichannel pipet to expedite the transfer.
6. Quickly place the plate in a pre-warmed luminometer (37 °C) programmed to measure light production in each well every 2 min for 2 h with gentle agitation between cycles. The luminescence signal is expressed as relative light units (RLU). Each stimulus will display different kinetics that may depend on its dose. If only a single stimulus is used, it may be useful to calculate area under the curve to facilitate comparison of different donors (*see* **Note 21**).

3.3 Immunoblot Analysis of NADPH Oxidase Protein Subunits

The NADPH oxidase consists of two membrane components, $p22^{phox}$ and $gp91^{phox}$; three cytosolic components, $p47^{phox}$, $p67^{phox}$, and $p40^{phox}$; and several guanosine triphosphate (GTP)-binding proteins. CGD is characterized by defects in any one of the four of these components. The severity of CGD can be related to the specific protein defect. Determination of the specific protein defect in CGD by western blot analysis also provides direction for determination of the specific genetic defect as well as enables appropriate genetic counseling for the extended family.

1. Add 90 μL of 1× sample buffer and 10 μL reducing agent to 5×10^6 DFP-treated neutrophil pellet, sonicate 80 rapid pulses (<1 s) at low setting (to disrupt DNA but prevent sample loss), cool, and repeat sonication. Spin at 10,600 × *g* for 5 min.
2. Repeat sonication with 40 rapid pulses at low setting, if necessary.
3. Prepare the NuPAGE gel (10% for $p40^{phox}$, $p47^{phox}$, and $p67^{phox}$, 4–12% for $p22^{phox}$ and $gp91^{phox}$) by removing the comb and rinsing the wells three times with 1× running buffer. Then fill the wells with 1× running buffer.
4. Assemble the gel electrophoresis apparatus and fill upper chamber with 600 mL of running buffer and 1.5 mL antioxidant. Fill the lower chamber with 600 mL of running buffer (without antioxidant) until each chamber is full.
5. Load 10 μL of pre-stained molecular weight standards in lane 10 for molecular weight determinations and orientation purposes. Load 20 μL (1×10^6 cell equivalents) of a neutrophil lysate from a normal subject in lane 1 and load a neutrophil

lysate from a known gp91^{phox}-deficient patient in lane 9. Load 20 μL of each patient sample (1×10^6 cell equivalents) in lanes 2–8.

6. Resolve the proteins on the gel at a constant voltage of 125 volts for 1.5 h or 100 volts for 2.5 h at 4 °C.
7. Prepare Immobilon-P transfer membrane by soaking in methanol for 30 s. Rinse with distilled H_2O and place in staining dish with 50 mL transfer buffer.
8. Transfer the resolved proteins on the gel to the Immobilon-P membrane using either a Trans-Blot unit or other transfer instrumentation.
9. Disassemble the blot from the gel, carefully removing the membrane from the transfer apparatus. To ensure adequate protein transfer, check to see if there are any pre-stained standards remaining in the gel. Mark the leading edge of each band of the molecular weight standards on the membrane.
10. Place the membrane in the membrane-blocking solution (5% nonfat milk solution) and incubate for 6 h at 4 °C with gentle rocking.
11. Remove the membrane from the blocking solution and wash the membrane 3× for 5 min with TBS/0.5% Tween 20.
12. Add the appropriate primary antibody solution to the membrane and incubate overnight at 4 °C with gentle rocking.
13. Remove the membrane from the primary antibody solution, reserving it for future use.
14. Wash the membrane 3× for 5 min with TBS/0.5% Tween 20.
15. Add the secondary antibody solution to the membrane and incubate with gentle agitation for 2 h at room temperature.
16. Repeat **step 15**.
17. Add either a color development peroxidase substrate or an enhanced chemiluminescence peroxidase substrate to the membrane to develop the specific protein band.
18. Carefully monitor development of the membrane to ensure adequate staining. Staining for p22^{phox} should be extended to ensure detection of the protein. A faint p22^{phox} band is generally observed in the gp91^{phox}-deficient CGD patient.
19. To stop the development, quickly wash the membrane twice with distilled H_2O.
20. Evaluate the expression of the individual phox proteins by either direct comparison to the normal control samples analyzed on the same blot or scanning densitometry (*see* **Note 22**).

3.4 7D5/Flavocytochrome b_{558} Staining of Neutrophils

The expression of cell surface antigens is determined on neutrophils stained with specific fluorescent monoclonal antibodies and analyzed by flow cytometric analysis. To determine the expression of a specific antigen on circulating neutrophils that avoid artifacts induced by neutrophil isolation, an aliquot of whole blood can be stained with the appropriate antibody prior to fixation and lysis of the erythrocytes. The antibody 7D5 recognizes an extracellular epitope of gp91^{phox} and can be used to identify surface expression of gp91^{phox} as well as mobilization of latent pools of gp91^{phox} stored in the specific granules [16]. A nonspecific isotype antibody is used to determine the nonspecific background staining. During flow cytometric analysis, the neutrophils are easily distinguished from other cell types using forward and right angle light scattering (FSxSS) to gate on the neutrophil population. Since very little blood is needed (100 μL) for each antigen, neutrophils can be stained with a panel of antibodies to many relevant surface antigens so that a more complete representation of surface antigen expression on neutrophils can be obtained.

1. Add 4 μL of FITC-labeled 7D5 to one tube and 20 μL of FITC-labeled IgG1 to a second tube.
2. Add 100 μL of whole blood to each tube and gently vortex.
3. Cover with foil and incubate at 4 °C for 30 min.
4. Add 500 μL of Optilyse C to each tube and vortex immediately. Cover with foil and incubate for 10 min.
5. Add 2 mL of PBS to each tube.
6. Spin at 400 × *g* for 5 min.
7. Aspirate the supernatant fluid and wash with 2 mL of PBS.
8. Spin at 400 × *g* for 5 min.
9. Resuspend the cells in 200 μL of flow buffer and store at 4 °C in the dark until analysis is performed.
10. Analyze by flow cytometry, using FSxSS to gate on the neutrophil population (*see* **Note 23**).

3.5 Flow Cytometry Analysis of p47^{phox} Protein Expression

Defects in p47^{phox} account for 25% of patients with CGD. p47^{phox} is a cytosolic component of NOX2 that becomes associated with the membranous cytochrome b_{558} during activation of the oxidase. The intracellular level of p47^{phox} can be assessed by permeabilizing neutrophils and staining with a specific anti-p47^{phox} antibody [17]. A fluorescent secondary antibody is then added and then the cells are analyzed by flow cytometry. A nonspecific primary antibody is added to cells treated similarly to determine the nonspecific background fluorescence (*see* **Note 24**).

1. Transfer two aliquots (25 μL) of EDTA-treated whole blood into separate polypropylene tubes. If the WBC is >5.0, dilute 1:1 with PBS.

2. Add Reagent 1 (100 μL) of IntraPrep Permeabilization kit to each tube. Vortex the tubes and incubate at room temperature for 15 min.
3. Wash the cells with PBS (4.0 mL) and spin for 5 min at 300 × *g*.
4. Aspirate the supernatant fluid and then gently add Reagent 2 (100 μL) from IntraPrep kit without vortexing. Incubate the cells for an additional 5 min at room temperature.
5. Shake the tubes slowly by hand and then add 10 μL of anti-$p47^{phox}$ antibody to one tube and 0.5 μL of the nonspecific IgG_1 isotype antibody to the other tube.
6. Incubate the tubes for 30 min at room temperature, wash once with PBS, and spin at 600 × *g* for 5 min. Aspirate the supernatant fluid, add AlexaFluor488-labeled goat anti-mouse IgG (1 μL), and incubate at room temp for 30 min.
7. Wash cells twice with PBS, pelleting the cells at 600 × *g* for 5 min.
8. Resuspend the cells in 200 μL of PBS with 0.5% BSA and read on a flow cytometer, gating on the neutrophil population using the FSxSS light scattering. Program the data collection so that at least 10,000 cells are analyzed within the neutrophil gate.

4 Notes

1. In general, blood is typically drawn using either citrate or heparin as anticoagulant and maintained at 20–25 °C in polypropylene containers. Most isolation protocols require 1–2 h to obtain relatively pure neutrophils and use differences in cell buoyant densities as the basis for the separation. The relative densities of blood cells are as follows: erythrocytes > neutrophils and eosinophils > monocytes, lymphocytes, and basophils > platelets.
2. Given the relatively brief life span of the neutrophils in vivo, it is essential to begin experiments without delay following their isolation from blood. Nevertheless, neutrophils can generally remain viable in anticoagulated blood (preferably acid citrate dextrose as an anticoagulant) for up to 24 h after collection, permitting analysis of patient blood samples shipped overnight. However, it is important to include a blood sample from a healthy, unrelated subject in the shipment to control for sample handling, temperature extremes, etc. and control for effects of shipment and delayed processing on the final neutrophil preparation.

3. All procedures are performed at room temperature and the isolated cells are maintained in a balanced salt solution without divalent cations. Hanks' balanced salt solution without Ca^{2+} and Mg^{2+} (HBSS[−]) or PBS supplemented with glucose is recommended for storage. Buffers containing divalent cations should be avoided during purification to reduce clumping and spontaneous activation of neutrophils; however, these cations should be present at physiologic levels in functional assays. Sterile technique helps to reduce contamination of neutrophil preparations with pyrogens that are known to alter their responsiveness in downstream assays. Avoid vigorous mixing of leukocytes as this tends to damage cells and may promote release of granules.
4. Ficoll-Paque is a solution of sodium diatrizoate (a dense, triiodinated compound) and Ficoll (a polysaccharide) with a density (1.077 g/cc) that falls between the density of neutrophils and that of the mononuclear cells [18].
5. Dextran promotes the formation of rouleaux (coin stacks) by the erythrocytes, causing them to sediment more rapidly than neutrophils at $1 \times g$.
6. Contaminating erythrocytes are removed by a brief (30 s) lysis with hypotonic saline solution. The isotonicity is quickly restored with an equal volume of hypertonic saline. A second hypotonic lysis removes many of the red cell ghosts. For most neutrophil isolation protocols, hypotonic lysis buffers are generally preferred compared to isotonic NH_4Cl/potassium lysis buffer because NH_4^+ can alter phagosomal pH.
7. In general, $1–2 \times 10^6$ neutrophils can be isolated per milliliter of whole blood from a normal subject with a purity and viability of >95%. To assess purity of isolated neutrophils, differential stains (e.g., Diff-Quik) should be used to differentiate eosinophils, neutrophils, and other cell types. The most common cell contaminants of the neutrophils preparation are eosinophils; however, some lymphocytes and monocytes may also be present. Contamination with eosinophils can also vary considerably with season and donor. In some cases, up to 20–30% of recovered polymorphonuclear leukocytes can be eosinophils. Because of their unique morphology, microscopic examination of neutrophil preparations with a differential stain such as a Wright's stain or a histochemical stain such as Kaplow's stain for myeloperoxidase [19] can provide valuable insight into some genetic immunodeficiencies. Typically, a neutrophil on a blood smear (Fig. 1A) or an isolated neutrophil on a cytospin slide (Fig. 1B), stained with Wright's stain, is multi-lobed (usually 2–5 lobes) with each lobe connected by a narrow nuclear filament. The nuclear chromatin is coarsely clumped

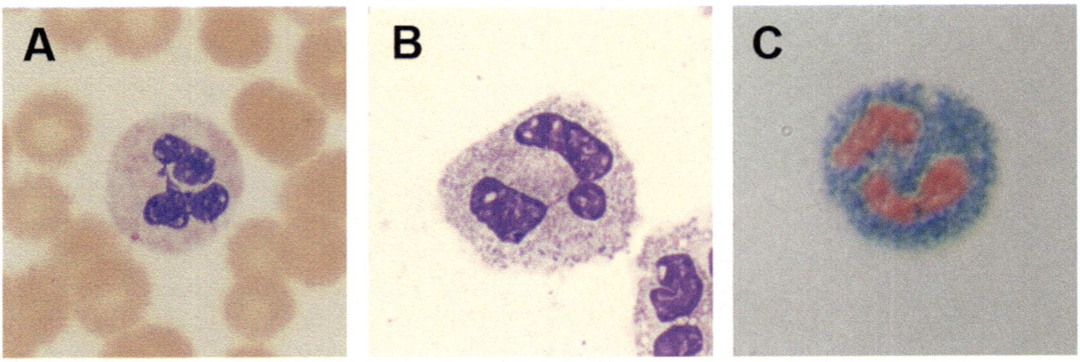

Fig. 1 Histochemical characterization of nuclear morphology. (**A**) Diff-Quik stain on neutrophil in a blood smear, (**B**) Diff-Quik stain of an isolated neutrophil, and (**C**) Kaplow's staining of myeloperoxidase in an isolated neutrophil

with purple staining. Nucleoli are generally not present. In band neutrophils, the nucleus is horseshoe shaped with no indication of constriction into lobes. The pink-violet staining of the cytosol is associated with numerous, evenly distributed, specific granules; occasionally a dark staining primary granule may be present. Kaplow's staining of a neutrophil (Fig. 1C) identifies the myeloperoxidase-containing primary granules as dark blue granules uniformly distributed throughout the cytosol with a pink, safranin O-counterstained nucleus.

8. An alternative neutrophil isolation protocol that uses a discontinuous gradient of plasma/Percoll has often been used to minimize exposure of neutrophils to trace contamination by bacterial LPS and reduce neutrophil priming [20].

9. For immunoblot studies, isolated neutrophils can be frozen in aliquots of 5×10^6 cells/vial. Neutrophils (1×10^6 cells/mL of buffer) can be pretreated for 20 min with the cell-permeant, irreversible serine protease inhibitor, diisopropylfluorophosphate (DFP, 1–5 mM), to inactivate endogenous proteases. DFP is a volatile, potent neurotoxin that can irreversibly bind to and inactivate acetylcholinesterase and should be used with extreme caution. The DFP-treated neutrophil cells can be pelleted and frozen as a cell pellet. These frozen cell pellets can be directly solubilized in sample buffer for electrophoretic analysis. In addition, these pellets can also be used as a source of DNA for genetic analyses.

10. In some disease states (lupus, cancer, neutrophil-specific granule deficiency), and often in blood samples held overnight, some neutrophils have decreased density and fail to penetrate the Ficoll-Paque density cushion. Instead these low-density neutrophils co-localize with the peripheral blood mononuclear cells atop the density cushion. Because of this, it is important to

realize that normo-dense neutrophils isolated from below the Ficoll-Paque density cushion may represent a subpopulation of the total neutrophil pool.

11. Subjects with myeloperoxidase deficiency, found in 1 per 2000 individuals, as well as several other disease states can give a false-negative DHR response, depending on the extent of their deficiency [21, 22]. Therefore, a negative DHR should be confirmed with a negative nitroblue tetrazolium test and/or abnormal $O_2^{\cdot -}$ generation using superoxide dismutase-inhibitable ferricytochrome c reduction. Analysis of $O_2^{\cdot -}$ production by cytochrome c reduction in neutrophils from a patient with myeloperoxidase deficiency often will yield normal/supranormal production of $O_2^{\cdot -}$. Myeloperoxidase deficiency can be confirmed by either histochemical staining [19] or quantitation of myeloperoxidase by ELISA.

12. Treatment of neutrophils from a normal subject with PMA results in a two-log shift in the mean fluorescence intensity (MFI) in more than 90% of the cells (Fig. 2, row A, "DHR" column); in addition, there is a redistribution of the neutrophil population within the FSxSS plot, with a shift to the left and overall reduction in the variance of the forward light scattering in PMA-stimulated neutrophils compared to basal neutrophils (Fig. 2, row A, "Buffer" and "PMA" columns). Note the distribution of the DHR(+) neutrophils (shown in blue in "Overlay") compared to the distribution of unstimulated neutrophils (shown in black). Neutrophils from patients with X-linked CGD exhibit little increase in fluorescence after stimulation with PMA (Fig. 2, row B, "DHR" column), and very little redistribution of the DHR(−) neutrophils (shown in red) within the FSxSS plot (Fig. 2, row B, "Buffer," "PMA," and "Overlay" columns) compared to the basal neutrophils (shown in black). Neutrophils from a female X-linked carrier of CGD exhibit two populations—an abnormal DHR(−) population expressing the mutant X allele and a bright DHR(+) population, expressing the functional X allele (Fig. 2, row C, "DHR" column). Moreover, the distribution of neutrophils within the FSxSS region is not symmetric; the DHR(+) neutrophils (shown in blue) shift and cluster along the lower left region of the gated population while the DHR(−) population (shown in red) parallels the distribution of the basal neutrophils (shown in black) (Fig. 2, row C, "Overlay" column). Neutrophils from patients with $p47^{phox}$ CGD and some CGD patients with hypomorphic mutations in $gp91^{phox}$, $p22^{phox}$, or $p67^{phox}$ can exhibit significant increases (up to a log) in the MFI but still much less fluorescence than observed in neutrophils from a normal subject (Fig. 2, row D, "DHR" column). There is very little redistribution of the DHR(−) neutrophils (shown in red)

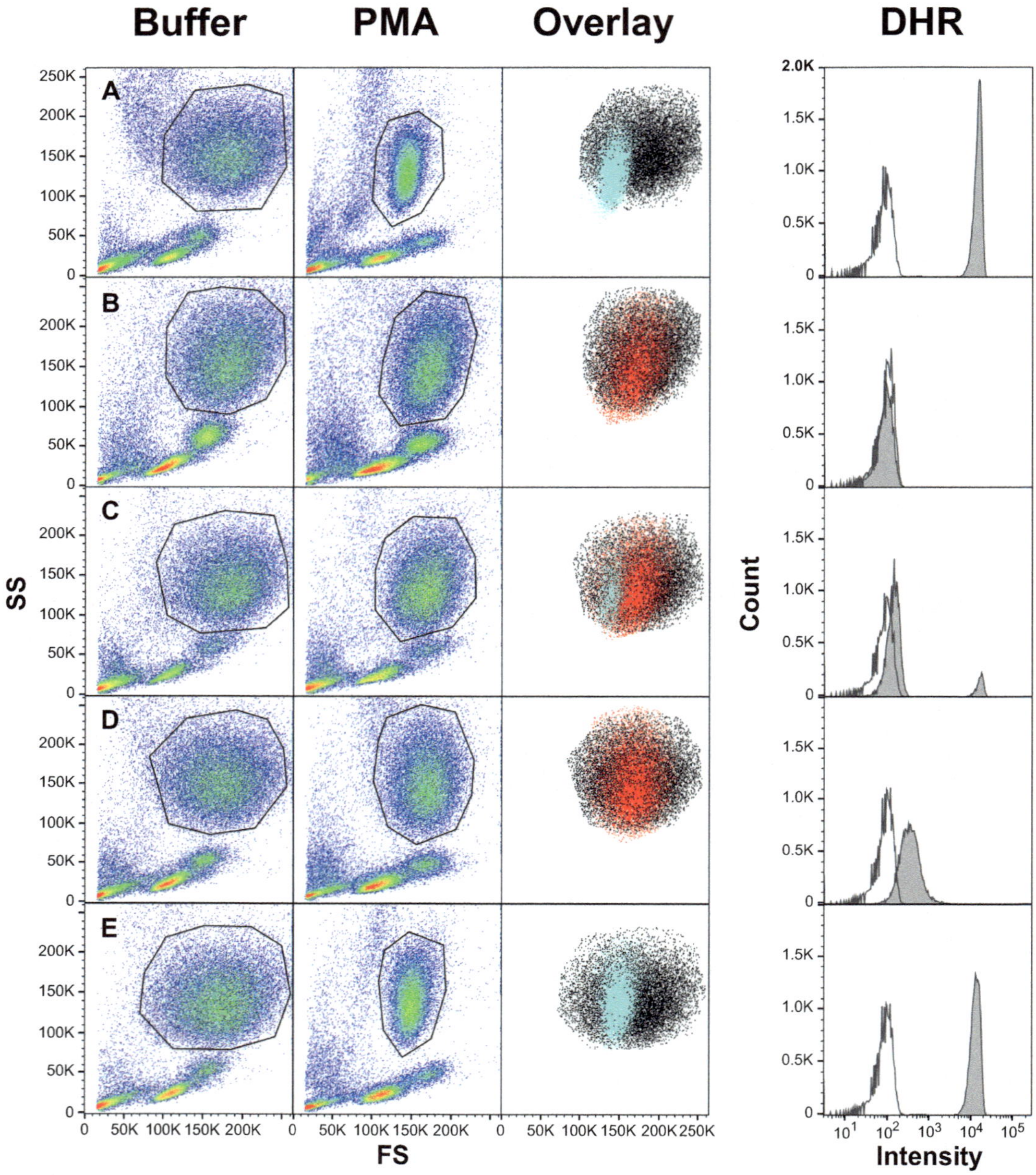

Fig. 2 Flow cytometric analysis of neutrophil ROS production by DHR. Presented are DHR analyses of blood samples from a normal subject (row **A**), a male X-linked CGD patient (row **B**), a female X-linked carrier (row **C**), a female autosomal p47^{phox} CGD patient (row **D**), and a female autosomal carrier of p47^{phox} CGD (row **E**). In column 1 (Basal) and column 2 (PMA) are the FSxSS plots. Note that a substantial shift in the distribution of the neutrophil populations after stimulation with PMA is more apparent in the back-gated presentation of the neutrophil distribution presented in column 3 (Overlay). The basal cells are represented in black; after stimulation with PMA, the DHR(+) neutrophils are in blue, and the DHR(−) are in red

within the FSxSS plot (Fig. 2, row D, "Buffer," "PMA," and "Overlay" columns) compared to the basal neutrophils (shown in black). Treatment of neutrophils from a p47^{phox} carrier with PMA results in a response comparable to that observed in neutrophils from a normal subject (Fig. 2, row E, "Buffer," "PMA," "Overlay," and "PMA" columns). Patients with p40^{phox} CGD exhibit normal responses to PMA and cannot be used to identify the status of those patients.

13. The major advantages of the DHR assay are the sensitivity, the signal-to-noise ratio, and the ease of counting large numbers of cells. Moreover, it has been shown that the DHR assay can yield reliable results on EDTA- or heparin-treated blood samples that have been stored overnight. A blood sample collected from a normal subject should be subjected to the same conditions, and the integrity of these samples should be assessed. Nonviable neutrophils in the analysis region can be misinterpreted as DHR(−) cells. Hence, a normal subject can be misdiagnosed as an X-linked CGD carrier. For the same reason, overnight samples should not be used to definitively rule out X-linked heterozygosity since a highly lyonized CGD carrier (>90% DHR(+)) could yield a DHR test result that mimics an aged sample from a normal subject. In these cases, sequence analysis of *CYBB* is needed to substantiate the final diagnosis.
14. CGD patients with mutations in the PU.1-binding domain of the promoter region of *CYBB* may express DHR(+) eosinophils, while neutrophils remain DHR(−) [23]. In addition, in several atypical CGD patients, we have noted small populations (2–7%) of DHR(+) neutrophils that may result from genetic reversion. It is often necessary to increase the number of events in the neutrophil gate to accurately assess these minor populations of cells.
15. Treatment of neutrophils from a normal subject with PMA results in the production of ROS, and the conversion of NBT to a precipitate of blue-black formazan crystals in the cytosol of the neutrophil with a pink counterstained nucleus (Fig. 3B). Under basal conditions, neutrophils from a normal subject typically do not generate ROS and do not convert NBT to formazan; the cytosol and nucleus of the neutrophil appear pink because of the safranin counterstain (Fig. 3A). However, spontaneous activation of neutrophils occasionally occurs after attachment of neutrophils to a plastic surface. Addition of the NADPH oxidase inhibitor diphenyleneiodonium chloride (DPI) blocks ROS production and ensures a negative control. In general, an NBT test on neutrophils from a patient with CGD exhibits only the safranin-counterstained nucleus with no formazan precipitate in the cytosol (Fig. 3C). As observed with the DHR assay, the NBT test can also be used to diagnose

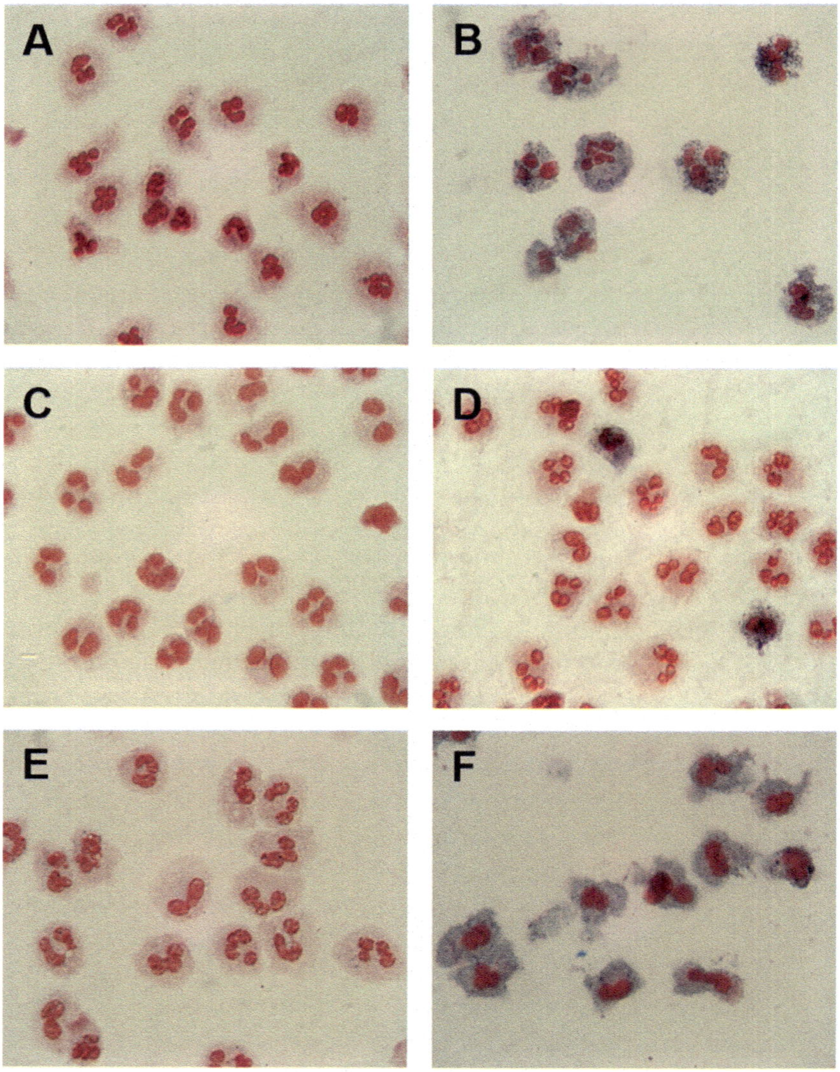

Fig. 3 Histochemical analysis of ROS production using NBT. (**A** and **B**) represent neutrophils isolated from a normal subject under basal conditions and after stimulation with PMA, respectively; (**C**) neutrophils from a gp91^{phox}, X-linked CGD patient after stimulation with PMA; (**D**) neutrophils from a gp91^{phox}, X-linked CGD carrier after stimulation with PMA; (**E**) neutrophils from p47^{phox} CGD patient after stimulation with PMA (occasionally a few blue granules can be found within the cytosol); and (**F**) neutrophils isolated from a p47^{phox} CGD carrier after stimulation with PMA

X-linked carriers of CGD. Treatment of neutrophils from a female X-linked carrier of CGD will exhibit both NBT(+) and NBT(−) staining cells (Fig. 3D). Because of inactivation of one of the X-chromosomes, an NBT test on neutrophils from an X-linked carrier of CGD reveals two populations of neutrophils—one in which cells are functioning correctly and able to generate ROS contain blue-black precipitate in the cytosol and a distinct, separate population unable to generate ROS with only a pink, safranin O-counterstained cytosol. The NBT test can result in a wide distribution in cell staining positivity among X-linked carriers of CGD, from as little as 5% NBT(+) to as much as 95% NBT(+). Neutrophils from patients with $p47^{phox}$ CGD are also NBT(−) but often can exhibit weak-staining NBT(+) crystals in the cytosol (Fig. 3E). Neutrophils from an autosomal carrier of CGD will exhibit only NBT(+) neutrophils, matching the appearance of neutrophils from a normal subject (Fig. 3F). Hence, the NBT test cannot be used to diagnose carriers of autosomal CGD.

16. The drawback of the NBT test is the visual examination and manual counting required to enumerate sufficient individual cells to obtain an accurate reflection of the percentage of positive cells.

17. Since cytochrome c is not cell permeant, the data obtained using this assay represent only a portion of the total ROS production, i.e., $O_2^{\cdot -}$ released from the cell and detected in the extracellular milieu. In addition, $O_2^{\cdot -}$ generation using ferricytochrome c reduction reflects the average response of all the cells in the population. For example, a CGD patient with 100% of cells functioning at 10% cannot be distinguished from an X-linked CGD carrier with 10% of cells functioning at 100% capacity. Hence, detection of $O_2^{\cdot -}$ generation by ferricytochrome c reduction alone is insufficient to diagnose X-linked carriers and should be used in conjunction with the DHR or NBT assay.

18. Neutrophils isolated from normal subjects produce 0.4 ± 0.3 nmol/10^6 cells/10 min and 2.1 ± 1.4 nmol/10^6 cells/60 min under basal conditions; treatment with PMA results in 45.0 ± 12.7 nmol/10^6 cells/10 min. An estimate of $O_2^{\cdot -}$ production in 60 min in neutrophils from a normal subject can be obtained by reducing the number of neutrophils in the assay to 2×10^5/mL. Treatment of normal neutrophils with PMA results in 238.8 ± 53.9 nmol/10^6 cells/60 min (Fig. 4, grey region represents the normal range, mean $\pm$ 2 SD). Although PMA-induced NOX2 function is impaired in all patients with CGD (the one exception being patients with $p40^{phox}$ CGD), there is significant variability (0.1–27% of normal) in PMA-stimulated ROS production by patient neutrophils. Patients with defects in $gp91^{phox}$, $p67^{phox}$, and $p22^{phox}$ exhibit a wide spectrum of residual ROS production depending on the specific mutation. In contrast, patients with defects in $p47^{phox}$ have less variability in their ROS production. Although the detection of $O_2^{\cdot -}$ by reduction of cytochrome c is useful in the diagnosis of patients with CGD, it cannot be used in the diagnosis of carriers because of the wide spectrum of responses that result from

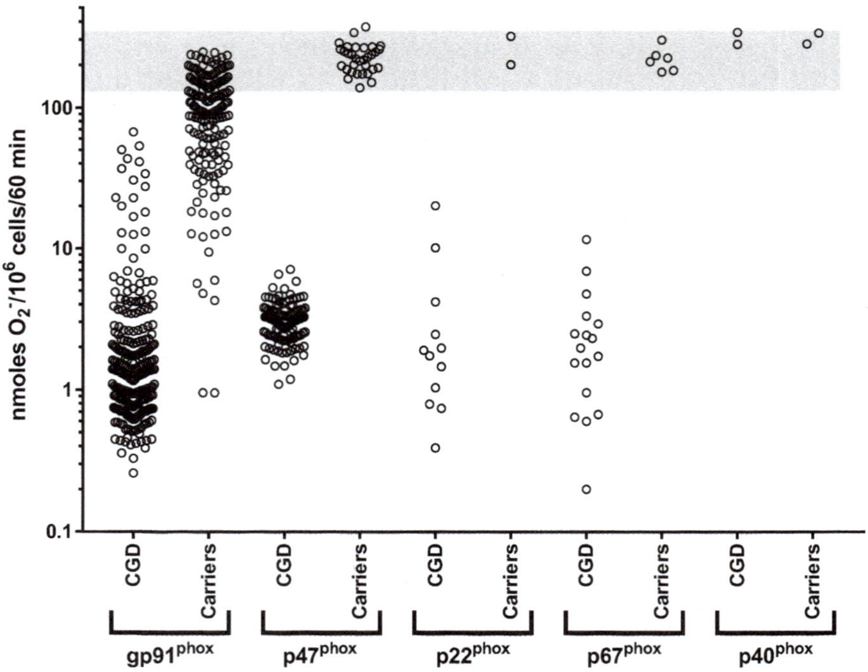

Fig. 4 $O_2^{\cdot -}$ production by ferricytochrome c reduction in CGD patients and carriers. Neutrophils (1×10^6/mL HBSS) were incubated in the presence of 100 μM cytochrome c with phorbol myristate acetate (PMA: 100 ng/mL) for 60 min at 37 °C. The reaction was terminated by centrifugation at 4 °C and the absorbance determined. An identical tube containing superoxide dismutase (100 μg/mL) served as a blank

the degree of X chromosome lyonization. Moreover, $O_2^{\cdot -}$ production in neutrophils from autosomal recessive carriers of CGD generally falls within the normal range. Patients with p40^{phox} CGD exhibit a normal response to PMA. To identify these patients, a particulate stimulus (such as opsonized *Staph. aureus)* is needed to identify these patients [9].

19. Studies have shown that $O_2^{\cdot -}$ determinations sufficiently reliable to diagnose CGD can be obtained from neutrophils isolated from heparinized whole blood that has been stored overnight. Hence, analyses can be performed on blood samples shipped by overnight express. A normal control blood sample should accompany the sample to ensure adequate shipment handling. By 48 h of storage, however, there are marked reductions in the PMA-induced response.

20. Analysis of this residual ROS production allowed discrimination of CGD into subgroups with different rates of mortality. The production of ROS has become an important tool to perform risk assessment in patients with CGD. Patients with the lowest ROS generation (<1% of normal generation) have lower predicted survival than patients with higher ROS generation (3–10% of normal) [7].

21. Luminol-enhanced chemiluminescence is a measure of both intracellular and extracellular ROS production, though it may not detect them with equivalent efficiency. Addition of PMN (100 ng/mL) to neutrophils from a normal subject results in a characteristic rapid increase in chemiluminescence that peaks at 15 min, followed by a gradual decline over 2 h (Fig. 5, grey line). Addition of superoxide dismutase significantly reduces both peak height and AUC after stimulation with PMA, suggesting that at least a portion of the response can be attributed to $O_2^{\cdot -}$. Typically, treatment of neutrophils from patients with gp91*phox*-deficient CGD with PMA results in little to no detectable luminescence in this assay (blue circles, Fig. 5); however, as observed with the ferricytochrome c assay, neutrophils from p47*phox*-deficient CGD patients have detectable luminescence that becomes most evident at later readings (40–80 min) (red circles, Fig. 5).
22. A lysate of a DFP-treated neutrophil pellet isolated from a normal subject is included on each gel as a reference for both migration position and intensity. Figure 6 presents composite western blots depicting the typical phox protein expression in neutrophils from a normal subject in comparison with those from seven different CGD patients. The expression of each phox component in the lysate of a neutrophil pellet from a patient is compared to the expression observed in the neutrophil lysate from a normal subject. The intensity of the protein band corresponding to each of the phox components can be scanned by densitometry or visually assessed and assigned a value based on an arbitrary, semiquantitative scale. In most cases, CGD patients with a defect in $p40^{phox}$ or $p47^{phox}$ exhibit undetectable expression of the corresponding protein in the lysate of the neutrophils while the remaining phox proteins exhibit normal expression. CGD patients with a defect in $p67^{phox}$ can exhibit variable expression of $p67^{phox}$ in the lysate of the neutrophil proteins, depending on the specific genetic mutation. In addition, because $p40^{phox}$ is stabilized in 1:1 complex with $p67^{phox}$, CGD patients who lack $p67^{phox}$ CGD also lack $p40^{phox}$ [24]. The phox components of NOX2 found in the plasma membrane, $gp91^{phox}$ and $p22^{phox}$, are expressed as a complex. CGD patients with defects in $gp91^{phox}$ generally exhibit undetectable expression of $gp91^{phox}$ but exhibit faint expression of $p22^{phox}$. In contrast, CGD patients with defects in $p22^{phox}$ disrupt the expression of both $p22^{phox}$ and $gp91^{phox}$. Neutrophils isolated from overnight samples can be used to diagnose $p47^{phox}$ deficiency because of the stability of the protein. However, detection of other phox protein defects in overnight samples can become more problematic because of proteolysis of $p67^{phox}$ and of the $gp91^{phox}$:$p22^{phox}$ complex.

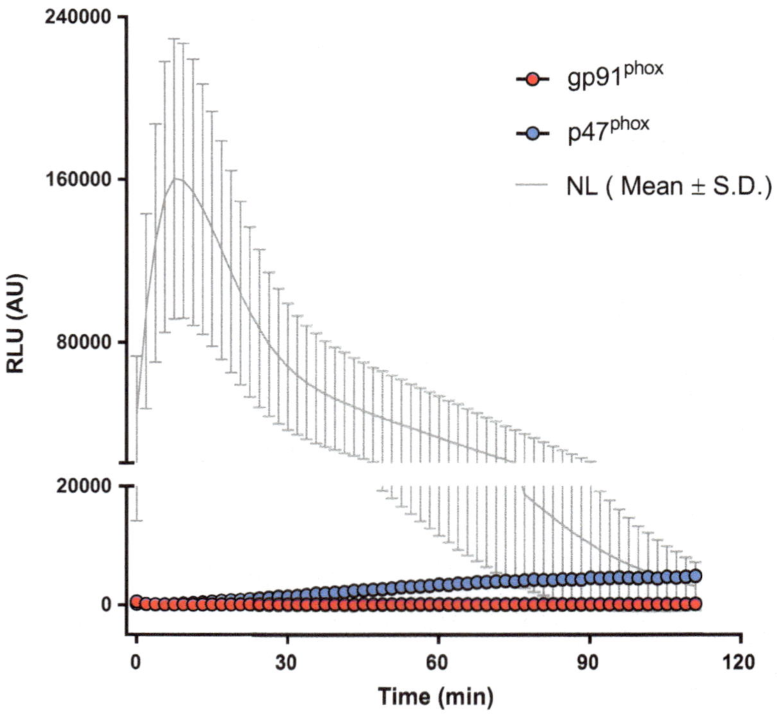

Fig. 5 Luminol-enhanced chemiluminescence. Neutrophils (1 × 10^5/200 μL) were pre-loaded with luminol (100 μM) for 10 min at 37 °C. At $t = 0$, either buffer or PMA (100 ng/mL) was added and luminescence was monitored for 2 h, with readings recorded every 2 min. Note that at later readings, neutrophils from a patient with p47^{phox} deficiency had increased luminescence compared to neutrophils from a CGD patient with gp91^{phox} deficiency

23. In general, neutrophils from a healthy subject exhibit 7D5 expression, though there may be some variability in the expression level among different individuals (Fig. 7A). Neutrophils from an X-linked CGD carrier with a protein(−) defect in gp91^{phox} exhibit both 7D5(+) and 7D5(−) populations (Fig. 7B). Neutrophils from a CGD patient with a protein(−) defect in either gp91^{phox} or p22^{phox} fail to exhibit any 7D5 expression (Fig. 7C). The percentage of 7D5(+) neutrophils should correlate with the percentage of DHR(+) neutrophils. Flow cytometric analysis of neutrophils stained with 7D5 can often be used to identify patients with X-linked gp91^{phox} or p22^{phox} CGD (little or no 7D5 staining) and X-linked carriers of CGD (mosaic pattern of staining), particularly in patients where the number of cells available for testing is limited.
24. While in most p47^{phox} CGD patients the expression of p47^{phox} is clearly absent (Fig. 8c), there is overlap of p47^{phox} expression by flow cytometry between healthy subjects and p47^{phox} CGD carriers (Fig. 8a, b), indicating that additional testing may be

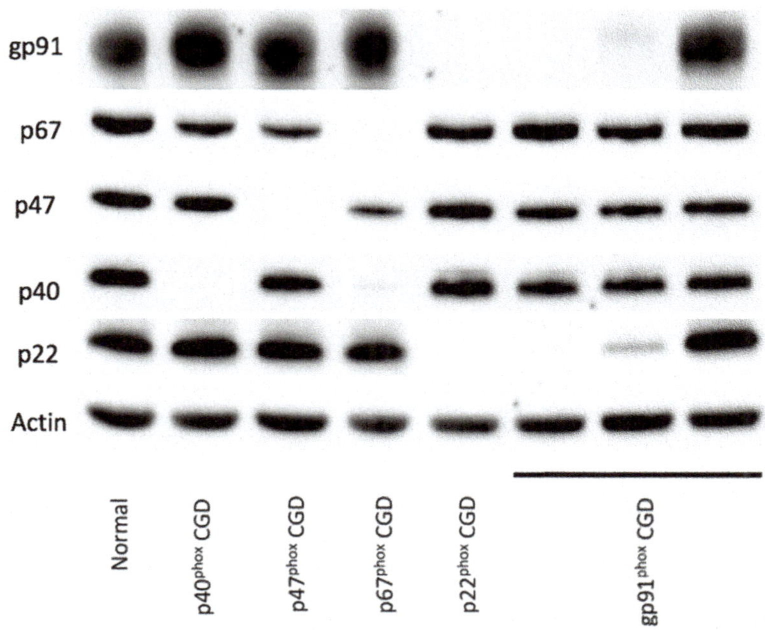

Fig. 6 Determination of NADPH oxidase protein defect by Western blot analysis. Cellular lysates (20 μL or 1 × 10^6 cell equivalents) were loaded into each lane (10% PAGE gels for $p40^{phox}$, $p47^{phox}$, and $p67^{phox}$ and actin; 4–12% gradient PAGE gels for $p22^{phox}$ and $gp91^{phox}$). A validated normal control was run on each gel for quality control. The gels were transferred to nitrocellulose, blocked with 5% powdered milk, and incubated overnight with specific antibodies to the phox proteins. The blots were washed, incubated with peroxidase-labeled secondary antibody, and developed using enhanced chemiluminescence detection reagents. A composite figure was created using the relevant bands from each blot

required to evaluate the status of a potential carrier of $p47^{phox}$ CGD. A similar approach is viable for detection of $p67^{phox}$ [17].

Acknowledgments

This project has been funded in whole or in part with federal funds from the National Cancer Institute, National Institutes of Health, under Contract HHSN261200800001E. The content of this publication does not necessarily reflect the views or policies of the Department of Health and Human Services, nor does the mention of trade names, commercial products, or organizations imply endorsement by the US Government.

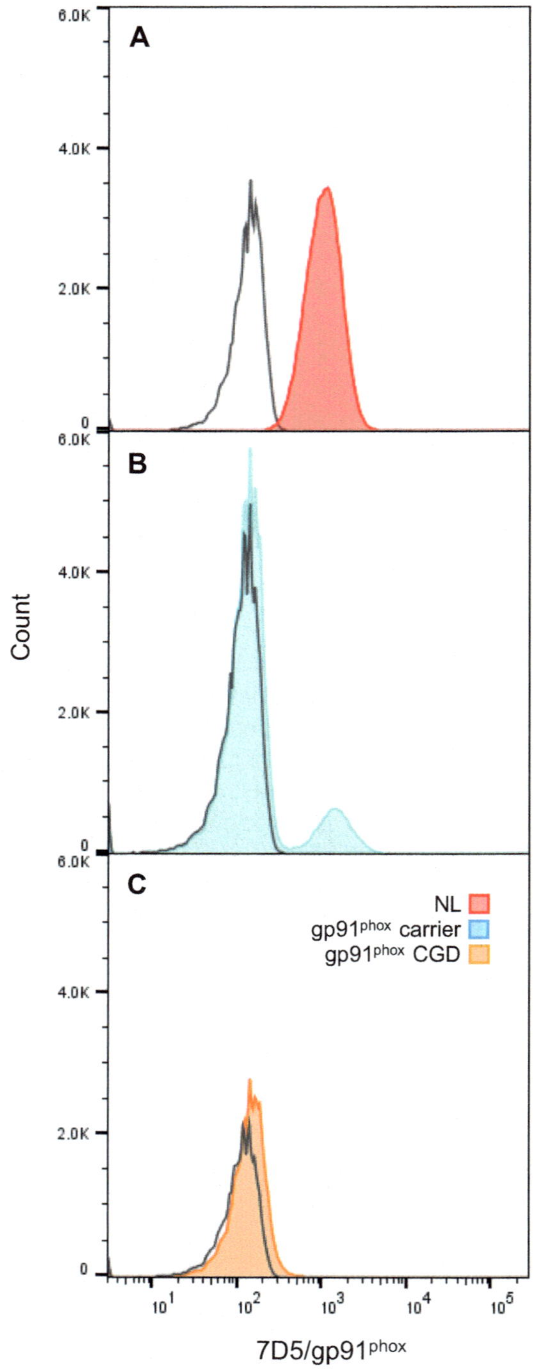

Fig. 7 Surface expression of gp91^{phox} using antibody 7D5. To determine gp91^{phox} expression in neutrophils, whole blood was stained with antibody 7D5, then fixed, lysed, and analyzed by flow cytometry. The neutrophil population was gated using FSxSS. The histograms represent (**A**) a normal subject, (**B**) an X-linked gp91^{phox} CGD carrier, and (**C**) a patient with gp91^{phox} CGD. The black histograms represent the binding of a nonspecific isotype control antibody

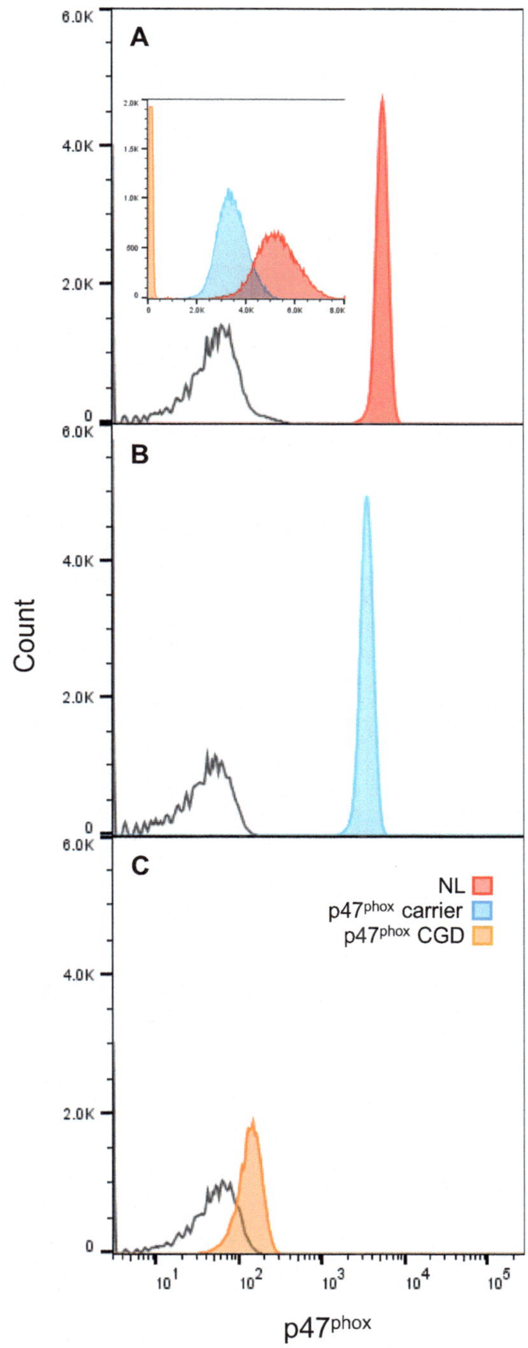

Fig. 8 Expression of p47^{phox} by flow cytometry. To determine p47^{phox} expression, whole blood was permeabilized/fixed and then stained with anti-p47^{phox} antibody. The neutrophil population was gated using FSxSS. The histograms represent (**A**) a normal subject, (**B**) a p47^{phox} CGD carrier, and (**C**) a patient with p47^{phox} CGD. The black histograms represent the binding of a nonspecific isotype control antibody. In the inset figure, the *x*-axis is plotted on a linear scale and illustrates more clearly the differences between the three groups

References

1. Nunoi H, Rotrosen D, Gallin JI, Malech HL (1988) Two forms of autosomal chronic granulomatous disease lack distinct neutrophil cytosol factors. Science 242:1298–1301
2. Volpp BD, Nauseef WM, Clark RA (1988) Two cytosolic neutrophil oxidase components absent in autosomal chronic granulomatous disease. Science 242:1295–1297
3. Wientjes FB, Hsuan JJ, Totty N, Segal AW (1993) p40^{phox}, a third cytosolic component of the activation complex of the NADPH oxidase to contain src homology 3 domains. Biochem J 296:557–561
4. Parkos CA, Dinauer MC, Walker LE, Allen RA, Jesaitis AJ, Orkin SH (1988) Primary structure and unique expression of the 22-kilodalton light chain of human neutrophil cytochrome *b*. Proc Natl Acad Sci U S A 85:3319–3323
5. Segal AW, Cross AR, Garcia RC, Borregaard N, Valerius NH et al (1983) Absence of cytochrome b_{-245} in chronic granulomatous disease—a multicenter European evaluation of its incidence and relevance. N Engl J Med 308:245–251
6. Berendes H, Bridges RA, Good RA (1957) A fatal granulomatosus of childhood: the clinical study of a new syndrome. Minn Med 40:309–312
7. Kuhns DB, Alvord WG, Heller T, Feld JJ, Pike KM et al (2010) Residual NADPH oxidase and survival in chronic granulomatous disease. N Engl J Med 363:2600–2610
8. Matute JD, Arias AA, Wright NAM, Wrobel I, Waterhouse CCM et al (2009) A new genetic subgroup of chronic granulomatous disease with autosomal recessive mutations in p40^{phox} and selective defects in neutrophil NADPH oxidase activity. Blood 114:3309–3315
9. van de Geer A, Nieto-Patlán A, Kuhns DB, Tool ATJ, Arias AA et al (2018) Inherited p40phox deficiency differs from classic chronic granulomatous disease. J Clin Invest 128:3957–3975
10. Gu Y, Jia B, Yang F-C, D'Souza M, Harris CE et al (2001) Biochemical and biological characterization of a human Rac2 GTPase mutant associated with phagocytic immunodeficiency. J Biol Chem 276:15929–15938
11. Kurkchubasche AG, Panepinto JA, Tracy J, Thomas F, Thurman GW, Ambruso DR (2001) Clinical features of a human Rac2 mutation: a complex neutrophil dysfunction disease. J Pediatr 139:141–147
12. Cooper MR, DeChatelet LR, McCall CE, Lavia MF, Spurr CL, Baehner RL (1972) Complete deficiency of leukocyte glucose-6-phosphate dehydrogenase with defective bactericidal activity. J Clin Investig 51:769–778
13. Kuijpers TW, Tool ATJ, van der Schoot CE, Ginsel LA, Onderwater JJM et al (1991) Membrane surface antigen expression on neutrophils: a reappraisal of the use of surface markers for neutrophil activation. Blood 78:1105–1111
14. Emmendörffer A, Hecht M, Lohmann-Matthes M-L, Roesler J (1990) A fast and easy method to determine the production of reactive oxygen intermediates by human and murine phagocytes using dihydrorhodamine 123. J Immunol Methods 131:269–275
15. Van Gelder BF, Slater EC (1962) The extinction coefficient of cytochrome c. Biochim Biophys Acta 58:593–595
16. Nakamura M, Murakami M, Koga T, Tanaka Y, Minakami S (1987) Monoclonal antibody 7D5 raised to cytochrome b_{558} of human neutrophils: immunocytochemical detection of the antigen in peripheral phagocytes of normal subjects, patients with chronic granulomatous disease, and their carrier mothers. Blood 69:1404–1408
17. Wada T, Muraoka M, Toma T, Imai T, Shigemura T et al (2013) Rapid detection of intracellular p47phox and p67phox by flow cytometry; useful screening tests for chronic granulomatous disease. J Clin Immunol 33:857–864
18. Böyum A (1968) Isolation of mononuclear cells and granulocytes from human blood. Isolation of mononuclear cells by one centrifugation, and of granulocytes by combining centrifugation and sedimentation at 1 g. Scand J Clin Invest Suppl 97:77–89
19. Kaplow LS (1965) Simplified myeloperoxidase stain using benzidine dihydrochloride. Blood 26:215–219
20. Haslett C, Guthrie LA, Kopaniak MM, Johnston RBJ, Henson PM (1985) Modulation of multiple neutrophil functions by preparative methods or trace concentrations of bacterial lipopolysaccharide. Am J Pathol 119:101–110
21. Mauch L, Lun A, O'Gorman MRG, Harris JS, Schulze I et al (2007) Chronic granulomatous disease (CGD) and complete myeloperoxidase deficiency both yield strongly reduced dihydrorhodamine 123 test signals but can be easily discerned in routine testing for CGD. Clin Chem 53:890–896
22. Milligan KL, Mann D, Rump A, Anderson VL, Hsu AP et al (2016) Complete

myeloperoxidase deficiency: beware the "false-positive" dihydrorhodamine oxidation. J Pediatr 176:204–206

23. Weening RS, De Boer M, Kuijpers TW, Neefjes VME, Hack WWM, Roos D (2000) Point mutations in the promoter region of the CYBB gene leading to mild chronic granulomatous disease. Clin Exp Immunol 122:410–417

24. Tsunawaki S, Mizunari H, Nagata M, Tatsuzawa O, Kuratsuji T (1994) A novel cytosolic component, $p40^{phox}$, of respiratory burst oxidase associates with $p67^{phox}$ and is absent in patients with chronic granulomatous disease who lack $p67^{phox}$. Biochem Biophys Res Commun 199:1378–1387

Chapter 34

Gastrointestinal Complications in Chronic Granulomatous Disease

E. Liana Falcone and Steven M. Holland

Abstract

Almost half of patients with chronic granulomatous disease (CGD) suffer from gastrointestinal (GI) inflammation, the pathogenesis of which is complex and multifactorial. As a result, the management of CGD-associated GI inflammation remains challenging due to its chronicity and difficulty in managing the simultaneous need for immunomodulation with increased susceptibility to infection. In order to contextualize prospective treatment interventions for CGD-associated GI inflammation, we have reviewed the clinical presentation, pathogenesis and current management of this disease. Increased understanding of the role of nicotinamide adenine dinucleotide phosphate (NADPH) oxidase complex 2 (NOX2)-derived reactive oxygen species (ROS) in inflammatory bowel disease (IBD) will likely reveal novel targets for therapeutic intervention.

Key words Chronic granulomatous disease, Inflammatory bowel disease, NADPH oxidase, NOX2, Reactive oxygen species, Microbiome

1 Introduction

Defects in any one of the nicotinamide adenine dinucleotide phosphate (NADPH) oxidase complex 2 (NOX2) subunits (Nox2 (a.k.a. gp91^{phox}), p22^{phox}, p47^{phox}, p67^{phox}, p40^{phox}), largely present in phagocytes, lead to defective reactive oxygen species (ROS) production and the development of chronic granulomatous disease (CGD). CGD is a rare inherited immune deficiency characterized by recurrent infections due to a relatively characteristic set of pathogens (e.g., *Staphylococcus aureus* sp., *Serratia marcescens* sp., *Burkholderia* spp., *Nocardia* spp., *Aspergillus* spp., etc.) and inflammatory dysregulation especially affecting the hollow viscera, such as the gastrointestinal (GI) tract (reviewed in [1]).

Almost 50% of patients with CGD suffer from some form of GI inflammation, which is independent of residual ROS production in neutrophils [2] and appears to equally affect patients with CGD of all genotypes. Although inflammatory bowel disease (IBD)

Ulla G. Knaus and Thomas L. Leto (eds.), *NADPH Oxidases: Methods and Protocols*, Methods in Molecular Biology, vol. 1982, https://doi.org/10.1007/978-1-4939-9424-3_34,

associated with CGD is often compared to Crohn disease (CD) [3], it is in many ways its own disease entity, as it shares features with CD and ulcerative colitis (UC) in addition to some distinct characteristics [4, 5]. The unique features of CGD-associated IBD (CGD-IBD) combined with inherent susceptibility to infections make CGD-associated intestinal inflammation difficult to treat. Thus, gaining a better understanding of the pathogenesis of CGD-IBD may improve treatment interventions while further elucidating the role of the NOX2 complex in intestinal inflammation in general. In order to contextualize prospective treatment interventions for CGD-IBD, the following chapter will review the **Clinical Presentation**, **Pathogenesis** and current **Management** of this disease.

2 Clinical Presentation

GI inflammation affects almost 50% of patients with CGD. A study of the cohort of patients with CGD followed at the National Institutes of Health (NIH) reported that out of 140 patients, 33% had GI involvement. More specifically, 43% of patients with X-linked disease and 11% of patients with autosomal recessive (AR) CGD had some form of IBD [6]. A follow-up evaluation of this cohort performed 10 years later showed that 44% of all patients with CGD ($n = 215$) were found to have a history of GI inflammation. Interestingly, the current survey, which included patients from the original study as well as new patients, demonstrated that patients with $p47^{phox}$-deficient CGD had almost similar rates of GI inflammation as patients with X-linked disease ($gp91^{phox}$-deficient 48%; $p47^{phox}$-deficient 38%), as opposed to the previously reported 11% rate in the $p47^{phox}$-deficient population [6] (Table 1). The increased rate of CGD-IBD in $p47^{phox}$-deficient patients may be due to a later age of diagnosis, as the mean age of IBD diagnosis in $gp91^{phox}$-deficient patients was 7 years versus 22 years in $p47^{phox}$-deficient patients (Table 2). In a report of inflammatory complications in a cohort of patients with CGD followed in France, 61% of patients had at least one episode of GI inflammation. More specifically, 66% of patients with X-linked disease and 50% of patients with AR disease had GI inflammation [7]. The median age for the first GI manifestation was 2 years for X-linked disease and 3 years for AR disease [7]. Together, these data from both the NIH and French cohorts support the notion that GI inflammation affects about half of patients with CGD independent of genotype. Interestingly but reproducibly, there does not appear to be an association between residual ROS production and prevalence or severity of infections and CGD-associated IBD [2, 8]. GI manifestations may occur at any age (even antenatally [9]) and can precede the diagnosis of CGD and related

Table 1
Distribution by chronic granulomatous disease genotype of patients with gastrointestinal complications followed at the National Institutes of Health [70]

Affected gene, deficient protein	Total patients, *n*	Patients with GI complications, *n* (%)
CYBB, gp91^{phox}	144	69 (48%)
NCF1, p47^{phox}	56	21 (38%)
CYBA, p22^{phox}	8	3 (38%)
NCF2, p67^{phox}	6	0 (0%)
NCF4, p40^{phox}	1	1 (100%)
All patients with CGD	215	94 (44%)

Table 2
Clinical characteristics of patients with chronic granulomatous disease and gastrointestinal disease followed at the National Institutes of Health [70]

Variables	p47^{phox}-deficient	gp91^{phox}-deficient
Total patients with GI disease	$n = 21$	$n = 69$
Mean age of living patients (years)	34	22
Mean age at diagnosis of GI disease (years)	22	7
GI manifestations (past or present), *n* (%)		
Esophagitis/gastritis/gastric or duodenal ulcers	11 (52%)	20 (29%)
Hypomotility and/or obstruction	3 (14%)	18 (26%)
Fistulae/fissures or Crohn disease diagnosis	9 (43%)	16 (23%)
Colitis	17 (81%)	54 (78%)
Perianal abscesses	18 (86%)	23 (33%)
Liver disease, *n* (%)		
History of liver abscess	5 (24%)	14 (20%)
Other	1 (5%)	6 (9%)

infectious complications. The difference in age of onset between the two geographically distinct cohorts, especially in AR CGD, may be related to differences in population-specific single nucleotide polymorphisms, environmental exposures, diet, and perhaps even in the composition of the intestinal microbiota.

Most patients with CGD and GI complications complain of crampy, diffuse, or lower abdominal pain which may be the main symptom in up to one third of patients. Other symptoms include non-infectious diarrhea (39% [6], with or without blood and/or mucus), oral aphthae, anal fistulae, nausea and vomiting (24% [6]),

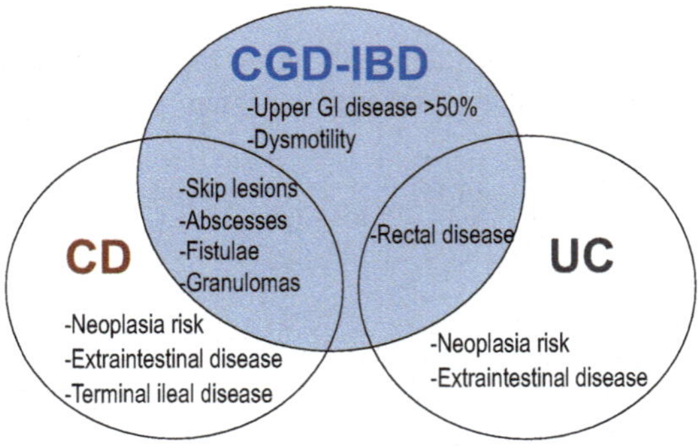

Fig. 1 Comparison of clinical findings between CGD-IBD and conventional IBD

anorexia, gastritis, constipation (2% [6]), intestinal obstruction, and growth delay or failure to thrive (11% [6]) [6, 7, 10]. Characterization of the GI manifestations in the CGD population reveals that independent of genotype, the majority of patients have colitis. There also appear to be higher rates of Crohn-like manifestations such as fistulae, fissures, and perianal abscesses (Table 2, [3, 4, 6, 10]).

Endoscopically, patients with CGD may present with discontinuous inflammation of the upper and/or lower GI tracts, with features characteristic of CD, as well as UC [6]. In a survey of 78 patients with CGD followed at the NIH, patients who underwent upper endoscopy were found to have esophagitis (21%), esophageal ulcers (7%), esophageal dysmotility (26%), gastritis (74%), gastric ulcers/erosions (19%), duodenitis (37%), and duodenal ulcers/erosions (11%). All patients were negative for *Helicobacter pylori* infection, and esophageal ulcers were negative for fungal or viral etiologies. Esophageal inflammation varied from mild to severe with deep linear ulcers, while the areas of chronic inflammation observed were similar to those seen in the colon. Of all patients who underwent colonoscopies, 74% had colitis (mild 17%, moderate 35%, severe 48%). Anorectal involvement (erythema, ulcers, stenosis, fissures and fistulae), affecting 93% of patients with CGD-IBD, was a remarkable feature. Given that CGD-IBD is often transmural, enteric fistulae were noted in 18% of patients with GI inflammation. Interestingly, no patients were found to have dysplasia on any of their biopsies. Figure 1 compares the clinical and endoscopic findings in CGD-IBD with those reported in CD and UC.

A study of patients with CGD colitis reported the following histopathological findings: the left colon was the most commonly involved site, acute colitis was usually associated with cryptitis/crypt abscess, and chronic colitis was associated with lymphoplasmacytic infiltrates in the lamina propria and architectural distortion. Other findings included ulceration (10% of all biopsies), eosinophilic microabscesses (26% of all biopsies), poorly to well--formed microgranulomas (46% of all biopsies) and pigmented macrophages (65% of all biopsies) [4].

3 Pathogenesis

3.1 Human Studies

The lack of effective treatment options for CGD-IBD is partly due to the limited understanding of the mechanisms underlying this disease process. Studies from patients with CGD suggest that although residual neutrophil-derived ROS production is associated with increased survival, it is not associated with CGD-IBD frequency or severity [2]. This suggests that NOX2-derived ROS has distinct roles in hematopoietic and non-hematopoietic cell types. The specific contributions of the mucosal non-hematopoietic cells in human CGD-IBD have yet to be elucidated, and the roles of ROS-deficient phagocytes in the development and persistence of CGD-IBD in humans are also unclear. T cells express NOX2, and reduced T-cell numbers have been reported in patients with CGD [11]. Although the underlying mechanisms for this are unclear, it is possible that the persistent inflammatory state in patients with CGD may contribute to T-cell exhaustion [12]. Defects in T-cell autophagy [13] and increased susceptibility to apoptosis [14] have also been described in the setting of CGD. In addition, decreased T-cell-intrinsic ROS production is associated with disturbances in T-cell effector differentiation and regulatory function [15–17]. In fact, aging normal individuals (without CGD) have $CD8^{+}CCR7^{+}$ regulatory T-cell (T_{REG}) failure due to deficient NOX2 function [18]. In addition, patients with CGD have an expansion of interleukin (IL)17-producing T helper (T_{H}) cells, which may contribute to a state of heightened immune activation [19]. Thus, it is possible that in addition to having expanded $T_{H}17$ compartments due to infection-mediated IL23 production [20], patients with CGD-IBD may also have aberrant T_{REG} function.

A study examining immunohistochemical staining (CD3, CD4, CD8, CD68, CD79, FoxP3, CD163) in inflamed colon sections from patients with CGD versus CD or controls showed decreased CD68 staining in patients with CGD compared to CD and healthy subjects. CD68 is a monocyte/macrophage marker primarily found in endosomes and lysosomes. However, based on the evaluation of other macrophage markers such as CD163, there was no actual decrease in monocytes/macrophages in CGD colons [21]. Thus, the mechanism underlying decreased CD68 expression in CGD-IBD remains unclear.

Unlike in CD in the normal population, CD-associated antimicrobial antibodies were increased indiscriminately in patients with CGD (i.e., ASCA IgA and IgG, anti-OmpC, anti-I2, and anti-CBir1) independent of CGD genotype or the presence of active colitis [22]. These findings suggest an underlying and pervasive defect in innate immunity in CGD that is not directly linked to CGD-IBD.

3.2 Mouse Studies

gp91^{phox}-, p47^{phox}-, p22^{phox}-and p40^{phox}-deficient mice have been used to examine how the lack of phagocyte-derived ROS affects susceptibility to chemical and infectious colitis. gp91^{phox} deficiency may be protective in both acute dextran sodium sulfate (DSS) colitis [23] and *Citrobacter rodentium*-induced colitis [24]. However, gp91^{phox}-deficient mice do succumb to 2,4,6-trinitrobenzenesulfonic acid (TNBS)-induced colitis [25]. Similarly, p22^{phox}-deficiency (including selective intestinal p22^{phox} deficiency) was protective in intestinal infection with *C. rodentium* and *Listeria monocytogenes* through a mechanism involving intestinal ROS-mediated enrichment of H_2O_2-producing taxa in the intestinal flora [24]. In contrast, colitis susceptibility was increased in p40^{phox}-deficient mice using both acute DSS and anti-CD40-induced colitis models [26]. Studies using DSS in p47^{phox}-deficient mice have been conflicting. In an early study using the acute DSS model, there was no difference in colitis severity between p47$^{phox-/-}$ and control mice [27], whereas in two subsequent studies, p47^{phox}-deficient mice were more susceptible to both chemical and infectious colitis. In a study using a modified DSS colitis model, p47^{phox}-deficient mice had more severe colitis than controls, which was associated with increased weight loss, colitis severity, infiltrating intestinal neutrophils, macrophages and lymphocytes, as well as increased systemic and localized cytokine (IL6, IL10, TNFα, IFNγ, and IL17A) production [28]. In a second study, p47$^{phox-/-}$ mice were more susceptible to both acute DSS and *C. rodentium*-induced colitis. However, colitis susceptibility was not reversed by restoring ROS production to the hematopoietic compartment but was largely mediated by the intestinal microbiome established at birth [29]. Interestingly, *Rac2*$^{-/-}$ mice have been found to be more susceptible to *C. rodentium*-induced colitis [30]. Rac2 is a GTPase involved in the electron transfer required for NADPH-derived ROS production [31] as well as cell motility. This cytosolic subunit predominates in phagocytes and functions through its interaction with p67^{phox}, which in turn binds to p47^{phox} [32].

3.3 Role of NADPH Oxidase in the Gut

A fine balance in the cell-specific production of ROS is necessary for the maintenance of intestinal homeostasis (reviewed in [33]). Disruption of the mucosal barrier leads to recruitment of intestinal phagocytes, where NOX2-derived ROS is needed for antimicrobial defense and to create a hypoxic environment, which supports tissue repair [25]. Meanwhile, tightly controlled ROS production by non-hematopoietic cells such as intestinal epithelial cells and myofibroblasts highlight the significant role of low physiological levels of ROS in various signaling pathways and possibly in modulating the microbiota in favor of colonic restitution [34–36]. Thus, a dysregulated redox balance compromises antimicrobial defense, alters innate and adaptive immune responses, including autophagy, and prolongs immune activation [33].

ROSs include oxygen radicals (e.g., $O_2^{\bullet-}$) and non-radicals (e.g., HOCl, $^1O_2$, H_2O_2). Prolonged high concentrations of ROS can cause DNA damage, lipid peroxidation and protein oxidation, and irreversibly destroy cells [33]. The main sources of ROS in the GI tract are NADPH oxidases and mitochondria, while the NADPH oxidases reportedly present in the gut are NOX1 (ileum, cecum and colon epithelium), DUOX2 (epithelium of entire gut), NOX2 (mostly phagocytes), and NOX4 (intestinal epithelium, fibroblasts and smooth muscle) [37]. NOX1 shares about 60% homology with NOX2. The NOX1 complex is generally quiescent until it is stimulated by proinflammatory cytokines and microbial products. NOX1 requires heterodimerization with $p22^{phox}$, and the associating cytosolic components are termed NOXO1 (NOX1 organizer; homologue to $p47^{phox}$) and NOXA1 (NOX1 activator; homologue to $p67^{phox}$). Similar to NOX2, activation of NOX1 results in $O_2^{\bullet-}$ production [38]. Studies in *Nox1*$^{-/-}$ mice using either DSS or TNBS models have not revealed a severe colitis phenotype unless the mice are crossed onto *Il10*$^{-/-}$ backgrounds [39]. Nevertheless, a growing body of evidence supports a role for NOX1-derived ROS in regulating intestinal epithelial proliferation and differentiation [40], as well as colonic restitution [35]. Moreover, signals derived from intestinal pathogens and commensals, such as *Lactobacilli* spp., may drive NOX1-mediated ROS production via stimulation of formyl peptide receptor 1 (FPR1) (present on intestinal epithelial cells), thereby promoting tissue repair [35, 36, 41]. NOX4 is largely found in the kidney cortex, as well as in fibroblasts, smooth muscle cells and the intestinal epithelium, where it is constitutively active and produces H_2O_2 [42]. Its role in GI inflammation is not yet clear. DUOX2 forms a membrane-bound heterodimer with DUOX2A and generates H_2O_2 in a calcium-dependent manner. Intestinal commensals and pathogens have been shown to induce intestinal DUOX2 expression in mice [43, 44], while increased DUOX2 expression has been detected in colon sections from patients with UC in association with relative increases in Proteobacteria [45].

Loss of function genetic variants in *NOX1*, *NOX2* (in the absence of CGD), and *DUOX2* [46–48], as well as biallelic inherited *DUOX2* inactivating mutations [49] have been associated with very-early-onset and early-onset IBD in the pediatric population. Clearly, intestinal NOX complexes play important roles in mucosal redox balance and susceptibility to IBD. However, it is still unclear exactly how NOX2 defects, such as those associated with CGD, promote intestinal inflammation. While the neutrophil-driven hypoxic mucosal milieu needed to promote intestinal repair is lacking in CGD, the NOX2 complex in non-hematopoietic cells of the intestinal mucosa may also be important. The NOX2 complex in non-hematopoietic cells likely directly impacts the net ROS balance in the intestinal milieu. Precisely how these different levels and expressions of NOX complexes affect IBD susceptibility remain to be determined but could include shifting the balance and function of T-cell subsets as well as the composition of the intestinal microbiota.

4 Management

The treatment of CGD-IBD is challenging as the disease is often persistent and prone to relapse. Glucocorticoids are the most common treatment for CGD-IBD. While they are effective, they can be problematic in terms of susceptibility to infections and, with long-term use, may be associated with growth impairment and osteoporosis. Therefore, steroid-sparing agents have been used extensively [50]. Similar to the treatment of conventional IBD, the therapeutic backbone for CGD-IBD consists of steroids followed by the addition of a salicylic acid derivative and/or a purine antimetabolite (e.g., azathioprine, 6-mercaptopurine). This treatment regimen is relatively effective in moderately severe CGD-IBD, but patients often complain of residual abdominal pain, persistent noninfectious diarrhea and relapses. In severe CGD-IBD cases, including those involving strictures and fistulizing disease, patients may eventually undergo bowel surgery or hematopoietic stem cell transplantation. The following section will review standard and more experimental CGD-IBD therapies.

4.1 Immune Modulators

Azathioprine is a purine analogue (prodrug of 6-mercaptopurine) that is incorporated into replicating DNA, blocking de novo purine synthesis. As a result, it mainly hinders B- and T-cell proliferation [51]. It is often second-line treatment for moderate to severe or steroid-dependent CGD colitis. However, it can take several months before a clinical response is observed [52].

Hydroxychloroquine, an antimalarial drug, has also been used to treat CGD-IBD based on its ability to reduce inflammation by decreasing tumor necrosis factor (TNF)-α and IL1β release. More specifically, it has been reported to improve gastric outlet obstruction [53]. Other immune modulators that have been considered but not extensively studied include methotrexate, sulfasalazine and cyclosporine [50, 54].

Thalidomide blocks nuclear localization of nuclear factor (NF)-κB, thereby inhibiting production of inflammatory cytokines in the setting of inflammatory conditions such as rheumatoid arthritis and CD. Its use was reported in a cohort of eight patients with refractory inflammatory complications (including CGD-IBD) with good response. However, thalidomide use is associated with an increased risk of deep vein thrombosis and peripheral neuropathy [55].

Peroxisome proliferator-activated receptor γ (PPARγ) is a central mediator of metabolic responses including fatty acid metabolism and gluconeogenesis. In the context of CGD, PPARγ agonists such as pioglitazone have been shown to increase efferocytosis and mitochondrial ROS production [56, 57]. Moreover, it may also regulate IL17, which is elevated in CGD. A single case of severe infection in CGD was treated with pioglitazone in addition to standard therapy with an improvement in superoxide production [58].

Previous studies have shown that NOX2-deficient phagocytes have defective autophagy and ROS-independent inflammasome activation [59, 60]. Rapamycin, an inhibitor of mammalian target of rapamycin complex 1 (mTORC1), has been shown to induce autophagy [61]. In a study of 15 patients with CGD (7 of whom had colitis) treated with rapamycin, the authors reported an inflammatory phenotype consisting of (1) increased peripheral nonclassical and intermediate monocytes, (2) mononuclear phagocytes producing increased IL1β and TNFα, (3) increased T_H17 $CD4^+$ cell subsets and (4) increased IL17A-producing neutrophils. This inflammatory phenotype was reversed with treatment with rapamycin [19]. However, it is unclear whether these effects translated directly into clinical improvement in the seven patients with colitis.

4.2 Biologics

Given that patients with CGD have dysregulated responses to inflammation, TNFα inhibitors have been used to treat CGD-IBD with some success. However, they confer a high risk of infection and death in patients with CGD and should be used with extreme caution in this population, if at all [62].

Recent studies have shown that cells from patients with CGD have defects in autophagy resulting in increased IL1β release and inflammation [13]. Use of anakinra to inhibit IL1 activity has been reported in patients with CGD and refractory CGD-IBD. Although few patients have been treated with this agent, it appears that therapeutic benefits may be modest and not sustained over time [13, 63].

Vedolizumab, an anti-$\alpha_{4:}\beta_7$ integrin monoclonal antibody, inhibits the gut-specific homing of lymphocytes, including activated effector mucosal lymphocytes. Thus, it may allow for relatively gut-specific immune suppression. Its use in CGD-IBD appears promising [64, 65], but larger studies are needed to evaluate safety and efficacy.

IL17 is an important mediator of inflammation, especially at epithelial surfaces. It is induced in $CD4^+$ T cells by IL23 and in turn leads to the production of G-CSF and IL22. IL17 inhibition with agents such as brodalumab (IL17 receptor A inhibitor) and secukinumab has been shown to improve psoriasis. While these drugs should also theoretically improve IBD, cases of drug-associated onset or exacerbation of IBD have been reported [66]. Therefore, despite increased IL17-producing $CD4^+$ cell and neutrophil subsets in CGD, these agents have not been formally evaluated for the treatment of CGD-IBD. The expansion of T_H17 cells in CGD may reflect increased IL23 production from chronic infections [20]. IL23 is formed by the heterodimerization of IL23p19 and IL12p40, while IL12 is formed by the heterodimerization of IL-12p35 and IL12p40. In mice, neutralization of IL23 using an anti-p19 antibody significantly improved colitis through the downstream inhibition of IL17 expression, leading to diminished neutrophil infiltration [67]. Ustekinumab, an anti-p40 monoclonal antibody which blocks both IL12 and IL23 that has been used to treat psoriasis and CD, was used in a single case of refractory CGD-IBD. Although the colitis improved, the patient developed severe infectious complications [68], reinforcing the difficulties and risks of immune modulation in CGD.

4.3 Microbial Modulation

As is sometimes the case in the treatment of conventional IBD, antimicrobials (e.g., ciprofloxacin and metronidazole) have been anecdotally used for the treatment of CGD-IBD. Studies in mice with CGD [29] and some preliminary studies in patients with CGD (unpublished data) suggest that the microbiome plays a significant role in CGD-IBD susceptibility and pathogenesis. The ideal constitution and way to manipulate the intestinal flora in CGD-IBD remains unclear.

Rare patients with CGD who have undergone hematopoietic cell transplantation (HCT), but later lost their grafts, have had resolution of their CGD colitis, suggesting that conditioning regimens and/or prolonged antibiotic treatment during HCT may affect CGD-IBD. Examination of the microbiota pre- and post-HCT in CGD patients is underway. These studies will help determine whether interventions targeting the microbiota through diet and/or bacteriotherapy are worthy of pursuit.

4.4 Hematopoietic Cell Transplantation

Although HCT dramatically and reliably improves CGD-IBD, formal evaluation of CGD-IBD outcomes post-HCT has only recently been performed. Data from the Primary Immune Deficiency Treatment Consortium (PIDTC) show that pre-HCT colitis does not decrease post-HCT survival, nor does it increase the risk of acute grade III-IV graft-versus-host disease. In almost all patients, HCT improved CGD-associated colitis. In those rare patients in whom GI symptoms returned, it remains unclear whether they are due to recurrent CGD colitis or gut graft-versus-host disease [69].

5 Conclusions

CGD-IBD is a unique disease entity, the pathogenesis of which is complex and multifactorial. The management of CGD-IBD remains challenging due to its chronicity and the difficulty of managing the simultaneous need for immunomodulation and the increased susceptibility to infection that it often brings. ROS production in phagocytes, lymphocytes and non-hematopoietic cells; the expanded T_H17 compartment in the CGD circulation and gut; abnormalities of autophagy; and the role of the intestinal microbiome are all fertile subjects for the understanding of CGD, as well as the role of superoxide in mammalian and microbial physiology. Exciting opportunities for novel understandings and new therapies await those studying NOX2-derived ROS in IBD.

References

1. Arnold DE, Heimall JR (2017) A review of chronic granulomatous disease. Adv Ther 34 (12):2543–2557
2. Kuhns DB, Alvord WG, Heller T, Feld JJ, Pike KM, Marciano BE et al (2010) Residual NADPH oxidase and survival in chronic granulomatous disease. N Engl J Med 363 (27):2600–2610
3. Marks DJ, Miyagi K, Rahman FZ, Novelli M, Bloom SL, Segal AW (2009) Inflammatory bowel disease in CGD reproduces the clinicopathological features of Crohn's disease. Am J Gastroenterol 104(1):117–124
4. Alimchandani M, Lai JP, Aung PP, Khangura S, Kamal N, Gallin JI et al (2013) Gastrointestinal histopathology in chronic granulomatous disease: a study of 87 patients. Am J Surg Pathol 37(9):1365–1372
5. Khangura SK, Kamal N, Ho N, Quezado M, Zhao X, Marciano B et al (2016) Gastrointestinal features of chronic granulomatous disease found during endoscopy. Clin Gastroenterol Hepatol 14(3):395–402 e5
6. Marciano BE, Rosenzweig SD, Kleiner DE, Anderson VL, Darnell DN, Anaya-O'Brien S et al (2004) Gastrointestinal involvement in chronic granulomatous disease. Pediatrics 114 (2):462–468
7. Magnani A, Brosselin P, Beaute J, de Vergnes N, Mouy R, Debre M et al (2014) Inflammatory manifestations in a single-center cohort of patients with chronic granulomatous disease. J Allergy Clin Immunol 134 (3):655–62 e8
8. Marciano BE, Spalding C, Fitzgerald A, Mann D, Brown T, Osgood S et al (2015) Common severe infections in chronic granulomatous disease. Clin Infect Dis 60 (8):1176–1183
9. Labrosse R, Abou-Diab J, Blincoe A, Cros G, Luu TM, Deslandres C et al (2017) Very early-onset inflammatory manifestations of X-linked chronic granulomatous disease. Front Immunol 8:1167
10. Damen GM, van Krieken JH, Hoppenreijs E, van Os E, Tolboom JJ, Warris A et al (2010) Overlap, common features, and essential differences in pediatric granulomatous inflammatory bowel disease. J Pediatr Gastroenterol Nutr 51 (6):690–697
11. Heltzer M, Jawad AF, Rae J, Curnutte JT, Sullivan KE, Diminished T (2002) cell numbers in patients with chronic granulomatous disease. Clin Immunol 105(3):273–278
12. Albuquerque AS, Fernandes SM, Tendeiro R, Cheynier R, Lucas M, Silva SL et al (2017) Major CD4 T-cell depletion and immune senescence in a patient with chronic granulomatous disease. Front Immunol 8:543
13. de Luca A, Smeekens SP, Casagrande A, Iannitti R, Conway KL, Gresnigt MS et al (2014) IL-1 receptor blockade restores autophagy and reduces inflammation in chronic granulomatous disease in mice and in humans. Proc Natl Acad Sci U S A 111(9):3526–3531
14. Montes-Berrueta D, Ramirez L, Salmen S, Berrueta L (2012) Fas and FasL expression in leukocytes from chronic granulomatous disease patients. Invest Clin 53(2):157–167
15. Jackson SH, Devadas S, Kwon J, Pinto LA, Williams MS (2004) T cells express a phagocyte-type NADPH oxidase that is

activated after T cell receptor stimulation. Nat Immunol 5(8):818–827
16. Padgett LE (2016) Tse HM. NADPH oxidase-derived superoxide provides a third signal for CD4 T cell effector responses. J Immunol 197 (5):1733–1742
17. Kwon BI, Kim TW, Shin K, Kim YH, Yuk CM, Yuk JM et al (2017) Enhanced Th2 cell differentiation and function in the absence of Nox2. Allergy 72(2):252–265
18. Wen Z, Shimojima Y, Shirai T, Li Y, Ju J, Yang Z et al (2016) NADPH oxidase deficiency underlies dysfunction of aged CD8+ Tregs. J Clin Invest 126(5):1953–1967
19. Gabrion A, Hmitou I, Moshous D, Neven B, Lefevre-Utile A, Diana JS et al (2017) Mammalian target of rapamycin inhibition counterbalances the inflammatory status of immune cells in patients with chronic granulomatous disease. J Allergy Clin Immunol 139 (5):1641–9 e6
20. Horvath R, Rozkova D, Lastovicka J, Polouckova A, Sedlacek P, Sediva A et al (2011) Expansion of T helper type 17 lymphocytes in patients with chronic granulomatous disease. Clin Exp Immunol 166(1):26–33
21. Liu S, Russo PA, Baldassano RN, Sullivan KE (2009) CD68 expression is markedly different in Crohn's disease and the colitis associated with chronic granulomatous disease. Inflamm Bowel Dis 15(8):1213–1217
22. De Ravin SS, Naumann N, Cowen EW, Friend J, Hilligoss D, Marquesen M et al (2008) Chronic granulomatous disease as a risk factor for autoimmune disease. J Allergy Clin Immunol 122(6):1097–1103
23. Bao S, Carr ED, Xu YH, Hunt NH (2011) Gp91(phox) contributes to the development of experimental inflammatory bowel disease. Immunol Cell Biol 89(8):853–860
24. Pircalabioru G, Aviello G, Kubica M, Zhdanov A, Paclet MH, Brennan L et al (2016) Defensive mutualism rescues NADPH oxidase inactivation in gut infection. Cell Host Microbe 19(5):651–663
25. Campbell EL, Bruyninckx WJ, Kelly CJ, Glover LE, McNamee EN, Bowers BE et al (2014) Transmigrating neutrophils shape the mucosal microenvironment through localized oxygen depletion to influence resolution of inflammation. Immunity 40(1):66–77
26. Conway KL, Goel G, Sokol H, Manocha M, Mizoguchi E, Terhorst C et al (2012) p40phox expression regulates neutrophil recruitment and function during the resolution phase of intestinal inflammation. J Immunol 189 (7):3631–3640
27. Krieglstein CF, Cerwinka WH, Laroux FS, Salter JW, Russell JM, Schuermann G et al (2001) Regulation of murine intestinal inflammation by reactive metabolites of oxygen and nitrogen: divergent roles of superoxide and nitric oxide. J Exp Med 194(9):1207–1218
28. Rodrigues-Sousa T, Ladeirinha AF, Santiago AR, Carvalheiro H, Raposo B, Alarcao A et al (2014) Deficient production of reactive oxygen species leads to severe chronic DSS-induced colitis in Ncf1/p47phox-mutant mice. PLoS One 9(5):e97532
29. Falcone EL, Abusleme L, Swamydas M, Lionakis MS, Ding L, Hsu AP et al (2016) Colitis susceptibility in p47(phox−/−) mice is mediated by the microbiome. Microbiome 4:13
30. Fattouh R, Guo CH, Lam GY, Gareau MG, Ngan BY, Glogauer M et al (2013) Rac2-deficiency leads to exacerbated and protracted colitis in response to Citrobacter rodentium infection. PLoS One 8(4):e61629
31. Diebold BA, Bokoch GM (2001) Molecular basis for Rac2 regulation of phagocyte NADPH oxidase. Nat Immunol 2(3):211–215
32. Dorseuil O, Reibel L, Bokoch GM, Camonis J, Gacon G (1996) The Rac target NADPH oxidase p67phox interacts preferentially with Rac2 rather than Rac1. J Biol Chem 271(1):83–88
33. Aviello G, Knaus UG (2017) ROS in gastrointestinal inflammation: rescue or Sabotage? Br J Pharmacol 174:1704–1718
34. Holmstrom KM, Finkel T (2014) Cellular mechanisms and physiological consequences of redox-dependent signalling. Nat Rev Mol Cell Biol 15(6):411–421
35. Leoni G, Alam A, Neumann PA, Lambeth JD, Cheng G, McCoy J et al (2013) Annexin A1, formyl peptide receptor, and NOX1 orchestrate epithelial repair. J Clin Invest 123 (1):443–454
36. Jones RM, Luo L, Ardita CS, Richardson AN, Kwon YM, Mercante JW et al (2013) Symbiotic lactobacilli stimulate gut epithelial proliferation via Nox-mediated generation of reactive oxygen species. EMBO J 32 (23):3017–3028
37. Bedard K, Krause KH (2007) The NOX family of ROS-generating NADPH oxidases: physiology and pathophysiology. Physiol Rev 87 (1):245–313
38. Rada B, Leto TL (2008) Oxidative innate immune defenses by Nox/Duox family NADPH oxidases. Contrib Microbiol 15:164–187
39. Treton X, Pedruzzi E, Guichard C, Ladeiro Y, Sedghi S, Vallee M et al (2014) Combined

NADPH oxidase 1 and interleukin 10 deficiency induces chronic endoplasmic reticulum stress and causes ulcerative colitis-like disease in mice. PLoS One 9(7):e101669

40. Coant N, Ben Mkaddem S, Pedruzzi E, Guichard C, Treton X, Ducroc R et al (2010) NADPH oxidase 1 modulates WNT and NOTCH1 signaling to control the fate of proliferative progenitor cells in the colon. Mol Cell Biol 30(11):2636–2650
41. Swanson PA 2nd, Kumar A, Samarin S, Vijay-Kumar M, Kundu K, Murthy N et al (2011) Enteric commensal bacteria potentiate epithelial restitution via reactive oxygen species-mediated inactivation of focal adhesion kinase phosphatases. Proc Natl Acad Sci U S A 108 (21):8803–8808
42. Geiszt M, Kopp JB, Varnai P, Leto TL (2000) Identification of renox, an NAD(P)H oxidase in kidney. Proc Natl Acad Sci U S A 97 (14):8010–8014
43. Sommer F, Backhed F (2015) The gut microbiota engages different signaling pathways to induce Duox2 expression in the ileum and colon epithelium. Mucosal Immunol 8 (2):372–379
44. Grasberger H, Gao J, Nagao-Kitamoto H, Kitamoto S, Zhang M, Kamada N et al (2015) Increased expression of DUOX2 is an epithelial response to mucosal dysbiosis required for immune homeostasis in mouse intestine. Gastroenterology 149 (7):1849–1859
45. MacFie TS, Poulsom R, Parker A, Warnes G, Boitsova T, Nijhuis A et al (2014) DUOX2 and DUOXA2 form the predominant enzyme system capable of producing the reactive oxygen species H2O2 in active ulcerative colitis and are modulated by 5-aminosalicylic acid. Inflamm Bowel Dis 20(3):514–524
46. Dhillon SS, Fattouh R, Elkadri A, Xu W, Murchie R, Walters T et al (2014) Variants in nicotinamide adenine dinucleotide phosphate oxidase complex components determine susceptibility to very early onset inflammatory bowel disease. Gastroenterology 147 (3):680–9 e2
47. Hayes P, Dhillon S, O'Neill K, Thoeni C, Hui KY, Elkadri A et al (2015) Defects in NADPH oxidase genes and in very early onset inflammatory bowel disease. Cell Mol Gastroenterol Hepatol 1(5):489–502
48. Schwerd T, Bryant RV, Pandey S, Capitani M, Meran L, Cazier JB et al (2018) NOX1 loss-of-function genetic variants in patients with inflammatory bowel disease. Mucosal Immunol 11:562–574
49. Parlato M, Charbit-Henrion F, Hayes P, Tiberti A, Aloi M, Cucchiara S et al (2017) First identification of biallelic inherited DUOX2 inactivating mutations as a cause of very early onset inflammatory bowel disease. Gastroenterology 153(2):609–11 e3
50. Magnani A, Mahlaoui N (2016) Managing inflammatory manifestations in patients with chronic granulomatous disease. Paediatr Drugs 18(5):335–345
51. Maltzman JS, Koretzky GA (2003) Azathioprine: old drug, new actions. J Clin Invest 111(8):1122–1124
52. Seger RA (2010) Chronic granulomatous disease: recent advances in pathophysiology and treatment. Neth J Med 68(11):334–340
53. Arlet JB, Aouba A, Suarez F, Blanche S, Valeyre D, Fischer A et al (2008) Efficiency of hydroxychloroquine in the treatment of granulomatous complications in chronic granulomatous disease. Eur J Gastroenterol Hepatol 20 (2):142–144
54. Rosh JR, Tang HB, Mayer L, Groisman G, Abraham SK, Prince A (1995) Treatment of intractable gastrointestinal manifestations of chronic granulomatous disease with cyclosporine. J Pediatr 126(1):143–145
55. Noel N, Mahlaoui N, Blanche S, Suarez F, Coignard-Biehler H, Durieu I et al (2013) Efficacy and safety of thalidomide in patients with inflammatory manifestations of chronic granulomatous disease: a retrospective case series. J Allergy Clin Immunol 132 (4):997–1000 e1–4
56. Fernandez-Boyanapalli RF, Frasch SC, Thomas SM, Malcolm KC, Nicks M, Harbeck RJ et al (2015) Pioglitazone restores phagocyte mitochondrial oxidants and bactericidal capacity in chronic granulomatous disease. J Allergy Clin Immunol 135(2):517–27 e12
57. Fernandez-Boyanapalli RF, Falcone EL, Zerbe CS, Marciano BE, Frasch SC, Henson PM et al (2015) Impaired efferocytosis in human chronic granulomatous disease is reversed by pioglitazone treatment. J Allergy Clin Immunol 136(5):1399–401 e1–3
58. Migliavacca M, Assanelli A, Ferrua F, Cicalese MP, Biffi A, Frittoli M et al (2016) Pioglitazone as a novel therapeutic approach in chronic granulomatous disease. J Allergy Clin Immunol 137(6):1913–5 e2
59. van de Veerdonk FL, Smeekens SP, Joosten LA, Kullberg BJ, Dinarello CA, van der Meer JW et al (2010) Reactive oxygen species-independent activation of the IL-1beta inflammasome in cells from patients with chronic

granulomatous disease. Proc Natl Acad Sci U S A 107(7):3030–3033
60. Meissner F, Seger RA, Moshous D, Fischer A, Reichenbach J, Zychlinsky A (2010) Inflammasome activation in NADPH oxidase defective mononuclear phagocytes from patients with chronic granulomatous disease. Blood 116 (9):1570–1573
61. Kim YC, Guan KL (2015) mTOR: a pharmacologic target for autophagy regulation. J Clin Invest 125(1):25–32
62. Uzel G, Orange JS, Poliak N, Marciano BE, Heller T, Holland SM (2010) Complications of tumor necrosis factor-alpha blockade in chronic granulomatous disease-related colitis. Clin Infect Dis 51(12):1429–1434
63. Hahn KJ, Ho N, Yockey L, Kreuzberg S, Daub J, Rump A et al (2015) Treatment with anakinra, a recombinant IL-1 receptor antagonist, unlikely to induce lasting remission in patients with CGD colitis. Am J Gastroenterol 110(6):938–939
64. Campbell N, Chapdelaine H (2017) Treatment of chronic granulomatous disease-associated fistulising colitis with vedolizumab. J Allergy Clin Immunol Pract 5(6):1748–1749
65. Zerbe CS, Kreuzburg SA, Daub J, Marciano BE, Strongin A, Holland S et al (2017) Vedolizumab in chronic granulomatous disease: a safe and promising bridge therapy for CGD related colitis. J Clin Immunol 37(2):235
66. Hohenberger M, Cardwell LA, Oussedik E, Feldman SR (2018) Interleukin-17 inhibition: role in psoriasis and inflammatory bowel disease. J Dermatolog Treat 29(1):13–18
67. Wang R, Hasnain SZ, Tong H, Das I, Che-Hao Chen A, Oancea I et al (2015) Neutralizing IL-23 is superior to blocking IL-17 in suppressing intestinal inflammation in a spontaneous murine colitis model. Inflamm Bowel Dis 21 (5):973–984
68. Butte MJ, Park KT, Lewis DB (2016) Treatment of CGD-associated Colitis with the IL-23 blocker ustekinumab. J Clin Immunol 36 (7):619–620
69. Leiding JW, Logan BR, Yin Z, Arbuckle E, Bleesing JJ, Sullivan KE et al (2018) Resolution of CGD related colitis after allogeneic hematopoietic stem cell transplantation in patients with chronic granulomatous disease-early results from the 6903 study of the Primary Immune Deficiency Treatment Consortium (PIDTC). Biol Blood Marrow Transplant 24(3):S53–SS4
70. Falcone EL (2016) Intestinal inflammation in chronic granulomatous disease: reactive oxygen species interact with the microbiome at the intestinal barrier (unpublished doctoral dissertation). University of Cambridge, Cambridge, United Kingdom

Chapter 35

Ex Vivo Models of Chronic Granulomatous Disease

Julie Brault, Bénédicte Vigne, and Marie José Stasia

Abstract

Induced pluripotent stem cells (iPSCs) are pluripotent stem cells that can be established from dedifferentiation of all somatic cell types by epigenetic phenomena. iPSCs can be differentiated into any mature cells like neurons, hepatocytes, or pancreatic cells that have not been easily available to date. Thus, iPSCs are widely used for disease modeling, drug discovery, and cell therapy development. Here, we describe a protocol to obtain human mature and functional neutrophils and macrophages as ex vivo models of X-linked chronic granulomatous disease (X-CGD). This method can be applied to model the other genetic forms of CGD. We also describe methods for testing the characteristics and functions of neutrophils and macrophages by morphology, phagocytosis assay, release of granule markers or cytokines, cell surface markers, and NADPH oxidase activity.

Key words iPSCs, CGD, Hematopoietic differentiation, Neutrophils, Macrophages, NADPH oxidase, Phagocytosis, Exocytosis, Cytokines

1 Introduction

Chronic granulomatous disease (CGD) is a rare primary immunodeficiency, mainly affecting phagocytes, which is characterized by an increased susceptibility to severe and recurrent bacterial and fungal infections, along with the development of granulomas. The average worldwide birth prevalence is estimated at 1/200,000 to 1/250,000 [1, 2]. CGD is caused by mutations in one of the five genes encoding subunits of the nicotinamide adenine dinucleotide phosphate (NADPH) oxidase complex expressed in phagocytic cells (neutrophils, monocytes/macrophages, eosinophils) [3, 4]. The NADPH oxidase complex consists of a membrane redox element named cytochrome b_{558}, composed of gp91^{phox} (also named NOX2) and p22^{phox} and of cytosolic proteins p47^{phox}, p67^{phox}, and p40^{phox}. This complex becomes assembled upon stimulation of phagocytes after opsonized pathogen recognition. The main genetic form present in humans is X-linked CGD (X-CGD), caused by mutations in *CYBB* and leading to the absence or dysfunction of cytochrome b_{558}, the redox element of the NADPH

Ulla G. Knaus and Thomas L. Leto (eds.), *NADPH Oxidases: Methods and Protocols*, Methods in Molecular Biology, vol. 1982, https://doi.org/10.1007/978-1-4939-9424-3_35, © Springer Science+Business Media, LLC, part of Springer Nature 2019

oxidase complex. X-CGD represents about 70% of the total cases reported to date. The other forms of CGD are autosomal recessive (AR), characterized by mutations in *NCF1*, *NCF2*, and *CYBA* encoding $p47^{phox}$, $p67^{phox}$, and $p22^{phox}$, respectively. Whereas AR-CGD22^0 and AR-CGD67^0 are extremely rare (less than 5% of cases), AR-CGD47^0 represents about 25% of CGD cases due to the presence of two *NCF1* pseudogenes carrying the main mutation of this genetic form. Twenty-four cases of Inherited $p40^{phox}$ deficiency was recently reported in the literature. Clinical characteristics of this innate immunodeficiency are reminiscent of a mild, atypical form of CGD [5].

Antibiotic and antifungal prophylaxis is the current treatment used for CGD patients throughout life. Although allogeneic hematopoietic stem cell (HSC) transplantation is potentially a curative strategy, less than 30% of patients have access to an HLA genetic match. Furthermore, the risk of graft versus host disease following HSC transplantation is a significant deterrent for patients and clinicians alike. However, recent advancements in patient conditioning have made this treatment safer in high-risk CGD patients [6]. Encouraging results were obtained using the gene therapy approach, and a new generation of viral vectors is currently under investigation. Indeed, gene therapy is an attractive alternative, as it enables a genetic defect in the patient's own cells to be corrected [7]. However, to develop new therapy, ex vivo cellular models are needed. Up to 2011, the only cell-based model mimicking the X-CGD form available for this purpose was the knockout *CYBB* PLB-985 cells [8]. Since the advent of induced pluripotent stem cell (iPSC) technology a decade ago, enormous progress has been made in stem cell biology and regenerative medicine [9]. Human iPSCs have been widely used for disease modeling, drug discovery, and cell therapy development [10]. Indeed, iPSCs provide a unique opportunity in that they can be generated from patient cells and differentiated to a desired cell type for disease modeling and therapeutic purposes. Functional neutrophils modeling X-linked chronic granulomatous disorder were made using induced pluripotent stem cells derived from fibroblasts [11–13], mesenchymal cells [14], and peripheral blood $CD34^+$ cells [15]. These ex vivo CGD models are useful to develop and test new therapeutic methods such as genome editing [14–17], gene therapy [18], BAC transgenesis and gene targeting [19], or protein therapy [20].

Non-integrating methods using mRNA, Sendaï virus, or plasmid constructs are preferred to transduce the transcription factors needed for epigenetic changes in the somatic cells leading to the generation of iPSCs [13, 21, 22]. This approach minimizes the appearance of deleterious mutations in the genome of the cells. In addition, feeder-free and xeno-free culture tends to replace the classical iPSC co-culture on irradiated mouse embryonic fibroblasts (MEFs), especially for a clinical use [23–25]. Although mature phagocytic cells can be obtained by feeder-free culture of iPSCs [17, 18, 26–28], the classical method for obtaining human neutrophils or macrophages is using a co-culture of iPSCs or embryoid bodies (EBs) on mouse OP9 bone marrow stromal cells [13–15, 19, 27, 29, 30].

Here we describe an original method to obtain an ex vivo model of X-CGD using iPSCs derived from patients' fibroblasts [13]. Both mature and functional X-CGD neutrophils and macrophages are generated from hematopoietic differentiation of iPSCs co-cultured with the OP9 cell line. Neutrophils and macrophages are characterized by their morphological aspect, phagocytosis function, the expression of NADPH oxidase components (to assess the absence of NOX2 and $p22^{phox}$), and the absence of NADPH oxidase activity compared to that of human control iPSC-derived neutrophils or macrophages. Neutrophils are specifically characterized by the presence of azurophil and specific granules and secretory vesicle markers. For macrophages, the expression of specific membrane markers has also been carried out. We have used mature X^0-CGD macrophages generated by this protocol to test a protein therapy approach using NOX2/$p22^{phox}$ liposomes [20].

2 Materials

Prepare all solutions and media using deionized water and analytical grade reagents. All media and solutions need to be sterilized by filtration with a 0.22-μm membrane filter and stored at 2–8 °C for up to 2 weeks.

2.1 Fibroblast Reprogramming in Feeder-Free Conditions with Episomal Vectors

1. Skin biopsies from X^0-linked patients were obtained after informed consent (Ethical permission Ref CPP-Sud-Est V: 09-CHUG-36, Ref Study: FibroCGD, N_ AFSSAPS:2009-A00944–53).
2. Human fibroblasts are obtained from skin biopsy of CGD patients or healthy donors and are amplified until passages 2–3.
3. Fibroblast medium: Dulbecco's Modified Eagle Medium (DMEM), 10% (v/v) fetal bovine serum (FBS), 1% penicillin/streptomycin, 1% nonessential amino acids (NEAA).
4. Phosphate-buffered saline, pH 7.4 (PBS).
5. 0.05% (w/v) trypsin-Ethylene Diamine Tetraacetic Acid (EDTA).
6. 0.1% (w/v) Porcine C gelatin solution diluted in deionized water and autoclaved.
7. Human dermal fibroblast (NHDF) Nucleofector® kit VPD 1001 and transfection cuvettes (Lonza).
8. Nucleofector II™ and transfection cuvettes (Lonza).

9. Episomal vectors: pCXLE-hOCT3/4-shp53, pCXLE-hSOX2-hKLF4, and pCXLE-hLIN28-hL-MYC (Addgene, references 27077, 27078, 27080, respectively).
10. Truncated recombinant human vitronectin (VTN).
11. TrypLE Select Enzyme (1×) (Gibco™).
12. StemMACS™ iPS-Brew XF medium (Miltenyi Biotec) supplemented with 5 μM PD 0325901, 1 μM CHIR 99021, 2 μM SB 4315435, 0.4 μM sodium butyrate, 500 μM thiazovivin, and 50 μg/mL ascorbic acid. An appropriate amount of this medium must be prepared extemporaneously because it cannot be stored (*see* **Note 1**). Store stock solutions of 10 mM PD 0325901, 10 mM CHIR 99021, and 10 mM SB431542 at −80 °C for 6 months. Store stock solutions of 100 mg/mL sodium butyrate and 50 mg/mL ascorbic acid at −20 °C for 6 months. Warning: sodium butyrate is toxic, and ascorbic acid must be protected from light.
13. 100-mm, 6-well, and 24-well tissue culture-treated dishes and plates, tissue culture-treated flasks 25 cm^2 (T25) or 75 cm^2 (T75).
14. 15 mL conical tubes.
15. 1.5 mL microtubes.

2.2 Feeder-Free iPS Cell Culture

1. iPSC medium for feeder-free culture: StemMACS™ iPS-Brew XF xeno- and serum-free medium (Miltenyi Biotec).
2. Truncated recombinant human vitronectin (VTN).
3. 0.5 mM EDTA.
4. PBS, pH 7.4.
5. 12-well and 6-well tissue culture-treated plates.

2.3 Hematopoietic Progenitor Differentiation and Characterization

1. OP9 mouse bone marrow stromal cell line (ATCC® CRL-2749™) (*see* **Note 2**).
2. OP9 medium: Minimum Essential Media (α-MEM), 20% FBS.
3. 0.1% (w/v) porcine C gelatin solution diluted in deionized water and autoclaved.
4. PBS, pH 7.4.
5. 0.05% trypsin-EDTA.
6. T75 tissue culture-treated flasks.
7. Hematopoietic differentiation induction medium: α-MEM medium, 10% FBS, 100 μM monothioglycerol (MTG), 50 ng/mL ascorbic acid. Add the ascorbic acid solution just prior to using the medium.
8. 0.5 mM EDTA.

9. Collagenase IV 1 mg/mL in Dulbecco's Modified Eagle Medium/Ham's Nutrient Mixture F-12 (DMEM/F-12).
10. 50 mL conical tubes.
11. Cell scrapers.
12. Cell strainers: 40 μm and 100 μm.
13. Magnetic cell sorting (MACS) buffer: 5% FBS, 2 mM EDTA in PBS, filter-sterilized. Keep buffer cold (4–8 °C) (*see* **Note 3**).
14. CD34 MicroBead kit, human (Miltenyi Biotec).
15. Magnetic separation: columns and separator (manual or automatized device).
16. Optional: CD34 monoclonal antibodies to evaluate the purity of sorted cells by flow cytometry (*see* **Note 4**).
17. Iscove's modified Dulbecco's medium (IMDM), 20% FBS.
18. Methylcellulose medium containing hematopoietic cytokines (StemMACS HSC-CFU media 130-091-277, Miltenyi Biotec): thaw the bottle at 4 °C overnight and then shake the bottle vigorously 3–5 s to mix the medium. Leave the bottle for 30 min at room temperature before use to let air bubbles disappear. Make aliquots of 3 mL of medium in 5-mL sterile propylene tubes with cap and store at −20 °C.
19. 3 mL syringe and 18-gauge needles.
20. 35 mm and 100-mm Petri dishes.

2.4 Neutrophil Differentiation Culture

Add cytokines to the medium directly before use. Prepare an appropriate quantity of this medium as it cannot be stored.

1. Stage 1 neutrophil differentiation medium (for day 10 to day 14 of differentiation): IMDM, 20% FBS, 100 ng/mL stem cell factor (SCF), 100 ng/mL Fms-related tyrosine kinase 3 Ligand (Flt3L), 100 ng/mL interleukin (IL)-6, 10 ng/mL thrombopoietin (TPO), and 10 ng/mL IL-3.
2. Stage 2 neutrophil differentiation medium (for day 14 to day 25 of differentiation): IMDM, 20% FBS, 100 ng/mL granulocyte colony-stimulating factor (G-CSF).
3. Poly(2-hydroxyethyl methacrylate) (pHEMA) (Sigma-Aldrich) solution: add 4 g of pHEMA to 40 mL of 95% ethanol and 10 mM NaOH to a 50 mL conical tube, and dissolve by continuous rotation overnight at room temperature.
4. pHEMA-coated culture flask: add 3 mL of pHEMA solution to a T25 tissue culture-treated flask and rotate to cover the entire surface of the flask with the pHEMA solution. Then, tilt the flask and remove the excess pHEMA solution (put the excess pHEMA solution back in the tube to use within the next few months). Let the flask dry overnight at room temperature under a sterile cell culture hood (*see* **Note 5**).

5. 15 mL conical tubes.
6. Gelatinized 6-well tissue culture-treated plates are prepared by adding 1 mL of 0.1% gelatin in each well and incubated for 3 h at 37 °C.

2.5 Macrophage Differentiation Culture

Add cytokines to the medium right before use. Prepare the right quantity of this medium as it cannot be stored.

1. Stage 1 macrophage differentiation medium (for day 10 to day 18 of differentiation): α-MEM medium, 10% FBS, 100 μM MTG, and 200 ng/mL granulocyte-macrophage colony-stimulating factor (GM-CSF).
2. Stage 2 macrophage differentiation medium (for day 18 to day 28 of differentiation): IMDM, 10% FBS, and 20 ng/mL macrophage colony-stimulating factor (M-CSF).
3. pHEMA solution: add 4 g of pHEMA to 40 mL of 95% ethanol and 10 mM NaOH, and dissolve by continuous rotation overnight at room temperature.
4. pHEMA-coated culture flask: add 3 mL of pHEMA solution to a T25 tissue culture-treated flask and rotate to cover the entire surface of the flask with the pHEMA solution. Then, tilt the flask and remove the excess pHEMA solution (put the excess pHEMA solution back in the tube to use within the next few months). Let the flask dry overnight at room temperature under a sterile cell culture hood (*see* **Note 5**).
5. 15 mL conical tubes.
6. Cell strainer: 40 μm.

2.6 May-Grünwald-Giemsa Staining

1. Glass microscope slides.
2. Cytocentrifuge and cytospin slide chambers.
3. May-Grünwald-Giemsa (MGG) staining solution and staining tank.

2.7 Flow Cytometry Analysis

1. Fluorescence-Activated Cell Sorting (FACS) buffer: 2% FBS, 2 mM EDTA, 0.05% NaN_3 in PBS.
2. Permeabilization buffer: 0.2% (w/v) bovine serum albumin (BSA), 0.01% (w/v) saponin in PBS.
3. Fluorescent-conjugated antibodies directed against human CD11b, CD14, CD16, CD45, and CD68.
4. Primary and secondary antibodies directed against human components of the NADPH oxidase complex (*see* Table 1). All the antibodies are prepared using a 1:200 dilution in the appropriate buffer extemporaneously.
5. 5 mL polypropylene round-bottom FACS tubes.
6. Flow cytometer.

Table 1
List of antibodies used for detection of the NADPH oxidase subunits by flow cytometry

NADPH oxidase subunit	Fixation/ permeabilization	Primary antibody	Irrelevant antibody	Secondary antibody
gp91^{phox}	No	IgG1 mouse 7D5 (D162–3, Clinisciences)	mouse IgG1	goat anti-mouse IgG1-FITC
p22 ^{phox}	Yes	IgG2a mouse anti-p22^{phox} clone 44.1 (sc-130550, Santa Cruz)	mouse IgG2a	goat anti-mouse IgG-PE
p47 ^{phox}	Yes	IgG1 mouse clone 1/p47^{phox} (610354, Beckton Dickinson)	mouse IgG1	goat anti-mouse IgG-PE
p67 ^{phox}	Yes	rabbit monoclonal antibody anti-p67^{phox}, clone EPR5065 (109523, Abcam)	rabbit monoclonal IgG	goat anti-rabbit IgG-AF488

2.8 Myeloperoxidase Activity in Neutrophils

1. Benzidine/nitroprusside solution: first, dilute 0.25 g of benzidine powder (4,4′-diaminobiphenyl; MW = 184.24 g/mol) in 100 mL of absolute ethanol. Then, prepare the nitroprusside solution by diluting 25 g of sodium nitroprusside (MW = 297.97 g/mol) in 100 mL of distilled water. Add 2 mL of this solution to 100 mL of benzidine solution. Store this preparation at 4 °C in an amber glass bottle for up to 2 months (*see* **Note 6**).
2. Hydrogen peroxide (H_2O_2) solution: extemporaneously dilute 0.3 μL of a hydrogen peroxide 30% stock solution in 50 mL of distilled water.
3. 10% Giemsa staining solution.
4. Chemical fume hood.

2.9 Exocytosis Experiments

1. PBS$^+$: dilute 25 μL of 1 M $MgCl_2$ and 45 μL of 1 M $CaCl_2$ in 50 mL PBS. Filter the solution with a 0.22-μm membrane filter.
2. Dimethyl sulfoxide (DMSO).
3. 0.25 mg/mL cytochalasin B: prepare a stock solution by dissolving 1 mg of powder (MW = 479.61 g/mol) in 200 μL of DMSO. Store in 20 μL aliquots at −20 °C. For the exocytosis experiment, prepare a 1:20 dilution in PBS.
4. 5 μM f-Met-Leu-Phe (fMLF): prepare a stock solution of 10^{-2} M fMLF by dissolving 10 mg of fMLF powder (MW = 437.6 g/mol) in 2 mL of DMSO, and store at

−20 °C. Then, prepare 10^{-3} M fMLF aliquots by diluting 50 μL of 10^{-2} M fMLF in 450 μL of DMSO. Store these aliquots at −20 °C. For the exocytosis experiment, prepare a 1:10 dilution of a 10^{-3} M fMLF aliquot in DMSO, and then perform a 1:20 dilution in PBS^{+}.

5. 0.2% (w/v) Triton® X-100: perform a 1:50 dilution of 10% (w/v) Triton® X-100 in PBS. Keep on ice until use.
6. Complete, EDTA-free protease inhibitor cocktail (25×): dissolve 1 tablet of complete, EDTA-free protease inhibitor in 2 mL PBS. This stock solution can be stored at −20 °C for up to 12 weeks.
7. 1.5 mL microtubes.

2.10 Lactoferrin Concentration

1. Lactoferrin ELISA kit (Calbiochem #427275).
2. Microplate spectrophotometer.

2.11 Matrix Metalloprotéinase 9 (MMP9) Concentration

1. Separating gel of 10% SDS-polyacrylamide gel containing 0.5 mg/mL gelatin: dissolve 15 mg of gelatin in 7.5 mL of 1.5 M Tris–HCl, pH 8.9, and 14.6 mL distilled water and heat the solution at 80 °C while shaking for at least 10 min. Then, let the solution cool to room temperature. Add 7.5 mL of 40% (w/v) acrylamide solution, 0.3 mL of 10% (w/v) sodium dodecyl sulfate (SDS), 75 μL of 10% (w/v) ammonium persulfate (APS), and 10 μL of *N,N,N′,N′*-tetramethylethylenediamine (TEMED).
2. Stacking gel: mix 6.32 mL of distilled water with 2.5 mL of 0.5 M Tris–HCl pH 6.8, 1 mL of 40% (w/v) acrylamide solution, 100 μL of 10% (w/v) SDS, 70 μL of 10% (w/v) APS, and 10 μL of TEMED.
3. MMP9 gelatinase standard: use gelatinase purified from exocytosis supernatants of stimulated neutrophils stored at −20 °C or from commercial sources [31].
4. Gelatinase buffer (1×): prepare 1 L of 0.05 M Tris–HCl, 0.05 M NaCl, 0.05% (w/v) Brij 35 in distilled water, pH 7.6. Store at 4 °C for up to 14 days.
5. 10× Sample buffer (10×SB) without reducing agents.
6. Renaturation solution: prepare 1 L of 0.05 M Tris–HCl, 5 mM $CaCl_2$, 1 μM $ZnCl_2$ in distilled water, pH 7.6, 2.5% (w/v) Triton® X-100.
7. Incubation buffer 2× pH 7.6: prepare 1 L of 100 mM Tris–HCl, 10 mM $CaCl_2$, 2 μM $ZnCl_2$, 2% (w/v) Triton® X-100, 0.04% (w/v) NaN_3, and adjust the pH to 7.6.
8. 10 mM *p*-aminophenylmercuric acetate (APMA) solution: dissolve 35.18 mg of APMA (MW = 351.8 g/mol) with 0.75 mL 1 N NaOH, and then add 2.5 mL of distilled water. Mix until

complete dissolution. Add 0.6 mL of HCl 1 N (the solution becomes cloudy), and add 1 drop of 1 N NaOH (the solution becomes clear). Top up the volume to 10 mL with distilled water. Store at 4 °C for up to 1 week.

9. 1 mM APMA incubation solution: dilute 10 mL of 10 mM APMA in 50 mL of incubation buffer 2× pH 7.6, and top up to a final volume of 100 mL with distilled water. Prepare fresh just before use.
10. 0.25% (w/v) Coomassie blue staining solution: dissolve 2.5 g of Coomassie Blue R250 in 100 mL of a mix solution of 10% acetic acid, 50% methanol, and 40% distilled water.
11. Destaining solution: mix of 40% methanol, 10% glacial acetic acid, and 50% distilled water.
12. Storage solution: mix of 1% glycerol, 10% glacial acetic acid, and 89% distilled water.
13. Densitometer: CD 60 Desaga.

2.12 Cytokine Release Experiments

1. 0.1 mg/mL lipopolysaccharide (LPS from *Pseudomonas aeruginosa)*: perform a 1:10 dilution of 1 mg/mL of LPS (stock solution) in IMDM.
2. Human Inflammatory Cytokines Multi-Analyte Profiler ELISArray™ kit.
3. Microplate spectrophotometer: Multiskan™ Spectrum.

2.13 Phagocytosis Assay by Flow Cytometry

1. FACS buffer: 2% FBS, 2 mM EDTA, 0.05% NaN_3 in PBS.
2. AF488 fluorescent heat-killed particles of *Staphylococcus aureus*: resuspend the lyophilized powder (2 mg, amount of bacteria depends on the batch ~ 6×10^8) with PBS, and adjust the concentration to 2×10^9 bacteria/mL. Make aliquots of 20 μL and store at −20 °C.
3. AF488 fluorescent heat-killed Zymosan A particles isolated from *Saccharomyces cerevisiae*: resuspend the lyophilized powder (2 mg, amount of yeast depends on the batch ~ 4×10^7) with PBS, and adjust the concentration to 2×10^9 yeast/mL. Make aliquots of 20 μL, and store at −20 °C.
4. Human AB serum: this serum is collected from plasma donations of healthy AB male donors at EFS-licensed facilities located in France and then pooled, filtered, bottled, and tested. In the lab, the unit is aliquoted in 15 mL conical tubes and frozen at −20 °C. Sterile-filtered human AB serum from several other suppliers can be also used (Sigma-Aldrich™, Innovative Research™).
5. Opsonization of fluorescently labeled *S. aureus* and Zymosan A: add 500 μL of human AB serum to a 20 μL aliquot of bacteria or yeast, and transfer the bacterial or yeast suspension

to a 1.5 mL microtube. Incubate for 1 h at 37 °C while shaking at 350 rpm in the dark, and then centrifuge for 8 min at 20,000 × *g*. Discard the supernatant, and resuspend with 200 μL of PBS. Count the bacteria or yeast suspension using a fluorescent microscope, and adjust the concentration to 2×10^8/mL with PBS. Store 20 μL aliquots at −20 °C for up to 3 months.

6. 0.8 M glucose: dilute 1.44 g of glucose powder (MW = 180.16 g/mol) in 10 mL of PBS.
7. PBS^+: dilute 25 μL of 1 M $MgCl_2$ and 45 μL of 1 M $CaCl_2$ in 50 mL of PBS. Filter the solution with a 0.22-μm membrane filter.
8. PBS^+/glucose buffer: mix 250 μL of 0.8 M glucose with 9.75 mL of PBS^+, and filter using a 0.22-μm membrane filter.
9. 2 mg/mL trypan blue: dissolve 100 mg of trypan blue powder in 50 mL of PBS. Shake for 30 min and then filter using a Whatman paper to remove aggregates.
10. 5 mL polypropylene round-bottom FACS tubes.
11. Flow cytometer.

2.14 ROS Experiments: Nitro Blue Tetrazolium (NBT) Test

1. 0.1 mg/mL lipopolysaccharide (LPS from *Pseudomonas aeruginosa*): perform a 1:10 dilution of 1 mg/mL of LPS (stock solution) in IMDM.
2. 53 mM phosphate buffer, pH 7.5: dilute separately 3.8 g Na_2HPO_4, $12H_2O$ (MW = 358.14 g/mol) in 200 mL of distilled water, and 1.44 g KH_2PO_4 (MW = 136.09 g/mol) in 200 mL of distilled water. Add the KH_2PO_4 solution to the Na_2HPO_4 solution until pH 7.5 is reached.
3. 7 mM NBT: dissolve 6 mg of NBT powder in 1 mL of 53 mM phosphate buffer pH 7.5 just before use. Do not store it.
4. Dimethyl sulfoxide (DMSO).
5. Human AB serum: *see* Subheading 2.13.
6. 3.9% (w/v) glucose: dissolve 1.95 g of glucose powder (MW = 180.2 g/mol) in 50 mL of distilled water.
7. 4 μg/mL phorbol-myristate-acetate (PMA): perform a 1/50 dilution of a 2 mg/mL stock solution in DMSO and then a 1/10 dilution in PBS.
8. 100 mM Tris, pH 8.5: dissolve 605.5 mg of Trizma base (MW = 121.1 g/mol) in 50 mL of distilled water, and adjust the pH to 8.5 with 1 N HCl.
9. Immunoglobulin G (IgG) from human serum, lyophilized powder.
10. Opsonized latex beads: incubate 100 μL of latex beads (LB8 Ø 0.8 μm, Sigma-Aldrich) with 120 μL of 100 mM Tris,

pH 8.5, and 3 mg of human IgG for 30 min at 37 °C with shaking at 130 rpm. Then, wash three times with 1 mL of PBS, and centrifuge for 2 min at 10,000 × *g* at room temperature. Resuspend in 500 μL PBS.

11. Safranin O staining solution 0.5% (w/v): add 5 g of safranin O powder to 1000 mL of distilled water. Store at room temperature for 1 year.
12. Mayer's hemalum solution (v/v): dilute 1 mL of Mayer's hemalum solution in 40 mL of distilled water.
13. Glass microscope slides.
14. 5 mL round-bottom tubes.
15. Chemical fume hood.

2.15 ROS Experiments: Dihydrorhodamine (DHR) Assay

1. 500 μM DHR 123: dilute 5 μL of a stock solution 5 mM DHR in 45 μL of DMSO, extemporaneously. Keep cold and in the dark before use. Store 50 μL aliquots of 5 mM DHR solution at −20 °C for 6 months.
2. PBS^+: dilute 25 μL of 1 M $MgCl_2$ and 45 μL of 1 M $CaCl_2$ in 50 mL of PBS. Filter the solution with a 0.22 μm membrane filter.
3. Dimethyl sulfoxide (DMSO).
4. 0.8 M glucose: dilute 1.44 g of glucose powder (MW = 180.16 g/mol) in 10 mL of PBS. Store aliquots at −20 °C.
5. 10% (w/v) NaN_3: dissolve 0.1 g of sodium azide powder (MW = 65.01 g/mol) in 1 mL of distilled water. Store 50 μL aliquots at −20 °C.
6. PBS^+ containing 20 mM glucose, 3 mM NaN_3, and 0.5 μM DHR: mix 250 μL of 0.8 M glucose with 20 μL of 10% NaN3 in 9.75 ml of PBS^+. Filter using a 0.22-μm membrane filter. Mix 5 mL of this buffer with 5 μL of 500 μM DHR, and keep in the dark.
7. 20 μg/mL phorbol-myristate-acetate (PMA): perform a 1:10 dilution of a 2 mg/mL stock solution in DMSO and then a 1:10 dilution in PBS.
8. Human AB serum: *see* Subheading 2.13.
9. 16 mg/mL opsonized Zymosan A: dissolve 16 mg of Zymosan A from *Saccharomyces cerevisiae* in 1.6 mL of sterile water in a 5 mL round-bottom tube, and incubate for 15 min in boiling water. Centrifuge for 15 s at 10,000 × *g* at room temperature, and discard the supernatant. Slowly add 4 mL of human AB serum while transferring to a large round-bottom tube. Incubate for 30 min at 37 °C while shaking at 130 rpm, and then centrifuge for 5 min at 600 × *g*. Wash the pellet three times

with 1 mL of PBS, and resuspend in 1 mL of PBS. Store 50 μl aliquots at −20 °C (*see* **Note 7**).

10. 1.5 mL microtubes.
11. 5 mL polypropylene round-bottom FACS tubes.
12. Flow cytometer.

2.16 ROS Experiments: Chemiluminescence Assay

1. PBS⁺: dilute 25 μL of 1 M $MgCl_2$ and 45 μL of 1 M $CaCl_2$ in 50 mL of PBS. Filter the solution with a 0.22-μm membrane filter.
2. 100 mM luminol: dissolve 17.7 mg of powder (MW = 177.16 g/mol) in 1 mL of DMSO. This solution can be stored protected from light at 4 °C for 1 week.
3. 2 mM luminol: perform extemporaneously before use a 1:50 dilution of the 100 mM luminol solution in PBS^+, vortex well, and dilute 1:5 in PBS^+. Keep this solution in the dark. Do not store this solution.
4. 0.8 M glucose: dilute 1.44 g of glucose powder (MW = 180.16 g/mol) in 10 mL of PBS. Store aliquots at −20 °C.
5. 500 U/mL horseradish peroxidase (HRP): dissolve 2.5 mg of HRP powder in 1 mL of PBS^+. Keep this solution in the dark. This solution can be stored protected from light at 4 °C for 1 week.
6. Chemiluminescence medium: prepare 200 μL/well. For 1 mL, mix 945 μL of PBS^+, 25 μL of 0.8 M glucose, 10 μL of 2 mM luminol, and 20 μL of 500 U/mL HRP. Prepare fresh and keep in the dark until use.
7. Dimethyl sulfoxide (DMSO).
8. 2 μg/mL PMA: perform a 1:10 dilution of the 2 mg/mL stock solution in DMSO. This solution can be stored at −20 °C for up to 1 month. Then, perform a 1:100 dilution in PBS^+.
9. Human AB serum: *see* Subheading 2.13.
10. 16 mg/mL opsonized Zymosan A: dissolve 16 mg of Zymosan A from *Saccharomyces cerevisiae* in 1.6 mL of sterile water in a 5 mL round-bottom tube, and incubate for 15 min in boiling water. Centrifuge at 10,000 × *g* for 15 s at room temperature, and discard the supernatant. Slowly add 4 mL of human AB serum while transferring to a large round-bottom tube. Incubate for 30 min at 37 °C while shaking at 130 rpm, and then centrifuge for 5 min at 600 × *g*. Wash the pellet three times with 1 mL of PBS, and resuspend in 1 mL of PBS. Store 50 μl aliquots at −20 °C (*see* **Note 7**).
11. Chemiluminometer.

3 Methods

For all the methods, centrifugations and experiments are done at room temperature when not specified otherwise.

3.1 Generation of iPSCs

Reprogramming of X^0-CGD human dermal fibroblasts was performed using episomal vectors as previously described, with minor modifications [13, 22, 23]. Other integration-free methods like mRNA or Sendaï virus are also pertinent in order to produce clinical grade iPSCs [21–23]. Alternatively, iPSCs can be cultured on MEFs, Matrigel or vitronectin-coated plates before hematopoietic differentiation into neutrophils or macrophages [24–30].

3.1.1 Culture of Human Dermal Fibroblasts (HDFs) (Day 3 to Day 0)

1. Grow HDFs passages 2–3 to confluence in fibroblast medium (*see* **Note 8**).
2. Wash the culture with PBS and incubate with 0.05% trypsin-EDTA for 3 min at 37 °C in the incubator (adjust volumes according to the size of the culture flask).
3. Add fresh fibroblast medium to the flask, detach the cells by flushing, and then transfer the cells into a 15 mL conical tube.
4. Centrifuge at 500 × *g* for 5 min and aspirate the supernatant.
5. Resuspend the cell pellet with 5 mL of fibroblast medium and count the cells.
6. Plate cells at a density of $2 \times 10^4/\text{cm}^2$ cells in a new T25 flask, and culture them in fibroblast medium for 3 days at 37 °C, in normoxic conditions under 5% CO_2 (*see* **Note 9**).

3.1.2 Reprogramming of Fibroblasts into iPS Cells by Episomal Technique

Transfection Day 0

1. Coat two wells of a 6-well plate by adding 1 mL/well of 0.1% gelatin, and incubate for at least 3 h.
2. Add 2 mL/well of fibroblast medium, and incubate 2 h at 37 °C to equilibrate the medium.
3. Wash the T25 flask containing the HDFs with 5 mL of PBS, and repeat this step once.
4. Incubate with 1 mL of 0.05% trypsin-EDTA for 3 min at 37 °C in the incubator.
5. Add 5 mL of fibroblast medium and detach the HDFs by flushing, then transfer the cells to a 15 mL conical tube, count the cells, and adjust the cell suspension to 10^6 cells/mL (*see* **Note 8**).
6. Transfer 1×10^6 cells into a new 15 mL conical tube, centrifuge at 300 × *g* for 10 min, and aspirate the supernatant (*see* **Note 9**).

7. Prepare the solution of Nucleofactor™ with Nucleofactor™ supplement (thaw at room temperature) in a 1.5 mL microtube according to the manufacturer's instructions and flush.
8. Add 2.5 μg of each episomal vector: pCXLE-hOct3/4-shp53, pCXLE-hSox2-hKlf4, and pCXLE-hLin28-hL-Myc to the microtube (adjust the volume of each plasmid according to the results of the DNA concentration) (*see* **Note 10**).
9. Resuspend the HDFs pellet (**step 6**) with this solution by gently flushing it, and transfer the whole suspension into the transfection cuvette. Perform the nucleofection immediately after adding the nucleofection solution to the cells.
10. Use the U023 program of the Nucleofector II™ for high transfection efficiency or P022 program for high cell viability according to the manufacturer's instructions.
11. After transfection, add 500 μL of the pre-equilibrated fibroblast medium to the transfection cuvette, and use the pipettes provided in the kit to transfer to the culture well (final volume 2 mL).
12. Incubate the cells at 37 °C, 5% CO_2.
13. Culture the cells in the fibroblast medium for 7 days, changing the medium every 2–3 days.

Day 7

1. Prepare VTN-coated 100-mm culture dishes 1 h before (around 5 dishes for 2 wells of a 6-well plate).
2. Aspirate the culture supernatant, wash with 2 mL/well of PBS, and incubate with 500 μL of pre-warmed TrypLE solution for 3 min at 37 °C.
3. Add 2 mL of fibroblast medium and detach cells by flushing, and then transfer the cells into a 15 mL conical tube.
4. Count the cells and transfer 1×10^5 cells to each VTN-coated 100-mm culture dish. Adjust the volume to 5 mL with fibroblast medium, and add 5 mL of iPS Brew XF medium.
5. Incubate at 37 °C, 5% CO_2, overnight.

Day 8

1. Aspirate the culture medium, and add 10 mL/100-mm dish of iPS Brew XF medium containing pharmacological inhibitors (*see* Subheading 2.1 and **Note 1**).
2. Culture for 7 days and replace the medium with fresh medium every 2 days.

Day 15

Change the culture medium for iPS Brew XF medium without inhibitors (*see* **Note 1**) every day until a change in the morphology of the cells similar to human embryonic stem cells (hESC) is observed. Undifferentiated IPS cells are small, round cells with a clear margin (Fig. 1a).

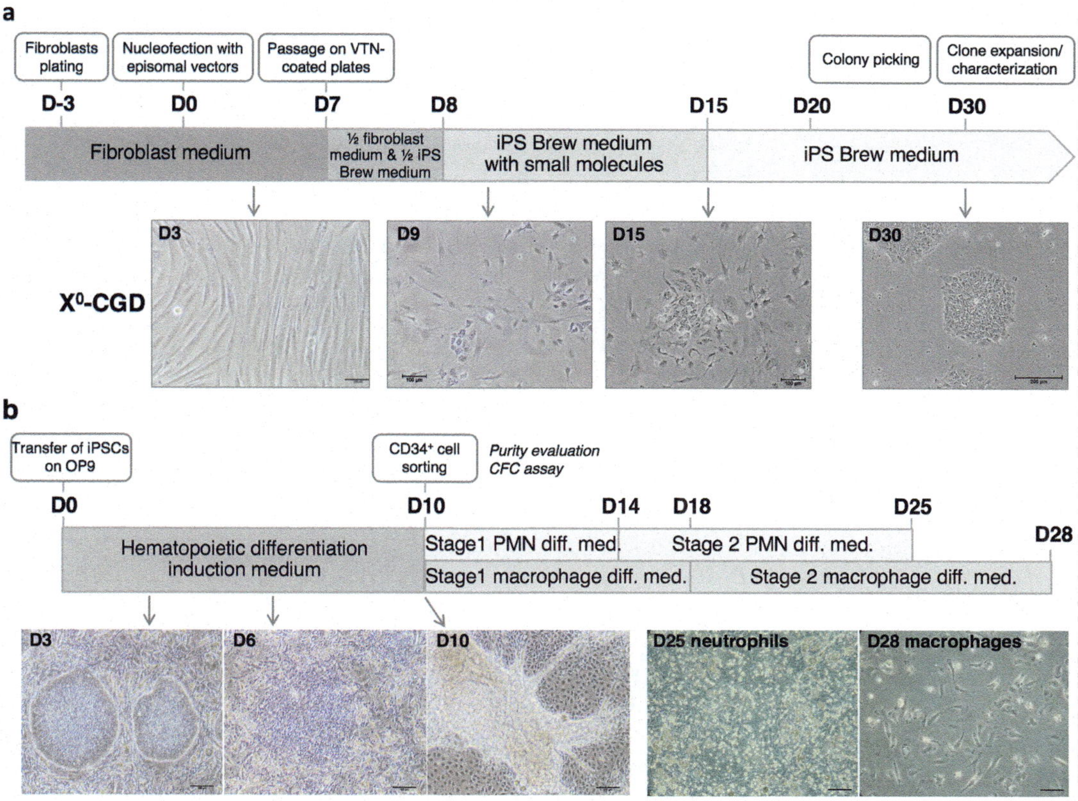

Fig. 1 iPS cells generation from human fibroblasts and hematopoietic differentiation into neutrophils and macrophages. (**a**) Timeline of the non-integrating reprogramming from fibroblasts cultured in feeder-free conditions illustrated by photos of the morphology of fibroblasts (D3), the appearance of small iPS colonies surrounded by non-reprogrammed fibroblasts (D9 and 15), and a picked iPS clone cultured on VTN-coated plates (D30) (scale bar 100 and 200 μm). (**b**) Timeline of the hematopoietic differentiation of iPSCs into CD34$^+$ progenitors (D0 to D10) and then into mature neutrophils in suspension (D25) or into adherent macrophages (D28) (scale bar 100 μm)

Days 20–25

1. Pick at least 12–24 iPS colonies: aspirate carefully using a 200 μL tip, and transfer one colony/well to a VTN-coated 24-well plate (*see* **Note 11**).
2. Replace the medium with fresh iPS Brew XF medium every other day, and check the growth of iPS colonies (Fig. 1a).

Expansion and Characterization of iPS Cell Lines

1. When iPS colonies have reached a sufficient size for passaging, aspirate the culture medium.
2. Wash the well once with 2 mL of PBS and aspirate.
3. Add 500 μL of 0.5 mM EDTA and incubate for 3 min at 37 °C.
4. Aspirate and add 1 mL of fresh iPS Brew XF medium. Harvest colonies by flushing gently using a 5 mL serological pipet.

Transfer colonies into a new VTN-coated well (12-well plates and then 6-well plates).

5. Expand several clones and freeze cells in liquid nitrogen.

After several passages, perform a complete characterization of each iPS cell line: all the X^0-CGD iPSC lines cultured in an undifferentiated state are characterized by a specific hESC-like morphology, a high alkaline phosphatase activity, and the expression of pluripotency markers Oct4, Sox2, Nanog, SSEA-3, SSEA-4, and Tra-1-81 by immunocytochemistry and/or flow cytometry [13]. Furthermore, these cell lines form embryoid bodies in vitro, containing cells originating from the three germ layers (ectoderm, mesoderm, and endoderm) as shown by expression of lineage-specific markers [13]. The genetic integrity of iPSC clones must be controlled to discard aberrant genetic clones after chromosome karyotyping and array-comparative genomic hybridization analysis [13]. We recommend completing the phenotypic characterization of iPSCs with a genetic analysis to confirm that they retain the mutation found in the original patient cells (not described here).

3.2 Hematopoietic Progenitor Differentiation and Characterization

From among all the published protocols demonstrating the production of hematopoietic progenitors from pluripotent stem cells, we selected the iPS/OP9 co-culture method [31–33]. This approach is reproducible, scalable, and easy to handle.

3.2.1 Preparation of the OP9 Feeder Layer for Differentiation

Four days before the differentiation of human X^0-CGD iPSCs, the OP9 feeder layer is prepared. The number of flasks to prepare depends on the number of progenitors or neutrophils/macrophages required for the experiments.

1. Prepare gelatinized T75 flasks by adding 5 mL of 0.1% gelatin, and incubate for 3 h at 37 °C.
2. Wash a confluent OP9 cell culture with 5 mL of PBS, and aspirate and repeat this step once.
3. Incubate with 1.5 mL of 0.05% trypsin-EDTA for 3 min at 37 °C in the incubator (*see* **Note 12**).
4. Add 5 mL of OP9 medium to the flask and detach cells by flushing, and then transfer the cells to a 15 mL conical tube.
5. Centrifuge at 500 × *g* for 5 min and aspirate the supernatant.
6. Resuspend the cell pellet with 5 mL of OP9 medium and count the cells.
7. Plate cells at a density of 8×10^3 cells/cm^2 in the gelatinized flask, and culture in OP9 medium for 4 days at 37 °C, in normoxic conditions under 5% CO_2 (*see* **Note 13**).

3.2.2 Induction of Hematopoietic Differentiation (for Days 0 Through 10)

1. Aspirate the medium of OP9 cell cultures, and add 10 mL of freshly prepared hematopoietic differentiation induction medium.
2. Remove the iPSC culture plates from the incubator. Aspirate the supernatant, and wash the well with 1 mL of PBS.
3. Add 1 mL of 0.5 mM EDTA/well and incubate for 3 min at 37 °C. Then, aspirate the EDTA solution, and add 2 mL of differentiation medium.
4. Gently dissociate the cells by flushing using a 5 mL serological pipet, and transfer the small clumps to a 15 mL conical tube (*see* **Note 14**).
5. Sediment the clumps for 5–10 min and aspirate the supernatant.
6. Resuspend the small clumps in 5 mL of hematopoietic differentiation induction medium, and transfer onto the confluent OP9 cell layer at a density of 1×10^6 cells/T75 (*see* **Note 15**). The final volume must be around 15 mL/T75 flask.
7. Spread the clumps homogeneously onto the entire surface, and incubate at 37 °C in normoxic conditions and 5% CO_2 (*see* **Note 16**).
8. At days 4 and 7 of differentiation, observe the morphology of the cultures (Fig. 1b) under a light microscope to ensure that differentiation is progressing normally. Remove half of the medium (7–8 mL), and replace it with fresh differentiation medium.
9. Incubate the cultures until day 10. At this step of differentiation, areas with a cobblestone-like morphology can be observed (Fig. 1b).

3.2.3 Isolation of $CD34^+$ Cells (Day 10)

In comparison to previously published protocols, the timing of co-culture was increased. Indeed, in our hands, a peak in $CD34^+$ cell production was observed after 10 days of co-culture for all the iPS cell lines tested (WT and CGD) ([13] and unpublished data). As the hematopoietic differentiation efficiency can differ among different cell lines, we recommend performing a kinetic of hematopoietic differentiation for each cell line to choose the day of co-culture showing the maximum $CD34^+$ cell production.

1. Aspirate the supernatant of day 10 co-cultures and add 5 mL of 1 mg/mL collagenase IV pre-warmed at 37 °C to each T75 flask.
2. After 25 min of incubation at 37 °C, transfer the 5 mL of cell suspension with collagenase IV into a 50 mL conical tube.
3. Add 2.5 mL of 0.05% trypsin-EDTA to the T75 flask, and then incubate for 10 min at 37 °C.

4. Use a cell scraper to detach all the cells from the surface. Then, dissociate cell aggregates carefully by mechanical dissociation, and transfer into the same 50 mL conical tube (*see* **Note 17**).
5. Add 10 mL of cold MACS buffer to the flask (*see* **Note 3**), rinse the surface, transfer into the conical tube used previously, and then flush vigorously to continue the mechanical dissociation.
6. Pass the cell suspension through a 100-μm cell strainer and then through a 40-μm cell strainer in order to remove aggregates.
7. Centrifuge cells at 400 × *g* for 5 min at 4 °C and aspirate the supernatant.
8. Wash the cell pellet with 20 mL of cold MACS buffer, centrifuge again at 400 × *g* for 5 min at 4 °C, and aspirate the supernatant.
9. Perform the sorting protocol according to the manufacturer's instructions (Miltenyi Biotec). Add 300 μL of cold MACS buffer, and resuspend the cell pellet by flushing gently. Add 100 μL of FcR blocking reagent and mix. Add 100 μL of CD34 microbeads and mix. Incubate for 25 min at 4 °C while mixing every 10 min.
10. Wash cells twice with 20 mL of cold MACS buffer and resuspend in 500 μL.
11. Perform magnetic cell isolation according to the manufacturer's instructions.
12. After isolation resuspend the cells in the MACS buffer and count cells (*see* **Note 15**).
13. *Optional*: The viability and purity of the sorted $CD34^+$ cells can be checked by flow cytometry (*see* **Note 18**). This also enables the yield of $CD34^+$ production to be calculated (*see* **Note 19**).

3.2.4 Characterization of the Hematopoietic Potential of the $CD34^+$ Progenitors [Colony-Forming Cell (CFC) Assay]

1. Thaw a 3 mL aliquot of methylcellulose medium at room temperature.
2. Resuspend the appropriate number of $CD34^+$ cells to plate in 300 μL of IMDM-20% FBS, and transfer into a 5 mL tube containing 3 mL of methylcellulose. It is advisable to test two dilutions of cells (1000 and 10,000 cells) (*see* **Note 20**).
3. Vigorously mix the cells with the medium using a 3 mL syringe fitted with a 18G needle, and allow air bubbles to escape for at least 5–10 min.
4. Transfer 1.5 mL of methylcellulose containing cells to each 35-mm Petri dish using the same 3 mL syringe fitted with a 18G needle, and cover with the lid.
5. Homogeneously spread the medium by rotating the dishes, and place them into a 100-mm Petri dish. Place an uncovered

35-mm dish filled with PBS in the same 100-mm Petri dish to maintain humidity in the dish and to stop the methylcellulose from drying out.

6. Incubate for 14 days in the incubator at 37 °C, 5% CO_2.
7. Observe the dishes under a light microscope, and identify colonies according to their specific size and morphology (*see* **Note 21**).
8. Calculate the average number of colony-forming units (CFU) obtained between duplicates (or by considering the two different plating densities), and estimate the CFU frequency (%) = 100 × number of CFU/input number of $CD34^+$ cells (*see* **Note 20**).

3.3 Production of Mature Neutrophils

Here we describe the protocol of terminal differentiation of $CD34^+$ cells into mature neutrophils (days 10 through 25). The first step of the first 4 days is entirely performed in suspension to trigger the differentiation toward the myeloid lineage using a cocktail of hematopoietic cytokines [34–36]. Then, terminal maturation into neutrophils is obtained by co-culture with OP9 cells and addition of G-CSF only for 11 days [28, 33, 34, 36].

1. Prepare pHEMA-coated T25 flasks at least 3 h before the differentiation step.
2. Wash $CD34^+$ cells obtained at Subheading 3.2.3 "Isolation of $CD34^+$ cells" **step 12**, with 10 mL of IMDM, 20% FBS, and centrifuge at 400 × *g* for 5 min.
3. Aspirate supernatant and resuspend at a density of 3–6 × 10^5 $CD34^+$ cells/mL in the stage 1 neutrophil differentiation medium (*see* **Note 22**).
4. Transfer to pHEMA-coated flasks and culture cells for 4 days at 37 °C in normoxic conditions and under 5% CO_2 (*see* **Note 23**).
5. The same day, prepare a gelatinized 6-well plate with OP9 cells, and place it into the incubator.
6. After 4 days, collect the suspension cells in a 15 mL conical tube. Rinse the flask with IMDM, 20% FBS medium, and transfer to the same conical tube. Centrifuge at 400 × *g* for 5 min.
7. Aspirate the supernatant and resuspend the cell pellet with 3 mL of stage 2 neutrophil differentiation medium. Count and transfer 1 × 10^5 cells/well to the 6-well plate pre-plated with OP9 cells (*see* **Note 24**).
8. Incubate for 11 days at 37 °C under 5% CO_2. Every 3–4 days, gently pipet half of the supernatant containing floating cells (1.5 mL/ well), and transfer into a 15 mL conical tube.

Centrifuge at 400 *g* for 5 min and then remove the supernatant. Add 1.5 mL fresh stage 2 neutrophil differentiation medium, and replate the cells.

9. At the end of the differentiation stage, collect floating neutrophils (Fig. 1b), and gently wash the well with 3 mL of PBS to collect the remaining cells (*see* **Note 25**).
10. Count the cells. A quick assessment of the purity of the neutrophil differentiation can be performed by MGG staining based on the specific neutrophil morphology (Fig. 2a) (*see* **Note 19**).

3.4 Production of Mature Macrophages

As for neutrophil differentiation, the first step is performed in suspension but only with addition of GM-CSF for 8 days, as previously published [29]. However, we extended this suspension culture by three additional days in the presence of M-CSF to induce the differentiation of monocytes into macrophages. The terminal maturation of macrophages is generally obtained after a last step of adherent culture for 5–7 days with M-CSF alone [16, 28] or with other cytokines [29].

1. Prepare pHEMA-coated T25 flasks at least 3 h before the differentiation step.
2. Wash $CD34^+$ cells obtained at Subheading 3.2.3 "Isolation of $CD34^+$ cells" **step 12**, in α-MEM supplemented with 10% FBS, 100 μM MTG and centrifuge at 400 × *g* for 5 min.
3. Aspirate supernatant and resuspend at a density of $3–6 \times 10^5$ $CD34^+$ cells/mL in the stage 1 macrophage differentiation medium.
4. Transfer to pHEMA-coated flasks, and culture cells for 8 days at 37 °C in normoxic conditions and 5% CO_2 with half-medium change every 3 days (*see* **Note 23**).
5. At day 18, transfer the cell suspension into a 15 mL conical tube, and centrifuge at 400 × *g* for 5 min.
6. Aspirate supernatant, and resuspend the cells with the same volume of stage 2 macrophage differentiation medium, and transfer into the same flask.
7. After 3 days, gently dissociate aggregates by pipetting up and down and filter the cell suspension through a 40-μm cell strainer to eliminate remaining aggregates.
8. Count cells and transfer $4–5 \times 10^4$ cells/cm^2 to the appropriate culture plate depending on the further analysis (*see* **Note 26**).
9. Allow cells to adhere in the same medium for 7 days with half-medium change after 3–4 days (Fig. 1b).
10. A quick assessment of the purity of the macrophage differentiation can be performed after detaching the adherent macrophages and MGG staining (Fig. 3a) (*see* **Note 19**).

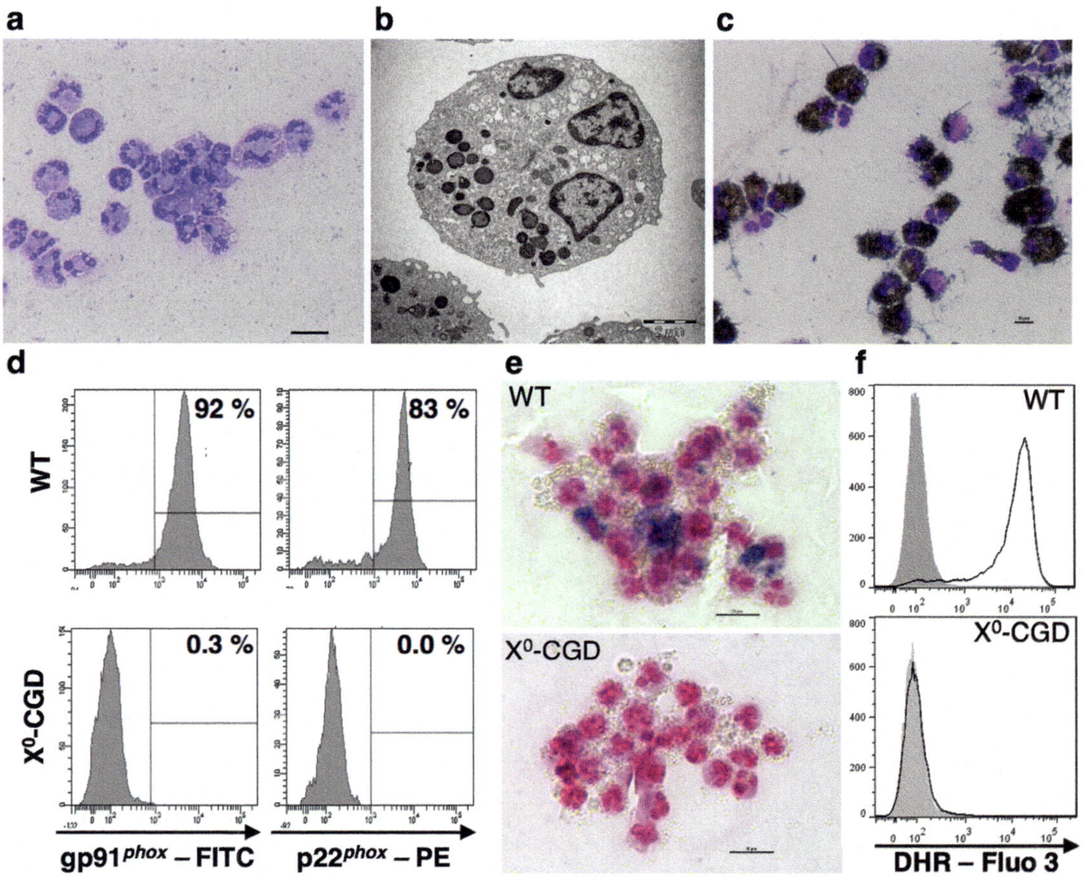

Fig. 2 Characterization of mature WT and X^0-CGD iPS-derived neutrophils. (**a**) Morphological aspect after MGG staining (scale bar 20 μm). (**b**) Presence of different types of cytoplasmic granules in electron transmission microscopy (scale bar 2 μm). (**c**) Presence of azurophil granules containing MPO (scale bar 10 μm). (**d**) Analysis of the expression of gp91^{phox} and p22^{phox} subunits by flow cytometry. (**e**) Analysis of the ability of iPS-derived neutrophils to produce ROS upon PMA stimulation by the NBT test (scale bar 10 μm) and (**f**) by DHR flow cytometry assay (filled gray curves represents unstimulated neutrophils, and black curves represent PMA-stimulated neutrophils)

3.5 Morphologic and Phenotypic Characterization of Mature Neutrophils and Macrophages

Morphologic and phenotypic characterization of generated phagocytic cells is essential to validate the efficiency of the established protocol of differentiation. Morphological analysis is generally performed by MGG staining showing the cell size, the presence of different populations of granules, and the form of the nucleus (Figs. 2a and 3a). However, phenotypic characterization of phagocytes by flow cytometry provides for the most precise quantification of the purity and maturation stage of the desired cells. We used CD16 and CD45, and CD11b, CD14, CD45 and CD68 markers to characterize neutrophils and macrophages, respectively (Fig. 3b). The expression of other surface markers such as CD13 or CD33 can also be analyzed for neutrophils [29, 36] as well as more specific markers to determine the subtype of macrophages

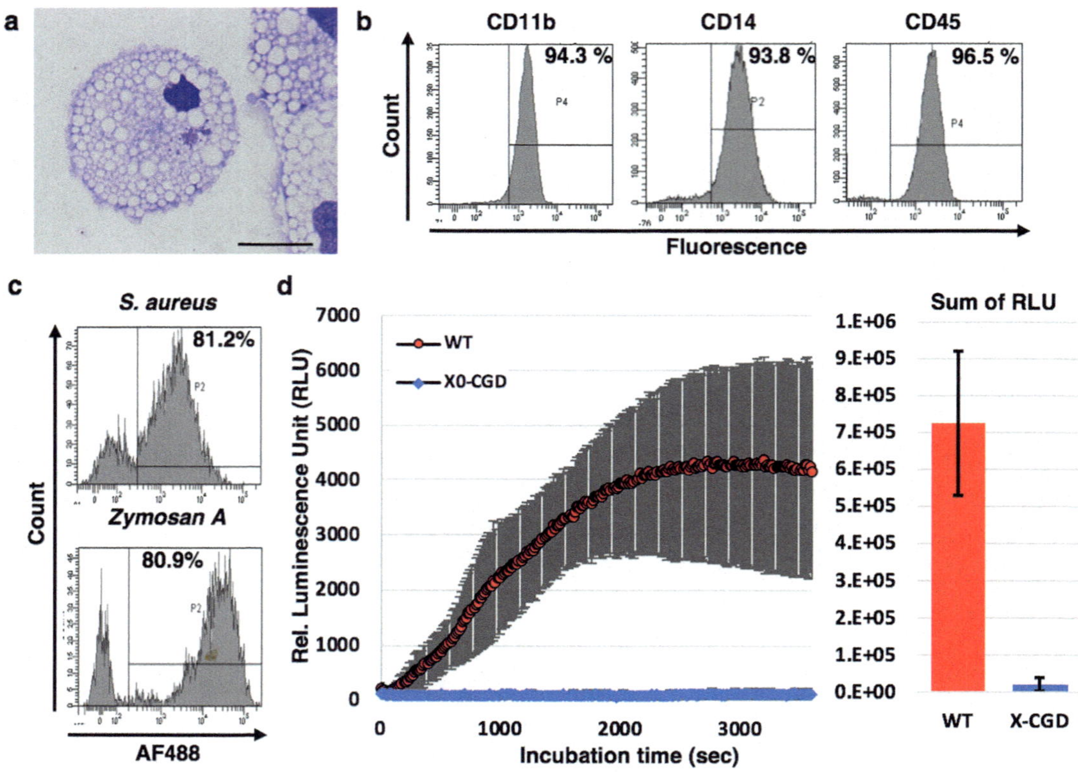

Fig. 3 Characterization of mature WT and X^0-CGD iPS-derived macrophages. (**a**) Morphological aspect after MGG staining (scale bar 20 μm). (**b**) Expression of specific markers analyzed by flow cytometry. (**c**) Ability of the iPS-derived macrophages to phagocytose AF488-labeled bacteria and yeast analyzed by flow cytometry. (**d**) ROS production by iPS-derived macrophages upon PMA stimulation measured by chemiluminescence ($n = 3$)

[13, 27]. Additional assays should be performed to complete this characterization. The complete maturation of neutrophils is established by the presence of the three types of granules (azurophilic, specific, and tertiary granules) as shown by electron microscopy (Fig. 2b), MPO staining (Fig. 2c), and an exocytosis experiment [13]. The ability of macrophages to be involved in an immune reaction can also be explored using cytokine release assays [12, 13, 28]. Finally, phagocytic ability is a fundamental function for both cells to be analyzed [13] (Fig. 3c). Readers can also refer to well-described protocols to explore the killing capacity of neutrophils or macrophages [37, 38].

3.5.1 May-Grünwald-Giemsa (MGG) Staining

1. Wash 10,000–20,000 collected cells with FACS buffer, and centrifuge at 400 × *g* for 5 min.
2. Resuspend in 100 μL of FACS buffer and transfer into the cytospin slide chamber.

3. Centrifuge at 400 × *g* for 8 min using a cytocentrifuge. Remove slides from the cytocentrifuge and air dry.
4. Put slides into a staining tank containing MGG stain solution.
5. Observe the cells using a light microscope equipped with a camera (*see* **Note 27**).

3.5.2 Phenotypic Analysis by Flow Cytometry

1. Wash at least 300,000 collected cells with FACS buffer, centrifuge at 500 × *g* for 5 min, and resuspend the pellet in 200 μL of FACS buffer.
2. Split into two tubes: The first for labeling with matched control isotypes and the second for labeling with specific antibodies.
3. Stain with the following conjugated antibodies according to manufacturer's instructions: CD16 and CD45 for neutrophil characterization and CD11b, CD14, CD45, and CD68 for macrophage characterization. Keep in the dark during incubation.
4. Wash the cells twice with 500 μL of FACS buffer and resuspend in FACS buffer.
5. Transfer cells to 5 mL polypropylene round-bottom FACS tubes.
6. Acquire at least 10,000 events using a flow cytometer, and analyze data with software.

3.5.3 Myeloperoxidase Activity in Neutrophils

1. Prepare a glass microscope slide of cytospin neutrophils as described for the MGG staining **steps 1–3** (*see* **Note 28**).
2. Cover the slide with benzidine/sodium nitroprusside solution for 3 min.
3. Carefully remove the solution and cover with a 1:1 mix of benzidine/sodium nitroprusside solution and a freshly prepared diluted H_2O_2 solution, and then incubate for 15 min (*see* **Note 6**).
4. Wash the slide with distilled water, air dry and stain with 10% Giemsa staining solution for 20 min.
5. Observe the slide. A blue-black precipitate in the cytoplasm (Fig. 2c) is evidence of neutrophils with myeloperoxidase-containing granules. Acquire images using a microscope equipped with a camera.

3.5.4 Exocytosis Experiment of Neutrophils

1. Resuspend a minimum of 5 × 10^5 collected neutrophils in 270 μL PBS$^+$, and split into three 5 mL tubes: tube #1 = control (no stimulation), tube #2 = f-Met-Leu-Phe (fMLF) stimulation, and tube #3 = cytochalasin B (CB) pre-treated cells plus fMLF stimulation (*see* **Note 29**).

2. Incubate neutrophils of tube #3 with 2 μL of 0.25 mg/mL cytochalasin B for 7 min at 37 °C in a shaker bath (300 rpm). Put tubes #1 and #2 in the shaker bath at the same time at 37 °C.
3. Add 2 μL of 5 μM fMLF to tubes #2 and #3, and incubate for 15 min at 37 °C in a shaker bath (300 rpm) to induce degranulation.
4. Stop the reaction by putting the tubes on ice, and transfer the cell suspensions into 1.5 mL microtubes. Keep them on ice during the transfer (*see* **Note 30**).
5. Centrifuge at 500 × *g* for 10 min at 4 °C.
6. Collect the supernatant (S) in separate microtubes S1, S2, and S3, and keep them on ice.
7. Resuspend the pellet (P) with 100 μL 0.2% Triton X-100, and vortex rapidly every 3 min for 20 min. Always keep on ice.
8. Centrifuge at 1500 × *g* for 10 min at 4 °C.
9. Collect the supernatant of each pellet (P) in separate microtubes P1, P2, and P3, and keep them on ice.
10. For each supernatant and pellet collected (S1–3 and P1–3), prepare an aliquot of 60 μL for the quantification of lactoferrin and an aliquot of 38.5 μL for the quantification of gelatinase (MMP9) assay (*see* **Note 31**). For the quantification of MMP9, add 1.5 μL of complete, EDTA-free protease inhibitor cocktail (25×).
11. Perform the quantification of lactoferrin and/or MMP9 immediately, or, if not possible, store these aliquots at −20 °C until analysis.

Lactoferrin Quantification

The release of lactoferrin by neutrophils was quantified by using an enzyme-linked immunosorbent assay (ELISA) kit according to the manufacturer's instructions.

1. For "**n**" different iPS-derived neutrophil cell lines, use "**n + 1**" racks of 8-precoated wells. Each well of the microplate is coated with a primary monoclonal antibody directed against human lactoferrin.
2. Perform the calibration curve of lactoferrin in column 1 as described in the manufacturer's protocol.
3. In column 2, distribute 60 μL of the supernatants obtained at **step 10** of Subheading 3.5.4 exocytosis experiment (S1 to 3 and P1 to 3) or use 30 μL for duplicates. Top each well up to 100 μL with diluting buffer.
4. Perform the quantification of lactoferrin according to the manufacturer's protocol.

5. Read the absorbance in each well at 450 nm using a microplate spectrophotometer.
6. Calculate the mean absorbance for each set of duplicate standards and samples, and subtract the mean optical density of the negative standard well. Use the standard curve to calculate the concentration of lactoferrin in each well of supernatant, taking into account the dilution factor.

MMP9 Quantification

MMP9 (or gelatinase B) release is quantified by gelatin zymography, an electrophoretic method for measuring proteolytic activity [30]. Indeed, the gelatin inside the gel is digested in the presence of MMP9, so that the degradation of gelatin is visible as white bands after Coomassie blue staining.

1. Prepare a 10% SDS-polyacrylamide separating gel with 0.5 mg/mL gelatin, and immediately let it polymerize completely into a cassette at room temperature.
2. Prepare the stacking gel solution, immediately pipet it on top of the polymerized separating gel, and insert a 12-well comb. Let the stacking gel polymerize at room temperature.
3. Prepare the MMP9 standard: dilute the MMP9 gelatinase standard in gelatinase buffer and 10×SB in order to load a quantity of 150 pg of MMP9 in 40 μL final volume.
4. Prepare the samples: add 4 μL of 10×SB to the 40 μL of supernatant obtained at **step 10** of Subheading 3.5.4
5. Load the wells of the stacking gel with 40 μL of MMP9 gelatinase standard in lane 1 and the exocytosis samples in the subsequent lanes (*see* **Note 32**).
6. Run the gel at 30 mA for approximatively 4 h until the bromophenol blue tracking dye reaches the bottom of the gel.
7. After migration, transfer the gel to the renaturation solution, and perform two washes of 15 min with shaking.
8. Incubate overnight at 37 °C in 1 mM APMA incubation solution.
9. The day after, stain the gel with a Coomassie staining solution for at least 1 h.
10. Pour off the Coomassie staining solution, and destain the gel in several baths of destaining solution. At this step, the gelatinolytic activity is revealed by the presence of clear bands over the blue background.
11. Transfer and leave the gel in a storage bath until analysis.
12. A semiquantitative analysis can be done using a densitometer evaluating the gelatinolytic area of the supernatants compared to that of the MMP9 standard band.

3.5.5 Assay of Cytokine Profile Release from Activated Macrophages

The detection of 12 inflammatory cytokines is performed using a Human Inflammatory Cytokines Multi-Analyte Profiler ELISArray™ kit according to the manufacturer's instructions. One 96-well microplate enables the analysis of six different samples compared to negative and positive controls (*see* **Note 33**). Indeed, each column of the microplate is coated with a primary monoclonal antibody directed against one of the twelve cytokines (IL-1A, IL-1B, IL-2, IL-4, IL-6, IL-8, IL-10, IL-12, IL-17A, IFNγ, TNFα, GM-CSF).

1. At **step 8** of Subheading 3.4 "Production of mature macrophages" (day 21), plate 2 wells of a 12-well plate with 2×10^5 cells/well, and culture for 7 days as indicated.
2. At day 28 of macrophage differentiation, aspirate the culture medium of the 2 wells, and add 1 mL of fresh stage 2 macrophage differentiation medium.
3. Pre-stimulate 1 well of macrophages by adding 10 μL of 0.1 mg/mL LPS (final concentration 1 ng/mL) directly to the well, and incubate for 24 h at 37 °C in the incubator. The other well is used as the unstimulated control.
4. After 24 h, collect the supernatant of each well in a 1.5 microtube, and centrifuge at 1000 × *g* for 10 min to remove cell debris. The supernatant can be used immediately or stored at −20 °C to be analyzed later.
5. For each cell line, distribute 50 μL of supernatant from the stimulated or unstimulated macrophages in each column, and then perform the experiment as described by the manufacturer.
6. Read the absorbance in each well at 450 nm using a microplate spectrophotometer.
7. Subtract the absorbance of the negative control from the absorbance measured for the positive control and each sample.

3.5.6 Phagocytosis Assay by Flow Cytometry

1. Wash at least 450,000 collected cells with FACS buffer, centrifuge at 500 × *g* for 5 min, and resuspend the pellet with 1450 μL of PBS^+/glucose buffer.
2. Transfer 482.5 μL of the cell suspension to three 5 mL tubes: tube #1 = control (no stimulation), tube #2 = *S. aureus* stimulation, and tube #3 = Zymosan A stimulation.
3. Add the appropriate volume of opsonized *S. aureus* or Zymosan A in order to obtain a multiplicity Of infection (MOI) of 5:1, and incubate for 30 min at 37 °C while shaking (300 rpm) in the dark.
4. Stop the reaction by transferring the suspension to a 5 mL polypropylene round-bottom FACS tubes on ice, and add the same volume of 2 mg/mL trypan blue.
5. Immediately (<1 min) acquire at least 10,000–30,000 events (*see* **Note 34**). Use cells not incubated with the particles to establish the threshold for positive staining (Fig. 3c).

3.6 Phenotypic and Functional Characterization of X-CGD iPS-Derived Neutrophils and Macrophages

Several assays are essential to fully characterize the NADPH oxidase-deficient phenotype of neutrophils and macrophages. These are the same as those used for the routine diagnosis of CGD patients [38–40] or in research [41] with minor modifications [13]. Expression of the NADPH oxidase subunits is quickly performed by flow cytometry (Fig. 2d). Western blotting could also be used, but it is time consuming and requires a higher number of cells than flow cytometry [39]. Among all the functional assays available, the NBT reduction slide test which measures superoxide generation by phagocytes is probably the oldest and most robust assay (Fig. 2e). Reactive oxygen species can also be measured by flow cytometry using the DHR probe (Fig. 2f). Although this last assay is probably the most sensitive and easy-to-do method, it needs the presence of a peroxidase [39]. We generally use both particulate (opsonized latex beads and Zymosan A) and soluble (PMA) stimulatory agents to trigger the respiratory burst of phagocytic cells after phagocytosis or not, respectively [13].

3.6.1 NADPH Oxidase Component Expression by Flow Cytometry

1. Use at least 200,000 collected cells in a 1.5-mL microtube for each staining (specific antibody or isotype-matched control). Centrifuge at 500 × *g* for 5 min, and aspirate the supernatant.
2. For staining the membrane $p22^{phox}$ subunit and the intracellular NADPH oxidase subunits $p47^{phox}$ and $p67^{phox}$, fix cells with 100 μL of a 2% PFA solution for 15 min, wash twice with PBS, and incubate for 20 min with permeabilization buffer.
3. As indicated in Table 1, incubate cells with 100 μL of the appropriate diluted primary antibody for 15 min, and then wash the cells twice with 500 μL of the appropriate buffer. Include a control staining with isotype-matched control in order to establish thresholds for positive staining. Use the permeabilization buffer to dilute primary and secondary antibodies directed against $p22^{phox}$ subunit and the cytoplasmic subunits and for the washing steps. In the case of $gp91^{phox}$ staining, use FACS buffer for antibody dilution and washing.
4. Wash twice with 500 μL of the appropriate buffer, and centrifuge at 500 × *g* for 5 min.
5. Incubate cells with 100 μL of diluted secondary antibody for 15 min, and then wash the cells twice with 500 μL of the appropriate buffer, and centrifuge at 500 × *g* for 5 min.
6. Resuspend in 300–500 μL of FACS buffer, and transfer to a 5 mL polypropylene round-bottom FACS tube.
7. Acquire at least 10,000 events using a flow cytometer, and analyze data with software (Fig. 2d).

3.6.2 NBT Assay: PMA and Opsonized Latex Beads Stimulation on Adherent Macrophages

1. Pre-stimulate 3 wells containing macrophages (cultured in 12-well plates with or without a glass slide as described at **step 8** of Subheading 3.4 "Production of mature macrophages") by adding 10 μL of 0.1 mg/mL LPS (final concentration 1 ng/mL) directly to 1 mL of fresh culture medium. Incubate for 3 h at 37 °C in the incubator (*see* **Note 35**).
2. Prepare the solutions as indicated below (volumes are indicated for 1 well). Negative control, 350 μL of human AB serum added to 50 μL of 3.9% glucose and 100 μL of PBS; for determination of superoxide, 350 μL of human AB serum added to 50 μL of 3.9% glucose and 100 μL 7 mM NBT.
3. Aspirate the culture medium and add the solutions to dedicated wells. Incubate for 2 min at 37 °C in a rotator at 250 rpm.
4. Add the stimulating agent to the NBT wells: 27 μL of 4 μg/mL PMA (200 ng/mL final) or 15 μL of opsonized latex beads; and incubate for 1 h at 37 °C while shaking (250 rpm in a rotary shaker).
5. Aspirate the supernatant and gently wash with 1 mL of PBS.
6. Directly observe plates under a light microscope and take images using a camera.
7. *Optional*: In order to conserve slides, plate macrophages in a 12-well plate previously covered with a glass slide, and, after performing the NBT test, gently remove the glass slide from the wells, and transfer onto microscope slides with mounting medium.

3.6.3 NBT Assay: PMA and Opsonized Latex Beads Stimulation of Neutrophils or Detached Macrophages

1. Transfer to a 5 mL round-bottom plastic tube at least 200,000 cells/tube of suspension neutrophils or harvested macrophages and then wash gently with 1 mL of PBS (*see* **Note 36**). Centrifuge at 400 × g for 5 min and aspirate the supernatant.
2. Add the reagent suspension as follows: control, 175 μL of human AB serum added to 25 μL of 3.9% glucose and 50 μL of PBS; stimulation tubes, 175 μL of human AB serum added to 25 μL of 3.9% glucose and 50 μL 7 mM NBT.
3. Incubate for 2 min at 37 °C in a water bath while shaking (160 rpm).
4. Add the stimulating agent to the NBT tubes: 13.5 μL of 4 μg/mL PMA (200 ng/mL final) or 7 μL of opsonized latex beads, and incubate for 10–20 min at 37 °C while shaking (250 rpm) for PMA stimulation and 20–30 min at 37 °C (130 rpm) for latex beads stimulation of neutrophils. Extend to 1 h for macrophages.
5. Centrifuge the tubes at 250 × g for 10 min at room temperature, and aspirate 2/3 of the supernatant.

6. Carefully resuspend the neutrophils with the leftover supernatant (around 80–100 μL), and prepare a smear on a glass microscope slide (15 μL). Then, let the slides air dry for around 1 h.
7. Fix slides with methanol under a chemical fume hood until the methanol has completely evaporated.
8. Stain slides in a 0.5% safranin O staining solution for 5 min to color cytoplasm pink to red, and then remove the staining solution.
9. Stain with 1:40 diluted Mayer's hemalum solution for 5 min to color nuclei blue, and then remove the solution.
10. Gently rinse the slide with distilled water and allow to air dry.
11. Optional: For long-term storage, put a cover glass on the slide, add mounting medium, and store at 4 °C.
12. Observe slides under a light microscope. Count the percentage of cells (neutrophils or macrophages) showing NBT reduction (blue formazan precipitates) upon PMA or opsonized latex beads stimulation (Fig. 2e).

3.6.4 DHR Assay: PMA and Opsonized Zymosan Stimulation

1. Collect at least 600,000 cells (neutrophils in suspension or detached adherent macrophages) in a 15 mL conical tube, and then gently wash with 5 mL of PBS (*see* **Note 36**). Centrifuge at 400 × *g* for 5 min and discard the supernatant.
2. Resuspend in PBS and separate into four 1.5 mL microtubes: two control tubes (without stimulation) and two stimulated tubes (PMA and opsonized Zymosan).
3. Centrifuge at 400 × *g* for 5 min and aspirate the supernatant.
4. Add 490 μL of PBS^+ containing 20 mM glucose, 3 mM NaN_3, and 0.5 μM DHR solution to each microtube. Resuspend the cells carefully and incubate for 5 min at 37 °C while shaking (300 rpm), and keep in the dark.
5. Then, add 5 μL of 20 μg/mL PMA or 5 μL of 16 mg/mL opsonized Zymosan to the stimulated tubes and mix gently.
6. Incubate in the dark at 37 °C while shaking (300 rpm) for 15 and 20 min after PMA or opsonized Zymosan stimulation and their controls, respectively.
7. Directly transfer into FACS tubes, and keep them on ice and in the dark until analysis by a flow cytometer (no longer than 30 min).
8. Acquire at least 10,000–30,000 events using a flow cytometer, and analyze data with software (Diva, FlowJo) (Fig. 2f).

3.6.5 Chemiluminescence Assay

1. Collect the suspension of neutrophils at the end of the differentiation step or detach adherent macrophages (*see* **Notes 35** and **37**), and centrifuge at 400 × *g* for 5 min. Discard the supernatant.
2. Resuspend the pellet in 1 mL of PBS (*see* **Note 36**) to wash the cells. Centrifuge at 400 × *g* for 5 min, and aspirate the supernatant.
3. Resuspend the pellet in 150 μL of PBS (volume required to perform 3 wells: cell blank, PMA, and opsonized Zymosan stimulation).
4. Prepare a 96-well plate: put 50 μL of PBS in the first well (medium blank) and then 50 μL of cells in the next 3 wells.
5. Add 200 μL of chemiluminescence medium to each well, and incubate for 5 min at 37 °C in the chemiluminometer in the dark.
6. Carefully vortex the 2 μg/mL PMA and 16 mg/mL opsonized Zymosan solutions for 1 min and 30 s, respectively, and then add 10 μL of PMA or Zymosan solutions to the dedicated wells.
7. Immediately start the kinetic (with shaking just before the first measurement), and acquire data every 1 min for 1 h at 37 °C.
8. Results of the NADPH oxidase activity can be expressed as the maximal value of relative unit of luminescence (RLU) corresponding to the peak of the kinetic curve. Otherwise, results can be expressed as the sum of RLU produced in a determined time by calculating the AUC (area under the curve) (Fig. 3d). This last result expression implies that the shape of the different kinetics is of the same type.

4 Notes

1. PD0325901 is a selective and non-ATP competitive MEK inhibitor; CHIR99021 is a selective small molecule GSK3 inhibitor that activates Wnt signaling; SB431542 is a potent inhibitor of TGF-β signaling, and thiazovivin is a RHO/R-OCK pathway inhibitor. These molecules are added to favor the reprograming of somatic cells [23].
2. Other murine stromal cell lines such as S17 and MS-5 demonstrated an ability to direct the hematopoietic differentiation of human ESCs toward the hematopoietic lineage [31]. We also compared the efficacy of the hematopoietic differentiation of human iPSCs in co-culture with OP9 and MS-5, and we observed a superiority of the OP9 cells in terms of production of CD34-positive cells and of hematopoietic potential of these progenitors (unpublished data).

3. Keeping the MACS buffer cold decreases the non-specific binding of CD34 beads to CD34 negative cells and thus leads to increased purity after sorting. All steps of the cell sorting should be performed at cold temperature whenever possible.
4. It is important to assess the purity when the $CD34^+$ progenitors are tested for their hematopoietic potential using in vitro tests like the colony-forming cell (CFC) assay or the long-term colony-initiating cell (LTC-IC) assays. Indeed, it enables the accurate calculation of the number of colonies (CFU) arising from a specific number of input $CD34^+$ cells. However, this step can be skipped when the $CD34^+$ progenitors are used for further differentiation.
5. Be sure to let the flasks dry for long enough so that all the ethanol has completely evaporated. Flasks can be prepared a few days before and stored at room temperature.
6. Prepare all the reagents and perform the experiment under a fume chemical hood. Because of the toxicity of benzidine and sodium nitroprusside, they must always be handled very carefully with gloves. Discard all the reagents in a dedicated waste bottle containing sodium hydrochloride solution.
7. During the preparation of the opsonized Zymosan A, close the tube with a parafilm to prevent splashing of boiling water into the tube. However, make a small hole (with a needle) in the parafilm to prevent pressure buildup in the tube. When the human serum is added to the Zymosan A suspension, be careful to avoid aggregates.
8. If the reprogramming is not performed directly after the establishment of the fibroblast cell line, freeze cells at early passages (2–3). We recommend performing at least one passaging of the fibroblasts before the transfection.
9. The ideal quantity for the transfection experiment is one million fibroblasts, so prepare at least one T25 flask, but the minimum is $4–5 \times 10^5$ cells for one transfection well. Excess cells can be frozen.
10. Plasmids: shRNAp53 is added to plasmid vectors containing the transcription factors Oct3/4, Sox2, Klf4, Lin28, and Myc to prevent apoptosis [22].
11. Do not break the colonies or they may not adhere to the culture plate. We recommend picking at least 12–24 colonies in order to obtain several fully reprogrammed iPS clones. Indeed, after picking, some clones differentiate immediately and loose the iPS specific morphology. If very few clones are obtained after a first round of picking, pick new colonies.
12. The use of OP9 cells after passage 60 is not recommended due to the risk of a decreased efficacy of the hematopoietic

differentiation [34]. Additionally, spontaneous adipogenesis can occur after overconfluence or extended times in culture. This can be observed by the presence of cells with adipocyte morphology, i.e., lipid droplets in the cytoplasm. Thus, it is extremely important to detect signs of adipocyte differentiation which could explain low efficiency of the progenitor's production.

13. Be careful that the cell layer is homogenous and confluent. If not, adjust the cell plating density of OP9 in order to obtain a confluent cell layer 4 days after plating.
14. This step is particularly critical. It is essential to take care not to produce too small clumps. Indeed, too small colonies will not adhere to the OP9 cell layer, and thus it decreases the yield of the hematopoietic progenitors produced.
15. You can use a separate well to perform the cell counting using trypan blue staining. For that, after incubation with EDTA and addition of medium, completely dissociate the colonies by vigorous pipetting in order to obtain a single cell suspension. Some fluorescent automated cell counters are now able to accurately count cells and estimate the viability even in the presence of aggregates by using a combination of acridine orange and propidium iodide, for example.
16. Be careful when transferring the plates to the incubator. If aggregates concentrate only in the center of the flask, the differentiation will not be efficient.
17. Sometimes, day 10 co-cultures are difficult to dissociate, so you may need to use a cell scraper to detach them from the flasks. If the mechanical dissociation is ineffective, you can also increase the incubation time with collagenase (to 40–50 min) and with trypsin/EDTA (to 20 min) [34]. However, with digestive enzymes, this extended incubation time can decrease the cell viability of the cell suspension and decrease the yield of sorted $CD34^+$ cells.
18. The purity can be determined after staining with an antibody directed against the cell marker CD34 and analysis by flow cytometry. In parallel, viability of the cell suspension can be assessed using fluorescent dyes by flow cytometry, for example, 7-actinoaminomycin (7AAD), DAPI, propidium iodide, or other commercial probes.
19. We previously determined that an average number of $1–2.5 \times 10^6$ $CD34^+$ progenitors can be obtained from 10×10^6 iPSCs, which requires preparing 10 T75 flasks. This quantity of progenitors enabled the production of around 3×10^5 neutrophils or 5×10^5 macrophages [13].

20. Measure the purity of the $CD34^+$ cell suspension after sorting. The number of $CD34^+$ cells to plate is dependent on their progenitor ability. We recommend, for the first experiment, testing two different plating densities in order to obtain dishes which are easy to read (enough colonies but well separated). Based on our experience, we usually plate between 100,000 and 250,000 cells harvested from the whole co-culture dish and between 10,000 and 50,000 sorted $CD34^+$ cells per a 35-mm dish. Because the total number of colony-forming units (CFU) increases over the time of co-culture, you may need to plate less number of cells at day 10 or 13 than at day 7. In that case, resuspend cells in 150 μL IMDM-20% FBS, and transfer into 1.5 mL methylcellulose.
21. To accurately identify a colony, pick the colony using a plastic pipette, wash the cells in FACS buffer, and perform a MGG staining as described.
22. Use one T25 pHEMA-coated flask for $1.5–5 \times 10^6$ cells. Do not resuspend cells in less than 5 mL. Thus, if the number of $CD34^+$ obtained after cell sorting is below 1.5×10^6 cells, culture them in one pHEMA-coated well of a 6-well plate.
23. Regularly observe the culture health. Cells must form small aggregates with single cells in suspension as well. In case of the presence of adherent cells, check your pHEMA solution and/or coating.
24. The number of cells obtained after the first differentiation step decreases so that one T25 generally enables the plating of $CD34^+$ into a maximum of a 6-well plate. Prepare more 6-well plates pre-plated with OP9 cells or use T25/T75 flasks if you produce a higher number of differentiated cells.
25. Vigorous washing of the well may detach OP9 cells or other differentiated cells like macrophages and will thus decrease the purity of the neutrophil suspension.
26. This plating density should be sufficient to cover the surface to 90% confluency. We recommend testing different plating densities to choose a density which enables a 90% coverage of the surface.
27. We routinely obtain purity of approximately 50–70% for neutrophils (contamination mainly with myeloid progenitor cells, monocytes/macrophages, and eosinophils) and >99% for macrophages [13].
28. Do not store the slides, and immediately perform the myeloperoxidase activity assay.
29. We recommend using at least 150,000 neutrophils in each tube in order to release a large enough amount of granule content to be detectable by the assays.

30. From this step on, it is important to keep the tubes and reagents on ice during the entire procedure to limit the degradation of enzymes by proteases.
31. Granule enzymes other than lactoferrin or gelatinase can also be analyzed. Be sure to store aliquots in specific conditions depending on the enzyme to analyze.
32. If you are loading a molecular weight marker in lane 1, leave lane 2 empty, because it may contain β-mercaptoethanol which diffuses to lane 2 and inhibits the enzymatic activity.
33. We recommend comparing a maximum of three different iPS-derived macrophage cell lines, each one with a stimulated and unstimulated well. 1 mL of LPS-induced macrophage supernatant is obtained, so the experiment can easily be performed in duplicate or triplicate.
34. You also can acquire cells before the addition of trypan blue. The fluorescence measured will be due to the fluorescent particles internalized but also to the non-specific binding of fluorescent particles to the surface of phagocytic cells. The addition of blue trypan enables the extracellular fluorescence to be suppressed and reflects the phagocytic ability of the neutrophils and macrophages [42].
35. We pre-activated macrophages with 200 ng/mL IFNγ for 24 h [43], which worked well (unpublished data).
36. It is important to carefully manipulate phagocytic cells in order to avoid activating them before performing the functional assays.
37. We recommend using at least 50,000–100,000 cells for each well and performing a duplicate or triplicate depending on the number of available neutrophils or macrophages. Alternatively, monocytes (day 21 of macrophage differentiation) can also be plated in 96-well plates, then day 28 macrophages are pre-stimulated with LPS or IFNγ as described, and the medium is added directly to the 96-well plate for the chemiluminescence measurement.

Acknowledgments

MJS is grateful for the support from the University Grenoble Alpes (AGIR program 2014), the Faculty of Medicine and the Pole Recherche, University Hospital Grenoble Alpes, and Interreg France-Suisse (Programme de Cooperation Territoriale Europeenne, Fond Europeen de Developpement Regional (FEDER), 2017–2019). This work was also supported by the Delegation for Clinical Research and Innovations (DRCI, Rementips project 2014). We also thank Sylvain Beaumel and Michèle Mollin for their helpful and valuable work at the Centre Diagnostic et Recherche sur la CGD (CDiReC), Grenoble France. This article is dedicated to the memory of Cécile Martel, an outstanding technician at the CDiReC, who passed away recently. We miss you.

References

1. Winkelstein JA, Marino MC, Johnston RB Jr et al (2000) Chronic granulomatous disease. Report on a national registry of 368 patients. Medicine 79:155–169
2. Van den Berg JM, van Koppen E, Ahlin A et al (2009) Chronic granulomatous disease: the European experience. PLoS One 4:e5234
3. Roos D, Kuhns DB, Maddalena A et al (2010) Hematologically important mutations: X-linked chronic granulomatous disease (third update). Blood Cells Mol Dis 45:246–265
4. Roos D, Kuhns DB, Maddalena A et al (2010) Hematologically important mutations: the autosomal recessive forms of chronic granulomatous disease (second update). Blood Cells Mol Dis 44:291–299
5. Van de Geer A, Nieto-Patlán A, Kuhns DB, et al (2018) Inherited p40^{phox} deficiency differs from classic chronic granulomatous disease. J Clin Invest 128:3957-3975
6. Gungor T, Teira P, Slatter M et al (2014) Reduced-intensity conditioning and HLA-matched haemopoietic stem-cell transplantation in patients with chronic granulomatous disease: a prospective multicentre study. Lancet 383:436–448
7. Kang EM, Malech HL (2012) Gene therapy for chronic granulomatous disease. Methods Enzymol 507:125–154
8. Zhen L, King AA, Xiao Y et al (1993) Gene targeting of X chromosome-linked chronic granulomatous disease locus in a human myeloid leukemia cell line and rescue by expression of recombinant gp91phox. Proc Natl Acad Sci U S A 90:9832–9983
9. Takahashi K, Tanabe K, Ohnuki M et al (2007) Induction of pluripotent stem cells from adult human fibroblasts by defined factors. Cell 131:861–872
10. Scudellari M (2016) A decade of iPSCs. Nature 534:310–312
11. Mukherjee S, Santilli G, Blundell MP et al (2011) Generation of functional neutrophils from a mouse model of X-linked chronic granulomatous disorder using induced pluripotent stem cells. PLoS One 6:e17565
12. Jiang Y, Cowley SA, Siler U et al (2012) Derivation and functional analysis of patient-specific induced pluripotent stem cells as an in vitro model of chronic granulomatous disease. Stem Cells 30:599–611
13. Brault J, Goutagny E, Telugu N et al (2014) Optimized generation of functional neutrophils and macrophages from patient-specific induced pluripotent stem cells: ex vivo models of X(0)-linked, AR22(0)- and AR47(0)-chronic granulomatous diseases. Biores Open Access 3:311–326
14. Zou J, Sweeney CL, Chou BK et al (2011) Oxidase-deficient neutrophils from X-linked chronic granulomatous disease iPSCs: functional correction by zinc finger nuclease-mediated safe harbor targeting. Blood 117:5561–5572
15. Merling RK, Sweeney CL, Choi U et al (2013) Transgene-free iPSCs generated from small volume peripheral blood non-mobilized CD34+ cells. Blood 121:98–107
16. Flynn R, Grundmann A, Renz P et al (2015) CRISPR-mediated genotypic and phenotypic correction of a chronic granulomatous disease mutation in human iPSCs. Exp Hematol 43:838–848
17. Dreyer AK, Hoffmann D, Lachmann N et al (2015) TALEN-mediated functional correction of X-linked chronic granulomatous disease in patient-derived induced pluripotent stem cells. Biomaterials 69:191–200
18. Merling RK, Sweeney CL, Chu J et al (2015) An AAVS1-targeted minigene platform for correction of iPSCs from all five types of chronic granulomatous disease. Mol Ther 23:147–157
19. Laugsch M, Rostovskaya M, Velychko S et al (2016) Functional restoration of gp91phox-oxidase activity by BAC transgenesis and gene targeting in X-linked chronic granulomatous disease iPSCs. Mol Ther 24:812–822
20. Brault J, Vaganay G, Le Roy A, Lenormand JL, Cortes S, Stasia MJ (2017) Therapeutic effects of proteoliposomes on X-linked chronic granulomatous disease: proof of concept using macrophages differentiated from patient-specific induced pluripotent stem cells. Int J Nanomedicine 12:2161–2177

21. Zhou YY, Zeng F (2013) Integration-free methods for generating induced pluripotent stem cells. Genomics Proteomics Bioinformatics 11:284–287
22. Okita K, Matsumura Y, Sato Y et al (2011) A more efficient method to generate integration-free human iPS cells. Nat Methods 8:409–412
23. Yu J, Chau KF, Vodyanik MA, Jiang J, Jiang Y (2011) Efficient feeder-free episomal reprogramming with small molecules. PLoS One 6: e17557
24. Villa-Diaz LG, Ross AM, Lahann J et al (2013) Concise review: the evolution of human pluripotent stem cell culture: from feeder cells to synthetic coatings. Stem Cells 31:1–7
25. Fan Y, Wu J, Ashok P et al (2015) Production of human pluripotent stem cell therapeutics under defined xeno-free conditions: progress and challenges. Stem Cell Rev 11:96–109
26. Salvagiotto G, Burton S, Daigh CA et al (2011) A defined, feeder-free, serum-free system to generate in vitro hematopoietic progenitors and differentiated blood cells from hESCs and hiPSCs. PLoS One 6:e17829
27. Senju S, Haruta M, Matsumura K et al (2011) Generation of dendritic cells and macrophages from human induced pluripotent stem cells aiming at cell therapy. Gene Ther 18:874–883
28. Yanagimachi MD, Niwa A, Tanaka T et al (2013) Robust and highly-efficient differentiation of functional monocytic cells from human pluripotent stem cells under serum- and feeder cell-free conditions. PLoS One 8: e59243
29. Choi KD, Vodyanik MA, Slukvin II (2009) Generation of mature human myelomonocytic cells through expansion and differentiation of pluripotent stem cell-derived lin-CD34+CD43+CD45+ progenitors. J Clin Invest 119:2818–2829
30. Morishima T, Watanabe K, Niwa A et al (2011) Neutrophil differentiation from human-induced pluripotent stem cells. J Cell Physiol 226:1283–1291
31. Trocmé C, Gaudin P, Berthier S et al (1998) Human B lymphocytes synthesize the 92-kDa gelatinase, matrix metalloproteinase-9. J Biol Chem 273:20677–20684
32. Vodyanik MA, Bork JA, Thomson JA (2005) Human embryonic stem cell-derived $CD34^+$ cells: efficient production in the coculture with OP9 stromal cells and analysis of lymphohematopoietic potential. Blood 105:617–626
33. Vodyanik MA, Slukvin II (2007) Hematoendothelial differentiation of human embryonic stem cells. Curr Protoc Cell Biol Chapter 23: Unit 23:6
34. Choi K-D, Vodyanik M, Slukvin II (2011) Hematopoietic differentiation and production of mature myeloid cells from human pluripotent stem cells. Nat Protoc 6:296–313
35. Yokoyama Y, Suzuki T, Sakata-Yanagimoto M et al (2009) Derivation of functional mature neutrophils from human embryonic stem cells. Blood 113:6584–6592
36. Niwa A, Heike T, Umeda K et al (2011) A novel serum-free monolayer culture for orderly hematopoietic differentiation of human pluripotent cells via mesodermal progenitors. PLoS One 6:e22261
37. Sweeney CL, Merling RK, Choi U et al (2014) Generation of functionally mature neutrophils from induced pluripotent stem cells. Methods Mol Biol 1124:189–206
38. Kuhns DB, Long Priel DA, Chu J et al (2015) Isolation and functional analysis of human neutrophils. Curr Protoc Immunol 111:7.23.1–7.2316
39. Stasia MJ, Li XJ (2008) Genetics and immunopathology of chronic granulomatous disease. Semin Immunopathol 30:209–235
40. Roos D, de Boer M (2014) Molecular diagnosis of chronic granulomatous disease. Clin Exp Immunol 175:139–149
41. Beaumel S, Picciocchi A, Debeurme F et al (2017) Down-regulation of NOX2 activity in phagocytes mediated by ATM-kinase dependent phosphorylation. Free Radic Biol Med 113:1–15
42. Nuutila J, Lilius E-M (2005) Flow cytometric quantitative determination of ingestion by phagocytes needs the distinguishing of overlapping population of binding and ingesting cells. Cytometry A 65A:93–102
43. Greenlee-Wacker MC, Nauseef WM (2017) IFN-γ targets macrophage-mediated immune responses toward *Staphylococcus aureus*. J Leukoc Biol 101:751–758

Chapter 36

Gene Editing in Chronic Granulomatous Disease

Colin L. Sweeney, Randall K. Merling, Suk See De Ravin, Uimook Choi, and Harry L. Malech

Abstract

Chronic granulomatous disease (CGD) is an immune deficiency characterized by defects in the production of microbicidal reactive oxygen species (ROS) by the phagocytic oxidase (phox) enzyme complex in neutrophils. We have previously described targeted gene editing strategies using zinc finger nucleases (ZFNs), transcription activator-like effector nucleases (TALENs), or clustered regularly interspaced short palindromic repeats (CRISPR)/Cas9 nucleases for gene targeting with homology-directed repair in CGD patient stem cells to achieve functional restoration of expression of phox genes and NADPH oxidase activity in differentiated neutrophils. In this chapter, we describe detailed protocols for targeted gene editing in human-induced pluripotent stem cells and hematopoietic stem cells and for subsequent differentiation of these stem cells into mature neutrophils, as well as assays to characterize neutrophil identity and function including flow cytometry analysis of neutrophil surface markers, intracellular staining for phox proteins, and analysis of ROS generation.

Key words Chronic granulomatous disease, Gene editing, CRISPR/Cas9, Neutrophil differentiation

1 Introduction

Chronic granulomatous disease (CGD) is an immune deficiency caused by defects in the production of reactive oxygen species (ROS) by neutrophils and other phagocytes for microbial killing, resulting in recurrent, life-threatening microbial infections and granulomatous inflammation in CGD patients. There are five forms of CGD, depending on which subunit of the phagocytic oxidase (phox) enzymatic complex is affected: X-linked CGD (involving the *CYBB* gene on the X-chromosome encoding gp91^{phox}, also known as NOX2) and the four autosomal recessive forms resulting from mutations in either *CYBA* (encoding p22^{phox}), *NCF1* (encoding p47^{phox}), *NCF2* (encoding p67^{phox}), or *NCF4* (encoding p40^{phox}). Stem cell or gene therapies for CGD include allogeneic transplantation of donor hematopoietic stem cells (HSCs), which is complicated by graft-versus-host disease

Ulla G. Knaus and Thomas L. Leto (eds.), *NADPH Oxidases: Methods and Protocols*, Methods in Molecular Biology, vol. 1982, https://doi.org/10.1007/978-1-4939-9424-3_36, © Springer Science+Business Media, LLC, part of Springer Nature 2019

and limited availability of suitable matched donors or autologous transplant of patient HSCs after phox gene transfer ex vivo using lentiviral or retroviral vectors [1], which carries a risk of insertional mutagenesis from undirected viral integration into the genome.

Targeted editing of a specific genomic sequence can be mediated using zinc finger nucleases (ZFNs) [2, 3], transcription activator-like effector nucleases (TALENs) [4], or clustered regularly interspaced short palindromic repeats (CRISPR)/Cas9 nucleases [5–7]. For CRISPR/Cas9, nuclease activity is provided by an assembled ribonucleoprotein (RNP) containing a guide RNA corresponding to a target genomic sequence (typically 20 nucleotides in length) adjacent to a suitable protospacer-adjacent motif (PAM) site; while the mechanism of target site recognition and binding differs for ZFNs and TALENs, in each case nuclease-mediated cutting results in a double-strand DNA break that recruits cellular DNA repair machinery to the target site. This can result in the induction of DNA *in*sertions or *del*etions (indels) at that site by inaccurate repair from error-prone microhomology-mediated end-joining or nonhomologous end-joining (NHEJ) repair pathways, potentially resulting in gene knockout. However, when nuclease-targeted cells are supplied with an abundance of an exogenous DNA donor containing homology to DNA sequences flanking the nuclease cut site, homology-directed repair (HDR) can occur instead, resulting in insertion of the DNA donor sequence at the target site.

We have previously reported on the use of ZFNs, TALENs, or CRISPR/Cas9 to develop targeted gene editing approaches for treatment of CGD, initially using patient-derived induced pluripotent stem cells (iPSCs) to model the effectiveness of the gene editing strategies [8–11], and then in patient $CD34^+$ HSCs as the more relevant cell type for eventual clinical gene therapy [11–13]. These gene editing strategies included targeted insertion of minigenes for each of the phox proteins into the safe harbor AAVS1 locus under the control of a constitutive promoter or targeted gene insertion or repair at the endogenous *CYBB* or *NCF1* gene locus to restore NADPH oxidase activity while maintaining normal physiological regulation of $gp91^{phox}$ or $p47^{phox}$ expression, respectively. Donor DNA constructs commonly used for HDR in such strategies include short (100–200 bp) single-stranded DNA oligonucleotides (ssODNs) for repair of individual mutations or larger DNA constructs for replacement of whole exons or insertion of larger cDNAs or minigenes (Fig. 1). These larger donor constructs may be introduced into cells either as plasmid DNA (which is cytotoxic in cell transfections, and consequently its effective use is limited to iPSCs or other highly expandable cell lines) or as adeno-associated virus (AAV) vectors (a less cytotoxic alternative suitable for use in either iPSCs or HSCs, but with an upper limit on vector packaging sizes which complicates insertion of donor constructs larger than ~4.4-kb). It is important for efficient targeted gene insertion that the donor construct does not contain the nuclease target sequence, in order to prevent cutting of the donor DNA or the inserted sequence after HDR; to this end, mutations can be introduced in the donor construct (including silent mutations in exonic sequences) at the nuclease target sequence or the PAM site for CRISPR/Cas9.

Larger donor constructs may also be designed to contain a drug-resistance gene expression cassette that enables subsequent selection for targeted DNA insertion; if desired, the drug-resistance cassette may be flanked by excisable elements for removal after targeted insertion (e.g., using loxP sites for Cre recombinase-mediated excision or piggyBac transposon elements for excision by piggyBac transposase). For drug selection in iPSCs, we typically utilize a constitutively expressed puromycin-resistance (Puro) gene cassette to select and screen for iPSC clones containing targeted insertion of the donor construct; this strategy may also be combined with negative selection by fusion of the puromycin-resistance gene with a herpes simplex virus thymidine kinase (HSV-tk) gene [14], which allows for positive selection with puromycin for donor insertion, followed by excision of the Puro/HSV-tk cassette and negative selection with ganciclovir against HSV-tk expression to eliminate cells in which excision was unsuccessful [10]. However, these drug selection strategies are not suitable for selection of HSCs in vitro, due to the limited capacity of these stem cells for self-renewal in vitro without the loss of their potential for multi-lineage differentiation and in vivo engraftment potential; consequently, we exclude the drug selection cassettes from the donor constructs used in HSCs (Fig. 1).

In this chapter, we provide protocols for the design of CRISPR/Cas9 guide RNAs and cloning of CRISPR/Cas9 expression constructs. Additionally, we include protocols for assessing nuclease activity and performing gene editing that may be used with ZFNs, TALENs, or CRISPR/Cas9 nucleases. The strategies and protocols presented are applicable to gene editing in iPSCs and HSCs and include examples from our previous publications on gene correction in CGD patient cells [8–13]. We also present detailed protocols for the production of AAV vector donor constructs, differentiation of iPSCs and HSCs into mature neutrophils, and neutrophil assays for expression of phox proteins and for assessment of ROS generation, which may be used to confirm successful gene editing for CGD, with examples from corrected neutrophils obtained using our previously described approach for seamless correction of *CYBB* exon 5 mutations in iPSCs [10].

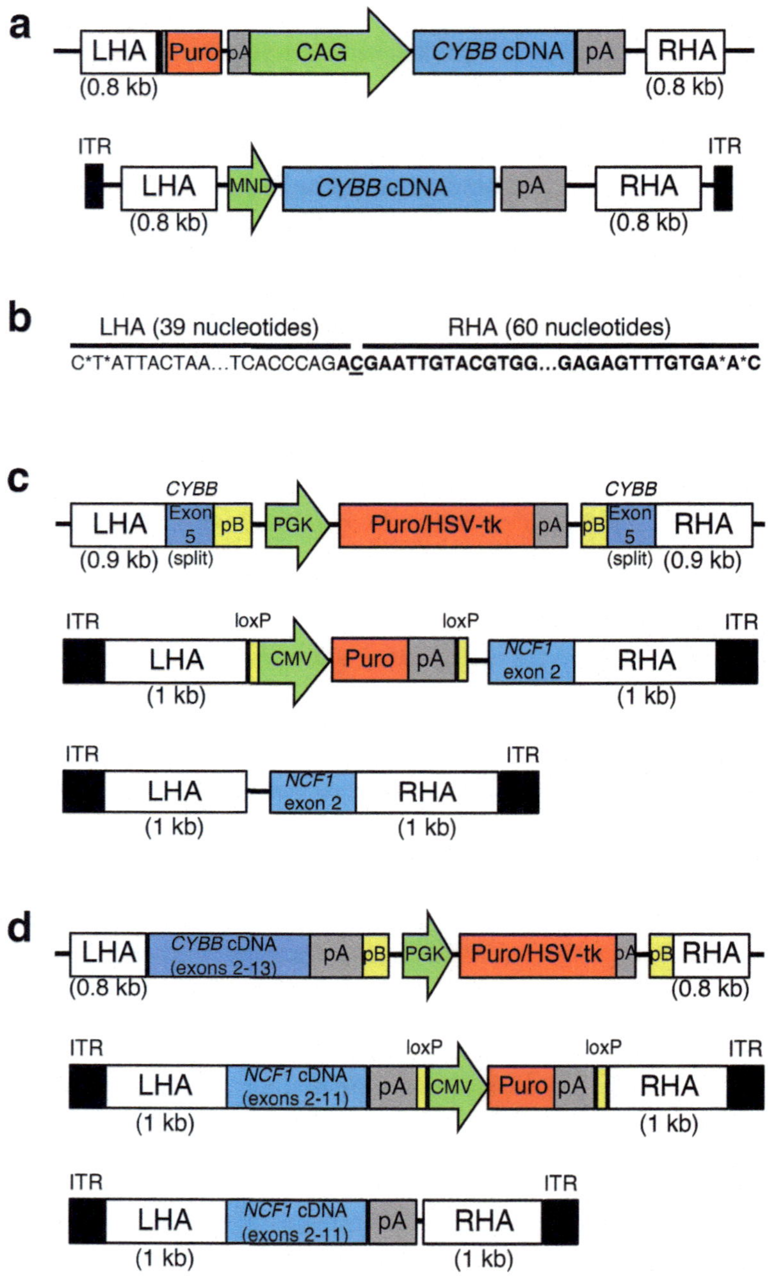

Fig. 1 Donor constructs for targeted insertion by HDR for CGD correction. Locations of polyadenylation signals (pA) from either rabbit b-globin or bovine growth hormone are indicated; arrows denote promoters. Also shown are the left and right homology arms (LHA and RHA) of indicated sizes, which contain homologous sequences to regions flanking the nuclease cut site for HDR. (**a**) Donors for minigene insertion at the safe harbor AAVS1 locus (top, plasmid donor for use in iPSCs, with puromycin selection gene expressed from the endogenous promoter at the AAVS1 locus; bottom, AAV donor for use in HSCs, lacking a drug selection cassette). A synthetic CAG or MND promoter is included for high levels

2 Materials

2.1 General Lab, Cell Culture, and CRISPR/Cas9 Equipment and Reagents

1. Sterile DNase-free/RNase-free 1.5-mL microcentrifuge tubes.
2. Vortexer.
3. P20, P200, and P1000 Pipetman with compatible pipet tips.
4. Pipet-Aid (or equivalent) with 5-mL, 10-mL, and 25-mL pipets.
5. NanoDrop 2000 or equivalent spectrophotometer.
6. Laboratory balance.
7. Deionized water.
8. Molecular biology grade water (double-distilled, sterilized, nuclease-free).
9. TE buffer (pH 8.0): 10 mM Tris–Cl (pH 8.0), 1 mM EDTA, prepared in distilled water and filter-sterilized with 0.22 μM filter.
10. Gel electrophoresis apparatus suitable for agarose gels, with power supply.

Fig. 1 (continued) of constitutive expression of the *CYBB* cDNA. (**b**) 100-nucleotide ssODN donor for targeted editing of a specific C > T mutation in *CYBB* exon 7. Phosphorothioate modifications to the ssODN to inhibit DNA degradation by exonucleases are denoted by "*"; Exon 7 sequences are in bold, while intron 6 sequences are in plain text. The location of the specific C nucleotide that is mutated in the targeted X-CGD patients is underlined; this nucleotide is flanked on the left side by 39 nucleotides of homology and on the right side by 60 nucleotides of homology to the target region. Although this single nucleotide change is the only difference between the ssODN donor and the patient mutation, the CRISPR used in this prior study was highly specific for the patient sequence compared to the corrected sequence, so that the CRISPR did not appreciably target the donor construct before or after HDR. (**c**) Donors for exon replacement of either *CYBB* exon 5 (top: plasmid donor for use in iPSCs, with exon 5 sequences interrupted by piggyBac transposon "pB" elements flanking a puromycin/HSV-tk dual-selection cassette for seamless excision by piggyBac transposase) or *NCF1* exon 2 (middle, AAV donor with Cre-excisable puromycin selection cassette for use in iPSCs; bottom, AAV donor without drug selection cassette, for use in HSCs). Constitutive expression of Puro or Puro/HSV-tk drug selection cassettes is driven by a phosphoglycerate kinase (PGK) or cytomegalovirus (CMV) promoter. (**d**) Donors for insertion of large multi-exon cDNAs for targeted gene editing of either *CYBB* (top, exon 2–13 donor plasmid with piggyBac transposon-excisable puromycin/HSV-tk selection cassette for use in iPSCs) or *NCF1* (middle, AAV with Cre-excisable puromycin selection cassette for use in iPSCs; bottom, AAV for use in HSCs). Constitutive expression of Puro or Puro/HSV-tk drug selection cassettes is driven by a PGK or CMV promoter

11. Agarose gel electroporation buffer: 1× Tris-acetate-EDTA (TAE) or 0.5× Tris-borate-EDTA (TBE) buffer.
12. 1% to 3% agarose gels prepared with ethidium bromide (or an alternative less-toxic/less-mutagenic DNA stain): use commercial pre-cast gels, or prepare gels using 1–3 g of agarose powder in 100 mL of TAE or TBE buffer (dissolved by heating in a microwave for ~2 min, and then cooled for several minutes before adding 5 μL of 10 mg/mL ethidium bromide, followed by mixing and pouring into a horizontal gel casting tray containing a gel comb).
13. Ultraviolet (UV) transilluminator for visualizing ethidium bromide-stained DNA, with 365-nm wavelength UV for gel band excision (or an appropriate light source for visualizing alternatives to ethidium bromide).
14. PCR thermocycler.
15. 0.2 mL PCR tubes (or appropriate size tubes for PCR thermocycler).
16. NucleoSpin Gel and PCR Clean-Up Kit (Macherey-Nagel) or equivalent kits for PCR production purification and DNA gel extraction.
17. Heat blocks set to 37 and 98 °C.
18. 100% ethanol.
19. 1× phosphate-buffered saline (PBS; pH 7.4) without calcium or magnesium.
20. Water bath set to 37 °C.
21. Tissue culture incubator set to 37 °C and 5% CO_2.
22. Tabletop centrifuge capable of centrifugation of 15-mL or 50-mL conical centrifuge tubes at 300–700 × *g* (~1200–1800 rpm) and 250-mL conical centrifuge tubes at 500 × *g* (~1500 rpm); equivalent rpm information is provided for a Sorvall Legend XTR centrifuge equipped with a TX-750 swinging-bucket rotor (with a 195-mm maximum radius).
23. 0.05% trypsin-EDTA (for passaging of HEK 293T cells).
24. Polyethyleneimine (PEI), branched, M.W. 25,000 (for HEK 293T cell transfection): 1 mg/mL stock solution (*see* **Note 1** regarding preparation of PEI stock solution).
25. Optional: Opti-MEM reduced serum medium (for HEK 293T cell transfection).
26. Puromycin dihydrochloride (for drug selection of puromycin-resistant cells): prepare 1 mg/mL stock solution in molecular grade water; filter-sterilize using a 0.22 μm filter, and store aliquots at −20 or −80 °C.

27. Genomic DNA purification kit: QIAGEN DNeasy Blood and Tissue kit (or equivalent).
28. Flow cytometer.

2.1.1 CRISPR/Cas9 Cloning and Synthesis of Single-Guide RNA (sgRNA) and Cas9 mRNA

1. Expression plasmid for sgRNA and Cas9 in mammalian cells (LentiCRISPRv2, Addgene plasmid # 52961) [15] or equivalent.
2. FastDigest BsmBI restriction endonuclease.
3. Single-stranded forward and reverse strand DNA oligos for cloning sgRNAs into LentiCRISPRv2 plasmid (*see* Subheading 3.1.2); the examples listed here are for cloning an sgRNA targeting the sequence CACTCTCTGAACTTGGAGAC<u>AGG</u> (in which the PAM sequence is underlined) in exon 5 of the *CYBB* gene:

   CYBB5 Forward oligo: 5′-caccgCACTCTCTGAACTTGGAGAC-3′

   CYBB5 Reverse oligo: 5′-aaacGTCTCCAAGTTCAGAGAGTGc-3′

   When the forward and reverse strand oligos for the target site are annealed, the nucleotides shown in uppercase letters (corresponding to the guide sequence) hybridize to each other to form double-stranded DNA, while the nucleotides in lower-case letters provide sticky ends for cloning into BsmBI-digested LentiCRISPRv2. It is important to note that the oligos do not contain the PAM sequence of the target site, since this sequence should not be present in the expressed sgRNA.
4. FastAP alkaline phosphatase.
5. T4 DNA ligase with T4 DNA ligase buffer or DNA Quick Ligation kit.
6. T4 polynucleotide kinase (PNK).
7. Competent *E. coli* bacteria for transformation with plasmid DNA (for cloning of LentiCRISPRv2 plasmid, use Stbl3 recombination-deficient bacteria to maintain the plasmid long-terminal repeat sequences without recombination).
8. Luria-Bertani (LB) medium.
9. Ampicillin (100 mg/mL for 1000× stock).
10. LB agar plates prepared with 100 μg/mL ampicillin.
11. 37 °C bacterial shaker incubator.
12. Plasmid DNA miniprep and maxiprep kits; an endotoxin-free maxiprep kit is preferred for preparing plasmid DNA to be used in iPSC transfections.

13. Optional: polymyxin B-agarose suspension, to remove endotoxins from plasmid DNA prepared using a standard maxiprep kit.
14. DNA polymerase for PCR: ThermoFisher DreamTaq Green PCR 2× Master Mix (or equivalent).
15. pLenti-2277-FWD primer: 5′-ATTATCGTTTCAGACCCAC-3′ (for screening LentiCRISPRv2 plasmid to confirm sgRNA cloning).
16. Optional: hU6-FWD primer – 5′-GAGGGCCTATTTCCCATGATT-3′ (for sequencing LentiCRISPRv2 plasmid to confirm sgRNA cloning).
17. T7-sgRNA PCR primers for appending the T7 promoter sequence to the beginning of the sgRNA for in vitro transcription, as an alternative to using commercially synthesized sgRNA (T7-sgRNA forward PCR primer – 5′-TTAATACGACTCACTATAGGN$_{20}$–3′, where the underlined sequence is the T7 promoter and N$_{20}$ is the 20-nucleotide target sequence of the sgRNA; T7-sgRNA reverse PCR primer – 5-′-AAAAGCACCGACTCGGTGCC-3′).
18. RNase AWAY or other RNase-decontaminating reagent.
19. Invitrogen MEGAshortscript T7 transcription kit, for in vitro transcription to produce sgRNA, as an alternative to using commercially synthesized sgRNA.
20. Cas9 expression plasmid for T7 in vitro transcription (MLM3613; Addgene plasmid # 42251) [16] or equivalent, as an alternative using commercial Cas9 mRNA or Cas9 protein.
21. PmeI restriction endonuclease for linearizing MLM3613 plasmid.
22. Invitrogen mMESSAGE mMACHINE T7 Ultra Transcription Kit (for in vitro transcription to produce Cas9 mRNA, as an alternative to using commercially synthesized Cas9 mRNA or Cas9 protein).
23. Ambion MEGAclear transcription clean-up kit, to purify RNA from MEGAshortscript or mMESSAGE mMACHINE transcription kits.

2.1.2 T7 Endonuclease I Assay of Nuclease Activity

1. HEK 293T cells.
2. HEK 293T cell culture medium: Dulbecco's Modified Eagle's Medium (DMEM, 4.5 g/L glucose formulation), 10% fetal bovine serum (heat-inactivated), 2 mM L-glutamine, 1% penicillin-streptomycin.
3. 6-well tissue culture plates.

4. DNA polymerase for PCR: Q5 high-fidelity DNA polymerase (or equivalent).
5. PCR primers to amplify a 300–1000 base pair region encompassing the genomic target for gene editing (with primers located preferably at least 100 base pairs away from the genomic cut site, but preferably not equidistant from the target site); for indel analysis of *CYBB* exon 5 CRISPRs, the primers are CYBB5-FWD: 5′-TGGCTTATCATAGAGTCAGAGGCTG-3’ and CYBB5-BWD: 5′-GGCAAAGGAGAGGTCTTCACTCAC-3′, generating a 332-bp PCR product.
6. T7 endonuclease I (T7E1) with 10× buffer.
7. Bio-Rad ChemiDoc XRS+ gel imaging system (or equivalent) with software for densitometric gel band analysis.

2.2 AAV Production

1. HEK 293T cells (*see* **Note 2**).
2. HEK 293T cell culture medium (*see* Subheading 2.1.2, **step 2**).
3. 150-mm tissue culture plates.
4. Adenovirus helper plasmid: pRS449B [17] or equivalent, containing adenovirus E2a, E4, and VA RNA genes necessary for AAV replication.
5. AAV Rep-Cap plasmid: AAV2 RepCap (AAV2-Y730+500+444F) [18], AAV6 RepCap (pACGr2c6-Y705+731F) [19], or equivalent Rep-Cap plasmids without these mutations (*see* **Note 3**).
6. rAAV donor plasmid containing the DNA sequence to be inserted at the genomic target site, surrounded on each side by homologous sequences (termed homology arms; typically ~400–1000 bp each) to the regions flanking target site for HDR. The entire construct to be packaged into AAV should be flanked by inverted terminal repeat (ITR) sequences that are typically derived from the AAV2 genome (*see* **Note 4** regarding size limits for rAAV packaging).
7. Sterile cell scrapers.
8. TD buffer: 140 mM NaCl, 5 mM KCl, 0.7 mM K_2HPO_4, 25 mM Tris (pH 7.4).
9. 250-mL conical centrifuge tubes, with suitable rotor buckets for centrifugation in a compatible tabletop centrifuge.
10. Sodium deoxycholate.
11. Benzonase nuclease, ultrapure.
12. Cesium chloride (CsCl), molecular biology grade.
13. Thinwall polypropylene ultracentrifugation tubes (13.2 mL; Beckman, 9/16×31/2) or equivalent ultracentrifugation tubes.

14. SW 41 Ti ultracentrifuge rotor (Beckman) or equivalent rotor capable of centrifugation at $178,000 \times g$.
15. Optima L-90 K Ultracentrifuge (Beckman) or equivalent ultracentrifuge compatible with the above rotor.
16. Sterile butterfly needle (18–21 gauge) with clamp.
17. Refractometer.
18. ThermoFisher Slide-A-Lyzer Dialysis Cassette, 10 K MWCO, 3 mL (or equivalent).
19. Takara AAVpro Titration Kit for Real-Time PCR (or equivalent SYBR green-based real-time PCR reagents with suitable primers specific for your rAAV donor construct).
20. Applied Biosystems 7500 Real-Time PCR System (or equivalent thermocycler for quantitative PCR).

2.3 Targeted Gene Modification of Human iPSCs

1. Human iPSCs. Our previous studies on targeted gene editing or safe harbor minigene insertion for CGD described herein have utilized iPSCs from normal healthy donors or from patients with gp91^{phox}-deficient X-linked CGD [8–10] and p22^{phox}-, p40^{phox}-, p47^{phox}-, or p67^{phox}-deficient autosomal recessive CGD [9, 11]. iPSCs in these studies were reprogrammed from CD34^{+} HSCs purified from small volumes of patient peripheral blood [20], which was obtained after written informed consent under the auspices of the National Institute of Allergy and Infectious Diseases (NIAID) Institutional Review Board-approved protocols 05-I-0213 and 94-I-0073. The conduct of these studies conforms to the Declaration of Helsinki protocols and all US federal regulations required for protection of human subjects.
2. Medium for feeder-free culture of human iPSC: NutriStem, mTeSR1, or Essential 8 (E8) medium (or equivalent).
3. Complete iPSC medium for feeder co-culture: KnockOut DMEM/F12 medium, supplemented with 20% KnockOut Serum Replacement, 1% nonessential amino acids, 2 mM L-glutamine, 100 U/mL penicillin, 100 μg/mL streptomycin, 10 ng/mL basic fibroblast growth factor, and 0.1 mM ß-mercaptoethanol.
4. Mouse embryonic fibroblast (MEF) feeder cells: DR4 MEFs (puromycin, neomycin, hygromycin, and 6-thioguanine resistant) or other puromycin-resistant MEFs for drug selection in iPSC cloning strategies or CF-1 MEFs for iPSC cloning without drug selection.
5. 96-well, 24-well, and 6-well tissue culture plates.
6. hESC-qualified Matrigel (or equivalent extracellular matrix for iPSC culture): coat wells of tissue culture plates according to manufacturer's protocol; store pre-coated plates at 4 °C per manufacturer's protocol.

7. Optional: Hank's Balanced Salt Solution (HBSS) without calcium or magnesium, as an alternative to PBS for washing cells.
8. Accutase cell detachment solution: for complete dissociation of iPSCs for electroporation and cloning.
9. Y-27632 ROCK inhibitor: enhances survival of dissociated iPSCs by inhibiting anoikis-induced apoptosis; prepared as 5 mM stock in sterile ultrapure water (1 mg Y-27632 in 624 μL water); store aliquots at −20 °C; store at 4 °C after thaw.
10. Amaxa mouse embryonic stem cell Nucleofector kit (or P3 primary cell 4D-Nucleofector kit).
11. Nucleofector I or II/2b device (or 4D-Nucleofector device).
12. Optional: Commercially synthesized sgRNA (preferably with 2′-O-methyl 3′ phosphorothioate modifications in the first three and last three nucleotides for increased stability and activity), as an alternative to LentiCRISPRv2 expression plasmid (*see* Subheading 3.1.2) or T7 in vitro transcribed sgRNA (*see* Subheading 3.1.3).
13. Optional: Cas9 mRNA or Cas9 protein or from a commercial source, as an alternative to LentiCRISPRv2 expression plasmid (*see* Subheading 3.1.2) or T7 in vitro transcribed Cas9 mRNA (*see* Subheading 3.1.4).

2.4 Neutrophil Differentiation Culture from iPSCs (#1–4) or HSCs (#4)

1. Stage-1 APEL differentiation medium (for day 0 through day 4 of neutrophil differentiation from human iPSCs; *see* **Note 5**): STEMdiff APEL medium (Stem Cell Technologies) supplemented with the following human cytokines – 30 ng/mL vascular endothelial growth factor (VEGF), 30 ng/mL bone morphogenetic protein-4 (BMP-4), and 40 ng/mL stem cell factor (SCF) and 50 ng/mL Activin A.
2. Stage-2 APEL differentiation medium (for day 4 through day 13 of neutrophil differentiation from human iPSCs; *see* **Note 5**): STEMdiff APEL medium supplemented with the following human cytokines – 25 ng/mL BMP-4, 300 ng/mL SCF, 300 ng/mL Flt-3 ligand (Flt-3 L), 10 ng/mL interleukin-6 (IL-6), and 10 ng/mL interleukin-3 (IL-3).
3. Stage-3 hematopoietic stem/progenitor cell medium (for day 13 through day 20 of neutrophil differentiation from human iPSCs; *see* **Note 5**): IMDM supplemented with 20% FBS, 5% protein-free hybridoma medium (PFHM-II, Life Technologies), 0.1 mM ß-mercaptoethanol, plus the following human cytokines – 100 ng/mL SCF, 100 ng/mL Flt3-L, 100 ng/mL IL-6, 10 ng/mL IL-3, and 10 ng/mL thrombopoietin (TPO).
4. Neutrophil differentiation medium (Stage-4 differentiation medium for day 20 through day 26 or 27 of neutrophil

differentiation from human iPSCs or for 2 weeks of neutrophil differentiation from somatic human HSCs; *see* **Note 5**): IMDM supplemented with 10% FBS, 0.1 mM β-mercaptoethanol, and 50 ng/mL human granulocyte colony-stimulating factor (G-CSF).

2.5 Targeted Gene Modification of Human HSCs

1. Human CD34[+] HSCs purified from peripheral blood mobilized with G-CSF (supplemented with plerixafor in some donors). For our prior studies described herein utilizing HSCs obtained from normal healthy donors or CGD patients [11–13], cells were obtained after written informed consent under the auspices of NIAID Institutional Review Board-approved protocols 05-I-0213 and 94-I-0073. The conduct of these studies conforms to the Declaration of Helsinki protocols and all US federal regulations required for protection of human subjects.
2. HSC medium: StemSpan SFEM II serum-free medium containing 0.75 μM of StemRegenin 1, 35 nM of UM171, and 100 ng/mL each of human SCF, Flt3-L, TPO, and IL-6.
3. 6-well tissue culture plates, T-75 or T-175 tissue culture flasks.
4. Amaxa P3 primary cell 4D-Nucleofector kit or 96-well Nucleofector kit.
5. Amaxa 4D-Nucleofector device.
6. Optional: Commercially synthesized sgRNA (preferably with 2′-O-methyl 3′ phosphorothioate modifications in the first three and last three nucleotides for increased stability and activity), as an alternative to T7 in vitro transcribed sgRNA (*see* Subheading 3.1.3).
7. Optional: Cas9 mRNA or Cas9 protein or from a commercial source, as an alternative to T7 in vitro transcribed Cas9 mRNA (*see* Subheading 3.1.4).

2.6 Neutrophil Assays

2.6.1 Giemsa Staining for Neutrophil Morphology

1. Cytocentrifuge (Cytospin 3 or equivalent) with compatible cytology funnels and microscope slides.
2. Methanol.
3. Giemsa stain.
4. Optional: Permount and coverslips.

2.6.2 Flow Cytometry Analysis of Cell Surface Markers and Phox Proteins

1. FACS buffer: 0.5% bovine serum albumin (BSA) and 0.1% sodium azide in PBS.
2. Fixing solution: 2% paraformaldehyde in PBS.
3. Permeabilization/staining solution: water containing 0.1% saponin, 0.1% BSA, 75 mM sodium acetate (5.1 g for 500 mL total), 25 mM HEPES; adjust pH to 7.2.

4. Antibodies to human phox protein subunits: $gp91^{phox}$ (unconjugated clone 7D5 mouse IgG1 antibody) [21], $p47^{phox}$ (unconjugated clone 1/p47Phox mouse IgG1 antibody), $p67^{phox}$ (unconjugated polyclonal rabbit IgG anti-NCF2 Ab2 from Sigma-Aldrich), $p40^{phox}$ (unconjugated clone EP2142Y rabbit IgG anti-NCF4 antibody), and $p22^{phox}$ (PE-conjugated clone 44.1 mouse IgG2 antibody) or equivalent antibodies.
5. Fluorochrome-conjugated secondary antibodies to detect unconjugated anti-phox primary antibodies (e.g., FITC-conjugated rat anti-mouse IgG for detection of $gp91^{phox}$ or $p47^{phox}$ primary antibody).
6. Fluorochrome-conjugated antibodies to human CD34, CD43, and CD45 (hematopoietic stem/progenitor cell markers for assessing hematopoietic differentiation from iPSCs).
7. Fluorochrome-conjugated antibodies to human CD45, CD13, CD15, or CD16 (neutrophil surface markers).

2.6.3 Dihydrorhodamine (DHR) Assay of ROS Production

1. Hank's balanced salt solution without calcium or magnesium (HBSS−).
2. Hank's balanced salt solution with calcium and magnesium (HBSS+).
3. Dimethyl sulfoxide (DMSO).
4. Dihydrorhodamine 123: dissolve 10 mg in 1 mL of DMSO for 29 mM stock. Protect from light. Store at −20 °C in 25 μL aliquots.
5. Catalase: dissolve 560,000 units in 400 μL of HBSS−. Store at −20 °C in 10 μL aliquots.
6. Phorbol 12-myristate 13-acetate (PMA): dissolve 1 mg in 500 μL of DMSO. Store at −20 °C in aliquots of at least 5 μL.
7. Optional: ACK lysis buffer for red blood cell lysis of peripheral blood samples (lyse red blood cells prior to use of normal healthy donor blood as a positive control for DHR assay).

3 Methods

3.1 CRISPR Design, Production, and Qualification

3.1.1 General Considerations for Designing CRISPR sgRNAs

In order to target a specific DNA sequence in the genome for cutting by a CRISPR nuclease, a suitable PAM sequence must be present adjacent to the target DNA sequence that corresponds to the (typically) 20-nucleotide sequence of the CRISPR sgRNA (*see* **Note 6** regarding shorter sgRNAs). For the standard *Streptococcus pyogenes* Cas9 (SpCas9), the required PAM sequence is NGG (any nucleotide followed by two guanines) immediately 3′ of the genomic target sequence, where the nuclease cuts within the target sequence at 3–4 bp upstream of the PAM site. Although this

PAM sequence must be present at the genomic target site, it should not be included in the sgRNA.

Online CRISPR sgRNA design tools have been developed for identification of suitable target sequences (based on the presence of a PAM site) and prediction of potential off-target sites for CRISPR cutting in the genome (based on one or more nucleotide mismatches with the target sequence), including Feng Zheng lab's CRISPR design tool (http://crispr.mit.edu) and Jin-Soo Kim lab's Cas-Designer (http://www.rgenome.net/cas-designer/). Use of these online tools simply requires selecting the particular species of CRISPR/Cas-derived nuclease desired (if other options besides SpCas9 are available), pasting a target DNA sequence corresponding to the genomic region to be targeted (which can include 250–1000 nucleotides, depending on the tool) and selecting the organism of the genome to be targeted; the resulting output includes a scored list of guide RNA sequences present in the target DNA sequence and their potential off-targets. These design tools cannot predict the actual DNA cutting activity of a particular CRISPR/sgRNA toward a target sequence, which could be affected by a number of potential factors including binding efficiencies, secondary structure effects, or chromatin accessibility of the target sequence. Consequently, we recommend designing at least two sgRNAs aimed at different sequences within the target gene, to increase the likelihood of finding at least one sgRNA with high cutting activity to achieve efficient gene editing.

The protocols in this chapter assume the use of the standard SpCas9; if using an alternative CRISPR to target another PAM sequence, different plasmids would be needed in Subheadings 3.1.2, 3.1.3, and 3.1.4 to provide the appropriate sgRNA scaffold and CRISPR nuclease.

3.1.2 Cloning the sgRNA Sequence into pLentiCRISPRv2 Vector

Once a desired guide has been designed, the appropriate 20-nucleotide guide sequence can be cloned into a plasmid vector such as LentiCRISPRv2, which contains the remaining non-specific scaffold portion of the guide RNA (corresponding to the CRISPR tracrRNA and a portion of the crRNA sequence) to reconstitute a complete sgRNA. LentiCRISPRv2 can be used for co-expression of the sgRNA and Cas9 nuclease in mammalian cells, either as plasmid DNA (suitable for HEK 293T or iPSC transfections) or after packaging as a lentiviral vector. In addition, the sgRNA/Cas9 expression cassette in LentiCRISPRv2 also co-expresses a puromycin-resistance gene, allowing for in vitro drug selection to eliminate untransfected cells.

1. Order commercial synthesis of both the forward and reverse strand oligonucleotides (*see* Subheading 2.1.1, **step 3** for examples of oligos to clone an sgRNA targeting *CYBB* exon 5):

sgRNA forward oligonucleotide: 5′-caccgNNNNNNNNNNNNNNNNNNNN-3′

sgRNA reverse oligonucleotide: 3′-cNNNNNNNNNNNNNNNNNNNNcaaa-5′

In the forward strand oligonucleotide, the string of N's corresponds to the guide sequence matching the genomic target (without the PAM) and corresponds to the complementary sequence of this guide sequence in the reverse strand oligonucleotide. Note that the reverse strand oligonucleotide is shown above in 3′ to 5′ orientation (for purposes of showing DNA strand alignment with the forward strand oligonucleotides), so care should be taken to change the reverse strand oligonucleotide sequence to 5′ to 3′ orientation before synthesizing. The PAM sequence (NGG) should not be included in these oligonucleotides.

2. Dissolve the lyophilized oligonucleotides in TE buffer at a concentration of 100 μM.
3. Digest 5 μg of LentiCRISPRv2 plasmid with 1 μL of FastDigest BsmBI restriction enzyme plus 1 μL of FastAP for 1 h at 37 °C.
4. Electrophorese the digested plasmid on a 1% agarose gel, and visualize the DNA by UV (or use an equivalent method for visualizing DNA stained with alternatives to ethidium bromide). Cut out the upper band (~12.8-kb in size), and transfer it to a pre-weighed 1.5-mL microcentrifuge tube (*see* **Note 7** regarding UV gel band excision and safety issues). Weigh the excised gel piece in the tube to determine the gel weight (subtracting the total weight from the weight of the tube). Purify the digested plasmid DNA from the gel band using the NucleoSpin Gel and PCR Clean-Up Kit (or equivalent) according to manufacturer's protocol. Determine concentration of purified DNA using NanoDrop spectrophotometer.
5. Add 1 μL forward oligo, 1 μL of reverse oligo, 1 μL of 10× T4 DNA ligase buffer, 6.5 μL water (molecular biology grade), and 0.5 μL of T4 PNK enzyme (T4 DNA ligase buffer is used instead of T4 PNK buffer, which does not include ATP; alternatively, use PNK buffer supplemented with 1 mM ATP). Mix and then incubate at 37 °C for 30 min to phosphorylate the oligos. Incubate at 98 °C for 5 min in a heat block, then remove the block from heat (leaving the tubes inside the block), and allow to cool for 40–60 min to anneal the oligos. Place the annealed oligos on ice until use.
6. Dilute annealed oligos 1:200 in water or TE buffer. Mix 1 μL of annealed oligos, 50 ng of BsmBI-digested LentiCRISPRv2 DNA, 2 μL T4 DNA ligase buffer (standard T4 ligase buffer or from DNA Quick Ligation kit), 14 μL water, and 1 μL T4

ligase (standard T4 ligase or from DNA Quick Ligation kit). Incubate at 16 °C overnight (using standard T4 DNA ligase/ buffer) or at room temperature for 10 min (using DNA Quick Ligation kit).

7. Transform 1–5 μL of the ligation mix into Stabl3 *E. coli* competent bacteria, and plate onto an LB agar plate containing 100 μg/mL ampicillin, following manufacturer's supplied protocols. Culture overnight at 37 °C in a bacteria incubator.
8. Pick 4–6 bacterial colonies and transfer each into 1–5 mL of LB medium containing 100 μg/mL ampicillin. Culture bacteria 8 h or overnight in a bacterial shaker incubator for plasmid minipreps (performed using manufacturer's supplied protocol), setting aside 0.25–1 mL of the bacterial culture to inoculate subsequent maxiprep cultures. Screen for correct clones by performing PCR on the miniprep plasmid DNA samples using pLenti-2277-FWD and the sgRNA reverse oligo (from **step 1**) as primers, resulting in a PCR product of ~600-bp. Alternatively, perform DNA sequencing on the plasmid DNA using hU6-FWD primer to identify clones containing the correct sgRNA sequences.
9. Inoculate 250–500 mL of LB medium containing 100 μg/mL ampicillin with 1:500 to 1:1000 dilution of remaining bacterial culture from one of the correct bacterial clones, and culture overnight in a bacterial shaker incubator for plasmid DNA maxiprep (using manufacturer's supplied protocol). This plasmid DNA is suitable for testing CRISPR activity in HEK 293T cells or other mammalian cell lines (*see* Subheading 3.1.5) to identify potent CRISPR/sgRNAs for further use in gene editing. For preparing suitable plasmid DNA for use in iPSC transfections, an endotoxin-free maxiprep kit is preferred. Alternatively, treat plasmid DNA with polymyxin B-agarose to remove endotoxins (*see* **Note 8**).

3.1.3 T7 Promoter-Driven sgRNA Synthesis

While LentiCRISPRv2 plasmid is suitable for CRISPR/Cas9 expression in mammalian cell lines including iPSCs, the use of sgRNAs with Cas9 mRNA or protein can provide highly efficient gene targeting with reduced cytotoxicity compared to plasmid DNA transfection (particularly for use in HSC electroporation). For some CRISPR/Cas9 gene targeting strategies, we have found that commercially synthesized sgRNAs containing 2′-O-methyl 3′ phosphorothioate modifications in the first three and last three nucleotides (for increased stability and activity) can result in a much greater efficiency of targeted gene editing than using unmodified sgRNAs. Chemical modification, however, will likely alter the specificity of the sgRNAs. As a lower-cost alternative to using commercial modified sgRNAs, we provide the following protocol for in-house production of sgRNAs using the MEGAshortscript T7 Transcription Kit.

Efficient transcription from the T7 promoter requires that the sgRNA sequence starts with two G nucleotides. If the sgRNA does not already start with one or two G nucleotides, these G's can be inserted in front of the guide sequence using the forward primer in **step 1** below, which will result in the addition of GG to the start of the sgRNA during transcription, generally without substantially affecting CRISPR targeting specificity or activity. If the sgRNA sequence does start with one or two G's, you can shorten the 20-nucleotide guide sequence below by the same number of G's in the forward primer in **step 1**, if desired. All reagents should be RNase-free (*see* **Note 9**).

1. Prepare DNA template for in vitro transcription by performing PCR using primers to append the T7 promoter sequence to the beginning of the sgRNA sequence present in the Lenti-CRISPRv2 plasmid. The forward PCR primer sequence is 5′-TTAATACGACTCACTATAGGN_{20}-3′, where the underlined sequence is the T7 promoter and N_{20} is the 20-nucleotide target sequence of the sgRNA. The reverse PCR primer sequence is 5′-AAAAGCACCGACTCGGTG CC-3′, which is complementary to the 3′ end of sgRNA scaffold.
2. Purify PCR product using NucleoSpin Gel and PCR Clean-Up Kit (or equivalent) according to manufacturer's protocol. Determine concentration of purified DNA using NanoDrop spectrophotometer.
3. In an RNase-free microfuge tube at room temperature, gently mix 1 μg of purified PCR product in 8 μL of nuclease-free water with 2 μL of T7 reaction buffer, 2 μL of each NTP, and 2 μL of T7 enzyme mix (from MEGAshortscript T7 Transcription kit). Incubate at 37 °C for 2–4 h.
4. Add 1 μL of TURBO DNase to the reaction and incubate at 37 °C for 15 min.
5. Purify sgRNA transcript using MEGAclear transcription clean-up kit, following manufacturer's protocol (alternatively, purify RNA with phenol/chloroform and again with chloroform followed by ethanol precipitation according to manufacturer's protocol for MEGAshortscript T7 Transcription kit).
6. Measure the sgRNA concentration by NanoDrop spectrophotometer (using setting for single-stranded RNA—i.e., 1 A_{260} unit = 40 μg/mL RNA).
7. Aliquot sgRNA and store at −80 °C.

3.1.4 T7 Promoter-Driven Cas9 mRNA Synthesis

As an alternative to using commercial Cas9 mRNA or protein, a protocol for Cas9 mRNA synthesis using the mMESSAGE mMACHINE T7 Ultra Transcription Kit follows. All reagents should be RNase-free (*see* **Note 9**).

1. Linearize 12 μg of MLM3613 T7-Cas9 plasmid by digesting with 100 units of PmeI restriction enzyme in 20 μL of 10× enzyme digestion buffer and 180 μL of nuclease-free water at 37 °C for 1 h.
2. Purify digested DNA (purify the entire digestion reaction; gel band excision is not necessary) using NucleoSpin Gel and PCR Clean-Up Kit (or equivalent) according to manufacturer's protocol; elute with 30 μL of nuclease-free water.
3. At room temperature, add 14 μL of nuclease-free water, 40 μL of T7 2× NTP/ARCA, 8 μL of 10× T7 reaction buffer, 10 μL of linearized MLM3613, and 8 μL of T7 enzyme mix (from mMESSAGE mMACHINE T7 Ultra Transcription kit). Incubate at 37 °C for 2 h.
4. Add 4 μL of TURBO DNase; mix well and incubate at 37 °C for 15 min.
5. Add 144 μL of nuclease-free water, 80 μL of 5× E-PAP buffer, 40 μL of 25 mM $MnCl_2$, and 40 μL of ATP solution. Add 16 μL of E-PAP; incubate at 37 °C for 45 min.
6. Purify RNA transcript using MEGAclear transcription clean-up kit, following the manufacturer's protocol (alternatively, purify RNA with phenol/chloroform and again with chloroform followed by isopropanol precipitation according to manufacturer's protocol for mMESSAGE mMACHINE T7 Ultra Transcription kit). If desired, check mRNA integrity by agarose gel (1% agarose) electrophoresis (intact, high-quality mRNA should appear as a distinct band without smearing).
7. Measure the Cas9 mRNA concentration by NanoDrop spectrophotometer (using setting for single-stranded RNA—i.e., 1 A_{260} unit = 40 μg RNA).
8. Aliquot mRNA and store at −80 °C.

3.1.5 T7 Endonuclease I Assay of Targeted Nuclease Cutting Activity

The cutting activity of CRISPR/Cas9 or other nucleases (including ZFNs, TALENs, and alternative CRISPRs) for a target DNA sequence can be assessed by the formation of indels at that site resulting from NHEJ or other error-prone DNA repair pathways; however, this may underestimate the actual cutting activity of the nuclease, since any error-free DNA repair that occurs would not result in indels. A protocol for T7 endonuclease I assay of indel activity follows, suitable for testing the cutting activity of target-specific nucleases by transfection of HEK 293T cells with a plasmid nuclease-expression construct such as pLentiCRISPRv2;

alternatively, nuclease cutting activity can be tested in patient-derived iPSCs or HSCs by replacing **steps 1–4** of this protocol with the nucleofection protocols in Subheadings 3.3 or 3.5, and replacing the pLentiCRISPRv2 expression plasmid with sgRNA plus Cas9 mRNA or Cas9 protein, or with expression plasmids or mRNAs for ZFNs or TALENs, as needed. As an alternative to analyzing indels using T7E1 enzyme (**steps 7–10** of the protocol), the purified PCR products from **step 6** (from both naïve and transfected cells) can be sequenced using one of the PCR primers, and the resulting sequence files can instead be analyzed for indel activity using the Tracking of Indels by Decomposition (TIDE) webtool [22] at https://tide.deskgen.com.

1. Plate 0.2–0.4 million HEK 293T cells per well of a 6-well tissue culture plate in 2 mL of HEK 293T cell culture medium (*see* **Note 10** regarding culture of HEK 293T cells).
2. Culture at 37 °C, 5% CO_2 for 1–2 days (until cells are ~40–60% confluent). Change medium on wells to 2 mL of fresh HEK 293T cell culture medium.
3. For each well of HEK 293T cells to be transfected, bring 3–10 μg of pLentiCRISPRv2 (or other nuclease expression plasmid) to 50 μL total with Opti-MEM (or DMEM without serum) in a sterile 1.5-mL tube; mix by pipetting or vortexing. In a separate sterile 1.5-mL tube, bring 9–30 μL of 1 mg/mL PEI (for a 3:1 ratio of PEI:DNA) to 50 μL total with Opti-MEM (or DMEM without serum); mix by pipetting or vortexing. Transfer contents of first tube to second tube, mix by pipetting or vortexing, and then incubate at room temperature for 5–15 min. Add the transfection mixture to a well of HEK 293T cells, dropwise. Leave one well untransfected as a naïve negative control.
4. Optional: starting at 1 day after transfection, treat transfected HEK 293T cells with 3–4 μg/mL puromycin for 3 days for selection of pLentiCRISPRv2-transfected cells prior to DNA extraction, in order to eliminate variations in transfection efficiencies when comparing Cas9 cutting activity of different sgRNAs.
5. At 3 days after transfection (or after completing puromycin selection), collect cells and extract genomic DNA using Qiagen DNeasy Blood and Tissue kit (or equivalent genomic DNA purification method) according to manufacturer's protocol. Determine DNA concentration using NanoDrop spectrophotometer.
6. PCR across the nuclease target site using 200 ng of template DNA (preferably to obtain a PCR product of 300–1000 bp, using primers that are each at least 100-bp from the nuclease target site but preferably not equidistant from the target site, so

as to allow optimal visualization of all digested bands by gel electrophoresis; *see* Subheading 2.1.2, **step 5** for examples of PCR primers for analysis of *CYBB* exon 5 indels). Purify the PCR product and reconstitute in 25 μL water. Determine DNA concentration using NanoDrop spectrophotometer.

7. Dilute 200 ng of purified PCR product to 16 μL in water. Add 2 μL of 10× T7E1 buffer. Incubate in 98 °C heat block for 5 min to DNA double-stranded DNA. Remove the block from heat, leaving samples in block, and allow to cool to room temperature in heat block for 40–60 min.
8. Add 2 μL (2 U) of T7E1 to denatured DNA. Include an uncut control sample (either without T7E1 enzyme or from an untransfected naïve DNA sample).
9. Incubate at 37 °C for 1 h in a heat block.
10. Electrophorese samples on a 1–3% agarose gel for 30–60 min at 100–125 V. Visualize gel DNA bands using UV (or use an equivalent method for visualizing DNA stained with alternatives to ethidium bromide). Photograph gel using gel imaging software and perform densitometry analysis of DNA bands; indel activity of the nuclease is equal to the density of the cut bands (added together) divided by the uncut band (Fig. 2).

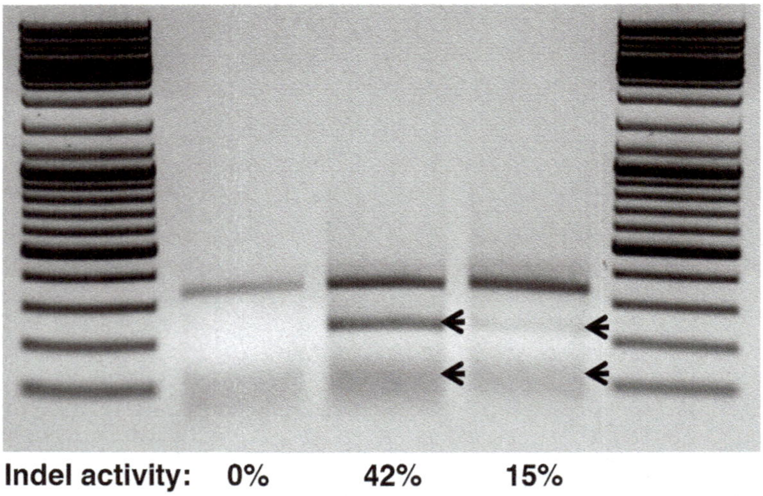

Fig. 2 T7E1 assay of indel formation at *CYBB* exon 5 CRISPRs in HEK 293T cells. HEK 293T cells were transiently transfected with one of two pLentiCRISPRv2 expression plasmids expressing different sgRNAs targeting *CYBB* exon 5, to determine which sgRNA has greater targeted cutting activity. Gel lanes 1 and 5 are 100-bp DNA ladders; lane 2 is a negative control showing the uncut 332-bp PCR product, and lanes 3 and 4 are from cells transfected with one of the two pLentiCRISPRv2 plasmids then selected with puromycin. Arrows indicate T7E1 digestion products of the PCR product; densitometry analysis of these bands determines the indel activity of that sgRNA. The sgRNA sequence for the lane 3 CRISPR with the higher rate of indel formation is listed in Subheading 2.1.1, **step 3**

3.2 AAV Donor Production and Quantification

A protocol follows for production of AAV with capsid from serotype 2 or serotype 6 (with or without tyrosine to phenylalanine mutations; *see* **Note 3**), for use in generating large DNA donors for homology-directed repair in iPSCs or HSCs in combination with CRISPR/Cas9 or other nucleases (*see* Subheading 1 and Fig. 1 for further information on donor design). As an alternative to this protocol, custom AAV production services can be purchased from commercial sources to obtain high titer AAV with a desired capsid serotype. We have observed better transduction of iPSCs in the absence of electroporation when using AAV generated with capsid from AAV serotype 2 and better transduction of HSCs in the absence of electroporation when using AAV serotype 6. However, since electroporation of the target cells with CRISPR/Cas9 or other nucleases prior to transduction with AAV appears to make cells more permissive to AAV entry, the use of different AAV serotypes for iPSCs versus HSCs may not be strictly necessary for efficient gene correction when transducing electroporated cells.

1. Plate 2–4 million HEK 293T cells on 150-mm plates in 18 mL of HEK 293T cell culture medium (*see* **Note 10** regarding culture of HEK 293T cells). Culture at 37 °C, 5% CO_2 overnight or until cells are 40–60% confluent. Prepare ten 150-mm plates of cells for a regular scale AAV preparation. or adjust the scale as desired (*see* **Note 11**).
2. Adjust plasmid concentrations to 1 μg/μL in TE buffer for each of the rAAV donor plasmid, Rep-Cap plasmid (for the desired AAV serotype), and adenovirus helper plasmid. Prepare the transfection reaction (amounts listed are for ten 150-mm plates of cells; adjust scale as desired) in a 50-mL tube by adding 75 μg of rAAV donor plasmid, 75 μg of Rep-Cap plasmid, and 150 μg of adenovirus helper plasmid (300 μg of plasmids in 300 μL total) to 1.4 mL of Opti-MEM (or DMEM without serum). Mix well and then add 300 μL of 1 mg/mL PEI. Mix by vortexing and then incubate at room temperature for 5–15 min. Add 18 mL of HEK 293T medium and mix again by pipetting.
3. Add 2 mL of the transfection reaction to each plate of HEK 293T cells, dropwise. Gently swirl medium in plates to mix. Return cells to 37 °C, 5% CO_2 for 48–72 h.
4. At 48 to 72 h after transfection, harvest cells by scraping (since the majority of AAV will be intracellular) and transfer cells into a 250-mL conical centrifuge tube. Centrifuge cells at 500 × *g* (~1500 rpm) for 15 min and wash with 1× PBS. Resuspend in 11 mL TD buffer. Freeze/thaw cells by freezing on dry ice (5–10 min) and then thaw in 37 °C water bath.
5. To disrupt cells, add 5 mg of sodium deoxycholate powder per mL of lysate (for a final concentration of 0.5% weight/volume,

i.e., 55 mg for 11 mL of cells in TD buffer). Then add benzonase (for a final concentration of 20 U/mL), and incubate at 37 °C for 30 min to degrade all genomic DNA and DNA not encapsulated in virions. After incubation, centrifuge at 500 × *g* for 5 min to pellet cellular debris. Transfer supernatant to a new 50 mL tube.

6. Add 0.55 g of CsCl powder per mL of lysate. Mix and then incubate for 10 min at room temperature. Transfer to ultracentrifuge tubes, filling tubes to the top (insufficient filling will result in collapse of the tubes and loss of the sample). If necessary, use TD buffer to supplement the lysate volume to fill the tubes. Balance and load ultracentrifuge tubes in a SW 41 Ti rotor (*see* **Note 12**) and then centrifuge for 72 h at 178,000 × *g* (~38,000 rpm) at room temperature with the brakes turned off.
7. Prepare approximately 20 1.5-mL microcentrifuge tubes in a tube rack for collection of AAV fractions. Puncture a hole 1-cm from the bottom of the ultracentrifuge tube using a clamped butterfly needle. Slowly open the clamp and collect 0.5 mL fractions in microcentrifuge tubes. Determine the refractive index for each fraction using a refractometer. The expected peak rAAV fraction is a refractive index of 1.371–1.372; pool fractions matching this desired refractive index.
8. Pre-wet a Slide-A-Lyzer Dialysis Cassette by immersing in DMEM for 1–2 min and then inject the pooled AAV fractions into the cassette according to manufacturer's instructions. Dialyze the cassette at 4 °C against 500 mL of DMEM two times for at least 2 h each and against 1000 mL of DMEM overnight in order to remove the CsCl from the AAV preparation.
9. Remove the AAV from the cassette according to manufacturer's protocol, and aliquot as desired into sterile 1.5-mL microcentrifuge tubes. Aliquots may be stored at −80 °C for at least a year.
10. Prepare serial dilutions of the AAV preparation (*see* **Note 13**), and quantitate the AAV vector titer of the diluted samples by SYBR green-based real-time PCR (using AAVpro Titration Kit containing ITR-specific primers or using equivalent reagents with primers specific for your rAAV donor construct) against a standard curve generated by serially diluting a positive control plasmid (diluted rAAV donor plasmid or AAVpro Titration Kit Positive Control, according to manufacturer's protocol).

3.3 Targeted Gene Modification of Human iPSCs

Due to the potential of iPSCs for unlimited self-renewal, low-efficiency targeted gene editing events can be selected for in clonal populations (using drug-resistance gene selection strategies, if desired), and desired iPSC clones can be expanded for further characterization. This self-renewal capacity also allows for the use of plasmid DNAs in iPSC transfections to express either as Cas9/sgRNA (or other nucleases) or as plasmid donor constructs for HDR (*see* Subheading 1 and Fig. 1 for further information on donor design), despite the cytotoxicity of transfected plasmid DNAs. Alternatively, replacing nuclease and donor plasmid DNAs with nuclease mRNA or RNP and ssODN or AAV donors can result in increased iPSC survival and potentially greater gene targeting efficiencies. For drug-resistance strategies, we have observed that puromycin provides superior clonal selection of undifferentiated iPSCs compared to neomycin, hygromycin, or ganciclovir. A protocol for iPSC transfection by electroporation (nucleofection) follows. All iPSC culture is performed at 37 °C, 5% CO_2.

1. Add 100 μL of Amaxa mouse ES cell Nucleofector solution (*see* **Notes 9**, **14**, and **15**) to a sterile, DNase-free/RNase-free 1.5-mL microcentrifuge tube, and allow solution to reach room temperature.
2. Add 5 μg of nuclease (ZFN, TALEN, or Cas9/sgRNA)-expression plasmid DNA or 4–10 μg of nuclease (ZFN, TALEN, or Cas9) mRNA to the Nucleofector solution (*see* **Notes 16** and **17**). When using Cas9 mRNA, include 5–20 μg of sgRNA (*see* **Note 18** regarding the use of in vitro-assembled Cas9 RNP complexes instead of Cas9 mRNA or Cas9/sgRNA expression plasmids). For targeted insertion by HDR, add 5 μg of donor plasmid DNA (*see* **Note 16**) or 5–30 μg of donor ssODN (*see* **Note 19**) to the Nucleofector solution (*see* **Note 20** regarding the use of an AAV donor for HDR).
3. Harvest 1–5 × 10^6 iPSCs (equivalent to ~1–3 confluent wells of a Matrigel-coated 6-well plate for iPSCs grown in E8, NutriStem, or mTeSR1 medium). For iPSCs cultured in feeder-free conditions, wash plates or wells with HBSS or PBS and then aspirate. Add 1 mL Accutase per well and then incubate at 37 °C for 5 min (*see* **Note 21**). Dissociate iPSC colonies into individual cells by pipetting several times with a P1000 Pipetman. Pool and then count an aliquot of the cells to determine cell number.
4. Add 5× volume of HBSS or PBS to cells to dilute Accutase. Centrifuge at 300 × *g* (~1200 rpm) for 4 min.
5. Resuspend cells in the prepared Amaxa/nuclease solution from **step 2**. Transfer cells to a Nucleocuvette vessel and electroporate with an Amaxa Nucleofector I or II/2b device using program A-023 (*see* **Note 15**).
6. Replate 0.5–2 × 10^6 iPSCs in each well of a 6-well Matrigel-coated plate in 3 mL/well of prewarmed (37 °C) NutriStem medium containing 10 μM Y-27632 ROCK

inhibitor (*see* **Note 20** regarding the use of AAV donors for HDR), in the presence or absence of MEF feeder cells (*see* **Notes 22–25**). If drug selection for transfection or targeted insertion will be performed, plate the iPSCs onto Matrigel-coated wells containing drug-resistant MEF feeder cells (e.g., puromycin-resistant MEFs or DR4 MEFs for puromycin selection) in NutriStem medium; alternatively, plate iPSCs on Matrigel for feeder-free culture, and then add MEFs in NutriStem medium on the next day.

7. At 1 day after nucleofection, replace medium with 2.5 mL/well fresh iPSC medium (without Y-27632; *see* **Note 25** regarding the choice of iPSC medium for MEF co-culture). Change medium (2.5 mL/well) daily.
8. When selecting for cells transfected with a nuclease expression construct coexpressing a puromycin-resistance gene (such as pLentiCRISPRv2), treat cells on day 2 after nucleofection with iPSC medium containing 0.25–0.5 μg/mL of puromycin. The next day, visually assess iPSC survival by microscopy. If puromycin selection was effective (as evidenced by marked cytotoxicity but with individual cells or small colonies persisting), stop selection on day 3 post-nucleofection by replacing with fresh iPSC medium without puromycin. Otherwise, if a single day of puromycin selection was ineffective (as evidenced by little or no cytotoxicity), continue selection for one more day using medium containing puromycin (but not longer, since the nuclease/puromycin-resistance plasmid DNA will be rapidly lost from the cells after this time point).
9. When performing puromycin selection for stable integration of a donor construct co-expressing a puromycin-resistance gene, wait until small colonies of 10–20 cells form and then treat cells with iPSC medium containing 0.25–0.4 μg/mL of puromycin. Continue selection for 3–7 days with daily media changes. Afterward, continue culture without puromycin to allow for robust colony growth.
10. If desired, harvest polyclonal DNA from a parallel culture well for DNA extraction and molecular analysis of the frequency of indel formation (by T7E1 assay) and/or targeted insertion (i.e., by quantitative PCR, sequence analysis, or restriction enzyme analysis for an altered sequence, if applicable). Assessment of the frequency of desired gene targeting events in the polyclonal population will help to determine the appropriate number of colonies to be plucked and expanded in order to obtain iPSC clones containing the desired alterations.
11. When iPSC colonies are of sufficient size (≥100–200 cells), pluck colonies using a p20 Pipetman by carefully scraping any MEFs (if present) away from the colony using the pipet and

then scraping underneath the colony to detach it from the plate surface. Draw the colony up into the p20 pipet tip, and transfer the colony into a sterile 1.5-mL microcentrifuge tube containing 50 μL of feeder-free iPSC medium with 10 μM Y-27632 (*see* **Notes 22** and **26**). For each colony, pipet a few times to break it apart into smaller clumps and then transfer the entire volume into a well of a Matrigel-coated 96-well plate. Change feeder-free medium (without Y-27632) each day thereafter, until the well is at least 50% confluent (or when the individual colonies within the well begin to come into contact with each other).

12. Thereafter, for each iPSC clone to be passaged, treat cells in well with 0.5 mM EDTA in HBSS or PBS for 2–5 min at room temperature. Remove EDTA and add 200 μL fresh feeder-free medium. Scrape cells using p200 Pipetman to detach cell clumps, and transfer to a well of a Matrigel-coated 24-well plate (equivalent to a 1:4 split). Supplement with 0.5 mL of fresh medium. Add Y-27632 to a final concentration of 10 μM. Change medium (without Y-27632) each day thereafter, until cells are nearly confluent and ready to passage (or when the individual colonies within the well begin to come into contact with each other).
13. Passage clones from the 24-well plate (as described above) into individual wells of a Matrigel-coated 6-well plate (equivalent to a 1:4 split) in a final volume of 2.5 mL/well. Change feeder-free medium (without Y-27632 each day thereafter) until iPSCs are nearly confluent and ready to passage (or when the individual colonies within the well begin to come into contact with each other).
14. Upon reaching confluence in a well of a 6-well plate, harvest the iPSC clone as small clumps of cells (as described above), setting aside half of the cells for DNA analysis to screen for desired gene targeting events and off-target events (by PCR-based screening, DNA sequencing, or other desired methods). The other half of the cells may be cryopreserved in 1 mL of iPSC medium containing 10% DMSO and 10 μM Y-27632 in a single cryovial for later use (*see* **Note 27**); this reduces laborious and unnecessary expansion of clones that lack the desired gene targeting and conserves cell culture reagents (*see* **Notes 28** and **29**).

3.4 In Vitro Differentiation of Human iPSCs into Neutrophils

To generate hematopoietic stem/progenitor cells from iPSCs, we differentiate iPSCs in adherent cell cultures on Matrigel-coated plates for 13 days consisting of 4 days of Stage-1 APEL medium followed by 9 days of Stage-2 APEL medium. The day 13 cells are then harvested and replated in Stage-3 medium for 1 week for hematopoietic stem/progenitor cell expansion and myeloid differentiation, followed by a final week of differentiation in Stage-4 neutrophil medium. If desired, cells from replicate wells may be periodically harvested during the differentiation and stained with antibodies to hematopoietic stem/progenitor cell surface markers (CD34/CD43/CD45) or myeloid markers (CD15) for flow cytometry analysis of differentiation status and efficiency (*see* Subheading 3.6.2 for an antibody staining protocol). At the end of Stage-4 differentiation, neutrophils can be identified and characterized using the methods described in Subheading 3.6; this differentiation protocol typically results in a mixture of both mature and immature neutrophils as well as monocytes/macrophages.

1. Day -2 or -1: Passage cells onto Matrigel-coated wells of a 6-well tissue culture plate in desired feeder-free iPSC medium (*see* **Note 30**). Since iPSC differentiation can vary based on the density and size of the initial iPSC colonies, we recommend setting up one or more wells at each of two or three different passaging dilutions (usually between 1:6 and 1:10) for each iPSC line to be differentiated. Culture iPSCs for 1–2 days, changing feeder-free iPSC medium as normally, until iPSC colonies enlarge.
2. Day 0: Change medium to 3 mL/well of freshly prepared Stage-1 APEL differentiation medium. Culture for 4 days at 37 °C, 5% CO_2. If the culture medium turns orange or yellow before day 4, replace partially or completely with additional Stage-1 medium as needed (*see* **Note 31**).
3. Day 4: Change medium to 3 mL/well of Stage-2 APEL differentiation medium. Culture for 9 additional days at 37 °C, 5% CO_2, changing the entire volume of medium every 3 days. If the culture medium turns orange or yellow before the next scheduled media change, replace partially or completely with additional Stage-2 medium as needed (*see* **Note 31**). If necessary, the Stage-2 culture can be extended by up to 3 days for iPSC lines that are differentiating slowly. Robust hematopoietic differentiation is indicated by the presence of cobblestone-like regions (typically arising around day 6–10) surrounding endothelial-enclosed "hematopoietic zones" containing adherent, round, putative hematopoietic stem/progenitor cells (typically arising around day 10–13; Fig. 3a,b). By day 7 (overall) of iPSC differentiation, $CD34^+$ $CD43^-$ endothelial cells are typically present in large numbers, and $CD34^+$ $CD43^+$ adherent cells (hemogenic endothelial cells or early hematopoietic stem cells) may be evident; by day 10–13 of differentiation, $CD34^+CD45^+$ hematopoietic stem/progenitor cells should be present among the adherent cells (Fig. 4a,b).
4. Day 13 overall or 1–3 days later for slowly differentiating iPSC lines: At the end of Stage-2 APEL culture, collect and transfer

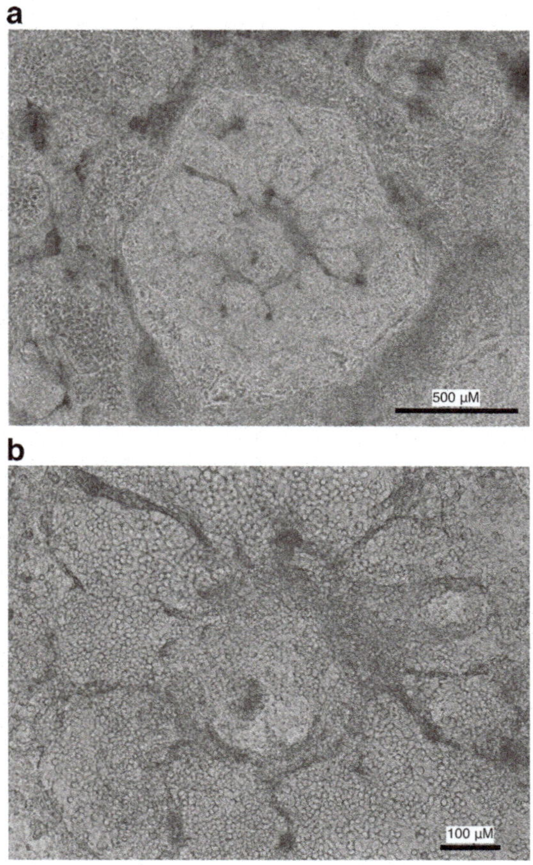

Fig. 3 Hematopoietic differentiation from iPSCs. Shown are images of a hematopoietic zone (containing round putative hematopoietic stem/progenitor cells) surrounded by cobblestone-like cells, arising by day 13 total of iPSC differentiation (late Stage-2 APEL culture), at (**a**) 4× magnification (scale bar = 500 μm) and (**b**) 10× magnification (scale bar = 100 μm). Images were collected by brightfield microscopy using a Nikon Eclipse Ti microscope with a Nikon DS-Qi1MC camera and NIS Elements BR software; contrast and brightness adjustments were performed on whole images without other processing

Stage-2 supernatant to 15-mL or 50-mL conical tubes (to collect any non-adherent cells). Wash wells containing adherent cells with 1× with HBSS or PBS, and treat with 1 mL/well Accutase for 10–20 min at 37 °C in incubator to detach cells from adherent culture. Dislodge cells with a P1000 pipet and repeated pipetting, or use a cell scraper if necessary. Transfer the cells to 15-mL or 50-mL conical tube containing the Stage-2 supernatant that was previously set aside, pooling the non-adherent and adherent cells. Wash the wells with HBSS or PBS or with the previously set-aside Stage-2 supernatant to collect any residual cells, and add them to the same conical tube. Centrifuge at 300 × *g* (~1200 rpm) for 10 min.

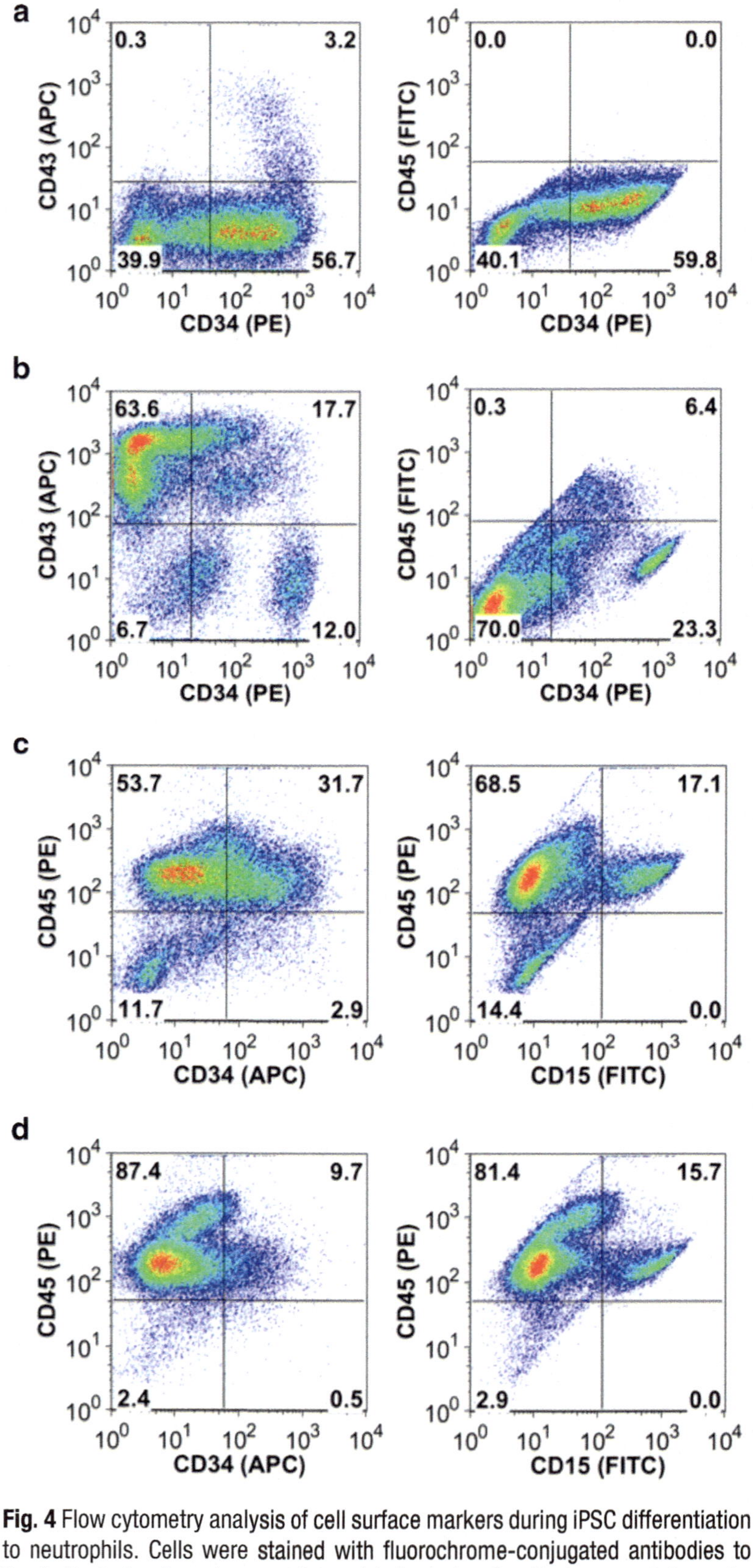

Fig. 4 Flow cytometry analysis of cell surface markers during iPSC differentiation to neutrophils. Cells were stained with fluorochrome-conjugated antibodies to the indicated cell surface markers. (**a**) Analysis of cells harvested from the

5. Aspirate and resuspend the cell pellet in Stage-3 differentiation medium (3 mL/well), replating into tissue-culture plates in the same number of wells as were harvested from Stage-2 culture. Culture for 7 days at 37 °C, 5% CO_2, changing medium every 3–4 days (*see* **Note 32**). Stage-3 culture may result in substantial expansion of the hematopoietic stem/progenitor cells within the non-adherent cell population, but these non-adherent cells rapidly lose CD34 expression and begin expressing myeloid markers by 2–4 days of culture (Fig. 4c, d), differentiating into myeloid-committed progenitor cells.
6. Day 20 overall: Collect the culture medium containing non-adherent hematopoietic cells, and transfer to a 15-mL or 50-mL conical tube (the adherent cells may be discarded or cultured further in Stage-3 medium to induce further hematopoiesis). Wash the wells with PBS or HBSS and pool into the conical tube. Centrifuge at 300 × *g* (~1200 rpm) for 10 min. Aspirate and resuspend the cell pellet in Stage-4 neutrophil medium (3 mL/well), replating into tissue-culture plates in the same number of wells as were harvested from Stage-3 culture. Culture for 6–7 days at 37 °C, 5% CO_2, changing Stage-4 neutrophil medium every 3–4 days by removing the old medium and centrifuging it at 300 × *g* for 10 min to collect non-adherent cells for replating; neutrophils should remain non-adherent or loosely adherent throughout Stage-4 culture (although a population of adherent or non-adherent monocytes and macrophages are also typically present in neutrophil differentiation cultures; *see* **Note 33** regarding directed monocyte/macrophage differentiation).

Due to their cytotoxicity, plasmid DNAs for nuclease expression or for donor constructs are not suitable for HSC transfections for

Fig. 4 (continued) adherent population on day 7 overall (early Stage-2 APEL culture) demonstrates the presence of $CD34^+CD43^-CD45^-$ endothelial cells and $CD34^+CD43^+CD45^-$ cells representing hemogenic endothelium or early hematopoietic stem cells. (**b**) Cells harvested from the adherent population on day 13 overall (late Stage-2 APEL culture) demonstrate the emergence of $CD34^+CD45^+$ hematopoietic stem/progenitor cells. (**c, d**) During Stage-3 culture, analysis of cells harvested from the non-adherent population shows the presence of $CD34^+CD45^+$ hematopoietic stem/progenitor cells which peak at around day 15 overall (**c**; early Stage-3 culture) and declined by day 17 overall (**d**; mid Stage-3 culture). Also present at days 15 and 17 are $CD15^+CD45^+$ myeloid-committed cells, which emerge prior to neutrophil maturation during Stage-4 culture

3.5 Targeted Gene Modification of Human HSCs and Differentiation into Neutrophils

efficient gene targeting. Consequently, we utilize nuclease mRNAs or Cas9 ribonucleoproteins for transfection of HSCs and either ssODN or AAV (capsid serotype 6, with or without tyrosine to phenylalanine mutations; *see* **Note 3**) donors for HDR (*see* Subheading 1 and Fig. 1 for further information on donor design). Our previous studies on targeted HSC gene therapy [11–13] have mainly utilized a clinically scalable cGMP-compliant MaxCyte GT electroporation system; however, we have also observed efficient gene editing using a protocol for nucleofection of HSCs with the more commonly used Amaxa 4D-Nucleofector system, which we present here. This protocol assumes the use of cryopreserved HSCs or other initially quiescent sources of HSCs and consequently utilizes 2 days of cytokine stimulation to induce HSC proliferation for enhanced HDR, since HDR pathways are active during G_2/S phase of the cell cycle [23, 24].

1. (Day -2) Thaw a cryovial containing 1–5 × 10^6 cryopreserved human HSCs in a 37 °C water bath with occasional gentle swirling or inverting of the cryovial, until a sliver of ice remains. Spray the outside of the cryovial with 70% ethanol or another disinfectant. In a sterile tissue culture hood, transfer the cells to a 15-mL centrifuge tube. Add 9 mL of hematopoietic stem cell medium dropwise while stirring. Centrifuge at 300 × *g* (~1200 rpm) for 10 min at room temperature. Resuspend cells in fresh HSC medium at a density of 0.2–0.5 × 10^6 cells per mL. Plate into an appropriate-sized tissue culture dish or flask, and culture for 2 days at 37 °C, 5% CO_2 to induce HSC proliferation (*see* **Note 34**).
2. (Day 0) Aliquot 100 μL of Amaxa P3 Nucleofector solution (*see* **Note 9**) into a sterile, DNase-free/RNase-free 1.5-mL microcentrifuge tube, and allow solution to reach room temperature. This assumes the use of a 100 μL Nucleocuvette vessel for nucleofection. If using 16-well Nucleocuvette strips instead, reduce the volume of Nucleofector solution to 20 μL, and scale down the amounts of HSCs, RNA, RNP, or donor ssODN in subsequent steps accordingly.
3. Add 4–10 μg of nuclease (ZFN, TALEN, or Cas9) mRNA to the Nucleofector solution (*see* **Notes 16** and **17**). When using Cas9 mRNA, include 5–20 μg of sgRNA (*see* **Note 18** regarding the use of in vitro-assembled Cas9 RNP complexes instead of Cas9 mRNA). For targeted insertion of donor ssODN by HDR, add 5–30 μg of ssODN (*see* **Note 19**) to the Nucleofector solution.
4. Harvest the non-adherent HSCs from the culture flask. Centrifuge at 300 × *g* (~1200 rpm) for 10 min. Wash with 1–5 mL of PBS to remove any RNases present in the culture medium.

Count an aliquot of the HSCs to determine cell number. Centrifuge 1–5 × 10^6 cells at 300× for 10 min.

5. Resuspend 1–5 × 10^6 HSCs in the prepared Amaxa/nuclease solution from **step 3**. Transfer cells to a Nucleocuvette vessel (or strips) and electroporate using an Amaxa 4D-Nucleofector device with program DZ-100.
6. Gently transfer nucleofected cells into fresh HSC medium for a density of 0.2–0.5 × 10^6 HSCs per mL. Transfer to an appropriate-size tissue culture dish or flask for this volume. For targeted insertion of AAV6 donor by HDR, add AAV to the electroporated HSCs at a multiplicity of infection of 10^4–10^6 viral genomes (vg) per cell (*see* **Note 35**) within 30 min of nucleofection (since electroporation appears to temporarily enhance AAV entry into cells); culture cells at 37 °C, 5% CO_2 for 12–24 h for AAV transduction; remove AAV after 12–24 h by centrifuging HSCs at 300 × *g* for 10 min and resuspending HSCs in fresh medium at a density of 0.2–0.5 × 10^6 HSCs per mL in an appropriate-size tissue culture dish or flask.
7. Continue culturing HSCs for 1–2 additional days (*see* **Note 34**) prior to harvesting for molecular analysis of gene editing efficiency (by indel assay, digital PCR or real-time PCR, high-throughput sequencing, or other desired methods).
8. For neutrophil differentiation of HSCs, replate cells in neutrophil differentiation medium at a density of 0.2–0.5 × 10^6 cells per mL. Change medium every 2–3 days for 14 days total to obtain mature neutrophils for characterization (*see* Subheading 3.6). Expression of phox proteins in a substantial portion of the HSC-derived population may be evident within as little as 4–5 days of neutrophil differentiation culture, but full neutrophil maturation (resulting in mature neutrophil morphology and ROS production) peaks at around 14 days of in vitro differentiation under these conditions.

3.6 Neutrophil Characterization and Functional Assays

The following protocols for characterization of neutrophil identity and function are included for the confirmation of successful in vitro differentiation of iPSCs/HSCs into mature neutrophils and may be used for assessment of targeted gene correction for restoration of phox gene expression and NADPH oxidase activity in CGD patient cells.

3.6.1 Cytospin and Giemsa Stain

Differentiated mature neutrophils can be identified morphologically by the presence of granules and a characteristic multilobed (typically into 3–5 segments) nucleus (Fig. 5a). For both iPSC and HSC differentiations, mature and immature neutrophils together typically comprise between 30 and 80% of the total differentiated

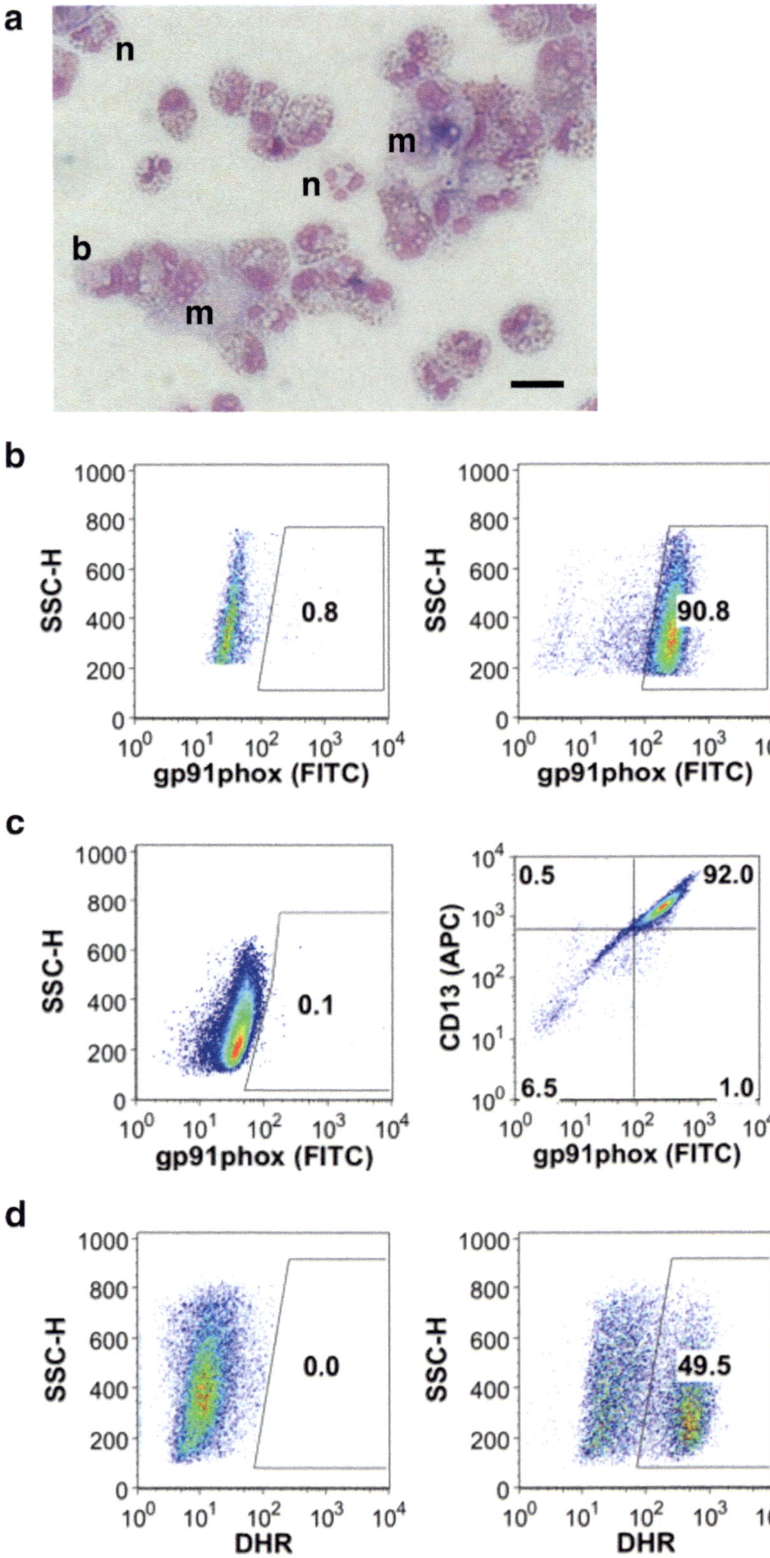

Fig. 5 Characterization of neutrophils differentiated from an X-CGD patient iPSC clone corrected by seamless targeted replacement of *CYBB* exon 5. (**a**) Giemsa-

cell population, with macrophages and monocytes comprising the majority of the remaining cells.

Recommended cell numbers for analysis: 1000–10,000 cells.

1. Assemble slide, filter, and cytofunnel according to manufacturer's instructions.
2. Resuspend cells in 20–200 μL of PBS. Apply cells to cytofunnel.
3. Cytospin cells in cytocentrifuge at 70 × *g* for 5 min. Remove slide from assembly.
4. Add methanol to cover cells on slide and then incubate for 5–7 min to fix cells. Air-dry.
5. Add Giemsa stain to cover cells on slide. Stain for 15–30 min.
6. Rinse slide in deionized water. Air dry.
7. Optional: For permanent storage, mount coverslip on slide with Permount.
8. Visualize by light microscopy.

3.6.2 Flow Cytometry Analysis of Phox Protein Expression and Cell Surface Markers

Flow cytometry analysis to detect restoration of phox subunit expression following gene correction in CGD patient cells can be performed by fixing, permeabilizing, and staining differentiated neutrophils with an antibody to the appropriate targeted phox protein subunit (Fig. 5b). For additional analysis of neutrophil surface markers (Fig. 5c), we typically stain (either separately or after intracellular staining for phox protein expression) with antibodies to detect expression of one or more of the following: CD45

Fig. 5 (continued) stained cytospins of differentiated cells (scale bar = 20 μm). Representative mature neutrophils possessing the characteristic polymorphonuclear morphology with intracellular granules are indicated (labeled "n"), as are immature band cells ("b") and larger monocytes ("m"). Images were collected by brightfield microscopy using an EVOS XL Core imaging system; contrast and brightness adjustments were performed on whole images without other processing. (**b**) Flow cytometry analysis of gp91^{phox} protein expression in neutrophils differentiated from uncorrected (left) or corrected (right) X-CGD patient iPSCs, demonstrating restoration of gp91^{phox} expression following gene correction. (**c**) Flow cytometry analysis demonstrating co-expression of gp91^{phox} protein with CD13 myeloid surface marker in neutrophils differentiated from corrected X-CGD patient iPSCs (right), but no expression of gp91^{phox} in undifferentiated iPSCs (left), as expected for normal regulation of gp91^{phox} expression from the *CYBB* promoter after targeted *CYBB* gene correction. (**d**) Flow cytometry analysis of ROS generation by DHR assay of neutrophils differentiated from uncorrected (left) or corrected (right) X-CGD patient iPSCs, demonstrating restoration of NADPH oxidase activity following targeted *CYBB* gene correction (a pan-leukocyte marker), CD13 (myeloid cell marker), CD15 (granulocyte/monocyte marker), or CD16 (granulocyte/macrophage marker).

Recommended cell numbers for analysis: 100,000–500,000 cells.

1. Wash cells with PBS, centrifuging at 500–700 × *g* (~1500–1800 rpm) for 5 min.
2. For staining cells for antibodies to phox subunit proteins, resuspend cells in 100 μL of fixing solution, and incubate for 10 min. Wash cells with 3 mL PBS, centrifuging as above. Resuspend cells in 100 μL of permeabilization/staining solution. Add 1 μL of unconjugated primary antibody. Incubate at room temperature for 20 min and then wash cells with 3 mL PBS as above. For detection of unconjugated primary antibodies, resuspend cells in 100 μL of PBS, add 2 μL of fluorochrome-conjugated secondary antibody specific to the primary antibody isotype, incubate in the dark at room temperature for 15–20 min, and then wash cells with 3 mL PBS as above.
3. If performing surface marker analysis, resuspend cells in 100 μL of FACS buffer, and stain cells in the dark with fluorochrome-conjugated antibodies to surface markers according to manufacturer's recommendations (e.g., incubate with 20 μL of antibody at room temperature for 20–30 min). Wash cells with 3 mL PBS, centrifuging as above.
4. Resuspend cells in 300–500 μL of FACS buffer. Analyze by flow cytometry.

3.6.3 DHR Assay of ROS Production

The DHR assay [25] detects the ROS-mediated oxidation of non-fluorescent DHR to fluorescent rhodamine-123, which can be used to measure NADPH oxidase activity in mature neutrophils on a per-cell basis by flow cytometry (Fig. 5d; *see* **Note 36**). DHR assay can be performed with as few as 5×10^4 cells, although at least 10^5 cells are preferred for optimal analysis. ROS production in this assay is stimulated by addition of PMA; if sufficient numbers of neutrophils are available for analysis, additional negative controls for DHR activity may be prepared in replicate samples without PMA stimulation. As a positive control, either 400 μL of peripheral blood from a healthy donor (*see* **Note 37**) or neutrophils differentiated in vitro from normal control cells (either iPSCs or HSCs) can be used.

Recommended cell numbers for analysis: 100,000–500,000 cells.

1. Thaw DHR, catalase, and PMA aliquots to room temperature while protecting DHR from light.

2. Prepare 100 U/μL catalase solution by adding 130 μL of HBSS− to 10 μL catalase aliquot.
3. Prepare 5 μg/mL PMA solution by adding 5 μL of 2 μg/μL aliquot to 2 mL of HBSS+.
4. Resuspend neutrophils in 400 μL of HBSS− in a flow cytometry tube.
5. Add 1.8 μL of 29 mM DHR to all cell samples.
6. Quickly add 5 μL of 100 U/μL catalase to all cell samples.
7. Vortex cells and incubate at 37 °C for 5 min in a heat block or water bath.
8. Add 100 μL of 5 μg/mL PMA to cells to stimulate ROS production (or add 100 μL of HBSS− without PMA, for unstimulated negative controls).
9. Incubate at 37 °C for 14–30 min in a heat block or water bath (*see* **Note 38**).
10. Analyze by flow cytometry for DHR fluorescence in the FL1 or FL2 channel (*see* **Note 39**).

4 Notes

1. Prepare 1 mg/mL stock of PEI by adding 100 mg of PEI to 90 mL of molecular grade water and then adjust pH to 2.0 by adding hydrochloric acid dropwise while stirring. Cover and continue stirring until PEI completely dissolves (this may take several hours). Adjust pH to 7.0 by adding sodium hydroxide dropwise as necessary, and then adjust total volume to 100 mL with molecular grade water. Filter-sterilize (0.22 μm filter). Store aliquots at −20 or −80 °C. After thawing, store aliquot at 4 °C for up to 2 months.
2. As an alternative to HEK293T cells, the parental HEK293 cell line (without SV40 large T antigen) may be used. Certain clones of HEK293 cells, such as AAV-293, have been reported to exhibit improved cell attachment during transfection and to result in higher titers of AAV production.
3. Packaging of AAV vectors using AAV2 and AAV6 Rep-Cap plasmids containing these specific tyrosine to phenylalanine mutations protects AAV capsids from ubiquitination and proteasome-mediated degradation [18, 19], thereby resulting in increased transduction (by >4-fold) of human iPSCs and HSCs. However, the use of these mutant capsids may not be strictly necessary for efficient AAV transduction of iPSCs and HSCs when using prior electroporation of cells for delivery of CRISPR/Cas9 or other nucleases (as in the gene targeting protocols described in this chapter), since electroporation is

thought to temporarily make cells more permissive to AAV transduction regardless of capsid serotype or capsid mutations.

4. The packaging capacity of rAAV is considered to be ~4.7-kb; packaging of larger vector constructs may result in substantially decreased viral titers. This size limit includes the flanking ITRs (typically 145-bp each), reducing the maximum vector packaging size limit to ~4.4-kb. In AAV donor constructs for HDR, this vector size limit includes the two homology arms of ~400–1000 bp each, further reducing the size of the intervening donor sequence that can be inserted at the target site without compromising AAV titer.
5. If desired, differentiation media may also be supplemented with preferred antibiotics and/or antimycotic (i.e., penicillin, streptomycin, amphotericin B) at normal concentrations for tissue culture usage.
6. The targeting sequence of a CRISPR sgRNA is typically 20 nucleotides in length, but it has been reported that this can be shortened to 17 or 18 nucleotides and still retain efficient targeted cutting activity by Cas9 nuclease while reducing undesirable off-target cutting activity [26]. The sgRNA design, cloning, and synthesis protocols presented here assume the use of 20-nucleotide sgRNAs and should be altered accordingly if utilizing shorter sgRNA sequences.
7. When visualizing DNA stained with ethidium bromide during gel band excision, minimize UV exposure time during gel band excision to avoid DNA damage. A 365-nm UV transilluminator produces weaker fluorescence but will cause less DNA damage than a 302-/312-nm transilluminator. Use a lab coat, gloves, and a UV blocking face shield when performing gel band excision to protect from UV damage to the skin and eyes.
8. Vortex polymyxin B-agarose suspension, and transfer 300 μL to a 1.5-mL microcentrifuge tube. Add 1 mL of TE buffer. Centrifuge in a tabletop microcentrifuge at maximum speed for 1 min. Remove and discard liquid. Add maxiprep plasmid DNA. Incubate at 4 °C with rotation for 20–30 min. Centrifuge at maximum speed for 1 min. Transfer liquid containing endotoxin-free plasmid DNA to a sterile 1.5-mL microcentrifuge tube.
9. When working with RNA, wear gloves and use RNase-free tubes and pipet tips. To prevent contamination of materials with RNases, spray gloves and surfaces with RNase AWAY or other RNase-decontaminating solution before handling materials, and never allow materials to come into contact with the skin.

10. When expanding and maintaining HEK 293T cells, do not allow cells to reach 100% confluence prior to passaging (ideally, passage at 80–90% confluence). Passage cells 1:3 to 1:8 as follows: aspirate medium, rinse the cell layer with PBS, add 0.05% trypsin-EDTA to cover cells, incubate for 5 min at 37 °C, then add HEK 293T cell culture medium to inactivate trypsin, and gently pipet to dislodge cells from plate. Change medium every 2–3 days as needed.
11. AAV production with capsid 2 or capsid 6 at this scale typically results in AAV titers of approximately 10^{11} viral genomes (vg)/mL.
12. Proper balancing of ultracentrifuge tubes within the rotor is *absolutely critical* to prevent damage to the centrifuge/rotor or serious physical injury.
13. Serial dilutions of the AAV preparation may be used directly for real-time PCR. Alternatively, AAV vector DNA may extracted from the AAV preparation using the AAVpro Titration Kit (according to manufacturer's protocol) and then diluted at least 50-fold prior to real-time PCR analysis.
14. The Amaxa human stem cell Nucleofector kits 1 and 2 are designed for use with human pluripotent stem cells and may also be used with Nucleofector I or II/2b program A-023. However, we have achieved comparable or better results with human iPSCs using the mouse embryonic stem cell kit.
15. As an alternative to the Nucleofector I or II/2b device, the 4D-Nucleofector system can be used for iPSC transfection with P3 primary cell Nucleofector solution using program CA-137 or CB-150, with either a 100 μL Nucleocuvette vessel or 16-well Nucleocuvette strips. If using 16-well Nucleocuvette strips, reduce the volume of Nucleofector solution to 20 μL, and scale down the amount of cells, DNA, RNA, or RNP accordingly.
16. The total volume of the nuclease and donor DNA components added to 100 μL of Amaxa solution should ideally be no more than 20 μL (preferably $\leq$10 μL). Consequently, highly concentrated nuclease and donor reagents are preferred ($\geq$1 μg/μL for DNA and mRNA, or $\geq$ 2 μg/μL for Cas9 protein).
17. If using ZFNs or TALENs, include DNA or mRNA for expression of both halves of the nuclease pair.
18. If using in vitro-assembled Cas9 RNP complexes for nucleofection (instead of Cas9/sgRNA DNA expression vectors or Cas9 mRNA and sgRNA), add 16 μg (approximately 100 pmol) of Cas9 protein and 250–500 pmol of sgRNA to 100 μL of Nucleofector solution. Incubate for 10–15 min at room temperature to form RNP complexes (prior to adding

cells or donor constructs and prior to nucleofection). When using RNPs, we have observed a much higher degree of gene editing with sgRNAs that have been commercially synthesized to contain 2′-O-methyl 3′ phosphorothioate modifications in the first three and last three nucleotides, compared to using unmodified sgRNAs.

19. Using chemically modified ssODN with one to three phosphorothioate-modified bases at each end improves stability of the donor DNA by inhibiting its degradation by exonucleases, increasing the efficiency of targeted insertion.

20. If using AAV vector instead of either plasmid or ssODN for the homologous donor, do not add AAV prior to nucleofection. Instead, add AAV to the culture medium containing the nucleofected cells within 30 min after nucleofection (since electroporation appears to temporarily make the cells more permissive to AAV entry). For iPSCs, use AAV at a multiplicity of infection (MOI) of 200–10,000 viruses per cell (higher MOIs may improve HDR, albeit with greater cytotoxicity); if transducing iPSCs in the presence of MEFs, an increased AAV titer may be required to compensate for viral loss due to transduction of MEFs. Change the culture medium the next day (after 12–24 h) to remove AAV for reduced cytotoxicity.

21. For iPSCs cultured on MEFs, treat with 1 mg/mL dispase at 37 °C for 15–30 min to detach iPSC colonies from fibroblasts and then centrifuge at 200 × *g* for 2 min. Aspirate to remove dispase, then add 1–3 mL Accutase to cell pellet, and incubate at 37 °C for 5 min.

22. Inclusion of Y-27632 ROCK inhibitor greatly improves iPSC survival and subsequent clonal growth after dissociation of iPSCs into individual cells or small clumps of cells. Y-27632 should only be included on the day of passaging iPSCs and should be removed with the next change of culture medium.

23. We have observed that survival of dissociated iPSCs following is generally improved when iPSCs are plated onto Matrigel-coated dishes (with or without MEFs) in NutriStem medium (which is suitable for both feeder-free and MEF co-culture), compared to plating iPSCs onto gelatin-coated dishes containing MEFs in complete iPSC medium for MEF co-culture.

24. After plating, dissociated individual iPSCs grown in feeder-free culture may migrate within the well to form larger polyclonal colonies, which complicates the establishment of clonal iPSC lines containing only the desired gene targeting events. For this reason, we recommend culturing the dissociated and electroporated iPSCs on MEF feeders to limit iPSC migration. This can be done either by initially plating the dissociated iPSCs onto MEFs (in Matrigel-coated wells) immediately following

iPSC electroporation, or by plating the iPSCs onto Matrigel without MEFs in feeder-free medium, then adding MEFs to the culture on the following day.

25. NutriStem medium can be used for subsequent MEF co-culture, or it can be replaced with complete iPSC medium for MEF co-culture, as desired. We prefer to use NutriStem medium instead of E8 or mTeSR1 media for co-culture of dissociated iPSCs on MEFs, since in our experience NutriStem appears to perform better at maintaining and expanding undifferentiated iPSCs when used in MEF co-cultures.
26. For subsequent expansion of iPSC clones without MEFs, any feeder-free iPSC medium may be used, but we generally observe a faster and more robust expansion of undifferentiated iPSCs in Essential-8 medium.
27. Use of a Nalgene Mr. Frosty freezing chamber is recommended for gradual controlled freezing of cells in cryovials to −80 °C overnight, followed by transfer of cryovials to liquid nitrogen for long-term storage.
28. Alternatively, each iPSC clone may be further expanded prior to cryopreservation by passaging 1:4 to 1:8 into additional wells of Matrigel-coated 6-well plates in feeder-free iPSC medium to establish a larger bank of cryovials for each clone.
29. Long-term adaptation of iPSCs to culture in Essential-8-medium can result in a persistent and greatly increased proliferation rate and may adversely affect subsequent attempts at directed differentiation of iPSCs to some cell lineages. Consequently, our group prefers to use NutriStem or mTeSR1 medium for long-term feeder-free maintenance of iPSCs and typically only uses Essential-8 medium for robust expansion of iPSC clones from individual colonies until they are sufficiently expanded for passaging into a 6-well plate.
30. Rapidly growing iPSC lines can be problematic for APEL differentiation, as they often become confluent quickly and can start to die off at around the end of the Stage-1 culture, unless the culture media is changed every 1–2 days. Since iPSCs cultured in E8 medium generally exhibit faster growth rates than in other feeder-free medium, we prefer to perform differentiations using iPSCs that have been adapted to slower growth in mTeSR1 or NutriStem media for at least several passages.
31. A major issue for hematopoietic differentiation in Stage-1 and Stage-2 APEL media is that the culture medium may begin to turn yellow before the next scheduled media change. If so, supplement with fresh medium, or cell death can occur within the next day. If necessary, remove some of the old medium before supplementing with fresh medium.

32. During Stage-3 culture, non-adherent round hematopoietic cells should be present in the culture, although dead cells or debris may also be present floating in the medium within the first 1–2 days after beginning Stage-3 culture. For media changes, collect the medium (replacing the medium in the wells with fresh Stage-3 medium to keep the adherent cells from drying out), and centrifuge the old medium at 300 × *g* for 10 min to collect the non-adherent hematopoietic cells. Resuspend the resulting cell pellet in fresh Stage-3 medium, and replate back into the same wells.
33. For directed differentiation of monocytes/macrophages instead of neutrophils, replace the G-CSF in Stage-4 neutrophil medium with 100 ng/mL macrophage colony-stimulating factor (M-CSF), and culture the day 20 cells for 10–14 days at 37 °C, 5% CO_2, changing medium every 3–4 days by centrifuging at 300 × *g* for 10 min to collect non-adherent cells for replating; most monocytes/macrophages should become adherent in the wells within the first 7 days of Stage-4. The cells present at day 10–14 are typically nearly all monocytes and macrophages.
34. The duration of HSC culture may be increased for greater cell proliferation, if desired. However, prolonged HSC culture results in substantial loss of stem cell potential as the HSC differentiate into hematopoietic progenitor cells. While the use of StemRegenin 1 and UM171 in HSC culture medium can help to maintain stem cell potential during HSC expansion [27], culturing for more than 5–6 days total may still result in substantial loss of HSC capacity for repopulation and multilineage differentiation.
35. For using AAV6 donor constructs in HSC gene editing, we find that an MOI of 10^5 vg/cell results in a good trade-off between HDR efficiency and HSC survival, while higher MOIs can result in increased HDR with a substantial loss of HSC viability.
36. ROS production requires a greater degree of neutrophil maturation than is generally required for expression of phox proteins. Consequently, since in vitro differentiation of iPSCs and HSCs typically results in varying degrees of maturation within the resulting neutrophil population, the proportion of DHR^+ cells is typically lower than the proportion of $phox^+$ cells in the same differentiation culture (*see* Fig. 5b, c versus Fig. 5d).
37. Before analyzing a peripheral blood sample for DHR, first lyse the red blood cells by adding 4 mL of prewarmed (37 °C) ACK lysis buffer to 400 μL of peripheral blood and incubating at 37 °C for 5 min. Centrifuge at 220 × *g* (~1000 rpm) for 5 min, aspirate, and then wash with 4 mL of HBSS−. Avoid

overtreating with ACK lysis buffer so as not to lyse the neutrophils as well.

38. Longer incubations with DHR reagent result in a stronger fluorescent signal for better separation of positive and negative cells. However, this can also eventually result in a gradual false-positive fluorescent signal in neutrophils lacking NADPH oxidase activity, due to other oxidative activities present in cells. Consequently, flow cytometry analysis should ideally be performed within 45 min of the initial addition of DHR reagent.

39. DHR fluorescence encompasses both FL1 and FL2 channels, precluding co-staining with other fluorochromes in those channels. If co-staining for expression of a cell surface marker is desired, live unfixed cells may be stained with an allophycocyanin (APC)-conjugated or Alexa Fluor 647-conjugated antibody to the desired surface marker (*see* Subheading 3.6.2 for an antibody staining protocol) prior to performing the DHR assay.

Acknowledgments

This work was supported by the Intramural Research Program of the NIH and NIAID.

References

1. Ott MG, Schmidt M, Schwarzwaelder K, Stein S, Siler FU, Koehl U, Glimm H, Kuhlcke K, Schilz A, Kunkel H, Naundorf S, Brinkmann A, Deichmann A, Fischer M, Ball C, Pilz I, Dunbar C, Du Y, Jenkins NA, Copeland NG, Luthi U, Hassan M, Thrasher AJ, Hoelzer D, von Kalle C, Seger R, Grez M (2006) Correction of X-linked chronic granulomatous disease by gene therapy, augmented by insertional activation of *MDS1-EVI1*, *PRDM16* or *SETBP1*. Nat Med 12 (4):401–409
2. Porteus MH, Baltimore D (2003) Chimeric nucleases stimulate gene targeting in human cells. Science 300(5620):763
3. Urnov FD, Miller JC, Lee Y-L, Beausejour CM, Rock JM, Augustus S, Jamieson AC, Porteus MH, Gregory PD, Holmes MC (2005) Highly efficient endogenous human gene correction using designed zinc-finger nucleases. Nature 435:646–651
4. Miller JC, Tan S, Qiao G, Barlow KA, Wang J, Xia DF, Meng X, Paschon DE, Leung E, Hinkley SJ, Dulay GP, Hua KL, Ankoudinova I, Cost GJ, Urnov FD, Zhang HS, Holmes MC, Zhang L, Gregory PD, Rebar EJ (2010) A TALE nuclease architecture for efficient genome editing. Nat Biotechnol 29:143–148
5. Jinek M, Chylinski K, Fonfara I, Hauer M, Doudna JA, Charpentier E (2012) A programmable dual-RNA–guided DNA endonuclease in adaptive bacterial immunity. Science 337 (6096):816–821
6. Cong L, Ran FA, Cox D, Lin S, Barretto R, Habib N, Hsu PD, Wu X, Jiang W, Marraffini LA, Zhang F (2013) Multiplex genome engineering using CRISPR/Cas systems. Science 339(6121):819–823
7. Mali P, Yang L, Esvelt KM, Aach J, Guell M, DiCarlo JE, Norville JE, Church GM (2013) RNA-guided human genome engineering via Cas9. Science 339(6121):823–826
8. Zou J, Sweeney CL, Chou BK, Choi U, Pan J, Wang H, Dowey SN, Cheng L, Malech HL (2011) Oxidase-deficient neutrophils from X-linked chronic granulomatous disease iPS cells: functional correction by zinc finger nuclease-mediated safe harbor targeting. Blood 117(21):5561–5572

9. Merling RK, Sweeney CL, Chu J, Bodansky A, Choi U, Priel DL, Kuhns DB, Wang H, Vasilevsky S, De Ravin SS, Winkler T, Dunbar CE, Zou J, Zarember KA, Gallin JI, Holland SM, Malech HL (2015) An AAVS1-targeted minigene platform for correction of iPSCs from all five types of chronic granulomatous disease. Mol Ther 23(1):147–157
10. Sweeney CL, Zou J, Choi U, Merling RK, Liu A, Bodansky A, Burkett S, Kim J-W, De Ravin SS, Malech HL (2017) Targeted repair of *CYBB* in X-CGD iPSCs requires retention of intronic sequences for expression and functional correction. Mol Ther 25(2):321–330
11. Merling RK, Kuhns DB, Sweeney CL, Wu X, Burkett S, Chu J, Lee J, Koontz S, Di Pasquale G, Afione SA, Chiorini JA, Kang EM, Choi U, De Ravin SS, Malech HL (2017) Gene-edited pseudogene resurrection corrects $p47^{phox}$-deficient chronic granulomatous disease. Blood Advances 1(4):270–278
12. De Ravin SS, Reik A, Liu P-Q, Li L, Wu X, Su L, Raley C, Theobald N, Choi U, Song AH, Chan A, Pearl JR, Paschon DE, Lee J, Newcombe H, Koontz S, Sweeney C, Shivak DA, Zarember KA, Peshwa MV, Gregory PD, Urnov FD, Malech HL (2016) Targeted gene addition in human $CD34^+$ hematopoietic cells for correction of X-linked chronic granulomatous disease. Nat Biotechnol 34:424–429
13. De Ravin SS, Li L, Wu X, Choi U, Allen C, Koontz S, Lee J, Theobald-Whiting N, Chu J, Garofalo M, Sweeney C, Kardava L, Moir S, Viley A, Natarajan P, Su L, Kuhns D, Zarember KA, Peshwa MV, Malech HL (2017) CRISPR-Cas9 gene repair of hematopoietic stem cells from patients with X-linked chronic granulomatous disease. Sci Transl Med 9(372): eaah3480
14. Yusa K, Rashid ST, Strick-Marchand H, Varela I, Liu P-Q, Paschon DE, Miranda E, Ordonez A, Hannan NRF, Rouhani FJ, Darche S, Alexander G, Marciniak SJ, Fusaki N, Hasegawa M, Holmes MC, Di Santo JP, Lomas DA, Bradley A, Vallier L (2011) Targeted gene correction of α_1-antitrypsin deficiency in induced pluripotent stem cells. Nature 478(7369):391–394
15. Sanjana NE, Shalem O, Zhang F (2014) Improved vectors and genome-wide libraries for CRISPR screening. Nat Methods 11:783–784
16. Hwang WY, Fu Y, Reyon D, Maeder ML, Tsai SQ, Sander JD, Peterson RT, Yeh JRJ, Joung JK (2013) Efficient genome editing in zebrafish using a CRISPR-Cas system. Nat Biotechnol 31(3):227–229
17. Smith RH, Afione SA, Kotin RM (2002) Transposase-mediated construction of an integrated adeno-associated virus type 5 helper plasmid. BioTechniques 33(1):204–206. 208, 210–211
18. Markusic DM, Herzog RW, Aslanidi GV, Hoffman BE, Li B, Li M, Jayandharan GR, Ling C, Zolotukhin I, Ma W, Zolotukhin S, Srivastava A, Zhong L (2010) High-efficiency transduction and correction of murine hemophilia B using AAV2 vectors devoid of multiple surface-exposed tyrosines. Mol Ther 18 (12):2048–2056
19. Song L, Li X, Jayandharan GR, Wang Y, Aslanidi GV, Ling C, Zhong L, Gao G, Yoder MC, Ling C, Tan M, Srivastava A (2013) High-efficiency transduction of primary human hematopoietic stem cells and erythroid lineage-restricted expression by optimized AAV6 serotype vectors *in vitro* and in a murine xenograft model *in vivo*. PLoS One 8(3): e58757
20. Merling RK, Sweeney CL, Choi U, De Ravin SS, Myers TG, Otaizo-Carrasquero F, Pan J, Linton G, Chen L, Koontz S, Theobald NL, Malech HL (2013) Transgene-free iPSCs generated from small volume peripheral blood nonmobilized $CD34^+$ cells. Blood 121(14): e98–e107
21. Yamauchi A, Yu L, Pötgens Andy JG, Kuribayashi F, Nunoi H, Kanegasaki S, Roos D, Malech Harry L, Dinauer Mary C, Nakamura M (2013) Location of the epitope for 7D5, a monoclonal antibody raised against human flavocytochrome b_{558}, to the extracellular peptide portion of primate $gp91^{phox}$. Microbiol Immunol 45(3):249–257
22. Brinkman EK, Chen T, Amendola M, van Steensel B (2014) Easy quantitative assessment of genome editing by sequence trace decomposition. Nucleic Acids Res 42(22):e168–e168
23. Heyer W-D, Ehmsen KT, Liu J (2010) Regulation of homologous recombination in eukaryotes. Annu Rev Genet 44:113–139
24. Gutschner T, Haemmerle M, Genovese G, Draetta Giulio F, Chin L (2016) Post-translational regulation of Cas9 during G1 enhances homology-directed repair. Cell Rep 14(6):1555–1566
25. Henderson LM, Chappell JB (1993) Dihydrorhodamine 123: a fluorescent probe for superoxide generation? Eur J Biochem 217 (3):973–980
26. Fu Y, Sander JD, Reyon D, Cascio VM, Joung JK (2014) Improving CRISPR-Cas nuclease specificity using truncated guide RNAs. Nat Biotechnol 32:279–284

27. Fares I, Chagraoui J, Gareau Y, Gingras S, Ruel R, Mayotte N, Csaszar E, Knapp DJHF, Miller P, Ngom M, Imren S, Roy D-C, Watts KL, Kiem H-P, Herrington R, Iscove NN, Humphries RK, Eaves CJ, Cohen S, Marinier A, Zandstra PW, Sauvageau G (2014) Pyrimidoindole derivatives are agonists of human hematopoietic stem cell self-renewal. Science 345(6203):1509–1512

Chapter 37

DUOX Defects and Their Roles in Congenital Hypothyroidism

Xavier De Deken and Françoise Miot

Abstract

Extracellular hydrogen peroxide is required for thyroperoxidase-mediated thyroid hormone synthesis in the follicular lumen of the thyroid gland. Among the NADPH oxidases, dual oxidases, DUOX1 and DUOX2, constitute a distinct subfamily initially identified as thyroid oxidases, based on their level of expression in the thyroid. Despite their high sequence similarity, the two isoforms present distinct regulations, tissue expression, and catalytic functions. Inactivating mutations in many of the genes involved in thyroid hormone synthesis cause thyroid dyshormonogenesis associated with iodide organification defect. This chapter provides an overview of the genetic alterations in *DUOX2* and its maturation factor, *DUOXA2*, causing inherited severe hypothyroidism that clearly demonstrate the physiological implication of this oxidase in thyroid hormonogenesis. Mutations in the *DUOX2* gene have been described in permanent but also in transient forms of congenital hypothyroidism. Moreover, accumulating evidence demonstrates that the high phenotypic variability associated with altered DUOX2 function is not directly related to the number of inactivated *DUOX2* alleles, suggesting the existence of other pathophysiological factors. The presence of two DUOX isoforms and their corresponding maturation factors in the same organ could certainly constitute an efficient redundant mechanism to maintain sufficient H_2O_2 supply for iodide organification. Many of the reported *DUOX2* missense variants have not been functionally characterized, their clinical impact in the observed phenotype remaining unresolved, especially in mild transient congenital hypothyroidism. DUOX2 function should be carefully evaluated using an in vitro assay wherein (1) DUOXA2 is co-expressed, (2) H_2O_2 production is activated, (3) and DUOX2 membrane expression is precisely analyzed.

Key words NADPH oxidase, Thyroid, Congenital hypothyroidism, DUOX, DUOXA, H_2O_2, Dual oxidase, DUOX maturation factor, Inherited disease

1 Introduction

In 1908, a huge oxidative burst was reported upon sea urchin egg fertilization [1]. Hydrogen peroxide (H_2O_2) produced during this process has been shown later to mediate the formation of covalent dityrosine bounds by the ovoperoxidase in the extracellular matrix protecting the egg from polyspermy [2]. The discovery in 1986 of the molecular nature of the phagocyte oxidase, NOX2/gp91^{phox}, responsible for the "respiratory burst" [3] revealed that molecular complexes present from fungi to mammals have been selected to

Ulla G. Knaus and Thomas L. Leto (eds.), *NADPH Oxidases: Methods and Protocols*, Methods in Molecular Biology, vol. 1982,
https://doi.org/10.1007/978-1-4939-9424-3_37,

produce ROS dedicated to specialized cellular functions in a variety of subcellular compartments [4]. The existence of a thyroid H_2O_2-generating system was postulated in the early seventies to be necessary for thyroid hormone (TH) synthesis by the enzyme thyroperoxidase (TPO) [5]. Further biochemical studies demonstrated that H_2O_2 was produced in the thyroid follicular space by a membrane-bound NADPH-dependent flavoprotein using calcium ions to be fully activated [6, 7]. Thirty years later, the molecular nature of the entity responsible of the thyroid H_2O_2-generating system was discovered by two independent groups following different strategies. Starting from purified membranes from pig thyroid follicles, the group of C. Dupuy isolated the $p138^{Tox}$ corresponding to the dual oxidase DUOX2 lacking the first 338 amino acids [8]. Meanwhile, based on functional similarities with NOX2, our group cloned two cDNAs coding for the complete sequences of DUOX1 and DUOX2 after the molecular screening of cDNA libraries generated from human thyrocytes in primary culture [9]. These new members of the NOX family present, respectively, 53% and 47% similarities with the NOX2 catalytic core, including the six transmembrane regions harboring the four histidines and the arginine involved in the heme binding and the COOH intracellular extremity containing the FAD- and NADPH-binding sites characteristic of the NADPH oxidase enzymes. The primary structure of the DUOX proteins (1551 and 1548 amino acids for DUOX1 and DUOX2) is extended at the NH_2-extremity of the conserved catalytic domain by an intracellular loop containing two EF-hand motifs and an additional transmembrane segment followed by an extracellular peroxidase homology domain (PHD) presenting 43% similarities with TPO. The final demonstration of DUOX NADPH oxidase activity has to wait six more additional years and the discovery of the DUOX maturation factors, DUOXA1 and DUOXA2, allowing the reconstitution of a DUOX-based functional H_2O_2-generating complex in heterologous cell systems [10]. Up to now, DUOX/DUOXA expression has been documented in a growing list of non-thyroid tissues, among which the salivary glands, the airways, and the intestinal tract, revealing additional cellular functions associated with DUOX-related H_2O_2 generation [11, 12]. In the present review, we will mainly focus on the role of hydrogen peroxide and DUOXs in thyroid function and their implication in inherited congenital hypothyroidism (CH).

2 The Thyroid Gland

2.1 The Follicular Structure

During evolution, the thyroid function emerges with the capacity to concentrate iodide and to synthesize iodoproteins. Multiple tissues of insects can accumulate radioiodide, but instead of TH synthesis, these iodo-compounds mostly reflect by-products generated during the cuticle formation [13]. An important evolutionary event was the development of iodination units in the endostyle of protochordate species presenting a peroxidase activity [14]. Most vertebrates possess a thyroid able to generate iodothyronines. The thyroid architecture is characterized by follicular units dispersed along the ventral aorta (between the first gill arch and the bulbus arteriosus) like in the zebrafish [15] or encapsulated in a compact glandular structure often divided in two lobes like in mammals. These thyroid functional units are composed of a monolayer of cuboidal cells, the thyrocytes, surrounding a colloidal lumen full of thyroglobulin (TG), the backbone of THs. The polarized organization of the thyroid follicles is critical for iodide concentration and TH storage as iodinated TG. These ovoid 3D structures are embedded in a dense network of blood capillaries allowing intense metabolite exchanges with the thyrocytes: iodide uptake and TH secretion [16]. Thyroid hormones play major roles in the regulation of multiple biological processes including development, growth, and metabolism [17]. Their critical role in embryogenesis is conserved in all vertebrates, especially for neuronal and skeleton development.

Dietary iodine is reduced into iodide before its absorption in the small intestine. About 20% of the iodide perfusing the thyroid is removed at each passage by the basolateral sodium/iodide symporter protein (NIS) allowing iodide to be concentrated 20–50 times in the thyroid (Fig. 1) [18]. However, under sufficient iodine intake, 90% of ingested iodide is lost in urine excretion [19]. To maintain a normal thyroid function, the recommended daily intake of iodine is around 150 μg in human [20]. However, iodine deficiency still remains a worldwide health problem promoting the development of multinodular goiters and thyroid nodules and, in case of severe iodine deficiency, causing hypothyroidism with mental retardation and cretinism. The main sources of iodine are seafood products, iodized salts, and bakery products [21, 22]. After NIS-mediated active transport from the blood, iodide is passively transported across the apical membrane in the follicular lumen. The transmembrane protein anion exchanger pendrin that exchanges chloride for bicarbonate, iodide, or thiocyanate (SCN^-) has been suggested to be involved in this transport [23]. Pendrin protein is expressed not only in the thyroid but also in other tissues, including the kidney, the airways, the mammary gland, and the inner ear. Pendred syndrome, an autosomal recessive disorder, is mainly characterized by deafness, but some patients also suffer from hypothyroidism. However, no thyroid phenotype has been reported in the pendrin knockout mice [24]. Recently, the calcium-activated chloride channel, anoctamin-1, has been demonstrated to mediate iodide efflux across the apical membrane of thyrocytes [25].

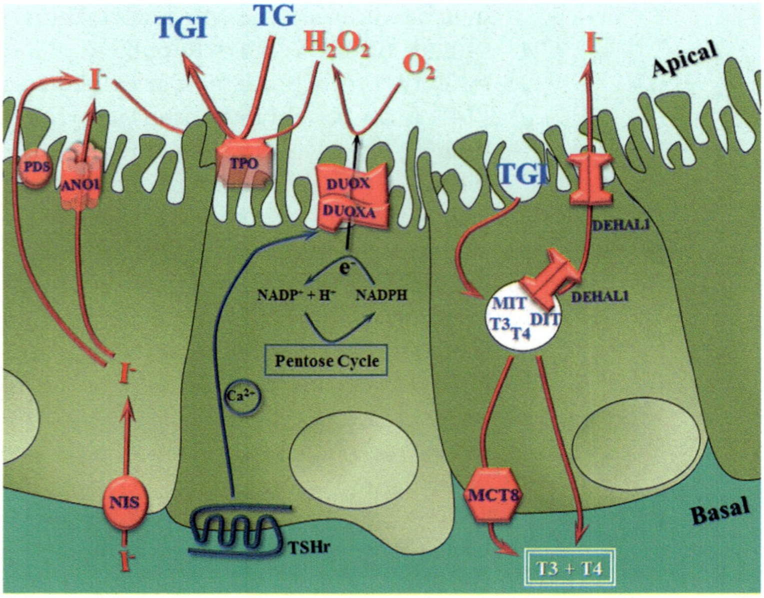

Fig. 1 Thyroid hormone synthesis. At the basal pole of thyrocytes, iodide uptake from the blood is mediated by the symporter NIS. Iodide is transported in the follicular lumen via the iodide channel anoctamin-1 (ANO1) and the anion exchanger pendrin (PDS). At the apex, thyroperoxidase (TPO) catalyzes iodide oxidation and coupling to tyrosine residues of thyroglobulin (TG) in the presence of H_2O_2 generated by DUOX/DUOXA complex. After endocytosis, thyroid hormones (T3 and T4) are released from iodinated TG (TGI) by proteolytic cleavage. Iodotyrosines (MITs and DITs) are deiodinated to recycle iodine by the iodotyrosine dehalogenase (DEHAL1). Thyroid hormones are secreted in the blood via dedicated transporters like the monocarboxylate transporter 8 (MCT8). The thyroid function is under the control of thyrotropin via its protein G-coupled receptor (TSHr)

2.2 Thyroid Hormone Biosynthesis

In the follicular lumen, iodide is oxidized and covalently linked to tyrosine residues of the macromolecule thyroglobulin [5]. This so-called organification process is catalyzed by the transmembrane thyroperoxidase at the apex of the thyrocytes in the presence of H_2O_2 generated by DUOX (Fig. 1). A close proximity between DUOX and TPO has been demonstrated at the plasma membrane of the follicular cells [26, 27]. This apical membrane complex named thyroxisome favors the hormonogenesis and limits H_2O_2 leakage [28]. Under sufficient iodine supply, the amount of H_2O_2 produced constitutes the limiting factor for TH synthesis [29]. The 933 residues of TPO present 44% sequence similarity with the myeloperoxidase and is composed of a short intracellular COOH-extremity, one transmembrane region, and a long catalytic ectodomain containing the heme moiety [30]. The final step of TH biosynthesis consists of the TPO-mediated coupling reactions. The assembly of two diiodotyrosines (DIT) forms the

3,5,3′,5′-tetraiodothyronine or thyroxin (T4), while the coupling of one monoiodotyrosine (MIT) with one DIT generates the 3,5,3′-triiodothyronine (T3), the active form of THs [31].

TG is the most abundant protein in the thyroid at a concentration of 200–300 mg/mL in the follicular lumen. Its main function is to provide the polypeptide backbone for TH synthesis, as well as TH storage and iodine depot when iodine availability is limited. The TG transcript encodes a protein of 2767 amino acids containing numerous proline and cysteine residues in constant position participating in the secondary structure of the protein via intramolecular disulfide bonds [32]. Mature TG is mainly found as homodimers with a molecular weight of 660 kDa, the carbohydrates comprising about 10% of its weight. Among the 132 tyrosyl residues of TG dimers, only 25–30 participate in the iodination reactions. These hormonogenic residues will be further used for the coupling reactions. One iodophenoxyl group from a MIT or a DIT residue called the "donor" is transferred onto a DIT residue called the "acceptor" [33]. Only 5–16 can be associated to generate 2–8 molecules of T4 and T3 [34]. A typical distribution for a TG containing 0.5% iodine is 5 residues of MIT, 5 of DIT, 2.5 of T4, and 0.7 of T3 [35].

Before being delivered into the bloodstream, THs must be released from TG after its internalization by endocytosis and fusion in lysosomal compartments [36]. Proteolytic cleavage is mediated by multiple glycohydrolases, phosphatases, sulfatases, and proteases including the cathepsins D, H, and L [37]. Nevertheless, about 70% of the TG iodine contents are in the forms of MITs and DITs which are deiodinated to recycle iodine in the intrathyroidal iodide pool. An iodotyrosine dehalogenase (DEHAL1) present at the apical plasma membrane and in endocytic vesicles of the thyrocytes has been characterized [38]. Patients carrying biallelic loss-of-function mutations in the corresponding gene present high levels of MIT/DIT in the urines and may develop a goiter under iodine deficiency [39]. For decades, the lipophilic nature of T3/T4 suggested that TH secretion was mediated mainly by simple diffusion across the basal plasma membrane. However, recent data clearly demonstrated that TH efflux from the thyroid in the blood and their uptake by the targeted cells are mediated by dedicated transporters like the monocarboxylate transporter 8 (MCT8) [40].

Thyroid physiology is mainly controlled by iodide availability and the plasma level of THs. A negative feedback loop controls TH synthesis and secretion via the thyrotropin (TSH) secreted by the pituitary. The TSH receptor (TSHr), a G-protein-coupled receptor, activates in humans two signaling pathways [41]: (1) The cAMP cascade stimulates TH secretion as well as the expression of genes involved in TH synthesis (*NIS*, *TPO*, *TG*); and (2) The Gq/phospholipase C cascade activates TH synthesis mainly via the activation of the thyroid H_2O_2-generating system. Under intense and chronic thyroid stimulation by TSH, thyroid cell proliferation is increased leading to the classical goiter formation. Iodine excess can also rapidly induce an inhibition of thyroid function [42] called the Wolff-Chaikoff effect mediated by an iodo-compound, 2-iodo-hexadecanal, that is able to inhibit H_2O_2 generation, blocking iodide organification [43].

3 The Thyroid H_2O_2-Generating Complex: DUOX/DUOXA

3.1 The DUOX/DUOXA Gene Locus

DUOX and *DUOXA* genes are oriented head to head in an operon-like unit, each couple of genes being located in tandem on the long arm of chromosome 15 (Fig. 2A) and sharing the same bidirectional promoter [10, 44, 45]. *DUOX1* spans 36 kb and is composed of 35 exons, *DUOX2* spans 21.5 kb containing 34 exons, and both genes are composed of 33 coding exons [46]. Interestingly, the length and position of the exons coding for the functional domains (EF-hands, FAD- and NADPH-binding sites) are well conserved between the two genes as well as with *NOX2*, reflecting their common molecular evolution (Fig. 2B) [4]. The DUOXA2 open reading frame spans 6 exons encoding a 320 amino acid protein composed of five transmembrane segments, the first extracellular loop presenting N-glycosylation sites, and a C-terminal cytoplasmic region. Four alternative DUOXA1 splicing variants have been identified. DUOXA1α (343 residues) corresponds to the closest homolog of DUOXA2 and is the major

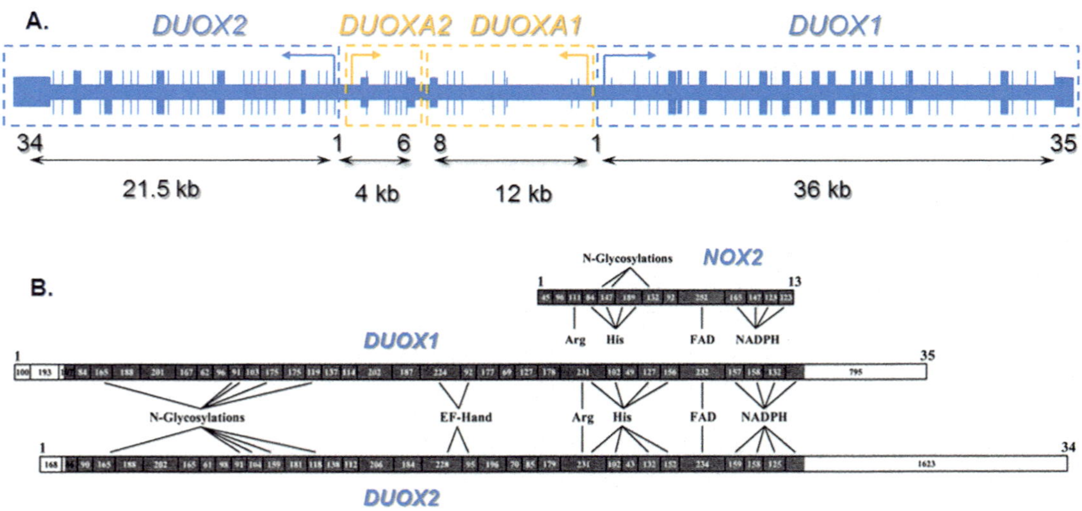

Fig. 2 A. Schematic structure of the genomic organization of the *DUOX/DUOXA* gene locus on chromosome 15q15.3 with the number of exons represented as vertical bars. B. Comparison of the transcripts encoding DUOX1, DUOX2, and NOX2. The length of each exon is mentioned and the corresponding coding region is represented in gray. Exons coding for N-glycosylation sites, EF-hand motifs, and heme-, FAD-, and NADPH-binding sites are indicated

variant in DUOX1/DUOXA1 expressing tissues [47, 48]. Due to their extra-thyroid expression, *DUOX/DUOXA* could not be defined as thyroid-specific genes. However, their expression appears only at the late stages of cell differentiation during thyroid embryogenesis in mice and fish, when the follicular structure is functionally specified making them important thyroid differentiation markers [15, 49]. In human thyroid, the *DUOX2* transcript is 2–5 times more expressed than *DUOX1* [46]. Transcriptional regulation by TSH seems to be species dependent. In dog, rat, and pig, *DUOX2* mRNA expression is positively controlled through the activation of the cAMP pathway [50–52], whereas in mouse and human, no significant modulation of *DUOX* transcription is observed [46, 49, 53].

3.2 Maturation of DUOX Proteins

The oxidases are fully functional when properly addressed at the apical membrane of the thyroid cell. When traveling to the apex, DUOX proteins undergo N-linked glycosylation in the Golgi apparatus to adopt the active 190 kDa form [50, 51]. Complete glycosyl-defective DUOX mutants generated by site-directed mutagenesis demonstrate impairment of cell surface expression and ROS production in reconstituted cellular system [54]. However, inhibition of the Golgi complex α-mannosidase II by swainsonine results in a fully active enzyme targeted at the membrane, demonstrating that the maturation of the N-glycan moieties in the Golgi is dispensable for the function of the oxidases [55]. In the absence of DUOX maturation factors, the oxidases are retained in the endoplasmic reticulum (ER) compartment where only low levels of superoxide ($O_2^{\cdot -}$) are detected [56]. Their critical role for DUOX function has been clearly demonstrated in DUOXA1/DUOXA2 double knockout mice [57] showing a hypothyroid phenotype characterized by the impairment of T4 production in thyroid follicles caused by the absence of DUOX cell surface expression and loss of H_2O_2 generation (Fig. 3; personal communication and [58]). Initially, DUOXA proteins were characterized as ER-resident proteins allowing ER-to-Golgi transition of mature DUOX enzymes [10]. However, accumulating evidence suggests now that they most probably act as organizing elements required for surface expression but also regulation of DUOX activity [47, 48, 59]. Their role can be related to the $p22^{phox}$ function for the NOX enzymes.

3.3 Control of DUOX Catalytic Activity

As NOX5, DUOX isoenzymes are obligate Ca^{2+}-dependent NADPH oxidases via the two EF-hand Ca^{2+}-binding motifs [60]. Contrary to the other NOX proteins, Rac1 activation is not required for DUOX-mediated thyroid H_2O_2 generation [61]. The intrinsic activity of DUOX enzymes can be also modulated via direct serine/threonine phosphorylation. In DUOX/DUOXA reconstituted Cos-7 cells, protein kinase A-mediated

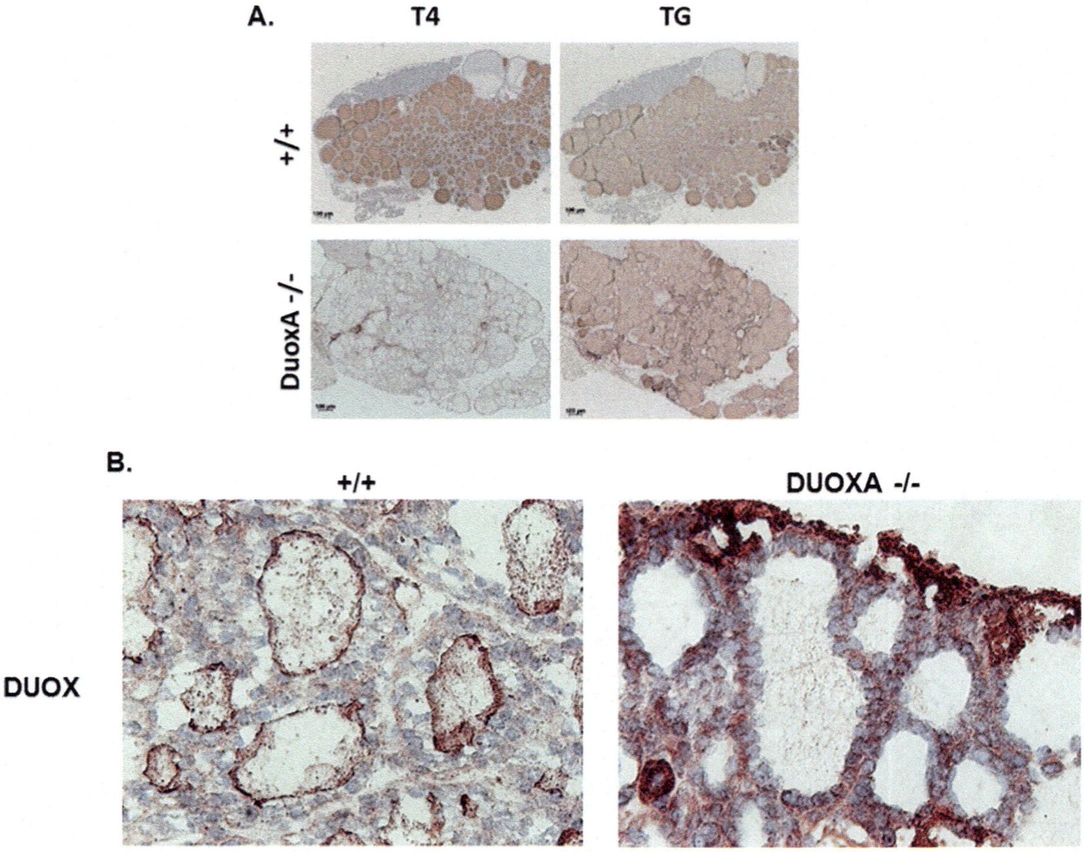

Fig. 3 A. T4 and TG immunostaining on serial thyroid sections (5 μm thick) from wild-type (+/+) and DUOXA-deficient (DUOXA−/−) mice. B. DUOX immunodetection in thyroid sections (5 μm thick) from wild-type and DUOXA −/− mice. DUOX immunostaining was localized at the apical membrane of wild-type mice and in the cytoplasm for DUOXA −/− mice

phosphorylation on serine 955 activates DUOX1, while DUOX2 is stimulated by nanomolar concentrations of phorbol 12-myristate 13-acetate (PMA), associated with protein kinase C-dependent phosphorylation [60]. Primary cultured human thyrocytes and bronchial epithelial cells show an increase in H_2O_2 generation after PMA treatment [60, 62, 63]. Finally, micromolar concentrations of iodide are also able to trigger H_2O_2 generation in human, pig, and dog thyroid slices [64], whereas higher concentrations inhibit H_2O_2 production via the Wolff-Chaikoff effect that represses iodide metabolism, preventing thyrotoxicosis [42].

Based on their sequence homology with NOX2 and the obligate one-electron transfer from the heme, the dual oxidases should primarily produce superoxide [65]. However, DUOX1 and DUOX2 co-expressed with their corresponding partner produce mainly hydrogen peroxide. Structure/function studies have demonstrated that the second intracellular loop and the COOH-terminal tail of DUOXA1 are required for H_2O_2 production by

DUOX1, while active DUOX2 depends on the integrity of the NH_2-terminal extremity of its maturation factor [66]. Moreover, exchanging the first N-linked glycosylated extracellular loop between DUOXA1 and DUOXA2 does not alter DUOX maturation, suggesting that this region could rather be involved in DUOX cell surface expression, a common feature shared by the two maturation factors. The existence of cysteine disulfide bridges for intermolecular protein-protein interactions between DUOX1 and DUOXA1 has been postulated [67]. In addition, recent studies by the group of C. Dupuy demonstrated the implication of two cysteine residues (Cys-124 and Cys-1162) in the formation of an intramolecular disulfide bound that stabilizes the conformation of DUOX2 supporting its interaction with DUOXA2 [68].

DUOX2, but not DUOX1, generates superoxide when co-expressed with the DUOXA1 maturation factor [47, 69]. Moreover, alterations of the NH_2-terminal end of DUOXA2 by deletion or exchange with the NH_2-extremity of DUOXA1 are sufficient to turn DUOX2 to a superoxide-generating enzyme [66]. Likewise, addition of a small unrelated sequence in front of this region in wild-type DUOXA2 converts DUOX2 to a dual-generating oxidase producing H_2O_2 and superoxide. A similar switch to $O_2\bullet^-$ production has been reported for NOX4 after the replacement of its signal peptide with the corresponding NOX1 sequence [70]. An elegant structure/function study has been conducted to delineate the domain in DUOX1 that constrains H_2O_2 production [54]. Using chimeric constructs between DUOX1 and DUOX2, Ueyama et al. identified the first extracellular loop of DUOX1 responsible for the reduction of $O_2\bullet^-$ leakage when transferred in DUOX2. However, the purified corresponding peptides did not show any superoxide dismutase activity. NOX4 also possesses a unique third extracellular loop involved in its hydrogen peroxide-generating capacity [71]. Mutational analysis identified an essential residue, His^{222}, involved in this process, suggesting that it could be an important proton donor to facilitate the formation of H_2O_2 [72]. Interestingly, Ueyama et al. identified in the first extracellular loop of DUOX1 two critical histidines, His^{1017} and His^{1072}, for the reduction of superoxide release.

4 Congenital Hypothyroidism

4.1 Genetic Causes of Congenital Hypothyroidism

Congenital hypothyroidism, characterized by high TSH and low T4 serum levels, is one of the most frequent inherited endocrine disorders affecting one in 3000 newborns [73]. The most frequent cause of sporadic CH is iodine deficiency. During human embryogenesis, iodide trapping and TH synthesis begin only at 10–12 days of gestation [74]. Before, transplacental passages of maternal THs ensure the normal fetal development, especially the maturation of the central nervous system. The implementation of neonatal screening for CH with blood spot assays in the early seventies allowed early treatment of newborn babies with thyroxin supplementation reducing the risk of mental retardation [75].

Thyroid dysgenesis (TD) is another major cause (85%) of CH resulting in thyroid ectopy, athyreosis, or hypoplasia. Loss-of-function mutations in the TSHr encoding gene have been shown to cause thyroid hypoplasia and TSH resistance in humans and in the hyt/hyt mouse model [76, 77]. Germline mutations in thyroid-related transcription factors have also been described in patients suffering from TD with athyreosis associated with cleft palate and spiky hair (*FOXE1*) [78] or thyroid gland hypoplasia sometimes mislocalized (*NKX2.*1 and *PAX8*) [79, 80]. However, the genetic causes of the majority of TD familial cases are still unknown. High prevalence of congenital heart diseases co-occurring with TD suggested that cardiovascular and thyroid developments could be linked [81]. Recent studies performed in the zebrafish have beautifully demonstrated the relationship between cardiovascular development and thyroid morphogenesis, the former probably being used as tissue guidance for correct thyroid migration [82, 83].

About 15% of CH cases are due to defects in thyroid hormone synthesis causing thyroid dyshormonogenesis (TDH), a group of disorders often inherited in an autosomal recessive manner [84]. Inactivating mutations in many of the genes involved in TH synthesis cause TDH associated with iodide organification defect (IOD). After thyroid trapping, free iodide is rapidly covalently bound to TG tyrosyl residues, remaining in the thyroid gland even after the blocking of its transport by the NIS-competitive inhibitor perchlorate. In the perchlorate discharge test, the amount of radioiodide in the neck is followed using a gamma camera after its uptake by the thyroid was blocked by perchlorate, 2 hours after radioisotope administration [85]. In healthy patients, less than 10% of the isotope initially present in the thyroid is washed out 1 hour after perchlorate injection [86]. A discharge value between 10 and 90% reflects a partial IOD. A summary of the etiologic classification of primary CH is presented in Fig. 4 [75, 85]. When ultrasonography revealed an eutopic thyroid gland, high TSH serum levels associated with low serum TG concentrations often reflect inactivating mutations in the *TG* gene. An absence or low iodide uptake detected by scintigraphy most probably suggests a NIS defect. Goiter is not always present in the affected patients, and the severity of hypothyroidism will be dependent on the dietary iodine intake [87]. The most prevalent cause of TDH is TPO deficiency usually associated with a total IOD (perchlorate discharge value >90%) [88]. Alterations of genes encoding TG, NIS, DUOX2, DUOXA2, or pendrin have been identified in patients with partial IOD [69, 86, 89–91].

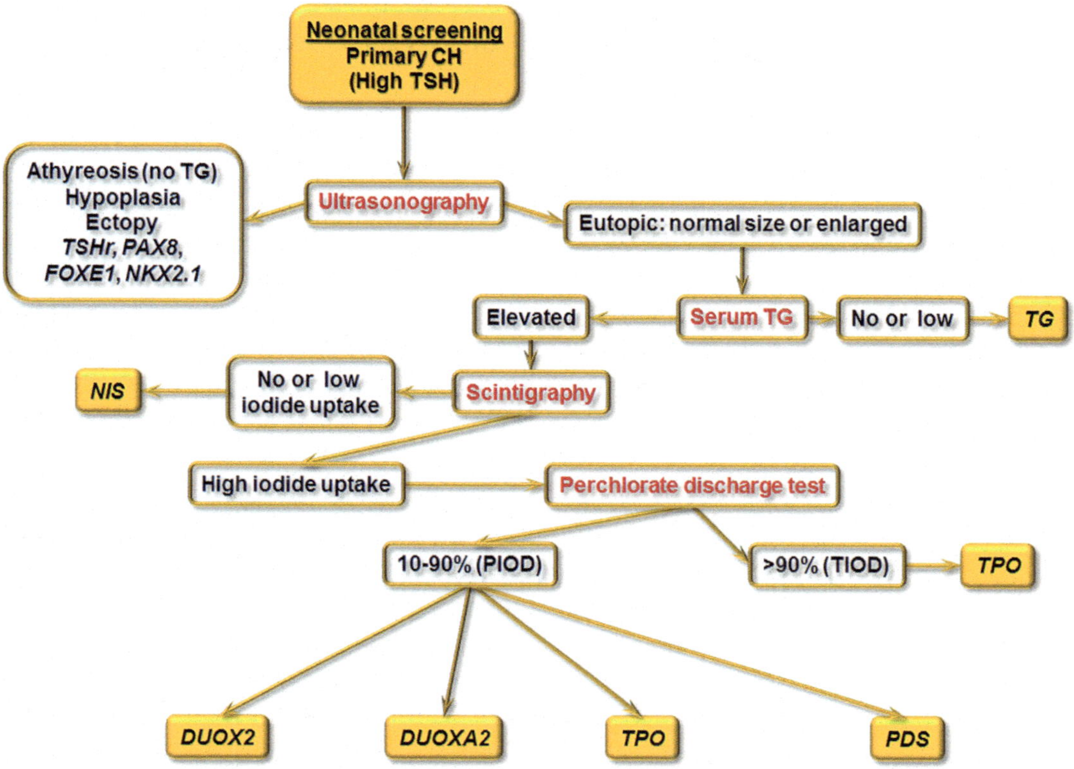

Fig. 4 Etiologic classification of primary congenital hypothyroidism (adapted from [75, 85]). *PIOD* partial iodide organification defect, *TIOD* total iodide organification defect

4.2 DUOX Defects in Congenital Hypothyroidism

Biallelic inactivation of *TPO*, *NIS*, *TG*, or *pendrin* causes permanent CH. Transient CH is frequently associated with a temporally limited exposure to external factors during pregnancy such as a lack or an excess of iodide intake [92], transplacental antibodies [93], or antithyroid drug treatment [94]. In 2002, the characterization of the first inactivating *DUOX2* mutations causing IOD undoubtedly demonstrates the essential role played by DUOX2 in TH synthesis [86]. Furthermore, the authors established the first genetic cause for transient CH with mono-allelic inactivation of *DUOX2* and postulated that *DUOX2* biallelic mutations would be associated with a permanent form of CH. However, numerous subsequent studies provide further evidence that the permanent or transient nature of congenital hypothyroidism is not directly related to the number of inactivated *DUOX2* alleles, suggesting the existence of other pathophysiological factors [95–98].

Multiple interesting reviews about *DUOX2/DUOXA2* genetic disorders have been published in the last 10 years [85, 99, 100], including the recent publication by Muzza et al. [101] analyzing the numerous described DUOX2/DUOXA2 variants. We have completed their extensive analysis of the literature with additional reported cases of novel *DUOX2/DUOXA2* deficient patients [102–111]. In summary and to the best of our knowledge, about 105 *DUOX2* variants including in-frame deletions, missense, nonsense, splice site, and frameshift mutations have been described in more than 200 unrelated CH patients. One third of the mutations are found in the ectodomain, one third in the first intracellular loop, and one third in the NOX catalytic domain (Fig. 5A). Interestingly, very few nonsense or frameshift mutations are present in the catalytic core of the protein. The p.S965PfsX29 variant, a frameshift mutation localized in the first intracellular loop of the protein, is the most prevalent *DUOX2* mutation. Multiple reported cases are not compatible with the initial hypothesis, showing transient CH with homozygous *DUOX2* inactivation and permanent CH with heterozygous mutations. Intrafamilial variabilities have also been reported in siblings presenting the same genetic defects. For example, four affected siblings of a family carrying the same compound heterozygous *DUOX2* mutations, p.L479SfsX2 and p. K628RfsX10, present permanent or transient CH [96].

The prevalence of *DUOX2* mutations among CH patients is quite variable but generally high with 29–83% in China [112–115], 43% in Japan [116], 30–45% in Italy [98, 117], 44% in Netherlands [86], and 35% in Korea [118]. In case of suspicion of an inherited congenital hypothyroidism, the best criteria for a *DUOX2* genetic screening are the presence of a goiter, a partial IOD, a low T4/TSH serum ratio, high serum TG levels, and a transient phenotype [119]. The prevalence of *DUOXA2* variants is much lower with less than 1%. Six missense, two nonsense, and two splice site mutations have been described (Fig. 5B) [69, 105, 107, 109, 120–125]. The first homozygous missense p.Y246X mutation was reported in 2008 in a Chinese patient suffering from a mild permanent CH [69]. This mutation has been reported in four additional cases but with various severities in the clinical outcome [107, 121, 122, 124]. Intrafamilial variabilities have also been reported for *DUOXA2*-affected patients. Among two dizygotic twins with a mono-allelic p.Y246X *DUOXA2* mutation and a heterozygous p.R885Q *DUOX2* mutation, the girl presented a more severe hypothyroid phenotype than the brother [124]. Another case of two siblings with biallelic *DUOXA2* p. Y138X mutation showed again that the girl suffered from a permanent CH while her brother was not clinically affected [105].

What would be the possible mechanisms to explain this high phenotypic variability? The presence of two DUOX isoforms and their corresponding maturation factors in the same organ could certainly constitute an efficient redundant mechanism to maintain sufficient H_2O_2 supply for iodide organification. Furthermore, the transient nature of CH could be related to the different requirement of THs with age, from the neonatal period (10–15 μg/kg/day) to adulthood (2 μg/kg/day) [126]. In the presence of

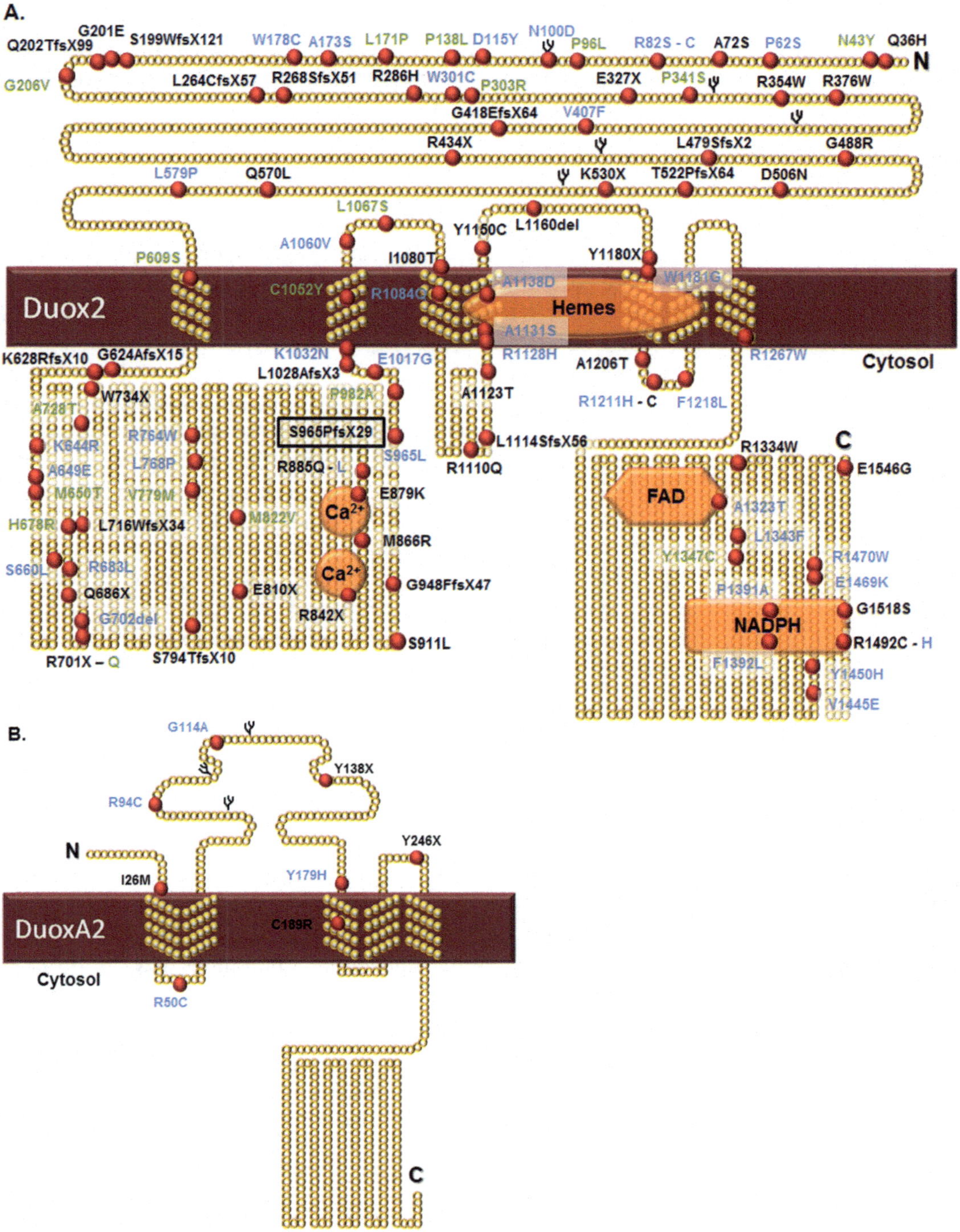

Fig. 5 Schematic structures of DUOX2 (A) and DUOXA2 (B) proteins with the genetic alterations (red dots) identified in congenital hypothyroid-affected patients. For the clarity of the figure, only the missense, nonsense, frameshift mutations and in-frame deletions have been localized. Black, damaging mutations (<60% residual activity); green, functional missense mutants (>60% residual activity); blue, not tested mutations

complete *DUOX2* deficiency, H_2O_2 supply by DUOX1 would be sufficient only after the infantile period when the need of TH decreases. Two independent studies performed in Italian and Japanese populations with CH-affected patients carrying *DUOX2* defects showed a majority of cases where thyroxin supplementation could be stopped after puberty [117, 127]. However, the reduction of T3/T4 production will be permanent in these affected children implying their continuous follow-up throughout their life, especially during pregnancy [99]. Recently, the first biallelic *DUOX1* splice site mutation c.1823-1G>C resulting in a truncated protein (p.V607DfsX43) has been reported in two siblings suffering from a particular severe form of permanent CH [108]. These children presented an additional homozygous p.R434X *DUOX2* mutation that has been originally associated with total IOD [86]. The severity of the phenotype could reflect the absence of compensatory mechanism played by DUOX1 in these affected patients. A mild congenital hypothyroid phenotype in a patient with only one *DUOX2* allele, two *DUOX1* alleles, and one remaining *DUOXA1* functional allele supports also the existence of a compensatory mechanism with DUOXA1 [120]. Another possible source of hormonogenic H_2O_2 could be NOX4. Its expression has been shown to be positively controlled by TSH, but its localization, mainly found in intracellular vesicles, is obviously incompatible with TH synthesis [128]. Finally, the dietary iodide intake was clearly demonstrated to be a disease modifier retarding the appearance of the hypothyroid phenotype [129, 130]. The percentage of households having access to iodide salts is higher in North America and Japan than in Europe where the hypothyroid phenotype seems to be more severe [131].

4.3 DUOX Functional Characterization in Heterologous Cell Systems

With the development of next-generation sequencing techniques, analyses of multiple genetic alterations associated with CH-affected patients have been largely facilitated. The coexistence of multiple genetic alterations in the *DUOX2* gene such as tri-allelic mutations has been associated with an increase in the severity of the disease [106, 112, 113]. In addition, increasing number of clinical case studies report *DUOX2* pathogenic variants concomitant with genetic alterations in other genes involved in TH synthesis including *TG* [125, 127], *TSHr* [102, 109, 118, 132], *TPO* [133], *Pendrin* [115], and *DUOXA2* [107, 123]. Additional studies would clarify their functional relevance in the evolution of the pathology. However, many of the reported *DUOX2* missense variants have not been functionally characterized raising the issue of their real functional impact in the observed phenotype, especially in mild transient CH.

To date, about 78 *DUOX2* missense mutations have been described. Only 41 have been functionally characterized in various heterologous systems, 23 of them showing a reduced catalytic activity by more than 60% (summarized in Table 1). However, an

Table 1
Summary of published functional DUOX2/DUOXA2 assays performed in heterologous cell systems to test the activity of *DUOX2/DUOXA2* missense genetic variants identified in congenital hypothyroid-affected patients

	Enzymatic activity					Surface expression			
Date	DUOX2/DUOXA2 constructs	Cells	Activators	Tests	DUOX2 mutants with <60% activity	Cells	FACS/IF	DUOX2 mutant classification	References
2007	HA-DUOX2 DUOXA2-Myc	Hela/Cos-7	Ionomycin	Amplex Red Homovanillic acid	p.Q36H (0%) p.R376W (0%) p.D506N (50%)	Hela/Cos-7	FACS/IF	p.Q36H (0) p.R376W (0) p.D506N (−)	[55]
2008	HA-DUOX2 Myc-DUOXA2	Hela	NM	Amplex Red Diogenes Reagent		Hela	IF		[69][a]
2009	HA-DUOX2 DUOXA2-Myc	Hela	NM	Amplex Red	p.S911 L (50%)	Hela	FACS	p.S911 L (−) **p.C1052Y (−)**	[138]
2010	HA-DUOX2 DUOXA2-Myc	CHO	Ionomycin PMA	Homovanillic acid Diogenes Reagent	p.G1518S (0%)	CHO	FACS	p.G1518S (+)	[135]
2011	HA-DUOX2 DUOXA2-Myc	Hek293	Ionomycin	Amplex Red	p.I1080T (50%) p.R1110Q (15%)	NT	NT		[116]
2011	HA-DUOX2 Myc-DUOXA2	Hela	NM	Amplex Red		NT	NT		[120][a]
2011	HA-DUOX2 DUOXA2-Myc	Hela	NM	Amplex Red	p.Y1150C (10%) **p.A728T (10%)**	Hela	FACS	p.Y1150C (+) **p.A728T (+)**	[98]

(continued)

Table 1
(continued)

	Enzymatic activity					Surface expression			
Date	**DUOX2/DUOXA2 constructs**	**Cells**	**Activators**	**Tests**	**DUOX2 mutants with <60% activity**	**Cells**	**FACS/IF**	**DUOX2 mutant classification**	**References**
2014	GFP-DUOX2 V5-DUOXA2	A549	NM	Amplex Red	p.A72S (25%) p.G488R (0%) p.E879K (5%) p.R885Q (5%) p.R1110Q (10%) p.A1123T (10%) p.R1334W (40%)	NT	NT		[118]
2014	HA-DUOX2 DUOXA2-Myc	Hela	NM	Amplex Red	p.Q570L (20%) p.M866R (0%) **p.C1052Y (20%)** p.E1546G (50%)	NT	NT		[119]
2015	HA-DUOX2 DUOXA2-Myc	Hek293	Ionomycin	Amplex Red		Hek293	IF	p.Y1347C (−)	[134]
2015	HA-DUOX2 DUOXA2	H661 clone DUOXA2	Thapsigargin	Homovanillic Ac	p.R1211C (55%) p.R1492C (15%)	H661 clone DUOXA2	FACS/IF	p.R1211C (+) p.R1492C (+)	[111]
2016	HA-DUOX2 DUOXA2-Myc	Hek293	Ionomycin	Amplex Red	p.R1110Q (5%) p.R1334W (25%)	NT	NT		[133]
2016	HA-DUOX2 Flag-DUOX2 DUOXA2-GFP	Hek293/ Hela	Ionomycin + PMA	Amplex Red Diogenes Reagent	**p.P303R (40%)**	Hek293	FACS	**p.P303R** (−)	[110]

2017	HA-DUOX2 DUOXA2	H661 clone DUOXA2	Thapsigargin	Homovanillic acid	p.R286H (0%)	H661 clone DUOXA2	FACS	p.R286H (0) **p.P609S** (−)	[104]
2017	HA-DUOX2 DUOXA2-Myc	Hek293	Ionomycin	Amplex Red	p.G201E (0%)	Hek293	FACS/IF	p.G201E (0)	[106]
2018	HA-DUOX2 DUOXA2-Myc	Hek293	Ionomycin	Amplex Red		NT	NT		[102]
2018	DUOX2 DUOXA2	Hela/ Hek293	NM	Amplex Red	p.R354W (0%) p.A1206T (0%)	NT	NT		[103]

Proposed *DUOX2* mutant classification: DUOX2^0: Absence of H_2O_2 generation associated with no DUOX2 cell surface expression; DUOX2$^-$: Low H_2O_2 generation associated with reduced DUOX2 cell surface expression; DUOX2$^+$: No or reduced H_2O_2 generation associated with DUOX2 cell surface expression. *FACS* flow cytometry, *IF* immunofluorescence, *NT* not tested, *NM* not mentioned. DUOX2 variants with conflicting reported data are presented in bold

[a]Only DUOXA2 missense variants tested

ideal test to carefully assess DUOX2 function must include specific quantification of H_2O_2 production but also evaluate the expression of the enzyme at the cell surface by flow cytometry using NH_2-terminally epitope-tagged proteins. From the 17 publications including DUOX2 functional assays, only 10 analyze the membrane expression of the protein. Reagents specific for H_2O_2 detection must be systematically used in the assays as the Amplex Red or the homovanillic acid. Moreover, as the basal activity of the oxidase is very low, agents increasing the intracellular Ca^{2+} concentration must be added in the assay. The calcium ionophore ionomycin [55, 106, 110, 116, 133–135] and the Ca^{2+}-ATPase inhibitor thapsigargin [104, 111] have been successfully used in different assays. The cell type could also modify the final outcome of the assay in terms of activity or maturation of the protein. The Hek293 and Hela cells are the two main cell lines used in the published assays.

A reduction of H_2O_2 generation for a particular DUOX2 variant has to be normalized to the level of protein expression but foremost compared to the cell surface expression of the oxidase. Based on this assay, we have characterized the functional consequences of the first DUOX2 variant in the NOX catalytic core [135]. The DUOX2 p.G1518S mutant was nonfunctional but correctly processed at the cell surface. A NOX2 p.C537R variant affecting a highly conserved cysteine localized in the fourth NADPH-binding site, next to the DUOX2 mutated glycine, presented similar functional consequences [136]. We already proposed to apply a comparable classification used for NOX2 mutants depending on the behavior of the protein in heterologous system (DUOX2 mutant classification mentioned in Table 1): $DUOX2^0$ variants showing a loss of H_2O_2 generation associated with no surface expression, $DUOX2^-$ characterized by a low oxidase activity and a reduced membrane expression, and the $DUOX2^+$ mutants which are expressed at the cell surface but less or not active [135, 137]. Additional mutations affecting functional domains of DUOX2 have been described: in the second EF-hand (p.E879K [96, 112, 114, 118, 127]) and in the FAD- (p.A1323T [113, 132]) and NADPH-binding sites (p.P1391A, p.F1392L, p.R1492C, p.R1492H [109, 111, 113, 114, 127]). Interestingly, the three mutants in DUOX2 functional domains that have been fully characterized (activity and membrane expression) to date (p.G1518S, p.R1492C, and p.E879Q generated by site-directed mutagenesis [60]) all belong to the $DUOX2^+$class.

About 18 tested missense DUOX2 variants show residual activity with more than 60% of the wild-type protein (displayed in green in Fig. 5A), raising the question of their clinical relevance in the pathology of the affected CH patient. The following mutants could be considered as functional single-nucleotide polymorphism (SNP) with limited damaging impact: p.P138L, p.P341S [119], p.H678R

[116, 118, 119, 127], p.R701Q [119], p.P982A [98, 119], p.L1067S [119], p.A728T [98, 106], p.M650T, p.M822V, p.P609S [104, 106], p.C1052Y [138], p.Y1347C [83], p.V779M [102], p.L171P, p.P303R [110, 119], p.G206V, p.N43Y, and p.P96L [118]. The p.H678R mutant has been functionally characterized as a non-deleterious variant, even if the mutation was combined with two other DUOX2 variants, p.R701Q and p.P982A, in the same expressed protein [98].

Conflicting data about the functional analysis of some DUOX2 variants have also been reported. The p.A728T mutant has been shown to be a DUOX2$^+$ nonfunctional variant when expressed in Hela cells [98]. But the same mutant showed unaffected H_2O_2 production and cell surface expression in Hek293 cells [106]. The possible absence of stimulating agents (not mentioned) in the first study could explain this discrepancy as well as the different cell types used. The p.C1052Y DUOX2 variant expressed in Hela cells presented a residual activity of more than 60% in one study [138] but only 20% in another [119]. The p.P303R mutant presented about 75% enzymatic activity of the wild type in one study [119] but less than 50% with reduced cell surface expression in another study (DUOX2$^-$) [110]. Additional conflicting data concern a DUOX2 variant reported in a boy suffering from very early-onset inflammatory bowel disease (IBD) with compound heterozygous mutations [104]. After expression in the large cell lung carcinoma H661 cell line and stimulation with thapsigargin, the p.R286H mutant was undetectable at the cell surface (DUOX2^0), while p.P609S expressing cells showed a limited reduction of H_2O_2 generation with lower protein surface expression (DUOX2$^-$). When expressed in Hek293 cells, the p.P609S DUOX2 mutant presented unaffected H_2O_2 generation as well as normal cell surface expression compared to the wild-type protein [106].

5 Conclusions

The implication of DUOX-mediated H_2O_2 in thyroid function is undoubtedly established. The DUOX2/DUOXA2 system represents the major H_2O_2 provider for TPO in thyroid hormonogenesis, but it could be supplemented and compensated by DUOX1/DUOXA1 under pathological circumstances. Recently, additional DUOX-related diseases have been reported in familial cases of IBD with mono- or biallelic *DUOX2* mutations. Even if no strong thyroid phenotype was identified in these patients, a recent meta-analysis-based study suggested that patients with transient CH with partial defects in DUOX2 are at higher risk of developing IBD [139]. The phenotypic characterization of new tissue-targeted DUOX- and DUOXA-deficient animal models will be very useful to study the DUOX function in extra-thyroid tissues. Better functional characterization of the novel *DUOX2* missense mutations using standardized DUOX functional assays combining quantifications of extracellular H_2O_2 generation and DUOX cell surface expression will be required to better evaluate their functional consequences in the clinical outcome of the affected patients. However, the different cell types used in the assays are not of thyroid origin and would therefore not reflect the physiological behavior of the protein in a real thyrocyte.

Acknowledgments

The authors acknowledge the support of the "Fonds de la Recherche Scientifique" (FRS-FNRS), the "Fonds Docteur J.P. Naets" managed by the "Fondation Roi Baudouin," and the "Fondation Tournay-Solvay."

References

1. Warburg O (1908) Beobachtungen über die Oxydationsprozesse im Seeigelei. Z Physiol Chem 57:1–11
2. Wong JL, Creton R, Wessel GM (2004) The oxidative burst at fertilization is dependent upon activation of the dual oxidase Udx1. Dev Cell 7(6):801–814
3. Royer-Pokora B, Kunkel LM, Monaco AP, Goff SC, Newburger PE, Baehner RL, Cole FS, Curnutte JT, Orkin SH (1986) Cloning the gene for an inherited human disorder-chronic granulomatous disease-on the basis of its chromosomal location. Nature 322:32–38
4. Kawahara T, Quinn MT, Lambeth JD (2007) Molecular evolution of the reactive oxygen-generating NADPH oxidase (Nox/Duox) family of enzymes. BMC Evol Biol 7:109
5. Dumont JE (1971) The action of thyrotropin on thyroid metabolism. Vitam Horm 29:287–412
6. Carvalho DP, Dupuy C, Gorin Y, Legue O, Pommier J, Haye B, Virion A (1996) The Ca^{2+}- and reduced nicotinamide adenine dinucleotide phosphate-dependent hydrogen peroxide generating system is induced by thyrotropin in porcine thyroid cells. Endocrinology 137(3):1007–1012
7. Gorin Y, Ohayon R, Carvalho DP, Dème D, Leseney AM, Haye B, Kaniewski J, Pommier J, Virion A, Dupuy C (1996) Solubilization and characterization of a thyroid Ca^{2+}-dependent and NADPH-dependent $K_3Fe(CN)_6$ reductase. Relationship with the NADPH-dependent H_2O_2-generating system. Eur J Biochem 240:807–814
8. Dupuy C, Ohayon R, Valent A, Noël-Hudson MS, Dème D, Virion A (1999) Purification of a novel flavoprotein involved in the thyroid NADPH oxidase. J Biol Chem 274 (52):37265–37269
9. De Deken X, Wang D, Costagliola S, Libert F, Vassart G, Dumont JE, Miot F (2000) Cloning of two human thyroid cDNAs encoding new members of the NADPH oxidase family. J Biol Chem 275(30):23227–23233
10. Grasberger H, Refetoff S (2006) Identification of the maturation factor for dual oxidase: evolution of an eukariotic operon equivalent. J Biol Chem 281(27):18269–18272
11. Geiszt M, Witta J, Baffi J, Lekstrom K, Leto TL (2003) Dual oxidases represent novel hydrogen peroxide sources supporting mucosal surface host defense. FASEB J 17 (11):1502–1504
12. De Deken X, Corvilain B, Dumont JE, Miot F (2014) Roles of DUOX-mediated hydrogen peroxide in metabolism, host defense, and signaling. Antioxid Redox Signal 20 (17):2776–2793
13. Eales JG (1997) Iodine metabolism and thyroid-related functions in organisms lacking thyroid follicles: are thyroid hormones also vitamins? Proc Soc Exp Biol Med 214 (4):302–317
14. Tong W, Kerkof P, Chaikoff IL (1962) Identification of labeled thyroxine and

triiodothyronine in amphioxus treated with 131-I. Biochim Biophys Acta 56:326–331
15. Opitz R, Maquet E, Zoenen M, Dadhich R, Costagliola S (2011) TSH receptor function is required for normal thyroid differentiation in zebrafish. Mol Endocrinol 25 (9):1579–1599
16. Imada M, Kurosumi M, Fujita H (1986) Three-dimensional aspects of blood vessels in thyroids from normal, low iodine diet-treated, TSH-treated, and PTU-treated rats. Cell Tissue Res 245:291–296
17. Boelaert K, Franklyn JA (2005) Thyroid hormone in health and disease. J Endocrinol 187 (1):1–15
18. Pochin EE (1950) Investigation of thyroid function and disease with radioactive iodine. Lancet 2(6620):84–91
19. Dunn JT (1998) What's happening to our iodine? J Clin Endocrinol Metab 83 (10):3398–3400
20. Soldin OP, Soldin SJ, Pezzullo JC (2003) Urinary iodine percentile ranges in the United States. Clin Chim Acta 328(1–2):185–190
21. Colin IM, Denef JF, Lengelé B, Many MC, Gérard AC (2013) Recent insights into the cell biology of thyroid angiofollicular units. Endocrine Rev 34(2):1–30
22. Vandevijvere S, Amsalkhir S, Oyen HV, Ines E, Moreno-Reyes R (2013) Iron status and its determinants in a nationally representative sample of pregnant women. J Acad Nutr Diet 113(5):659–666
23. Kopp P, Pesce L, Solis-S JC (2008) Pendred syndrome and iodide transport in the thyroid. Trends Endocrinol Metab 19(7):260–268
24. Wangemann P, Kim HM, Billings S, Nakaya K, Li X, Singh R, Sharlin DS, Forrest D, Marcus DC, Fong P (2009) Developmental delays consistent with cochlear hypothyroidism contribute to failure to develop hearing in mice lacking Slc26a4/pendrin expression. Am J Physiol Renal Physiol 297(5):F1435–F1447
25. Twyffels L, Strickaert A, Virreira M, Massart C, Van Sande J, Wauquier C, Beauwens R, Dumont JE, Galietta LJ, Boom A, Kruys V (2014) Anoctamin-1/TMEM16A is the major apical iodide channel of the thyrocyte. Am J Physiol Cell Physiol 307(12):C1102–C1112
26. Song Y, Ruf J, Lothaire P, Dequanter D, Andry G, Willemse E, Dumont JE, Van Sande J, De Deken X (2010) Association of duoxes with thyroid peroxidase and its regulation in thyrocytes. J Clin Endocrinol Metab 95(1):375–382
27. Fortunato RS, Lima de Souza EC, Ameziane-el Hassani R, Boufraqech M, Weyemi U, Talbot M, Lagente-Chevallier O, Carvalho DP, Bidart JM, Schlumberger M, Dupuy C, Ameziane-El-Hassani R (2010) Functional consequences of dual oxidase-thyroperoxidase interaction at the plasma membrane. J Clin Endocrinol Metab 95 (12):5403–5411
28. Song Y, Driessens N, Costa M, De Deken X, Detours V, Corvilain B, Maenhaut C, Miot F, Van Sande J, Many MC, Dumont JE (2007) Roles of hydrogen peroxide in thyroid physiology and disease. J Clin Endocrinol Metab 92(10):3764–3773
29. Corvilain B, Van Sande J, Laurent E, Dumont JE (1991) The H_2O_2-generating system modulates protein iodination and the activity of the pentose phosphate pathway in dog thyroid. Endocrinology 128:779–785
30. Libert F, Ruel J, Ludgate M, Swillens S, Alexander N, Vassart G, Dinsart C (1987) Thyroperoxidase, an auto-antigen with mosaic structure made of nuclear and mitochondrial gene modules. EMBO J 6:4193–4196
31. Taurog A, Dorris ML, Doerge DR (1996) Mechanism of simultaneous iodination and coupling catalyzed by thyroid peroxidase. Arch Biochem Biophys 330(1):24–32
32. Van de Graaf SAR, Ris-Stalpers C, Pauws E, Mendive FM, Targovnik HM, De Vijlder JJM (2001) Up to date with human thyroglobulin. J Endocrinol 170:307–321
33. Carvalho DP, Dupuy C (2017) Thyroid hormone biosynthesis and release. Mol Cell Endocrinol 458:6–15
34. Cahnmann HJ, Pommier J, Nunez J (1977) Spatial requirement for coupling of iodotyrosine residues to form thyroid hormones. Proc Natl Acad Sci U S A 74(12):5333–5335
35. Dunn JT, Dunn AD (2001) Update on intrathyroidal iodine metabolism. Thyroid 11(5):407–414
36. Bernier-Valentin F, Kostrouch Z, Rabilloud R, Munari-Silem Y, Rousset B (1990) Coated vesicles from thyroid cells carry iodinated thyroglobulin molecules. First indication for an internalization of the thyroid prohormone via a mechanism of receptor-mediated endocytosis. J Biol Chem 265(28):17373–17380
37. Dunn AD, Dunn JT (1982) Thyroglobulin degradation by thyroidal proteases: action of purified cathepsin D. Endocrinology 111 (1):280–289

38. Gnidehou S, Caillou B, Talbot M, Ohayon R, Kaniewski J, Noel-Hudson MS, Morand S, Agnangji D, Sezan A, Courtin F, Virion A, Dupuy C (2004) Iodotyrosine dehalogenase 1 (DEHAL1) is a transmembrane protein involved in the recycling of iodide close to the thyroglobulin iodination site. FASEB J 18(13):1574–1576
39. Moreno JC, Klootwijk W, van Toor H, Pinto G, D'Alessandro M, Léger A, Goudie D, Polak M, Grüters A, Visser TJ (2008) Mutations in the iodotyrosine deiodinase gene and hypothyroidism. N Engl J Med 358(17):1811–1818
40. Cosmo CD, Liao XH, Dumitrescu AM, Philp NJ, Weiss RE, Refetoff S (2010) Mice deficient in MCT8 reveal a mechanism regulating thyroid hormone secretion. J Clin Invest 120 (9):3377–3388
41. Song Y, Massart C, Chico-Galdo V, Jin L, De Maertelaer V, Decoster C, Dumont JE, Van Sande J (2010) Species specific thyroid signal transduction: conserved physiology, divergent mechanisms. Mol Cell Endocrinol 319 (1–2):56–62
42. Corvilain B, Van Sande J, Dumont JE (1988) Inhibition by iodide of iodide binding to proteins: the "Wolff-Chaikoff" effect is caused by inhibition of H_2O_2 generation. Bioch Biophys Res Commun 154(3):1287–1292
43. Panneels V, Van den Bergen H, Jacoby C, Braekman JC, Van Sande J, Dumont JE, Boeynaems JM (1994) Inhibition of H_2O_2 production by iodoaldehydes in cultured dog thyroid cells. Mol Cell Endocrinol 102:167–176
44. Christophe-Hobertus C, Christophe D (2010) Delimitation and functional characterization of the bidirectional THOX-DUOXA promoter regions in thyrocytes. Mol Cell Endocrinol 317(1–2):161–167
45. Xu C, Linderholm A, Grasberger H, Harper RW (2012) Dual oxidase 2 bidirectional promoter polymorphisms confer differential immune responses in airway epithelia. Am J Respir Cell Mol Biol 47(4):484–490
46. Pachucki J, Wang D, Christophe D, Miot F (2004) Structural and functional characterization of the two human *ThOX/Duox* genes and their 5′-flanking regions. Mol Cell Endocrinol 214(1–2):53–62
47. Morand S, Ueyama T, Tsujibe S, Saito N, Korzeniowska A, Leto TL (2009) Duox maturation factors form cell surface complexes with Duox affecting the specificity of reactive oxygen species generation. FASEB J 23 (4):1205–1218
48. Luxen S, Noack D, Frausto M, Davanture S, Torbett BE, Knaus UG (2009) Heterodimerization controls localization of Duox-DuoxA NADPH oxidases in airway cells. J Cell Sci 122(Pt 8):1238–1247
49. Milenkovic M, De Deken X, Jin L, De Felice M, Di Lauro R, Dumont JE, Corvilain B, Miot F (2007) Duox expression and related H_2O_2 measurement in mouse thyroid: onset in embryonic development and regulation by TSH in adult. J Endocrinol 192(3):615–626
50. De Deken X, Wang D, Dumont JE, Miot F (2002) Characterization of ThOX proteins as components of the thyroid H_2O_2-generating system. Exp Cell Res 273(2):187–196
51. Morand S, Chaaraoui M, Kaniewski J, Deme D, Ohayon R, Noel-Hudson MS, Virion A, Dupuy C (2003) Effect of iodide on nicotinamide adenine dinucleotide phosphate oxidase activity and Duox2 protein expression in isolated porcine thyroid follicles. Endocrinology 144(4):1241–1248
52. Morand S, Dos Santos OF, Ohayon R, Kaniewski J, Noel-Hudson MS, Virion A, Dupuy C (2003) Identification of a truncated dual oxidase 2 (DUOX2) messenger ribonucleic acid (mRNA) in two rat thyroid cell lines. Insulin and forskolin regulation of DUOX2 mRNA levels in FRTL-5 cells and porcine thyrocytes. Endocrinology 144(2):567–574
53. Raad H, Eskalli Z, Corvilain B, Miot F, De Deken X (2013) Thyroid hydrogen peroxide production is enhanced by the Th2 cytokines, IL-4 and IL-13, through increased expression of the dual oxidase 2 and its maturation factor DUOXA2. Free Radic Biol Med 56:216–225
54. Ueyama T, Sakuma M, Ninoyu Y, Hamada T, Dupuy C, Geiszt M, Leto TL, Saito N (2015) The extracellular A-loop of dual oxidases affects the specificity of reactive oxygen species release. J Biol Chem 290(10):6495–6506
55. Grasberger H, De Deken X, Miot F, Pohlenz J, Refetoff S (2007) Missense mutations of dual oxidase 2 (DUOX2) implicated in congenital hypothyroidism have impaired trafficking in cells reconstituted with DUOX2 maturation factor. Mol Endocrinol 21 (6):1408–1421
56. Ameziane-El-Hassani R, Morand S, Boucher JL, Frapart YM, Apostolou D, Agnandji D, Gnidehou S, Ohayon R, Noel-Hudson MS, Francon J, Lalaoui K, Virion A, Dupuy C (2005) Dual oxidase-2 has an intrinsic Ca^{2+}–dependent H_2O_2-generating activity. J Biol Chem 280(34):30046–30054
57. Grasberger H, De Deken X, Mayo OB, Raad H, Weiss M, Liao XH, Refetoff S

(2012) Mice deficient in dual oxidase maturation factors are severely hypothyroid. Mol Endocrinol 26(3):481–492

58. Eskalli Z, Achouri Y, Hahn S, Many MC, Craps J, Refetoff S, Liao XH, Dumont JE, Van Sande J, Corvilain B, Miot F, De Deken X (2016) Overexpression of IL-4 in the thyroid of transgenic mice upregulates the expression of Duox1 and the anion transporter Pendrin. Thyroid 26(10):1499–1512
59. Leto TL, Morand S, Hurt D, Ueyama T (2009) Targeting and regulation of reactive oxygen species generation by Nox family NADPH oxidases. Antioxid Redox Signal 11 (10):2607–2619
60. Rigutto S, Hoste C, Grasberger H, Milenkovic M, Communi D, Dumont JE, Corvilain B, Miot F, De Deken X (2009) Activation of dual oxidases Duox1 and Duox2: differential regulation mediated by cAMP-dependent protein kinase and protein kinase C-dependent phosphorylation. J Biol Chem 284(11):6725–6734
61. Fortemaison N, Miot F, Dumont JE, Dremier S (2005) Regulation of H_2O_2 generation in thyroid cells does not involve Rac1 activation. Eur J Endocrinol 152(1):127–133
62. Nakanaga T, Nadel J, Ueki IF, Koff JL, Shao MXG (2007) Regulation of interleukin-8 via an airway epithelial signaling cascade. Am J Physiol Lung Cell Mol Physiol 292(5): L1289–L1296
63. Shao MX, Nadel JA (2005) Dual oxidase 1-dependent MUC5AC mucin expression in cultured human airway epithelial cells. Proc Natl Acad Sci U S A 102(3):767–772
64. Corvilain B, Collyn L, Van Sande J, Dumont JE (2000) Stimulation by iodide of H_2O_2 generation in thyroid slices from several species. Am J Physiol Endocrinol Metab 278:692–699
65. Lambeth JD (2007) Nox enzymes, ROS, and chronic disease: an example of antagonistic pleiotropy. Free Radic Biol Med 43 (3):332–347
66. Hoste C, Dumont JE, Miot F, De Deken X (2012) The type of DUOX-dependent ROS production is dictated by defined sequences in DUOXA. Exp Cell Res 318(18):2353–2364
67. Meitzler JL, Hinde S, Banfi B, Nauseef WM, Ortiz de Montellano PR (2013) Conserved cysteine residues provide a protein-protein interaction surface in dual oxidase (DUOX) proteins. J Biol Chem 288(10):7147–7157
68. Carre A, Louzada RA, Fortunato RS, Ameziane-El-Hassani R, Morand S, Ogryzko V, Carvalho DP, Grasberger H, Leto TL, Dupuy C (2015) When an intramolecular disulfide bridge governs the interaction of DUOX2 with its partner DUOXA2. Antioxid Redox Signal 23(9):724–733
69. Zamproni I, Grasberger H, Cortinovis F, Vigone MC, Chiumello G, Mora S, Onigata K, Fugazzola L, Refetoff S, Persani L, Weber G (2008) Biallelic inactivation of the dual oxidase maturation factor 2 (DUOXA2) gene as a novel cause of congenital hypothyroidism. J Clin Endocrinol Metab 93(2):605–610
70. Helmcke I, Heumüller S, Tikkanen R, Schröder K, Brandes RP (2009) Identification of structural elements in Nox1 and Nox4 controlling localization and activity. Antioxid Redox Signal 11(6):1279–1287
71. Takac I, Schroder K, Zhang L, Lardy B, Anilkumar N, Lambeth JD, Shah AM, Morel F, Brandes RP (2011) The E-loop is involved in hydrogen peroxide formation by the NADPH oxidase Nox4. J Biol Chem 286 (15):13304–13313
72. Nisimoto Y, Diebold B, Consentino-Gomes-D, Lambeth JD (2014) Nox4: a hydrogen peroxide-generating oxygen sensor. Biochemistry 53(31):5111–5120
73. Van Vliet G (2004) Hypothyroidism in infants and children. In: Braverman LE, Utiger RD (eds) Werner & Ingbar's the thyroid: a fundamental and clinical text, 9th edn. Lippincott Williams & Wilkins, Philadelphia, pp 1029–1048
74. Williams ED, Toyn CE, Harach HR (1989) The ultimobranchial gland and congenital thyroid abnormalities in man. J Pathol 159 (2):135–141
75. Park SM, Chatterjee VK (2005) Genetics of congenital hypothyroidism. J Med Genet 42 (5):379–389
76. de Roux N, Misrahi M, Brauner R, Houang M, Carel JC, Granier M, Le Bouc Y, Ghinea N, Boumedienne A, Toublanc JE, Milgrom E (1996) Four families with loss of function mutations of the thyrotropin receptor. J Clin Endocrinol Metab 81 (12):4229–4235
77. Stein SA, Oates EL, Hall CR, Grumbles RM, Fernandez LM, Taylor NA, Puett D, Jin S (1994) Identification of a point mutation in the thyrotropin receptor of the hyt/hyt hypothyroid mouse. Mol Endocrinol 8 (2):129–138
78. Clifton-Bligh RJ, Wentworth JM, Heinz P, Crisp MS, John R, Lazarus JH, Ludgate M, Chatterjee VK (1998) Mutation of the gene encoding human TTF-2 associated with

thyroid agenesis, cleft palate and choanal atresia. Nat Genet 19(4):399–401

79. Devriendt K, Vanhole C, Matthijs G, de Zegher F (1998) Deletion of thyroid transcription factor-1 gene in an infant with neonatal thyroid dysfunction and respiratory failure. N Engl J Med 338(18):1317–1318

80. Macchia PE, Lapi P, Krude H, Pirro MT, Missero C, Chiovato L, Souabni A, Baserga M, Tassi V, Pinchera A, Fenzi G, Gruters A, Busslinger M, Di Lauro R (1998) *PAX8* mutations associated with congenital hypothyroidism caused by thyroid dysgenesis. Nat Genet 19(1):83–86

81. Devos H, Rodd C, Gagne N, Laframboise R, Van Vliet G (1999) A search for the possible molecular mechanisms of thyroid dysgenesis: sex ratios and associated malformations. J Clin Endocrinol Metab 84(7):2502–2506

82. Opitz R, Maquet E, Huisken J, Antonica F, Trubiroha A, Pottier G, Janssens V, Costagliola S (2012) Transgenic zebrafish illuminate the dynamics of thyroid morphogenesis and its relationship to cardiovascular development. Dev Biol 372(2):203–216

83. Opitz R, Hitz MP, Vandernoot I, Trubiroha A, Abu-Khudir R, Samuels M, Desilets V, Costagliola S, Andelfinger G, Deladoey J (2015) Functional zebrafish studies based on human genotyping point to netrin-1 as a link between aberrant cardiovascular development and thyroid dysgenesis. Endocrinology 156(1):377–388

84. Moreno JC, de Vijlder JJM, Vulsma T, Ris-Stalpers C (2003) Genetic basis of hypothyroidism: recent advances, gaps and strategies for future research. Trends Endocrinol Metab 14(7):318–326

85. Grasberger H (2010) Defects of thyroidal hydrogen peroxide generation in congenital hypothyroidism. Mol Cell Endocrinol 322 (1–2):99–106

86. Moreno JC, Bikker H, Kempers MJ, van Trotsenburg P, Baas F, De Vijlder JJM, Vulsma T, Ris-Stalpers C (2002) Inactivating mutations in the gene for thyroid oxidase 2 (*THOX2*) and congenital hypothyroidism. N Engl J Med 347(2):95–102

87. Kosugi S, Sato Y, Matsuda A, Ohyama Y, Fujieda K, Inomata H, Kameya T, Isozaki O, Jhiang SM (1998) High prevalence of T354P sodium/iodide symporter gene mutation in Japanese patients with iodide transport defect who have heterogeneous clinical pictures. J Clin Endocrinol Metab 83(11):4123–4129

88. Bikker H, Vulsma T, Baas F, De Vijlder JJ (1995) Identification of five novel inactivating mutations in the human thyroid peroxidase gene by denaturing gradient gel electrophoresis. Hum Mutat 6(1):9–16

89. Targovnik HM, Frechtel GD, Mendive FM, Vono J, Cochaux P, Vassart G, Medeiros-Neto G (1998) Evidence for the segregation of three different mutated alleles of the thyroglobulin gene in a Brazilian family with congenital goiter and hypothyroidism. Thyroid 8 (4):291–297

90. Fujiwara H, Tatsumi K, Miki K, Harada T, Miyai K, Takai S, Amino N (1997) Congenital hypothyroidism caused by a mutation in the Na^+/I^- symporter. Nat Genet 16(2):124–125

91. Everett LA, Glaser B, Beck JC, Idol JR, Buchs A, Heyman M, Adawi F, Hazani E, Nassir E, Baxevanis AD, Sheffield VC, Green ED (1997) Pendred syndrome is caused by mutation in a putative sulphate transporter gene (*PDS*). Nat Genet 17:411–422

92. Delange F, Dodion J, Wolter R, Bourdoux P, Dalhem A, Glinoer D, Ermans AM (1978) Transient hypothyroidism in the newborn infant. J Pediatr 92(6):974–976

93. Ordookhani A, Mirmiran P, Walfish PG, Azizi F (2007) Transient neonatal hypothyroidism is associated with elevated serum antithyroglobulin antibody levels in newborns and their mothers. J Pediatr 150 (3):315–317. 317

94. Bhavani N (2011) Transient congenital hypothyroidism. Indian J Endocrinol Metab 15 (Suppl 2):S117–S120

95. Ohye H, Fukata S, Hishinuma A, Kudo T, Nishihara E, Ito M, Kubota S, Amino N, Ieiri T, Kuma K, Miyauchi A (2008) A novel homozygous missense mutation of the dual oxidase 2 (*DUOX2*) gene in an adult patient with large goiter. Thyroid 18(5):561–566

96. Maruo Y, Takahashi H, Soeda I, Nishikura N, Matsui K, Ota Y, Mimura Y, Mori A, Sato H, Takeuchi Y (2008) Transient congenital hypothyroidism caused by biallelic mutations of the dual oxidase 2 gene in Japanese patients detected by a neonatal screening program. J Clin Endocrinol Metab 93(11):4261–4267

97. Yoshizawa-Ogasawara A, Ogikubo S, Satoh M, Narumi S, Hasegawa T (2013) Congenital hypothyroidism caused by a novel mutation of the dual oxidase 2 (*DUOX2*) gene. J Pediatr Endocrinol Metab 26:45–52

98. De Marco G, Agretti P, Montanelli L, Di Cosmo C, Bagattini B, De Servi M, Ferrarini E, Dimida A, Freitas Ferreira AC, Molinaro A, Ceccarelli C, Brozzi F, Pinchera A, Vitti P, Tonacchera M (2011)

Identification and functional analysis of novel dual oxidase 2 (*DUOX2*) mutations in children with congenital or subclinical hypothyroidism. J Clin Endocrinol Metab 96(8): E1335–E1339

99. Moreno JC, Visser TJ (2007) New phenotypes in thyroid dyshormonogenesis: hypothyroidism due to *DUOX2* mutations. Endocrine Dev 10:99–117
100. Ohye H, Sugawara M (2010) Dual oxidase, hydrogen peroxide and thyroid diseases. Exp Biol Med 235(4):424–433
101. Muzza M, Fugazzola L (2017) Disorders of H_2O_2 generation. Best Pract Res Clin Endocrinol Metab 31:225–240
102. Abe K, Narumi S, Suwanai AS, Adachi M, Muroya K, Asakura Y (2018) Association between monoallelic *TSHR* mutations and congenital hypothyroidism: a Statistical approach. Eur J Endocrinol 178:137–144
103. Liu S, Zhang W, Zhang L, Zou H, Lu K, Li Q, Xia H, Yan S, Ma X (2018) Genetic and functional analysis of two missense *DUOX2* mutations in congenital hypothyroidism and goiter. Oncotarget 9 (4):4366–4374
104. Parlato M, Charbit-Henrion F, Hayes P, Tiberti A, Aloi M, Cucchiara S, Bègue B, Bras M, Pouliet A, Rakotobe S, Ruemmele F, Knaus UG, Cerf-Bensussan N (2017) First identification of biallelic inherited *DUOX2* inactivating mutations as a cause of very early onset inflammatory bowel disease. Gastroenterology 153(2):609–611
105. Sugisawa C, Higuchi S, Takagi M, Hasegawa Y, Taniyama M, Abe K, Hasegawa T, Narumi S (2017) Homozygous *DUOXA2* mutation (p.Tyr138*) in a girl with congenital hypothyroidism and her apparently unaffected brother: case report and review of the literature. Endocr J 64 (1):1–6
106. Kizys MML, Louzada RA, Mitne-Neto M, Jara JR, Furuzawa GK, de Carvalho DP, Dias-da-Silva MR, Nesi-França S, Dupuy C, RMB M (2017) *DUOX2* mutations are associated with congenital hypothyroidism with ectopic thyroid gland. J Clin Endocrinol Metabol 102(11):4060–4071
107. Zheng X, Ma S, Guo M, Yl Q, Yang L (2017) Compound heterozygous mutations in the *DUOX2/DUOXA2* genes cause congenital hypothyroidism. Yonsei Med J 58 (4):888–890
108. Aycan Z, Cangul H, Muzza M, Bas VN, Fugazzola L, Chatterjee VK, Persani L, Schoenmakers N (2017) Digenic *DUOX1* and *DUOX2* mutations in cases with congenital hypothyroidism. J Clin Endocrinol Metabol 102(September):3085–3090
109. Park KJ, Park HK, Kim YJ, Lee KR, Park JH, Park JH, Park HD, Lee SY, Kim JW (2016) *DUOX2* mutations are frequently associated with congenital hypothyroidism in the Korean population. Ann Lab Med 36 (2):145–153
110. Levine AP, Pontikos N, Schiff ER, Jostins L, Speed D, Lovat LB, Barrett JC, Grasberger H, Plagnol V, Segal AW (2016) Genetic complexity of Crohn's disease in two large ashkenazi jewish families. Gastroenterology 151(4):698–709
111. Hayes P, Dhillon S, O'Neill K, Thoeni C, Hui KY, Elkadri A, Guo CH, Kovacic L, Aviello G, Alvarez L, Griffiths AM, Snapper SB, Brant SR, Doroshow JH, Silverberg MS, Peter I, McGovern DPB, Cho J, Brumell JH, Uhlig HH, Bourke B, Muise AM, Knaus UG (2015) Defects in nicotinamide-adenine dinucleotide phosphate oxidase genes *NOX1* and *DUOX2* in very early onset inflammatory bowel disease. Cell Mol Gastroenterol Hepatol 1 (5):489–502
112. Fu C, Zhang S, Su J, Luo S, Zheng H, Wang J, Qin H, Chen Y, Shen Y, Hu X, Fan X, Luo J, Xie B, Chen R, Chen S (2015) Mutation screening of *DUOX2* in Chinese patients with congenital hypothyroidism. J Endocrinol Investig 38(11):1219–1224
113. Fu C, Luo S, Zhang S, Wang J, Zheng H, Yang Q, Xie B, Hu X, Fan X, Luo J, Chen R, Su J, Shen Y, Gu X, Chen S (2016) Next-generation sequencing analysis of DUOX2 in 192 Chinese subclinical congenital hypothyroidism (SCH) and CH patients. Clin Chim Acta 458:30–34
114. Tan M, Huang Y, Jiang X, Li P, Tang C, Jia X, Chen Q, Chen W, Sheng H, Feng Y, Wu D, Liu L (2016) The prevalence, clinical, and molecular characteristics of congenital hypothyroidism caused by *DUOX2* mutations: a population-based cohort study in Guangzhou. Horm Metab Res 48(9):581–588
115. Jiang H, Wu J, Ke S, Hu Y, Fei A, Zhen Y, Yu J, Zhu K (2016) High prevalence of *DUOX2* gene mutations among children with congenital hypothyroidism in central China. Eur J Med Genet 59(10):526–531
116. Narumi S, Muroya K, Asakura Y, Aachi M, Hasegawa T (2011) Molecular basis of thyroid dyshormonogenesis: genetic screening in population-based Japanese patients. J Clin Endocrinol Metab 96(11):E1838–E1842
117. Rabbiosi S, Vigone MC, Cortinovis F, Zamproni I, Fugazzola L, Persani L,

Corbetta C, Chiumello G, Weber G (2013) Congenital hypothyroidism with eutopic thyroid gland: analysis of clinical and biochemical features at diagnosis and after re-evaluation. J Clin Endocrinol Metab 98(4):1395–1402

118. Jin HY, Heo SH, Kim YM, Kim GH, Choi JH, Lee BH, Yoo HW (2014) High frequency of *DUOX2* mutations in transient or permanent congenital hypothyroidism with eutopic thyroid glands. Horm Res Paediatr 82:252–260

119. Muzza M, Rabbiosi S, Vigone MC, Zamproni I, Cirello V, Maffini M, Maruca K, Schoenmakers N, Beccaria L, Gallo F, Park SM, Beck-Peccoz P, Persani L, Weber G, Fugazzola L (2014) The clinical and molecular characterization of patients with dyshormonogenic congenital hypothyroidism reveals specific diagnostic clues for Duox2 defects. J Clin Endocrinol Metab 99: E544–E553

120. Hulur I, Hermanns P, Nestoris C, Heger S, Refetoff S, Pohlenz J, Grasberger H (2011) A single copy of the recently identified dual oxidase maturation factor (*DUOXA*) 1 gene produces only mild transient hypothyroidism in a patient with a novel biallelic *DUOXA2* mutation and monoallelic *DUOXA1* deletion. J Clin Endocrinol Metab 96(5): E841–E845

121. Yi RH, Zhu WB, Yang LY, Lan L, Chen Y, Zhou JF, Wang J, Su YQ (2013) A novel dual oxidase maturation factor 2 gene mutation for congenital hypothyroidism. Int J Mol Med 31 (2):467–470

122. Liu S, Liu L, Niu X, Lu D, Xia H, Yan S (2015) A novel missense mutation (I26M) in *DUOXA2* causing congenital goiter hypothyroidism impairs NADPH oxidase activity but not protein expression. J Clin Endocrinol Metabol 100:1225–1229

123. Zheng X, Ma SG, Qiu YL, Guo ML, Shao XJ (2016) A novel c.554+5C>T mutation in the *DUOXA2* gene combined with p.R885Q mutation in the *DUOX2* gene causing congenital hypothyroidism. J Clin Res Pediatr Endocrinol 8(2):224–227

124. Lx Y, Sg M, Yl Q, Zheng X (2016) Heterozygous mutations of the *DUOXA2* and *DUOX2* genes in dizygotic twins with congenital hypothyroidism. Clin Lab 62 (5):849–854

125. Nicholas AK, Serra EG, Cangul H, Alyaarubi S, Ullah I, Schoenmakers E, Deeb A, Habeb AM, Almaghamsi M, Peters C, Nathwani N, Aycan Z, Saglam H, Bober E, Dattani M, Shenoy S, Murray PG, Babiker A, Willemsen R, Thankamony A, Lyons G, Irwin R, Padidela R, Tharian K, Davies JH, Puthi V, Park SM, Massoud AF, Gregory JW, Albanese A, Pease-Gevers E, Martin H, Brugger K, Maher ER, Chatterjee VK, Anderson CA, Schoenmakers N (2016) Comprehensive screening of eight known causative genes in congenital hypothyroidism with gland-in-situ. J Clin Endocrinol Metab 101(12):4521–4531

126. LaFranchi S (1999) Congenital hypothyroidism: etiologies, diagnosis, and management. Thyroid 9(7):735–740

127. Maruo Y, Nagasaki K, Matsui K, Mimura Y, Mori A, Fukami M, Takeuchi Y (2016) Natural course of congenital hypothyroidism by dual oxidase 2 mutations from the neonatal period through puberty. Eur J Endocrinol 174(4):453–463

128. Weyemi U, Caillou B, Talbot M, Ameziane-El-Hassani R, Lacroix L, Lagent-Chevallier O, Al Ghuzlan A, Roos D, Bidart JM, Virion A, Schlumberger M, Dupuy C (2010) Intracellular expression of reactive oxygen species-generating NADPH oxidase NOX4 in normal and cancer thyroid tissues. Endocr Relat Cancer 17(1):27–37

129. Vigone MC, Fugazzola L, Zamproni I, Passoni A, Di Candia S, Chiumello G, Persani L, Weber G (2005) Persistent mild hypothyroidism associated with novel sequence variants of the *DUOX2* gene in two siblings. Hum Mutat 26(4):395

130. Kasahara T, Narumi S, Okasora K, Takaya R, Tamai H, Hasegawa T (2013) Delayed onset congenital hypothyroidism in a patient with *DUOX2* mutations and maternal iodine excess. Am J Med Genet A 161A(1):214–217

131. Zimmermann MB, Jooste PL, Pandav CS (2008) Iodine-deficiency disorders. Lancet 372(9645):1251–1262

132. Satoh M, Aso K, Ogikubo S, Yoshizawa-Ogasawara A, Saji T (2015) Hypothyroidism caused by the combination of two heterozygous mutations: one in the TSH receptor gene the other in the *DUOX2* gene. J Pediatr Endocrinol Metab 28(5–6):657–661

133. Yoshizawa-Ogasawara A, Abe K, Ogikubo S, Narumi S, Hasegawa T, Satoh M (2016) Transient congenital hypothyroidism caused by compound heterozygous mutations affecting the NADPH-oxidase domain of DUOX2. J Pediatr Endocrinol Metab 29(3):363–371

134. Abe K, Narumi S, Suwanai AS, Hamajima T, Hasegawa T (2015) Pseudodominant inheritance in a family with nonautoimmune hypothyroidism due to biallelic *DUOX2* mutations. Clin Endocrinol 83(3):394–398

135. Hoste C, Rigutto S, Van Vliet G, Miot F, De Deken X (2010) Compound heterozygosity for a novel hemizygous missense mutation and a partial deletion affecting the catalytic core of the H_2O_2-generating enzyme DUOX2 associated with transient congenital hypothyroidism. Hum Mutat 31(4): E1304–E1319
136. Jirapongsananuruk O, Malech HL, Kuhns DB, Niemela JE, Brown MR, Anderson-Cohen M, Fleisher TA (2003) Diagnostic paradigm for evaluation of male patients with chronic granulomatous disease, based on the dihydrorhodamine 123 assay. J Allergy Clin Immunol 111(2):374–379
137. Stasia MJ (2007) The X^+ chronic granulomatous disease as a fabulous model to study the NADPH oxidase complex activation. Med Sci (Paris) 23(5):526–532
138. Tonacchera M, De Marco G, Agretti P, Montanelli L, Di Cosmo C, Freitas Ferreira AC, Dimida A, Ferrarini E, Ramos HE, Ceccarelli C, Brozzi F, Pinchera A, Vitti P (2009) Identification and functional studies of two new dual-oxidase 2 (*DUOX2*) mutations in a child with congenital hypothyroidism and a eutopic normal-size thyroid gland. J Clin Endocrinol Metab 94(11):4309–4314
139. Grasberger H, Noureldin M, Kao TD, Adler J, Lee JM, Bishu S, El Zaatari M, Kao JY, Waljee AK (2018) Increased risk for inflammatory bowel disease in congenital hypothyroidism supports the existence of a shared susceptibility factor. Sci Rep 8 (1):10158

Chapter 38

NAPDH Oxidases in Inflammatory Bowel Disease

Emily Stenke, Billy Bourke, and Ulla G. Knaus

Abstract

Inflammatory bowel diseases (IBD), categorized as ulcerative colitis (UC), Crohn's disease (CD), or IBD-undetermined (IBDU), are increasing in incidence. IBD is understood to result from environmental factors interacting with a pre-existing genetic susceptibility. Approximately 1% of all patients with inflammatory bowel disease (IBD) are diagnosed before the age of 6 years, designated as very-early-onset IBD (VEOIBD). This cohort of patients is distinguished from other age groups by differences in disease phenotype and by a higher burden of genetic mutations. Recent studies have linked mutations in NADPH oxidase function to VEOIBD and even pediatric IBD. Loss-of-function NOX2 variants expressed in phagocytes and NOX1/DUOX2 variants expressed in intestinal epithelial cells have been associated with VEOIBD and pediatric and adult IBD in patients. Cell and animal studies suggest a protective role for these reactive oxygen species (ROS)-producing enzymes in intestinal homeostasis—a paradigm that challenges the conventional concept that only increased ROS result in cell and tissue damage. Examining the role of NADPH oxidases in VEOIBD may improve our understanding of the pathophysiology of this disease and will uncover new therapeutic possibilities.

Key words Inflammatory bowel disease, VEOIBD, Intestine, Reactive oxygen species, NADPH oxidase, NOX, DUOX

1 Introduction

Inflammatory bowel disease (IBD) is a group of conditions manifesting as chronic intestinal inflammation that has become increasingly common in the last two decades [1–3]. IBD is classified based on clinical and histological phenotype as either Crohn's disease (CD), ulcerative colitis (UC), or IBD-undetermined (IBDU). Patients present with bloody or non-bloody diarrhea, abdominal cramps, weight loss, or extraintestinal symptoms such as skin rashes or arthritis. Diagnosis is based on clinical presentation and endoscopic/histological findings. CD is classically characterized by discontinuous, transmural inflammation occurring anywhere throughout the digestive tract, while UC manifests as continuous mucosal inflammation confined to the colon. Among patients with IBD, differences in disease presentation and behavior are observed

Ulla G. Knaus and Thomas L. Leto (eds.), *NADPH Oxidases: Methods and Protocols*, Methods in Molecular Biology, vol. 1982, https://doi.org/10.1007/978-1-4939-9424-3_38, © Springer Science+Business Media, LLC, part of Springer Nature 2019

based on age of onset. Subclassification of IBD into specific age groups has been codified internationally and is referred to as the Paris classification (actually a modification of the Montreal classification of IBD) [4]. In 2017 up to 25% of new diagnoses of IBD were in patients younger than 18 years. As many as 15% of these patients develop IBD under the age of 6 years, which is defined as very-early-onset IBD (VEOIBD) [4–6]. Children with VEOIBD are significantly more likely than older children or adults to have disease confined to the colon at diagnosis and a positive family history of IBD [3, 7, 8]. Patients with VEOIBD are also more likely to be classified as IBDU than older patients, likely reflecting not only the underlying pathophysiology, but also the constraints on a comprehensive diagnostic endoscopic workup in this age group [9]. Despite efforts to identify genetic markers for the development of IBD, as yet only a small percentage of the heritability determined by twin studies can be explained by gene variants identified in genome-wide association studies (GWAS), suggesting a polygenic model of genetic susceptibility in adult IBD [10, 11]. Nevertheless, genetic analysis including GWAS of populations of common ancestry and exome or whole-genome sequencing of index patients have proved useful tools to highlight pathways and processes where variation in gene expression, protein structure, or protein function leads to an increased risk of IBD. This knowledge can inform our understanding of the pathophysiology of IBD. Genetic variants that increase the risk of adult IBD and/or VEOIBD have been identified in both coding and noncoding DNA. Many of these mutations also confer an overlapping risk of other inflammatory conditions (e.g., type 1 diabetes, rheumatoid arthritis) highlighting the multifactorial pathogenesis of IBD, where environmental factors (e.g., smoking, diet, composition of the microbiome) combine with a pre-existing genetic susceptibility to produce a specific disease phenotype [10, 12].

2 VEOIBD: The Clinical Perspective

The incidence of adult and pediatric IBD has risen in the last decade and the rate of increase appears to be highest in the pediatric group [1–3]. However, VEOIBD remains rare, with an incidence of 0.36–4.37 per 100,000 children in 2009–2011 and a prevalence of 14 per 100,000 children in 2009 [3, 7, 13]. The clinical presentation varies depending on disease location. Colonic inflammation occurring in UC or CD typically manifests with bloody diarrhea, while symptoms of small intestinal CD can be subtle, including poor weight gain and growth, abdominal pain, poor appetite, and fatigue. Perianal disease associated with CD may present with perianal abscesses, ulcers, or draining fistulae. Patients are diagnosed with VEOIBD and subclassified as VEO-CD, -UC, or -IBDU following ileocolonoscopy and esophagogastroduodenoscopy [4, 14, 15]. UC is characterized by mucosal inflammation limited to the colon; inflammation starts in the rectum and extends proximally in a continuous fashion. The inflammation seen in CD can occur anywhere throughout the digestive tract. Skip lesions in the colon and/or ileum are common, inflammation can be transmural with granulomas present on microscopic examination, and fibrosis, stenosis, fistulae, and perianal disease can occur. Radiological imaging at diagnosis using magnetic resonance (MR) or computed tomography (CT) enterography or contrast-enhanced X-rays permits detection of inflammation, fistulae, or stenosis in the small intestine that is inaccessible to standard endoscopy. Phenotypic differences have been noted between VEOIBD patients and later onset pediatric or adult IBD. CD is more common than UC in adult and older pediatric patients with IBD, whereas the rate of UC is higher in patients with VEOIBD [3, 5, 8, 16]. UC is more likely to present as a pancolitis, while in CD perianal and colonic disease are common [6, 16, 17].

A healthy intestine requires immune tolerance to normal intestinal microbiota while simultaneously being able to mount a robust defense against pathogenic bacteria. This is provided by an intact epithelial and mucosal barrier and a functioning innate and adaptive immune system. Many primary immunodeficiencies are associated with an increased risk of colitis and patients presenting with VEOIBD are significantly more likely than adult IBD patients to have an identifiable genetic defect [7]. These include conditions characterized by epithelial barrier dysfunction, neutrophil dysfunction (e.g., chronic granulomatous disease (CGD), glycogen storage disease type 1b), T- and B-cell dysfunction (e.g., combined variable immunodeficiency (CVID), severe combined immunodeficiency (SCID)), and defects in immune response regulation (e.g., IL10 and IL10-R mutations) [7]. Most of these primary immunodeficiencies present at a young age and inflammation of the bowel may be an early or a dominant manifestation. Hence, it is important that pediatric gastroenterologists consider other disorders in young patients presenting with intestinal inflammation. Multiple affected family members, consanguinity, younger age of onset (<2 years), and severe, treatment-resistant disease are all associated with an increased likelihood of a monogenic disorder manifesting as VEOIBD. Although differences in endoscopic appearance and disease location have been described at a population level (e.g., CGD-IBD has a higher rate of perianal disease than VEOIBD [18]), at an individual patient level endoscopic and histological findings are not specific enough to rule in or rule out monogenic disorders. Laboratory tests to assess immune function including the neutrophil oxidative burst by either nitroblue tetrazolium test (NBT) or dihydrorhodamine fluorescence assay (DHR) to diagnose CGD, T and B lymphocyte counts and subsets, immunoglobulin levels, and responses to vaccines are routinely accessible and performed, while more specific tests including IL10-R function and targeted genetic testing may require input from specialist laboratories.

More recently, single-gene defects associated with VEOIBD and/or adult IBD in the absence of an identifiable immunodeficiency or coexisting disorder have been identified. These include defects in genes encoding the NADPH oxidase family of enzymes. Reactive oxygen species (ROS) have long been considered harmful in IBD by perpetuating inflammation and causing tissue injury [19]. In the last decade primary ROS (superoxide, hydrogen peroxide) have been recognized as critical second messengers transmitting outside-in signals and initiating redox relays [20, 21]. Among all ROS sources the NADPH oxidases are the only enzymes generating ROS as their sole known function. Some members of the NADPH oxidase family expressed in the innate immune system or in the intestinal epithelium have been implicated in the pathogenesis of inflammatory bowel diseases; they form the subject of this chapter.

3 NOX and Innate Immunity

The NADPH oxidase 2 complex (NOX2 complex) is a multimeric complex composed of the transmembrane NOX2 (gp91^{phox}) and p22^{phox} proteins (encoded by *CYBB* and *CYBA* genes, respectively) and the cytosolic proteins p47^{phox}, p67^{phox}, and p40^{phox} (encoded by *NCF1*, *NCF2*, and *NCF4* genes, respectively); additionally the GTP-binding protein RAC2 is required for NOX2 activation [22, 23]. The NOX2 catalytic unit consists of a C-terminal cytosolic half with binding sites for NADPH and FAD, and an N-terminal half with six transmembrane helices linked by three extracellular and two intracellular loops and a cytosolic N-terminus. Histidines located in helices 3 and 5 coordinate the two low-potential hemes required for shuttling electrons from NADPH/FAD to molecular oxygen (O_2) at the opposite face of the membrane, resulting in the one-electron reduction of O_2 to superoxide anion [23]. In the resting state the NOX2-p22^{phox} membrane-bound heterodimer and the cytosolic multimeric complex of p47^{phox}-p67^{phox}-p40^{phox} remain dissociated. Signaling-induced posttranslational modifications cause conformational changes, dissociation, and translocation of the cytosolic components and their assembly with the membrane-bound heterodimer. Stimulation of RAC2 translocation, GDP to GTP exchange, and binding of the active GTPase to the NOX2 complex lead to full activation of the oxidase [22, 23].

The importance of phagocyte NOX2 for human health is evident in CGD. CGD is caused by mutations in NOX2 complex components that result in significantly reduced or absent protein expression or catalytic activity, and consequently impaired superoxide production and pathogen killing by phagocytes. This condition typically presents in early childhood and is diagnosed based on a history of recurrent bacterial and fungal infections and significantly reduced phagocyte oxidative burst measured by NBT or DHR. The majority (70%) of cases are caused by mutations in the *CYBB* gene encoding the NOX2 protein (previously termed gp91^{phox}). A further 25% of cases are due to mutations in the *NCF1* gene encoding p47^{phox} and the remainder are due to rare mutations in *NCF2*, *NCF4* and *CYBA* (encoding p67^{phox}, p40^{phox} and p22^{phox}, respectively) [24]. Up to 40% of CGD patients develop chronic intestinal inflammation which is often clinically similar to CD [25]. In younger patients the colitis-predominant inflammation and perianal disease closely mimic VEOCD [26, 27]. This finding reinforces the current conceptual framework of IBD as a disturbance of the complex interplay between the gut microbiota, intestinal epithelial barrier, and immune system, where disturbance of any, or a combination, of these factors may contribute to the development of IBD.

Abnormalities in the NOX2 complex have also been associated with VEOIBD and pediatric IBD in the absence of CGD. Dhillon et al. studied 122 patients classified as VEOIBD without apparent immunodeficiency and used targeted exome sequencing and tagged SNP genotyping to identify mutations in the genes encoding the NOX2 complex or required for its activation (*CYBB*, *CYBA*, *NCF1*, *NCF2*, *NCF4*, *RAC1*, and *RAC2*) [28]. They identified eight previously described and three novel SNPs associated with VEOIBD in their cohort. Patients were heterozygous for these mutations and did not have CGD. Several identified SNPs did not alter protein expression but overexpression studies in transfected HEK293 cells revealed decreased protein-protein associations within the NOX2 complex. Two *NCF2* variants (encoding p67^{phox}; c.1256A>T; p.N419I; rs35012521, p.G501R) reduced the binding of p67^{phox} to p40^{phox} and p47^{phox}, respectively, and an *NCF4* variant (encoding p40^{phox}; c.923G>A; p.R308Q; rs141160114) showed weaker binding to p67^{phox} [28]. Additional 26 missense mutations in NOX2 complex components, all leading to low ROS generation, were detected in pediatric IBD (age 6–12) [29]. These ROS-low patients showed a more aggressive course of disease including perianal involvement, structuring, and abdominal surgery, linking decreased ROS generation firmly to IBD.

Translocation, activation, and binding of active GTP-RAC are essential for activation of the NOX2 complex [30, 31]. Muise et al. recently identified an *NCF2* variant (encoding p67^{phox}; c.113G>A; p.R38Q; rs147415774) present in 11/268 VEOIBD patients (9 heterozygous, 2 homozygous) that resulted in reduced binding of the mutant p67^{phox} protein to RAC2 [32]; 11 out of 122 patients were heterozygous for this variant in another cohort

[28]. Phagocyte superoxide production of the index patient was reduced. This variant was also identified in 1 out of 341 adult patients with IBD and in 1 out of 480 healthy controls. Since variants within the NOX2 complex are associated with both pediatric and adult-onset IBD, and have also been observed in healthy controls, this suggests a shared risk for VEOIBD and adult-onset IBD as well as incomplete penetrance of NOX2 defects and a complex interplay between different susceptibility genes and the environment.

Mutations in introns causing aberrant gene regulation are also associated with VEOIBD. SNPs leading to increased GATA1 binding at a *RAC1* promoter site (rs35761891) and loss of an SP1-binding site at the *CYBA* promoter (rs72550704) have been associated with VEOIBD/VEOUC and VEOCD, respectively. Although not confirmed experimentally, reduced RAC1 or $p22^{phox}$ expression and subsequently decreased superoxide anion production are predicted to occur with these variants [28]. Similarly, an intronic *NCF4* SNP (rs4821544, intron 1) has been associated with adult Crohn's disease in some GWAS [33, 34], although this was not replicated in other studies [35, 36]. Neutrophils from patients with this SNP showed reduced fMLF-induced superoxide production after GM-CSF priming compared to CD controls. No difference in superoxide production was observed following fMLF stimulation in the absence of GM-CSF priming, confirming that the importance of $p40^{phox}$ in superoxide production is dependent on the signaling context [37, 38].

The majority of mutations in the NOX2 complex leading to CGD are autosomal recessive (*NCF1*) or X-linked (*CYBB*) and result in absent or reduced catalytic activity, while mutations associated with VEOIBD tend to be dominant mutations with incomplete penetrance resulting in structural variants of proteins $p40^{phox}$, $p67^{phox}$, and RAC2. *CYBB* (NOX2) variants are rare in VEOIBD with only one SNP identified in a single patient (p.G364R; rs141756032). This missense mutation is located adjacent to a FAD-binding site and is predicted to be damaging, although confirmatory experimental or human patient data are not available [28]. Functionally, patients with CGD have very low levels of phagocytic superoxide production. There is some evidence that patients with VEOIBD as well as adult patients with Crohn's disease have reduced ROS levels [39–41], but those levels remain above the 5–10% threshold associated with clinical immunodeficiency [42–44]. For example, analyses by Dhillon et al. of several of the SNPs associated with VEOIBD predict reduced superoxide production and five out five VEOIBD patients generated less superoxide than healthy controls in the NBT test [28]. A study of CGD patients with and without IBD described a higher burden of IBD-risk alleles in CD patients compared to CGD-IBD patients, suggesting that defective superoxide production is a major

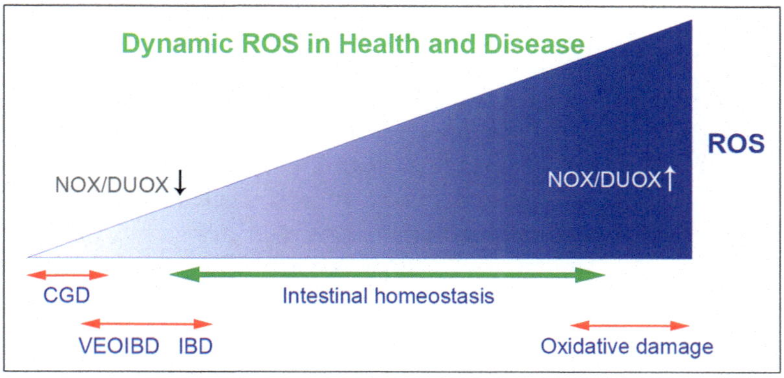

Fig. 1 Regulated production of NADPH oxidase-derived ROS is important for maintaining intestinal epithelial homeostasis. Low ROS generation due to genetic variants in *NOX1*, *NOX2*, or *DUOX2* is associated with VEOIBD and IBD. Constantly increased ROS production by hyperactivated signaling pathways or feedback loops may lead to oxidative damage. Thus, permanent over- or underproduction of ROS or the overresponsiveness or the lack of responsiveness of NADPH oxidases contributes to inflammatory disease

independent risk factor for IBD [45] (Fig. 1). Supporting this hypothesis, treatment with granulocyte-macrophage colony-stimulating factor (GM-CSF) to increase bone marrow production of neutrophils has been reported to induce remission in a patient with VEOIBD [46], and serum anti-GM-CSF antibody levels have been associated with risk of clinical relapse in IBD patients [47]. However, it is known that not all patients with CGD develop IBD and indeed Kuhns et al. showed that in patients with CGD maximal superoxide levels do not correlate with the presence of IBD [48]. Furthermore, variants of *NCF4* and *RAC2* have been associated with adult CD, while the identification of a dominant variant with incomplete penetrance (*NCF2* encoding $p67^{phox}$; c.113G>A; p.R38Q; rs147415774) in VEOIBD and adult IBD suggests that a second hit in addition to impaired NOX2-mediated superoxide production is required for the development of IBD [32].

Several SNPs in the NOX2 complex associated with VEOIBD have also been connected to other immune-mediated diseases such as systemic lupus erythematosus (SLE) and rheumatoid arthritis (RA). Epidemiological data have confirmed an association between IBD and other immune-mediated conditions [49]. It is likely that defects in phagocyte function due to NOX2 mutations described above play a role in this association. These observations highlight the complexity of the pathogenesis of IBD and VEOIBD, where an underlying genetic and/or immune predisposition leads to IBD in the presence of specific environmental and microbial triggers.

4 Epithelial NADPH Oxidases: NOX1 and DUOX2

The intestinal epithelium is responsible for nutrient absorption, mucus layer maintenance, and communication between commensal bacteria, pathogenic microorganisms, and immune system. The importance of epithelial integrity for normal intestinal function and homeostasis is illustrated by evidence from GWAS, from monogenic diseases with coexisting colitis, and from animal studies. Identified monogenic variants of epithelial genes are associated with IBD-like intestinal inflammation [50]. TTC7A is expressed by enterocytes and regulates PI4K signaling and hence development of enterocyte polarity. Genetic analysis revealed an association between VEOIBD or multiple intestinal atresia and *TTC7A* mutations, which cause reduced expression or mislocalization of TTC7A and reduced binding to PI4K [51]. Mutations in *IKBKG* (encoding NEMO) lead to impaired NF-κB signaling, epithelial apoptosis, and enterocolitis; a case report described the reversal of the associated immunodeficiency following bone marrow transplant, while the intestinal inflammation persisted, suggesting that the enterocolitis is due to specific epithelial defects rather than generalized immune dysfunction [50, 52]. IBD-like colitis has been described in epidermolysis bullosa due to congenital [53] or acquired [54] collagen type VII deficiency, presumably related to loss of intestinal epithelial integrity. Additionally, GWAS in adult IBD have identified several genes involved in the mucosal barrier that are more prevalent in IBD than in healthy controls, including *GNA12* (encoding the heterotrimeric G protein Gα12, involved in stabilization of tight junctions) [12] and *MUC19* (encoding mucin 19) [55]. NOX1 and DUOX2 are both expressed in intestinal epithelial cells and recent evidence suggests that mutations in these genes are associated with an increased risk of IBD (Fig. 1).

4.1 NOX1

The importance of the phagocyte NOX2 complex in host defense is well established, while interest in the role of epithelial ROS in intestinal homeostasis developed only recently. Two members of the NADPH oxidase family are highly expressed in the digestive tract. The NADPH oxidase 1 (NOX1) complex is expressed in intestinal epithelial cells (IECs), mainly in the colon and rectum, while dual oxidase 2 (DUOX2) can be found throughout the gastrointestinal tract with highest expression in the ileum [56–59]. NOX1 and DUOX2 are the primary sources of ROS in the intestinal epithelium, generating superoxide and hydrogen peroxide, respectively. NOX1's multimeric protein complex is comprised of NOX1, $p22^{phox}$, NOX organizer 1 (NOXO1), and NOX activator 1 (NOXA1), and is regulated by transcription, posttranslational modifications, and RAC1-GTP [60]. Inactivating *NOX1* variants have been associated with VEOIBD as well as adult UC. Hayes et al. used targeted exome sequencing to identify a novel *NOX1* missense mutation in 1 out of 59 patients with VEOIBD (c.988G>A; p.P330S) and a previously described *NOX1* variant (rs34688635; c.967G>A; p.D360N) in 2 out of 59 patients [61]. When transfected into NOX/DUOX-deficient COS7 cells both mutated NOX1 enzymes showed reduced catalytic activity (50% reduction of superoxide production) with preserved protein expression and correct intracellular localization. A similar loss of superoxide generation was observed ex vivo in colonic biopsies derived from $Nox1^{-/-}$ mice lentivirally transduced in vivo with *NOX1* variants [61]. Whole-genome sequencing of an index patient with VEOIBD and targeted exome sequencing of a cohort of other VEOIBD patients detected six *NOX1* variants [62]. Functional analysis of one of these mutants, NOX1 N122H, in colonic biopsies of the index patient showed reduced superoxide production when compared to IBD and healthy controls but a more pronounced reduction of NOX1 function predicted for this mutation may have been masked by stress-induced ROS generation due to other ROS sources. Colonic epithelial organoids expressing NOX1 N122H generated significantly less superoxide, both constitutively and following PMA stimulation, compared to organoids containing NOX1 wild type. Similarly, HCT116 colonic epithelial cells transfected with several *NOX1* variants showed compromised catalytic activity. Patient IBD phenotypes included CD, UC, and IBDU, with age of symptom onset ranging from 2 to 9.5 years. There was no obvious correlation between younger age of onset and more severe phenotype in loss-of-function versus hypomorphic mutations, albeit patient numbers were small [62]. The p.D360N *NOX1* variant was also associated with adult-onset UC in a cohort of Ashkenazi males [61]; however this association was not replicated in a population of non-Jewish patients of European descent [62], suggesting that a further genetic or environmental trigger is required.

The mechanisms by which NOX1 contributes to intestinal homeostasis are as yet incompletely understood. Although no specific intestinal pathogen or defined shift in the microbiome has yet been conclusively linked with IBD, intestinal dysbiosis with reduced bacterial diversity and an altered metabolic profile is regularly observed in colitis [63–66]. The gradient of NOX1 expression in human IECs alters from minimal/no expression in the small intestine to high expression in the distal colon [67], mirroring bacterial load, mucus layer thickness, and distribution of inflammation in UC. This suggests a role for NOX1 in bacterial defense and/or mucin production. Supporting this hypothesis, cells expressing hypomorphic *NOX1* variants (P330S and D360N) identified in VEOIBD patients showed increased susceptibility to *C. jejuni* adhesion and invasion, thereby linking intestinal epithelial superoxide production by NOX1 to immune defense [61]. Cell and animal studies showed superoxide- and NOX1-dependent epithelial regeneration and repair, modulated by the commensal *Lactobacillus* [68, 69]. These findings require further validation but together raise the possibility that NOX1 may play a role in cell differentiation, migration and microbial defense, mediated at least in part by the microbiome.

4.2 DUOX2

DUOX2 is expressed in the epithelium throughout the gastrointestinal tract with maximal expression in the ileum [56, 57, 59]. DUOX2 consists of an extracellular N-terminal peroxidase-like domain and a C-terminal Nox domain featuring six transmembrane helices and a long cytosolic extension containing FAD- and NADPH-binding sites. These two parts are linked by an additional transmembrane helix and cytosolic loop with two calcium-binding EF hands [23]. Heterodimerization with the transmembrane protein DUOXA2 is necessary for release of the heterodimer from the ER to the Golgi, transport to the plasma membrane, and H_2O_2 production. *DUOX2* upregulation has been observed during intestinal inflammation but the relevance of increased DUOX2 expression for severity or duration of colitis has not yet been addressed [70–72]. The identification of rare inactivating *DUOX2* variants in patients with VEOIBD [61, 73] and in adult patients with CD [74] suggests that DUOX2 function is generally protective for the host. Hayes et al. identified two novel variants of *DUOX2* through exome sequencing of 59 patients with VEOIBD [61]. One patient with VEOIBD, likely CD, was heterozygous for *DUOX2* p.R1211C (c.3631C>T) and a patient with VEOUC was heterozygous for *DUOX2* p.R1492C (c.4474G>A). Both are missense mutations with amino acid substitution in the last intracellular loop and the third NADPH-binding domain, respectively. H661 cells expressing either of these two mutant proteins showed conserved protein expression and localization but reduced H_2O_2 production when compared to wild-type DUOX2, and cells expressing these mutants were more susceptible to invasion by *C. jejuni* [61]. Two additional *DUOX2* variants were identified in a compound heterozygous patient with VEOIBD [73]. The parents were asymptomatic carriers of *DUOX2* p.P609S (c.1825G>A) and *DUOX2* p.R286H (c.857G>A). Biopsies of this patient displayed markedly reduced DUOX2 expression as compared to healthy and unrelated disease controls. Functional studies in H661 cells pointed to a failure in ER or Golgi exit of DUOX2 R286H, possibly due to impaired dimerization with DUOXA2, and markedly decreased H_2O_2 production. Exome sequencing in an Ashkenazi Jewish family with a high incidence of adult CD identified another rare missense variant in *DUOX2* (rs 151261408; c.908C>G; p.P303R) in 15 heterozygous individuals, although the association was not replicated in an independent cohort [74]. Functional studies in HEK293 cells showed preserved protein expression but reduced membrane localization and H_2O_2 production by DUOX2 P303R compared to wild-type DUOX2. Colon biopsies indicated upregulation of *DUOX2* mRNA and DUOX2 protein in patients with active compared to both quiescent UC and healthy controls. Ex vivo culture of these biopsies revealed enhanced upregulation of *DUOX2* expression in the presence of 5-ASA, an effective first-line treatment for both UC and CD in adults and children [75], suggesting that increased DUOX2 expression may not be linked to oxidative stress but somehow supports the anti-inflammatory effects of 5-ASA (Fig. 1).

Prior to its detection in the digestive tract the DUOX2/DUOXA2 complex was identified in thyroid tissue where thyroid hormone synthesis by thyroperoxidase (TPO) requires H_2O_2 [76]. Inactivating mutations of *DUOX2* and *DUOXA2* have been commonly identified in patients with congenital hypothyroidism (CH), either transient or permanent [22, 77]. *DUOX2* mutations have variable and incomplete penetrance in CH and the genotype is not always predictive of CH phenotype (permanent vs. transient, partial vs. total iodination defects) either in heterozygous or homozygous patients [22, 77]. The *DUOX2* variant identified as a CD risk factor by Levine et al. (rs 151261408; c.908C>G; p.P303R) [74] was detected in a patient with CH without intestinal inflammation and in healthy controls, and in contrast to Levine functional studies in HeLa cells showed normal H_2O_2 production [77]. A recent case report describes two siblings with unusually severe CH with homozygous mutations in both *DUOX2* and *DUOX1* [78]. On the other hand, a large retrospective CH cohort analysis indicated that CH patients had a significantly higher overall IBD prevalence with increased odds for patients with transient CH [79]. To date VEOIBD patients with DUOX2 missense mutations had normal thyroid function. Together these findings suggest that thyroid-expressed DUOX1 may partially compensate for DUOX2 defects and that the variable penetrance and clinical manifestation (CH, IBD, CH with IBD) may be explained by the characteristics of the *DUOX2* variant, and the yet unknown influence of other genetic and environmental disease modifiers.

Animal studies provide insight into the role of intestinal DUOX2 in inflammation and pathogen defense. *Duox* knockdown or deficiency resulted in increased susceptibility to bacterial infection in *Drosophila*, *C. elegans*, and zebrafish [80–83]. In mouse models, infection with *Citrobacter rodentium* and *Helicobacter felis* results in increased Duox2 and Duoxa2 expression, while Duoxa1/2 deficiency in mice, resulting in nonfunctional Duox1/2 and compromised intestinal H_2O_2 production, gave rise to increased mucosal colonization by *H. felis* and aggravated gastritis [84, 85]. Corcionivoschi et al. used intestinal biopsies and cellular models to demonstrate increased NOX1 and DUOX2 expression and augmented H_2O_2 production following infection with

C. jejuni, and an H_2O_2-dependent decrease in *C. jejuni* capsule formation through downregulation of bacterial phosphotyrosine signaling, further supporting a vital role for DUOX2 in epithelial defense [86]. Pircalabioru et al. observed for the first time in vivo cross talk between NOX1 and DUOX2: Nox1 inactivation in *Cyba*$^{-/-}$ and *Cyba*$^{fl/fl}$ Vil-cre mice or in *Nox1* knockout mice markedly decreased *C. rodentium*-induced upregulation of Duox2 at the height of infection [84].

In summary, patient data suggest a protective role for epithelial NOX1 and DUOX2 in intestinal inflammation. The complexity of NOX1's signaling mechanisms and co-regulation with other signaling molecules, including DUOX2 and IL10, may explain the incomplete penetrance of NOX1 and DUOX2 defects associated with adult-onset IBD and VEOIBD in humans.

5 NADPH Oxidases Within a Complex System

The intestine is a complex system which in health maintains the communication and balance between bacteria (pathogenic and commensal), the epithelial and mucus barrier, and the innate immune system. NOX enzymes are part of multiple signaling pathways and are regulated not only by each other [84] but also by other signaling molecules. NOD2 defects were identified as an independent genetic risk factor for adult CD in 2001 [87–89] and a DUOX2-dependent increase in ROS production following NOD2 stimulation by muramyl dipeptide and TNFα was observed in Caco-2 cells [90], suggesting a role in host defense that may lead to oxidant-induced inflammation. Deficiencies of the anti-inflammatory cytokine IL10 or its receptor IL10-R have been identified as VEOIBD risk factors [7]. IL-10 inhibited IFNγ- and TNFα-dependent superoxide production and upregulation of NOX1 protein and mRNA in cells, while *Nox1* mRNA expression was increased in the colon of *IL10*$^{-/-}$ mice [91]. Tréton et al. describe early onset of spontaneous colitis in Nox1/IL10 double-knockout mice with a reduction in goblet cell numbers and mucin production and a colitis phenotype resembling human UC in distribution and histology [92]. Together these findings point to co-regulation of IL10 and NOX1 signaling in triggering inflammation, but also to a protective role for NOX1 in barrier maintenance. In addition, the effect of the intestinal microbiome must be considered. Dysbiosis has been frequently described in patients with IBD [63–66]. Whether this dysbiosis leads to or results from intestinal inflammation remains unclear. There is some preliminary evidence that fecal transplantation may be beneficial in adult UC [93–95] but in published studies the numbers of patients were small, study follow-up durations were relatively short, and there was no consensus on mode, frequency, or duration of treatment.

While antibiotics are often used for the treatment of Crohn's related complications (perianal disease, fistulae), antibiotic therapy to induce remission in IBD is not yet supported by current evidence. Intriguingly, there is evidence from a handful of uncontrolled studies that antibiotics might be efficacious in pediatric IBD [96, 97], including VEOIBD [98], although there is a necessity for more robust data from randomized controlled trials before this approach can be adopted more widely.

6 Outlook

Patients with VEOIBD remain rare but carry a significant disease burden throughout their childhood and adult life. Recent epidemiological and genetic analyses discriminate between VEOIBD patients and later childhood or adult-onset IBD patients based on differences in disease behavior and higher burden of genetic risk factors. The identification of monogenic disease associated with an increased risk of developing intestinal inflammation (e.g., CGD, IL10 deficiency) as well as the detection of single-gene defects in patients with VEOIBD alone provide important clues to the pathogenesis of VEOIBD and, by extension, to adult-onset IBD. Our new understanding of VEOIBD as monogenic diseases, together with easier access to whole-genome and targeted gene analysis, is leading to increased genetic testing of VEOIBD patients [7, 99]. The identification of several reduced function or loss-of-function NADPH oxidase variants as risk factors for the development of VEOIBD and pediatric IBD makes a compelling case for protective roles of NADPH oxidases and NOX/DUOX-derived ROS production in intestinal barrier maintenance, homeostasis, and innate immune cell functions (Fig. 1). The complexity of NADPH oxidase signaling in addition to their co-regulation with other signaling molecules and with the intestinal microbiome is a likely explanation for the incomplete penetrance of NOX1 and DUOX2 defects associated with adult-onset IBD and VEOIBD in humans.

As genetic analysis of patients with VEOIBD becomes more routine [100], it is likely that more disease-causing variants in this enzyme family as well as in associated signaling molecules leading to a secondary reduction in NADPH oxidase-dependent ROS production will be identified. In fact, a VEOIBD patient with combined presence of *NOX1* (c.C721T, p.R241C) and *CYBA* (c.T214C, p.Y72H) variants was just identified [101]. Expanding our understanding of the role of ROS in VEOIBD has exciting implications not only for treatment of these young patients but also for the treatment of adult IBD and other diseases manifesting as intestinal inflammation. Current animal models are limited in their replication of human disease as they seem not to reproduce the complex interplay between NOX enzymes, immune system, and epithelial and bacterial signaling. The development of new models will be important for progress in defining how diminished ROS levels contribute to inflammation and to test the hypothesis that controlled stimulation of physiological ROS production might be beneficial to selected patients.

Acknowledgements

This work was supported by the Health Research Board (HPF 2016-1677 to ES), the National Children's Research Centre/ CMRF Crumlin (UGK, BB), and Science Foundation Ireland (UGK).

References

1. Hope B, Shahdadpuri R, Dunne C et al (2012) Rapid rise in incidence of Irish paediatric inflammatory bowel disease. Arch Dis Child 97(7):590–594. https://doi.org/10.1136/archdischild-2011-300651
2. Benchimol EI, Fortinsky KJ, Gozdyra P et al (2011) Epidemiology of pediatric inflammatory bowel disease: a systematic review of international trends. Inflamm Bowel Dis 17 (1):423–439. https://doi.org/10.1002/ibd.21349
3. Benchimol EI, Mack DR, Nguyen GC et al (2014) Incidence, outcomes, and health services burden of very early onset inflammatory Bowel disease. Gastroenterology 147 (4):803–813.e807. https://doi.org/10.1053/j.gastro.2014.06.023
4. Levine A, Griffiths A, Markowitz J et al (2011) Pediatric modification of the Montreal classification for inflammatory bowel disease: the Paris classification. Inflamm Bowel Dis 17 (6):1314–1321. https://doi.org/10.1002/ibd.21493
5. Heyman MB, Kirschner BS, Gold BD et al (2005) Children with early-onset inflammatory bowel disease (IBD): analysis of a pediatric IBD consortium registry. J Pediatr 146 (1):35–40. https://doi.org/10.1016/j.jpeds.2004.08.043
6. Aloi M, Lionetti P, Barabino A et al (2014) Phenotype and disease course of early-onset pediatric inflammatory bowel disease. Inflamm Bowel Dis 20(4):597–605. https://doi.org/10.1097/01.mib.0000442921.77945.09
7. Uhlig HH, Schwerd T, Koletzko S et al (2014) The diagnostic approach to monogenic very early onset inflammatory Bowel disease. Gastroenterology 147 (5):990–1007.e1003. https://doi.org/10.1053/j.gastro.2014.07.023
8. Paul T, Birnbaum A, Pal DK et al (2006) Distinct phenotype of early childhood inflammatory bowel disease. J Clin Gastroenterol 40 (7):583–586
9. de Bie CI, Buderus S, Sandhu BK et al (2012) Diagnostic workup of paediatric patients with inflammatory bowel disease in Europe: results of a 5-year audit of the EUROKIDS registry. J Pediatr Gastroenterol Nutr 54(3):374–380. https://doi.org/10.1097/MPG.0b013e318231d984
10. Jostins L, Ripke S, Weersma RK et al (2012) Host–microbe interactions have shaped the genetic architecture of inflammatory bowel disease. Nature 491(7422):119–124. https://doi.org/10.1038/nature11582
11. Spehlmann ME, Begun AZ, Burghardt J et al (2008) Epidemiology of inflammatory bowel disease in a German twin cohort: results of a nationwide study. Inflamm Bowel Dis 14 (7):968–976. https://doi.org/10.1002/ibd.20380
12. Khor B, Gardet A, Xavier RJ (2011) Genetics and pathogenesis of inflammatory bowel disease. Nature 474(7351):307–317. https://doi.org/10.1038/nature10209
13. Bequet E, Sarter H, Fumery M et al (2017) Incidence and phenotype at diagnosis of very-early-onset compared with later-onset paediatric inflammatory bowel disease: a population-based study [1988–2011]. J Crohns Colitis 11(5):519–526. https://doi.org/10.1093/ecco-jcc/jjw194

14. Levine A, Koletzko S, Turner D et al (2014) ESPGHAN revised Porto criteria for the diagnosis of inflammatory bowel disease in children and adolescents. J Pediatr Gastroenterol Nutr 58(6):795–806. https://doi.org/10.1097/mpg.0000000000000239
15. Castellaneta SP, Afzal NA, Greenberg M et al (2004) Diagnostic role of upper gastrointestinal endoscopy in pediatric inflammatory bowel disease. J Pediatr Gastroenterol Nutr 39(3):257–261
16. Mamula P, Telega GW, Markowitz JE et al (2002) Inflammatory bowel disease in children 5 years of age and younger. Am J Gastroenterol 97(8):2005–2010. https://doi.org/10.1111/j.1572-0241.2002.05915.x
17. Coughlan A, Wylde R, Lafferty L et al (2017) A rising incidence and poorer male outcomes characterise early onset paediatric inflammatory bowel disease. Aliment Pharmacol Ther 45(12):1534–1541. https://doi.org/10.1111/apt.14070
18. Khangura SK, Kamal N, Ho N et al (2016) Gastrointestinal features of chronic granulomatous disease found during endoscopy. Clin Gastroenterol Hepatol 14(3):395–402.e395. https://doi.org/10.1016/j.cgh.2015.10.030
19. Biasi F, Leonarduzzi G, Oteiza PI et al (2013) Inflammatory Bowel disease: mechanisms, redox considerations, and therapeutic targets. Antioxid Redox Signal 19(14):1711–1747. https://doi.org/10.1089/ars.2012.4530
20. Bedard K, Krause K-H (2007) The NOX family of ROS-generating NADPH oxidases: physiology and pathophysiology. Physiol Rev 87(1):245–313
21. Lambeth JD, Neish AS (2014) Nox enzymes and new thinking on reactive oxygen: a double-edged sword revisited. Annu Rev Pathol 9:119–145. https://doi.org/10.1146/annurev-pathol-012513-104651
22. O'Neill S, Brault J, Stasia M-J, Knaus UG (2015) Genetic disorders coupled to ROS deficiency. Redox Biol 6:135–156. https://doi.org/10.1016/j.redox.2015.07.009
23. Sumimoto H (2008) Structure, regulation and evolution of Nox-family NADPH oxidases that produce reactive oxygen species. FEBS J 275(13):3249–3277. https://doi.org/10.1111/j.1742-4658.2008.06488.x
24. Roos D, Boer M (2014) Molecular diagnosis of chronic granulomatous disease. Clin Exp Immunol 175(2):139–149. https://doi.org/10.1111/cei.12202
25. Holland SM (2013) Chronic granulomatous disease. Hematol Oncol Clin North Am 27 (1):89–99. https://doi.org/10.1016/j.hoc.2012.11.002
26. Marks DJ, Miyagi K, Rahman FZ et al (2009) Inflammatory bowel disease in CGD reproduces the clinicopathological features of Crohn's disease. Am J Gastroenterol 104 (1):117–124. https://doi.org/10.1038/ajg.2008.72
27. Schappi MG, Smith VV, Goldblatt D et al (2001) Colitis in chronic granulomatous disease. Arch Dis Child 84(2):147–151
28. Dhillon SS, Fattouh R, Elkadri A et al (2014) Variants in nicotinamide adenine dinucleotide phosphate oxidase complex components determine susceptibility to very early onset inflammatory Bowel disease. Gastroenterology 147(3):680–689.e682. https://doi.org/10.1053/j.gastro.2014.06.005
29. Denson LA, Jurickova I, Karns R, Shaw KA, Cutler DJ, Okou DT, Dodd A, Quinn K, Mondal K, Aronow BJ, Haberman Y, Linn A, Price A, Bezold R, Lake K, Jackson K, Walters TD, Griffiths A, Baldassano RN, Noe JD, Hyams JS, Crandall WV, Kirschner BS, Heyman MB, Snapper S, Guthery SL, Dubinsky MC, Leleiko NS, Otley AR, Xavier RJ, Stevens C, Daly MJ, Zwick ME, Kugathasan S (2018) Clinical and genomic correlates of neutrophil reactive oxygen species production in pediatric patients with Crohn's disease. Gastroenterology 154 (8):2097–2110. https://doi.org/10.1053/j.gastro.2018.02.016
30. Knaus UG, Heyworth PG, Evans T et al (1991) Regulation of phagocyte oxygen radical production by the GTP-binding protein Rac 2. Science 254(5037):1512–1515
31. Pick E (2014) Role of the Rho GTPase Rac in the activation of the phagocyte NADPH oxidase: outsourcing a key task. Small GTPases 5: e27952. https://doi.org/10.4161/sgtp.27952
32. Muise AM, Xu W, Guo CH et al (2012) NADPH oxidase complex and IBD candidate gene studies: identification of a rare variant in NCF2 that results in reduced binding to RAC2. Gut 61(7):1028–1035. https://doi.org/10.1136/gutjnl-2011-300078
33. Rioux JD, Xavier RJ, Taylor KD et al (2007) Genome-wide association study identifies new susceptibility loci for Crohn's disease and implicates autophagy in disease pathogenesis. Nat Genet 39:596. https://doi.org/10.1038/ng2032
34. Roberts RL, Hollis-Moffatt JE, Gearry RB et al (2008) Confirmation of association of IRGM and NCF4 with ileal Crohn's disease in a population-based cohort. Genes Immun

9:561. https://doi.org/10.1038/gene.2008.49

35. Glas J, Seiderer J, Pasciuto G et al (2009) rs224136 on chromosome 10q21.1 and variants in PHOX2B, NCF4, and FAM92B are not major genetic risk factors for susceptibility to Crohn's disease in the German population. Am J Gastroenterol 104(3):665–672. https://doi.org/10.1038/ajg.2008.65
36. Amre DK, Mack DR, Israel D et al (2012) NELL1, NCF4, and FAM92B genes are not major susceptibility genes for Crohn's disease in Canadian children and young adults. Inflamm Bowel Dis 18(3):529–535. https://doi.org/10.1002/ibd.21708
37. Somasundaram R, Deuring JJ, van der Woude CJ et al (2012) Linking risk conferring mutations in NCF4 to functional consequences in Crohn's disease. Gut 61(7):1097
38. Ellson CD, Davidson K, Ferguson GJ et al (2006) Neutrophils from p40phox−/− mice exhibit severe defects in NADPH oxidase regulation and oxidant-dependent bacterial killing. J Exp Med 203(8):1927–1937. https://doi.org/10.1084/jem.20052069
39. Verspaget HW, Mieremet-Ooms MA, Weterman IT et al (1984) Partial defect of neutrophil oxidative metabolism in Crohn's disease. Gut 25(8):849–853
40. Palmer CD, Rahman FZ, Sewell GW et al (2009) Diminished macrophage apoptosis and reactive oxygen species generation after phorbol ester stimulation in Crohn's disease. PLoS One 4(11):e7787. https://doi.org/10.1371/journal.pone.0007787
41. Worsaae N, Staehr Johansen K, Christensen KC (1982) Impaired in vitro function of neutrophils in Crohn's disease. Scand J Gastroenterol 17(1):91–96
42. Anderson-Cohen M, Holland SM, Kuhns DB et al (2003) Severe phenotype of chronic granulomatous disease presenting in a female with a de novo mutation in gp91-phox and a non familial, extremely skewed X chromosome inactivation. Clin Immunol 109(3):308–317
43. Lekstrom-Himes JA, Gallin JI (2000) Immunodeficiency diseases caused by defects in phagocytes. N Engl J Med 343(23):1703–1714. https://doi.org/10.1056/nejm200012073432307
44. Marciano BE, Zerbe CS, Falcone EL et al (2018) X-linked carriers of chronic granulomatous disease: Illness, lyonization, and stability. J Allergy Clin Immunol 141(1):365–371. https://doi.org/10.1016/j.jaci.2017.04.035
45. Huang C, De Ravin SS, Paul AR et al (2016) Genetic risk for inflammatory Bowel disease is a determinant of Crohn's disease development in chronic granulomatous disease. Inflamm Bowel Dis 22(12):2794–2801. https://doi.org/10.1097/mib.0000000000000966
46. Vaughan D, Drumm B (1999) Treatment of fistulas with granulocyte colony-stimulating factor in a patient with Crohn's disease. N Engl J Med 340(3):239–240. https://doi.org/10.1056/nejm199901213400317
47. Dabritz J, Bonkowski E, Chalk C (2013) Granulocyte macrophage colony-stimulating factor auto-antibodies and disease relapse in inflammatory bowel disease. Am J Gastroenterol 108(12):1901–1910. https://doi.org/10.1038/ajg.2013.360
48. Kuhns DB, Alvord WG, Heller T et al (2010) Residual NADPH oxidase and survival in chronic granulomatous disease. N Engl J Med 363(27):2600–2610. https://doi.org/10.1056/NEJMoa1007097
49. Halling ML, Kjeldsen J, Knudsen T et al (2017) Patients with inflammatory bowel disease have increased risk of autoimmune and inflammatory diseases. World J Gastroenterol 23(33):6137–6146. https://doi.org/10.3748/wjg.v23.i33.6137
50. Uhlig HH (2013) Monogenic diseases associated with intestinal inflammation: implications for the understanding of inflammatory bowel disease. Gut 62(12):1795–1805. https://doi.org/10.1136/gutjnl-2012-303956
51. Avitzur Y, Guo C, Mastropaolo LA et al (2014) Mutations in tetratricopeptide repeat domain 7A result in a severe form of very early onset inflammatory bowel disease. Gastroenterology 146(4):1028–1039. https://doi.org/10.1053/j.gastro.2014.01.015
52. Pai SY, Levy O, Jabara HH et al (2008) Allogeneic transplantation successfully corrects immune defects, but not susceptibility to colitis, in a patient with nuclear factor-kappaB essential modulator deficiency. J Allergy Clin Immunol 122(6):1113–1118.e1111. https://doi.org/10.1016/j.jaci.2008.08.026
53. Freeman EB, Koglmeier J, Martinez AE et al (2008) Gastrointestinal complications of epidermolysis bullosa in children. Br J Dermatol 158(6):1308–1314. https://doi.org/10.1111/j.1365-2133.2008.08507.x
54. Reddy H, Shipman AR, Wojnarowska F (2013) Epidermolysis bullosa acquisita and inflammatory bowel disease: a review of the literature. Clin Exp Dermatol 38

(3):225–229.; quiz 229-230. https://doi.org/10.1111/ced.12114
55. Kumar V, Mack DR, Marcil V et al (2013) Genome-wide association study signal at the 12q12 locus for Crohn's disease may represent associations with the MUC19 gene. Inflamm Bowel Dis 19(6):1254–1259. https://doi.org/10.1097/MIB.0b013e318281f454
56. Geiszt M, Witta J, Baffi J, Lekstrom K, Leto TL (2003) Dual oxidases represent novel hydrogen peroxide sources supporting mucosal surface host defense. FASEB J 17 (11):1502–1504. https://doi.org/10.1096/fj.02-1104fje
57. El Hassani RA, Benfares N, Caillou B et al (2005) Dual oxidase2 is expressed all along the digestive tract. Am J Physiol Gastrointest Liver Physiol 288(5):G933–G942. https://doi.org/10.1152/ajpgi.00198.2004
58. Rokutan K, Kawahara T, Kuwano Y et al (2008) Nox enzymes and oxidative stress in the immunopathology of the gastrointestinal tract. Semin Immunopathol 30(3):315–327. https://doi.org/10.1007/s00281-008-0124-5
59. Grasberger H, Gao J, Nagao-Kitamoto H et al (2015) Increased expression of DUOX2 is an epithelial response to mucosal dysbiosis required for immune homeostasis in mouse intestine. Gastroenterology 149 (7):1849–1859. https://doi.org/10.1053/j.gastro.2015.07.062
60. Ueyama T, Geiszt M, Leto TL (2006) Involvement of Rac1 in activation of multicomponent Nox1- and Nox3-based NADPH oxidases. Mol Cell Biol 26(6):2160–2174. https://doi.org/10.1128/MCB.26.6.2160-2174.2006
61. Hayes P, Dhillon S, O'Neill K et al (2015) Defects in NADPH oxidase genes NOX1 and DUOX2 in very early onset inflammatory Bowel disease. Cell Mol Gastroenterol Hepatol 1(5):489–502. https://doi.org/10.1016/j.jcmgh.2015.06.005
62. Schwerd T, Bryant RV, Pandey S et al (2018) NOX1 loss-of-function genetic variants in patients with inflammatory bowel disease. Mucosal Immunol 11(2):562–574. https://doi.org/10.1038/mi.2017.74
63. Gevers D, Kugathasan S, Denson LA et al (2014) The treatment-naive microbiome in new-onset Crohn's disease. Cell Host Microbe 15(3):382–392. https://doi.org/10.1016/j.chom.2014.02.005
64. Maukonen J, Kolho KL, Paasela M et al (2015) Altered fecal microbiota in paediatric inflammatory Bowel disease. J Crohns Colitis 9(12):1088–1095. https://doi.org/10.1093/ecco-jcc/jjv147
65. Oyri SF, Muzes G, Sipos F (2015) Dysbiotic gut microbiome: a key element of Crohn's disease. Comp Immunol Microbiol Infect Dis 43:36–49. https://doi.org/10.1016/j.cimid.2015.10.005
66. Shafquat A, Joice R, Simmons SL et al (2014) Functional and phylogenetic assembly of microbial communities in the human microbiome. Trends Microbiol 22(5):261–266. https://doi.org/10.1016/j.tim.2014.01.011
67. Szanto I, Rubbia-Brandt L, Kiss P et al (2005) Expression of NOX1, a superoxide-generating NADPH oxidase, in colon cancer and inflammatory bowel disease. J Pathol 207 (2):164–176. https://doi.org/10.1002/path.1824
68. Jones RM, Luo L, Ardita CS et al (2013) Symbiotic lactobacilli stimulate gut epithelial proliferation via Nox-mediated generation of reactive oxygen species. EMBO J 32 (23):3017–3028. https://doi.org/10.1038/emboj.2013.224
69. Leoni G, Alam A, Neumann PA et al (2013) Annexin A1, formyl peptide receptor, and NOX1 orchestrate epithelial repair. J Clin Invest 123(1):443–454. https://doi.org/10.1172/jci65831
70. Haberman Y, Tickle TL, Dexheimer PJ et al (2014) Pediatric Crohn disease patients exhibit specific ileal transcriptome and microbiome signature. J Clin Invest 124 (8):3617–3633. https://doi.org/10.1172/jci75436
71. Csillag C, Nielsen OH, Vainer B et al (2007) Expression of the genes dual oxidase 2, lipocalin 2 and regenerating islet-derived 1 alpha in Crohn's disease. Scand J Gastroenterol 42 (4):454–463. https://doi.org/10.1080/00365520600976266
72. Hamm CM, Reimers MA, McCullough CK et al (2010) NOD2 status and human ileal gene expression. Inflamm Bowel Dis 16 (10):1649–1657. https://doi.org/10.1002/ibd.21208
73. Parlato M, Charbit-Henrion F, Hayes P et al (2017) First identification of biallelic inherited DUOX2 inactivating mutations as a cause of very early onset inflammatory Bowel disease. Gastroenterology 153(2):609–611.e603. https://doi.org/10.1053/j.gastro.2016.12.053
74. Levine AP, Pontikos N, Schiff ER et al (2016) Genetic complexity of Crohn's disease in two

large Ashkenazi Jewish families. Gastroenterology 151(4):698–709. https://doi.org/10.1053/j.gastro.2016.06.040

75. MacFie TS, Poulsom R, Parker A et al (2014) DUOX2 and DUOXA2 form the predominant enzyme system capable of producing the reactive oxygen species H2O2 in active ulcerative colitis and are modulated by 5-aminosalicylic acid. Inflamm Bowel Dis 20 (3):514–524. https://doi.org/10.1097/01.MIB.0000442012.45038.0e
76. De Deken X, Wang D, Many MC et al (2000) Cloning of two human thyroid cDNAs encoding new members of the NADPH oxidase family. J Biol Chem 275 (30):23227–23233. https://doi.org/10.1074/jbc.M000916200
77. Muzza M, Rabbiosi S, Vigone MC et al (2014) The clinical and molecular characterization of patients with dyshormonogenic congenital hypothyroidism reveals specific diagnostic clues for DUOX2 defects. J Clin Endocrinol Metab 99(3):E544–E553. https://doi.org/10.1210/jc.2013-3618
78. Aycan Z, Cangul H, Muzza M et al (2017) Digenic DUOX1 and DUOX2 mutations in cases with congenital hypothyroidism. J Clin Endocrinol Metab 102(9):3085–3090. https://doi.org/10.1210/jc.2017-00529
79. Grasberger H, Noureldin M, Kao TD, Adler J, Lee JM, Bishu S, El-Zaatari M, Kao JY, Waljee AK (2018) Increased risk for inflammatory bowel disease in congenital hypothyroidism supports the existence of a shared susceptibility factor. Sci Rep 8 (1):10158. https://doi.org/10.1038/s41598-018-28586-5
80. Ha EM, Oh CT, Bae YS et al (2005) A direct role for dual oxidase in Drosophila gut immunity. Science 310(5749):847–850. https://doi.org/10.1126/science.1117311
81. Kim SH, Lee WJ (2014) Role of DUOX in gut inflammation: lessons from Drosophila model of gut-microbiota interactions. Front Cell Infect Microbiol 3:116. https://doi.org/10.3389/fcimb.2013.00116
82. Chavez V, Mohri-Shiomi A, Garsin DA (2009) Ce-Duox1/BLI-3 generates reactive oxygen species as a protective innate immune mechanism in Caenorhabditis elegans. Infect Immun 77(11):4983–4989. https://doi.org/10.1128/iai.00627-09
83. Flores MV, Crawford KC, Pullin LM et al (2010) Dual oxidase in the intestinal epithelium of zebrafish larvae has anti-bacterial properties. Biochem Biophys Res Commun 400(1):164–168. https://doi.org/10.1016/j.bbrc.2010.08.037
84. Pircalabioru G, Aviello G, Kubica M et al (2016) Defensive mutualism rescues NADPH oxidase inactivation in gut infection. Cell Host Microbe 19(5):651–663. https://doi.org/10.1016/j.chom.2016.04.007
85. Grasberger H, El-Zaatari M, Dang DT et al (2013) Dual oxidases control release of hydrogen peroxide by the gastric epithelium to prevent Helicobacter felis infection and inflammation in mice. Gastroenterology 145 (5):1045–1054. https://doi.org/10.1053/j.gastro.2013.07.011
86. Corcionivoschi N, Alvarez LA, Sharp TH et al (2012) Mucosal reactive oxygen species decrease virulence by disrupting Campylobacter jejuni phosphotyrosine signaling. Cell Host Microbe 12(1):47–59. https://doi.org/10.1016/j.chom.2012.05.018
87. Hampe J, Cuthbert A, Croucher PJ et al (2001) Association between insertion mutation in NOD2 gene and Crohn's disease in German and British populations. Lancet 357 (9272):1925–1928. https://doi.org/10.1016/s0140-6736(00)05063-7
88. Ogura Y, Bonen DK, Inohara N et al (2001) A frameshift mutation in NOD2 associated with susceptibility to Crohn's disease. Nature 411(6837):603–606. https://doi.org/10.1038/35079114
89. Hugot JP, Chamaillard M, Zouali H et al (2001) Association of NOD2 leucine-rich repeat variants with susceptibility to Crohn's disease. Nature 411(6837):599–603. https://doi.org/10.1038/35079107
90. Lipinski S, Till A, Sina C et al (2009) DUOX2-derived reactive oxygen species are effectors of NOD2-mediated antibacterial responses. J Cell Sci 122(Pt 19):3522–3530. https://doi.org/10.1242/jcs.050690
91. Kamizato M, Nishida K, Masuda K et al (2009) Interleukin 10 inhibits interferon gamma- and tumor necrosis factor alpha-stimulated activation of NADPH oxidase 1 in human colonic epithelial cells and the mouse colon. J Gastroenterol 44 (12):1172–1184. https://doi.org/10.1007/s00535-009-0119-6
92. Treton X, Pedruzzi E, Guichard C et al (2014) Combined NADPH oxidase 1 and interleukin 10 deficiency induces chronic endoplasmic reticulum stress and causes ulcerative colitis-like disease in mice. PLoS One 9 (7):e101669. https://doi.org/10.1371/journal.pone.0101669
93. Paramsothy S, Kamm MA, Kaakoush NO et al (2017) Multidonor intensive faecal microbiota transplantation for active ulcerative colitis: a randomised placebo-controlled trial.

Lancet 389(10075):1218–1228. https://doi.org/10.1016/s0140-6736(17)30182-4
94. Narula N, Kassam Z, Yuan Y et al (2017) Systematic review and meta-analysis: fecal microbiota transplantation for treatment of active ulcerative colitis. Inflamm Bowel Dis 23(10):1702–1709. https://doi.org/10.1097/mib.0000000000001228
95. Browne AS, Kelly CR (2017) Fecal transplant in inflammatory Bowel disease. Gastroenterol Clin North Am 46(4):825–837. https://doi.org/10.1016/j.gtc.2017.08.005
96. Levine A, Turner D (2011) Combined azithromycin and metronidazole therapy is effective in inducing remission in pediatric Crohn's disease. J Crohns Colitis 5(3):222–226. https://doi.org/10.1016/j.crohns.2011.01.006
97. Turner D, Levine A, Kolho KL et al (2014) Combination of oral antibiotics may be effective in severe pediatric ulcerative colitis: a preliminary report. J Crohns Colitis 8(11):1464–1470. https://doi.org/10.1016/j.crohns.2014.05.010
98. Lev-Tzion R, Ledder O, Shteyer E et al (2017) Oral vancomycin and gentamicin for treatment of very early onset inflammatory Bowel disease. Digestion 95(4):310–313. https://doi.org/10.1159/000475660
99. Muise AM, Snapper SB, Kugathasan S (2012) The age of gene discovery in very early onset inflammatory Bowel disease. Gastroenterology 143(2):285–288. https://doi.org/10.1053/j.gastro.2012.06.025
100. Uhlig HH, Muise AM (2017) Clinical genomics in inflammatory Bowel disease. Trends Genet 33(9):629–641. https://doi.org/10.1016/j.tig.2017.06.008
101. Lipinsky S, Petersen BS, Barann M et al (2019) Missense variants in NOX1 and p22phox in a case of very-early-onset inflammatory bowel disease are functionally linked to NOD2. Cold Spring Harb Mol Case Stud 5:a002428. https://doi.org/10.1101/mcs.a002428

INDEX

A

B

C

D

Ulla G. Knaus and Thomas L. Leto (eds.), *NADPH Oxidases: Methods and Protocols*, Methods in Molecular Biology, vol. 1982, https://doi.org/10.1007/978-1-4939-9424-3,

S

T

W

Z

Printed by Printforce, the Netherlands